PHENOTYPIC SWITCHING

PHENOTYPIC SWITCHING

Implications in Biology and Medicine

Edited by

HERBERT LEVINE

Center for Theoretical Biological Physics, Rice University, Houston, TX, United States; Department of Bioengineering, Northeastern University, Boston, MA, United States; Department of Physics, Northeastern University, Boston, MA, United States

MOHIT KUMAR JOLLY

Center for BioSystems Science and Engineering, Indian Institute of Science, Bangalore, India; Department of Medicine, Duke University Medical Center, Durham, NC, United States; Duke Cancer Institute, Durham, NC, United States

PRAKASH KULKARNI

Department of Medical Oncology and Therapeutics Research, City of Hope National Medical Center, Duarte, CA, United States

VIDYANAND NANJUNDIAH

Centre for Human Genetics, Bangalore, India

ELSEVIER

ACADEMIC PRESS

An imprint of Elsevier

Academic Press is an imprint of Elsevier
125 London Wall, London EC2Y 5AS, United Kingdom
525 B Street, Suite 1650, San Diego, CA 92101, United States
50 Hampshire Street, 5th Floor, Cambridge, MA 02139, United States
The Boulevard, Langford Lane, Kidlington, Oxford OX5 1GB, United Kingdom

British Library Cataloguing-in-Publication Data
A catalogue record for this book is available from the British Library

Library of Congress Cataloging-in-Publication Data
A catalog record for this book is available from the Library of Congress

ISBN: 978-0-12-817996-3

For Information on all Academic Press publications
visit our website at https://www.elsevier.com/books-and-journals

Publisher: Andre Gerhard Wolff
Acquisitions Editor: Peter B. Linsley
Editorial Project Manager: Samantha Allard
Production Project Manager: Stalin Viswanathan
Cover Designer: Mark Rogers

Typeset by MPS Limited, Chennai, India

Contents

List of contributors xv
Preface xxi
Summaries of contributions xli
Acknowledgments xlix

1 The fundamentals of phenotypic plasticity 1
Sarthak Sahoo, Ayalur Raghu Subbalakshmi and Mohit Kumar Jolly
Introduction 1
Phenotypic plasticity at an intracellular level: macromolecules, pathways, and organelles 3
Phenotypic plasticity at a cellular level: Implications in development, homeostasis, and disease 5
Phenotypic plasticity at the organismal level 13
Conclusion 15
Acknowledgment 15
References 15

2 Rethinking the role of chance in the explanation of cell differentiation 23
Marco Casali and Francesca Merlin
Introduction 23
Noise in gene expression: a descriptive analysis 27
What, actually, is noise? 29
From noise to chance as explanatory 30
Chance in immunology: an example to follow 33
A positive view of chance: main features and theoretical advantages 37
Three reasons for biological explanations in terms of chance 42
Conclusion: Chance and the reductionism/antireductionism debate 46
Acknowledgments 47
References 47

3 Random walk across the epigenetic landscape 53
András Páldi
Concepts 54
Historical origins 57
The epigenetic landscape 59
A new conceptual framework 63
Putting together the pieces of the puzzle to build a new operative model 68
Conclusion 72
Acknowledgments 73
References 73

4 Maneuvering protein functions and functional levels by structural excursions 77
Vasam Manjveekar Prabantu, Arangasamy Yazhini and Narayanaswamy Srinivasan
Moonlighting proteins 78
Functional switch mediated by protein–protein interactions 81
Modulation of protein function by oligomerization 84
Influence of domain association on protein function 85
Modulation of protein function by posttranslational modifications ... 89
Silent mutation tunes gene function 90
Conclusion 97
Acknowledgments 98
Abbreviations 98
References 99

5 Prion-mediated phenotypic diversity in fungi 105
Mick F. Tuite
Introduction 105
Prion formation and loss 108
Prion propagation and transmission in the fungal cell 110
Prion-mediated phenotypes 112
Conformational diversity generates phenotypic diversity 120
Concluding remarks 122
Acknowledgments 123
References 123

6 Bistability in virus–host interaction networks underlies the success of hepatitis C treatments 131
Pranesh Padmanabhan, Rubesh Raja and Narendra M. Dixit
Introduction 131
Bistability in the interferon signaling network 133
Phenotypic heterogeneity, viral kinetics, and treatment outcomes.. 137
Potential considerations and strategies for optimizing DAA treatments 144
Concluding remarks 150
Acknowledgments 151
References 152

7 Quantifying Waddington landscapes, paths, and kinetics of cell fate decision making of differentiation/development 157
Li Xu and Jin Wang
Introduction 157
Potential and flux landscape theory of cell fate decision of differentiation and reprograming 160
Quantifying Waddington landscape and paths for differentiation/development 168
Discussions on critical issues of cell fate decision making of differentiation and development 175
Acknowledgment 182
References 182

8 The physics of cell fate 189
Philip Greulich, Rosanna Smith and Ben D. MacArthur
Introduction 189
The "physical laws" of cell fate dynamics 192
Statistical mechanics of cell state dynamics 196
Universality in cell biology 200
Critique and outlook 203
Conclusions 203
References 203

9 Disentangling the environmentally induced and stochastic developmental components of phenotypic variation 207
Günter Vogt
Introduction 207
Considerations on phenotypic, genetic, environmentally induced, and stochastic developmental variations 209
Occurrence of environmentally induced variation and stochastic developmental variation in the kingdoms of life 215
Determination of stochastic developmental variation in laboratory experiments 218
Disentangling genetic variation, environmentally induced variation, and stochastic developmental variation in the laboratory 227
Disentangling genetic variation, environmentally induced variation, and stochastic developmental variation in field studies 229
Disentangling genetic variation plus environmentally induced variation from stochastic developmental variation by mathematical modeling 231
Identification of the molecular mechanisms underlying environmentally induced variation and stochastic developmental variation 232
The marbled crayfish as a promising model for investigating the nongenetic components of phenotypic variation 236
Conclusions 241
Acknowledgments 242
References 242

10 The evolution of cell differentiation in animals: biomolecular condensates as amplification hubs of inherent cell functions 253
Stuart A. Newman
Introduction 253
Metazoan-specific modes of transcriptional regulation 259
Beta-catenin, Grainyhead-like and the role of multicellularity in the evolution of differentiation 263
Inherent cell functions in the origin of differentiation 266
Conclusion: Prolific cell differentiation as a metazoan evolutionary innovation 269

Abbreviations 273
References 273

11 Phenotypic switching and its evolutionary consequences 281
Dragan Stajic and Claudia Bank
Evolution and the principle of inheritance 282
Molecular basis and examples of epigenetically determined phenotypic switching 288
Evolutionary consequences of phenotypic switching 295
Acknowledgments 301
References 301

12 Cell-state organization by exploratory sloppy dynamics 305
Erez Braun
Introduction 305
Experimental approach: cell adaptation to an unforeseen challenge 310
Implications of the yeast adaptation experiments 320
Summary and open issues 330
Acknowledgments 331
References 331

13 Emergence of metabolic heterogeneity in cell populations: lessons from budding yeast 335
Sandeep Krishna and Sunil Laxman
Introduction: cell state heterogeneity in isogenic microbial populations 335
Metabolic heterogeneity and spatial organization within yeast colonies 337
General considerations on the manifestation of multiple cellular states at the population level 341
Cell-state heterogeneity of yeast in a well-mixed chemostat 346
Discussion 353
Acknowledgments 355
Appendix 355
References 358

14 Stochastic phenotypic switching in endothelial cell heterogeneity 361
Erzsébet Ravasz Regan
Introduction 361
Noise-mediated endothelial cell heterogeneity in vivo 372
Discussion—implications of biological noise-driven bed hedging as an adaptive trait 388
References 394

15 Regulation of phenotypic plasticity from the perspective of evolutionary developmental biology 403
Yuichiro Suzuki, Kenneth Z. McKenna and H. Frederik Nijhout
Introduction to phenotypic plasticity 403
What are polyphenisms? 404
Regulation of developmental plasticity 404
Nonadaptive plasticity and the evolution of robustness 422
Plasticity as an adaptation 423
Reaction norms and allometries 424
Are polyphenisms reaction norms? 424
Evolution via phenotypic plasticity 425
The role of cryptic genetic variation in genetic accommodation 427
Model of genetic accommodation 430
Conclusions 434
References 435

16 Phenotypic plasticity and the origins of novelty 443
Nicholas A. Levis and David W. Pfennig
Introduction 443
Plasticity-led evolution 444
Empirically evaluating plasticity-led evolution 447
Plasticity-led evolution in spadefoot toads 449
Plasticity and macroevolution 452
Conclusions 454
References 454

17 Niche construction and the transition to herbivory: Phenotype switching and the organization of new nutritional modes 459
Lynn Chiu and Scott F. Gilbert
Introduction 459
How the bovine got its stomach 461
Perturbational and mediational niche construction 467
Niche construction, plasticity, and developmental scaffolding 470
Implications for holobiont individuality 472
Conclusion 475
References 475

18 Nature, nurture, and noise in bird song ontogeny as determinants of phenotypic and functional variation among dialects 483
Jill A. Soha
Introduction: Song function, ontogeny, and phenotypic variation... 483
How phenotypic variation arises: plasticity in song development... 489
Song dialects as functional polymorphisms 498
Conclusion 503
References 503

19 Domestication as a process generating phenotypic diversity 511
Lyudmila N. Trut and Anastasiya V. Kharlamova
Domestication in evolutionary terms 511
Traditional view on possible mechanisms arisen of phenotypic diversity 513
New insights into the nature of changes under domestication 514
Phenotypic changes of experimental foxes 516
Constraints of development and its focus 519
Phenotypic changes of domesticated foxes and developmental rate 521
Acknowledgment 524
References 524
Further reading 526

20 The glycobiology of ovarian cancer progression: phenotypic switches and microenvironmental influences 527
Jimpi Langthasa, Purba Sarkar and Ramray Bhat
Introduction 527
Glycobiology and cancer.. 529
Ovarian cancer progression 532
N- and O-linked glycoproteins 536
Proteoglycans in ovarian cancer.. 539
Lectins... 540
Cancer evolution and glycan-driven plasticity 541
Conclusion... 542
Acknowledgments 543
References 543

21 Epithelial-mesenchymal transition in cancer.................................... 553
Dongya Jia, Mohit Kumar Jolly, Herbert Levine and José N. Onuchic
Introduction 553
Epithelial-mesenchymal transition 554
Regulation of epithelial-mesenchymal transition............................. 556
Hybrid epithelial/mesenchymal phenotypes 557
Therapeutic strategies targeting epithelial-mesenchymal plasticity 561
Acknowledgments 563
References 564

22 Phenotypic switching and prostate diseases: a model proposing a causal link between benign prostatic hyperplasia and prostate cancer.. 569
Prakash Kulkarni, Anusha Nathan, Ravi Salgia, Govindan Rangarajan and Mohit Kumar Jolly
Introduction 569
From BPH to PCa via EMT/MET... 572
The Integrated Model 573
The key drivers of EMT/MET in PCa are IDPs................................. 578
Phenotypic plasticity is a manifestation of plasticity at the molecular level... 580
Conclusions and future directions.. 582
References 583

23 Phenotypic plasticity and lineage switching in prostate cancer 591
Jason A. Somarelli, Andrew J. Armstrong, Maya U. Sheth, Kathryn E. Ware and Mohit Kumar Jolly
Prostate development and the progression to prostate cancer 591
Targeting the androgen receptor signaling axis in prostate adenocarcinoma 592
Treatment-induced phenotypic plasticity overcomes androgen receptor blockade 594
EMT and NEPC—common molecular drivers with unique phenotypic outputs? 602
Phenotypic plasticity is a dynamic driver of prostate cancer aggression 605
References 606

24 Implications of non-genetic heterogeneity in cancer drug resistance and malignant progression 617
Kaitlyn Johnson, Eric Brenner and Amy Brock
Introduction 617
Experimental evidence for nongenetic heterogeneity 620
Future of cell state characterization: defining cell states via relevant heterogeneity 630
References 633

25 Phenotypic plasticity: the emergence of cancer stem cells and collective cell migration 639
Caterina A.M. La Porta and Stefano Zapperi
Phenotypic plasticity and cancer stem cells 639
The landscape of epithelial-mesenchymal plasticity 641
Senescence and cancer cell plasticity 643
Phenotypic transformations and collective cell migration 645
Concluding remarks 646
References 646

26 Adaptive phenotypic switching in breast cancer in response to matrix deprivation 651
Ananthalakshmy Sundararaman, Saurav Kumar and Annapoorni Rangarajan
Heterogeneity and phenotypic plasticity in cancer 652
Sensing matrix deprivation 653
Signaling pathways regulating metabolic plasticity 654
Role of metabolic plasticity in cancer cell survival under matrix deprivation 655
Bistable systems and plasticity 665
Matrix deprivation and EMT 667
Matrix deprivation and stemness 668
EMT and stemness are associated with metabolic reprogramming in cancer 669
Future directions 670
References 671

27 Phenotypic instability induced by tissue disruption at the origin of cancer 677
Jean-Pascal Capp
Introduction 677
Phenotypic plasticity in cancer growth and drug resistance: role of cancer stem cells and stochastic gene expression 678
Cellular plasticity and stemness conferred by high stochastic gene expression 681
Cell–cell interactions as constraints decreasing cellular plasticity and stemness during differentiation 683
Tissue disruption at the origin of phenotypic instability and consequently cancer 685
References 687

28 Evolutionary strategies to overcome cancer cell resistance to treatment 691
Pedro M. Enriquez-Navas and Robert A. Gatenby
Introduction 691
Conclusion 700
References 701

Index 705

List of contributors

Andrew J. Armstrong Department of Medicine, Duke University Medical Center, Durham, NC, United States; Duke Cancer Institute, Durham, NC, United States

Claudia Bank Gulbenkian Science Institute, Oeiras, Portugal

Ramray Bhat Department of Molecular Reproduction Development and Genetics, Indian Institute of Science, Bangalore, India

Erez Braun Department of Physics, Technion-Israel Institute of Technology, Haifa, Israel

Eric Brenner Institute for Cellular and Molecular Biology, The University of Texas at Austin, Austin, TX, United States

Amy Brock Department of Biomedical Engineering, The University of Texas at Austin, Austin, TX, United States; Institute for Cellular and Molecular Biology, The University of Texas at Austin, Austin, TX, United States

Jean-Pascal Capp INSA/University of Toulouse, Toulouse Biotechnology Institute, UMR CNRS 5504, UMR INRA 792, Toulouse, France

Marco Casali Department of Philosophy, Sapienza – University of Rome, Rome, Italy; Institut d'histoire et de philosophie des sciences et des techniques (IHPST), CNRS & University of Paris 1 Panthéon-Sorbonne, Paris, France; Institut d'histoire et de philosophie des sciences et des techniques (IHPST), University of Paris 1 Panthéon-Sorbonne, Paris, France

Lynn Chiu School of Biology, University of St. Andrews, Scotland, United Kingdom; ImmunoConcept Lab, University of Bordeaux, Bordeaux, France

Narendra M. Dixit Department of Chemical Engineering, Indian Institute of Science, Bangalore, India; Centre for Biosystems Science and Engineering, Indian Institute of Science, Bangalore, India

Pedro M. Enriquez-Navas Department of Cancer Physiology, Moffitt Cancer Center & Research Institute, Tampa, FL, United States

Robert A. Gatenby Department of Cancer Physiology, Moffitt Cancer Center & Research Institute, Tampa, FL, United States

Scott F. Gilbert Swarthmore College, Swarthmore, PA, United States

Philip Greulich Mathematical Sciences, University of Southampton, Southampton, United Kingdom; Institute for Life Sciences, University of Southampton, Southampton, United Kingdom

Dongya Jia Center for Theoretical Biological Physics, Rice University, Houston, TX, United States

Kaitlyn Johnson Department of Biomedical Engineering, The University of Texas at Austin, Austin, TX, United States

Mohit Kumar Jolly Center for BioSystems Science and Engineering, Indian Institute of Science, Bangalore, India; Department of Medicine, Duke University Medical Center, Durham, NC, United States; Duke Cancer Institute, Durham, NC, United States

Anastasiya V. Kharlamova The Federal Research Center Institute of Cytology and Genetics of Siberian Branch of the Russian Academy of Sciences (IC&G SB RAS), Novosibirsk, Russia

Sandeep Krishna Simons Centre for the Study of Living Machines, National Centre for Biological Sciences, Tata Institute of Fundamental Research, Bangalore, India

Prakash Kulkarni Department of Medical Oncology and Therapeutics Research, City of Hope National Medical Center, Duarte, CA, United States

Saurav Kumar Department of Molecular Reproduction, Development and Genetics, Indian Institute of Science, Bangalore, India

Caterina A.M. La Porta Center for Complexity and Biosystems, University of Milan, Milano, Italy; Department of Environmental Science and Policy, University of Milan, Milano, Italy

Jimpi Langthasa Department of Molecular Reproduction, Development and Genetics, Indian Institute of Science, Bangalore, India

Sunil Laxman Institute for Stem Cell Science and Regenerative Medicine (inStem), Bangalore, India

Herbert Levine Center for Theoretical Biological Physics, Rice University, Houston, TX, United States; Department of Bioengineering, Northeastern University, Boston, MA, United States; Department of Physics, Northeastern University, Boston, MA, United States

Nicholas A. Levis Department of Biology, University of North Carolina, Chapel Hill, NC, United States

Ben D. MacArthur Mathematical Sciences, University of Southampton, Southampton, United Kingdom; Institute for Life Sciences, University of Southampton, Southampton, United Kingdom; Centre for Human Development, Stem Cells and Regeneration, Faculty of Medicine, University of Southampton, Southampton, United Kingdom

Kenneth Z. McKenna Department of Biology, Duke University, Durham, NC, United States

Francesca Merlin Institut d'histoire et de philosophie des sciences et des techniques (IHPST), CNRS & University of Paris 1 Panthéon-Sorbonne, Paris, France

Vidyanand Nanjundiah Centre for Human Genetics, Bangalore, India

Anusha Nathan Department of Medical Oncology and Therapeutics Research, City of Hope National Medical Center, Duarte, CA, United States

Stuart A. Newman Department of Cell Biology and Anatomy, New York Medical College, Valhalla, NY, United States

H. Frederik Nijhout Department of Biology, Duke University, Durham, NC, United States

José N. Onuchic Center for Theoretical Biological Physics, Rice University, Houston, TX, United States; Department of Chemistry, Rice University, Houston, TX, United States; Department of Biosciences, Rice University, Houston, TX, United States; Department of Physics and Astronomy, Rice University, Houston, TX, United States

Pranesh Padmanabhan Clem Jones Centre for Ageing Dementia Research, Queensland Brain Institute, The University of Queensland, St. Lucia, Brisbane, Australia

András Páldi École Pratique des Hautes Études, Paris Sciences Lettres, Inserm U938, Paris, France

David W. Pfennig Department of Biology, University of North Carolina, Chapel Hill, NC, United States

Vasam Manjveekar Prabantu Molecular Biophysics Unit, Indian Institute of Science, Bangalore, India

Rubesh Raja Department of Chemical Engineering, Indian Institute of Science, Bangalore, India

Annapoorni Rangarajan Department of Molecular Reproduction, Development and Genetics, Indian Institute of Science, Bangalore, India

Govindan Rangarajan Department of Mathematics, Indian Institute of Science, Bangalore, India; Center for Neuroscience, Indian Institute of Science, Bangalore, India

Erzsébet Ravasz Regan Biochemistry and Molecular Biology, The College of Wooster, Wooster, OH, United States

Sarthak Sahoo Centre for BioSystems Science and Engineering, Indian Institute of Science, Bangalore, India

Ravi Salgia Department of Medical Oncology and Therapeutics Research, City of Hope National Medical Center, Duarte, CA, United States

Purba Sarkar Department of Molecular Reproduction, Development and Genetics, Indian Institute of Science, Bangalore, India

Maya U. Sheth Department of Medicine, Duke University Medical Center, Durham, NC, United States; Duke Cancer Institute, Durham, NC, United States

Rosanna Smith Cancer Sciences, Faculty of Medicine, University of Southampton, Southampton, United Kingdom

Jill A. Soha Duke University, Durham, NC, United States

Jason A. Somarelli Department of Medicine, Duke University Medical Center, Durham, NC, United States; Duke Cancer Institute, Durham, NC, United States

Narayanaswamy Srinivasan Molecular Biophysics Unit, Indian Institute of Science, Bangalore, India

Dragan Stajic Gulbenkian Science Institute, Oeiras, Portugal

Ayalur Raghu Subbalakshmi Centre for BioSystems Science and Engineering, Indian Institute of Science, Bangalore, India

Ananthalakshmy Sundararaman School of Biochemistry, Faculty of Life Sciences, University of Bristol, Bristol, United Kingdom

Yuichiro Suzuki Department of Biological Sciences, Wellesley College, Wellesley, MA, United States

Lyudmila N. Trut The Federal Research Center Institute of Cytology and Genetics of Siberian Branch of the Russian Academy of Sciences (IC&G SB RAS), Novosibirsk, Russia

Mick F. Tuite Kent Fungal Group, School of Biosciences, University of Kent, Canterbury, United Kingdom

Günter Vogt Faculty of Biosciences, University of Heidelberg, Heidelberg, Germany

Jin Wang Department of Chemistry and of Physics and Astronomy, State University of New York at Stony Brook, Stony Brook, NY, United States

Kathryn E. Ware Department of Medicine, Duke University Medical Center, Durham, NC, United States; Duke Cancer Institute, Durham, NC, United States

Li Xu State Key Laboratory of Electroanalytical Chemistry, Changchun Institute of Applied Chemistry, Chinese Academy of Sciences, Changchun, P.R. China

Arangasamy Yazhini Molecular Biophysics Unit, Indian Institute of Science, Bangalore, India

Stefano Zapperi Department of Physics, University of Milan, Milano, Italy; CNR - Consiglio Nazionale delle Ricerche, ICMATE, Milano, Italy

Preface

Early thoughts on phenotypic plasticity in development and evolution

Introduction

Few ideas are as entrenched in biologists' minds as the one that genotypic change *must* precede phenotypic change during evolution. The idea that phenotypes can change first, and genotypes later, is considered absurd. The contributors to this volume present a different viewpoint: the same genotype can be consistent with more than one phenotype, and nongenetic phenotypic variation can have significant implications for development and evolution. The new phenotype can appear spontaneously or be triggered by an environmental stimulus. It can be very similar to the previous phenotype, and appear to change smoothly with the external environment. Alternatively, it can be very different, in which case the change can be viewed as an all-or-none switch from one phenotype to another. The articles that follow deal with both situations. They discuss formal models and specific cases that encompass molecular biology, cell biology, development, behavior, evolution, and ecology. Those that come toward the end focus on cancer, which can be thought of as an unusual case of multicellular development; not unexpectedly, they have a strong applied flavor.

Most authors use a common conceptual framework: organisms are clonal assemblages of cells; protein properties are determined entirely by the DNA sequences that encode them; differential gene expression mediated by interacting networks of genes and proteins is the key to understanding how phenotypes change during development. In addition, on the whole, Darwinian natural selection is a satisfactory explanation for evolution.[1] Four articles step outside the framework to bring up prion-mediated phenotypes (Tuite), environment-based protein multifunctionality (Prabantu et al.), cell-type specificity based on a physical condensation process involving chromatin and proteins (Newman), and the necessity of thinking about phenotypes as joint productions of the organism and its associated microbiota (Chiu and Gilbert).

[1] In this view the physical properties of living matter play an ancillary role, as intermediaries that help in implementing the roles of genes.

It is of interest to look at the history behind the theoretical concepts on which the contents of this book are based; it turns out that earlier thinkers had come close to many supposedly new notions.[2] They involve phenotypic variations deriving mainly from two sources.[3] One is autonomous to the cell or multicellular unit (in a defined environmental context, of course), and is spontaneous. It lacks direction, meaning that its influence on the phenotype is unpredictable—it is "random." Referred to as phenotypic stochasticity or developmental noise, it implies that even in the same external environment, more than one phenotype is consistent with a given genotype. Another sort of variation can be traced to an altered physical or biotic environment, that is, it is triggered by an external stimulus. The changed phenotype that comes about is predictable, at least to the extent that a defined stimulus gives rise to a characteristic response (though the response may vary within a range). The ability of the phenotype to vary in this manner has been referred to as phenotypic plasticity. Its existence too shows that more than one phenotype is consistent with the same genotype, though in this case there appears to be an identifiable cause behind the altered phenotype, namely, the environmental change.

Both sorts of variation have long been recognized. In the context of multicellular development, it is no surprise that genotypes do not determine phenotypes uniquely: cell differentiation itself is an example. Besides, organisms exhibit different phenotypes at different stages of the life cycle. More recently, it has come to be realized that the altered phenotypes, whether reflective of noise or plasticity, can be transmitted not only through mitotic divisions (regularly), but also meiotic divisions (occasionally). That has sparked a resurgence of interest in the role of noise and phenotypic plasticity.

One reason for the interest is that the explanatory power of generic physical mechanisms for developmental patterning has begun to be appreciated.[4] Another is the attention being paid to

[2]"The only sure guide to right action is to follow the path set by great people" (Yudhishthira, Mahābhārata, Vana Parva). Where interpretation is involved, quotes are used so that readers can check for themselves. Citations to this book have largely been avoided.

[3]Genetic polymorphism that leads to phenotypic variation (not considered here) is a third source. Also see note 52.

[4]Newman, S.A., 1992. Generic physical mechanisms of morphogenesis and pattern formation as determinants in the evolution of multicellular organization. J. Biosci. 17, 193–215.

the interplay between development biology and evolution ("evo-devo"), especially with regard to how the physical and biotic environments together influence evolution ("eco-evo-devo").[5] That has motivated evolutionary thinking that goes beyond the Modern Synthesis, extending to a willingness to take "Lamarckian" viewpoints seriously.[6] Experimental techniques make it possible to quantitate noise in gene expression in single cells, and to simultaneously monitor gene activities in many cells. Together, these factors go some way to explain the attention being paid to epigenetic mechanisms.[7]

The study of variation, in particular the origin of perceived discontinuities in phenotype, has been central to biology. The causes and implications of variation have long engaged embryologists and evolutionary biologists. Discontinuities in space (among the members of a species and between species, or between the cells of an organism), and discontinuities in time (among the members of a lineage, or between differentiated cells, or over the life cycle) have both been of concern. So has been the relative importance of two seemingly opposed explanations for the discontinuities. Were they because of internal causes (epigenesis, spontaneous internal changes), or because of identifiable external causes (the environment, cytoplasmic determinants in the fertilized egg also falling under that head)? How did the leading thinkers of the past address these questions?

[5]Hall, B.K., 2012. Evolutionary developmental biology (evo–devo): past, present, and future, Evol. Educ. Outreach 5 (2), 184–193; Gilbert, S.F., Bosch, T., Ledón-Rettig, C.G., Eco-evo-devo: developmental symbiosis and developmental plasticity as evolutionary agents. Nat. Rev. Genet. 16, 611–622.

[6]Gissis, S.B., Jablonka, E. (Eds.), 2011. Transformations of Lamarckism, From Subtle Fluids to Molecular Biology. The MIT Press, Cambridge, MA, pp. 474.

[7]As Haig (2012) says, "We are in the midst of an epidemic of the words 'epigenetic' and 'epigenetics'." (Haig, D., 2012. Commentary: the epidemiology of epigenetics. Int. J. Epidemiol. 41, 13–16.) The origin of the word "epigenetics" has been much discussed. Historians have traced the idea of epigenesis, the gradual manifestation during development of capacities latent in the embryo, to Aristotle and Harvey [Lopez, Angel, "William Harvey (1578–1657)." Embryo Project Encyclopedia (2010-06-18)]. Also see Gayon, J., 2016. From Mendel to epigenetics: history of genetics. C. R. Biologies 339, 225–230. "Epigenetic" and "epigenetics" are usually attributed to Waddington (1940, 1942), but Wright (1934) was clear that development was an epigenetic process [Wright, S., 1934. Physiological and evolutionary theories of dominance. Am. Natural. 68, 25–53].

Background: Lamarck and Darwin

For Lamarck, the environment made demands of the organism, and the organism responded by making an appropriate structural and physiological response, "somewhat as a fluid fits a vessel."[8] Therefore variations were elicited by the action of the environment on the organism; and the tendency of the organism was to respond adaptively.[9] Two theoretical assumptions were at the core of Lamarck's thought. One was that the plasticity of organic forms is inherently adaptive; environmental change made the plasticity manifest. The other, which we will come to later, was that behavioral change drove morphological change.[10] His chief concern was with how different forms came about during evolution. Changes in external circumstances led to a "motion of fluids" inside tissues, that could modify organs and "*the influence of new circumstances* on animals as they spread throughout habitable areas, were the two general causes."[11]

Darwin took the occurrence of variation for granted, and also the capacity to express a trait to be inherited. As regards the origin of variation, he was open. Variations could be on account of the environment, use and disuse, or simply "fluctuating." Importantly, acquired variations could be inherited.[12]

However, he added significant caveats. First, innate factors were the most significant: "the most frequent cause of variability may be attributed to the male and female reproductive elements having been affected prior to the act of conception."[13] Second, "Lamarckian" causes were minor, and furthermore, differed with regard to how important they were: "...species have

[8]The phrase in quotes is from Bateson, W., 1894. Materials for the Study of Variation: Treated With Especial Regard to Discontinuity in the Origin of Species. Cambridge University Press, p. 4. Also see note 15.

[9]Such a tendency was evident from the way muscle tissue grew after exercise, for example. Niche construction theory inverts the argument, in the sense that the environment "adapts" to the organism.

[10]The inheritance of acquired traits, held up today as the essence of Lamarckism, appears to have been self-evident to Lamarck (as well as many others). It was not something he claimed as an original idea. See Burkhardt, R.W., Jr., 2013. Lamarck, evolution and the inheritance of acquired characters. Genetics 194, 793–805.

[11]Italics in Hodge's translation; Hodge, M.J.S., 1971. Lamarck's science of living bodies. Brit. J. Hist. Sci. 5 (20), 323–352.

[12]Many have drawn attention to Darwin's 'Lamarckian' leanings. See Liu, Y., 2016. Darwinian evolution includes Lamarckian inheritance of acquired characters. Int. J. Epidemiol. 45 (6), 2206–2207. As stated, with regard to this he was not unusual.

[13]Darwin, C.R., 1859. On the Origin of Species by Means of Natural Selection, or the Preservation of Favoured Races in the Struggle for Life, first ed. John Murray, London, Chap. I, p. 8. This formulation is retained until the 4th edition of 1866.

been modified chiefly through the natural selection of numerous successive, slight, favourable variations aided in an *important* manner by the inherited effects of the use and disuse of parts; and in an *unimportant* manner, that is in relation to adaptive structures, whether past or present, by the direct action of external conditions, and by variations which seem to us in our ignorance to arise spontaneously"[14] (italics added). Given the struggle for existence, those variations that improved adaptation were likely to spread in the population. However, "Variations neither useful nor injurious would not be affected by natural selection, and would be left either a fluctuating element, as perhaps we see in certain polymorphic species, or would ultimately become fixed, owing to the nature of the organism and the nature of the conditions," which may be the earliest recognition of undirected neutral phenotypic variation.[15]

For William Bateson, there was a "cardinal difficulty" in the views of Lamarck as well as Darwin. The two had seen evolution as a gradual process, and both viewed environmental differences as the ultimate cause of differences in form (albeit working through very different routes). However, "diverse environments often shade into each other insensibly and form a continuous series, whereas the Specific Forms of life which are subject to them on the whole form a Discontinuous Series."[16]

Woltereck and the reaction norm

Woltereck's work brings us nearer to the modern concept of phenotypic plasticity.[17] He wanted to counter two beliefs that

[14]Darwin, C.R., 1872. The Origin of Species by Means of Natural Selection, or the Preservation of Favoured Races in the Struggle for Life, sixth ed. John Murray, London; with additions and corrections, p. 421.

[15]Darwin 1872, p. 63; and "...as I am inclined to believe, morphological differences, which we consider as important [...] first appeared in many cases as fluctuating variations" (Darwin, 1872, p. 176). Also see Bonner, J.T., 2013. Randomness in Evolution. Princeton University Press.

[16]Bateson, P., 1894. Materials for the Study of Variation: Treated with Especial Regard to Discontinuity in the Origin of Species. Macmillan, London, p. 5. The rediscovery of Mendel's principles 6 years later confirmed Bateson in his view that evolution proceeded via mutations of major effect.

[17]Woltereck, R., 1909. Weitere experimentelle Untersuchungen über Artveränderung, spezielüber das Wesen quantitativer Artunterschiede bei Daphnien. Verhandl. der Deutsch. Zool. Gesell. 19, 110–173. For a clear discussion of the background and implications of the work, see Sarkar, S., 1999. From the Reaktionsnorm to the adaptive norm: the norm of reaction, 1909–1960. Biol. Philos. 14, 235–252.

were gaining ground. One was that the continuous variations that occurred in nature were not heritable (as Johannsen had found in his experiments on pure lines of the bean, Phaseolus); the other was that major changes in form, such as those that evolutionary theory was required to explain, could be the outcome of single mutational events (as de Vries had found in crosses of the evening primrose, Oenothera). To the extent that the beliefs based on the work of Johannsen and de Vries went against the ideas of Lamarck, Woltereck was in agreement. However, as they also went against his belief in Darwinian natural selection, he felt impelled to carry out an independent test of the relevance of environmentally induced variation for evolution. He worked with parthenogenetically propagated—therefore genetically identical—lines of the crustaceans Daphnia and Hyalodaphnia, and used the height of the head (relative to the body length) as the phenotypic trait of interest. He found that in a given environment, heights were distributed narrowly about a mean, but the mean varied significantly as environmental conditions were changed.[18]

He had something significant to say regarding the variation. It enlarged on what Johannsen had pointed out based on his work with pure lines of beans, namely that two sorts of factors accounted for their size differences. There were those common to the group as a whole, deriving from external conditions that were the same for all plants in a set. Then there was a "fluctuating variability" that was specific to each member of the group.[19] Woltereck expanded on Johannsen's conclusion. First, as far as the individual was concerned, it did not matter whether the variations stemmed from a common external cause or distinct individual-specific causes.[20] Second, the individual-specific causes could be internal.[21] Thus the attitudes of Johannsen and

[18]The environmental variables he used were level of nutrition, temperature, and age (number of parthenogenetic generations).

[19]Johannsen, W. Elemente der exakten Erblichkeitslehre, pp. 219–220: Die kollektive Variabilität umfaßt den Unterschied der Individuengruppen; die fluktuierende Variabilität umfaßt die Variation von Einzelindividuen innerhalb einer Gruppe.

[20]"...fur das betroffene Individuum ist es ganz gleich, ob es von einem speziell oder einem generell andersartigen Milieu beeinflußt ist" (Woltereck, 1909, p. 167): both causes had to work through the identical internal makeup.

[21]Wie alle andern, so hat auch dieses Merkmal zweierlei Ursachen für seine spezifische Ausprägung: äußere Milieufaktoren und die innere, erbliche, spezifische Potenz" (Woltereck, p. 171). The ideas expressed in notes 19–21 are reminiscent of the contemporary distinction between intrinsic and extrinsic noise. Without restricting himself to any particular trait, he also says the internal capacity is heritable (erblich). Given that he and Johannsen worked with pure lines, it is difficult to interpret whether he meant that to hold for sexually reproducing organisms too. In the present case, by "dieses Merkmal" (this character) Woltereck happens to be referring to the capacity to produce parthenogenetic female eggs, which—like all other traits ("alle andern")—varies on account of the stated two factors.

Woltereck overlapped partly. Both agreed that instead of a unique well-defined phenotype, the same genotype could give rise to an array of phenotypes. However, for Johannsen, the telling fact was that natural selection could not discriminate between the members of the array.

On the other hand, Woltereck contended that natural selection could indeed act: not on the variations uncovered by Johannsen, but on the vastly enlarged range of variation, induced by the environment, that he had discovered. His argument began from what he called the phenotype curve. For simplicity, he illustrated it as a broken line that showed how the means of phenotype distributions (in his case, of Daphnia head height) changed as a single environmental parameter was varied. He named the entire set of phenotypic curves that could be so generated the "reaction norm."[22] He reasoned that just as each phenotype curve, the reaction norm proved that phenotypes varied continuously, and significantly. However, the problem still remained. Since the reaction norm referred to a pure genotype, how could selection act on it? Woltereck showed how. In the course of doing so, he may have come close to, if he did not actually enunciate, Waddington's later explanation (that he named genetic assimilation).[23]

The experimental design resembled the one Waddington was to use, but there were important differences too.[24] Woltereck began with a Daphnia biotype with a markedly smaller head height than usual, and made use of the plasticity of head height (in response to increased nutrition) to select for height increase. Crucially, he took care to intersperse sexual generations in between parthenogenetic ones. Following several generations of selection, the distribution of heights shifted to one appropriate

[22]The reaction norm would be represented by a hypersurface in multiple-dimensional space. Woltereck mistakenly said the reaction norm was equivalent to the genotype, and was pulled up for it by Johannsen (Johannsen, W., 1911. The genotype conception of heredity. Am. Natural. 45, 129–159.

[23]As we will see, it is not clear whether Woltereck's reasoning was a forerunner of genetic assimilation or the Baldwin effect. To the extent that his experiments dealt with asexual organisms much of the time, the Baldwin effect is the more plausible candidate. He himself uses the term "degree of assimilation." M. Press has noted the similarity to genetic assimilation (<https://github.com/maximilianpress/reaktionsnorm/blob/master/woltereck_reaktionsnorm.pdf>, 2020 (accessed 04.14.20.).

[24]As the account is not sufficiently detailed, it is not possible to say for sure.

to a different biotype,[25] namely, to one with a larger mean head height. His explanation was that the physiological processes (Reaktionsablauf) responsible for the larger height progressively "got accustomed" (gewohnt) to increased nutrition, to the extent that eventually a new reaction norm could be elicited without the stimulus of enhanced nutrition. He described the extent to which the reaction norm shifted during the course of the experiment as "degree of assimilation" (Assimilationsgrade).[26]

Waddington, canalization and genetic assimilation

Because Waddington's thinking forms the central backdrop to many of the articles that follow, it requires only a brief mention here. In 1940 Waddington presented his "somewhat romantic conceit, the epigenetic landscape," in the form of a pictorial representation by the artist John Piper.[27] The landscape was a system of serially branching valleys down which a branching river flowed. It stood for the diverse pathways of differentiation followed by cells, and was meant to illustrate change and stability during embryonic development. Change was represented by variations in steepness, and stability by the depth of the valley. Branching indicated that as development proceeded, the potentialities of a cell were increasingly restricted. Either an altered topography (gene mutations) or an externally caused disturbance (environmental influences) could cause the flow to shift from one valley to another; the likelihood of it happening would depend on the strength of the perturbation. Time-dependent

[25]Which could mean a different genotype. However, if one may use an anachronism, it is more likely to refer to a different canalized phenotype (i.e., comprising individuals of genotypes with the same reaction norm). See the section that follows on Waddington.

[26]The inference drawn here rests on equating the biotypes to wild-type genotypes of the same species with two distinct reaction norms. Woltereck had succeeded in changing one reaction norm to another.

[27]Waddington, C.H., 1940. Organisers and Genes. University Press, Cambridge. A later, and more widely copied, illustration included adjustable ropes (standing for interacting gene products) that pulled on the landscape and modulated its topography. See Waddington, C.H., 1957. The Strategy of the Genes. Abingdon-on-Thames, Routledge.

phenotypic plasticity is implicit in the model of the epigenetic landscape; its mathematical exploration is beginning.[28]

Waddington later brought in the concept of switching along with two new terms, canalization, and genetic assimilation.[29] "Canalization" stood for developmental stability, the buffering of the phenotype with respect to genetic and environmental change.[30] The application of stress, either external (e.g., raised temperature) or internal (e.g., mutation), would remove the organism from the range of environments to which it was adapted in nature. If survival was not compromised, the normally canalized developmental system would get destabilized and exhibit any among a range of abnormal phenotypes. If one of the abnormal phenotypes happened to be advantageous (in Waddington's experiments, because he bred from them), indirect selection for gene combinations that favored the advantageous phenotype would lead to its developing constitutively—that is, without the stress. The necessary genetic variation—on which selection acts later—is already present in the original population, but is cryptic. It is uncovered by the shuffling effect of recombination during meiosis in every

[28]One motivation for trying to construct models is that phenotypic plasticity in time may have been a precursor of phenotypic plasticity in space. The evolutionary origin of division of labor can be visualized as the change from one cell (or multicellular organism) doing different things at different times, to members of a group of cells (or group of organisms) doing different things at the same time. Multirhythmicity, which is a slightly different concept, is the temporal equivalent of multistability. It can be viewed as a synchronic analogue of temporal phenotypes. For an illustration of multirhythmicity see Goldbeter, A., Martiel, J.L., 1985. Birhythmicity in a model for the cyclic AMP signaling system of the slime mold Dictyostelium discoideum. FEBS Lett. 191, 149–153.

[29]Waddington, C.H., 1942. Canalization of Development and the Inheritance of Acquired Characters. Nature 3811, 563–565; Waddington, C.H., 1953. Genetic assimilation of an acquired character. Evolution 7, 118–126. Mather and De Winton had earlier used the terms "switch mechanism" and "switching genes" while discussing a famous polymorphism that had engaged Darwin, namely heterostyly in the flowers of Primula. Their view was that the polymorphism was better viewed as a developmental outcome—two independent solutions to the same problem—rather than a consequence of natural selection for distinct outbreeding mechanisms that evolved side-by-side (Mather, K., De Winton, D., 1941. Adaptation and counter-adaptation of the breeding system in Primula: the nature of breeding systems. Ann. Botany 5 (2), 297–311).

[30]Other terms coined independently to describe the buffering of the phenotype are "stabilizing selection" (Schmalhausen, I.I., 1949. Russian original in 1947. In: Dordick, I. (Trans.), Dobzhansky, T. (Ed.), The Theory of Stabilizing Selection. The Blakiston Company, Philadelphia, pp. xiv + 327) and "genetic homeostasis" (Lerner, I.M., 1954. Genetic Hoemostasis. University of California Press, Berkeley).

generation. Subsequent experiments confirmed the hypothesis.[31] Waddington coined the term "genetic assimilation" for the phenomenon. In effect, over successive generations there would be a gradual lowering of the threshold of stimulus intensity required to elicit a phenotype that fell outside the normal range; the environment would have acted like a switch.

Behavior; the Baldwin effect

Phenotype switching via genetic assimilation is one route for a seemingly Lamarckian route to evolutionary change: an enviromentally-induced phenotype becomes constitutive. In a behavioral context, behavior that has to be acquired by learning in every generation is expressed without the need for learning; it becomes instinctive. However, it is not the only route. There is a long history of thinking on how behavior can drive the evolution of form.[32] The beginnings go back to Lamarck himself.[33] Darwin's book *The Expression of the Emotions* motivated many to attempt evolutionary explanations for behavior, in particular

[31]Waddington, C.H., 1956. Genetic assimilation of the bithorax phenotype. Evolution 10, 1–13; Gibson, G., Hogness, D.S., 1996. Effect of polymorphism in the *Drosophila* regulatory gene *Ultrabithorax* on homeotic stability. Science 271, 200–203. For a simple explanation see Nanjundiah, V., 2003. Phenotypic plasticity and evolution by genetic assimilation. In: Müller, G., Newman, S.A., (Eds.), Origins of Organismal Form. MIT Press, pp. 244–263.

[32]Woltereck had drawn an analogy between the evolution of head height in his experiments and associative learning (see the following section on behavior).

[33]"I could prove it is not the form, either of the body or of its parts, that gives rise to habits and way of life of animals, but it is to contrary the habits, the way of life, and all the other influential circumstances that have with time constituted the form of the body and the parts of animals" (from Burkhardt, 2013; Lamarck, J.-B., 1801. Système des animaux sans vertèbres . . .; précédé du Discours d'ouverture de l'an VIII de la République. Déterville, Paris). In at least one place Lamarck asserted that environment-driven evolutionary change required a large number of generations. His "Second Law" of 1815 ascribes evolutionary gains and losses to "circumstances in which their race has long been exposed" ("Tout ce que la nature a fait acquérir ou perdre aux individus par *l'influence des circonstances où leur race se trouve depuis longtemps exposée,* [. . .] elle le conserve par la génération aux nouveaux individus qui en proviennent" (Lamarck, Philosophie zoologique, 1809, p. 235; italics added). Stéphane Tirard (personal communication), referring to *Histoire naturelle des animaux sans vertèbres, 1815, pp. 181–182),* adds "It is not explicit in the fourth law in 1815, but there is no indication that Lamarck has abandoned his idea." As we will see later, the Baldwin effect too works by the "assimilation" of an environment-induced change in phenotype after many generations, with the route being via genetic change.

instinctive behavior.[34] Among the older workers, the names of D. Spalding, J. M. Baldwin, H. S. Jennings, and C. Lloyd Morgan are prominent.[35] Jennings gives an especially clear description of how environmentally caused change can become innate, in an explanation now known as the Baldwin effect. In one version behavior plays the principal role. For instance, consider a cylindrically shaped ciliate ancestor. It can respond to danger by swimming away, or by hiding in in the detritus at the bottom of the pond. Natural selection will lead to a different outcome depending on whether it exhibits a slight heritable preference for one behavior or the other. Ciliates that swim away will evolve a streamlined form and organs suited for rapid movement, while those that dig into mud will evolve a shape and organs that aid burrowing. The upshot is that starting from a behavioral response to the environment, morphological change evolves.[36] There is a conceptually simpler version of the Baldwin effect that operates in the absence of behavioral change. A new phenotype can develop through the influence of the environment. If the influence persists, the phenotype recurs generation after generation. A mutational change makes its development constitutive and so independent of the changed environment.

In a sense, the Baldwin effect too leads to the "assimilation" of an environment-induced modification of the phenotype. In

[34]Darwin, C., 1872. The Expression of the Emotions in Man and Animals. John Murray, London. In addition to the direct action of natural selection, Darwin leans heavily on correlated changes in muscle tension for the origin of signals that convey emotional states.

[35]They came up with similar explanations. See Spalding, D.A., 1873. Instinct, With original observations on young animals. Macmillan's Magazine 27, 282–293 (reprinted in Brit. J. Animal Behav., 1954); Jennings, H.S., 1905/1965. Behaviour of the Lower Organisms. Oxford and IBH, Calcutta, and Griffiths, P.E., 2004. Instinct in the '50s: the British reception of Konrad Lorenz's theory of instinctive behavior. Biol. Philos. 19, 609–631 (see pp. 13–16 in the online version available at <http://philsci-archive-dev.library.pitt.edu/1676/1/Instinct_in_the_50s.pdf>). Later, by initiating the study of evolutionary epistemology through his theory of how instinctive behavior and imprinting could evolve—in which he was anticipated by Spalding—, Lorenz extended Darwinian thinking in a different direction (Lorenz, K., 1941. Kant's Lehre vom Apriorischen im Lichte gegenwärtiger Biologie. Blätter für Deutsche Philosophie 15, 94–125). As Lorenz puts it, evolution explains how we are born prepared to anticipate features of the external world, just as the hoof of the horse forms in anticipation of the hard floor of the steppe (Unsere vor jeder individuellen Erfahrung festliegenden Anschauungsformen und Kategorien passen aus ganz denselben Gründen auf die Außenwelt, aus denen der Huf des Pferdes schon vor seiner Geburt auf den Steppenboden, die Flosse des Fisches, schon ehe er dem Ei entschlüpft, ins Wasser paßt, p. 99). This has been put pithily as "what is a priori for the individual is a posteriori for the species."

[36]See Jennings (1965), pp. 319–327.

the Baldwin effect, the environment biases the direction of the initial modification. In the case of Waddington's genetic assimilation, it does not; instead the environmental stimulus potentiates an increase in phenotypic variance (or developmental noise). In addition, unlike Waddington's scheme, which can work with standing genetic variation, the Baldwin effect requires genetic change. Therefore, it is liable to require many more generations than genetic assimilation. On the other hand, it can work on asexual organisms, and more than one individual can be influenced simultaneously. Both genetic assimilation and the Baldwin effect are natural selection-based explanations for apparently Lamarckian evolution. The starting point is environmental stress-induced phenotypic variation in the first case, and phenotypic variation termed as plasticity in the second.

Bird song, in particular the fact that there are song dialects, is an example of behavioral phenotypic plasticity within a species-specific range. Song develops from an imperfect beginning as the bird itself develops, and we know more about the ontogeny of song than its evolution. The mature song is based on a combination of innate tendencies and learning, and variations are shaped by both the environment and social cues.[37] It is generally agreed that songs function as interindividual signals that evolved from vocalizations, and that sexual selection by female choice played an important role. In many species, females appear to be more attracted to males with a large song repertoire than a small repertoire, which may point to a role for the handicap principle and signal selection in the evolution of song.[38]

Domestication

Domestication provides an example of phenotypic plasticity leading to evolutionary change with human involvement. It shows how unintentional selection can lead to similar outcomes across diverse species. The observation to be explained is that a common set of characters marks the difference between domes-

[37]Baker, M.C., 2001. Bird song research: the past 100 years. Bird Behav. 14, 3–50.
[38]Zahavi, A., 2007. Sexual selection, signal selection and the handicap principle. Reproduc. Biol. Phyl. Birds, 143–159.

ticated plants and animals from their wild relatives. It was remarked upon by Darwin, who had noticed that domestic animals with widely differing ancestries shared morphological traits that their wild relatives did not possess.[39] Dubbed the domestication syndrome, it constitutes a remarkable example of rapid evolution in large organisms taking place well within a human lifetime.[40] In plants the relevant traits include, among many others, loss of seed dispersal, larger grain size and reduced sensitivity to environmental cues for flowering.[41] In the case of animals, the common morphological and behavioral changes that accompany domestication include drooping ears, piebald pigmentation, and reproduction throughout the year.[42]

Purely by virtue of becoming accustomed to living in the proximity of humans, wild animals get selected for tameness. In addition to mere closeness, cultivation seems to have been required in the case of plants. Still, the circumstances that hold good in the case of animals should hold good here too: with animals, initially breeding under captive conditions is not required and selection is unintentional.[43] Spurway proposed a link between the (then unnamed) domestication syndrome and genetic assimilation.[44] Her idea was that under natural conditions, wild-type phenotypes are more or less uniform. The canalization or homeostasis that they exhibit is because "The environment imposes a pattern of development on the organism, therefore it winnows out of a population those genotypes which can modify their ontogeny to produce that pattern." Human intervention destabilizes the developmental system: "Allopatric cultivation usually alters the habitat, and captivity, by definition, removes the developmental systems from the range of environmental stimuli for which they have been stabilized." If the plant or animal survives, it develops traits that fall far outside the normal range, and if conscious or unconscious selection acts on them, they are propagated. As in Waddington's

[39]For a historical review see Bidau, C.J., 2009. Domestication through the centuries: Darwin's ideas and Dmitry Belyaev's long-term experiment in silver foxes. Gayana 73 (Suplemento), 55–72.

[40]Wilkins, A.S., Wrangham, R.W., Fitch, W.T., 2014. The "domestication syndrome" in mammals: a unified explanation based on neural crest cell behavior and genetics. Genetics 197 (3), 795–808.

[41]Brown, T.A., Jones, M.K., Powell, W., Allaby, R.G., 2009. The complex origins of domesticated crops in the fertile crescent. Trends Ecol. Evol. 24 (2), 103–109.

[42]Larson, G., Fuller, D.Q., 2014. The evolution of animal domestication. Ann. Rev. Ecol. Evol. System. 45 (1), 115–136.

[43]Larson and Fuller (2014).

[44]Spurway, H., 1955. The causes of domestication: an attempt to integrate some ideas of Konrad Lorenz with evolution theory. J. Genet. 53, 325–362.

explanation, genetic assimilation follows from the instability deriving from the environmental change. Belayev, who initiated a long-term experimental study of domestication in foxes, concluded that the changes in phenotype that accompanied domestication across species were the consequence of selection for tameness per se, and that the process was associated with mutations in regulatory genes that affected the development of the neuroendocrine system. He described the process as "destabilizing selection"; the similarity to Spurway's thinking stands out.[45]

One can relate the domestication syndrome to Vavilov's "Law of homologous series in variation." Vavilov's approach to constructing the law was inductive: he wanted to understand why the spectrum of phenotypes seen in diverse groups of plants overlapped to the extent it did. He thought the explanation would come through an understanding of the laws that guide variation. His view, as that of many others before and since, was that the same forms occur again and again because the operation of natural selection is strongly constrained by the limited number of outcomes permitted by the physics and chemistry of development. That was why "...the great majority of varietal characters, not only within the limits of single genera and families but even in distant families, are homologous from a morphological point of view (colour, shape, etc.)."[46] Though the notion is anachronistic, one can imagine Vavilov explaining the domestication syndrome as a form of winnowing by selection, intentional or otherwise, of the variation thrown up by the altered environment of domestication; he could have called it evolutionary canalization.[47] In passing, it is worth noting that

[45]Belyaev too viewed domestication as a form of stress—a destabilizing influence on development, and one that could be expected to affect human evolution as well. Domestication selected for "changes in the regulation of genes—that is, in the timing and the amount of gene expression rather than changes in individual structural genes ... The selection becomes destabilizing when it affects, directly or indirectly, the systems of neuroendocrine control of ontogenesis." Belyaev, D.K., 1979. Destabilizing selection as a factor in domestication. J. Heredity 70 (5), 301–308.

[46]"The regularities in polymorphism of plants, established by a minute examination of variation in different genera and families which we have examined, can be compared to homologous series of organic chemistry e.g. carbohydrogen (CH_4, C_2H_4, C_2H_2, ...)." Vavilov, N.I., 1922. The law of homologous series in variation. J. Genet. 12, 47–89.

[47]Both Vavilov (1922) and Spurway (1954) allude to the winnowing of grain as a likely route for the action of unconscious selection leading to the evolution of similar forms in unrelated plants.

the domestication syndrome may apply to bird song as well.[48] If so, it would be a novel case to be followed up.

Developmental noise

As implied by the expression "developmental noise," it was believed that fluctuations in the phenotype that occurred spontaneously were irrelevant to the organism. Benzer showed that noise-dependent phenotypic switching could be of functional relevance. A clone of *Escherichia coli* raised in the same environment contained cells with a bistable phenotype that was related to their metabolic state; spontaneous fluctuations caused the phenotype to alternate between the two states. It was possibly the earliest demonstration of noise-induced phenotypic switching.[49] Similar findings have been made since then in numerous studies. For instance, differentiation in the hematopoietic lineage works on a stochastic basis.[50] Protein energy landscapes contain multiple minima, and the environment biases the "conformational substate" that a protein adopts.[51] Noise can take other forms, including stress-induced, error-prone translation initiation leading to the synthesis of variant proteins from the same DNA sequence.[52] It has been argued that noise-induced phenotype switching is ubiquitous, and can be adaptive at more than one level of organization.[53]

[48]Güttinger, H.R., 1985. Consequences of domestication on the song structures in the canary. Behaviour 94 (3/4), 254–278.
[49]The approach was elegant: he used bacteriophage growth as an indicator of whether a bacterial cell contained beta galactosidase enzyme activity or not. Benzer, S., 1953. Induced synthesis of enzymes in bacteria analyzed at the cellular level. Biochem. Biophys. Acta. 11, 383–395. Subsequently Novick and Weiner's work made the induced and uninduced states of the lac operon a paradigm for genetic switching (Novick, A., Weiner, M., 1957. Enzyme induction as an all-or-none phenomenon. Proc. Natl. Acad. Sci. U.S.A. 43 (7), 553–566).
[50]Till, J.E., McCulloch, E.A., Siminovitch, L., 1964. A stochastic model of stem cell proliferation, based on the growth of spleen colony forming cells. Proc. Natl. Acad. Sci. U. S. A. 51, 29–36.
[51]Frauenfelder, H., Fenimore, P.W., Young, R.D., 2007. Protein dynamics and function: insights from the energy landscape and solvent slaving. IUBMB Life 59 (8–9), 506–512.
[52]Schwartz, M.H., Waldbauer, J.R., Zhang, L., Pan, T., 2016. Global tRNA misacylation induced by anaerobiosis and antibiotic exposure broadly increases stress resistance in *Escherichia coli*. Nucl. Acid. Res.
[53]Vogt, G. 2015. Stochastic developmental variation, an epigenetic source of phenotypic diversity with far-reaching biological consequences. J. Biosci. 40, 159–204; Samhita, L., Raval, P.K., Agashe, D., 2020. Global mistranslation increases cell survival under stress in *Escherichia coli*. PLoS Genet. 16 (3), e1008654. <https://doi.org/10.1371/journal.pgen.1008654>.

In clonal populations of *Dictyostelium discoideum*, cells can spontaneously diverge into two phenotypes (presumptive spore/stalk) in proportions that vary about a mean. A model that combines cell-autonomous stochastic differentiation with intercellular feedbacks satisfactorily accounts for the mean as well as the spectrum of fluctuations.[54] There is a special feature: the stochastic predisposition of cell states in *D. discoideum* leads to complementary phenotypes that foster cooperative group behavior.[55] Recent work shows a behavioral phenotype, the transition from solitary swimming to cooperative schooling in fish, that is driven by noise.[56] Similarly, the energetic benefit of metabolic division of labor between cells has been considered as providing a selective advantage to phenotype switching.[57] Kauffman suggested that cell-type differentiation, and switching between cell types, could be viewed as transitions between different states of a set of genes, each of which functions like a binary switch.[58] Then there is the proposal that embryonic development can take place through a Darwinian mechanism: stochastic gene expression leads to a range of

[54]Saran, S., Azhar, M., Manogaran, P.S., Pande, G., Nanjundiah, V. 1994. The level of sequestered calcium in vegetative amoebae of Dictyostelium discoideum is a predictor of post-aggregative cell fate. Differentiation 57, 163–169 and Nanjundiah, V., Bhogle, A.S., 1995. The precision of regulation in Dictyostelium discoideum: implications for cell-type proportioning in the absence of spatial pattern. Ind. J. Biochem. Biophys. 32, 404–416. The result has a general bearing: when gene states or phenotypes occur on a probabilistic basis, relative proportions are maintained on average. This means that a single genotype is sufficient for a stable phenotypic polymorphism to be present.

[55]Two other examples of switching-dependent complementary phenotypes are the AC/VU switch in vulval development in *Caenorhabditis elegans* (Wilkinson, H.A., Fitzgerald, K., Greenwald, I., 1994. Reciprocal changes in expression of the receptor lin-12 and its ligand lag-2 prior to commitment in a *C. elegans* cell fate decision. Cell 79 (7), 1187–1198) and the production of FGF2 or FGF2 receptor in human pancreatic cells (Hardikar, A.A., Marcus-Samuels, B., Geras-Raaka, E., Raaka, B.M., Gershengorn, M.C., 2003. Human pancreatic precursor cells secrete FGF2 to stimulate clustering into hormone-expressing islet-like cell aggregates. Proc. Natl. Acad. Sci. U.S.A. 100 (12), 7117–7122).

[56]Jhawar, J., Morris, R.G., Amith-Kumar, U.R. et al. 2020. Noise-induced schooling of fish. Nat. Phys. 16, 488–493. The transition bears comparison with the ordering role played by noise in phase transitions, for instance the onset of full magnetization in a ferromagnet; local domains of oriented magnetization form first and later coalesce.

[57]With modifications, the explanation can be extended to division of labor in general, whether within or between organisms of the same species, or even between species. However, it is not self-evident that energetic considerations will always work. For possible evolutionary explanations, see Borges, R., 2017. Co-niche construction between hosts and symbionts: ideas and evidence. J. Genet. 96 (3), 483–489.

[58]Kauffman, S.A., 1973. Control circuits for determination and transdetermination. Science 181 (4097), 310–318.

phenotypes, and selection results in the "correct" phenotypes getting stabilized via intercellular interactions.[59] These examples are restricted to the outcome of phenotypic switching, not to the underlying dynamics. Besides, switching is a time-dependent process, and there are interesting models of how cell types might change in the course of development—how the river of development flows down Waddington's epigenetic landscape.[60]

Phenotypic noise and phenotypic plasticity

It is customary to restrict the term "phenotypic plasticity" to those cases in which the phenotypes are responses to different environments. However, whether spontaneous or directed, a change in phenotype must be implemented in the same way: through internal processes that are the same, whatever be the cause of the variation. In other words, the developmental or evolutionary effects of noise and plasticity act through similar routes. For this reason, it may be useful to use the same term, plasticity, to describe a situation in which individuals of the same genotype can possess different phenotypes.[61] A potential advantage of doing so is that it may suggest the route to a common explanatory framework. In the case of a physical system, a powerful result known as the fluctuation-dissipation theorem relates the statistical properties of undirected internal fluctuations at equilibrium to the directed response of the system to an external disturbance. A well-known case has to do with diffusion and mobility. The diffusion coefficient of a fluid is a measure of the disorganized random movements made by

[59]Kupiec, J.-J., 1997. A Darwinian theory for the origin of cellular differentiation. Mol. Gen. Genet. 255 (2), 201–208. The process may be mediated via differences in lifetimes of different RNA species (Corre, G., Stockholm, D., Arnaud, O., Kaneko, G., Viñuelas, J., Yamagata, Y., et al., 2014. Stochastic fluctuations and distributed control of gene expression impact cellular memory. PLoS One 9 (12), e115574. <https://doi.org/10.1371/journal.pone.0115574>). Also see note 34.

[60]Kaneko, K., 2016. A scenario for the origin of multicellular organisms. Perspective from multilevel consistency dynamics. In: Niklas, K.J., Newman, S.A. (Eds.), Multicellularity: Origins and Evolution. MIT Press, Cambridge, MA, pp. 201–223. Also see Suzuki et al., and Greulich et al., this book.

[61]Nanjundiah, V., 2003. Phenotypic plasticity and evolution by genetic assimilation. In: Müller, G., Newman, S.A. (Eds.), Origins of Organismal Form. MIT Press, pp. 244–263). One could also take the interactionist viewpoint that that phenotypic plasticity is not just an expression of organismal potential in a given environment, but a (temporary) manifestation of the interaction between the genotypic potential expressed in a phenotype and its niche (Isabella Sarto-Jackson, personal communication).

spontaneously the molecules of the fluid. The mobility is a measure of the organized ordered movement performed by the fluid when it flows smoothly under the application of an external shearing force. The first demonstration of the theorem was Einstein's proof that the diffusion coefficient is proportional to the mobility. It meant that the two phenomena were in fact closely related.[62] A beginning has been made to show that a similar link exists in the case of near-linear chemical reactions.[63] It is worth exploring whether the link can usefully be extended, even if formally, to changes in phenotype that are triggered by internal noise (in some situations) and external perturbation (in other situations).

Cancer

The articles dealing with cancer, which have been clubbed together, are also concerned with nongenetic phenotypic variation and phenotypic switching. However, the discussions deal largely with cancer as a developmental problem. Cancer as an evolutionary phenomenon is touched on indirectly, in the sense that it involves fitness differences between cells within an organism. Most of the cancers discussed are caused by disruptions of epithelial tissue, which is interesting in the context of the general belief that epithelia are the oldest organized tissues in metazoans. The authors share some views about cancer. First, there is an implicit assumption that at some meaningful level "cancer" is a unitary phenomenon with diverse manifestations. Second, cancer is seen within an interactionist framework, where the interactants are DNA sequences and proteins within a cell, or cells and the environment. Third, changed gene expression and aberrant cell behavior, not always due to mutations, are the significant causal factors. It appears that a slow switch in thinking is under way, and the origin of cancer is increasingly being thought of as a problem of a society of cells, a context-dependent aberration of development.[64]

[62]<https://en.wikipedia.org/wiki/Einstein_relation_(kinetic_theory)>.

[63]Yan, C.-C.S., Hsu, C.-P., 2013. The fluctuation-dissipation theorem for stochastic kinetics—implications on genetic regulations. J. Chem. Phys. 139, 224109.

[64]This, in spite of the acknowledgment that cancers present with characteristic genetic signatures; see Bissell, M.J., Hines, W.C., 2011. Why don't we get more cancer? A proposed role of the microenvironment in restraining cancer progression. Nat. Med. 17 (3), 320–329. <https://doi.org/10.1038/nm.2328> and Sonnenschein, C., Soto, A.M., 2016. Carcinogenesis explained within the context of a theory of organisms. Prog. Biophys. Mol. Biol. 122 (1), 70–76.

Summing up

In a larger sense, the articles in this volume, including those whose content is largely made up of experimental findings, convey a common message. A "physics way" of thinking about biological problems, which has a long history, is now coming into its own.[65] Developmental and evolutionary interactions are being studied as parts of a single overlapping scheme. The view is gaining ground, that regulatory mechanisms that evolved early in evolution have been put to multiple uses since then.[66]

Acknowledgments

I am grateful to Morrel Cohen, S. Mahadevan, Isabella Sarto-Jackson, Laasya Samhita, H. Sharat Chandra, and Stéphane Tirard for making useful suggestions concerning the Preface.

Vidyanand Nanjundiah

Centre for Human Genetics, Bangalore, India

[65]The change is chiefly one of emphasis, in the way the all-pervasive "laws of nature" are seen to work out. The new approach is more ahistorical than historical; in the language of physics, it stresses boundary conditions more than than initial conditions. This tendency is not new. For instance, it is apparent in the work of Theodor Eimer, D'Arcy Thompson and others, or in the Darwin-Fisher theory of the sex ratio. However, conventional selectionist thinking, which is built on the (supposedly) overwhelming importance of initial conditions and undirected mutations, sees things differently. It views organisms largely as special-purpose devices that evolved to solve problems that were particular to the ancestors of their own lineage. The distinction being drawn here overlaps partly with the one between functional and evolutionary explanations (See Nanjundiah V. and Morange M. (2015). Aging, sex ratio, and genomic imprinting: functional and evolutionary explanations in biology. Biol. Theory 10: 125-133.). Also see notes 1, 4, 5 and Newman, S. A. (2002), Developmental mechanisms: Putting genes in their place. J. Biosci. 27: 97–104.

[66]Robertson, A., Cohen, M.H., 1972. Control of developing fields. Ann. Rev. Biophys. Bioeng. 1, 409–464.

Summaries of contributions

The chapters have been ordered approximately in the following sequence, though there is a great deal of overlap. A nuts-and-bolts introduction to phenotypic plasticity is followed by papers dealing with philosophical and historical issues, and thereafter, with discussions that deal with the theme at the level of macromolecules, genes, cells, organisms, species associations, and behavior. The last nine chapters deal with phenotypic plasticity and switching in the context of cancer biology.

Sahoo et al. (The fundamentals of phenotypic plasticity) initiate the reader into the themes treated in this volume. Starting with a reference to the importance of plasticity for generating variation, they touch on both concepts and applications. They explain the concepts of phenotype variation, multistability, switching and Waddington's epigenetic landscape, and the treat the issue of variation at the subcellular, cellular, and organismal levels, with particular attention to phenotype switching in cancer. The review ends with a return to the possibility that plasticity may be an adaptation at one level of organization but not another. The educational value of the illustrations is noteworthy.

Merlin (Rethinking the role of chance in the explanation of cell differentiation) deals with the history and philosophy of the concept of chance in cell and developmental biology. She points out that ontology and epistemology have been conflated in the relevant biological literature. One reason is that chance is defined in terms of physical causes but measured in terms of biological consequences. She says it is important to distinguish between chance as "error" and chance as a concept with biological, possibly functional, relevance.

Páldi (Random walk across the epigenetic landscape) argues for a nondeterministic view of multicellular development. The proposal is that reliable differentiation is implemented by molecular reactions whose role is regulatory rather than instructive. Two-way interactions with higher-order organization channelize the reactions, which are noisy to begin with, along pathways that lead to predictable trajectories on the epigenetic landscape. Processes involved in energy production and consumption play a central role in this.

Prabantu et al. (Maneuvering protein functions and functional levels by structural excursions) illustrate how proteins with the same primary (amino acid) sequence can have distinct effects on the phenotype depending on various factors including their three-dimensional structure, interactions with other proteins, and the cellular environment. Viral infection, inflammation, and melanoma are among the conditions in which this can be important.

Tuite (Prion-mediated phenotypic diversity in fungi) brings in an entirely new element, namely heredity and plasticity based on the propagation of protein structure, albeit in the specific context of yeast. Switching can be reduced to the transition between alternative conformational states of a prion. Besides prions, nontransmissible protein aggregates can carry a form of epigenetic memory, and transient overexpression of a set of proteins with large intrinsically disordered domains can lead to phenotypes that are stably inherited.

Padmanabhan et al. (Bistability in virus–host interaction networks underlies the success of hepatitis C treatments) touch on a theme that has been made very topical on account of the Covid-19 epidemic. Models of the interaction dynamics between the human host and hepatitis C virus can yield a bistable outcome with two phenotypes that differ in responsiveness to interferon therapy. They show how the models can be extended to take into account combined treatment with interferon and direct-acting antivirals, and to choose the optimal treatment regime for an individual.

Xu and Wang (Quantifying Waddington landscapes, paths, and kinetics of cell fate decision making of differentiation/development) enlarge the standard representation of Waddington's epigenetic landscape by the inclusion of substrate-dependent chemical reaction fluxes. The landscape and fluxes are both dependent on an underlying genetic network. In such a picture, flows are not necessarily directed toward potential minima in the landscape, because kinetic considerations come into play. They show how to quantify cell fate transitions with the help of a novel nonequilibrium transition state theory.

Greulich et al. (The physics of cell fate) explain that a statistical mechanics-based, "coarse-grained" consideration can cut through the forbidding complexity of the networks that underlie gene regulation and expression. The methods can be used to understand, in probabilistic terms, the relationships between molecular expression states, functional cell types, and phenotypic switching within the broad framework of Waddington's "epigenetic landscape."

Vogt (Disentangling environmentally induced variation and stochastic developmental variation, the two epigenetic components of phenotypic variation) places phenotypic plasticity (environmentally induced) and developmental noise (stochastic) on a par, in the sense that both depend on epigenetic mechanisms and contribute significantly to phenotypic variation. In addition, they are adaptive, by optimizing the phenotype in a particular environment (plasticity) or allowing for evolutionary bet hedging (noise). The marbled crayfish, which propagates itself clonally, is a good model for exploring the consequences for development and evolution of these two modes of variation.

Newman (The evolution of cell differentiation in animals: biomolecular condensates as amplification hubs of inherent cell functions) points out that existing models of gene regulation networks are ill-suited to explain differentiation hierarchies in multicellular development and the stability of cell types across evolutionary lineages. The qualitatively new modes of gene regulation dynamics seen in chromatin-based metazoan genomes require a better theory. He discusses the hypothesis that cell type specificity, a metazoan characteristic, is regulated by "function-amplifying centers" (FACs), which are localized physical condensates of enhancer regions, transcription factors, and target genes that appear to have been present before metazoans evolved.

Stajic and Bank (Phenotypic switching and its evolutionary consequences) discuss the history of views regarding the role of phenotypic switching in adaptive evolution. An environment-driven phenotypic switch can work as a bet-hedging strategy. If switching is successful, the proportion of individuals capable of dealing with the challenge posed by the new environment is higher than it would be following a genetic mutation. That gives rise to a correspondingly higher probability—at the population level—of a beneficial mutation arising subsequently. The article discusses genetical assimilation, the theoretical framework originally proposed by Waddington, that is often invoked as an evolutionary route to phenotype change.

Braun (Cell-state organization by exploratory sloppy dynamics) highlights a conflict that cells are compelled to resolve. Their internal biochemistry must be sufficiently robust to lead to a reliable outcome under varied conditions, and sufficiently flexible for them to respond to environmental vicissitudes. He discusses experiments that monitor what happens to gene expression, biochemistry, and morphology when yeast cells are forced to respond to a novel circumstance, perhaps never before encountered in the cell's ancestry. The conclusion is that the internal dynamics of a cell is not always based on prestructured interaction networks. By being "sloppy," it enables the cell to mount condition-dependent responses to the environment.

Krishna and Laxman (Emergence of metabolic heterogeneity in cell populations: lessons from budding yeast) discuss experimental and theoretical studies on clonal yeast populations that are propelled by a metabolic constraint to self-organize and display temporal and/or spatial order. The same cell does different things at different times, or different cells do different things at the same time, in a reversible manner. The metabolite in question is glucose; the differences reflect the modulation and partitioning of the relevant energy-creating and consuming biosynthetic pathways.

Regan (Stochastic phenotypic switching in endothelial cell heterogeneity) points out that phenotypic plasticity and noise-driven switching are examples of what can happen in response to changes in the cell's internal environment, whether autonomously or in response to changes in the external environment. Does the overlap in pathways indicate a functional overlap too? The example of dynamic mosaicism in von Willebrand Factor expression in the mouse suggests as much. Mosaic expression, which is a consequence of random fluctuations in cells' internal states, is modulated by intercellular interactions to give rise to a stable spatial pattern.

Suzuki et al. (Regulation of phenotypic plasticity from the perspective of evolutionary developmental biology) build on the concepts of cryptic genetic variation, polyphenism, reaction norm, and genetic accommodation to discuss ways in which developmental plasticity can lead to the evolution of robust adaptive phenotypes. They point out that genetic accommodation permits new developmental trajectories to arise via changes in environment, with genetic change occurring only subsequently, and construct an epigenetic watershed model in which, besides a Waddingtonian landscape, phenotypic stability is represented.

Levis and Pfennig (Phenotypic plasticity and the origins of novelty) show how a comparative study of reaction norms can enable one to infer that a canalized trait has evolved from an ancestral plastic trait. One compares reaction norms in the appropriate environments between a lineage that displays the trait of interest and a related lineage with an ancestral version of the trait. If individuals that normally display the ancestral trait utilize cryptic genetic variation to develop the derived trait, the hypothesis gains support. They describe how the test is carried out with the example of a feeding polyphenism in the spadefoot toad.

Chiu and Gilbert (Niche construction and the transition to herbivory: phenotype switching and the organization of new nutritional modes) point out that at one time, heredity was not viewed as being congruent with genetics. Given the ubiquity of metazoan-microbe symbiotic associations, with microbes mediating adaptive developmental plasticity, they suggest that it may be useful to reexamine that viewpoint. The host and its microbiota collaborate symbiotically to construct a niche for both. In a functional sense, which harks back to the old view, the genes of the combine (the holobiont) are replicators, not those of either component by itself.

Soha (Nature, nurture, and noise in bird song ontogeny as determinants of phenotypic and functional variation among dialects) introduces bird song as a behavioral phenotype with rich biological and cultural underpinnings that await deeper investigation. Songs are often canalized and so believed to be "species specific"; but they also exhibit plasticity and developmental stochasticity. A stochastic component ought to result in songbirds singing differently in the absence of meaningful genetic or environmental variation. The efficacy of a song depends on the phenotype of the listener, which implies that on top of stochastic effects, the developmental histories of singer and listener should lead to functional differences between song dialects.

Trut and Kharlamova (Domestication as a process generating phenotypic diversity) consider an astonishing example of phenotypic plasticity in action. An ongoing experiment shows that phenotypic changes accumulate rapidly when foxes are selected for tameness. Morphological and behavioral transformations, which parallel those that accompanied the evolution of a wolf-like ancestor to the domestic dog, can take place over a small number of generations. The outcome resembles what has happened in the case of other domestic animals as well. Abnormal timing of ontogenetic processes appears to play a key role in mediating the consequences of domestication. The hypothesis is that this results in the development of an ensemble of correlated traits, jointly constituting a domestication syndrome.

Langthasa et al. (The glycobiology of ovarian cancer progression: phenotypic switches and microenvironmental influences) direct attention on the glycocalyx, a sugar-rich cellular envelope that is an intermediary between cells and their environment. They discuss glycobiological changes observed during ovarian cancer progression and point out that as cells move through different microenvironments in the course of metastasis, they acquire different phenotypes. Morphology, mechanical forces, and glycan-lectin expression are involved in the process, which helps cancer cells to survive.

Jia et al. (Epithelial-mesenchymal transition in cancer) draw a link between the epithelial-mesenchyme transition and the acquisition of stem cell-like properties of cells via metabolic reprogramming. Unexpectedly, cancer cells can display metastable intermediate phenotypes, which combine epithelial and mesenchymal aspects. Since such cells can transit in either direction, being poised on the edge can favor long-term growth if the future environment is uncertain. More study of how such hybrid phenotypes arise can offer clues for better therapy.

Kulkarni et al. (Phenotypic switching and prostate diseases: a model proposing a causal link between benign prostatic hyperplasia and prostate cancer) point out that benign prostatic hyperplasia and prostate cancer, thought to affect old and young males, respectively, may coexist in the same person. That is because the two diseases are linked via the phenomenon of phenotypic switching. They present a model of how the link works and draw attention to its potential impact on treatment.

Somarelli et al. (Phenotypic plasticity and lineage switching in prostate cancer) discuss a different aspect of prostate cancers: they can exhibit very different phenotypes at the molecular level. The apparent phenotypic plasticity tends to show up during the emergence of resistance to drug treatment, and may be an outcome of genetic or epigenetic variation followed by selection. Observations indicate that from the cancer's point of view, plasticity is a route to favor survival, enable spreading to new locations, and counter the immune response.

Johnson et al. (Implications of nongenetic heterogeneity in cancer drug resistance and malignant progression) discuss methods of monitoring and quantifying cell states that are associated with nongenetic heterogeneity and phenotype switching in cancers. In particular, they pick epithelial-mesenchyme transitions, which are common to normal embryonic development and tissue repair, both well studied, and cancer formation. The evidence shows that exposure to cytotoxic or targeted therapeutic treatments can induce cells to activate resistance pathways in diverse ways. A detailed understanding of how phenotypic plasticity is manifested will be useful for improving treatment therapies.

LaPorta and Zapperia (Phenotypic plasticity: the emergence of cancer stem cells and collective cell migration) discuss the relevance of phenotypic plasticity in various situations: metastasis, the switching back of cancer cells into a cancer stem cell state, the induction of other tumor cells to become cancer stem cells if the stem cell population is depleted, the switching of senescent cells, and the transition between epithelial and mesenchymal types. Computational analysis, along with transcriptome data, shows a range of intermediate hybrid states that are partly epithelial and partly mesenchymal. Studies of these changing cell state compositions offer the possibility of exploiting phenotypic plasticity for developing new therapeutic strategies.

Sundararaman et al. (Adaptive phenotypic switching in breast cancer in response to matrix deprivation) explore the set of properties utilized by cancer cells to progress through the stages of metastasis. Matrix deprivation and consequent cytoskeletal rearrangement leads to cell death in normal cells but not a subset of cancer cells, which go on to metastasize under anchorage-deprived conditions. Epithelial-Mesenchyme hybrid cells are more likely to metastasize and survive in a matrix-deprived condition. Energy metabolism plays a role in this, and promotion of glucose metabolism induces an epithelial-mesenchyme transition via intercellular signaling; the metabolic phenotype of cells impinges on other forms of phenotypic plasticity. What leads to differences in the propensity for transition, is not fully understood.

Capp (Phenotypic instability induced by tissue disruption at the origin of cancer) discusses what we have learnt about phenotypic plasticity from single-cell studies of cancer cells. Nongenetic phenotypic switching accelerates tumor growth and leads to drug resistance. Tissue disruption may be the cause of phenotypic instability, because stem cells whose phenotype is unstable to begin with, acquire phenotypic stability concomitantly with the establishment of cell–cell interactions. It may be therapeutically beneficial to partially restore intercellular signaling networks by allowing cells to interact with molecules that would normally be contributed by partners in a healthy tissue. In support of the approach, microenvironmental defects can lead to cancer, and healthy tissue can "normalize" cancer cells.

Enriquez-Navas and Gatenby (Evolutionary strategies to overcome cancer cell resistance to treatment) suggest using the fact that cancer cells seem to evolve by natural selection—successfully, at least in the short run—to devise evolutionary approaches to counter their effectiveness. Several genetic signatures of cancers have been identified, as have genes that, when mutated, provide fitness advantages to cancer cells in the intratumoral environment. That has led to targeted approaches to treat cancer. Still, on the whole cancers have proven resistant to treatment. What is needed is a multidisciplinary approach to cancer treatment by oncologists, experimentalists, evolutionary biologists, and mathematicians, working together to build on evolutionary principles.

Acknowledgments

This book is the result of numerous discussions between ourselves, and between us and our colleagues and acquaintances. Many of these discussions took place rather informally, but some arose from formal meetings at which we presented our ideas. Although the potential list of individuals that must be thanked is too long to cite here, we do wish to thank them, especially our respective laboratory members who contributed so much in so many ways to the genesis of this book.

We are indebted to all the authors who so graciously contributed the outstanding chapters despite their hectic schedules, engagements, and commitments. Likewise, we wish to acknowledge the feedback provided by the three reviewers of the book proposal; their constructive criticism provided new ideas and directions that we did not originally conceive.

We would also like to thank members of our families for their unstinted support, encouragement, and love that made this endeavor worth the effort. We dedicate this book to our families as a token of our gratitude.

Finally, we are to acknowledge Mr. Peter Linsley, Senior Acquisitions Editor, Biochemistry, Genetics and Genomics, Elsevier Publishers for his unstinted support and encouragement; Ms. Samantha Allard for her dedication; and Mr. Stalin Viswanathan and his colleagues for their meticulous attention to detail. It was a joy to work with them; they were not only friendly, professional, and extremely helpful but also always went the extra mile to help. They were just superb. Last but not the least, we express our gratitude to Prof. Jin Wang, Prof. Günter Vogt, Prof. Guanghua Huang, and Prof. Rogier van Vugt for graciously agreeing to share the figures used in the cover art.

Editors

1

The fundamentals of phenotypic plasticity

Sarthak Sahoo, Ayalur Raghu Subbalakshmi and Mohit Kumar Jolly

Centre for BioSystems Science and Engineering, Indian Institute of Science, Bangalore, India

Introduction

Plasticity is a double-edged sword; the more flexible an organism is the greater the variety of maladaptive, as well as adaptive, behaviours it can develop; the more teachable it is the more fully it can profit from the experiences of its ancestors and associates and the more it risks being exploited by its ancestors and associates.

Donald Symons.

The ability of a living organism that carries a unique genome but gives rise to multiple distinct phenotypes at various biological levels of organization remains one of the fundamental questions that has puzzled humans for ages. The diversity and stability of such phenotypes clearly indicate that the mapping of genotype to phenotype is clearly not one-to-one (Fusco and Minelli, 2010). While genotypes are discrete representations of biological information, phenotypes can be observed at a continuous scale, i.e. at varying levels of biological form, structure, or function; for instance, the function of a protein, a gene regulatory network and a cell state are all phenotypes. Thus, the genotype-phenotype maps are of fundamental interest across scales of biological organization (Ahnert, 2017).

One genotype is capable of giving rise to more than one phenotype, particularly in response to different environmental conditions; this property is called as phenotypic plasticity (Fusco and Minelli, 2010). Phenotypic plasticity is universal across most, if not all, living organisms, because the environment around them is a dynamic, multifactorial, and complex variable. Phenotypic plasticity may enable the acquisition of different phenotypes of optimal

Phenotypic Switching. DOI: https://doi.org/10.1016/B978-0-12-817996-3.00001-3

fitness for a given environment. Consequently, phenotypic plasticity has been generally proposed to facilitate evolution, particularly for fluctuating environments (Espinosa-Soto et al., 2011; Fierst, 2011; Forsman, 2014; Xue and Leibler, 2018). Earlier thought of as a predominantly morphological phenomenon, phenotypic plasticity is now considered to encompass variations in biochemistry, physiology, behavior, or the life history of an organism in response to an external signal (Whitman and Agrawal, 2009). The idea of phenotypic plasticity is also closely tied to organismal development, where the conditions for emergence of specific distinct phenotypes are maintained in a robust spatio-temporal manner. Thus, phenotypic plasticity encompasses the idea of both the ability of an individual organism to alter its phenotypic state (say metabolism) in a different environment, and the ability to modify its developmental trajectories in response to specific environmental conditions (Fusco and Minelli, 2010).

The emergence of phenotypic plasticity is often conveyed visually and conceptually via a Waddington's landscape (Fig. 1.1), or some more recent abstractions of the same (Feinberg et al., 2016; Pisco et al., 2016). Similar landscapes are being constructed to encapsulate the dynamics of adaptive cellular reprogramming including cell differentiation and transdifferentiation as well as pattern formation, including efforts to quantify such landscape based on the dynamics of underlying regulatory networks (Enver et al., 2009; Jolly et al., 2015). From a cellular decision-making perspective, the concept of "landscape," viewed from the lens of dynamical systems theory, posits each cell phenotype as an "attractor" of a high-dimensional landscape. Using the Waddington's landscape, it can be easily visualized how a cell can roll down a network of hills and valleys signifying the various differentiation stages that it passes through to finally settle in a low-lying valley (a stable differentiated cell state) (Fig. 1.1A). This conceptual framework also aids to visualize the experimentally observed "transdifferentiation" of cell types by overexpression of specific transcription factors where the cell can jump over a hill, given a sufficiently large perturbation (Jia et al., 2017). The origin of such perturbations could be either intrinsic or extrinsic noise in gene expression levels in the system or externally induced overexpression of transcription factors, such as cellular reprogramming to induced pluripotent stem cells (iPSCs) (Takahashi, 2012). Such perturbations may alter the gene expression profile of cells, pushing them to a different phenotype, owing to underlying interconnected gene regulatory networks (Fig. 1.1B and C). Thus, phenotypic plasticity seems to play a fundamental role in establishing cellular identity, driving

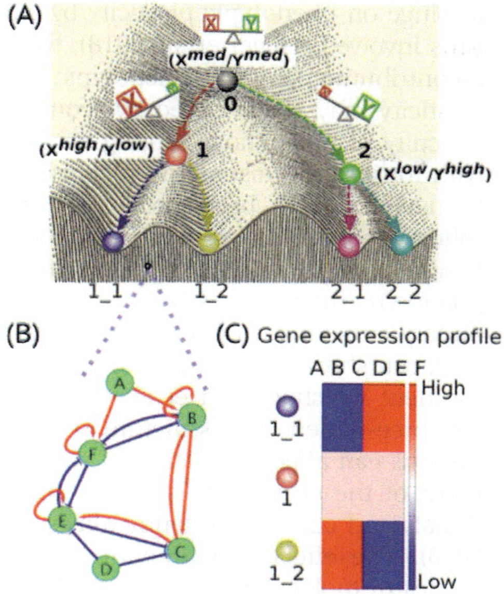

Figure 1.1 Waddington landscape as a metaphor for phenotypic plasticity and cellular differentiation and reprogramming. (A) A schematic for Waddington landscape, where each valley (1_1, 1_2, 2_1, 2_2) represents an "attractor" or cellular phenotype. Balls of different colors represent the possible different positions of cells as they roll down from an undifferentiated ("0") state to a terminally differentiated one through intermediate progenitor states ("1," "2"). (B) A schematic of a gene regulatory network, which may enable the existence of multiple steady states (phenotypes); red lines denote inhibition, blue ones denote activation. (C) Heatmap showing different possible gene expression profiles, each corresponding to a particular phenotype shown aside. *Source*: Courtesy Jia, D., Jolly, M.K., Kulkarni, P., Levine, H., 2017. Phenotypic plasticity and cell fate decisions in cancer: insights from dynamical systems theory. Cancers (Basel). 9, E70.

cellular responses, and in the evolution of traits over a larger timescale of generations that may have adapted to survive in a compendium of different environments. A common theme often connected with phenotypic plasticity is that of multistability, that is, existence of more than one stable states (phenotypes) which may interconvert stochastically (Guantes and Poyatos, 2008).

Phenotypic plasticity at an intracellular level: macromolecules, pathways, and organelles

Phenotypic plasticity can be manifested at levels from individual macromolecules to cellular. For instance, changes in protein configuration of G-protein coupled receptors in response to varying environmental conditions such as the binding of a given ligand may impinge on phenotypic plasticity by altering the signaling mechanisms involved (Hilger et al., 2018); thus, structural plasticity can be a contributing factor to phenotypic plasticity. Similarly, phenotypic plasticity may be furthered by "conformational noise" due to promiscuity of interactions among proteins, particularly among those with intrinsically disordered regions (Jolly et al., 2018). Conceptual frameworks have been proposed for mapping protein structure (phenotype) to amino acid sequence (genotype) (Ahnert, 2017). Similarly, some of the earliest theoretical works on genotype-phenotype maps treat RNA sequence as a genotype and RNA secondary structures as phenotypes. Thus, besides proteins, DNA and RNA can show different phenotypes depending on their local environment (Nechooshtan et al., 2009; Whelan et al., 2014). Further, the three-dimensional structure of the chromatin present inside the nucleus can affect regulation of various genes; this spatial arrangement of the chromatin is under the influence of diverse external signals and can consequently affect cellular phenotype (Virk et al., 2020). The effect of chromatin organization is often relayed via modes of molecular epigenetics: DNA methylation, histone modification, and nucleosome positioning (Pisco et al., 2016; Portela and Esteller, 2010). These changes facilitate diverse cellular responses, despite identical genomic content; for instance, a bivalent chromatin structure is reported for key genes in embryonic stem cells (ESCs) (Bernstein et al., 2006). Such bivalent or "poised" chromatin—the presence of both activating and repressive methylation marks—is thought of making cells more plastic and facilitate cellular reprogramming in many contexts. For instance, a poised chromatin at the promoter of ZEB1—a key regulator of phenotypic plasticity in cancer—can enhance cancer cell plasticity and tumorigenesis (Chaffer et al., 2013).

Plasticity has also been observed in the context of how various macromolecules interact within the cells to mediate different biological outcomes. Interaction networks among proteins, transcription factors, and metabolites are hubs of phenotypic plasticity as well as its evolution, and may offer mechanistic insights into environmental adaptation and mutational robustness (Harrison et al., 2007; Nijhout et al., 2017). The topology and/or parametric regime of interactions corresponding to these regulatory networks may be cell type specific; for instance, the regulation of ZEB1 may be different in cell types of epithelial versus mesenchymal origin (Martin et al., 2020; Somarelli et al., 2016). Similarly, the mode of respiration and metabolism a cell opts for under various cellular conditions can be traced back to the phenotypic plasticity enabled by the underlying networks. Perturbations to these gene regulatory networks can lead to various pathological states, but at the same time can also enable variability, adaptability, and consequent evolution of the biological system.

Phenotypic plasticity is also exhibited at the level of cellular organelles. Plants are prominent examples that show diverse adaptation mechanisms at an organelle level. Plastid specialization and multifunctional organellar protein features are central to enhance phenotypic plasticity in some plant species (MacKenzie and Kundariya, 2020). Plasticity in plants can be exhibited at physiological, anatomical, and morphological levels, and may allow the plants to reproduce and grow in environments varying in space and time (Gratani, 2014).

Phenotypic plasticity at a cellular level: Implications in development, homeostasis, and disease

Phenotypic plasticity at cellular level has been well studied in many biological systems. One of the simplest organisms, the bacterial cell, can exhibit varied phenotypes depending on its environment; for instance, bacterial cells can exhibit phenotypic plasticity in response to antibiotic treatments (Day, 2016). Such phenotypic plasticity may drive phenotypic (or nongenetic) heterogeneity in a genetically homogeneous population. It has been associated with multistability in bacterial cells, enabling them to switch their phenotype upon drug exposure (Ghosh et al., 2020). Such plasticity and heterogeneity has been proposed as a "bet-hedging" strategy opted by bacterial cells to better cope with unanticipated stresses that they may face (Veening et al., 2008; Xue and Leibler, 2018). This concept is central to antibiotic tolerance in various diseases that plague our world today.

The eukaryotic world is also abundant in instances where cells exhibiting phenotypic plasticity, starting from yeast to humans. Yeast has been shown to exhibit phenotypic plasticity related to its metabolism (Harrison et al., 2007). Such phenotypic plasticity can be genetically regulated over evolutionary timescales, thus setting up feedback loops between genotype and phenotype (Yadav et al., 2016). In the context of mammals, a well-characterized instance of phenotypic plasticity has been seen in the context of development of various cell/tissue types. The Waddington's landscape was strongly motivated by the puzzling question that how a single genome was able to produce multiple cell phenotypes that can maintain themselves over generations (Pisco et al., 2016). It has been believed that differentiation of most ESCs to stem cells and progenitor cells of different tissue lineages is unidirectional, that is, the ball (the state of a cell differentiating from being an ESC) in Waddington's landscape can only roll down the hill into valleys (representing terminal cell fates) at one or the other (developmental) stage following the "arrow of time" (Jia et al., 2017); for instance, a pluripotent cell giving rise to endoderm, mesoderm, or ectoderm (Fig. 1.2A) or generating a more terminally differentiated phenotype such as a mature neuron through various intermediates (Fig. 1.2B). However, recent investigations have questioned this irreversibility, and shown instances when cells can be dedifferentiated and/or transdifferentiated to adopt another phenotype. Such reprogramming implies some restructuring of the underlying phenotypic landscape, where a given phenotype is "destabilized" by lowering the "energy barrier" separating the two cell states (Fig. 1.2C). For instance, the overexpression of four Yamanaka factors OCT4, SOX2, KLF4, and NANOG was sufficient to push a subpopulation of somatic cells to be embryonic stem-like cells. These reprogrammed cells displayed incomplete pluripotency, subsequently underwent defective differentiation, and were labelled as iPSCs. These observations suggest that a cell phenotype is relatively robust, yet maintains the flexibility to be reprogrammed to varying degrees of efficiency and accuracy. The molecular mechanisms underlying iPSC generation, and the identification of factors that can help identify *a priori* which

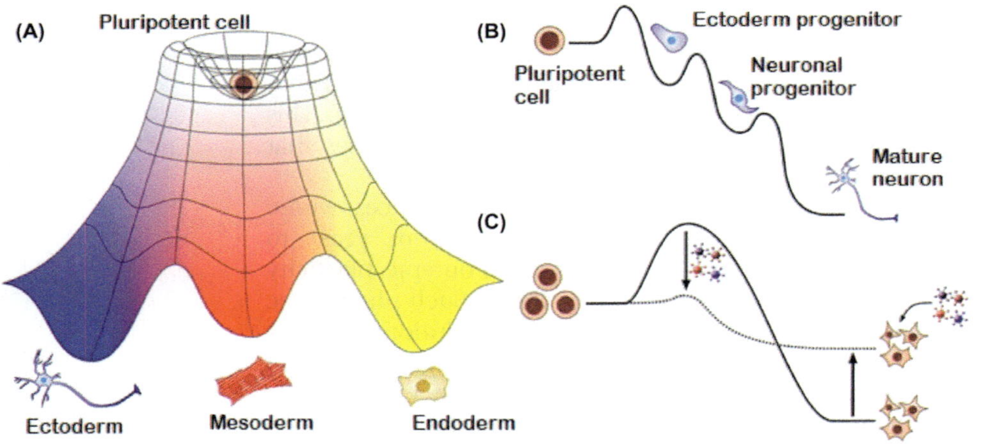

Figure 1.2 Epigenetic landscape and its implications in cellular differentiation. (A) A landscape inspired by Waddington's schematic. A pluripotent cell, shown at the top, can "roll" down into one of the three valleys, each representing a terminally differentiated phenotype. (B) A closer look at the putative trajectory of a pluripotent cell, which differentiates into a neuron through various intermediate progenitor states. (C) The process of cellular reprograming involves a change in the underlying epigenetic landscape (compare solid and dotted curves). *Source*: Courtesy Wikimedia Commons.

subpopulation of cells will undergo such reprogramming under the influence of Yamanaka factors yet remain elusive (Takahashi and Yamanaka, 2016).

Besides these evident examples of dedifferentiation, recent studies have shown how cells from one specific lineage can transition into a cell of a different lineage without necessarily converting to a stem-cell state *en route*. Such a "transdifferentiation" process is mediated by loss of morphological and functional traits of the former phenotype, and gaining that of the other (Wagers and Weissman, 2004). Many open questions remain in the field from a perspective of phenotypic plasticity: (1) what is the minimum amount of perturbation needed to push the cells to dedifferentiate? (2) How long do they stay in that dedifferentiated state before reversing? (3) What mechanisms mediate the mean residence times of cells in a dedifferentiated state?

One of the major events during embryonic development is the process of development of the three germ layers from the progenitor cells based on the myriad of signals these cells receive. Interestingly a general feature that has emerged which seems critical in governing the cell fate decisions at the critical branch points during the course of development are small core gene circuits—feedback loops with mutual inhibition and self-activations (Zhou and Huang, 2011). Such a circuit can enable bistability in the system, hence specifying two robust but interconvertible cell states. For instance, a bistable switch is known to exist between the proteins Ptf1a and Nkx6 during the development of pancreatic cells. The common pancreatic progenitor that gives rise to either the endocrine or the exocrine cells in the pancreas has a comparable level of both these proteins. During the differentiation of such cells, depending on the relative levels of the two proteins, the progenitors either give rise to exocrine cells (Ptf1a >> Nkx6) or to endocrine cells (Ptf1a << Nkx6) (Zhou and Huang, 2011). The decision of which progenitor cell adopts which cell state depends on various internal and external signals the cell receives in the given environment, including cell-cell communication through various biochemical and biomechanical signals (de Back et al., 2013). Various other cell-fate master regulators have been reported to engage in such feedback loops and drive distinct cellular phenotypes (Guantes and Poyatos, 2008).

Another key "transdifferentiation" processes that occur during embryonic development is epithelial to mesenchymal transition (EMT) and its reverse mesenchymal to epithelial transition (MET). EMT and MET are canonical example of phenotypic plasticity that can exist in cells where although the genetic composition of the cell remains the same, the phenotypes observed can be completely different at molecular and morphological levels, and can switch among one another (Nieto et al., 2016). EMT occurs during gastrulation phase where epithelial cells gain migratory properties; various consequent rounds of EMT and MET are implicated in formation of various organs (Thiery et al., 2009) and in cellular reprogramming (Pei et al., 2019). Mathematical modeling has revealed that many mutually inhibitory loops involved in EMT/MET can be potentially multistable, thus driving this phenotypic plasticity (Font-Clos et al., 2018; Lu et al., 2013; Mooney et al., 2017; Roca et al., 2013; Tian et al., 2013). Dynamic switching among epithelial, mesenchymal, and hybrid epithelial/mesenchymal phenotypes (Tripathi et al., 2020) has been found to play a crucial role in wound healing and tissue homeostasis as well (Kalluri and Weinberg, 2010). EMT and MET are also involved in fuelling cancer metastasis; EMT typically is involved in dissemination of cancer cells in circulation, while MET is believed to facilitate colonization of these cancer cells in distant organs (Jolly et al., 2017). EMT usually involves reduction in levels and/or membrane localization of cell-cell adhesion molecules such as E-cadherin, and a concomitant increase of intermediate filaments such as Vimentin (Nieto et al., 2016). An interconnected axis of cellular plasticity in the context of cancer cell invasion is mesenchymal to amoeboid transition (MAT) and amoeboid to mesenchymal transition (AMT)—mesenchymal cells can secrete matrix metalloproteinases to remodel the extracellular matrix (ECM) and act as "path generators," while amoeboid cells can squeeze through the gaps in ECM and act as "path finders." In a three-dimensional microenvironment, many cancer cells exhibit AMT and/or MAT in response to various signals in their local microenvironment (Pankova et al., 2010) (Fig. 1.3). Unlike EMT/MET where the intracellular feedback loops identified so far operate largely at transcriptional and posttranscriptional levels (Jolly and Levine, 2017), AMT/MAT is controlled by a similar feedback loop formed between two GTPases Rac1 and RhoA that control actin polymerization and actomyosin contraction (Huang et al., 2014). Such feedback loops may enable more than two phenotypes also (Zhou and Huang, 2011); consistently, hybrid cell phenotypes have been seen during both EMT/MET and MAT/AMT (Lecharpentier et al., 2011; Yoshida and Soldati, 2006). Cells in a partial EMT or hybrid E/M phenotype typically exhibit collective cell migration phenotype (Campbell and Casanova, 2016); these cells may directly switch to an amoeboid transition through a process known as collective to amoeboid transition (CAT) (Huang et al., 2015); thus, phenotypic plasticity is fundamental to cancer metastasis (Fig. 1.3).

Phenotypic plasticity can also be seen in macrophages—they can transdifferentiate into endothelial cells and endothelial

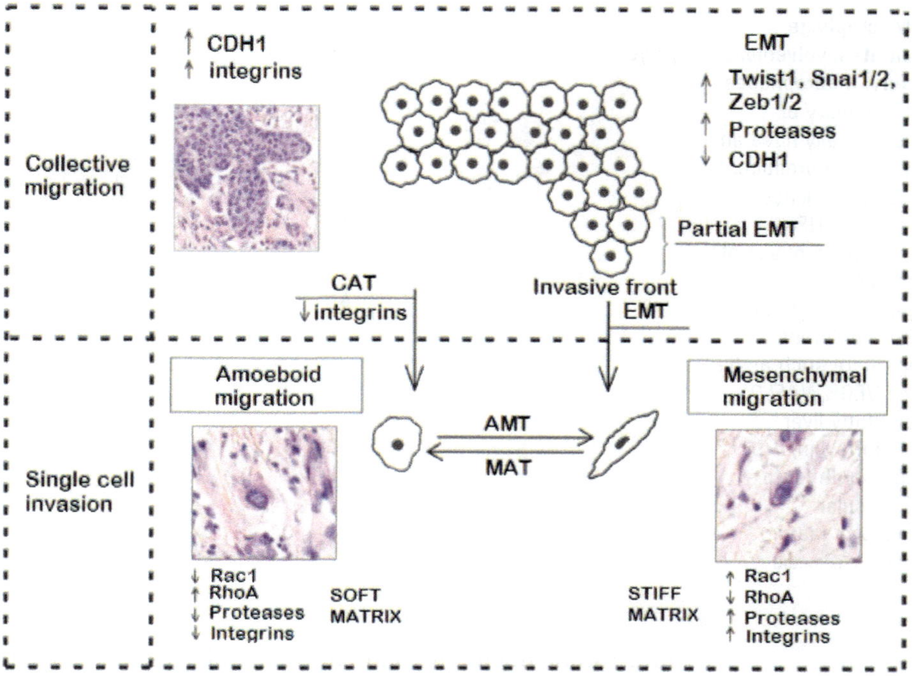

Figure 1.3 Patterns of cancer cell invasion: collective and individual. (Top) Collective migration involves a partial EMT with upregulation of some mesenchymal markers such as ZEB1/2, TWIST1, SNAI1/2, and decrease in CDH1 levels. These cells retain cell-cell adhesion yet gain migratory traits. Cells can undergo CAT where downregulation of integrins lead to cancer cell detachment and migration via amoeboid mode. Cells can also switch between amoeboid and mesenchymal states by altering the levels of Rac1 and RhoA. *Source*: Courtesy Wikimedia Commons.

progenitor cells in vitro and in vivo; this process may have subsequent roles in wound healing (Das et al., 2015; Yan et al., 2011). Macrophages help the body in maintaining immunological homeostasis, and are morphologically and phenotypically distinct in various tissues. The macrophage population originates from progenitors in the bone marrow and is highly heterogeneous and plastic in response to surrounding microenvironment. Macrophages are known to exhibit two common distinct polarization phenotypes: (1) classically activated M1 and (2) the alternatively activated M2. The uncommitted macrophages (M0) can be driven to adopt M1 and M2 states through LPS/IFN-γ and IL-4/IL-13 treatment, respectively (Tarique et al., 2015). These macrophages are characterized by different cell surface markers and transcriptional signatures that they express and can dynamically switch states. The switch from M1 to M2 macrophages is beneficial in scenarios such as wound healing, where a resolution of inflammation is needed (Fig. 1.4). However,

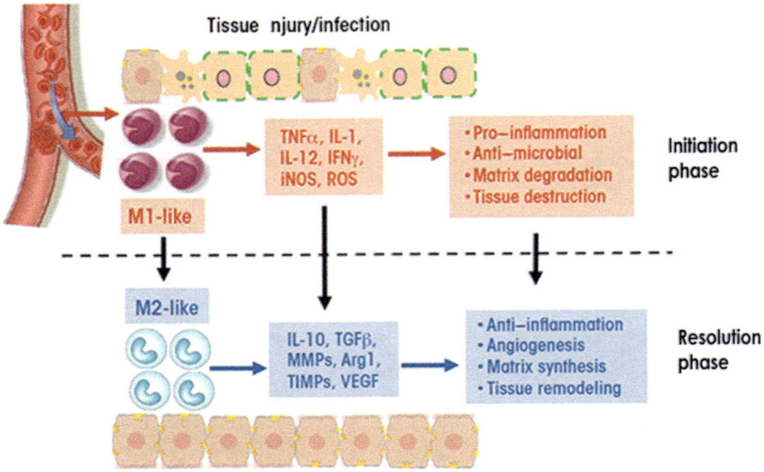

Figure 1.4 Macrophage plasticity and its involvement in tissue injury. Macrophages recruited at the injury or infection site typically have an M1 phenotype; they produce proinflammatory cytokines such as IL-12 and TNFα. During the resolution phase of the injury, the M1 macrophages convert to an M2 type with a different cytokine repertoire such as IL-10, TGFβ, and VEGF. *NAFLD*, Non-alcoholic fatty liver disease; *IL*, Interleukin; *TGFβ*, transforming growth factor-beta; *VEGF*, vascular endothelial growth factor. *Source*: Courtesy Wikimedia Commons.

in a tumor microenvironment, macrophages tend to exhibit more of a M2-like phenotype (Shapouri-Moghaddam et al., 2018). Similar to cases of hybrid E/M phenotypes discussed earlier, the binary view of M1-M2 classification has been now replaced with the concept of M1 and M2 activation states representing the extremes, and macrophages have been observed to exhibit phenotypes along the M1-M2 continuum in varying tissue microenvironments (Italiani and Boraschi, 2014).

Various putative mechanisms have been proposed to underlie phenotypic switching of macrophages from a proinflammatory M1 state to a proresolution M2 state and vice versa (Das et al., 2015). A major difference between M1 and M2 states is at a metabolic level—while the arginine metabolism is shifted to ornithine and polyamines in M2 macrophages, it is tilted toward citrulline and nitrix oxide (NO) in M1 macrophages. These metabolic differences are functionally relevant, and suggest that phenotypic plasticity can be enabled through metabolic networks too (Italiani and Boraschi, 2014). Thus, depending on the cytokine milieu present, the M1 and M2 macrophages can be reprogrammed (or repolarized) into one another, as well as dedifferentiated (or depolarized) to the uncommitted M0 state (Tarique et al., 2015). Similar to macrophages, neutrophils can also display phenotypic changes during inflammation resolution where an alteration in the production of lipid mediators is seen to transition from proinflammatory prostaglandins and leukotrienes to antiinflammatory lipoxins (Galli et al., 2011). Another set of immune cells can display phenotypic switching upon cytokine exposure: recent studies have highlighted that Th1/Th2 cells are not as terminally differentiated as originally thought of (Geginat et al., 2014). Thus, phenotypic plasticity permeates cell types.

Transdifferentiation of adult stem cells ushered in the idea of adult stem cell plasticity and questioned their linear and irreversible trajectories (Blau et al., 2001). Such plasticity can help the tissues to adapt and regenerate during stressed conditions (Wagers and Weissman, 2004). Similar plasticity has been observed in cancer stem cells (CSCs)—a subpopulation of cancer cells with the ability of repopulating an entire tumor after a therapeutic attack (Tang, 2012). CSCs can be quite heterogeneous, for instance, more epithelial-like CSCs (characterized by ALDH1 +) versus more mesenchymal-like CSCs (characterized by CD44 + CD24 −) can exist in different spatial positions in a tumor (Bocci et al., 2019). Both of them can give rise to one another and initiate a tumor (Liu et al., 2014). These CSC subsets have different metabolic profiles (Luo et al., 2018), reminiscent of observations for M1/M2 sets. Moreover, so-called non-CSCs can convert to being a CSC under hypoxia (Dhawan et al., 2016) and/or other programs such as EMT (Jolly et al., 2014). Such plasticity may help maintain a dynamic equilibrium of cells in varying phenotypes (Yang et al., 2012), facilitating symbiosis as has been observed through metabolic coupling (Hamada et al., 2018).

Metabolic reprogramming has been identified as a hallmark of cancer (Hanahan and Weinberg, 2011). It is generally believed that even in presence of oxygen, cancer cells generate energy via a glycolytic pathway to produce lactate. This phenomenon is referred to as the "Warburg Effect" or "aerobic glycolysis." Recent studies have indicated that cancer cells can also utilize oxidative phosphorylation (OXPHOS) pathway to generate energy when oxygen is present abundantly. Recent mathematical modelling studies have revealed the cancer cells can also exist in two additional phenotypic states in which both the glycolytic and the OXPHOS pathways are either active or inactive. AMP-activated protein kinase (AMPK) and hypoxia inducible factor-1 (HIF-1), master regulators of OXPHOS and glycolysis, respectively, are critical in maintaining the balance (Jia et al., 2019). Owing to multistability in the network, these phenotypic states can potentially switch among one another depending on stress conditions and oxygen availability faced by the cancer cells. Thus, cancer cells can be metabolically plastic and utilize different ways of energy production that may be advantageous as they face fluctuating environments during metastatic cascade (Celià-Terrassa and Kang, 2016). Fluctuating population dynamics has been suggested to aid the evolution of phenotypic plasticity (Svanbäck et al., 2009). Therefore, such observations of phenotypic plasticity in cancer cell invasion and cancer metabolism underscore that cancer is a highly adaptive system, where phenotypic plasticity enabled by the gene regulatory networks and protein–protein interactions within the cells can fuel cancer progression and aggressiveness (Jolly and Celia-Terrassa, 2019; Schwab and Pienta, 1996). Phenotypic plasticity can also influence the emergence of adaptive drug resistance in cancer in response to various therapeutic attacks (Hammerlindl and Schaider, 2018), and enable long-term evolution of resistance (Gunnarssson et al., 2020).

Phenotypic plasticity decisions need not be governed at an intracellular level, but can be coordinated at a population level as well. For instance, during collective cell invasion of breast cancer—the predominant mode of metastasis (Aceto et al., 2014)—leader and follower cells can dynamically exchange their spatial locations depending on metabolic coordination (Zhang et al., 2019). Similarly, cell-cell communication through juxtacrine and/or paracrine signaling may underlie spatially distinct locations of CSC subsets in a tumor microenvironment (Bocci et al., 2019) and can alter the EMT phenotype of cancer cells (Neelakantan et al., 2017). An in silico coculture setup of cancer cells of different phenotypes and macrophages of different polarizations suggested how phenotypic plasticity of different cell types (cancer cells and immune cells) may be coordinated in a given ecosystem (Li et al., 2019); thus, suggesting coevolution of cancer cells with their stromal counterparts.

In pathogenic infections, phenotypic plasticity plays a crucial role for both host and pathogen in determining the outcome. As mentioned earlier, macrophages can undergo phenotypic switching to different polarization states to control the infection and the resultant inflammation at the affected site. The pathogen *Mycobacterium tuberculosis (Mtb)* primarily attacks macrophages in the lungs and gives rise to nodular structures called granuloma (Ehlers and Schaible, 2012). *M. tuberculosis* has been shown to reprogram macrophages towards M2 for its better survival in the host; also, pathogens can undergo profound alterations in response to host-derived inputs (Dhar et al., 2016; Khan et al., 2019). Recent studies indicate that the *Mtb* cells that are phagocytosed by the macrophages can show drug tolerance as a virtue of phenotypic variations brought about by the acidic environment inside the macrophages (Mishra et al., 2019). In *Candida albicans*, a human fungal pathogen that can show differential phenotypes depending on the environmental conditions, it has been observed that the pathogen can switch from a "white" state to an "opaque" state (Mallick et al., 2016). These different states have different colony growth properties and interactions with the immune cells in vitro. Similarly, in vivo infection models of zebrafish showed differences in virulence rates in the fungus, again pointing to the fact that the fungi can dynamically change their states depending on the environmental conditions. The overall plasticity in the outcomes can be sometimes due to the complex interplay of the pathogen, host, and intervention strategy used. One prominent example is the treatment outcomes in the context of Hepatitis C treatment in humans. The emergent properties of the interferon signaling network due to its interactions with the viral dynamics and treatment strategies can dictate the outcomes of the infection and the chances of viral persistence inside the host (Padmanabhan et al., 2014).

Cardiac phenotypic plasticity or cardiac remodeling is yet another example where the tissue structure is remodeled both during the developmental origins and during diseased conditions due to various internal and external cues. The plasticity observed in this case is primarily adaptive in nature and depends on the mechanics of the local tissue, which, in turn, can be affected by the clinical conditions such as senescence, obesity, diabetes, ischemia, and neurohormonal reaction (Swynghedauw, 2006). The plasticity in this case is again controlled by the underlying gene regulatory networks that enable the reprogramming to take place. Across the world obesity and metabolic syndromes are on the rise due to unhealthy eating habits and sedentary lifestyles. During the progression of NAFLD, phenotypic alteration of the hepatocytes has been observed where they show an increased expression of the adipocytic features at the cost of the hepatic markers (Lakhani et al., 2018). These changes can be driven due to multistable nature of underlying gene regulatory networks (Sahoo et al., 2020), and may suggest therapies on reverting NAFLD (Parafati et al., 2018).

Phenotypic plasticity at the organismal level

Mathematical simulations and experiments across biological systems have provided insights into how a clonal (i.e., genetically identical) population of cells can generate a phenotypically heterogeneous population (Ruscetti et al., 2016; Stockholm et al., 2007; Tripathi et al., 2020). Conceptually, this process can incorporate two mutually exclusive phenomenon—one in which the cells exhibit phenotypic switching only under the influence of specific external environmental perturbations, and another in which the cells can switch states independent of environmental conditions just by virtue of its intrinsic state/noise. Simulations suggested that both these mechanisms can produce stable phenotypic heterogeneity; however, the population distribution can be different across the two scenarios considered. The in silico findings were later experimentally confirmed through a myogenic cell line showcasing the importance of both the external "ecological" context of the cells and the intrinsic phenotypic robustness of the cells (Stockholm et al., 2007).

Phenotypic variability evident through specific properties at a cellular level can give rise to larger group-level behaviors, and have major consequences for the ecology and evolution of populations (Vogel et al., 2015). For instance, in the population of a unicellular organism *Physarum polycephalum*, calcium signaling at the individual cell level dictates its social behavior, movement speed, and patterns. The phenotype of all cells in a population, taken together, finally translates into the alternative social strategies opted by the group as a whole for foraging (Vogel et al., 2015). Similarly, social behavior exhibited by the cells of the amoeba species *Dictyostelium discoideum* immediately after starvation may emerge from bistability and consequent phenotypic plasticity. The prestarved amoeba exhibit random movements that transition into a very coordinated set of movements of the cell colony as a single aggregate or slug if the cells are starved for a certain duration. The cells in the slug itself are largely of two types—one that form the spore forming body while the others form the stalk. Among the various phenotypic differences between the presumptive stalk and the spore cells, the cells show a distinctive bimodal behavior in terms of their movement speeds and differential amounts of calcium ions in the cells. This bimodality disappears when the nutrients are restored for cells, suggesting a reversible dynamical transition triggered by external signals (Goury-Sistla et al., 2012).

On the other hand, some systems may be required to be robust to external environmental fluctuations. What factors contribute to such environmental robustness, or in other words, the relative insensitivity of the phenotypic outcome? Such a kind of stabilizing selection can be brought about by specific properties of underlying gene regulatory networks involved such as redundancy in network structure, the network topology itself, modularity in the network, etc. (Braendle and Félix, 2009). This concept of how biological systems, particularly in embryonic development, have found robust ways of controlling cell fates and corresponding phenotypes despite extensive variations in environment has been intriguing. Recent studies have identified some design principles that may contribute to developmental robustness and maintenance of an invariant and precise phenotypic outcome (Bhat and Pally, 2017; Braendle and Félix, 2009).

Conclusion

Phenotypic plasticity is a central concept in generating functional heterogeneity in biological systems. Such heterogeneity and adaptability enabled by phenotypic plasticity may aid the evolution of a species as a whole, but the relationship of phenotypic plasticity to fitness across multiple levels—cell, organ, organism, and species—need not be a monotonic function. Thus, as Donald Symons pointed out, phenotypic plasticity can be thought of as a double-edged sword.

Acknowledgment

MKJ was supported by Ramanujan Fellowship (SB/S2/RJN-049/2018) provided by Science and Engineering Research Board (SERB), Department of Science and Technology (DST), Government of India.

References

Aceto, N., Bardia, A., Miyamoto, D.T., Donaldson, M.C., Wittner, B.S., Spencer, J. A., et al., 2014. Circulating tumor cell clusters are oligoclonal precursors of breast cancer metastasis. Cell 158, 1110–1122. Available from: https://doi.org/10.1016/j.cell.2014.07.013.

Ahnert, S.E., 2017. Structural properties of genotype-phenotype maps. J. R. Soc. Interface 14. Available from: https://doi.org/10.1098/rsif.2017.0275.

Bernstein, B.E., Mikkelsen, T.S., Xie, X., Kamal, M., Huebert, D.J., Cuff, J., et al., 2006. A bivalent chromatin structure marks key developmental genes in embryonic stem cells. Cell 125, 315–326. Available from: https://doi.org/10.1016/j.cell.2006.02.041.

Bhat, R., Pally, D., 2017. Complexity: the organizing principle at the interface of biological (dis) order. J. Genet. 96, 431–444. Available from: https://doi.org/10.1007/s12041-017-0793-8.

Blau, H.M., Brazelton, T.R., Weimann, J.M., 2001. The evolving concept of a stem cell: entity or function? Cell 105, 829–841. Available from: https://doi.org/10.1016/S0092-8674(01)00409-3.

Bocci, F., Gearhart-Serna, L., Boareto, M., Ribeiro, M., Ben-Jacob, E., Devi, G.R., et al., 2019. Toward understanding cancer stem cell heterogeneity in the tumor microenvironment. Proc. Natl. Acad. Sci. U. S. A. 116, 148–157. Available from: https://doi.org/10.1073/pnas.1815345116.

Braendle, C., Félix, M.A., 2009. The other side of phenotypic plasticity: a developmental system that generates an invariant phenotype despite

environmental variation. J. Biosci. 34, 543–551. Available from: https://doi.org/10.1007/s12038-009-0073-8.
Campbell, K., Casanova, J., 2016. A common framework for EMT and collective cell migration. Development 143, 4291–4300. Available from: https://doi.org/10.1242/dev.139071.
Celià-Terrassa, T., Kang, Y., 2016. Distinctive properties of metastasis-initiating cells. Genes. Dev. 30, 892–908. Available from: https://doi.org/10.1101/gad.277681.116.892.
Chaffer, C.L., Marjanovic, N.D., Lee, T., Bell, G., Kleer, C.G., Reinhardt, F., et al., 2013. Poised chromatin at the ZEB1 promoter enables breast cancer cell plasticity and enhances tumorigenicity. Cell 154, 61–74. Available from: https://doi.org/10.1016/j.cell.2013.06.005.
Das, A., Sinha, M., Datta, S., Abas, M., Chaffee, S., Sen, C.K., et al., 2015. Monocyte and macrophage plasticity in tissue repair and regeneration. Am. J. Pathol. 185, 2596–2606. Available from: https://doi.org/10.1016/j.ajpath.2015.06.001.
Day, T., 2016. Interpreting phenotypic antibiotic tolerance and persister cells as evolution via epigenetic inheritance. Mol. Ecol. 25, 1869–1882. Available from: https://doi.org/10.1111/mec.13603.
de Back, W., Zimm, R., Brusch, L., 2013. Transdifferentiation of pancreatic cells by loss of contact-mediated signaling. BMC Syst. Biol. 7, 77. Available from: https://doi.org/10.1186/1752-0509-7-77.
Dhar, N., McKinney, J., Manina, G., 2016. Phenotypic heterogeneity in mycobacterium tuberculosis. Microbiol. Spectr 4, 6. Available from: https://doi.org/10.1128/microbiolspec.tbtb2-0021-2016.
Dhawan, A., Madani Tonekaboni, S.A., Taube, J.H., Hu, S., Sphyris, N., Mani, S. A., et al., 2016. Mathematical modelling of phenotypic plasticity and conversion to a stem-cell state under hypoxia. Sci. Rep. 6, 1–10. Available from: https://doi.org/10.1038/srep18074.
Ehlers, S., Schaible, U.E., 2012. The granuloma in tuberculosis: dynamics of a host-pathogen collusion. Front. Immunol. 3, 411. Available from: https://doi.org/10.3389/fimmu.2012.00411.
Enver, T., Pera, M., Peterson, C., Andrews, P.W., 2009. Stem cell states, fates, and the rules of attraction. Cell Stem Cell 4, 387–397. Available from: https://doi.org/10.1016/j.stem.2009.04.011.
Espinosa-Soto, C., Martin, O.C., Wagner, A., 2011. Phenotypic plasticity can facilitate adaptive evolution in gene regulatory circuits. BMC Evol. Biol. 11, 5. Available from: https://doi.org/10.1186/1471-2148-11-5.
Feinberg, A.P., Koldobskiy, M.A., Gondor, A., 2016. Epigenetic modulators, modifiers and mediators in cancer aetiology and progression. Nat. Rev. Genet. 17, 284–299.
Fierst, J.L., 2011. A history of phenotypic plasticity accelerates adaptation to a new environment. J. Evol. Biol. 24, 1992–2011. Available from: https://doi.org/10.1111/j.1420-9101.2011.02333.x.
Font-Clos, F., Zapperi, S., La Porta, C.A.M., 2018. Topography of epithelial–mesenchymal plasticity. Proc. Natl. Acad. Sci. 115, 5902–5907. Available from: https://doi.org/10.1073/pnas.1722609115.
Forsman, A., 2014. Rethinking phenotypic plasticity and its consequences for individuals, populations and species. Heredity (Edinb.) 115, 276–284. Available from: https://doi.org/10.1038/hdy.2014.92.
Fusco, G., Minelli, A., 2010. Phenotypic plasticity in development and evolution: facts and concepts. Philos. Trans. R. Soc. B Biol. Sci. 365, 547–556. Available from: https://doi.org/10.1098/rstb.2009.0267.

Galli, S.J., Borregaard, N., Wynn, T.A., 2011. Phenotypic and functional plasticity of cells of innate immunity: macrophages, mast cells and neutrophils. Nat. Publ. Gr. 12. Available from: https://doi.org/10.1038/ni.2109.

Geginat, J., Paroni, M., Maglie, S., Alfen, J.S., Kastirr, I., Gruarin, P., et al., 2014. Plasticity of human CD4 T cell subsets. Front. Immunol. 5, 630. Available from: https://doi.org/10.3389/fimmu.2014.00630.

Ghosh, D., Veeraraghavan, B., Elangovan, R., Vivekanandan, P., 2020. Antibiotic resistance and epigenetics: more to it than meets the eye. Antimicrob. Agents Chemother. 64. Available from: https://doi.org/10.1128/AAC.02225-19.

Goury-Sistla, P., Nanjundiah, V., Pande, G., 2012. Bimodal distribution of motility and cell fate in dictyostelium discoideum. Int. J. Dev. Biol. 56, 263–272. Available from: https://doi.org/10.1387/ijdb.113384ps.

Gratani, L., 2014. Plant phenotypic plasticity in response to environmental factors. Adv. Bot. 208747. Available from: https://doi.org/10.1155/2014/208747.

Guantes, R., Poyatos, J.F., 2008. Multistable decision switches for flexible control of epigenetic differentiation. PLoS Comput. Biol. 4, e1000235. Available from: https://doi.org/10.1371/journal.pcbi.1000235.

Gunnarssson, E.B., De, S., Leder, K., Foo, J., 2020. Understanding the role of phenotypic switching in cancer drug resistance. J. Theor. Biol. 490, 110162.

Hamada, M., Schröder, K., Bathia, J., Kürn, U., Fraune, S., Khalturina, M., et al., 2018. Metabolic co-dependence drives the evolutionarily ancient Hydra–Chlorella symbiosis. Elife 35122. Available from: https://doi.org/10.7554/eLife.35122.

Hammerlindl, H., Schaider, H., 2018. Tumor cell-intrinsic phenotypic plasticity facilitates adaptive cellular reprogramming driving acquired drug resistance. J. Cell Commun. Signal. 12, 133–141. Available from: https://doi.org/10.1007/s12079-017-0435-1.

Hanahan, D., Weinberg, R.A., 2011. Hallmarks of cancer: the next generation. Cell 144, 646–674. Available from: https://doi.org/10.1016/j.cell.2011.02.013.

Harrison, R., Papp, B., Pál, C., Oliver, S.G., Delneri, D., 2007. Plasticity of genetic interactions in metabolic networks of yeast. Proc. Natl. Acad. Sci. U. S. A. 104, 2307–2312. Available from: https://doi.org/10.1073/pnas.0607153104.

Hilger, D., Masureel, M., Kobilka, B.K., 2018. Structure and dynamics of GPCR signaling complexes. Nat. Struct. Mol. Biol. 25, 4–12. Available from: https://doi.org/10.1038/s41594-017-0011-7.

Huang, B., Lu, M., Jolly, M.K., Tsarfaty, I., Onuchic, J., Ben-Jacob, E., 2014. The three-way switch operation of Rac1/RhoA GTPase-based circuit controlling amoeboid-hybrid-mesenchymal transition. Sci. Rep. 4, 6449. Available from: https://doi.org/10.1038/srep06449.

Huang, B., Jolly, M.K., Lu, M., Tsarfaty, I., Ben-Jacob, E., Onuchic, J.N., 2015. Modeling the transitions between collective and solitary migration phenotypes in cancer metastasis. Sci. Rep. 5. Available from: https://doi.org/10.1038/srep17379.

Italiani, P., Boraschi, D., 2014. From monocytes to M1/M2 macrophages: phenotypical vs. functional differentiation. Front. Immunol. 5, 514. Available from: https://doi.org/10.3389/fimmu.2014.00514.

Jia, D., Jolly, M.K., Kulkarni, P., Levine, H., 2017. Phenotypic plasticity and cell fate decisions in cancer: insights from dynamical systems theory. Cancers (Basel) 9, E70. Available from: https://doi.org/10.3390/cancers9070070.

Jia, D., Lu, M., Jung, K.H., Park, J.H., Yu, L., Onuchic, J.N., et al., 2019. Elucidating cancer metabolic plasticity by coupling gene regulation with metabolic pathways. Proc. Natl. Acad. Sci. U. S. A. 116, 3909–3918. Available from: https://doi.org/10.1073/pnas.1816391116.

Jolly, M.K., Levine, H., 2017. Computational systems biology of epithelial-hybrid-mesenchymal transitions. Curr. Opin. Syst. Biol. 3, 1–6. Available from: https://doi.org/10.1016/j.coisb.2017.02.004.

Jolly, M.K., Celia-Terrassa, T., 2019. Dynamics of phenotypic heterogeneity associated with EMT and stemness during cancer progression. J. Clin. Med. 8, 1542. Available from: https://doi.org/10.3390/jcm8101542.

Jolly, M.K., Huang, B., Lu, M., Mani, S.A., Levine, H., Ben-Jacob, E., 2014. Towards elucidating the connection between epithelial-mesenchymal transitions and stemness. J. R. Soc. Interface 11, 20140962. Available from: https://doi.org/10.1098/rsif.2014.0962.

Jolly, M.K., Boareto, M., Lu, M., Onuchic, J.N., Clementi, C., Ben-Jacob, E., 2015. Operating principles of Notch-Delta-Jagged module of cell-cell communication. N. J. Phys. 17, 55021. Available from: https://doi.org/10.1088/1367-2630/17/5/055021.

Jolly, M.K., Ware, K.E., Gilja, S., Somarelli, J.A., Levine, H., 2017. EMT and MET: necessary or permissive for metastasis? Mol. Oncol. 11, 755–769. Available from: https://doi.org/10.1002/1878-0261.12083.

Jolly, M.K., Kulkarni, P., Weninger, K., Orban, J., Levine, H., 2018. Phenotypic plasticity, bet-hedging, and androgen independence in prostate cancer: role of non-genetic heterogeneity. Front. Oncol. 8, 1–12. Available from: https://doi.org/10.3389/fonc.2018.00050.

Kalluri, R., Weinberg, R.A., 2010. The basics of epithelial-mesenchymal transition. J. Clin. Invest. 120, 1786. Available from: https://doi.org/10.1172/JCI39104C1.

Khan, A., Singh, V.K., Hunter, R.L., Jagannath, C., 2019. Macrophage heterogeneity and plasticity in tuberculosis. J. Leukoc. Biol. 106, 275–282. Available from: https://doi.org/10.1002/JLB.MR0318-095RR.

Lakhani, H.V., Sharma, D., Dodrill, M.W., Nawab, A., Sharma, N., Cottrill, C.L., et al., 2018. Phenotypic alteration of hepatocytes in non-alcoholic fatty liver disease. Int. J. Med. Sci. 15, 1591–1599. Available from: https://doi.org/10.7150/ijms.27953.

Lecharpentier, A., Vielh, P., Perez-Moreno, P., Planchard, D., Soria, J.C., Farace, F., 2011. Detection of circulating tumour cells with a hybrid (epithelial/mesenchymal) phenotype in patients with metastatic non-small cell lung cancer. Br. J. Cancer 105, 1338–1341. Available from: https://doi.org/10.1038/bjc.2011.405.

Li, X., Jolly, M.K., George, J.T., Pienta, K.J., Levine, H., 2019. Computational modeling of the crosstalk between macrophage polarization and tumor cell plasticity in the tumor microenvironment. Front. Oncol. 9, 1–12. Available from: https://doi.org/10.3389/fonc.2019.00010.

Liu, S., Cong, Y., Wang, D., Sun, Y., Deng, L., Liu, Y., et al., 2014. Breast cancer stem cells transition between epithelial and mesenchymal states reflective of their normal counterparts. Stem Cell Rep. 2, 78–91. Available from: https://doi.org/10.1016/j.stemcr.2013.11.009.

Lu, M., Jolly, M.K., Levine, H., Onuchic, J.N., Ben-Jacob, E., 2013. MicroRNA-based regulation of epithelial-hybrid-mesenchymal fate determination. Proc. Natl. Acad. Sci. U. S. A. 110, 18174–18179. Available from: https://doi.org/10.1073/pnas.1318192110.

Luo, M., Shang, L., Brooks, M.D., Jiagge, E., Zhu, Y., Buschhaus, J.M., et al., 2018. Targeting breast cancer stem cell state equilibrium through modulation of redox signaling. Cell Metab. 28, 69–86. Available from: https://doi.org/10.1016/j.cmet.2018.06.006.

MacKenzie, S.A., Kundariya, H., 2020. Organellar protein multi-functionality and phenotypic plasticity in plants. Philos. Trans. R. Soc. B Biol. Sci. 375, 20190182. Available from: https://doi.org/10.1098/rstb.2019.0182.

Mallick, E.M., Bergeron, A.C., Jones, S.K., Newman, Z.R., Brothers, K.M., Creton, R., et al., 2016. Phenotypic plasticity regulates *Candida albicans* interactions and virulence in the vertebrate host. Front. Microbiol. 7, 780. Available from: https://doi.org/10.3389/fmicb.2016.00780.

Martin, E.W., Pacholewska, A., Patel, H., Dashora, H., Sung, M.-H., 2020. Integrative analysis suggests cell type–specific decoding of NF-κB dynamics. Sci. Signal. 13, eaax7195. Available from: https://doi.org/10.1126/scisignal.aax7195.

Mishra, R., Kohli, S., Malhotra, N., Bandyopadhyay, P., Mehta, M., Munshi, M.H., et al., 2019. Targeting redox heterogeneity to counteract drug tolerance in replicating *Mycobacterium tuberculosis*. Sci. Transl. Med. 11. Available from: https://doi.org/10.1126/scitranslmed.aaw6635.

Mooney, S.M., Talebian, V., Jolly, M.K., Jia, D., Gromala, M., Levine, H., et al., 2017. The GRHL2/ZEB feedback loop—a key axis in the regulation of EMT in breast cancer. J. Cell. Biochem. 118, 2559–2570. Available from: https://doi.org/10.1002/jcb.25974.

Nechooshtan, G., Elgrably-Weiss, M., Sheaffer, A., Westhof, E., Altuvia, S., 2009. A pH-responsive riboregulator. Genes. Dev. 23, 2650–2662. Available from: https://doi.org/10.1101/gad.552209.

Neelakantan, D., Zhou, H., Oliphant, M.U.J., Zhang, X., Simon, L.M., Henke, D. M., et al., 2017. EMT cells increase breast cancer metastasis via paracrine GLI activation in neighbouring tumour cells. Nat. Commun. 8, 15773. Available from: https://doi.org/10.1038/ncomms15773.

Nieto, M.A., Huang, R.Y., Jackson, R.A., Thiery, J.P., 2016. EMT: 2016. Cell 166, 21–45. Available from: https://doi.org/10.1016/j.cell.2016.06.028.

Nijhout, H.F., Sadre-Marandi, F., Best, J., Reed, M.C., 2017. Systems biology of phenotypic robustness and plasticity. Integr. Comp. Biol. 57, 171–184. Available from: https://doi.org/10.1093/icb/icx076.

Padmanabhan, P., Garaigorta, U., Dixit, N.M., 2014. Emergent properties of the interferon-signalling network may underlie the success of hepatitis C treatment. Nat. Commun. 5, 1–9. Available from: https://doi.org/10.1038/ncomms4872.

Pankova, K., Rosel, D., Novotny, M., Brabek, J., Paňková, K., Rösel, D., et al., 2010. The molecular mechanisms of transition between mesenchymal and amoeboid invasiveness in tumor cells. Cell. Mol. Life Sci. 67, 63–71. Available from: https://doi.org/10.1007/s00018-009-0132-1.

Parafati, M., Kirby, R.J., Khorasanizadeh, S., Rastinejad, F., Malany, S., 2018. A nonalcoholic fatty liver disease model in human induced pluripotent stem cell-derived hepatocytes, created by endoplasmic reticulum stress-induced steatosis. DMM Dis. Model. Mech. 11. Available from: https://doi.org/10.1242/dmm.033530.

Pei, D., Shu, X., Gassama-Diagne, A., Thiery, J.P., 2019. Mesenchymal–epithelial transition in development and reprogramming. Nat. Cell Biol. 21, 44–53. Available from: https://doi.org/10.1038/s41556-018-0195-z.

Pisco, O.A., Fouquier D'hérouël, A., Huang, S., 2016. Conceptual confusion: the case of epigenetics. bioRxiv 053009. Available from: https://doi.org/10.1101/053009.

Portela, A., Esteller, M., 2010. Epigenetic modifications and human disease. Nat. Biotechnol. 28, 1057–1068. Available from: https://doi.org/10.1038/nbt.1685.

Roca, H., Hernandez, J., Weidner, S., McEachin, R.C., Fuller, D., Sud, S., et al., 2013. Transcription factors OVOL1 and OVOL2 induce the mesenchymal to

epithelial transition in human cancer. PLoS One 8, e76773. Available from: https://doi.org/10.1371/journal.pone.0076773.
Ruscetti, M., Dadashian, E.L., Guo, W., Quach, B., Mulholland, D.J., Park, J.W., et al., 2016. HDAC inhibition impedes epithelial-mesenchymal plasticity and suppresses metastatic, castration-resistant prostate cancer. Oncogene 35, 3781–3795. Available from: https://doi.org/10.1038/onc.2015.444.
Sahoo, S., Singh, D., Chakraborty, P., Jolly, M.K., 2020. Emergent properties of HNF4a-PPARg may drive consequent phenotypic plasticity in NAFLD. bioRxiv 9, 870. Available from: https://doi.org/10.3390/jcm9030870.
Schwab, E.D., Pienta, K.J., 1996. Cancer as a complex adaptive system. Med. Hypotheses 47, 235–241. Available from: https://doi.org/10.1016/S0306-9877(96)90086-9.
Shapouri-Moghaddam, A., Mohammadian, S., Vazini, H., Taghadosi, M., Esmaeili, S.-A., Mardani, F., et al., 2018. Macrophage plasticity, polarization and function in health and disease. J. Cell. Physiol. 233, 6425–6440. Available from: https://doi.org/10.1002/jcp.26429.
Somarelli, J.A., Shetler, S., Jolly, M.K., Wang, X., Dewitt, S.B., Hish, A.J., et al., 2016. Mesenchymal-epithelial transition in sarcomas is controlled by the combinatorial expression of microRNA 200s and GRHL2. Mol. Cell. Biol. 36, 2503–2513.
Stockholm, D., Benchaouir, R., Picot, J., Rameau, P., Neildez, T.M.A., Landini, G., et al., 2007. The origin of phenotypic heterogeneity in a clonal cell population in vitro. PLoS One 2, e394. Available from: https://doi.org/10.1371/journal.pone.0000394.
Svanbäck, R., Pineda-Krch, M., Doebeli, M., 2009. Fluctuating population dynamics promotes the evolution of phenotypic plasticity. Am. Nat. 174, 176–189. Available from: https://doi.org/10.1086/600112.
Swynghedauw, B., 2006. Phenotypic plasticity of adult myocardium: molecular mechanisms. J. Exp. Biol. 209, 2320–2327. Available from: https://doi.org/10.1242/jeb.02084.
Takahashi, K., 2012. Cellular reprogramming – lowering gravity on Waddington's epigenetic landscape. J. Cell Sci. 125, 2553–2560. Available from: https://doi.org/10.1242/JCS.084822.
Takahashi, K., Yamanaka, S., 2016. A decade of transcription factor-mediated reprogramming to pluripotency. Nat. Rev. Mol. Cell Biol. 17, 183–193. Available from: https://doi.org/10.1038/nrm.2016.8.
Tang, D.G., 2012. Understanding cancer stem cell heterogeneity and plasticity. Cell Res. 22, 457–472. Available from: https://doi.org/10.1038/cr.2012.13.
Tarique, A.A., Logan, J., Thomas, E., Holt, P.G., Sly, P.D., Fantino, E., 2015. Phenotypic, functional, and plasticity features of classical and alternatively activated human macrophages. Am. J. Respir. Cell Mol. Biol. 53, 676–688. Available from: https://doi.org/10.1165/rcmb.2015-0012OC.
Thiery, J.P., Acloque, H., Huang, R.Y.J., Nieto, M.A., 2009. Epithelial-mesenchymal transitions in development and disease. Cell 139, 871–890. Available from: https://doi.org/10.1016/j.cell.2009.11.007.
Tian, X.-J., Zhang, H., Xing, J., 2013. Coupled reversible and irreversible bistable switches underlying TGFβ-induced epithelial to mesenchymal transition. Biophys. J. 105, 1079–1089. Available from: https://doi.org/10.1016/j.bpj.2013.07.011.
Tripathi, S., Chakraborty, P., Levine, H., Jolly, M.K., 2020. A mechanism for epithelial-mesenchymal heterogeneity in a population of cancer cells. PLoS Comput. Biol. 16, e1007619. Available from: https://doi.org/10.1101/592691.

Veening, J.W., Stewart, E.J., Berngruber, T.W., Taddei, F., Kuipers, O.P., Hamoen, L. W., 2008. Bet-hedging and epigenetic inheritance in bacterial cell development. Proc. Natl. Acad. Sci. U. S. A. 105, 4393–4398. Available from: https://doi.org/10.1073/pnas.0700463105.

Virk, R.K.A., Wu, W., Almassalha, L.M., Bauer, G.M., Li, Y., VanDerway, D., et al., 2020. Disordered chromatin packing regulates phenotypic plasticity. Sci. Adv. 6, eaax6232. Available from: https://doi.org/10.1126/sciadv.aax6232.

Vogel, D., Nicolis, S.C., Perez-Escudero, A., Nanjundiah, V., Sumpter, D.J.T., Dussutour, A., 2015. Phenotypic variability in unicellular organisms: from calcium signalling to social behaviour. Proc. R. Soc. B Biol. Sci. 282, 20152322. Available from: https://doi.org/10.1098/rspb.2015.2322.

Wagers, A.J., Weissman, I.L., 2004. Plasticity of adult stem cells. Cell 116, 639–648. Available from: https://doi.org/10.1016/S0092-8674(04)00208-9.

Whelan, D.R., Hiscox, T.J., Rood, J.I., Bambery, K.R., McNaughton, D., Wood, B. R., 2014. Detection of an en masse and reversible B- to A-DNA conformational transition in prokaryotes in response to desiccation. J. R. Soc. Interface 11. Available from: https://doi.org/10.1098/rsif.2014.0454.

Whitman, D.A., Agrawal, A.A., 2009. What is phenotypic plasticity and why is it Important? In: Whitman, D.W., Ananthakrishna, T.N. (Eds.), Phenotypic Plasticity of Insects: Mechanisms and Consequences. Science Publishers, Inc: Enfield, N.H., pp. 1–63.

Xue, B.K., Leibler, S., 2018. Benefits of phenotypic plasticity for population growth in varying environments. Proc. Natl. Acad. Sci. U. S. A. 115, 12745–12750. Available from: https://doi.org/10.1073/pnas.1813447115.

Yadav, A., Dhole, K., Sinha, H., 2016. Genetic regulation of phenotypic plasticity and canalisation in yeast growth. PLoS One 11, e0162326. Available from: https://doi.org/10.1371/journal.pone.0162326.

Yan, D., Wang, X., Li, D., Qu, Z., Ruan, Q., 2011. Macrophages overexpressing VEGF, transdifferentiate into endothelial-like cells in vitro and in vivo. Biotechnol. Lett. 33, 1751–1758. Available from: https://doi.org/10.1007/s10529-011-0645-1.

Yang, G., Quan, Y., Wang, W., Fu, Q., Wu, J., Mei, T., et al., 2012. Dynamic equilibrium between cancer stem cells and non-stem cancer cells in human SW620 and MCF-7 cancer cell populations. Br. J. Cancer 106, 1512–1519. Available from: https://doi.org/10.1038/bjc.2012.126.

Yoshida, K., Soldati, T., 2006. Dissection of amoeboid movement into two mechanically distinct modes. J. Cell Sci. 119, 3833–3844. Available from: https://doi.org/10.1242/jcs.03152.

Zhang, J., Goliwas, K.F., Wang, W., Taufalele, P.V., Bordeleau, F., Reinhart-King, C. A., 2019. Energetic regulation of coordinated leader–follower dynamics during collective invasion of breast cancer cells. Proc. Natl. Acad. Sci. U. S. A. 116, 7867–7872. Available from: https://doi.org/10.1073/pnas.1809964116.

Zhou, J.X., Huang, S., 2011. Understanding gene circuits at cell-fate branch points for rational cell reprogramming. Trends Genet. 27, 55–62. Available from: https://doi.org/10.1016/j.tig.2010.11.002.

2

Rethinking the role of chance in the explanation of cell differentiation

Marco Casali[1,2] and Francesca Merlin[2]
[1]*Department of Philosophy, Sapienza – University of Rome, Rome, Italy*
[2]*Institut d'histoire et de philosophie des sciences et des techniques (IHPST), CNRS & University of Paris 1 Panthéon-Sorbonne, Paris, France*

Introduction

How does it happen that populations of clonal cells show cell-to-cell variability even when their environment is kept homogeneous and constant? Until the mid-twenty century, isogenic populations of cells were treated as ideal gases, which are composed by identical point particles, so that phenotypic variability among genetically identical cells that are placed in the same environment was fading away.[1] Was it a useful approximation to model and predict more easily the behavior of cell populations, as in the case of an ideal gas, or such variability went simply unnoticed by biologists until that time? The answer lies somewhere between these two scenarios. What we do know is that phenotypic variability among genetically identical cells had started to be observed and studied starting from the 1970s, thanks to the development of new technologies for single-cell analysis (in particular, the use of fluorescent biological markers like variants of fluorescent proteins, and the development of highly sensitive systems for signal light detection such as flux cytometry and fluorescence microscopy). However, the origin of

[1]Jacques Monod is a paradigmatic example of this viewpoint. In the 1950s, he explained that he was interested in *Escherichia coli* bacteria in a liquid medium because they represented an experimental model similar to the study of perfect gazes. See Monod's biography by Debré (1996: 188).

Phenotypic Switching. DOI: https://doi.org/10.1016/B978-0-12-817996-3.00005-0

phenotypic variability among genetically identical cells had remained a mystery for decades: if cells do share the same DNA and the same milieu, where their differences could come from? What does determine that they make different fate decisions, which have no apparent reason to take place? From the 1950s onward, chance[2] had been the main answer to these questions, as it is often the case when no other solution is available, and the mystery had remained for a long time. Some internal stochastic instability had been supposed to affect the normal course of intracellular processes, thus producing random variability in cell phenotype even in the absence of any genetic and environmental change. Starting from the 1990s, this hypothesis has been clearly established thanks to theoretical and experimental investigations: cell-to-cell nongenetic variability is shown to be due to stochastic fluctuations during the process of gene expression, what has started to be called "noise" (e.g., see McAdam and Arkin, 1997; McAdams and Arkin, 1999; Thattai and van Oudenaarden, 2001; Elowitz et al., 2002; Blake et al., 2003; Fraser et al., 2004; Thattai and Van Oudenaarden, 2004; Raser and O'Shea, 2005; for a review see Maheshri and O'Shea, 2007). Cell-to-cell phenotypic variability among isogenic cells is thus considered simply as an error, a random deviation with respect to what is expected when one knows

Table 2.1 Chance: terms and definitions.

Terms	Definitions
Chance, chancy	General terms used in this paper to designate probabilistic phenomena (processes, patterns), i.e., phenomena that can be described in terms of probability.
Stochasticity, stochastic	Terms used in this paper to refer to chancy processes. More precisely, starting from a given set of initial conditions, a stochastic process can produce various results according to a probability law or distribution.
Randomness, random	Terms used in this paper to refer to chancy patterns or results. More precisely, a random pattern or result is characterized by a certain number of statistical properties that can be described in terms of probability.
Noise, noisy	Terms used in the biological literature to refer to internal instability affecting a physico-chemical process (e.g., gene expression) and producing deviations in its expected result (in the absence of any other internal or external source of perturbation).

[2]See Table 2.1 for the specific way we use, in this paper, the terms "chance," "noise," "stochasticity," "randomness," and their respective adjectival forms.

their genotype and the environment they grow in. In other words, chance in this context is a mere nuisance to the regular, deterministic, and predictable progress of molecular processes taking place inside cells, in particular the process of gene expression (on this point see also Nanjundiah, 2003; Vogt, 2015)[3].

Nowadays, stochasticity in intracellular processes is still considered as an important source of isogenic cell-to-cell variability. Theoretical and experimental researches are often focused, at the molecular level, on noise in gene expression, but more and more studies have also dealt with stochasticity affecting various cellular and metabolic processes involved in the development of multicellular organisms, whose cells are genetically identical but follow various differentiation pathways during ontogenesis. Stochastic fluctuations affecting all these processes are still qualified in negative terms as "noise," in line with the deterministic view of development that, since the mid-20th century, conceives it as the unfolding of a genetic program. Thus, noise is not, in this context, a possible explanatory element of the developmental process.

Despite this predominant viewpoint, there is now more and more evidence that chance can also play a role in the development of biological systems, sometimes even a functional one, rather than always being a mere nuisance (e.g., Nanjundiah and Bhogle, 1995; Nanjundiah, 2003; Raser and O'Shea, 2005; Maamar et al., 2007; Losick and Desplan, 2008; Raj and van Oudenaarden, 2008; Cağatay et al., 2009; Eldar and Elowitz, 2010; Feinberg and Irizarry, 2010; Pilpel, 2011; Gandrillon et al., 2012; Pujadas and Feinberg, 2012; Meyer and Roeder, 2014; Vogt, 2015; Roberfroid et al., 2016; Holmes et al., 2017). This opens the possibility for it to enter into biological explanations alongside acknowledged deterministic processes. In this paper, we take these recent research results seriously. In their light, we argue that chance affecting various biological processes involved in cell functioning and organism development may actually play a positive, constructive role, and should thus be conceived as more than just noise (see also Huang, 2009). Rather than being a perturbation represented by an error term, chance can contribute to the explanation of these processes. More specifically, our objective is dual. First, even if today there is some degree of consensus recognizing the functional role of chance (see above), we aim at showing that the way it is currently conceived (as noise) in the biological literature about cell differentiation and development is still ambiguous because it

[3]For more details about the history of noise in biology, see Merlin (2015).

mixes up ontological and epistemological considerations. As we will show, this is due to the fact that at the same time noise is characterized by referring to its physical causes and is measured in terms of its biological consequences (see "What, actually, is noise?" section). Second, we argue that the notion of chance as noise is not in line with the most recent theoretical and experimental research results on its role in molecular and cellular processes involved in development, and thus is in need of being reexamined. We thus suggest a positive view of chance, which is intended to enrich and augment the mostly negative way it has been characterized in the biological discourse since the 1970s. We call it an "augmented epistemological view"[4] according to which chance can be perturbation with no biological relevance, thus represented by an error term, but can also be seen as a "theoretical operator" (i.e., as defined by Morizot (2012), a theoretical entity operating within a theory) with a biological, and sometimes even functional, significance. When it is shown to play a role in the functioning of the cell as well as in the development of individual organisms throughout time, chance can be conceived as describing the behavior of cellular processes, and integrated in models and explanations as such.

To achieve our two objectives, the paper is structured as follows. The first part is a brief presentation of current researches on noise in gene expression. Our focus on this molecular process is motivated by the fact that, during the last 20 years, theoretical and experimental studies on noise have been more numerous and extensive at this level. In the second part, we offer a short critical analysis of the notion of noise found in this kind of research, and show its ambiguities. The third part is the turning point of our paper. By relying on some recent results of empirical and theoretical researches on stochasticity in gene expression (in particular, in bacteria), we argue that the negative notion of chance as noise is not well suited in some cases because it prevents it from playing an explanatory role. The fourth part is meant to take advantage of the way chance is conceived in immunology in order to undermine the ongoing resistance, in the study of development at the molecular and cellular levels, to adopt a positive view of chance that is already

[4]By "augmented epistemological view" of chance we mean a wider perspective that brings about new ways of conceiving it. This expression has been inspired by Mitchell's "augmented epistemology," expression by which she refers to her own philosophical approach to the study of complex systems, that is, "a new approach that will extend the scope of what counts as reliable knowledge of our complex world," and will bring about "new ways of understanding" the phenomena under study (2009: 12).

present and operative in other biological fields studying molecular and cellular processes (namely, in immunology). We present, in the fifth and sixth parts, the main features of the augmented epistemological view of chance we argue for, and its theoretical advantages from the explanatory point of view that we illustrate, in more concrete terms, with an example taken from the study of cell differentiation and development in animals.

Noise in gene expression: a descriptive analysis

Theoretical and experimental researches on noise in biology have particularly developed in the 2000s and have been focused on its impact on gene expression (Elowitz et al., 2002; Klingenberg, 2005; Samoilov et al., 2006; Maheshri and O'Shea, 2007; Raj and van Oudenaarden, 2008; Pilpel, 2011). In this context, the term "noise" generally refers to various microscopic stochastic events taking place inside the cell and producing fluctuations in one or more steps of these intracellular processes (namely, during transcription and translation). Even though various verbal characterizations of this same phenomenon can be found in the literature, noise is always characterized by the following main features: it is an internal and irregular phenomenon, stochastic and unpredictable, both inherent and unavoidable, and a source of variation, which is not due to neither genetic nor environmental changes. Moreover, noise in gene expression is always described in the negative as a deviation or an error with respect to an average or expected result.[5] The way noise is quantified, as a random variation around the mean value of gene expression (i.e., the average amount of proteins produced) has become its more common definition in the literature. More precisely, noise in gene expression is statistically measured by calculating the coefficient of variation of protein or mRNA abundance (i.e., the standard deviation by the mean) in a clonal population of cells at some point in time (synchronically) or in a single cell over a certain duration of time

[5]More explicitly, the expected result corresponds to the one predicted by deterministic models of gene expression, which rely on deterministic chemical kinetics. As a matter of fact, researches on noise in gene expression have led biologists to acknowledge the limits of deterministic models of gene expression and to complement them with probabilistic ones (Gillespie, 2007), which are based on stochastic chemical kinetics, and with stochastic simulations.

(diachronically), in a stable and homogeneous environment. This is considered as "the most direct and unambiguous measure of gene-expression noise" (Kaern et al., 2005, p. 454).

Let us look at the causes and consequences of such stochastic and unpredictable fluctuations (i.e., noise). The origin of noise in gene expression is commonly attributed to various physical phenomena that can affect the concentration, the localization, and the state of different molecular species involved in this intracellular process. First, molecules inside the cell are constantly in thermal agitation, that is, they continuously move around and collide to each other. Second, the concentration of molecules involved in gene expression is low (sometimes a few copies per cell) and the intracellular environment is not homogeneous. Third, quantum events can affect biochemical bonds between molecules. All these physical phenomena can have an impact on the process of gene expression, being the source of fluctuations in some of its steps, that which produces random and unpredictable variation of its result (i.e., protein abundance). The consequences of noise in gene expression are not limited to the intracellular level in terms of random variation in the abundance of proteins produced. More broadly, noise can produce random variation in the phenotype, namely phenotypic differences in isolated individual organisms over time or between members of clonal populations, even in the absence of genetic and environmental changes. There are many biological examples of the effect of noise in gene expression on cell-fate decision, from virus and bacteria to humans. For instance, it produces (random) variation in bacteria's persistence to antibiotics (Kint et al., 2012) as well as in their ability to uptake DNA from the environment and incorporate it into their genome (i.e., bacterial competence state; more on it in "From noise to chance as explanatory" section); Maamar et al., 2007, Losick and Desplan., 2008); (random) variation in the cell units (called "ommatidia") of drosophila's compound eyes (Johnston and Desplan, 2010) as well as in the phenotype of cloned animals like in the case of the first cloned cats (see Raser and O'Shea, 2005). Noise in gene expression has also been shown to be involved in the development of various human pathologies like different cancer types (Han et al., 2016).

Up to now, we have briefly considered the scientific literature on noise in gene expression and have highlighted the way noise is defined and measured, its commonly acknowledged causes, and consequences. This gives us an idea of the way it is commonly conceived in the recent biological literature, and thus of the main features of the negative notion of chance which dominates the biological discourse in molecular and cellular biology. However, despite a rather broad consensus on these features, it seems difficult to understand what the term "noise" actually refers to. Is noise reducible to the physical phenomena affecting the various molecular processes taking place inside the cell? Does it correspond to the random variation around the average amount of protein produced by gene expression? Or is it something else? In the next section, we provide an answer to these questions by showing that, in the way noise is currently conceived, ontological and epistemological features are mixed up, which makes this notion ambiguous.

What, actually, is noise?

It is commonly considered that noise in gene expression is due to various physical phenomena (thermal agitation, time delays due to low concentration of molecules, and quantum effects) that can affect one or more steps of this process. Moreover, noise is usually defined as the random variation around the average result of gene expression, that is, the mean abundance of proteins (in a single cell over time or in a population of cells at some point in time, in the absence of any genetic and environmental change). Noise thus receives what we call an "operational" definition because it is defined by the way it can be observed and measured. However, it is pretty unclear what noise actually is. It is not supposed to be identified either with its physical sources or with its consequences because it is actually caused by various physical phenomena and variations in protein abundance are its effects. However, in the biological literature, in particular in the studies focused on gene expression, one can find various verbal characterizations with no precise ontological and epistemological "flesh." For instance, Kaern et al. (2005) writes that noise arise from some sort of "fluctuations in transcription and translation, despite constant environmental conditions" (p. 451); Pilpel (2011) characterizes it by saying that "stochastic variation in gene expression levels among genetically identical cells grown under the same conditions is often dubbed 'noise'"(p. 410). This sort of formulation does not allow one to better understand what noise actually is.

As already mentioned in the previous section, noise is an internal, intracellular source of variation, stochastic, and unpredictable. It is supposed to lie between its physical causes and its biological effects, but is often ontologically reduced to the first and epistemologically defined and measured in terms of the latter. This mix of ontological and epistemological features blurs the issue of what noise actually is. In the present biological literature on noise in gene expression, we thus have the following scenario. From an epistemological perspective, "noise" is just a verbal way to acknowledge our ignorance about the causes of cell-to-cell variability and, from an ontological one, "noise" does not refer to anything biological but is reducible to some physical phenomena.[6] We are not happy with this view because chance can play no biological and explanatory role, if any. As a matter of fact, neither lack of knowledge nor physical processes can allow taking into account chance as a behavior of certain biological processes, which could be integrated into their models and explanations. In order to escape this scenario, one should look into the black box that lies between the two, that is, the physical phenomena at the origin of noise and the result of gene expression. Our aim is to provide noise an unambiguous characterization that allows attributing it more than a nuisance role: chance as a behavior characterizing the processes of gene expression, of cell differentiation, and of development, and contributing to their progress.

From noise to chance as explanatory

Let us look at some paradigmatic research results in the study of noise in gene expression: it is about the way bacteria *Bacillus subtilis* become competent or not, that is, capable of taking up DNA fragments from the environment and integrating them in their own genome by recombination (Maamar et al., 2007, Çağatay et al., 2009, for a review see Johnston and Desplan, 2010). As we will show, this case study attests that we have both empirical and theoretical evidence that, in some biological systems, chance is not just a nuisance (i.e., noise) but plays a central role in the system's fate determination, and thus in its explanation. Indeed, biologists involved in this kind of research talk of the "bet-hedging" competence in bacteria. Why so? It has been shown that, in colonies of *B. subtilis*, each bacterium stochastically "choose" the competence or noncompetence state (i.e., a phenomenon of phenotypic switching). More

[6]At this point a specification is needed. On the one hand, by "physical phenomena" we mean phenomena that are described and explained in terms of the principles and theories of physics. On the other hand, "biological phenomena" are those described and explained in terms of biological principles and theories. However, we do not deny the possibility that, by reduction, some biological phenomena could be described and explained in terms of physical principles and theories.

precisely, the phenotypic state of each individual bacterium depends on the level of the expression of some gene, the *comK* gene, that randomly fluctuates over time. This gene is involved in the production of the binding protein *ComK* whose abundance in each cell, as a consequence, randomly varies. Thus, as far as the competent or noncompetent state of bacteria depends on the fluctuating abundance of *ComK* (above a certain threshold in terms of protein abundance, the cells become competent), their phenotypic state is not unequivocally determined but in a chancy way, that is, with some specific probability. This is why, in a colony of isogenic *B. subtilis* bacteria, we can observe both competent and noncompetent cells, even though the environment is kept stable and homogeneous (see Fig. 2.1).

Maamar and colleagues (2007) conclude their paper by suggesting that the rate at which bacterial cells become competent, and thus are able to integrate in their chromosome DNA taken from the environment, can have an impact in terms of their fitness. Thus, as far as stochasticity in *comK* gene expression defines this rate, we can deduce that the level of stochasticity at the origin of randomly fluctuating numbers of *ComK* proteins can have an impact in terms of fitness (both at the individual and population levels). This leads them to claim that "noise

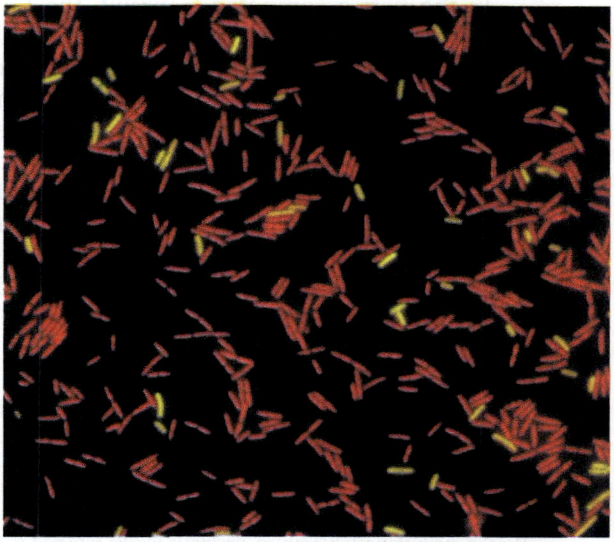

Figure 2.1 Random distribution of cell states (competent and noncompetent) in *Bacillus subtilis* bacteria. *Source*: © Jeanette Hahn and David Dubnau.

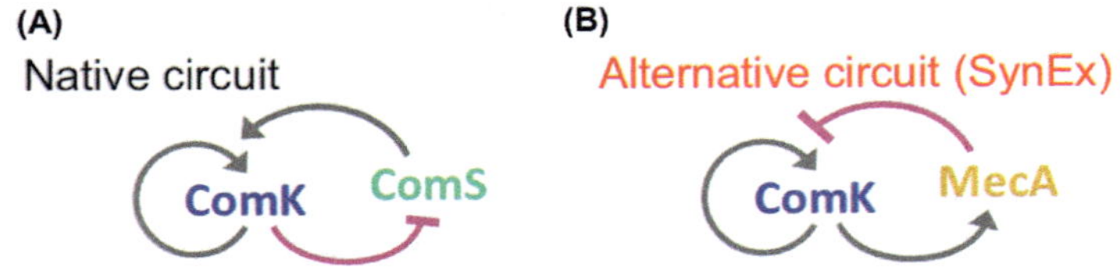

Figure 2.2 (A) The native *B. subtilis* competence circuit; (B) the alternative gene circuit architecture SynEx. *Source*: From Cağatay, T., Turcotte, M., Elowitz, M.B., Garcia-Ojalvo, J., Süel, G.M., 2009. Architecture-dependent noise discriminates functionally analogous differentiation circuits. Cell 139 (3), 512–522.

characteristics of particular genes may be subject to evolutionary pressures" (Maamar et al., 2007, p. 529), and to conclude: "Our results imply that noise properties are subject to evolutionary forces and suggest how cells might alter those rates to increase fitness. Because noise has been implicated in a variety of cellular behaviors, such knowledge can help both in the understanding of natural regulatory networks and in the synthesis of artificial networks" (Maamar et al., 2007, p. 529).[7]

Another group of researchers has taken up this challenge by performing synthetic biology experiments that allowed them to test this same hypothesis (Cağatay et al., 2009). They built a competence genetic circuit architecture, alternative to the native one observed in *B. subtilis* bacteria, in order to change the noise profile (see Fig. 2.2).

In the native competence circuit, the *ComK* protein positively regulates its own synthesis forming a positive feedback loop and, at the same time, represses the synthesis of the *ComS* protein, which in turn promotes *ComK* synthesis, thus establishing a net negative feedback loop. In the alternative circuit, called SynEx, the *ComK* protein also autoregulates its own synthesis through a positive feedback loop, but induces the expression of another factor, *MecA*, which in turn induces *ComK* degradation. Therefore, the difference with respect to the native circuit architecture is in the net negative feedback loop whose reactions, in the SynEx, are interchanged. Cağatay and colleagues have shown that the native circuit is highly noisy and thus advantageous under unpredictable environmental conditions while the SynEx, which exhibits less noisy and more precise dynamics, is advantageous under stable environmental conditions. On this basis, they argue that the level of noise

[7] In the evolutionary literature, previous works referred to the different consequences of natural selection acting on the mean and variance of a trait; in particular, see Haldane and Jayakar (1963) and Gillespie (1977).

matters, both from an evolutionary and a physiological point of view. In fact, as argued by Maamar and colleagues, it can make a difference in terms of fitness depending on the environmental conditions of the bacterial colony. More importantly for us, by building and comparing the two circuit architectures and the stochastic behavior they respectively generate, Cağatay and colleagues have also shown that the level of noise of each circuit "in turn dictates the system's physiological response range" (Cağatay et al., 2009, p. 520). More explicitly, stochasticity in gene regulation and expression can also be essential for the functioning of biological systems, like deterministic regulation of genetic circuits is. It can thus enter into its explanation alongside deterministic regulation.

The "bet-hedging" property of *B. subtilis* bacteria is particularly enlightening because it clearly shows that stochasticity (more generally, chance) can be more than just an error term (i.e., more than just noise) in the description of biological systems. Rather, it sometimes deserves a positive role, and even a functional one, both in the physiology of these systems and in the determination of their fitness value. Other research studies dealing with stochasticity in various processes involved in the development of multicellular organisms, from gene expression and cellular differentiation to tissue and organ formation, suggest this same conclusion: stochasticity is a relevant feature of these processes because it can contribute both to the rise of variation and stability in the developmental process (Meyer and Roeder, 2014). Actually, stochasticity is at the origin of heterogeneity among the isogenic cells of a developing multicellular body and, by generating cellular fates then stabilized by regulatory mechanisms such as feedback loops, it contributes to make developmental outcomes reproducible. We argue that an augmented epistemological view of chance is thus needed that takes into account both that chance can be a perturbation to the normal, deterministic functioning of biological systems, but that it can also positively, and even functionally, contribute to it. Such a positive view of chance is already present and operative in other biological fields studying molecular processes taking place inside the cell, namely immunology. This is why, in the following section, we briefly show the role that chance plays in some immunological explanation in order to take it as an example.

Chance in immunology: an example to follow

In this section, we briefly describe the way the synthesis of antibodies is accounted for in immunology, and focus our attention on the acknowledged role of chance in the explanation of this molecular process. Our motivation is to point at the fact that, in immunology, chance has a well-recognized and unproblematic explanatory role. This shows that the positive view of chance we put forward in order to make sense of the role stochasticity can have in gene expression is already present and acknowledged in other biological fields dealing with molecular and cellular processes. Immunology thus provides us with an example to follow in order to counter the widespread, longstanding resistance, in the study on noise in gene expression and cell differentiation, to consider chance more than just noise.

In any immunology textbook, one can learn that mature B-lymphocytes produce antibodies, a specific type of glycoproteins, which in turn bind to antigens and, through a cascade of molecular mechanisms, allow the organism to respond properly to the presence of these molecules. Each mature B-lymphocyte can produce only one type of antibody that can bind only one type of antigen. The entire repertory of antibodies cannot be linearly stocked in the DNA. Indeed, it would need too much space in the nucleus for such amount of information. So, how is this diversity possible? The answer is: by chance!

Present in a developing B lymphocyte are numerous gene segments categorized as variable (V), diversity (D), and joining (J), whose main function is to synthetize antibodies. During B-lymphocyte development, these regions are stochastically brought together in a process called V(D)J rearrangement. Each mature B-lymphocyte thereby contains a unique DNA V(D)J sequence which is then involved in the synthesis of a unique antibody (see Fig. 2.3).

Different pieces of V(D)J are rearranged in two steps: (1) two types of recombinase enzyme called RAG-1 and RAG-2 (Recombination Activating Gene) cut different parts of genes at different sites[8]; (2) the DNA repair machinery ligates the DNA strand back together.[9] What is called the "snip-and-fix process" is the fact that different pieces of DNA (VDJ gene segments) create a unique sequence, which can give rise to a single, unique antibody. The variation of these sequences results from: (1) the stochastic rearrangement of (V), (D), and (J) gene segments by recombinase enzymes (RAG); and (2) what is commonly expressed as "imprecise joining" between rearranged gene segments. This variation, which is thus considered to be due to partly

[8]The RAG enzymes recognize the recombination signal sequences (RSSs) to bind DNA.

[9]For further details, see any immunological textbook (e.g., Kindt et al., 2007; Mak and Saunders, 2008).

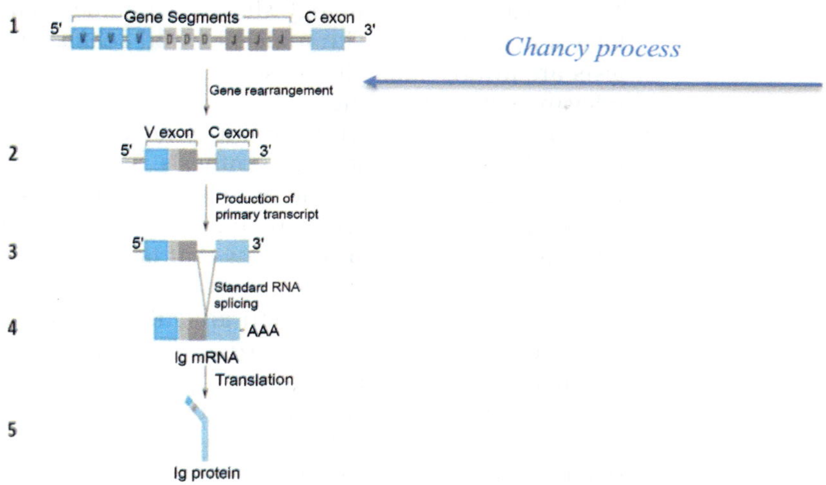

Figure 2.3 A qualitative overview of the process which transforms V(D)J genes segments into antibodies (Ig protein). (1) The DNA strand with segments V, D, and J (segment C is not relevant to our purpose) of the immature B-lymphocyte; (2) the DNA after the V(D)J rearrangement in the mature B-lymphocyte. The transition from (2) to (3) represents the first step of gene expression (transcription, DNA → immature RNA); the transition from (3) to (4) represents the maturation of RNA through the splicing process, a mechanism of splitting and lumping of RNA sequences; the transition from (4) to (5) represents the second mechanism of gene expression (translation, mature RNA → proteins). *Source*: Figure by Mak, T., Saunders, M., 2008. Primer to the Immune Response, first ed. Elsevier.

stochastic molecular processes, is recognized by immunologists as a fundamental part of the explanation of antibodies synthesis.

To better understand just how essential chance is in immunology explanation, let us consider two counterfactual scenarios in which no chancy processes are involved in the production of antibodies. Either the organism is able to produce just one type of antibody from all mature B-lymphocytes, or the rearrangement of VDJ sequences allowing the production of different types of antibodies happens in a deterministic way. In the first scenario, the immune system would be effectively useless for the organism as it could allow it to fight only against one type of antigen. In the latter, we can make the hypothesis that the rearrangement process would be far away from the optimal solution in terms of energy costs. Indeed, it would require a lot of energy, and could thus destabilize the general energy balance of the cell.[10]

[10]More general considerations about energy metabolism and optimization can be found in Paldi (2020), this volume. For more details about the relationship between the precision of a biological process and energy costs, see Lestas et al. (2010) also mentioned in Paldi's chapter.

The case of immunology shows, we argue, that in other biological fields dealing with molecular and cellular processes biologists are aware, since decades, of the functional and explanatory relevance of chance. They acknowledge that the chancy character of DNA sequence rearrangement is an essential component of the explanation of antibody synthesis. This is why we point at immunology as an example of how to think about chance when it has a constructive role in molecular and cellular processes, and thus to adopt a positive view that takes it into consideration, and even puts it forward. By contrast, research studies on what is called noise in gene expression are still embedded in the deterministic framework characterizing molecular biology since its origin in the 1950s. The idea of "genetic program" is a central feature of this view (Jacob and Monod, 1961; Mayr, 1961). It postulates that the relation between genotype and phenotype is deterministic (or nearly): the same genes give rise to the same phenotypes, and chance can play no more than a disruptive role in genetic processes. Research advances since the end of the 1970s have shown that this strictly deterministic view is no more tenable because of the complex set of processes involved in gene expression, protein synthesis and, more broadly, organism development (for historical and philosophical analyses of genetic determinism, see Sarkar, 1996; Kay, 2000; Keller Fox, 2000; Kitcher, 2003; Rheinberger and Müller-Wille, 2018). Nevertheless, the most common view of chance in the literature on gene expression still describes it as a mere "nuisance" for the deterministic process of gene expression,[11] which consists in running the genetic program. While immunology and molecular biology both study processes taking place at the same level of biological organization (namely, gene rearrangement and gene expression), the explanatory role of chance in perfectly acknowledged in the former, and tends to be denied in the latter. Why is there a widespread resistance to provide chance with any explanatory relevance in the molecular study of gene expression?

One could argue that chance is actually detrimental for gene expression whereas it is actually functional for immunological response. This is why it is differently conceived in these two

[11]Examples of the negative role of chance (i.e., noise as a nuisance, hence not explanatory of the functioning of biological systems) in the literature are: (1) the study of cell identity, where noise is conceived as an obstacle for the correct measurement of the identity of cells (e.g., see Birnbaum and Kussell, 2011; Reiter et al., 2011; Efroni et al., 2015; Grün et al., 2016); (2) the study of alternative splicing process (e.g., Melamud and Moult, 2009, p. 4873; Jin et al., 2017, p. 11), characterized by the (more or less explicit) assumption that if the alternative splicing is due to noise, *then* it is not functional.

areas of biological research studies. However, this counterargument fails in front of recent empirical evidence in molecular and cellular biology calling for the positive view of chance already present in immunology. Indeed, more and more studies of noise in gene expression (as those described in "What, actually, is noise?" section) clearly show the role of chance in various molecular processes, in particular those taking place inside the cell (e.g., genetic recombination and gene expression). We thus need an augmented epistemological view for molecular and cellular biology that admits the possibility of chance being explanatory of cell functioning. It is worth noting that this does not mean that all research fields in biology have to include the notion of chance in their theoretical apparatus or, put it differently, that chance is essential to account for all biological processes. However, we want to underline how, in certain cases, it is necessary to admit chance as an important and even necessary part of biological explanation. Moreover, such a positive view could allow to ask new questions, which cannot be addressed from the mainstream perspective (chance as a mere nuisance). For example, can chance play a functional role in development? Can it be selected and passed on across generations? How its functional role (if any) could be explained in an evolutionary framework?[12]

A positive view of chance: main features and theoretical advantages

Before entering in a more detailed description of the positive view of chance we suggest for the study of molecular and cellular processes, let us introduce one misleading direction we want to avoid, and which can stand in the way of research about chance.

The study of chance could present the temptation to settle research from the question: "what is chance, and where is it in the cell?" This kind of research question leads to the risk of reifying chance as if it was a thing, a physical or a biological object that can be detected and identified. To avoid it, in this paper we adopt a strictly epistemological perspective on chance, and consider it as a "theoretical operator" (Morizot, 2012) in biology. In other words, we suggest that chance should be seen as a theoretical entity that enters in biological

[12]It is noteworthy that similar questions are not completely absent in literature. See, e.g., Heams (2014, p. 5) and Pavé (2007, p. 191).

descriptions and explanations as a parameter,[13] in order to account, with other factors, for the behavior of biological systems.

Let us now present in further detail the main features of the view of chance we defend here for molecular and cellular biology, and highlight the theoretical advantages it brings about both for philosophical reflection and biological research. In this context, chance is not always seen as a perturbation, thus described with an error term, but can also be part of the explanation of biological systems. This does not imply, however, an ontological commitment for chance as a specific and irreducible biological phenomenon. Rather, chance characterizing biological systems can be in principle reduced to the physical processes at its origin, but still these processes are constrained and canalized by highly evolved and organized systems (namely, developing organisms) that bias their behavior and main features.[14] Thus, to sum up, ontologically speaking, chance in biological systems is physical because it is in principle reducible to physical phenomena; however, it has a biological value from an epistemological perspective because of its explanatory role in biological explanations, which would be lost if chance was reduced to physical descriptions epistemologically speaking too.[15]

Now, let us illustrate with a biological example why, if we do not embrace the view of chance we defend, and thus do not acknowledge the role it plays in certain biological systems, we leave a gap in the explanation of their functioning.

The example, taken from Meyer and Roeder (2014), is about the various steps of cell differentiation and development in

[13]More precisely, by "parameter" we mean a term that participates in the formal description and explanation of a given system, that can have different values (vs a constant), and that corresponds to a state of the system in the world.

[14]This is the main difference between our view of chance and, for instance, Kupiec's who, from our perspective, does not stress enough that organisms are highly evolved and organized biological systems constraining chance. Contrarily to us, Kupiec is metaphysically committed to chance and ascribes it a preponderant, and even leading role in the origin of biological organization (see Kupiec, 2019). More precisely, he argues that molecular interactions lack of any specificity, are intrinsically noisy, such that the idea of genetic program is no more valid. More broadly, according to him, all biological processes, and in particular cell differentiation, necessarily need chance in order to take place (see Kupiec, 2010; Kupiec and Sonigo, 2003). Unlike Kupiec, in this paper we provide epistemological arguments in favor of the role of chance in biological explanations, without any commitment as regards its ontology.

[15]More generally, Kaiser (2015) writes that "what holds for ontology need not also apply to epistemology" and that "you can have ontological reduction without, at the same time, having epistemic reduction" (p. 51).

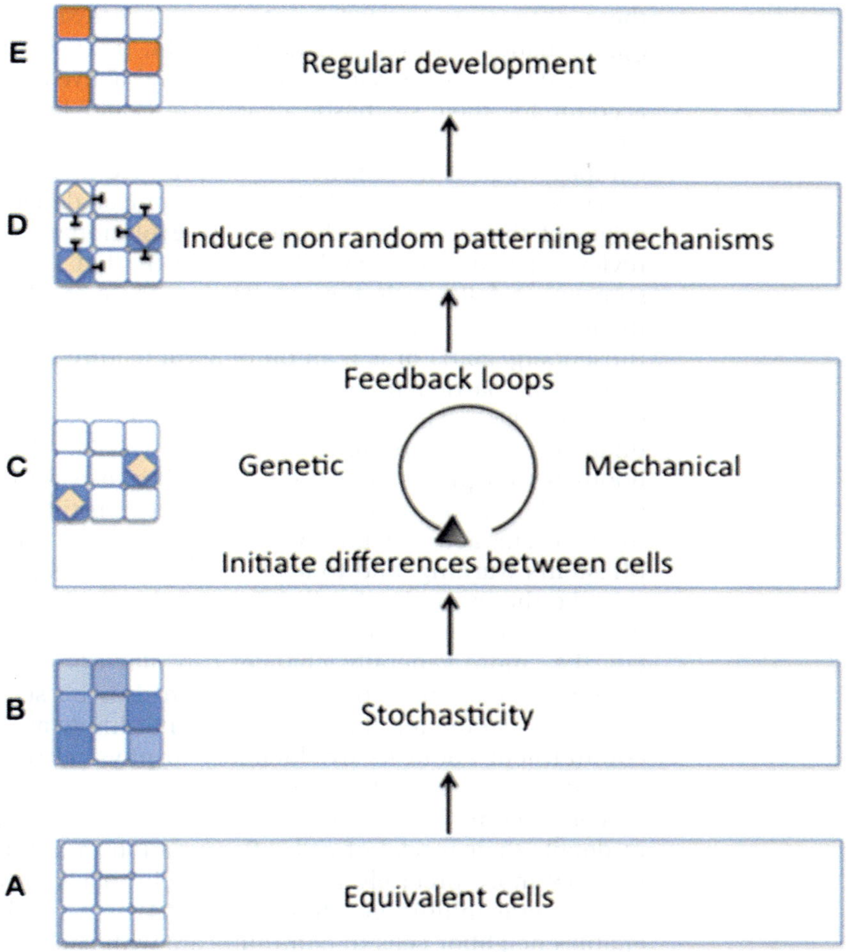

Figure 2.4 Schematic model of the importance of chance in promoting regular plant and animal development. (A) In early development, cells are morphologically equivalent. (B) Differences between morphologically equivalent cells arise through stochastic fluctuations in gene expression. (C) Differences between cells are stabilized by regulatory mechanisms such as feedback loops. (D) Once stabilized, lateral inhibition mechanisms take place. (E) Regular development takes place in a regular manner. *Source*: Taken from Meyer, M., Roeder, A., 2014. Stochasticity in plant cellular growth and patterning. Frontiers in Plant Science, Plant Evolution Dev. 5, 420, 1–14.

animals. The general schema proposed by these two authors is the following (see Fig. 2.4):

Looking at Fig. 2.4, one can notice that the second step of cell differentiation in animals is a "stochastic" one (step B). This is not the case of the following step (step C), during which "[g] enetic and mechanical feedback loops can enhance and solidify these differences [caused by stochastic processes] to begin cell differentiation" (Meyer and Roeder, 2014, p. 1). Let us focus on step B. What processes are stochastic? And above all, how can we account for their stochastic character? Could be the case that what we call "a stochastic step" is no chancy at all but rather a complex set of deterministic and unidentified mechanisms?

These questions recall the philosophical debate about the origin of the stochastic character of evolutionary theory (for a review, see Malaterre and Merlin, 2015) that we can roughly sum up as follows: is the stochastic character of evolutionary theory due to our ignorance of some "hidden variables" (Rosenberg, 2001) or is it an ontological feature of the biological world, in particular of biological evolution (Brandon and Carson, 1996)? This metaphysically ambitious question is still unanswered. According to Millstein (2000), the debate is doomed to a philosophical dead end because we have not yet enough knowledge to argue for the deterministic or the indeterministic character of the evolutionary process. In the light of Millstein's warning, and to avoid entering in this sort of debate, we take the example of cell differentiation and development in animals in order to show and argue for the fact that stochasticity (i.e., more generally, chance) can play an explanatory role in the study of these processes. It can thus be seen a specific biological parameter, that which does not imply any ontological commitment about its irreducibility to physical phenomena or its eliminability by hidden variables yet to be discovered.

Let us come back to Meyer and Roeder's paper and analyze in details their model of cell differentiation in animals in order to show more precisely how stochasticity plays an explanatory role. Lateral inhibition (step D, Fig. 2.4) is a mechanism that enables neighbor cells to communicate with each other in order to synchronize their differentiation during, for example, neural development (Morrison et al., 2000), that is to inhibit or to activate the expression of specific genes in a coordinated manner. In animals, two *trans*-membrane proteins, the receptor Notch and the ligand Delta, are involved in this mechanism. When the receptor Notch of one cell interacts with a *trans*-Delta ligand (i.e., the Delta of another cell), it triggers a phosphorylation cascade that activates the expression of Notch genes (*trans*-activation). Otherwise, the interaction between a Notch receptor and a *cis*-Delta (i.e., within the same cell) produces a *cis*-inhibition in the cell (i.e., no Notch genes are activated). Moreover, when a cell has a stabilized higher level of Notch receptors than Delta ligands, it is in a "receiving states," namely, it is "waiting for" an interaction with *trans*-Delta to activate its own Notch genes.

Instead, when a cell has a stabilized higher level of Delta ligands than Notch receptors, it is in a "sending state," namely, it is "waiting for" a cell with *trans*-Notch to activate. What is the mechanism allowing cells to be in a sending or waiting state? Regulatory mechanisms such as genetic or mechanical feedback loops (step C, Fig. 2.4). But what determines the concentrations of ligands and receptors in cells that, once stabilized, are responsible of the sending/waiting cell states? Sprinzak et al. (2010) show that small differences in the concentration of Delta and Notch factors are caused by stochastic fluctuations in the expression of genes involved in the production of these proteins. In the light of these molecular details, it is no longer possible to deny that stochasticity is a relevant feature of a precise step of cell differentiation (step B, Fig. 2.4), source of differential gene expression, and so of random variation in the state of cells. Put it differently, if we reject the explanatory role of stochasticity here, we cannot account for what step B gives rise to (namely, differences between isogenic cells of a developing organism), and we thus have a gap in the explanation of cell differentiation and development.

Let us examine this point closer in order to argue for it in a counterfactual manner. With no stochastic fluctuation, one should imagine the following scenario in order to explain what happens, at step B, during cell differentiation: cells can regulate in a fine-grained manner gene expression, and so the concentration of Notch and Delta proteins. Otherwise, with neither stochastic fluctuation nor a deterministic mechanism, cell differentiation could not take place. This counterfactual scenario shows why stochasticity should be seen as an explanatory feature of the developmental process because it contributes to explain how cell differentiation takes place—namely, because genes are expressed in a stochastic manner, thus producing random variation in protein concentration that are stabilized giving rise to various lineages of differentiated cells.

We argue that to consider chance as a biological parameter, which contributes to characterize the behavior of specific biological processes (i.e., fluctuations in gene expression and protein synthesis), allows providing a better explanation[16] of cellular differentiation and development than to build fine-grained (stochastic or

[16]We maintain that a better explanation is an explanation that better fits available theoretical and experimental evidence. In this specific context, a better explanation is an explanation that gives to chance the explanatory role it deserves based on recent research results on it.

deterministic) descriptions of physical mechanisms involved in these molecular processes. In the following section, we present three arguments in favor of this thesis that are, more precisely, two related reasons and a historical evidence.

Three reasons for biological explanations in terms of chance

The first general reason to favor explanations in which chance plays a role as a biological parameter is that a more detailed explanation, at a lower level, is not always a guarantee of higher explanatory power than a higher level, less detailed one. Think of Borges' story about an empire where cartography has reached an accuracy level where it only produces (very detailed) maps on the same scale as the empire itself. One can ask whether such maps would be useful to anybody. By analogy, would a complete description of the physical processes involved in gene expression and protein synthesis be explanatory of cell differentiation and development? The *hubris*[17] for complete reductionist explanations makes us forget that we can decide to abstract away from fine-grained details in order to produce better explanations of the phenomenon under study. Of course, in our case, an explanation mentioning all the physical details of gene expression and protein synthesis would be useful to explain the origin of variability in protein concentration. However, what is at stake here (i.e., our explanatory purpose) is not how it comes that the concentration of proteins varies in a random manner from cell to cell, and even in the same cell over time. Rather, we investigate how such variation in protein concentration, which turns out to be well described in a stochastic way (i.e., in terms of probability), contributes to the process of cell differentiation and, more broadly, to development. In other words, if the *explanandum* is the origin of stochastic fluctuations in gene expression ("noise"), the *explanans* should be formulated in terms of their physical causes (see "What, actually, is noise?" section). Otherwise, if the *explanandum* is how cell differentiation takes place, the *explanans* should be

[17]It is interesting to note how the reductionist tendency is not always due to a *hubris* for complete explanations. For example, Woodward (2010: 295) claims that adopting a "reductive direction" can come from the necessity to find a more "stable causal relationship" in a scenario of complex distal causal relationships. Nevertheless, he claims, even if the necessity of stability drives us toward a reductionist approach, it could be the case that macrorelations are highly more stable than micro ones, and that the former better satisfy other conditions (e.g., proportionality, see later in this section).

identified in stochasticity in gene expression and the consequent random variation in protein synthesis.[18]

All this means that the choice of the level of explanation is actually context dependent. In particular, it depends on the "investigator's purpose" (Woodward, 2010, p. 297), that is, what "the investigator wished to explain or understand." Sober (1999) argues against this thesis by claiming that, in a specific scenario (e.g., the explanation of cell differentiation), we do not have any objective argument to prefer one explanation (let us say, at the higher level of random patterns of protein concentration) to another (let us say, at the lower level of physical processes at the origin of stochasticity in gene expression). Which one to choose is just "a matter of taste." Therefore, according to Sober, there is no objective reason to choose one explanation rather than another in any explanatory context. We do not agree with Sober on this point. In line with Godfrey-Smith's contextualism (2003), we rather argue that the pieces of explanation one chooses, at some level, depend on the explanatory context in which the explanation is developed (i.e., what Woodward (2010) calls the "investigator's purposes").

The second reason we have for favoring higher level explanations in which chance intervenes as a biological parameter resides in what Woodward (2010) calls the "proportionality criteria." Following this author, proportionality is one relevant criterion we should use to choose an appropriate level of explanations. Roughly, a good level of explanation is a level in which causes and effects are proportional, so that no irrelevant detail at the level of causes is part of the explanation of the effects.[19] For example, as claimed by Woodward himself, if one's explanatory target is the generation of spike trains of individual neurons in response to incoming signals, it could be useful to study the detailed temporal features of this process for each neuron. Otherwise, if one wants to explain the neuronal response to external stimuli, these details may be explanatory irrelevant: only the overall firing rates should be mentioned in the explanation. In order to better understand what he means by proportionality between causes and effects, Woodward proposes an

[18]Note that our claim joins Garfinkel's argument (1981) for the autonomy of macro (higher level) explanations. They do not have the same object and explanatory target than micro (lower level) explanations, so that they should be preferred depending on the explanatory context.

[19]Note that Garfinkel (1981) also claims that we should favor macroexplanations because often microexplanations do not tell us what would have been otherwise, and this is due to the fact that different situations at the micro level can bring about the same result at the macro level (what he calls, "redundant causality").

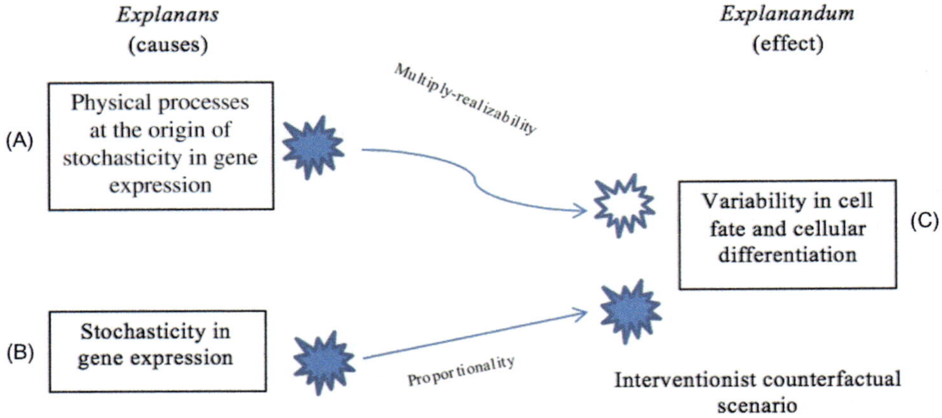

Figure 2.5 Schema of an interventionist counterfactual scenario. On the left, two possible *explanans*, A and B, of the *explanandum* C *(on the right)*. The arrows represent the causal connections between *explanans* and *explanandum*. In the case of A→C (multiple-realizable causality), if one intervenes on A *(blue star)*, C could fail to change *(white star)*; in the case of B→C (proportional causality), if one intervenes on B *(blue star)*, it is very likely that C changes too *(blue star)*. For the implication of this difference, see the main text.

interventionist counterfactual scenario. If I intervened on certain details at the level of causes and no change happened at the level of the effects, then these details should be considered as irrelevant for the explanation because they make no difference for the effects, and thus play no role in their explanation.[20]

Starting from our *explanandum*, which is cell differentiation (more specifically, variability in cell fates), let us know try to choose the appropriate *explanans* by using Woodward's proportionality criteria.[21] The *explanans* could be at different levels. It can correspond to the physical processes at the origin of stochasticity in gene expression (e.g., thermal motion; quantum effects, etc.) or it can be stochasticity in gene expression itself, which is measured as the random variation of protein concentration in a population of cells. Which one of these two *explanans* better account for our *explanandum*? In order to answer to this question, we can imagine the following counterfactual situations (see Fig. 2.5). First of all, what would happen if one

[20]The other way around works too: if one intervened on certain details at the level of causes and this change provoked corresponding changes at the level of the effects, then these details should be considered as relevant for the explanation because they make a difference for the effects, and they thus play a role in their explanation.

[21]It is worth noting that we are not just looking for the more suitable level of explanation but also arguing that this level is the one where chance plays an explanatory role.

intervened to change the physical processes at the origin of stochasticity in gene expression? Would this provoke some detectable change in the effect, that is, in the process of cell differentiation? In the light of the complexity of cell dynamics, a cautious answer could be "maybe yes." But this does not fit Woodward's proportionality criteria because it is likely that some changes in the physical processes at the origin of stochasticity in gene expression (the causes) "*fail* to be associated with changes in the effect" (Woodward, 2010, p. 298, italics in original), that is, in the distribution of various fates in a population of cells. In other words, because of the multirealizability, at the physical level, of the processes of gene expression and protein synthesis, there is no proportionality between these physical processes and variability in cell fates. We can thus conclude that the physical processes at the origin of stochasticity in gene expression can be considered causes of variability in cell fates but they are, if any, explanatorily irrelevant ones. Second, what would happen if one intervened to change stochasticity in gene expression, and thus the random patterns of protein concentration in a population of cells? Would this provoke some detectable change in the effect, that is, in the process of cell differentiation? The answer is yes. Actually, if one changed the temporal patterns of protein concentration (which is the way stochasticity in gene expression is measured), it is very likely that the distribution of various cell fates in a population of cells would change. Thus, in this case, the proportionality criteria between causes and effects is fulfilled, that which reveals which causes are relevant to account for our *explanandum*. To conclude, it seems reasonable to say that the explanation to choose in order to account for variability in cell fate and cellular differentiation is the one in terms of random patterns of protein concentration, in which chance is a biological parameter characterizing the behavior of biological processes, and playing a central explanatory role.

Let us finally look at the historical evidence in favor of our claim that a more detailed explanation, at the lower level, is not always a guarantee of a better explanation than the higher level, less detailed one. After the rise of molecular biology (1940s–1960s), cell biologists were afraid that their research would be reduced to molecular studies. This fear stemmed from the evidence that vesicles, transport proteins, organelles, and their activities (i.e., all research objects of cell biology) are made of the same macromolecules molecular biologists deal with. In this scenario, a molecular explanation of objects and activities studied by cell biology could have been considered as a better explanation than the cellular-level one, that which could have produced the end of cell biology. However, this did not happen. Indeed, in the 1980s cell biology had its "golden age" (Morange, 2000, p. 244) and has been a breeding ground of scientific research until today.[22] This event in the history of biology shows that sometimes lower level, reductionist approaches can fail to prevail over higher level, nonreductionist ones.[23] We think that it provides a good historical case going in the same direction as our specific epistemological antireductionism as regards the explanatory role of stochasticity in cell differentiation and development.

Conclusion: Chance and the reductionism/antireductionism debate

In this last section, we want to briefly clarify the relationship between the augmented epistemological view of chance suggested here and the reductionist-antireductionist debate in philosophy of science it is related to.

We have just argued that explanations of cellular differentiation and development in terms of chance as a biological parameter are to be preferred to lower level, physical (deterministic or stochastic) ones. One could object that, for our claim to be compelling, we need to argue against ontological reductionism, which means, in our case, to show that chance in gene expression cannot be reduced to physical chance or to deterministic physical phenomena. Otherwise, with no autonomous levels of causality, no autonomous levels of explanation can be defended. We resist this objection by claiming that no form of ontological antireductionism about chance in biology is needed in order to ground the autonomy of biological explanations (with respect to lower level, physical ones). More explicitly, we argue that, even though biological phenomena involved in cell differentiation and development could be reduced to physical ones, and their explanations in terms of biological chance could in principle be reduced to more detailed and complex physical ones, there are still reasons to prefer the former sort of explanation with respect to the latter

[22]Note that molecular approaches to cell biology coexist with strictly cellular ones; for example, see Alberts et al. (2002).

[23]Another historical example of an antireductionist development of biology can be seen in the relationship between classical genetics and biochemistry in the 1930s (see Morange, 2000).

(see "Three reasons for biological explanations in terms of chance" section). Contrary to the antireductionist thesis defended by authors like Kitcher (1984), these reasons are not ontological, but epistemological ones: they come from the difference in *explanandum* between lower and higher level explanations (in this case, physical and biological ones) and from the fact that the higher level phenomenon we could want to explain can be realized by many different lower level phenomena, that which is directly linked to the proportionality criteria suggested by Woodward. These two reasons, with the historical evidence just introduce above, allow us to argue for the need to prefer, in some cases, higher level explanations (namely, explanations of cell differentiation in terms of random patterns of gene expression and protein synthesis, rather than in terms of lower level physical processes) without the need of any strong commitment for both ontological and epistemological antireductionism.

To conclude, our augmented epistemological view of chance for the study of cell differentiation and development is coherent with our antireductionist position as regards explanation. In fact, it does not deny that chance can be a perturbation (described as an error term, i.e., noise), but opens the possibility of conceiving chance as a biological parameter with an explanatory role. As to the reductionism-antireductionism debate, we do not deny ontological reductionism as well as the possibility, in principle, to reduce biological explanations to more detailed physical ones; however, we argue for the epistemic advantage, in some cases, of preferring biological, less detailed but probabilistic explanations because of their higher explanatory power.

Acknowledgments

Many thanks to Jeanette Hahn and David Dubnau for kindly sharing with us the image for Fig. 2.1 and for permitting us to publish it in this chapter. We also thank Michel Morange and Philippe Huneman for their comments and suggestions.

References

Alberts, B., Johnson, A., Lewis, J., Raff, M., Roberts, K., Walter, P., 2002. Molecular Biology of the Cell. Garland Science Taylor & Francis Group, New York.

Birnbaum, D., Kussell, E., 2011. Measuring cell identity in noisy biological systems. Nucleic Acids Res. 39 (21), 9093–9107.

Blake, W.J., Kærn, M., Cantor, C.R., Collins, J.J., 2003. Noise in eukaryotic gene expression. Nature 422, 633–637.

Brandon, R., Carson, S., 1996. The indeterministic character of evolutionary theory: no 'no hidden variables' proof but no room for determinism either. Philosophy Sci. 63, 315–337.

Cağatay, T., Turcotte, M., Elowitz, M.B., Garcia-Ojalvo, J., Süel, G.M., 2009. Architecture-dependent noise discriminates functionally analogous differentiation circuits. Cell 139 (3), 512–522.

Debré, P., 1996. Jacques Monod. Flammarion, Paris.

Efroni, I., Ip, P., Nawy, T., Mello, A., Birnbaum, D., 2015. Quantification of cell identity from single-cell gene expression profiles. Genome Biol. 16 (9), 1–12. Available from: https://doi.org/10.1186/s13059-015-0580-x.

Eldar, A., Elowitz, M., 2010. Functional roles for noise in genetic circuits. Review, Nat. 467, 167–173. Available from: https://doi.org/10.1038/nature09326.

Elowitz, M.B., Levine, A.J., Siggia, E.D., Swain, P.S., 2002. Stochastic gene expression in a single cell. Reports. Science 297, 1183–1186.

Feinberg, A., Irizarry, R., 2010. Evolution in health and medicine Sackler colloquium: Stochastic epigenetic variation as a driving force of development, evolutionary adaptation, and disease. Proc. Natl. Acad. Sci. U S A 107 (Suppl. 1), 1757–1764. Available from: https://doi.org/10.1073/pnas.0906183107.

Fraser, H.B., Hirsh, A.E., Giaever, G., Kumm, J., Eisen, M.B., 2004. Noise minimization in eukaryotic gene expression. PLoS Biol. 2, 0834–0838.

Gandrillon, O., Kolesnik-Antoine, D., Kupiec, J.J., Beslon, G., 2012. Chance at the heart of the cell. Prog. Biophys. Mol. Biol. 110 (1), 1–150.

Garfinkel, A., 1981. Forms of Explanation. Yale University Press, New Haven, CT.

Gillespie, D., 2007. Stochastic simulation of chemical kinetics. Annu. Rev. Phys. Chem. 58, 35–55.

Gillespie, J.H., 1977. Natural selection for variance in offspring numbers: a new evolutionary principle. Am. Naturalist 111, 1010–1014.

Godfrey-Smith, P., 2003. *Theory and Reality. An Introduction to the Philosophy of Science*. The University of Chicago Press, Chicago and London.

Grün, D., Muraro, M.J., Boisset, J., Wiebrands, K., Lyubimova, A., Dharmadhikari, G., et al., 2016. De novo prediction of stem cell identity using single-cell transcriptome data. Cell Stem Cell 4 (19), 1–12.

Haldane, J.B.S., Jayakar, S.D., 1963. Polymorphism due to selection of varying direction. J. Genet. 58, 237–242.

Han, R., Huang, G., Wang, Y., Xu, Y., Hu, Y., Jiang, W., et al., 2016. Increased gene expression noise in human cancers is correlated with low p53 and immune activities as well as late stage cancer. Oncotarget 7 (44), 72011.

Heams, T., 2014. Randomness in biology. Mathematical Structures in Computer Science 24 (3), 1–24. Available from: https://doi.org/10.1017/S096012951200076X. e240308, Cambridge University Press.

Holmes, W.R., de Mochel, N.S.R., Wang, Q., Du, H., Peng, T., Chiang, M., et al., 2017. Gene expression noise enhances robust organization of the early mammalian blastocyst. PLoS Comput. Biol. 13 (1), e1005320.

Huang, S., 2009. Non-genetic heterogeneity of cells in development: more than just noise. Development 136, 3853–3862. Available from: https://doi.org/10.1242/dev.035139.

Jacob, F., Monod, J., 1961. Genetic regulatory mechanisms in the synthesis of proteins. J. Mol. Biol. 3, 318–356.

Jin, Y., Dong, H., Shi, Y., Bian, L., 2017. Mutually exclusive alternative splicing of pre-mRNAs. Wiley Interdiscip. Reviews: RNA 9.3 1–14. e1468.

Johnston, R., Desplan, C., 2010. Stochastic mechanisms of cell fate specification that yield random or robust outcome. Annu. Rev. Cell Dev. Biol. 26, 689–719.

Kaern, M., Elston, T.C., Blake, W.J., Collins, J.J., 2005. Stochasticity in gene expression: from theories to phenotypes. Nat. Reviews, Genet. 6, 451–464.
Kaiser, M.I., 2015. Reductive Explanation in the Biological Sciences. Springer International Publishing.
Kay, E.L., 2000. Who Wrote the Book of Life? A History of the Genetic Code. Stanford University Press, Stanford.
Keller Fox, E., 2000. The Century of the Gene. Harvard University Press, Cambridge MA.
Kitcher, P., 2003. Battling the undead: how and (how not) to resist genetic determinism. Mendel's Mirror: Philosophical Reflections on Biology. Oxford University Press, pp. 283–300.
Kindt, T.J., Goldsby, R.A., Osborne, B.A., Kuby, J., 2007. Kuby Immunology. Macmillan.
Kint, C.I., Verstraeten, N., Fauvart, M., Michiels, J., 2012. New-found fundamentals of bacterial persistence. Trends Microbiology 20 (12), 577–585.
Kitcher, P., 1984. 1953 and all that. A tale of two sciences. Philos. Rev. XCIII (3), 335–373.
Klingenberg, C., 2005. Developmental constraints, modules, and evolvability. In: Hallgrimsson, B., Hall, B.K. (Eds.), Variation. A Central Concept in Biology. pp. 219–247.
Kupiec, J.J., 2010. On the lack of specificity of proteins and its consequences for a theory of biological organization. Prog. Biophys. Mol. Biol. 102 (1), 45–52.
Kupiec, 2019. Et si le vivant était anarchique. La génétique est-elle une gigantesque arnaque? LLL, Les Liens Qui Libèrent.
Kupiec, J., Sonigo, P., 2003. Ni Dieu ni gène: Pour une autre théorie de l'hérédité. Paris, Seuil.
Lestas, I., Vinnicombe, G., Paulsson, J., 2010. Fundamental limits on the suppression of molecular fluctuations. Nature 467, 174–178.
Losick, R., Desplan, C., 2008. Stochasticity and cell fate. Science 320 (5872), 65–68. Available from: https://doi.org/10.1126/science.1147888.
Maamar, H., Raj, A., Dubnau, D., 2007. Noise in gene expression determines cell fate in Bacillus subtilis. Science 317, 526–529.
Maheshri, N., O'Shea, E., 2007. Living with noisy genes: How cells function reliably with inherent variability in gene expression. Annu. Rev. Biophys. Biomol. Struct. 36, 413–434.
Mak, T., Saunders, M., 2008. Primer to the Immune Response, first ed. Elsevier.
Malaterre, C., Merlin, F., 2015. The (in)determinism of biological evolution: where does the stochastic character of evolutionary theory come from? In: Heams, T., Huneman, P., Lecointre, G., Silberstein, M. (Eds.), Handbook of Evolutionary Thinking in the Science. Springer, Dordrecht, pp. 349–366.
Mayr, E., 1961. Cause and effect in biology. Science 134 (3489), 1501–1506.
McAdam, H., Arkin, A., 1997. Stochastic mechanisms in gene expression. Proc. Natl. Acad. Sci. U S A 94, 814–819.
McAdams, H.H., Arkin, A., 1999. It's a noisy business! Genetic regulation at the nanomolar scale. Trends Genet. 15 (2), 65–69.
Melamud, E., Moult, J., 2009. Stochastic noise in splicing machinery. Nucleic Acids Res. 37 (14), 4873–4886.
Merlin, F., 2015. Developmental noise: explaining the specific heterogeneity of individual organisms. In: Braillard, P., Malaterre, C. (Eds.), How does Biology Explain? An Enquiry into the Diversity of Explanatory Patterns in the Life Sciences. Springer, Dordrecht, pp. 91–110.
Meyer, M., Roeder, A., 2014. Stochasticity in plant cellular growth and patterning. Frontiers in plant science. Plant Evolution Dev. 5 (420), 1–14.

Millstein, R.L., 2000. Is Evolutionary Process Deterministic or Indeterministic? An Argument for Agnosticism. Presented at the Biennial Meeting of the Philosophy of Science Association Vancouver, Canada: <http://philsci-archive.pitt.edu/4544/1/Millstein2000Agnosticism.pdf>.

Mitchell, S., 2009. Unsimple Truths: Science, Complexity, and Policy. University of Chicago Press, Chicago.

Morange, M., 2000. A History of Molecular Biology. Harvard University Press, Cambridge, MA.

Morizot, B., 2012. Chance: from metaphysical principle to explanatory concept. The idea of uncertainty in a natural history of knowledge. Prog. Biophys. Mol. Biol. 110, 54–60.

Morrison, S., Perez, S., Qiao, Z., Verdi, J., Hicks, C., Weinmaster, G., et al., 2000. Transient Notch activation initiates an irreversible switch from neurogenesis to gliogenesis by neural crest stem cells. Cell 101, 499–510. Available from: https://doi.org/10.1016/S0092-8674(00)80860-0.

Nanjundiah, V., 2003. Phenotypic plasticity and evolution by genetic assimilation. In: Müller, G.B., Newman, S.A. (Eds.), Origination of Organismal Form: Beyond the Gene in Developmental and Evolutionary Biology. MIT Press, pp. 245–264.

Nanjundiah, V., Bhogle, A.S., 1995. The precision of regulation in Dictyostelium discoideum: Implications for cell-type proportioning in the absence of spatial pattern. Indian. J. Biochem. Biophys. 32, 404–416.

Paldi, A., 2020. Random walk across the epigenetic landscape. In: Levine, H., Kulkarni, P., Jolly, M., Nanjundiah, V. (Eds.), Phenotypic Switching: Implications in Biology and Medicine. Elsevier.

Pavé, A., 2007. Necessity of chance: biological roulettes and biodiversity. C. R. Biol. 330, 189–198.

Pilpel, Y., 2011. Noise in biological systems: pros, cons, and mechanisms of control. Yeast Systems Biology. Humana Press, pp. 407–425.

Pujadas, E., Feinberg, A., 2012. Regulated noise in the epigenetic landscape of development and disease, Cell, 148. Elsevier Inc., pp. 1123–1131.

Raj, A., van Oudenaarden, A., 2008. Nature, nurture, or chance: stochastic gene expression and its consequences. Cell 135, 216–226.

Raser, J.M., O'Shea, E.K., 2005. Noise in gene expression: origins, consequences, and control. Science 309 (5743), 2010–2013.

Reiter, M., Kirchner, B., Müller, H., Holzhauer, C., Mann, W., Pfaffl, M.W., 2011. Quantification noise in single cell experiments. Nucleic Acids Res. 39 (18), e124.

Rheinberger, H., Müller-Wille, S., 2018. The Gene: From Genetics to Postgenomics. University of Chicago Press, Chicago.

Roberfroid, S., Vanderleyden, J., Steenackers, H., 2016. Gene expression variability in clonal populations: causes and consequences. Crit. Rev. Microbiol. 42 (6), 969–984.

Rosenberg, A., 2001. Discussion note: Indeterminism, probability, and randomness in evolutionary theory. Philosophy Sci. 68 (4), 536–544.

Samoilov, M.S., Price, G., Arkin, A.P., 2006. From fluctuations to phenotypes: the physiology of noise. Science STKE 366 (re 17), 1–9.

Sarkar, S., 1996. Biological information: a skeptical look at some central dogmas of molecular biology. In: Sarkar, S. (Ed.), The Philosophy and History of Molecular Biology: New Perspectives. Kluwer, Dordrecht, pp. 187–231.

Sober, E., 1999. The multiple realizability argument against reductionism. Philosophy Sci. 66, 542–564.

Sprinzak, D., Lakhanpal, A., Lebon, L., Santat, L.A., Fontes, M.E., Anderson, G.A., et al., 2010. Cis interactions between Notch and Delta generate mutually exclusive signals states. Nature 465 (7294), 86–90.
Thattai, M., van Oudenaarden, A., 2001. Intrinsic noise in gene regulatory networks. Proc. Natl. Acad. Sciences, USA 98, 8614–8619.
Thattai, M., Van Oudenaarden, A., 2004. Stochastic gene expression in fluctuating environments. Genetics 167, 523–530.
Vogt, G., 2015. Stochastic developmental variation, an epigenetic source of phenotypic diversity with far-reaching biological consequences. J. Biosci. 40, 159–204.
Woodward, J., 2010. Causation in biology: stability, specificity, and the choice of level of explanation. Biol. Philosophy 25, 287–318.

3

Random walk across the epigenetic landscape

András Páldi
École Pratique des Hautes Études, Paris Sciences Lettres, Inserm U938, Paris, France

Despite an impressive number of studies, the problem of cell differentiation remains one of the greatest puzzles in modern biology (Newman, 2019). Two fundamentally different conceptual frameworks influence our way of thinking of how cells differentiate. The dominant paradigm supports a deterministic model. Here, cell differentiation as the "elementary" event of the development of multicellular organisms is part of the unfolding of a deterministic program encoded by the genes. Therefore, the phenotype of the cell is determined cell intrinsically and the change of the phenotype (differentiation) is induced by specific extrinsic signals. The perfect knowledge of the genome structure is sufficient to predict the exact timing and trajectory of the whole process of differentiation and development. According to an alternative theory, differentiation is a self-referencing process without a preset program or plan. It emerges gradually from the random interactions between the cell's own components (genes, for example) and between the cell and the environment. In this case, it is impossible to predict mechanically the final state of the process from the initial structure.

Should one subscribe to the dominant or the alternative theory, both come with the same set of interdependent conceptual difficulties. First, it is not clear how to define, or at least, how to provide a set of rules that allows unambiguous identification of what they consider as different cell phenotypes. To what extent does a cell have to be different from its precursor to be considered as acquiring a new phenotype? A second important issue concerns the time scale of the cell differentiation process: how is the life cycle of the cell related to the process of differentiation? How many times can a cell change phenotype during the cycle between two divisions and how long has a phenotype has to be maintained to consider it as differentiated? Last, but not least, a major issue is the understanding of the causes of differentiation. In biology, the identification of general

Phenotypic Switching. DOI: https://doi.org/10.1016/B978-0-12-817996-3.00008-6

patterns and rules underlying a phenomenon is usually not sufficient to consider that phenomenon as understood. It is a consensus among biologists that "understanding" means identification of the "mechanisms" and "mechanism" refers to a "step-by-step explanation of the mode of operation of a causal process that gives rise to a phenomenon of interest" (Nicholson, 2012). However, depending on how the problem is framed different "mechanisms" can satisfy this definition. The aim of this paper is to seek a mechanism for cell differentiation that is based on a nondeterministic vision of the cell. To do this, we will start from the historical roots of our fundamental concepts and then examine how on the basis of these theoretical foundations and recent experimental observations it is possible to build a synthetic model that explains cell differentiation in a step-by-step manner.

Concepts

The issues of cell identity and differentiation are interdepended and strongly related to the complementary concepts of "stability" and "change." One may see in many phenomena in biology change or process, or alternatively, stable categories. Indeed, when a cell divides, the two daughter cells can inherit the maternal cell's phenotype. This requires stability, the action of some kind of heredity or memory mechanisms. The term of epigenetic heredity or cellular memory is usually used to specify this phenomenon. However, cells also have the capacity to change their phenotype, that is, differentiate. Whatever is the underlying mechanism, it has to counteract the memory mechanisms. The concepts of "stability" and "change" cannot be separated from each other; one defines the other. Both concepts come from philosophical ontology and are frequently encountered in other scientific disciplines where they produce very similar difficulties. Typically, biologists claim, that biology, as an experimental science, is only concerned by facts, not with philosophical issues. This opinion is clearly detrimental for the biology itself. Ignoring the necessity to clarify the adequacy of the philosophical roots of a research project may lead to very practical difficulties. As the philosopher Daniel Dennett put it: "There is no such thing as philosophy-free science; there is only science whose philosophical baggage is taken on board without examination" (Dennett, 1995). A practical illustration of this point is a recent inquiry by the journal Cell Systems (Clevers et al., 2017). Fifteen prominent scientists were asked to give their conceptual definition of "cell type." The answers were all different. Although many agreed that the cells could be categorized on the basis of their "resemblance," there was no consensus about what "resemblance" is and how it can be defined. What kinds of features are important when two cells are compared: morphology, gene expression pattern or something else? It is also unclear how many features two cells have to share to be considered belonging to the same type and what is the minimal difference required to be classed as different. If very different cells are considered, it is relatively simple to answer these questions. The difference between, let us say, an epidermal cell and a neuron is obvious. However, closely related cells are much more difficult to demarcate from each other. When individual cells are analyzed for their gene expression profile or protein composition the emerging picture is a continuum, rather than discrete categories of cell phenotypes (Paldi, 2018). Although it is possible to measure the resemblance between the cells and cluster them into classes on the basis of shared characteristics, such classification remains a subjective dissection of a continuum into categories. The biologists working on the isolation of, for example, stem cells are confronted to this difficulty every day. The discussion about "stemness" as a distinguishing feature of stem cells is more reminiscent of a philosophical dissertation than an experimental paper (Blau et al., 2001; Zipori, 2004).

The discussion on cell types is reminiscent of another unsettled debate in biology: the definition of species. In both cases, the discussion is about whether cell types or species really exist or they represent a category of biological objects—cells in the first, individuals in the other case—created by the observers for their own convenience. Both of these discussions take their origin from the controversy over "essence" and "existence" and can be traced back to Greek philosophy and medieval theology. The dominant view of contemporary biology assigns the "essence" of the living to the genome. The essential characters of the individual are said to be determined by its genome. Individuals belong to the same species because they share similar genomes (=DNA sequence, but also chromosome number and structure). As a consequence, the usual narrative is about *the* human, *the* mouse etc. genome and not about the genome of individuals. This is a static vision. The ground state of the species is stability, because its essence is stable. Evolution of the species becomes possible only because of the random mutations of the genome cause the carrier to deviate from the "norm." Mutations therefore represent a mechanistic explanation of how spontaneous changes in the "essence" can occur. The identification of a robust mechanism of how evolutionary variation is generated led to a general acceptance to the essentialist view of the genetics.

The cells in an organism change. They differentiate. However, they also share the same genome [with some notable exceptions, as indicated by lymphocytes and chimeras (Fields and Levin, 2018)]. Hence, a mutation-based mechanism cannot account for the gradual change of the cell phenotype. How can a single set of genes determine so many different sets of characters and their ordered succession during the process of ontogenesis? According to the deterministic vision, this is only possible if the expression of the genes is precisely regulated by other genes coding for transcription factors (TFs). TFs are proteins that are capable of recognizing specifically some short DNA sequences and bind to them. These binding sites are located in the promoter regions of the genes. The binding of the TFs represents the first step in the initiation of transcription because all the other proteins and enzymes are recruited to the gene trough specific interactions. Since TFs are also coded by the genome, genetic determinism prevails. The genome encodes the construction bricks and also the plan to construct the organism using the bricks. In this way, cell differentiation is also conceptualized as the manifestation of the "essence." The metaphysical roots of this view are evident. It is not surprising, that Max Delbrück recommended Aristotle be symbolically awarded the Nobel prize in biology for his discovery of the principle underlying genetic determinism, that is, a plan as the causal principle for the creation of a new organism (Delbrück, 1971).

That the genetic determinism and developmental program paradigm carries inherent logical contradictions has been noticed by the founder of the classical genetics, Thomas Morgan and has been criticized many times by many authors (Kupiec, 2009; Oyama, 2000; Noble, 2006). Among others, it is clear that genome and phenotypic complexities are not related. Despite the explicit promises made by genetics, the knowledge of the genomic sequence is insufficient to deduce on the phenotype and developmental trajectory of an organism. Although this contradiction alone should be sufficient to reject it, the idea of genetically programmed cell differentiation remains dominant. Biologists typically claim that they believe in facts, not theories. Indeed, observations obtained on large cell populations are frequently compatible with deterministic models. However, now new experimental observations directly refute the deterministic view. First, the development of epigenetics has shown, that nonspecific interactions are more important in gene regulation than previously thought. Second, more important point is that the deterministic vision is directly contradicted by single-cell observations. It is now well acknowledged and clearly demonstrated by molecular and in vivo imaging tools that due to the low copy number of molecules within the cells biochemical reactions such as gene expression and vary probabilistically (Misteli, 2001). The ordered behavior at higher level of organization must therefore result from the collective action of a large number of individually stochastic molecular events.

This assertion is counterintuitive for most of us, because it is the exact opposition of the Cartesian view prevalent in biology. It is therefore necessary to examine in detail how the "order from disorder" rule is implemented by molecular mechanisms and why there is no contradiction between the observations and the assumption of stochasticity. To do this, it may be helpful to go back first to the historical roots of the deterministic and nondeterministic views of development.

Historical origins

From the beginning of human thought, the study of living organisms was challenged by the opposition between "stability" and "change." On the one hand, descendants resemble their parents. The apparent *stability* of characters across generations called for explanation. On the other hand, however, individuals can change substantially during their life and this capacity to *change* also required clarification.

The theory of preformation proposed an explicit explanation of the stable transgenerational heredity of characters. According to it, the characters of an adult organism are carried by a miniature individual (Fig. 3.1) already preformed in the gamete of the parent (sperm cell or egg). Ontogenesis is simply the growth of the individual. One of the main conceptual contradictions inherent in the theory of preformation is the fact that it leads to infinite regress. Indeed, each preformed individual must have performed individuals in their gametes that also have gametes with even smaller preformed individuals, etc. The only way out from the dilemma is to postulate the existence of a starting point—an explanation in perfect harmony with the religious view of that time.

In opposition to this view, the theory of epigenesis emphasized changes during ontogenesis by claiming that the new individual emerged gradually from homogenous plasma contained within the gametes. This theory conceptualized well the easily observable ontogenetic transformations, but it had difficulties to explain the apparent similarity of the characters between individuals of the same lineage in different generations.

By the end of the 19th century, the two concurrent theories contributed to the constitution of two distinct disciplines: classical genetics and embryology. As the American biologist William Morton Wheeler suggested in 1899, there were two different kinds of thinkers: some saw stability, while others saw change and

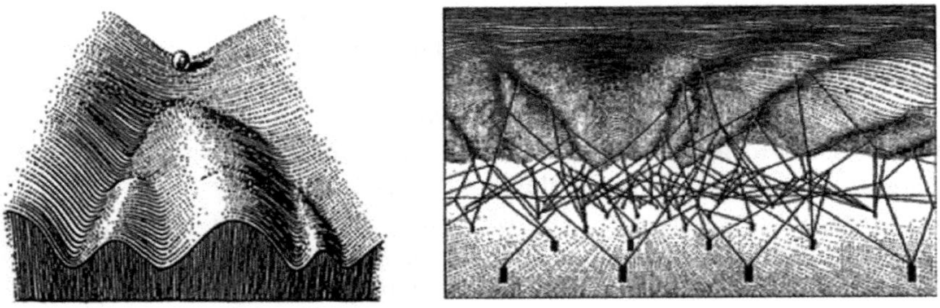

Figure 3.1 The epigenetic landscape as represented on the original drawing by Waddington. The left panel shows the top view of the landscape with the dice rolling down. The right panel is the view from below indicating the complex network of interactions (strings) between the genes (pegs) and the landscape.

process (Maieinschein, 2017). Indeed, while genetics focused on the stability of the form and function across generations, embryology emphasized the changes during ontogenesis. Genetics followed the path of preformation. As August Weismann, one of the founders of genetics put it: "the fate of the cells is determined by forces situated within them, and not by external influences" (cited in (Maieinschein, 2017)). The key conceptual innovation of the genetics was the separation of the genotype and the phenotype. In this setup, highly reminiscent of the "essence/existence" dualism, the organism is not preformed, only the "plan" for its construction is. The genotype is invariable and can be transmitted to the next generation; it contains the description of the organism. The phenotype corresponds to the realization of the plan. The idea of genetic determinism was born. The other important, but a frequently overlooked consequence of the gene/character separation is the exclusion of acquired characters from heredity. In fact, if the gene is the only to be transmitted invariably to the next generation, then the changes that may occur to a character cannot be inherited by definition. Thus, the impossibility of the so-called "Lamarckian" inheritance is part of the theoretical assumption and not a rule deduced from observations. This is why the dispute on Lamarckian inheritance was very passionate at the beginning of the last century (McLaren, 2000), but we also witness intense discussions since the recent re-emergence of epigenetics as a new discipline. The opposition of many scientists to transgenerational epigenetic inheritance remains strong (Heard and Martienssen, 2014). This is understandable, because accepting it would also mean accepting implicitly that the essence/existence duality at the heart of the genetic determinism paradigm cannot fully account for heredity. It is important to recall this here, because the possibility of transmission of the phenotype acquired by cells during their life, to daughter cells through mitosis is considered as self-evident despite the obvious conceptual similarity and the identity of known mechanisms involved in both mitotic and meiotic inheritance. Another challenge to the classical genetics was the difficulty to reconcile the stability of the genotype with the idea of evolutionary change. This contradiction was resolved successfully by the discovery of genetic mutations.

Following the basic idea of the theory of epigenesis, embryology focused on the capacity of change during the ontogenesis, especially during the initial stages of the life cycle. In order to understand how different structures, tissues and organs develop, embryology identified general patterns of morphogenetic changes (e.g., gastrulation, neurulation, segmentation, etc.), investigated the interactions between the different parts of the developing embryo that could explain those changes (induction, morphogenetic fields, etc.). Embryology, as a science of change, had no difficulties to accept the idea of evolution. There were attempts to identify invariant patterns in the two processes, as the theory of the "ontogenesis recapitulating phylogenesis" formulated by Ernest Haeckel. Nevertheless, genetics—whose practice dependent on a different viewpoint, as we have seen—had a strong influence on embryology. For example, the study of mutants with developmental defects became a standard approach.

The epigenetic landscape

By the middle of the last century, it became evident that some kind of synthesis between the two opposing views was not only necessary but also possible. The author of this synthesis was Conrad Waddington. By fusing "epigenesis" and "genetics," he created the new term of "epigenetics." What does this mean exactly? The use of this term in the current scientific literature is highly ambiguous (Greally, 2018). In this paper, we use "epigenetics" in Waddington's sense. The best is to quote how Waddington himself defined the term in his short correspondence sent to Nature (Waddington, 1956):

"The fact that the word 'epigenetics' is reminiscent of 'epigenesis' is to my mind one of the points in its favor. As Dr. Wightman points out, in the old controversy, now stilled, between the theories of epigenesis and preformationism, neither side emerged completely victorious. We all realize that, by the time development begins, the zygote contains certain 'preformed' characters, but that these must interact with one another, in processes of 'epigenesis', before the adult condition is attained. The study of the 'preformed' characters nowadays belongs to the discipline known as genetics; the name 'epigenetics' is suggested for the study of those processes which constitute the epigenesis which is also involved in development." (McLaren, 1999; Waddington, 1956) To illustrate his idea, Waddington sketched a visual metaphor: the epigenetic landscape (Waddington, 1957). In this metaphor, the cell is represented by a ball rolling down a landscape (Fig. 3.1). The slope makes the ball to roll from the top to the bottom, but the exact trajectory is uncertain. It may enter different valleys that direct the ball through a series of bifurcations to different end points. The valleys symbolize the differentiation pathways available to the cell and the endpoints represent the various terminally differentiated phenotypes. When the cell/ball is on the top, it is impossible to predict which final phenotype it will adopt. The final phenotype is the result of the cell's journey across the landscape. Each choice to enter a pathway depends on the actual interactions of the cell at the bifurcation with the surface and the environment in general. Since the ball is rolling down the slope and cannot come back, with each choice the cell loses all the other possibilities accessible from the other path. As a result, the cell's potential to differentiate gradually decreases. The actual state of a cell depends on its own past trajectory and on the actual forces acting on it. The surface of the landscape is determined by the organism's genetic makeup and environmental context, and remains unchanged. Ignored in his time, this illustration has now become very popular in our days. It is frequently reproduced in various publications.

A second illustration (Fig. 3.1) depicts how genes influence the shape of the landscape. This very important complementary drawing shows the underside of the landscape. It depicts how the genes, represented by pegs, are connected to the surface by strings. Importantly, there is no direct link between a gene/peg and a point on the surface. Rather, the action of each gene interacts with those of others, forming a network of complex relationships that determines *in fine* the shape of the surface in a given environment. This drawing is probably one of the first known illustrations representing a gene network.

Thus, epigenetics defined by Waddington, is a synthesis between genetics and embryology. It advocates the idea that cell differentiation is a process constrained by the genes but not programmed or determined by them. Waddington's proposition includes the changes in gene expression as a part of the process of differentiation, but makes no assumptions about the mechanisms. This original view is very different of the definition frequently used today that restricts epigenetics to a set of biochemical mechanisms acting on gene expression.

The scientific community, under the influence by the spectacular development of the molecular biology from the 1960s, retained the idea of the deterministic gene regulation. According to it, genes are essentially regulated at the transcriptional level. The synthesis of mRNAs is initiated by the specific interaction between one or several TFs with specific DNA sequences called enhancer or promoter. TFs are said to regulate gene expression in a specific way, because each gene has its own TF or combination of TFs required for the initiation of transcription. Since TFs are encoded by the so-called regulatory genes, the fine regulation of gene expression is supposed to ensure the ordered expression of the genes and implement in a strict deterministic way the program encoded by the genome.

This explanation can be satisfactory at the level of a single gene. The prototype of this model is based on the studies of the Lac operon in bacteria described by Monod and Jacob in the sixties. This approach can be successful in explaining simple on-off switches. However, if we consider the larger context, we have to explain how the regulators are regulated and the regulators of the regulators, etc. Here also, the logic of the linear causality leads to infinite regress. A rational way to reconcile causal relationships between the genes and take into account their finite number in the genome is to postulate some kind of circularity in the causal relationships. A feedback loop is the simplest representations of a circular causal relationship. A large interconnected network of genes is a system with multiple circles of causality. In this way, the network representation not only appears as a solution to save the deterministic concept of genome functioning, but it is in line with the observation that genes have multiple interactions.

The vision of genome function as a web of interactions became rapidly very popular. It opens new ways to conceptualize cell differentiation and allows the introduction of mathematical methods in the description of cellular phenotypes. Each individual cell can be characterized by a vector in a multidimensional space defined by expression level of every gene in the genome. In other terms, a gene expression pattern can be described by a single vector. Using multiparametric methods, it is then possible to class the cells on the basis of the proximity of their gene expression patterns and quantify the strength of interactions between the genes. In the simplest pragmatic interpretation, the network structure reflects measurable correlations between expression levels of the genes. According to the stronger hypothetical interpretation, only regulatory interactions between the expression of the regulator and the regulated genes are considered. In this second interpretation, the network structure is invariant among the genetically identical cells, because the regulatory interactions depend *in fine* only on the nucleotide sequence of the interacting genes. This is a strictly deterministic view and represents the network reformulation of the strict genetic determinism. The first, pragmatic interpretation leaves a room for interactions that are not encoded in and cannot be deduced from the genome sequence, such as alternative splicing of the mRNAs or posttranslational modifications of the proteins. In this sense, this is a "soft" genetic determinism.

In both the "hard" and "soft" interpretations, a "network state" is defined by the collective effect of all the expressed genes. Cellular phenotypes are defined by a particular network state. Due to interactions between the genes, different network states have different stability. The most stable network states behave as attractors and correspond to stable cell phenotypes. Differentiation then is conceptualized as the change of the network state in the cell due to the coordinated alteration of the expression levels of interacting genes. The network states can be formalized as a quasi-potential landscape where each point corresponds to a network state. Stable states are attractors and represented as wells, unstable equilibrium points between them are saddles. Cell differentiation can be represented as the trajectory between two attractors that goes through unstable intermediate states.

Some authors consider that this representation is a mathematical reformulation of Waddington's metaphoric landscape (Huang, 2009; Huang, 2012). Indeed, at the first glance the visual similarity between the simplified 3D representation of the quasi-potential landscape and Waddington's visual metaphor is compelling. However, there is a substantial difference that rules out a formal equivalence. In the case of Waddington's landscape, the inclination of the surface symbolizes the driving force that makes the cell differentiate. The quasi-potential landscape is flat and the attractors are potential wells. This model captures only the stability (determined genetically and proportional to the depth of the well), but misses the dynamic aspect of cell differentiation. While the differences between the possible phenotypes are precisely quantified by the distances between the attractors in the state space, it is unclear why a cell quits the attractor to move to another. To make the model complete, it is necessary to identify and include the force(s) that make the cell change. In other terms, it is necessary to identify the biological equivalent of gravitation that makes the cell rolling down on the metaphorical landscape (Paldi, 2012).

A new conceptual framework

We have witnessed an extraordinary development of our technology used in the study of living organisms over the last 50 years. Our experimental methodologies have higher sensitivity and resolution whenever. We are now able to sequence whole genomes and transcriptomes and even track individual molecules in cells. Massive amount of experimental observations has been accumulated and analyzed with highly efficient computational methods. Although new details can certainly be discovered in the future, it is unlikely that essential components or interactions playing key roles in cell differentiation could have completely escaped our attention. We may lack some pieces of the puzzle, but the contours of the overall picture are now emerging. This means that, rather than discover new facts, we need to change our point of view and frame the problem of differentiation in a different way. Following the path of the Newtonian tradition, it was generally considered that the cells remain phenotypically stable unless destabilized by an external force, usually a new signaling environment. The alternative view is to consider that a cell never remains stable spontaneously. It will fluctuate permanently and display different phenotypes unless constrained to remain more or less stable. The *explanandum* is not the change but the stability.

The idea that cell differentiation can be viewed as a process of selective stabilization of gene expression profiles generated by spontaneous variation of gene transcription was proposed long ago (Kupiec, 2009; Kupiec, 1996; Kupiec, 1997). The whole process is analogous to the process of Darwinian evolution of variation and selection. Today, the 25-year-old intuition is supported by direct and indirect experimental observations. Although usually not considered from this point of view, molecular mechanisms that empower the cells to change their phenotype are well known. They can be arranged in a coherent sequence of action that can provide mechanistic explanation of differentiation.

Single-cell studies draw attention to a previously overlooked aspect of cellular life. Most of the biochemical reactions in the cell involve a small number of molecules. If the copy number of the molecular species is low, the associated processes are noisy. Gene expression is a typical example. There are usually only two copies of a given promoter of a gene in the cell. The synthesis and processing of the mRNA requires the joint action of a large number of proteins and other molecules (RNA polymerase, TFs, etc.), all present in a very low copy number. As a result, gene expression is a stochastic event, leading to broad fluctuations in the gene product level (Chen and Larson, 2016).

The phenotypic heterogeneity observed even in clonal populations is essentially the consequence of the ubiquitous molecular fluctuations occurring during the individual steps of synthesis and degradation of molecules (Corre et al., 2014). Stochasticity of gene expression was suggested and experimentally detected long time ago but the phenomenon gained a significant interest only after the publication of a landmark paper in 2002 (Elowitz et al., 2002). Many consider these fluctuations as a simple molecular noise without real functional significance that the living organisms are able to tolerate. A recent study by Lestas et al. challenges the traditional view of high precision of cellular processes such as gene regulation. The study has demonstrated that molecular fluctuations are not only ubiquitous, but also impossible to suppress by specifically dedicated mechanisms (Lestas et al., 2010). Using methods of information theory and physical chemistry, the authors demonstrated that doubling the precision of a process requires 16 times more energy than keeping it at the initial level. Suppressing fluctuations is only possible if the number of chemical events integrated over the timescale of control is very high. When noise suppression is essential for the survival, the cells can use a brute-force strategy by synthesizing a very large number of molecules dedicated to the task. However, the energetic costs of such a strategy are too high to be applied to all but a few exceptional processes.

Single cell RNA sequencing studies fully confirm this view. Gene transcription usually operates with a high level of stochastic fluctuations. High and even level of transcription is restricted to a small number of genes as those coding for ribosomal RNA-s and proteins and some other housekeeping genes. Equally, if not more important, however, gene silencing is also "noisy." Indeed, RNA polymerase and TFs can bind to any DNA sequence other than to their target sites. Such interactions are weaker and last shorter; yet, they can occasionally initiate transcription with low probability. As early as in 1974, Von Hippel et al. has suggested on the basis of binding constants and concentrations that in bacteria nonspecific biding of regulatory proteins plays a crucial role in the activation and repression of genes (von Hippel et al., 1974; Kao-Huang et al., 1977). Recently, direct observations of individual molecules in living bacteria confirmed that a TF molecule spends 90% of the time bound to non-target sites (Elf et al., 2007). The most efficient way to suppress transcription is to occlude the access to the DNA by the RNA polymerase enzyme complex. In eukaryotes, this function is fulfilled by chromatin. Chromatin is essentially formed by histone octamers, nucleosomes, bound to the DNA in a nonspecific way independently of the nucleotide sequence. As a result, the gene promoters on the DNA molecule become relatively inaccessible to the transcription machinery (Zhu et al., 2018). The cell produces such a large amount of nucleosome-forming histone molecules, that they can fully saturate the genomic DNA and outcompete TF binding. The default state of the chromatin therefore is repressive. However, even under these conditions the complex formed by the DNA molecule and the nucleosomes is a highly dynamic structure. DNA wrapped around the nucleosome spontaneously unwraps and rewraps with high frequency of less than sec-1. These dynamic fluctuations provide opportunity to other proteins—for example, TFs—to access to their DNA target sites and initiate transcription (Tims et al., 2011). In this way, the impossibility to suppress chromatin fluctuations (=chromatin "noise") is the key to activation of repressed genes.

Chromatin fluctuations are not uniform along the genome. The strength of the interaction between the DNA molecule and the nucleosomes is strongly influenced by covalent posttranslational modifications of the histone molecules commonly called "epigenetic" (Zentner and Henikoff, 2013). (It is important to note that the term "epigenetic" is used here to design a set of molecular mechanisms, by contrast to the previously described use of the term.) Typically, a small molecule is attached covalently to one of the charged amino acid residues of the histone protein, like acetylation, methylation, phosphorylation, poly-ADP ribosylation, etc. These modifications are reversible and both the on and off reactions are catalyzed by dedicated enzymes. The histone molecules can be modified at multiple sites and a large number of different combinations may coexist. The different combinations of chromatin modifications can be roughly classed into two states: heterochromatin and euchromatin. The first corresponds to the transcriptionally repressed, the second to the transcriptionally permissive state (Misteli, 2001). In euchromatin, due to the covalent modifications of the histones, the interactions between the DNA and nucleosomes are weak and the local fluctuations of the nucleosome-DNA interactions are high. More fluctuations mean more opportunities for transcriptional initiation. In the heterochromatic regions, the histone modifications are different and the interaction between the nucleosomes and DNA are stronger (Misteli, 2001). Overall, this makes the DNA less frequently accessible; hence, the initiation of transcription remains a rare event.

The enzyme activities that act to establish euchromatin cooperate to mutually reinforce each other's effect. The enzymes bringing together the heterochromatin state also act together. Although individual modifications are short-lived, the overall profile of a genomic segment remains stable. This dynamic stability depends on the equilibrium of the "on" and "off" reactions. As a result, chromatin is able to conserve the transcriptionally permissive or repressive state and even to transmit it through cell division. This is the basis of cellular or epigenetic memory of gene expression patterns. A key feature of chromatin is the hysteretic nature of the on-off transitions between hetero- and euchromatin. This lends the chromatin the capacity to "register" earlier states. The transition between euchromatin and heterochromatin occurs only if the equilibrium between the two types of epigenetic modifications is perturbed. Therefore, chromatin can be considered as a classical bistable system (Dodd et al., 2007; Sneppen and Ringrose, 2019). Both stable states result from a dynamic equilibrium of opposing enzymatic reactions. The transition between the two states occurs when the equilibrium is perturbed.

What we call transcriptionally permissive euchromatin is in fact a highly mobile, a kind of "noisy" repressive chromatin. Epigenetic mechanisms appear therefore as modulators of chromatin fluctuations. Gaining control over epigenetic reactions means gaining control over the stability of gene expression patterns. A key issue therefore is to understand what determines the balance between the euchromatin- and heterochromatin-inducing histone modifications. The enzymes that catalyze the modification reactions or remove the modifications from the histones use small molecular substrates (some authors erroneously use the term "cofactor"). Without exception, these substrates are key intermediates of the basic energy metabolism of the cell (Etchegaray and Mostoslavsky, 2016; Lu and Thompson, 2012; Cyr and Domann, 2011; Nieborak and Schneider, 2018) and usually referred as "sentinel metabolites" (Walsh et al., 2018) key role in regulating bioenergetics pathways (Fig. 3.2).

These are, for example, acetyl-CoA for acetylation, NAD+ for poly-ADP ribosylation or histone deacetylation, ATP for phosphorylation, FAD+ and alpha-ketoglutarate for demethylation, etc. The universal methyl donor for DNA and histone methylation, S-adenosyl-methionine (SAM) in animal cells is synthesized from the essential amino-acid methionine taken up from the environment and a molecule of ATP. It is important to emphasize that these molecules are not cofactors, as some authors say, but substrates. They are consumed by the epigenetic reactions. Therefore, the different pathways using the same substrate are in competition for the same bioenergetics substrates. For example, full oxidation of an acetyl-CoA molecule through the Krebs cycle and terminal oxidation for example results in the synthesis of about 10–12 ATP molecules. If an acetyl-CoA molecule is used to acetylate a lysin residue of a histone molecule, the same molecule cannot be used for energy production. This means that acetylation

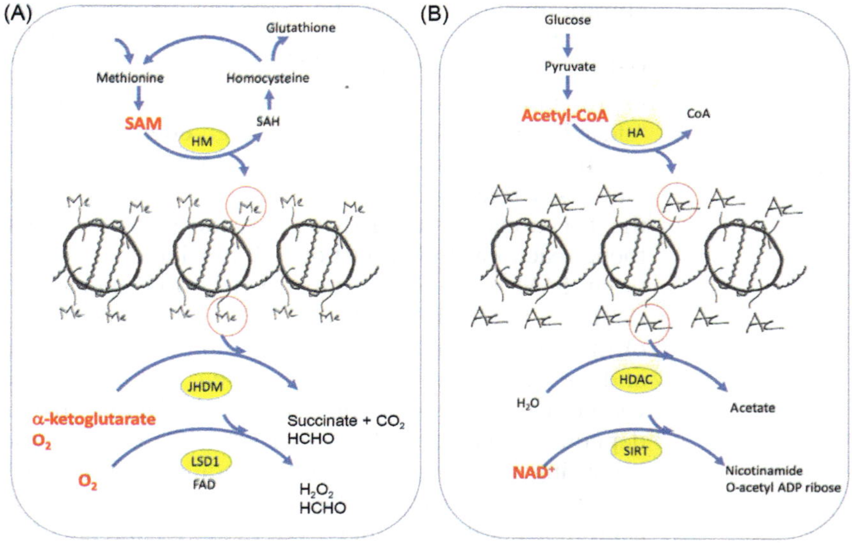

Figure 3.2 Link between posttranslational modifications of the chromatin and the energy metabolism of the cells through the use of "sentinel metabolites" as substrates (after Ye & Tu, 2018). Left panel: histone (but also DNA and any other) methylation uses SAM as substrate and a-ketoglutarate and oxygen for demethylation. These molecules are no more available for other energy-dependent processes. Thus, the energetic cost of each reaction can be calculated. Right panel: histone acetylation uses acetyl-CoA and deacetylation may be dependent on NAD+. Both are "sentinel metabolites."

has an associated energetic cost equivalent to this number of ATP molecules. The Krebs-cycle intermediate alpha-ketoglutarate is consumed during the demethylation reaction by 10–11 translocation hydroxylases involved in DNA demethylation and the Jumonji C domain containing lysine demethylases. If a single alpha-ketoglutarate is fully oxidized, about eight ATP molecules can be obtained. The energetic cost of each individual epigenetic modification can be calculated in the same way. Given the large number of nucleosomes, each containing eight histone molecules, each with numerous modifiable amino-acid residues and the rapid turnover of the modifications, the total energetic costs of the epigenetic modification could be much higher than we imagine. It is difficult to produce exact estimates at the present stage, because we do not know the absolute number of bioenergetics molecules produced and consumed by a single cell per unit of time. It is likely however, that the maintenance of the chromatin structure through epigenetic modifications represents a very important item on the energetic bill of the cell (Ye and Tu, 2018).

As shown by Lestas et al. "noise" reduction has a high associated energy cost (Lestas et al., 2010). The link between the chromatin states, epigenetic modifications and the energy metabolism of the cells is a particularly clear example of the implementation of this principle. Both stable repression and stable high expression can be seen as noise reduction. Both the "open" and "closed" chromatin states have a high energy cost.

The same molecules used as substrates for histone modification play a key role in a large number of other pathways. Acetyl-CoA or the Krebs cycle intermediates are the starting points for biosynthetic reactions and NADH or ATP provide the necessary energy for them, SAM is one of the most widely used metabolites in the cell (Walsh et al., 2018; Ye and Tu, 2018). The intracellular concentration of these metabolites reflects at each moment the overall metabolic and energetic state of the cell. Their abundance varies between wide ranges. In general, the rate of an individual enzymatic reaction depends critically on the actual substrate concentration. Enzymes catalyzing individual epigenetic modifications are not exceptions. The rate of the reactions they catalyze follows the intracellular concentrations of the substrates. For example, the rate of histone methylation in a given nuclear location is dependent on the local SAM concentration, while the rate of demethylation at the same location is a function of the alpha-ketoglutarate concentration. The concentration of these metabolites depends on the actual flux through the metabolic pathways that produce or consume them. As a result, epigenetic modifications act as a "sensor" of metabolic flux balance of the cell and depending on it tend to repress or derepress rapidly whole genomic regions without changing the overall concentration of the enzyme molecules involved. Repression (heterochromatin) and derepression (euchromatin) depend on different substrates produced by different pathways. As a consequence, the way the carbon and electron fluxes are organized in the cell also impacts the prevalence and distribution of the chromatin marks in the nucleus.

The link between bioenergetics and chromatin was predicted (Paldi, 2003) and experimentally confirmed (Wellen et al., 2009; Sivanand et al., 2018) long time ago. There is no doubt that this tight link is fundamental for the cell differentiation and development of multicellular organisms. Could this be more than a simple mechanistic link?

Putting together the pieces of the puzzle to build a new operative model

Maintaining specialized functions typical for differentiated cells in addition to the vital functions (ion channels, cytoskeleton, membrane trafficking, etc.) requires continuous flow of chemical energy. This energy is obtained by oxidizing the substrates taken up from the environment. The nature of these substrates and the pathways the electrons go through to reach the final electron acceptor and carbon-containing compounds are transformed are essential features of a given cellular phenotype. A cell phenotype remains stable only if the metabolism is able to provide the necessary chemical energy and material in the form of ATP and small metabolites required for the specific ongoing processes in the cell. In other terms, the cell phenotype is primarily determined by the absolute necessity to maintain the carbon and electron flow within a defined range. In this process the role of the "sentinel" metabolites carrying electrons and carbon atoms (acyl groups) as well as the availability of their acceptors, such as O_2 as the ultimate electron acceptor is determinant (Paldi, 2012). The strict stoichiometry of the metabolism makes that if the availability of any of the electron and carbon donors or acceptors changes, this overall flow can only be maintained if alternative pathways using different substrates are activated. This frequently requires the synthesis of new enzymes, hence, expression of previously silent genes. Defined as a change of phenotype, the process of cell differentiation therefore can be seen as a continuous metabolic adaptation, that requires changes in gene expression and its true driving force is the need to meet the energy requirements. However, the cell has no ability to anticipate which genes it has to express to acquire the capacity to metabolize the nutrient substrates it will be able to access in the future. The most efficient strategy that makes possible to meeting an unanticipated challenge is trial and error. This strategy is realized thanks to the mechanistic link between the energy-producing metabolism, chromatin stability, and gene expression.

The change of substrate availability results in modification of metabolic fluxes and leads to variation in the intracellular concentration of the sentinel metabolic intermediates. These variations are mechanistically relayed to the chromatin by the epigenetic mechanisms that use the same metabolites as substrates. As a result, fluctuations of the metabolite levels are transferred to the chromatin structure. It is important that the chromatin structure fluctuations are not targeted to specific genes or DNA sequences; their amplitude and frequency depend only on the preexisting epigenetic profile and the local availability of the substrates to maintain it. The chromatin becomes derepressed in many regions letting previously repressed genes transcribed in a disordered way leading to the rapid generation of a nonspecific mixed multilineage gene expression profile. This is a uniform reaction, analogous to the "alarm" stage of stress reaction observed in physiology, enables the cell to respond to any kind of challenge by expressing a large variety of previously repressed genes.

Although disordered, the pattern of gene expression is not entirely random; it is constrained by the epigenetic (cellular) memory that assigns lower probability of reactivation to genes suppressed long time ago than to those that were silenced recently. The consequence of the constraints imposed by cellular memory is that the cell can reach some new states easier than others depending on the nature of the starting point. While giving a direction to the whole process, cell memory does not exclude the reversibility of the process; it just makes it less likely. The fluctuations continue as trial-error, random walk process. This allows the acquisition in a nonpredictable fashion of new morphological and functional properties, such as membrane receptors able to take up new energetic substrates or metabolic enzymes able to metabolize them, etc. When the new configuration allows the cell to ensure or at least partially improve the energetic requirements of its own maintenance the fluctuations start to decrease and a new expression pattern is progressively set up. This is done by the stabilization of the new chromatin pattern by a feedback effect of the decreasing fluctuations of the metabolite concentrations and the emerging interactions between the components of the gene regulatory network. Selective interactions, for example between TFs and target promoters, play more a role of stabilizers of a new state than initiators of the change. The stable state can be conceptualized as an attractor state shaped by stabilizing gene–gene, cell–cell, or cell–environment interactions. At any moment, the phenotypic diversity of a cell population can be described as a quasi-potential landscape. Due to the dynamic nature of the process, many independent trajectories can lead to the same optimal phenotype. The differentiation of a cell is better described metaphorically as a random walk across the epigenetic landscape constrained by hills and valleys. It is as a process by which the cell optimizes the flow of energy and material by continuously adapting its own structure and functioning. It is based on the spontaneous fluctuation of the intracellular biochemical reactions and the opposing energy dissipating stabilizing forces that arise from the interactions between the cell's own components and the extracellular environment. The stochastic disorder at lower time- and structural scales results in a dynamically stable state at higher scales. The stable order in terms of a coherent gene expression profile, but also a coherent epigenetic pattern and a stable morphological phenotype is in fact a dynamic steady state. The whole process is dynamic, where the difference between the characteristic time scales of events at the molecular, cell organelle, or cellular levels encompasses several orders of magnitude going from microseconds to days, months, or even more. Stability observed at longer time scales is always emerging from the fluctuations of the components at shorter time scales. As a result, a cell can display a stable phenotype and maintain the same "identity" for a very long period of time, despite the rapid molecular turnover of its constituents.

This molecular scenario is a mechanistic implementation of previous theoretical models (Paldi, 2012; Kupiec, 1996; Paldi, 2003). It is in remarkable agreement with the experimental observations. One of the key predictions is that differentiating cells must necessarily go through a dynamically fluctuating phase with oscillating gene expression. This is systematically observed during the critical state transitory phase of the differentiation process (Hu et al., 1997; Mojtahedi et al., 2016; Moussy et al., 2017; Richard et al., 2016). The cells during the transitory period may display fluctuating morphological changes. These fluctuations disappear at the same time as the transcriptomic ones (Moussy et al., 2017). The central coordinating role of the metabolism is also well documented. Indeed, direct observations unambiguously established the dominant role of the basic energy metabolism in cell fate decisions (Xu et al., 2017; Oburoglu et al., 2014; Tischler et al., 2019; Moussaieff et al., 2015). In these examples, the cell fate decisions were directly dictated by the availability of the energetic substrates despite the high concentrations of signaling molecules supposed to instruct the cells to differentiate into a different phenotype.

The model presented here also has conceptual and practical advantages over deterministic scenarios. It allows to get rid of the cryptic finalism always hidden in determinism. It resolves the apparent conflict between the omnipresence of stochastic fluctuations and the ordered functioning of the cell by emphasizing the importance of the time scales. Fluctuations at a microscopic time scale produce ordered changes at a higher time scale. This simple rule can help to understand why cellular processes take the time they take. Deterministic explanations not only ignore the time scale of the biological processes but also have difficulties to incorporate it in the explanatory scheme. Ignoring the time-scales of cellular events lead other difficulties also. For example, short-lived forms of fluctuating cells can be easily confused with "rare" cell types. Last, but not least, our new model opens the way to new experimental strategies to investigate and influence normal and pathological cell differentiation.

Although we focused on the cellular level, it is evident that cells in a multicellular organism can hardly be considered individually. The surrounding cells are part of the extracellular environment. The interactions between the cells can contribute to the destabilization of the phenotype. The cells that compete for the same metabolic resources may destabilize each other and initiate phenotypic change. Alternatively, they may develop cooperation through metabolic complementarity. This type of cooperation has a stabilizing effect. Phenotypic differentiation and formation of stable spatial structures is induced by metabolic competition in yeast populations (Varahan et al., 2019). In social amoebae, and other group-forming microorganisms, it has been proposed that optimum resource utilization, for example, by metabolic cooperation, can be the driving force of the transition from the single cell to the social state (Hamant et al., 2019). This is a good illustration of how metabolic cooperation could contribute to the emergence of multicellular organization during evolution. A nice example of stabilizing metabolic cooperation through cross feeding can be observed in multicellular organisms, for example, between neurons and astrocytes in the mammalian brain (Belanger et al., 2011). This and many other examples illustrate that likewise in unicellular organisms, cell differentiation in multicellular organisms can also be seen as a process of spontaneous phenotypic diversification followed by constrained optimization through selective stabilization of cellular states.

Conclusion

Going back to the historical roots can be very useful for the understanding modern theories and interpretations. Interestingly, old theoretical approaches that thrive on historical philosophical tenets survive hidden behind a modern form and actively influence current thinking on fundamental biological issues. They give us the comfortable feeling that at least some fundamental issues have found their definitive explanation. The influence of preformationism on how cell differentiation and development are considered represents a clear example. The same causes lead to the same result. The deterministic vision of cellular processes implied by the traditional philosophical views contradicts the observed highly dynamic, apparently disordered, yet coordinated functioning of the cells. Biology needs to reframe the problem. The present paper proposes that we go back to epigenetics, as envisioned by Conrad Waddington, who made an attempt to create a synthesis between genetics inspired by preformation and the alternative vision advocated by the theory of epigenesis. The paper suggests a mechanism that can explain cell differentiation and development by integrating the deterministic and stochastic aspects of the living cells into an explanatory scheme. Each element of the proposed mechanism—core energy metabolism, chromatin modifications, gene transcription, etc.—are well known and fairly well characterized. The cornerstone of this model is the idea that differentiation is a process governed by the energy needs of cells to maintain their integrity and implemented by dynamic molecular processes. The apparently ordered functioning at the cellular level emerges from the dynamic disorder of molecular processes acting at a substantially shorter time scale. The higher level of organization feeds back as a causative agent to processes at lower organizational level by modulating reaction rates: not by direct transmission of specific "instructions" but by channeling them along certain pathways. Since substrates for energy production are provided by the environment or produced by neighboring cells, cell-environment, or cell–cell interactions are strong constraints that act as a selective force in the choice and stabilization of the differentiation pathway. Therefore, the model can further be developed to explain the emergence of phenotypic heterogeneity based on metabolic cooperation between cells in the multicellular organisms and the formation of tissues during ontogenesis.

Acknowledgments

The author is grateful to Romuald Parmentier, Laëtitia Racine, Alice Moussy, Olivier Gandrillon, and Jean-Jacques Kupiec for the stimulating discussions. The present work was supported by ANR-17-CE12–0031-002.

References

Belanger, M., Allaman, I., Magistretti, P.J., 2011. Brain energy metabolism: focus on astrocyte-neuron metabolic cooperation. Cell Metab. 14, 724–738.

Blau, H.M., Brazelton, T.R., Weimann, J.M., 2001. The evolving concept of a stem cell: entity or function? Cell 105, 829–841.

Chen, H., Larson, D.R., 2016. What have single-molecule studies taught us about gene expression? Genes. Dev. 30, 1796–1810.

Clevers, H., Rafelski, S., Elowitz, M., Klein, A., Shendure, J., et al., 2017. What is your conceptual definition of "cell type" in the context of a mature organism? Cell Syst. 4, 255–257.

Corre, G., Stockholm, D., Arnaud, O., Kaneko, G., Vinuelas, J., et al., 2014. Stochastic fluctuations and distributed control of gene expression impact cellular memory. PLoS One 9, e115574.

Cyr, A.R., Domann, F.E., 2011. The redox basis of epigenetic modifications: from mechanisms to functional consequences. Antioxid. Redox Signal. 15, 551–589.

Delbrück, M., 1971. Aristotle-totle-totle. In: Monod, J., Broek (Eds.), Of Microbes and Life. Columbia University Press.

Dennett, D., 1995. Darwin's Dangerous Idea. Simon & Schuster, New York.

Dodd, I.B., Micheelsen, M.A., Sneppen, K., Thon, G., 2007. Theoretical analysis of epigenetic cell memory by nucleosome modification. Cell 129, 813–822.

Elf, J., Li, G.W., Xie, X.S., 2007. Probing transcription factor dynamics at the single-molecule level in a living cell. Science 316, 1191–1194.

Elowitz, M.B., Levine, A.J., Siggia, E.D., Swain, P.S., 2002. Stochastic gene expression in a single cell. Science 297, 1183–1186.

Etchegaray, J.P., Mostoslavsky, R., 2016. Interplay between metabolism and epigenetics: a nuclear adaptation to environmental changes. Mol. Cell 62, 695–711.

Fields, C., Levin, M., 2018. Are Planaria individuals? What regenerative biology is telling us about the nature of multicellularity. Evolut. Biol. 45, 237–247.

Greally, J.M., 2018. A user's guide to the ambiguous word "epigenetics. Nat. Rev. Mol. Cell Biol. 19, 207–208.

Hamant, O., Bhat, R., Nanjundiah, V., Newman, S.A., 2019. Does resource availability help determine the evolutionary route to multicellularity? Evol. Dev. 21, 115–119.

Heard, E., Martienssen, R.A., 2014. Transgenerational epigenetic inheritance: myths and mechanisms. Cell 157, 95–109.

Hu, M., Krause, D., Greaves, M., Sharkis, S., Dexter, M., et al., 1997. Multilineage gene expression precedes commitment in the hemopoietic system. Genes. Dev. 11, 774–785.

Huang, S., 2009. Non-genetic heterogeneity of cells in development: more than just noise. Development 136, 3853–3862.

Huang, S., 2012. The molecular and mathematical basis of Waddington's epigenetic landscape: a framework for post-Darwinian biology? Bioessays 34, 149–157.

Kao-Huang, Y., Revzin, A., Butler, A.P., O'Conner, P., Noble, D.W., et al., 1977. Nonspecific DNA binding of genome-regulating proteins as a biological control mechanism: measurement of DNA-bound Escherichia coli lac repressor in vivo. Proc. Natl Acad. Sci. U S A 74, 4228–4232.

Kupiec, J.J., 1996. A chance-selection model for cell differentiation. Cell Death Differ. 3, 385–390.

Kupiec, J.J., 1997. A Darwinian theory for the origin of cellular differentiation. Mol. Gen. Genet. 255, 201–208.

Kupiec, J.-J., 2009. The origin of individuals. World Scientific, New Jersey, p. 253, xix.

Lestas, I., Vinnicombe, G., Paulsson, J., 2010. Fundamental limits on the suppression of molecular fluctuations. Nature 467, 174–178.

Lu, C., Thompson, C.B., 2012. Metabolic regulation of epigenetics. Cell Metab. 16, 9–17.

Maieinschein, J., 2017. Epigenesis and preformationism. Stanf. Encycl. Philosophy Spring Ed. 2017.

McLaren, A., 1999. Too late for the midwife toad: stress, variability and Hsp90. Trends Genet. 15, 169–171.

McLaren, A., 2000. Cloning: pathways to a pluripotent future. Science 288, 1775–1780.

Misteli, T., 2001. Protein dynamics: implications for nuclear architecture and gene expression. Science 291, 843–847.

Mojtahedi, M., Skupin, A., Zhou, J., Castano, I.G., Leong-Quong, R.Y., et al., 2016. Cell fate decision as high-dimensional critical state transition. PLoS Biol. 14, e2000640.

Moussaieff, A., Rouleau, M., Kitsberg, D., Cohen, M., Levy, G., et al., 2015. Glycolysis-mediated changes in acetyl-CoA and histone acetylation control the early differentiation of embryonic stem cells. Cell Metab. 21, 392–402.

Moussy A., Cosette J., Parmentier R., daSilva C., Corre G., et al. (2017) Integrated time-lapse and single-cell transcription studies highlight the variable and dynamic nature of human hematopoietic cell fate commitment. Available from: https://doi.org/10.1101/101428.

Newman S.A. (2019) Cell differentiation: what have we learned in 50 years? https://arxivorg/ftp/arxiv/papers/1907/190709551pdf.

Nicholson, D., 2012. The concept of mechanism in biology. Stud. History Philosophy Biol. Biochemical Sci. 43, 152–163.

Nieborak, A., Schneider, R., 2018. Metabolic intermediates - cellular messengers talking to chromatin modifiers. Mol. Metab. 14, 39–52.

Noble, D., 2006. The Music of Life: Biology Beyond the Genome. Oxford University Press.

Oburoglu, L., Tardito, S., Fritz, V., de Barros, S.C., Merida, P., et al., 2014. Glucose and glutamine metabolism regulate human hematopoietic stem cell lineage specification. Cell Stem Cell 15, 169–184.

Oyama, S., 2000. The Ontogeny of Information: Developmental Systems and evolution, 2nd edition Duke University Press, Durham, North Carolina.

Paldi, A., 2003. Stochastic gene expression during cell differentiation: order from disorder? Cell Mol. Life Sci. 60, 1775–1778.

Paldi, A., 2012. What makes the cell differentiate? Prog. Biophys. Mol. Biol. 110, 41–43.

Paldi, A., 2018. Conceptual challanges of the systematic approach in understanding cell differentiation. Methods Mol. Biol. 1702, 27–39.

Richard, A., Boullu, L., Herbach, U., Bonnafoux, A., Morin, V., et al., 2016. Single-cell-based analysis highlights a surge in cell-to-cell molecular variability preceding irreversible commitment in a differentiation process. PLoS Biol. e1002585.

Sivanand, S., Viney, I., Wellen, K.E., 2018. Spatiotemporal control of acetyl-CoA metabolism in chromatin regulation. Trends Biochem. Sci. 43, 61–74.

Sneppen, K., Ringrose, L., 2019. Theoretical analysis of Polycomb-Trithorax systems predicts that poised chromatin is bistable and not bivalent. Nat. Commun. 10, 2133.

Tims, H.S., Gurunathan, K., Levitus, M., Widom, J., 2011. Dynamics of nucleosome invasion by DNA binding proteins. J. Mol. Biol. 411, 430–448.

Tischler, J., Gruhn, W.H., Reid, J., Allgeyer, E., Buettner, F., et al., 2019. Metabolic regulation of pluripotency and germ cell fate through alpha-ketoglutarate. EMBO J. 38.

Varahan, S., Walvekar, A., Sinha, V., Krishna, S., Laxman, S., 2019. Metabolic constraints drive self-organization of specialized cell groups. Elife 8.

von Hippel, P.H., Revzin, A., Gross, C.A., Wang, A.C., 1974. Non-specific DNA binding of genome regulating proteins as a biological control mechanism: I. The lac operon: equilibrium aspects. Proc. Natl Acad. Sci. U S A 71, 4808–4812.

Waddington, C., 1956. Embryology, epigenetics and biogenetics. Nature 177, 1241.

Waddington, C.H., 1956. Correspondance. Nature 4522, 1241.

Waddington, C.H. (Ed.), 1957. The Strategy of the Genes. Allen & Unwin.

Walsh, C.T., Tu, B.P., Tang, Y., 2018. Eight kinetically stable but thermodynamically activated molecules that power cell metabolism. Chem. Rev. 118, 1460–1494.

Wellen, K.E., Hatzivassiliou, G., Sachdeva, U.M., Bui, T.V., Cross, J.R., et al., 2009. ATP-citrate lyase links cellular metabolism to histone acetylation. Science 324, 1076–1080.

Xu, T., Stewart, K.M., Wang, X., Liu, K., Xie, M., et al., 2017. Metabolic control of TH17 and induced Treg cell balance by an epigenetic mechanism. Nature 548, 228–233.

Ye, C., Tu, B.P., 2018. Sink into the epigenome: histones as repositories that influence cellular metabolism. Trends Endocrinol. Metab. 29, 626–637.

Zentner, G.E., Henikoff, S., 2013. Regulation of nucleosome dynamics by histone modifications. Nat. Struct. Mol. Biol. 20, 259–266.

Zhu, F., Farnung, L., Kaasinen, E., Sahu, B., Yin, Y., et al., 2018. The interaction landscape between transcription factors and the nucleosome. Nature 562, 76–81.

Zipori, D., 2004. The nature of stem cells: state rather than entity. Nat. Rev. Genet. 5, 873–878.

4

Manoeuvring protein functions and functional levels by structural excursions

Vasam Manjveekar Prabantu*, Arangasamy Yazhini* and Narayanaswamy Srinivasan

Molecular Biophysics Unit, Indian Institute of Science, Bangalore, India

Proteins are workhorses that perform almost all biological functions in the cell. These polypeptide chains are made from a meaningful sequential order of 20 amino acids that are connected by peptide bonds. Proteins assume globular fold by establishing interactions (such as hydrogen bonding, hydrophobic, ionic, and van der Waals) among residues that are proximal as well as distant in the amino acid sequence. These interresidue interactions mold a protein to adopt a stable three-dimensional (3D) structure and provide a shape to confer specific molecular functions. It is well established that the sequential order of amino acids determines 3D structure and function of a given protein (Anfinsen, 1973).

In the 3D structure of a protein, not all residues are involved in protein function. A localized region house functional residues with a specific spatial geometry and these execute molecular function while rest of the residues support in conferring right shape to the functional site and dynamics of functional residues.

Changes in amino acid sequence of a protein can have deleterious effects on the function (Panyasai et al., 2004). Divergent evolution results in the alteration of amino acid sequences that is conferred by mutations in protein coding regions of DNA. These genetic mutations influence protein functions at various levels such as mRNA transcription, splicing, translation, active site shape and its chemical property, allostery communication, and interactions between protein and with other biomolecules. Alteration in protein functions and/or regulation leading to disease conditions are often associated with genetic mutations. Nevertheless, genetic mutations are not the only influencer of

*Equal contribution.

Phenotypic Switching. DOI: https://doi.org/10.1016/B978-0-12-817996-3.00006-2

protein functions. Alterations in protein functions or functional levels are observed even without genetic mutations (Huberts and van der Klei, 2010; Nielsen et al., 2007). Cellular environment and inherent properties of proteins are also key factors involved in regulating the nature and levels of function.

In this chapter, the effect of nongenetic factors that modulate protein functions and functional levels are discussed. We focus mainly on the observed instances of nongenetic factors that lead to phenotypic changes associated with survival and abnormal cellular conditions. These nongenetic factors are categorized into moonlighting, protein–protein interactions, domain–domain interactions, and posttranslational modifications. Further, the effect of silent genetic mutations (synonymous codon substitutions) on protein functions is also discussed. Each section of the chapter discusses appropriate examples and explains how these cues assist proteins to switch their molecular functions under different circumstances and thus consequent in phenotypic variations.

Moonlighting proteins

The paradigm of "one protein for one function" has been questioned since the observation of multifunctional proteins. They are referred to as "moonlighting proteins" and are known to perform multiple functions under different cellular contexts. In the 1980s, *Piatigorsky* and *Wistow* reported the first observations of moonlighting proteins (Wistow et al., 1988). They found that crystallins present in the eye lens of vertebrates that are known to function as structural proteins, moonlight as enzymes. For example, τ-crystallin in eye lens of turtles moonlight as an α-enolase. Many such examples have been discovered since then and are attributed to "gene sharing", a phenomenon where a single protein performs distinct functions (Piatigorsky and Wistow, 1989). Moonlighting in proteins has been observed due to the pleiotropy in gene actions under identical genetic background. The pleiotropy may exist in a molecule as the different phenotypes or arise due to changes in the environment, interacting partners, cellular localization, or several other factors.

As many studies have focused on moonlighting proteins, the definition of moonlighting has been refined with time. Functions of proteins that are mechanistically different and are not a consequence of individual domain functions or gene fusion events are termed as moonlighting proteins. With the discovery of several moonlighting proteins, many databases have been developed to specifically provide information about moonlighting proteins (Franco-Serrano et al., 2018). For example, Moonprot is a database that has comprehensive details of nearly 400 moonlighting proteins from different organisms (Chen et al., 2018). Identification of such proteins is not a trivial task because moonlighting functions are generally not inferred by homology nor by the presence of sequence motifs that resemble multiple interaction sites (Jeffery, 2015). However, it has been reported that certain orthologs retain their moonlighting nature, but the kind of the moonlighting function may not be the same (Arnér and Holmgren, 2000).

Certain enzymes are multispecific, that is, they bind to multiple substrates, using the same binding site and perform similar functions. Likewise, a moonlighting function can be a result of a new use for an existing binding site or modifications of otherwise unused regions of the protein structure. It has been a common observation that moonlighting functions are associated with enzymes (Huberts and van der Klei, 2010). Either a protein that is known to perform a nonenzymatic function has been identified later to function as an enzyme or an enzyme has been identified subsequently to have a nonenzymatic role. The choice about which is the canonical function, and which is the moonlighting function is often arbitrary and challenging to predict. However, phylogenetic analysis of moonlighting proteins may shed light on the canonical function by sequence conservation (Hernández et al., 2015). In this section we discuss a few examples of such moonlighting proteins.

Moonlighting proteins have been recognized in all kingdoms of life. Thioredoxin that is present in most organisms generally functions as a molecular antioxidant. Additionally, it can perform different functions depending upon the organism in which it is expressed (Arnér and Holmgren, 2000). For example, upon bacteriophage T7 infection in *Escherichia coli*, thioredoxin, known as an antioxidant, binds to a loop region of T7 DNA polymerase and forms a sliding clamp. It allows the polymerase to bind DNA more strongly and enhances DNA replication. The enhanced DNA replication mediated by thioredoxin is a crucial step for bacteriophage to infect *E. coli* (Bedford et al., 2002). Hence, thioredoxin has dual function as being antioxidant as well as T7 DNA polymerase enhancer in *E. coli*. These two functions are completely independent of each other. Similarly, thioredoxin exhibits several other functions in different organisms and has been recognized as a well-known example of moonlighting proteins.

The θ subunit, encoded by *holE* gene, is a structural component of the DNA polymerase III core complex that is commonly found in prokaryotes (Kelman, 1995; McHenry, 2003). In the core complex, it acts as accessory component and associates with α subunit (DNA polymerization) and ε subunit (proofreader exonuclease) (Keniry et al., 2006). It was believed that it has no biochemical function until the recognition of sequence and structural similarity with a family of nucleoid-associated regulatory proteins referred as Hha/YdgT-like proteins. YdgT is a transcription factor and known to be involved in the transcriptional termination of *tna* operon (Dietrich et al., 2014). Microarray gene expression experiments on the YdgT and θ subunit mutants have revealed that mutations on both genes influence the expression of several other genes including *tna* operon genes. In addition, the transcriptional profiles of YdgT mutant are similar to that of θ subunit mutant. From the high similarity in sequence, structure, and expression profiles of θ subunit with the YdgT gene, it has been suggested that θ subunit can additionally function as a transcription terminator like YdgT protein apart from being an accessory component in the core complex of DNA polymerase III.

The ribosomal protein S4 from *Disctyostelium discoideum* (DdS4) is another example of a protein with pleiotropic roles (Amarnath et al., 2012). As a part of ribosomal complex, it participates in protein synthesis and contributes to cell viability. DdS4 also acts a scaffold to form a multiprotein complex, an analogous to a complex of cell cycle genes (*cdc42*, *cdc24*, and *bem1*) involved in morphogenesis of yeast. This multiprotein complex is involved in the morphogenesis of *D. discoideum*. Thus, DdS4 has a dual function of having role in protein synthesis and morphogenesis.

Several moonlighting proteins have been identified to be associated with disease processes (Sriram et al., 2005). One such example is dihydrolipoamide dehydrogenase (DLD), a mitochondrial enzyme. It associates with many enzyme complexes to participate in energy metabolism and to maintain redox balance. Impairment of DLD activity is linked to severe disorders in infancy such as inability to thrive, hypotonia, and metabolic disorders. This enzyme is functionally active as homodimer, which exists in equilibrium with inactive monomer (Klyachko et al., 2005). Acidification of the mitochondrial matrix shifts equilibrium toward inactive monomeric form of DLD and leads to reduced enzymatic activity. However, it has also been observed that the disassociation of an active DLD homodimer into monomers leads to the acquisition of protease activity upon exposing catalytic dyad present at the dimeric interface

(Fig. 4.1) (Babady et al., 2007). Thus, the DLD enzyme possesses dual enzymatic functions.

The nature of moonlighting in proteins has been mostly identified by accidents. It is possible that there are many more proteins with moonlighting functions that are yet to be recognized.

Functional switch mediated by protein–protein interactions

Given the crowded milieu of biological molecules (proteins, nucleic acids, polysaccharides, ions, etc.) inside the cell, interactions among biomolecules are unavoidable and influence protein functions (Kuznetsova et al., 2014). Interaction among proteins can influence the molecular functions of interacting protein partners. For example, Ras protein interacts with Raf kinases and activates them to cascade signals for cell division (Drugan et al., 1996). Similarly, β-catenin interacts with transcription factors (of T cell factor/lymphoid enhancer factor family) and coactivates them to regulate the expression of genes involved in Wnt pathway mediated cell proliferation and cell migration (MacDonald et al., 2009). Notably, interactions among proteins not only modulate molecular functions but also switch the functioning of interacting protein partners in certain circumstances. In the following paragraphs, a few examples of protein–protein interactions mediating functional switch are explained.

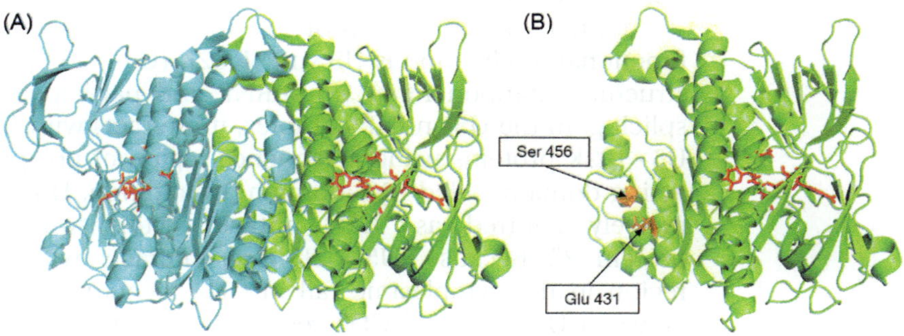

Figure 4.1 The crystal structure of human DLD. (A) Cartoon representation of DLD dimer and individual chains are shown in *cyan* and *green* color with cofactor flavin adenine dinucleotides in red [PDB code: 5NHG]. (B) Cartoon representation of one of the interacting chains and interface residues Ser456 and Glu431 are shown as sticks in *orange* color that form a catalytic dyad at the dimeric interface. *PDB*, Protein Data Bank.

Histones are a family of alkaline proteins that wrap double-stranded DNA into bundles of nucleosomes and compact them into chromatin. The histone family contains core and linker histones. Core histones comprise of H2A, H2B, H3, and H4 proteins. Each of these proteins functions as dimers and interacts among themselves to organize into octamer and wraps DNA around the octameric complex to form nucleosome. Linker histones comprise of H1 protein and its isoforms (histone H5). Two molecules of histone H1 bind to an entry and exit sites of DNA in the nucleosomes, lock the DNA in place and help in the formation of higher order nucleosome structures. Both core and linker histones interact among themselves and form homo- as well as hetero-oligomeric complexes to pack DNA compact. Since histones are the integral part of nucleosomes, they are highly localized in the nucleus (Mariño-Ramírez et al., 2005). However, few studies have evidenced that histone molecules are also present at the plasma membrane. In macrophages, histone H2B is located at the periphery of plasma membrane and act as a plasminogen receptor. Molecular signals mediated by the interactions between histone H2B and plasminogen are associated with cell adhesion and migration under inflammatory condition (Das et al., 2007). Similarly, histone H1 was observed to be located at the plasma membrane of macrophages where it binds to thyroglobulin and helps in the clearance of thyroglobulin from the blood circulation through endocytosis (Brix et al., 1998). Hence, under the circumstance of protein complex in the nucleus, histone H2B and H1 are functioning as structural proteins and contribute to DNA packing. Whereas, the secreted molecules of these histones are embedded into the plasma membrane of macrophages, act as receptors for other secretory proteins and regulate inflammatory responses.

Another example of protein–protein interaction mediated functional modulation is U5 snRNP 52K (U5–52K) protein. It is a structural component of U5 snRNP and participates in mRNA splicing. In the U5 snRNP complex, it interacts with U5–15K and U5–102K proteins using a sequence region of polyproline interacting domain called GYF-domain. In addition, U5–52K is alternatively referred as "CD2 antigen cytoplasmic tail-binding protein 2". Because, U5–52K also interacts with proline-rich region in the cytoplasmic tail of CD2 receptor. The interaction between U5–52K and CD2 receptor, mediated by GYF-domain, is involved in the production of interleukin 2. Although, U2–52K uses GYF-domain to interact with U5–15K/U5–102K and CD2 receptor, the interfacial regions in the GYF-domain are different between these two interactions (Fig. 4.2) (Nielsen et al., 2007). It means U5–52K uses different parts of the same domain to participate in diverge protein–protein interactions and play a role in mRNA splicing as well as interleukin 2 production.

Ski8 protein is an integral part of Ski complex, which is essential for the cytoplasmic functions of exosome such as RNA turnover, interference, and surveillance. Ski8 has seven WD40 repeats and adopts β-propeller conformation. In Ski complex, it interacts with Ski2 and Ski3 proteins in a 2:1:1 stoichiometry using WD40 repeats. Presence of Ski8 in the Ski complex helps in central positioning of Ski2 where it threads single-stranded RNA to pass through via Ski complex (Halbach et al., 2013). During meiotic division, Ski8 relocalizes into chromosomes in the nucleus and associates with Spo11 protein. It helps Spo11 to recruit other proteins into meiotic chromosomes for the formation of Spo11-dependent multiprotein complex. The Ski8 aided multiprotein complex catalyzes DNA double-strand breaks (DSBs) and initiates meiotic recombination (Arora et al., 2004). In this case, Ski8 acts as a scaffold protein to promote the assembly of multiprotein complex essential for meiotic recombination. Hence, Ski8 functions as an integral component of Ski complex in cytoplasm and participates in RNA metabolism, while it works as a scaffold protein for DSBs complex in the nuclear compartment during meiosis.

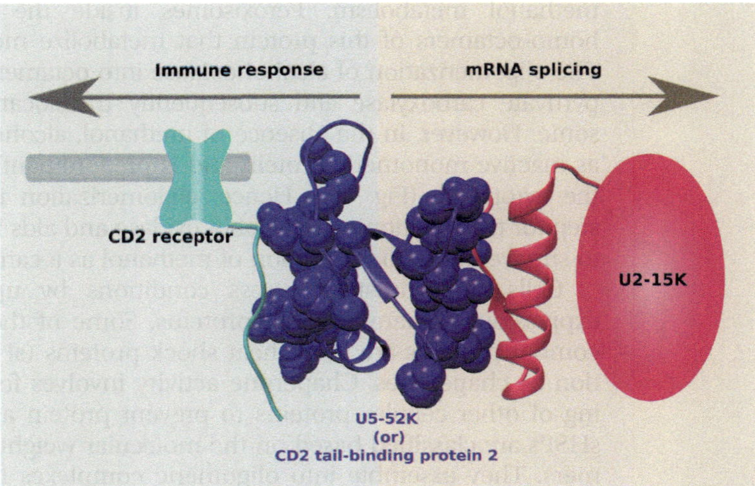

Figure 4.2 Protein-protein interaction mediated functional switch. This is an illustration of U5–52K (or CD2 tail-binding protein 2 protein, *blue*) interacting with CD2 receptor (left, *cyan*) and U2–15K protein (right, *magenta*). Residues at the interfacial regions in the GYF-domain of U5–52K are shown as spheres. 3D structure information was inferred from available structures in protein data bank (PDB codes: 1L2Z and 1SYX for U5–52K/CD2 receptor complex and U5–52K/U5–15K complex, respectively).

In addition to interactions involved between proteins of different sequences, interactions between identical proteins also have shown to modulate protein functions. The influence of self-interactions among proteins on their molecular functions is explained in the following section.

Modulation of protein function by oligomerization

Self-interactions among proteins of identical sequence form homo-oligomers. Oligomerization by self-interactions is essential for certain protein functions. It provides 3D topology for the active site located at the interfaces, structural stability, allosteric regulations, etc. (Miller et al., 1988; Goodsell and Olson, 2000; Cornish-Bowden and Koshland, 1971; Torshin, 1999). An alteration in the oligomeric status of self-interacting proteins may leads to changes in the molecular functions (like moonlighting proteins) or functional levels. Here, a few cases of oligomerization mediated functional modulation are discussed.

Methylotrophic yeast such as *Hansenula polymorpha* and *Pichia pastoris* can survive in methanol medium by metabolizing methanol as a carbon source. Alcohol oxidase is a key enzyme in methanol metabolism. Peroxisomes inside the cells assemble homo-octamers of this protein that metabolize methanol actively. The oligomerization of alcohol oxidase into octamer is mediated by pyruvate carboxylase and subsequently translocated into peroxisome. However, in the absence of methanol, alcohol oxidase exists as inactive monomeric protein and is predominantly distributed in the cytoplasm (Fig. 4.3). Hence, oligomerization is an important step for the functioning of alcohol oxidase and aids these organisms to survive under the condition of methanol as a carbon source.

Cells can withstand stress conditions by upregulating the expression of many combat proteins. Some of these well-known combat proteins are small heat shock proteins (sHSPs) that function as chaperones. Chaperone activity involves folding or refolding of other cellular proteins to prevent protein aggregation. The sHSPs are classified based on the molecular weight of their monomers. They assemble into oligomeric complexes in order to perform chaperone activity. The oligomeric complexes are tightly packed and hence the binding of sHSPs with substrates requires disassembly of large oligomers (Van Montfort et al., 2001; Lee and Vierling, 2000; Giese and Vierling, 2002). The suboligomeric complexes of sHSP are required in higher concentration for efficient chaperone activity (Feil et al., 2001). Large oligomeric complexes

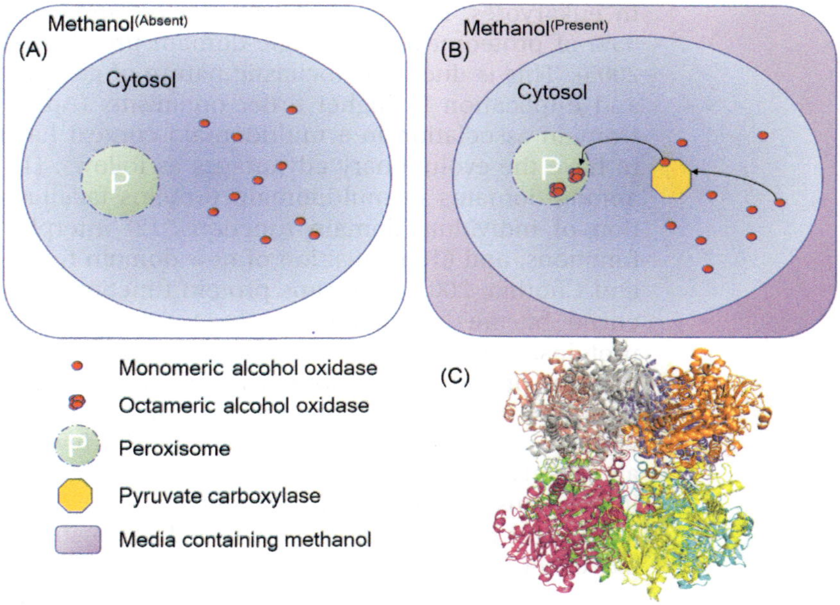

Figure 4.3 A schematic representation on the regulation of alcohol oxidase oligomeric state in methylotrophic organisms. (A) Inactive monomers of alcohol oxidase freely distributed in the cytosol under the condition of no methanol in the medium. (B) Pyruvate carboxylase catalyzes the oligomerization of alcohol oxidase into octamers and translocates to peroxisomes under the condition of methanol as a carbon source. (C) Shown is a cartoon representation of octameric form of alcohol oxidase from *Pichia pastoris* (PDB code: 5HSA). Colors highlight individual chains.

may still have chaperone activity but it is inefficient than those sHSP that can easily assemble and disassemble (Giese and Vierling, 2002). Hence, the dynamic nature of the oligomeric state is necessary for sHSPs to function efficiently.

These examples emphasize an essential role of protein–protein interactions in mediating context dependent protein functions. In addition, it highlights the robust mechanism of cells to use the same protein for several biological activities depending on the nature of interacting partners.

Influence of domain association on protein function

Many proteins are made of multiple domains that represent evolutionary units within a protein and can fold independently.

In eukaryotes, multidomain proteins are more common (over 65% of proteomes) than single-domain proteins (Ekman et al., 2005). This is due to a stochastic nature of gene recombination and duplication in higher order organisms (Apic et al., 2001). Domain association in a multidomain context has been shown to have the evolutionary advantages as follows: (1) interactions among domains in multidomain proteins facilitate the regulation of individual domain functions, (2) interplay of domain functions, and (3) acquisition of new domain function (Bashton and Chothia, 2007). It means protein function is strongly influenced by the interactions between domains in multidomain proteins. Thus, protein function is not only influenced by protein–protein interactions (as explained in the previous section) but also by the interactions among domains present within a protein. This section exemplifies a few cases in which alterations in protein functions are driven by domain–domain interactions in multidomain proteins.

A sugar hydrolyzing enzyme called β-glucanase breaks 1→3 glucosidic linkage in β-D-glucans. It has a domain belong to 'trans-glycosidase' superfamily (Fig. 4.4A) (Varghese et al., 1994). Homolog of this domain is present at the C-terminus of β-glucuronidase (domain 3) in the context of multidomain association. The β-glucuronidase has two additional domains located at the N-terminus to the domain 3. The first domain at the N-terminus (domain 1) belongs to "galactose binding" superfamily. Domain in the middle (domain 2) belongs to "glucuronidase domain" superfamily and act as a linker between domain 1 and domain 3 (Fig. 4.4B). Domain 2 does not participate in enzyme catalysis whereas domain 1 contributes to the catalytic function of domain 3 by taking part in the active site. An extended loop from domain 1 occludes the active site of domain 3 and makes it different from the active site of single-domain β-glucanase. The alteration in the active site provides substrate specificity to β-glucuronidase to act upon β-D-glucuronoside (Jain et al., 1996; Juers et al., 1999). Hence, additional domains in β-glucuronidase regulate hydrolyzing function of domain 3. Previous studies have shown that interactions among domains can influence 3D structure, dynamics and energetics of individual domains in multidomain proteins (Bhaskara and Srinivasan, 2011; Vishwanath et al., 2018). The influence of domain 1 and domain 2 on the structure as well as dynamics of domain 3 in β-glucuronidase was examined upon comparison with its homologous "trans-glycosidase" domain present as a single entity in β-glucanase. Notable structural differences are observed (root mean square deviation of Cα atoms is 3.3 Å)

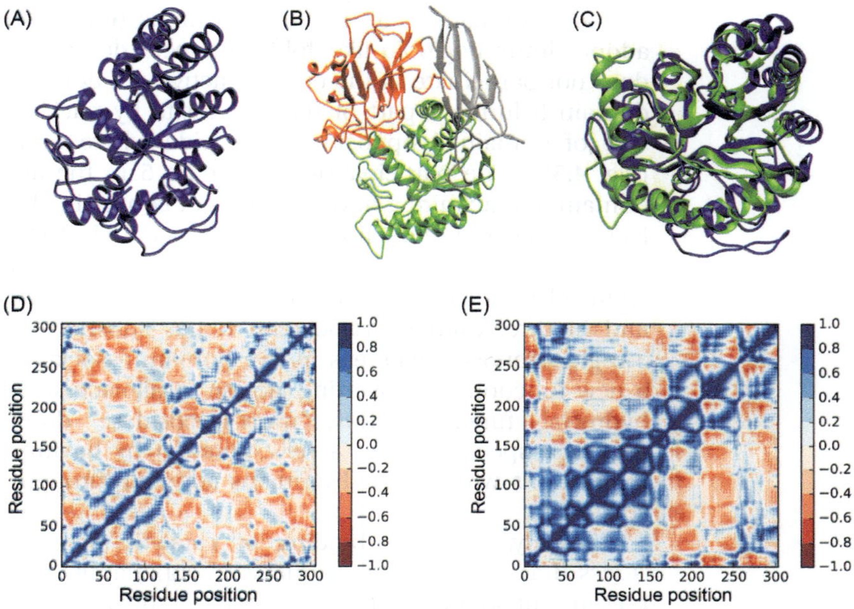

Figure 4.4 Changes in 3D structure and dynamics of "trans-glycosidase" domain upon interactions with other domains in multidomain context. (A) Cartoon representation of β-glucanase structure in which a domain of "trans-glycosidase" superfamily present in isolation (PDB code: 1GHS). (B) Cartoon representation of β-glucuronidase where domain 1, 2, and 3 are colored as *orange, gray,* and *green,* respectively (PDB code: 1BHG). (C) Structural superposition of "trans-glycosidase" domain of β-glucanase (*navy blue*) and β-glucuronidase (*green*) with Cα root mean square deviation of 3.3 Å. (D) Dynamics cross-correlation matrix (represents concerted interresidue motions) of "trans-glycosidase" domain from β-glucanase. (E) Dynamics cross-correlation matrix of "trans-glycosidase" domain (domain 3) from β-glucuronidase. Data on cross-correlations in residue motions were obtained from all-atom normal mode analysis (Skjaerven et al., 2014).

between "trans-glycosidase" domains of β-glucanase and β-glucuronidase (Fig. 4.4C). Dynamics analysis shows that inter-residue concerted motions are stronger in the context of multi-domain as compared to the domain being in isolation (Fig. 4.4D and E). Hence, 3D structure and dynamics of "trans-glycosidase" domain is strongly influenced by the interactions made with additional domains in β-glucuronidase.

Interestingly, the three-domain association of β-glucuronidase is also seen in β-galactosidase along with two other domains. This five-domain combination adds new function to domain 3, which is not observed when it occurs alone (in β-glucanase) or in association with two other domains (in β-glucuronidase). In β-galactosidase, the first three N-terminus domains (domain 1–3) are homologous to β-glucuronidase along with two additional domains (domain 4–5) at the C-terminus (Fig. 4.5A). Domain 4 adopts immunoglobin-like fold, connects domain 3 and 5 but does not participate in the enzyme activity of domain 3. However, domain 5 forms a part of the active site by packing against the side of domain 3 opposite to the side where domain 1 packs (Fig. 4.5B). The participation of domain 5 in the active site aids domain 3 to acquire a novel function of converting lactose to allolactose (Bashton and Chothia, 2007; Juers et al., 1999; Todd et al., 2001; Juers et al., 2000). These observations indicate that when "trans-glycosidase" domain functions alone, it hydrolyzes D-glucan and in three-domain association it has substrate specificity for β-D-glucuronoside whereas in five-domain association it acquires a new function of converting lactose to allolactose in addition to hydrolysis function. Thus, "trans-glycosidase" domain acquires substrate specificity and a new function by having interdomain interactions with tethered domains adopted through the event of speciation.

Similarly, substrate binding of catalytic domain belongs to "Class II aaRS and biotin synthetases" superfamily is altered by the presence of additional domain in aspartyl-tRNA synthetase as compared to that of its homologous catalytic domain present as a single entity in asparaginase synthetase (Nakatsu et al., 1998; Ruff et al., 1991). Substrate affinity of hexokinase-1 differs between single-domain and multidomain orthologs (Vishwanath et al., 2018; Miller

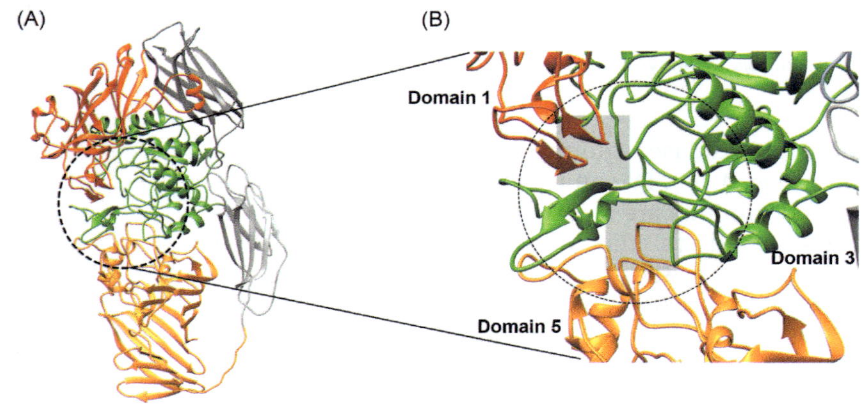

Figure 4.5 Domain–domain association alters the active site in a multidomain protein. (A) Cartoon representation of β-galactosidase structure (PDB code: 1JZ7). Colors discriminate individual domains where domains 1–3 are colored same as their homologs in Fig. 4.1B. (B) Occlusion of active site in "trans-glycosidase" domain (or domain 3, encircled by black dashed lines) by extended loops of domain 1 and 5 is highlighted in gray-colored boxes.

et al., 2007; Liu et al., 1991). In addition, domain association can lead to an acquisition of dual functions to the protein as observed in 6-phosphofructo-2-kinase under different genetic background (Bashton and Chothia, 2007). Hence, domain–domain interaction is yet another factor influencing protein functions.

Modulation of protein function by posttranslational modifications

Proteins are synthesized from mRNA by ribosomes in the process of translation. Upon translation, proteins may undergo chemical modification such as phosphorylation, acetylation, formation of disulfide bridges, carbonylation, and glycosylation (Duan and Walther, 2015). Phosphorylation is the common posttranslational modification that is catalyzed by kinases. It is known to play an important role in the regulation of enzyme activity and most cellular functions (Nestler and Greengard, 1983). In this section, two examples on how phosphorylation causes differences in protein function or functional levels are discussed.

Prostate associated gene-4 (PAGE4) is an intrinsically disordered protein (IDP) (Zeng et al., 2011). IDPs are extremely dynamic and do not fold into globular structures in isolation. Through dynamics, an IDP explores several nonstable states, which allow it to engage in interactions with other substrates. Homeodomain-interacting protein kinase-1 (HIPK1) phosphorylates PAGE4 at serine and threonine residues at positions 9 and 51 (Rajagopalan et al., 2014; Mooney et al., 2014). As a result, PAGE4 samples only compact folded conformations as compared to unphosphorylated protein. On the contrary, CDC-like kinase 2 (CLK2) hyperphosphorylates PAGE4 at multiple sites and allows the protein to acquire an elongated random loop-like conformations (Fig. 4.6) (Atala, 2017). Interestingly, these two conformational ensembles exhibit two distinct functions. HIPK1 phosphorylated PAGE4 binds to the canonical substrate and potentiates its activity whereas CLK2 phosphorylated PAGE4 attenuates the substrate activity.

Alpha B-crystallins oligomerise to function as chaperones. Under mitosis or stress conditions, αB-crystallins undergo phosphorylation at three serine residues 19, 45, and 59 (Ito et al., 1997; Kato et al., 1998). The phosphorylation is mediated by cAMP-dependent protein kinase. Mutational studies on phosphorylation sites has demonstrated that phosphorylation is essential for αB-crystallins to perform chaperone-like activity

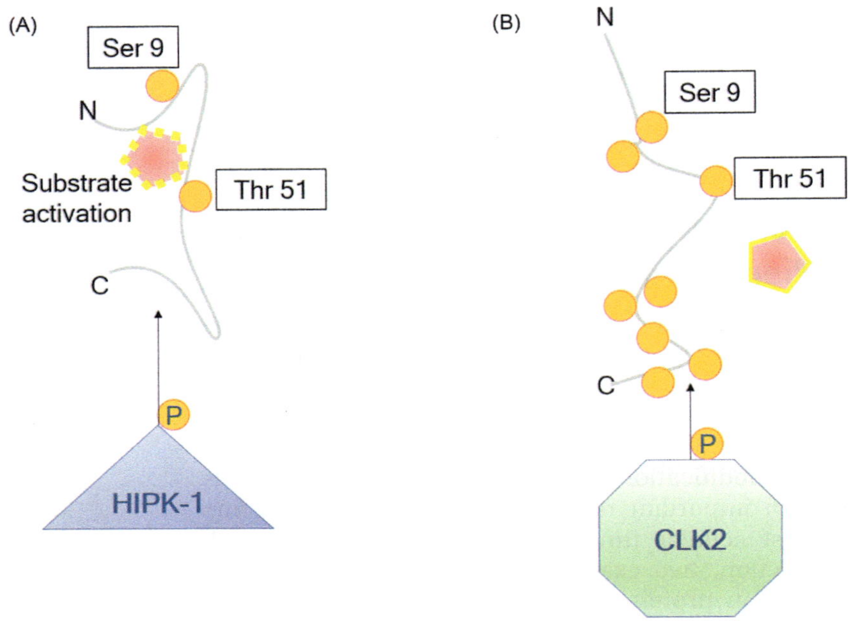

Figure 4.6 Differential phosphorylation of prostate associated gene-4 (PAGE4) protein results in differential functional levels. (A) The stress–response kinase homeodomain-interacting protein kinase-1 phosphorylates PAGE4 at two positions (Ser9 and Thr51) and result in a compact conformation that can bind to the substrate. (B) CDC-like kinase hyperphosphorylates PAGE4 at eight different Ser/Thr residues that leads to an elongated conformation of PAGE4 and cannot bind to the substrate.

(Spector et al., 1985; Voorter et al., 1989). Upon phosphorylation, the higher oligomers of αB-crystallins dissociate into lower order oligomers that consequently affects chaperone-like activity (Ito et al., 2001). Similar observations in sHSPs (HSP27 and HSP20) emphasize that phosphorylation alters functionally relevant oligomeric states (Kato et al., 1994; Lavoie et al., 1995; Kato et al., 1994). Hence, functional activities of αB-crystallin and sHSP are regulated by posttranslational phosphorylation.

Silent mutation tunes gene function

Codons are made up of nucleotide triplets, corresponding to 64 combinations of four bases. Of these, 61 combinations encode for 20 amino acids. According to Anfinsen theory, amino acid sequence is a determinant of protein 3D structure and hence having more than one codon for an amino acid (codon degeneracy) was believed to be redundant until the observation of organism-specific codon distributions. Codon degeneracy accounts for multiple codons for a single amino acid that differ in the nature of nucleotide often at the third position and hence they are synonymous. Preference of one codon over other synonymous codons varies among organisms indicating that codon distributions are biased in an organism-specific manner. A point mutation that converts one synonymous codon to another is referred as synonymous mutation. Since synonymous mutation does not affect encoded amino acid, they are considered to be "silent" and thus referred as "silent mutation". It is generally believed that "silent" mutation does not cause phenotypic effect. However, it has been observed that the occurrence of synonymous mutations is under selective pressure and they are associated with more than 50 human diseases (Hurst and Pál, 2001; Sauna and Kimchi-Sarfaty, 2013). Genetic studies on human cancer cells revealed that synonymous mutations in oncogenes often act as driver mutations for cancer conditions (Supek et al., 2014).

Hence, the impact of synonymous mutations has been studies in various disease conditions. From these studies, it was realized that synonymous mutations regulate gene function by various mechanisms at multiple levels starting from processing of mRNAs to cotranslational folding of proteins. The following sections give an overview of the regulatory mechanisms by which synonymous mutations regulate gene expression with the examples of alterations in protein functions.

Synonymous mutations dictate gene splicing

In eukaryotes, splicing is an essential mechanism that holds the responsibility of making mature mRNA from the nascent product of gene transcription to synthesize functional protein. Evidences from functional consequences of synonymous mutations in disease conditions indicate that substitution of synonymous codons modulate gene splicing. Synonymous mutations with most apparent impact are those located near the splice junctions. Another molecular mechanism by which synonymous mutations alter splicing is the activation of exonic cryptic or de novo splice sites. Synonymous mutations in exonic regulatory sequences located far from splice junctions have also been reported to be linked to the aberration of gene splicing. The earliest example of synonymous mutations associated with altered protein functional levels is β-thalassemia phenotype. A synonymous mutation (GGT → GGA for glycine) in codon 24 in β-globin gene activates abnormal 5′ exonic donor splice site at codon 25.

Activation of cryptic splice site truncates 16 nucleotides upstream of exon 1-intron 1 boundary. These truncated variant of mRNA are subsequently degraded by nonsense-mediated decay and consequent in 75% decreased expression of normal β-globin mRNA. This condition leads to the development of β-thalassemia (Goldsmith et al., 1983). Similarly, other incidences of synonymous mutations associated with various human diseases by affecting gene splicing are listed in Table 4.1.

However, synonymous codon substitution need not always results in aberrant splicing. In medium-chain acyl-CoA dehydrogenase (MCAD) deficiency, the presence of nonsynonymous mutation (c.362C→T) in exon 5 of MCAD gene inactivates a crucial exonic splicing enhancer, causes skipping of exon 5 and thus leads to the loss of functional proteins. Another mutation causing synonymous codon substitution (c.351A→C) at exon 5 nullifies the deleterious effect of c.363C→T mutation by disrupting its corresponding exonic splicing silencer. Hence, synonymous mutations may also be beneficial to gene splicing (Nielsen et al., 2007). These observations highlight the complex interplay between splicing regulatory elements and synonymous codons in order to dictate gene splicing.

Synonymous mutations regulate folding of mRNA secondary structure

Synonymous mutations affect gene expression by regulating mature mRNA folding. In Dopamine receptor D2 (DRD2) gene, C957T substitution alters mRNA secondary structure due to which mRNA stability decreases, the rate of translation reduces and eventually weakens dopamine-induced upregulation of DRD2 gene expression (Duan et al., 2003). This synonymous mutation is associated with schizophrenia and alcoholism. Human catechol-o-methyltransferase (COMT) is a key regulator of pain perception, cognitive function, and affected the mood by controlling catecholamine levels in the cell. Synonymous mutations in COMT gene (C/T; rs4633 and C/G; rs4818) increases mRNA stability by aiding the formation of varied stem-loop structure. This condition reduces translation rate and thus lower COMT enzymatic activity (Nackley et al., 2006). On the contrary, an another synonymous mutation (C166T) at the 5' end of COMT mRNA has shown to destabilize mRNA secondary structure which consequent in increased activity of COMT enzyme (Tsao et al., 2011). In both cases, synonymous mutations affect pain sensitivity.

Table 4.1 List of synonymous codons that influence gene splicing and associated with human disease conditions.

Disease condition	Gene (protein name)	Mutation	Mechanism	Effect on gene expression	Reference
Acute intermittent porphyria	PBGD or HMBS (porphobilinogen deaminase)	Exon 12 (position − 1): G → A	Alter splice junctions	Exon skipping	Grandchamp et al. (1989)
Tay–Sachs disease	HEXA (α subunit of β-hexosaminidase A)	Exon 5 (position −1): G → A	Alter splice junctions	Exon skipping	Akli et al. (1990)
Marfan syndrome	FBN1 (fibrillin-1)	Exon 51 (position c.6354): C → T	Disrupt exonic splicing enhancer	Exon skipping	Liu et al. (1997)
Frontotemporal dementia with parkinsonism	Tau (Microtubule Associated Protein Tau)	Exon 10 (codon position 284): T → C	Disrupt exonic splicing silencer	Exon inclusion	D'Souza et al. (1999)
Phenylketonuria	PAH (phenylalanine hydroxylase)	Exon 11 (position -3): A → T	Alter splice junctions	Exon skipping	Chao et al. (2001)
Spinal muscular atrophy	SMN2 (survival of motor neuron 2)	Exon 7 (position +6): C → T	Disrupt exonic splicing enhancer	Exon skipping	Cartegni and Krainer (2002)
Cystic fibrosis	CFTR (cystic fibrosis transmembrane regulator)	Exon 12: (position 19 and 40): T → A and T → C	Alter splice junctions	Exon skipping	Pagani et al. (2005)
von Hippel–Lindau disease	VHL (von Hippel-Lindau tumor suppressor)	Exon 2 (position c.462): A → C	Alter splice junctions	Exon skipping	Martella et al. (2006)
Congenital molecular dystrophy	POMGNT1 (O-mannose β-1,2-N-acetylglucosaminyl-transferase)	Exon 7 (position c.636): C → T	Activate exonic cryptic splice silencer	Exon skipping	Oliveira et al. (2008)
Laron syndrome	GHR (growth hormone receptor)	Exon 6 (codon position 180): A → G	Activate exonic cryptic splice sites	Exon truncation (24 nt)	Berg et al. (1992)
Treacher Collins syndrome	TCOF1 (treacle ribosome biogenesis factor 1)	Exon 22 (position c.3612): A → C	Disrupt exonic splicing enhancer	Exon skipping	Macaya et al. (2009)

(Continued)

Table 4.1 (Continued)

Disease condition	Gene (protein name)	Mutation	Mechanism	Effect on gene expression	Reference
Rett syndrome	MECP2 (Methyl-CpG binding protein 2)	Exon 1 (position c.48): C → T	Activate exonic cryptic splice sites	Exon deletion (16 nt)	Sheikh et al. (2013)
Crouzon syndrome	FGFR2 (fibroblast growth factor receptor 2)	Exon IIIc (position c.1032): G → A	Activate exonic cryptic splice sites	Exon truncation (51 nt)	Li et al. (1995); Del Gatto and Breathnach (1995)

In cystic fibrosis, synonymous mutations in CFTR gene affect not only mRNA splicing (as mentioned in the previous section) but also drastically alter secondary structural folding of its mRNA. Synonymous mutation (that converts a codon of isoleucine from ATC to ATT) in ΔF508 variant of CFTR mRNA alters mRNA folding and results in ribosome stalling. A delay in translation causes misfolding of CFTR protein. Subsequently, the misfolded protein is subjected to rapid endoplasmic reticulum-associated degradation (Bartoszewski et al., 2010). Thus, synonymous codon substitutions affect mRNA stability by altering its secondary structure folding.

Synonymous mutations impair the interactions of mRNA with RNA-binding proteins and miRNAs

Interactions between mRNA and RNA-binding proteins are yet another target for synonymous codons to regulate gene expression. In 15% of amyotrophic lateral sclerosis patients, synonymous mutations (G10G, S59S, T116T, N139N, A140A, and Q153Q) in copper/zinc superoxide dismutase (SOD1) gene have shown to affect the interaction of SOD1 mRNA with RNA-binding protein and thereby impair the formation of neuronal tissue-specific RNP complex (Ge et al., 2006).

Synonymous mutation (CDSN*971T) in corneodesmin (CDSN) decreases mRNA affinity for 39 kDa cytoplasmic RNA-binding protein. This results in increased CDSN mRNA stability and decreased rate of RNA decay which then consequent in the pathogenicity of psoriasis (Capon et al., 2004). Genome-wide protein coding sequence analysis has shown that the rate of evolution of synonymous sites in the motifs that interact with RNA-binding proteins is 2-3% lower than the same in regions that lack those motifs (Savisaar and Hurst, 2017). This observation indicates the need for evolutionary pressure on synonymous sites to conserve such motifs in order to retain mRNA-protein interactions.

Similarly, Hurst (2006) has showed that miRNA binding region in mRNA are under 35% reduced rate of evolution (Hurst, 2006). Subsequently, an evidence from immunity-related GTPase M (IRGM) gene has shown the association of synonymous mutation with Crohn's disease (Brest et al., 2011). A synonymous mutation of c.313C > T in IRGM mRNA is located in miR-196s (miR-196A and miR-196B) binding site. This codon substitution disrupts interaction between miR-196s and IRGM mRNA and leads to the loss of miR-196 regulation in IRGM-dependent xenophagy. It eventually causes abnormal persistence of Crohn's disease-associated intracellular bacteria in intestinal epithelial cells and thus induces intestinal inflammation. Another example is the association of C51T synonymous mutation in BCL2L12 (BCL2-like 12) gene with melanoma (Gartner et al., 2013). This mutation interrupts the binding of BCL2L12 mRNA with hsa-miR-671-5p. The loss of hsa-miR-671-5p mediated mRNA inhibition leads to elevated BCL2L12 protein concentration inside the cell. Interaction between BCL2L12 protein translated from the variant mRNA and p53 inhibits UV-induced apoptosis more efficiently than normal condition and reduces gene transcription of p53 targeted genes. As a consequence, antiapoptotic signals are enhanced and contribute to cell malignancy.

Synonymous mutations modulate cotranslational folding

Synonymous mutation in the coding region of a gene has shown to affect protein function by altering cotranslational folding (Yu et al., 2015). This phenomenon is exemplified in bovine gamma-B crystallin (Buhr et al., 2016). Gamma-B crystallin is a mammalian eye lens protein. It consists of an N-terminal domain and a C-terminal domain. In order to study the effect of synonymous mutations, Buhr et al. (2016) have introduced synonymous codons in the bovine gamma-B crystallin gene to have most similar frequencies as seen in the host expression system *E. coli* (harmonized, H). The expression of H variant was compared with the expression of native unharmonized *E. coli* gamma-B crystallin (U). Experimental results showed that an average translation rate of H variant (2.2 aa/s) was higher than U variant (1.8 aa/s). N-terminal domain synthesis took 30 seconds for U variant while H variant took only 20 seconds. The onset of folding was significantly different between U variant and H variant (50 seconds vs 35 seconds, respectively). An average folding time for N-terminal domain was 59 seconds for U variant and 39 seconds for H variant. Changes in folding rate leads to different conformations of H variant. These results had clearly indicated that synonymous mutations modulate real-time kinetics of translation and cotranslational protein folding.

The plausible mechanism by which synonymous codons gives rise to these alternative conformations of gamma-B crystallin is the modulation of translation elongation. Modulation of translation elongation would affect the interaction of N-terminal domain with ribosome, the rate at which N-terminal domain moves away from the ribosome surface and its interaction with emerging C-terminal domain. These effects interfere with the folding of N-terminal domain as well as conformational sampling of N- and C-terminal domains. Hence, synonymous mutations can produce proteins of different conformations with altered sensitivity to proteolytic cleavage (Buhr et al., 2016).

In human, modulation of cotranslation folding by synonymous mutation is implied in multidrug resistance of cancer cells. A synonymous mutation (C3435T) in exon 26 of multidrug resistance 1 (MDR1) gene changes frequent codon to rare codon and alters the cotranslational folding of P-glycoprotein by ribosome stalling. As a consequence, protein folds at a different rate, thereby affecting the insertion of P-glycoprotein into plasma membrane. Thus, it leads to an altered conformation of P-glycoprotein that has a different susbstrate specificity as compared to a protein with canonical frequent codon. The effect is directly linked to the phenotype of multidrug resistance in cancers (Kimchi-Sarfaty et al., 2007).

All these examples underpin the functional importance of codon degeneracy and biased distribution of synonymous codons. The mechanisms by which synonymous mutations fine-tune gene functions are of multiple levels such as RNA splicing, mRNA stability, interactions between mRNA and other biomolecules and cotranslational protein folding (Fig. 4.7). With this realization, efforts have been undertaken to recognize synonymous mutations associated with phenotypic switching using computational approaches such as Genomic Evolutionary Rate Profiling and Silent Variant Analyzer (Buske et al., 2013;

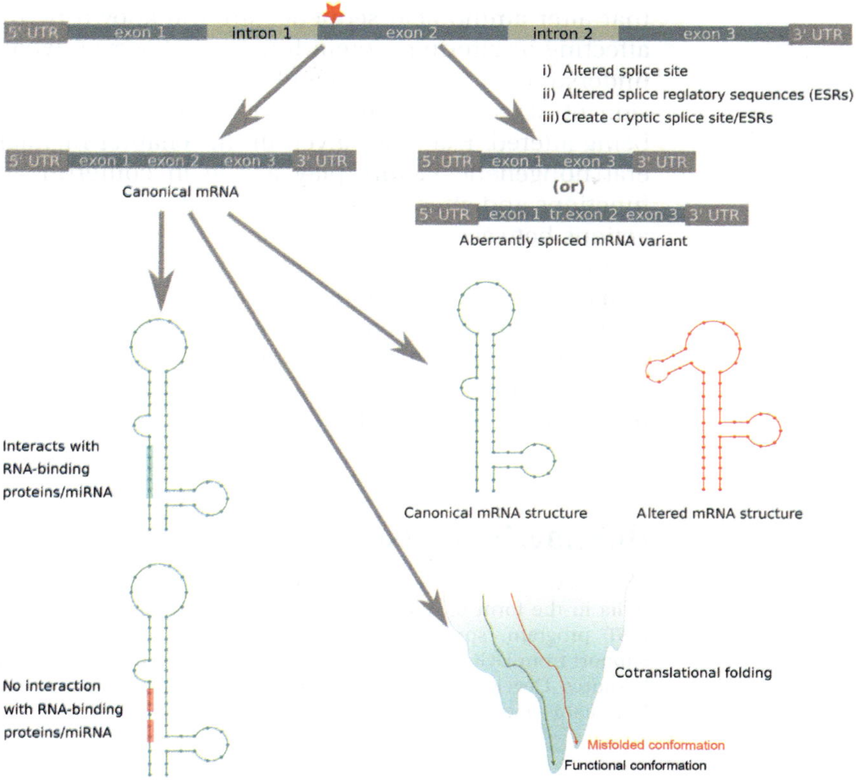

Figure 4.7 Regulation of gene expression by synonymous mutations. The effect of synonymous mutation (indicated as *red star*) on alternative splicing, mRNA secondary structure stability, mRNA and RNA-binding protein/miRNA interactions, and cotranslational protein folding are depicted. Abbreviation tr.exon indicates truncated exon and ESR corresponds to exonic splicing regulatory element. This image has been generated on the basis of a previous study (Gotea et al., 2015).

Davydov et al., 2010). Thus, growing evidences on the gene regulatory role may restrict the use of a tag "silent" while referring synonymous mutations.

Conclusion

Proteins are under selection pressure to preserve their function in evolution by conserving 3D structure. Although, only a localized part of the protein acts as a functional site, the chemical nature of remaining part and their interaction with functional site is essential to provide structural stability, shape, and regulatory cues to the functional site. Hence, a perturbation to protein structure may influence its function. Genetic mutations that alter amino acid sequence are the known perturbations in affecting or altering protein function. However, what is not well understood is the influence of cellular environment and excursion of 3D structure on protein function without its sequence being altered. Examples given in this chapter highlight that several nongenetic factors play a role in conferring moolighting functions and alter functional levels of proteins. These are interactions between proteins and other biomolecules, inter-domain interactions within a protein and posttranslational modifications. In addition, though synonymous mutation does not change amino acid sequence, it influences protein functions by modulating various mechanisms starting from mRNA processing to cotranslational folding. Thus, protein function is tightly regulated by diverse nongenetic factors.

Acknowledgments

This work is supported by the Department of Biotechnology, Government of India in the form of IISc-DBT partnership program, Mathematical Biology and FIST program, sponsored by the Department of Science and Technology. Support from UGC, India—Centre for Advanced Studies and Ministry of Human Resource Development, India, is gratefully acknowledged. NS is a J. C. Bose National Fellow.

Abbreviations

BCL2L12 BCL2-like 12
CDSN Corneodesmin
CLK2 CDC-like Kinase-2
COMT Catechol-o-methyltransferase
DdS4 Dictyostelium discoideum S4 ribosomal protein
DLD Dihydrolipoamide dehydrogenase
DRD2 Dopamine receptor D2
DSB DNA double strand breaks
HIPK1 Homeodomain-interacting protein kinase-1
IDP Intrinsically disordered proteins
IRGM Immunity-related GTPase M
MCAD Medium-chain acyl-CoA
MDR1 Multidrug resistance 1
PAGE4 Prostate associated gene-4
sHSP Small heat shock proteins
SOD1 Superdioxide dismutase
tna Tryptophanase

References

Akli, S., Chelly, J., Mezard, C., Gandy, S., Kahn, A., Poenaru, L., 1990. Akli, S., Chelly, J., Mezard, C., Gandy, S., Kahn, A., Poenaru, L., 1990, A "G" to "A" mutation at position -1 of a 5′ splice site in a late infantile form of Tay-Sachs disease. J. Biol. Chem. 265 (13), 7324–7330.

Amarnath, S., Kawli, T., Mohanty, S., Srinivasan, N., Nanjundiah, V., 2012. Pleiotropic roles of a ribosomal protein in dictyostelium discoideum. PLoS One.

Anfinsen, C.B., 1973. Principles that govern the folding of protein chains. Science 181, 223–230.

Apic, G., Gough, J., Teichmann, S.A., 2001. Domain combinations in archaeal, eubacterial and eukaryotic proteomes. J. Mol. Biol. 310 (2), 311–325.

Arnér, E.S.J., Holmgren, A., 2000. Physiological functions of thioredoxin and thioredoxin reductase. Eur. J. Biochem. 267, 6102–6109.

Arora, C., Kee, K., Maleki, S., Keeney, S., 2004. Antiviral protein Ski8 is a direct partner of Spo11 in meiotic DNA break formation, independent of its cytoplasmic role in RNA metabolism. Mol. Cell. 13 (4), 549–559.

Atala, A., 2017. Re: Phosphorylation-induced conformational dynamics in an intrinsically disordered protein and potential role in phenotypic heterogeneity. J. Urol. 198, 486–487.

Babady, N.E., Pang, Y.P., Elpeleg, O., Isaya, G., 2007. Cryptic proteolytic activity of dihydrolipoamide dehydrogenase. Proc. Natl Acad. Sci. U S A 104 (15), 6158–6163.

Bartoszewski, R.A., Jablonsky, M., Bartoszewska, S., Stevenson, L., Dai, Q., Kappes, J., et al., 2010. A synonymous single nucleotide polymorphism in ΔF508 CFTR alters the secondary structure of the mRNA and the expression of the mutant protein. J. Biol. Chem. 285 (37), 28741–28748.

Bashton, M., Chothia, C., 2007. The generation of new protein functions by the combination of domains. Structure. 15 (1), 85–99.

Bedford, E., Tabor, S., Richardson, C.C., 2002. The thioredoxin binding domain of bacteriophage T7 DNA polymerase confers processivity on Escherichia coli DNA polymerase I. Proc. Natl Acad. Sci. *U S A* 94 (2), 479–484.

Berg, M.A., Guevara-Aguirre, J., Rosenbloom, A.L., Rosenfeld, R.G., Francke, U., 1992. Mutation creating a new splice site in the growth hormone receptor genes of 37 Ecuadorean patients with Laron syndrome. Hum. Mutat. 1 (1), 24–34.

Bhaskara, R.M., Srinivasan, N., 2011. Stability of domain structures in multi-domain proteins. Sci. Rep. 1.

Brest, P., Lapaquette, P., Souidi, M., Lebrigand, K., Cesaro, A., Vouret-Craviari, V., et al., 2011. A synonymous variant in IRGM alters a binding site for miR-196 and causes deregulation of IRGM-dependent xenophagy in Crohn's disease. Nat. Genetics. 43, 242–245.

Brix, K., Summa, W., Lottspeich, F., Herzog, V., 1998. Extracellularly occurring histone H1 mediates the binding of thyroglobulin to the cell surface of mouse macrophages. J. Clin. Invest. 102 (2), 283–293.

Buhr, F., Jha, S., Thommen, M., Mittelstaet, J., Kutz, F., Schwalbe, H., et al., 2016. Synonymous codons direct cotranslational folding toward different protein conformations. Mol. Cell. 61 (3), 341–351.

Buske, O.J., Manickaraj, A., Mital, S., Ray, P.N., Brudno, M., 2013. Identification of deleterious synonymous variants in human genomes. Bioinformatics 29 (15), 1843–1850.

Capon, F., Allen, M.H., Ameen, M., Burden, A.D., Tillman, D., Barker, J.N., et al., 2004. A synonymous SNP of the corneodesmosin gene leads to increased mRNA stability and demonstrates association with psoriasis across diverse ethnic groups. Hum. Mol. Genet. 13 (20), 2361–2368.

Cartegni, L., Krainer, A.R., 2002. Disruption of an SF2/ASF-dependent exonic splicing enhancer in SMN2 causes spinal muscular atrophy in the absence of SMN. Nat. Genet. 30 (4), 377–384.

Chao, H.K., Hsiao, K.J., Su, T.S., 2001. A silent mutation induces exon skipping in the phenylalanine hydroxylase gene in phenylketonuria. Hum. Genet. 108 (1), 4–9.

Chen, C., Zabad, S., Liu, H., Wang, W., Jeffery, C., 2018. MoonProt 2.0: an expansion and update of the moonlighting proteins database. Nucleic Acids Res. 46 (D1), D640–D644.

Cornish-Bowden, A.J., Koshland, D.E., 1971. The quaternary structure of proteins composed of identical subunits. J. Biol. Chem. 246 (10), 3092–3102.

D'Souza, I., Poorkaj, P., Hong, M., Nochlin, D., Lee, V.M.Y., Bird, T.D., et al., 1999. Missense and silent tau gene mutations cause fr, ontotemporal dementia with parkinsonism-chromosome 17 type, by affecting multiple alternative RNA splicing regulatory elements. Proc. Natl. Acad. Sci. U S A 96 (10), 5598–5603.

Das, R., Burke, T., Plow, E.F., 2007. Histone H2B as a functionally important plasminogen receptor on macrophages. Blood. 110 (10), 3763–3772.

Davydov, E.V., Goode, D.L., Sirota, M., Cooper, G.M., Sidow, A., Batzoglou, S., 2010. Identifying a high fraction of the human genome to be under selective constraint using GERP++. PLoS Comput. Biol. 6 (12).

Del Gatto, F., Breathnach, R., 1995. A crouzon syndrome synonymous mutation activates a 5' splice site within the iiic exon of the fgfr2 gene. Genomics 27 (3), 558–559.

Dietrich, M., Pedró, L., García, J., Pons, M., Hüttener, M., Paytubi, S., et al., 2014. Evidence for moonlighting functions of the θ subunit of Escherichia coli DNA polymerase III. J. Bacteriol. 196 (5), 1102–1112.

Drugan, J.K., Khosravi-Far, R., White, M.A., Der, C.J., Sung, Y.J., Hwan, Y.W., et al., 1996. Ras interaction with two distinct binding domains in Raf-1 may be required for Ras transformation. J. Biol. Chem. 271 (1), 233–237.

Duan, G., Walther, D., 2015. The roles of post-translational modifications in the context of protein interaction networks. PLoS Comput. Biol. 11 (2).

Duan, J., Wainwright, M.S., Comeron, J.M., Saitou, N., Sanders, A.R., Gelernter, J., et al., 2003. Synonymous mutations in the human dopamine receptor D2 (DRD2) affect mRNA stability and synthesis of the receptor. Hum. Mol. Genet. 12, 205–216.

Ekman, D., Björklund, Å.K., Frey-Skött, J., Elofsson, A., 2005. Multi-domain proteins in the three kingdoms of life: orphan domains and other unassigned regions. J. Mol. Biol. 348 (1), 231–243.

Feil, I.K., Malfois, M., Hendle, J., Van der Zandt, H., Svergun, D.I., 2001. A novel quaternary structure of the dimeric α-crystallin domain with chaperone-like activity. J. Biol. Chem. 276 (15), 12024–12029.

Franco-Serrano, L., Hernández, S., Calvo, A., Severi, M.A., Ferragut, G., Pérez-Pons, J.A., et al., 2018. MultitaskProtDB-II: An update of a database of multitasking/moonlighting proteins. Nucleic Acids Res. 46 (D1), D645–D648.

Gartner, J.J., Parker, S.C.J., Prickett, T.D., Dutton-Regester, K., Stitzel, M.L., Lin, J.C., et al., 2013. Whole-genome sequencing identifies a recurrent functional synonymous mutation in melanoma. Proc. Natl Acad. Sci. U S A 110 (33), 13481–13486.

Ge, W.W., Leystra-Lantz, C., Sanelli, T.R., McLean, J., Wen, W., Strong, W., et al., 2006. Neuronal tissue-specific ribonucleoprotein complex formation on SOD1 mRNA: Alterations by ALS SOD1 mutations. Neurobiol. Dis. 23 (2), 342–350.
Giese, K.C., Vierling, E., 2002. Changes in oligomerization are essential for the chaperone activity of a small heat shock protein in vivo and in vitro. J. Biol. Chem. 277 (48), 46310–46318.
Goldsmith, M.E., Humphries, R.K., Ley, T., Cline, A., Kantor, J.A., Nienhuis, A.W., 1983. "Silent" nucleotide substitution in a β + -thalassemia globin gene activates splice site in coding sequence RNA. *Proc. Natl Acad. Sci. U S A* 80 (8 I), 2318–2322.
Goodsell, D.S., Olson, A.J., 2000. Structural symmetry and protein function. Annu. Rev. Biophys. Biomol. Struct. 29 (1), 105–153.
Gotea, V., Gartner, J.J., Qutob, N., Elnitski, L., Samuels, Y., 2015. The functional relevance of somatic synonymous mutations in melanoma and other cancers. Pigment. Cell Melanoma Res. 28, 673–684.
Grandchamp, B., Picat, C., de Rooij, F., Beaumont, C., Wilson, P., Deybach, J.C., et al., 1989. A point mutation G - A in exon 12 of the porphobilinogen deaminase gene results in exon skipping and is responsible for acute intermittent porphyria. Nucleic Acids Res. 17 (16), 6637–6649.
Halbach, F., Reichelt, P., Rode, M., Conti, E., 2013. The yeast ski complex: crystal structure and rna channeling to the exosome complex. Cell. 154 (4), 814–826.
Hernández, S., Franco, L., Calvo, A., Ferragut, G., Hermoso, A., Amela, I., et al., 2015. Bioinformatics and moonlighting proteins. Front. Bioeng. Biotechnol. 3, (Jun).
Huberts, D.H.E.W., van der Klei, I.J., 2010. Moonlighting proteins: an intriguing mode of multitasking. Biochimica et. Biophysica Acta-Mol. Cell Res. 1803, 520–525.
Hurst, L.D., Pál, C., 2001. Evidence for purifying selection acting on silent sites in BRCA1. Trends Genet. 17, 62–65.
Hurst, L.D., 2006. Preliminary assessment of the impact of microRNA-mediated regulation on coding sequence evolution in mammals. J. Mol. Evol. 63 (2), 174–182.
Ito, H., Kamei, K., Iwamoto, I., Inaguma, Y., Nohara, D., Kato, K., 2001. Phosphorylation-induced change of the oligomerization state of αB-crystallin. J. Biol. Chem. 276 (7), 5346–5352.
Ito, H., Okamoto, K., Nakayama, H., Isobe, T., Kato, K., 1997. Phosphorylation of αB-crystallin in response to various types of stress. J. Biol. Chem. 272 (47), 29934–29941.
Jain, S., Drendel, W.B., Chen, Z.W., Mathews, F.S., Sly, W.S., Grubb, J.H., 1996. Structure of human β-glucuronidase reveals candidate lysosomal targeting and active-site motifs. Nat. Struct. Biol. 3 (4), 375–381.
Jeffery, C.J., 2015. Why study moonlighting proteins? Front. Genet. 6, (Jun).
Juers, D.H., Huber, R.E., Matthews, B.W., 1999. Structural comparisons of TIM barrel proteins suggest functional and evolutionary relationships between beta-galactosidase and other glycohydrolases. Protein Sci. 8 (1), 122–136.
Juers, D.H., Jacobson, R.H., Wigley, D., Zhang, X., Huber, R.E., Tronrud, D.E., et al., 2000. High resolution refinement of β-galactosidase in a new crystal form reveals multiple metal-binding sites and provides a structural basis for α-complementation. Protein Sci. 9 (9), 1685–1699.
Kato, K., Goto, S., Inaguma, Y., Hasegawa, K., Morishita, R., Asano, T., 1994. Purification and characterization of a 20-kDa protein that is. Biochemistry 269 (21), 15302–15309.

Kato, K., Hasegawa, K., Goto, S., Inaguma, Y., 1994. Dissociation as a result of phosphorylation of an aggregated form of the small stress protein, hsp27. J. Biol. Chem. 269 (15), 11274–11278.
Kato, K., Ito, H., Kamei, K., Inaguma, Y., Iwamototo, I., Sagal, S., 1998. Phosphorylation of αB-crystallin in mitotic cells and identification of enzymatic activities responsible for phosphorylation. J. Biol. Chem. 273 (43), 28346–28354.
Kelman, Z., 1995. DNA polymerase III holoenzyme: structure and function of a chromosomal replicating machine. Annu. Rev. Biochem. 64 (1), 171–200.
Keniry, M.A., Park, A.Y., Owen, E.A., Hamdan, S.M., Pintacuda, G., Otting, G., et al., 2006. Structure of the θ subunit of Escherichia coli DNA polymerase III in complex with the ε subunit. J. Bacteriol. 188 (12), 4464–4473.
Kimchi-Sarfaty, C., Oh, J.M., Kim, I.W., Sauna, Z.E., Calcagno, A.M., Ambudkar, S.V., et al., 2007. A "silent" polymorphism in the MDR1 gene changes substrate specificity. Sci. 315 (5811), 525–528.
Klyachko, N.L., Shchedrina, V.A., Efimov, A.V., Kazakov, S.V., Gazaryan, I.G., Kristal, B.S., et al., 2005. pH-dependent substrate preference of pig heart lipoamide dehydrogenase varies with oligomeric state: Response to mitochondrial matrix acidification. J. Biol. Chem. 280 (16), 16106–16114.
Kuznetsova, I.M., Turoverov, K.K., Uversky, V.N., 2014. What macromolecular crowding can do to a protein. Int. J. Mol. Sci. 15 (12), 23090–23140.
Lavoie, J.N., Lambert, H., Hickey, E., Weber, L.A., Landry, J., 1995. Modulation of cellular thermoresistance and actin filament stability accompanies phosphorylation-induced changes in the oligomeric structure of heat shock protein 27. Mol. Cell Biol. 15 (1), 505–516.
Lee, G.J., Vierling, E., 2000. A small heat shock protein cooperates with heat shock protein 70 systems to reactivate a heat-denatured protein. Plant. Physiol. 122 (1), 189–197.
Li, X., Park, W.J., Pyeritz, R.E., Jabs, E.W., 1995. Effect on splicing of a silent FGFR2 mutation in Crouzon syndrome. Nat. Genet. 9 (3), 232–233.
Liu, F., Dong, Q., Myers, A.M., Fromm, H.J., 1991. Expression of human brain hexokinase in Escherichiacoli: purification and characterization of the expressed enzyme. Biochem. Biophys. Res. Commun. 177 (1), 305–311.
Liu, W., Qian, C., Francke, U., 1997. Silent mutation induces exon skipping of fibrillin-1 gene in marfan syndrome. Nat Genet. 16 (4), 328–329.
Macaya D, Katsanis SH, Hefferon TW, Audlin S, Mendelsohn NJ, Roggenbuck J, et al., A synonymous mutation in TCOF1 causes treacher collins syndrome due to mis-splicing of a constitutive exon, Am. J. Med. Genet. Part A 2009.
MacDonald, B.T., Tamai, K., He, X., 2009. Wnt/β-catenin signaling: components, mechanisms, and diseases. Developmental Cell. 9–26.
Mariño-Ramírez, L., Kann, M.G., Shoemaker, B.A., Landsman, D., 2005. Histone structure and nucleosome stability. Expert. Rev. Proteom. 719–729.
Martella, M., et al., 2006. Molecular analysis of two uncharacterized sequence variants of the VHL gene. J. Hum. Genet. 51 (11), 964–968.
McHenry, C.S., 2003. Chromosomal replicases as asymmetric dimers: Studies of subunit arrangement and functional consequences. Mol. Microbiology. 49, 1157–1165.
Miller, S., Lesk, A.M., Janin, J., Chothia, C., 1988. The accessible surface area and stability of oligomeric proteins. Nature. 328 (6133), 834–836.
Miller, S., Ross-Inta, C., Giulivi, C., 2007. Kinetic and proteomic analyses of S-nitrosoglutathione-treated hexokinase A: consequences for cancer energy metabolism. Amino Acids. 32 (4), 593–602.

Mooney, S.M., Qiu, R., Kim, J.J., Sacho, E.J., Rajagopalan, K., Johng, D., et al., 2014. Cancer/testis antigen PAGE4, a regulator of c-jun transactivation, is phosphorylated by homeodomain-interacting protein kinase 1, a component of the stress-response pathway. Biochemistry 53 (10), 1670–1679.

Nackley, A.G., Shabalina, S.A., Tchivileva, I.E., Satterfield, K., Korchynskyi, O., Makarov, S.S., et al., 2006. Human catechol-O-methyltransferase haplotypes modulate protein expression by altering mRNA secondary structure. Sci. 314 (5807), 1930–1933.

Nakatsu, T., Kato, H., Oda, J., 1998. Crystal structure of asparagine synthetase reveals a close evolutionary relationship to class II aminoacyl-tRNA synthetase. Nat. Struct. Biol. 5 (1), 15–19.

Nestler, E.J., Greengard, P., 1983. Protein phosphorylation in the brain. Nature.

Nielsen, K.B., Sørensen, S., Cartegni, L., Corydon, T.J., Doktor, T.K., Schroeder, L.D., et al., 2007. Seemingly neutral polymorphic variants may confer immunity to splicing-inactivating mutations: A synonymous SNP in exon 5 of MCAD protects from deleterious mutations in a flanking exonic splicing enhancer. Am. J. Hum. Genet. 80 (3), 416–432.

Nielsen, T.K., Liu, S., Lührmann, R., Ficner, R., 2007. Structural basis for the bifunctionality of the U5 snRNP 52K protein (CD2BP2). J. Mol. Biol. 369 (4), 902–908.

Oliveira, J., Soares-Silva, I., Fokkema, I., Gonçalves, A., Cabral, A., Diogo, L., et al., 2008. Novel synonymous substitution in POMGNT1 promotes exon skipping in a patient with congenital muscular dystrophy. J. Hum. Genet. 53 (6), 565–572.

Pagani, F., Raponi, M., Baralle, F.E., 2005. Synonymous mutations in CFTR exon 12 affect splicing and are not neutral in evolution. Proc. Natl. Acad. Sci. USA 102 (18), 6368–6372.

Panyasai, S., Fucharoen, S., Surapot, S., Fucharoen, G., Sanchaisuriya, K., 2004. Molecular basis and hematologic characterization of δβ-thalassemia and hereditary persistence of fetal hemoglobin in Thailand. Haematologica. 89 (7), 777–781.

Piatigorsky, J., Wistow, G.J., 1989. Enzyme/crystallins: gene sharing as an evolutionary strategy. Cell. 197–199.

Rajagopalan, K., Qiu, R., Mooney, S.M., Rao, S., Shiraishi, T., Sacho, E., et al., 2014. The Stress-response protein prostate-associated gene 4, interacts with c-Jun and potentiates its transactivation. Biochim. Biophys. Acta - Mol. Basis Dis. 1842 (2), 154–163.

Ruff, M., Krishnaswamy, S., Boeglin, M., Poterszman, A., Mitschler, A., Podjarny, A., et al., 1991. Class II aminoacyl transfer RNA synthetases: crystal structure of yeast aspartyl-tRNA synthetase complexed with tRNAAsp. Sci. (80-) 252 (5013), 1682–1689.

Sauna, Z.E., Kimchi-Sarfaty, C., 2013. Synonymous mutations as a cause of human genetic disease. eLS.

Savisaar, R., Hurst, L.D., 2017. Both maintenance and avoidance of RNA-binding protein interactions constrain coding sequence evolution. Mol. Biol. Evol. 34 (5), 1110–1126.

Sheikh, T.I., Mittal, K., Willis, M.J., Vincent, J.B., 2013. A synonymous change, p. Gly16Gly in MECP2 Exon 1, causes a cryptic splice event in a Rett syndrome patient. Orphanet J. Rare Dis. 8 (1), 108.

Skjaerven, L., Yao, X.Q., Scarabelli, G., Grant, B.J., 2014. Integrating protein structural dynamics and evolutionary analysis with Bio3D. BMC Bioinforma. 15 (1), 399.

Spector, A., Chiesa, R., Sredy, J., Garner, W., 1985. cAMP-dependent phosphorylation of bovine lens alpha-crystallin. Proc. Natl Acad. Sci. 82 (14), 4712–4716.

Sriram, G., Martinez, J.A., McCabe, E.R.B., Liao, J.C., Dipple, K.M., 2005. Single-gene disorders: What role could moonlighting enzymes play? Am. J. Hum. Genet. 76 (6), 911–924.

Supek, F., Miñana, B., Valcárcel, J., Gabaldón, T., Lehner, B., 2014. Synonymous mutations frequently act as driver mutations in human cancers. Cell. 156 (6), 1324–1335.

Todd, A.E., Orengo, C.A., Thornton, J.M., 2001. Evolution of function in protein superfamilies, from a structural perspective. J. Mol. Biol. 307 (4), 1113–1143.

Torshin, I., 1999. Activating oligomerization as intermediate level of signal transduction: analysis of protein-protein contacts and active sites in several glycolytic enzymes. Front. Biosci. 4, D557–D570.

Tsao, D., Shabalina, S.A., Gauthier, J., Dokholyan, N.V., Diatchenko, L., 2011. Disruptive mRNA folding increases translational efficiency of catechol-O-methyltransferase variant. Nucleic Acids Res. 39 (14), 6201–6212.

Van Montfort, R.L.M., Basha, E., Friedrich, K.L., Slingsby, C., Vierling, E., 2001. Crystal structure and assembly of a eukaryotic small heat shock protein. Nat. Struct. Biol. 8 (12), 1025–1030.

Varghese, J.N., Garrett, T.P., Colman, P.M., Chen, L., Høj, P.B., Fincher, G.B., 1994. Three-dimensional structures of two plant beta-glucan endohydrolases with distinct substrate specificities. Proc. Natl Acad. Sci. 91 (7), 2785–2789.

Vishwanath, S., de Brevern, A.G., Srinivasan, N., 2018. Same but not alike: Structure, flexibility and energetics of domains in multi-domain proteins are influenced by the presence of other domains. PLoS Comput. Biol. 14 (2), e1006008.

Voorter, C.E.M., de Haard-Hoekman, W.A., Roersma, E.S., Meyer, H.E., Bloemendal, H., de Jong, W.W., 1989. The in vivo phosphorylation sites of bovine αB-crystallin. FEBS Lett. 259 (1), 51–52.

Wistow, G.J., Lietman, T., Williams, L.A., Stapel, S.O., de Jong, W.W., Horwitz, J., et al., 1988. Tau-crystallin/alpha-enolase: one gene encodes both an enzyme and a lens structural protein. J. Cell Biol. 107 (6 Pt 2), 2729–2736.

Yu, C.H., Dang, Y., Zhou, Z., Wu, C., Zhao, F., Sachs, M.S., et al., 2015. Codon usage influences the local rate of translation elongation to regulate co-translational protein folding. Mol. Cell. 59 (5), 744–754.

Zeng, Y., He, Y., Yang, F., Mooney, S.M., Getzenberg, R.H., Orban, J., et al., 2011. The cancer/testis antigen prostate-associated gene 4 (PAGE4) is a highly intrinsically disordered protein. J. Biol. Chem.

Chapter 5

Prion-mediated phenotypic diversity in fungi

Mick F. Tuite
Kent Fungal Group, School of Biosciences, University of Kent, Canterbury, United Kingdom

Introduction

Members of the fungal kingdom show considerable genetic and phenotypic diversity both between and within species. The causes of this genetic diversity encompass changes at the nucleotide level (e.g., single nucleotide polymorphisms), at the gene level (e.g., horizontal gene transfer) and whole chromosome rearrangements both in terms of their structure and number (Taylor et al., 2017; review). The ever-growing volume of genomic data available for members of the fungal kingdom has also revealed the extent of genetic variation *between* strains of a single fungal species. This is exemplified by an analysis of the genomes of over 1000 laboratory-bred and wild strains of the budding yeast *Saccharomyces cerevisiae,* which revealed that only 85% of the known genes can be considered as "core" components of the genome of this model organism, that is, are conserved and present in all strains of the species (Peter et al., 2018; McCarthy and Fitzpatrick, 2019). The remaining 15% of the genes of *S. cerevisiae* must be considered as "accessory" being present only in the genomes of some of the strains. The repertoire of such accessory genes define in part the phenotypic diversity seen among strains of *S. cerevisiae* (and undoubtedly other fungal species) and would allow for rapid of the organism adaptation to changing environments in the diversity of ecological niches where these species are found.

Yet genetic variation and the consequent phenotypic variation in fungi is not restricted to changes in the nuclear and extranuclear (i.e., mitochondrial DNA)-based genomes. For example, RNA-based mycoviruses are present in many species of both filamentous fungi and yeasts (Kotta-Loizou and Coutts, 2017; Sato et al., 2018; Wickner et al., 2013; reviews). Mycoviruses typically have either a single-stranded (ss) or a double-stranded (ds) RNA

Phenotypic Switching. DOI: https://doi.org/10.1016/B978-0-12-817996-3.00027-X

genome packaged into virus-like capsids. While the impact of mycoviruses on the host phenotype is generally minimal, even neutral, there are exceptions to this generalization. For example, the dsRNA "killer" virus found in strains of *S. cerevisiae* produces a toxin that is secreted by the infected host cell and kills other yeast cells lacking a mycovirus-encoded toxin immunity function (Schmitt and Breinig, 2006; Becker and Schmitt, 2017; reviews). Some fungal species also contain multiple copies of small circular DNA plasmids as exemplified by the 2 μ plasmid of *S. cerevisiae* which is efficiently maintained in growing cells (Rizvi et al., 2018). The presence of the 2 μ plasmid has no discernible impact on the host unless the copy number of the plasmid is significantly elevated in which case cell death is observed.

Much of the phenotypic variability seen between different strains of a given fungal species can therefore be accounted for by changes in nuclear or extranuclear nucleic acid-based genomes. But not all. In the 1960s and 1970s three "mutant" phenotypes were identified in *S. cerevisiae* that were controlled by then unknown cytoplasmic "genetic" determinants that could not be linked to any known nucleic acid genome. That they showed a non-Mendelian inheritance pattern in genetic crosses was consistent with a cytoplasmic location for the underlying "genetic" determinant. The three phenotypes in question were a defect in translation termination leading to the suppression of nonsense mutations ([*PSI*$^+$], Cox, 1965), a defect in the regulation of nitrogen catabolism ([*URE3*]; Aigle and Lacroute, 1975) and resistance to glucosamine, a glucose analogue (Kunz and Ball, 1977). While these three stable and heritable phenotypes were not connected by a shared cellular or biochemical pathway, it subsequently emerged that they were connected by the nature of the underlying "genetic" determinants; they were prion based.

The term "prion" (*pro*teinaceous *in*fectious particle; Prusiner, 1982) is the now universally accepted term used to describe a class of unusual infectious agents that were first uncovered in studies of the pathology of certain brain degenerative diseases of humans (e.g., Creutzfeldt Jakob Disease; CJD) and domesticated animals typified by scrapie in sheep and bovine spongiform encephalopathy in cattle (Prusiner, 2013; review). What made prions unusual was that even though they could be transmitted between hosts where they multiplied, that is, they were infectious; the lack of any prion-associated nucleic acid component to facilitate this multiplication proved enigmatic. The mystery was eventually solved by the realization and ultimately experimental validation that the prion linked to these diseases was a normal cellular protein (PrP) with an atypical three-dimensional structure (designated PrP^{Sc}).

The PrP^{Sc} form, which is identical in amino acid sequence to the native PrP^{c} form, serves as a "conformational template" that captures and converts the native PrP^{c} form into the infectious PrP^{Sc} prion conformation leading to the formation of PrP^{Sc} aggregates with the physico-chemical characteristics of an amyloid (Fig. 5.1). Amyloids are robust proteinaceous fibrils that are self-templating polymers of an alternative conformational form of a protein. The presence of amyloid deposits in the brain is a diagnostic feature of a number of neurodenegerative diseases including Alzheimer's and Huntington's disease (Chiti and Dobson, 2017). Over a long time period—often years in the case of human prion diseases such as CJD and Kuru—the emergence of this alternative amyloid form in the infected individual leads to brain degeneration and ultimately death. Exactly how prions trigger these events in the brain remains to be established (Ironside et al., 2017; Aguzzi et al., 2018; reviews).

A decade after the discovery of the infectious mammalian PrP^{Sc} prion, it emerged that prions were not only associated with these rare yet fatal brain disease of animals. This came initially from a proposal that a prion-based mechanism could account for the non-Mendelian inheritance of the enigmatic [*URE3*] and [*PSI*$^+$] traits in *S. cerevisiae* (Wickner, 1994) and which was subsequently proposed to explain the third enigmatic non-Mendelian trait in *S. cerevisiae*, namely, glucosamine resistance (Brown and Lindquist, 2009). Following the revelation that certain proteins in *S. cerevisiae*

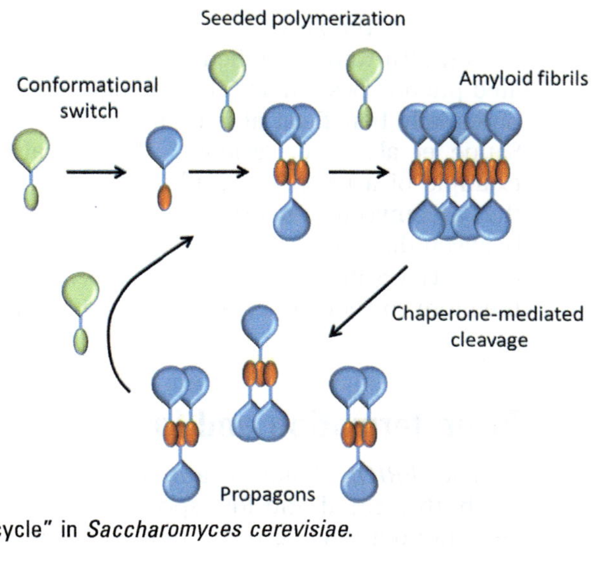

Figure 5.1 The prion "life cycle" in *Saccharomyces cerevisiae*.

Table 5.1 ***Saccharomyces cerevisiae*** **prions and their associated proteins.**

Prion	Prion protein (PFP)	Function of PFP	Variants reported	References
[*URE3*]	Ure2	Transcriptional regulation	Yes	Aigle and Lacroute (1975), Wickner (1994)
[PSI^+]	Sup35	Translation termination factor eRF3	Yes	Cox (1965), Wickner (1994)
[PIN^+]	Rnq1, Pin1 etc.	Various	Yes	Derkatch et al. (2001), Osherovich and Weissman (2001)
[RNQ^+]	Rnq1	Unknown function	Yes	Sondheimer and Lindquist (2000)
[$MOT3^+$]	Mot3	Transcriptional repressor/activator	No	Alberti et al. (2009), Holmes et al. (2013)
[SWI^+]	Swi1	Chromatin-remodeling factor	No	Du et al. (2008)
[GAR^+]	Pma1/Std1	Plasma membrane proton pump/ transcriptional regulator	Yes	Brown and Lindquist (2009)
[OCT^+]	Cyc8	Transcriptional corepressor	No	Patel et al. (2009)
[MOD^+]	Mod5	tRNA modifying enzyme	No	Suzuki et al. (2012)
[ISP^+]	Sfp1	Transcriptional activator	No	Rogoza et al. (2010)
[$NUP100^+$]	Nup100	Nucleoporin	No	Halfmann et al. (2012b)
[*KIL-D*]	ND	ND	No	Suzuki et al. (2015)

can form heritable prions, a number of other prion-forming proteins (PFPs) were identified in *S. cerevisiae* with a variety of associated phenotypes (Table 5.1) ranging from detrimental to neutral to potentially beneficial at least in some environments (Tuite, 2013; Saupe et al., 2016; reviews). There is also one well-established example of a prion-based phenotype in a filamentous fungus. In the ascomycete *Podospora anserina* vegetative incompatibility between different strains of the species is linked to the prion form of the Het-S protein (Riek and Saupe, 2016; a review). Prions can therefore considered as protein-based "genes."

Prion formation and loss

The [$PRION^+$] state is a metastable one in fungi since prions are both formed and lost spontaneously at measurable rates in cell populations grown under laboratory conditions. The degree of metastability is prion dependent with some prions such as [PSI^+] showing rates of formation and loss of $<10^{-7}$ (Lancaster et al., 2010) whereas other prions such as [LSB^+] are only maintained in a small percentage of a cell population (Chernova et al., 2017). Changes in a cell's environment can also transiently alter the rate of loss or acquisition of the [$PRION^+$] state thereby modulating the degree of metastability that would facilitate rapid adaptation to a constantly changing environment.

The rate at which the [$PRION^+$] state can arise spontaneously in *S. cerevisiae* this epigenetic state emerges varies depending on the prion. The reported rates of prionogenesis range from $\sim 10^{-7}$ as seen for the [PSI^+] (Lancaster et al., 2010) and [$MPH1^+$] prions (Chakrabortee et al., 2016) to 10^{-2} for the [PIN^+] prion (N. Koloteva-Levine and M.F. Tuite, unpublished). There are also a number of genetic, epigenetic and environmental factors that can increase or suppress the rate of de novo formation of a prion (e.g., Derkatch et al., 1997; Tyedmers et al., 2008; Manogaran et al., 2011) although not all may be impactful in nature. For example, the overexpression of a PFP or just its prion-forming domain, typically leads to significant elevation in the rate of formation of the prion form of that PFP. This was first demonstrated by overexpression of Sup35 and de novo formation of the [PSI^+] prion (Chernoff et al., 1993). A variety of stress-inducing conditions such as oxidative stress and high salt stress can also give rise to an increase in the rate of prion formation (Tyedmers et al., 2008; Doronina et al., 2015) possibly as a consequence of resulting protein damage triggering stochastic misfolding into the infectious prion form.

While de novo formation is likely to be triggered by stochastic protein misfolding event in which the PFP takes up the prion templating conformation (Halfmann et al., 2010; review), it is evident that a number of cellular structures facilitate the assembly of the misfolded PFP and thus the stable establishment of the [$PRION^+$] state. Foremost amongst these are interactions of the misfolded PFP with the actin cytoskeleton (Ganusova et al., 2006) and the vacuole-associated insoluble protein deposit, a cellular compartment that sequesters misfolded proteins that irreversibly aggregate and that may be the site of de novo formation (Tyedmers et al., 2010). There may also be at least two independent pathways of de novo prion formation (Speldewinde et al., 2017).

An additional factor that promotes or indeed regulates de novo formation of yeast prions is the [PIN^+] prion (Serio, 2018, review). [PIN^+] was first identified as a non-Mendelian factor necessary for efficient [PSI^+] *in*duction de novo, a property that gave rise to its name (Derkatch et al., 1997). Subsequently it emerged that in most laboratory strains of *S. cerevisiae*, [PIN^+] is the prion form of Rnq1, a nonessential, low abundance cytoplasmic protein that is rich is Asn (N) and Gln (Q) residues (Sondheimer and Lindquist, 2000). [PIN^+] is also the most commonly found prions in nonlaboratory strains of *S. cerevisiae* (Resende et al., 2003; Halfmann et al., 2012a; Kelly et al., 2012).

Precisely how the amyloid form of the Rnq1 protein enhances the rate of spontaneous formation of the [$PRION^+$] state has not been established but there are two models. The first and perhaps most plausible is that the [PIN^+] prion acts as a heterologous nucleus, interacting directly with the PFP to stimulate its nucleation into the prion form (Derkatch et al., 2001) and there are data from both in vivo and in vitro studies that are consistent with this model (Serio, 2018; review). The alternative model proposes that the amyloid form of Rnq1 titrates out an anti-aggregation factor that normally prevents templated aggregation of the PFP (Osherovich and Weissman, 2001), but the supporting data for this model are much weaker. In addition one has to consider that there are other amyloid-forming proteins, both endogenous and heterologous in origin, that establish a [PIN^+]-like state (Derkatch et al., 2001; Osherovich and Weissman, 2001).

Prion propagation and transmission in the fungal cell

Once established, fungal prions are generally stably maintained by their host over many generations thereby ensuring the longevity of the biological impact of any prion-mediated phenotype. Prions are efficiently vertically transmitted via mitosis and meiosis, but unlike their mammalian counterparts, fungal prions do not appear to be transmitted horizontally, that is, between individual cells. However, under laboratory conditions, prion aggregates either generated in vitro using purified a PFP or purified directly from [$PRION^+$] cells, can be introduced into prion-free cells to establish the [$PRION^+$] state in prion-free cells (Tanaka et al., 2004; King and Diaz-Avalos, 2004). This suggests that fungal prions can potentially be transmitted horizontally and it has been proposed that such horizontal transmission may occur via packaging of prion aggregates into extracellular vesicles (Kabani and Melki, 2016).

Once the [$PRION^+$] state has been triggered, the basic prion "life cycle" ensures that the state is maintained even in actively dividing cells. The life cycle (outlined in Fig. 5.1) involves the following sequence of events:

1. Initiating switch in conformation of the PFP to the "infectious" prion form.
2. Binding of the prion form to a native PFP molecular to template its conformational conversion to the prion form.
3. Continued polymerization of native PFP molecules, seeded by the newly emerged prion form of the PFP, to give rise to amyloid fibrils.
4. Fragmentation of the amyloid fibrils by the endogenous molecular chaperone network with the key role being played by the protein disaggregase Hsp104 and two cochaperones, Hsp40 (Sis1) and Hsp70.
5. Transmission of the resulting fragments of the amyloid fibrils which constitute new prion seeds (also called propagons), to the daughter cell.
6. Conformational conversion and polymerization of native PFP molecules, now seeded by the newly generated propagons.

This cyclical process occurs within the time frame of a single cell division (which for *S. cerevisiae* can be as short as 90 minutes). This ensures that the [*PRION*$^+$] state is retained and passed on to the emerging daughter cell before cell separation. The inheritance of a single propagon by the daughter cell is sufficient to maintain the [*PRION*$^+$] state in that cell with the propagon number doubling every ~20 minutes in exponentially growing cells (Ness et al., 2002).

There is no evidence of an active mechanism by which propagons are transmitted to the daughter cell during normal cell growth (Byrne et al., 2009). However, specialized spatial quality control systems exist in *S. cerevisiae* and other fungi that allow a mother cell to retain damaged, misfolded proteins (Schneider et al., 2018, review). Such a retention mechanism may impact on propagon transmission under conditions of heat stress (Newnam et al., 2011; Ali et al., 2014) or elevated levels of Hsp104 in nonheat-stressed cells (Ness et al., 2017) or in aged cells (Derdowski et al., 2010). There is also evidence supporting a size threshold for efficient transmission of Sup35-based propagons to daughter cells during cell division (Derdowski et al., 2010). Any defects in amyloid fibril fragmentation can lead to the accumulation of higher molecular weight fibrils that are physically constrained from passing through the mother-daughter link by cytoplasmic streaming. Any modulation in the size of the propagons generated by fragmentation of the fibrils can lead to changes in the number of propagons that are transmitted to other cells. Such heterogeneity can potentially impact on the prion-mediated phenotype as has been shown for the [*PSI*$^+$] prion (Derdowski et al., 2010).

A relatively small subset of proteins in fungi (and presumably other organisms) has the potential to switch to a heritable [*PRION*$^+$] form. In *S. cerevisiae,* for example, approximately 1%–2% of the proteome has that potential (Alberti et al., 2009; Chakrabortee et al., 2016; An et al., 2016). The fungal PFPs have characteristic structural features that are required for the establishment and/or maintenance of the [*PRION*$^+$] state. These features have been largely established by the analysis of several different PFPs, but in particular the Sup35 translation termination factor that gives rise to the [*PSI*$^+$] prion (Ross et al., 2013; Lancaster et al., 2014). The critical element in most PFPs including Sup35 is a region of the protein called the prion-forming domain (PrD). This is an intrinsically disordered region, typically located at the N-terminus of the PFP and that is characteristically rich in Asn and Gln residues (Harbi and Harrison, 2014). Mutations in the PrD, not necessarily affecting the Gln or Asn residues, can block the propagation of the [*PRION*$^+$] form of a PFP [e.g., (DePace et al., 1998; Marchante et al., 2013)] and do so either by impacting on the amyloidogenic properties of the protein (DePace et al., 1998) or by constraining the intrinsic disorder of the domain (Marchante et al., 2013). The Gln and Asn residues make different contributions to the conformational switch linked to prion formation (Halfmann et al., 2011) although neither mammalian PrP or the *P. anserina* Het-S PFPs have Asn/Gln-rich regions showing that this amino acid bias is not an essential feature of a PFP. The *S. cerevisiae* Mod5 protein that gives rise to the [MOD +] prion (Table 5.1) also lacks such a region (Suzuki et al., 2012).

Prion-mediated phenotypes

One of the most remarkable features of fungal prion biology is the diversity and range of phenotypes prions can generate as a direct consequence of a switch in the conformation of a single protein without any underlying change in the encoded primary protein sequence; a classic epigenetic phenomenon. Yet the link between the prion form(s) of the protein and the resulting changes in the biology of the host cell still remain relatively poorly understood. As is made clear in Table 5.1, those proteins that have behave unambiguously as PFPs in *S. cerevisiae,* span a wide range of cellular functions with an evident bias toward proteins that interact with either DNA (transcription factors) or RNA (translation factors). Consequently, affecting the function of proteins that play a key role in the flow of information from DNA-to-RNA-to-protein would be expected to generate a diversity of phenotypes linked to inheritance of the [*PRION*$^+$] state. This is indeed what we see.

What has also become evident as the catalogue of fungal prions expands is that prion-mediated phenotypes do not necessarily arise as a consequence of a loss of the cellular function of the PFP. One might expect this when a protein loses its native structure and rapidly forms high molecular weight aggregates. Yet, there are a number of prion-mediated phenotypes that do mimic the phenotypes linked to either mutated, defective forms of the PFP or, in some cases, even mimic the complete loss of function associated with deletion of the PFP gene. There are also examples of prions that are linked to gain-of-function phenotypes and, in one case, the [*MOT3*$^+$] prion, to both (Holmes et al., 2013). In some extreme cases, this can be detrimental as exemplified by the mammalian prion PrP^{Sc} which is linked to fatal neurodegeneration in humans and other mammals (Prusiner, 2013; a review) although few fungal prions have such a catastrophic effect on their host. There are several examples of gain-of-function phenotypes that are beneficial to the host and such phenotypes have been described in both fungi (Newby and Lindquist, 2013; review) and other species including insects, mollusks, and mammals (Si, 2015; a review).

The following are examples of four different fungal PFPs and the phenotypic outcomes of a switch to their [*PRION*$^+$] conformation. These illustrate the diversity of phenotypes linked to prion formation in fungi, and particularly in *S. cerevisiae*. For information on the phenotypes linked to the fungal prions not discussed in the succeeding text, the reader is referred to the references given in Table 5.1.

[*PSI*$^+$]/Sup35: regulating the decoding of stop codons and more

The [*PSI*$^+$] prion has been the most widely studied yeast prion and much of our knowledge of prion formation, propagation, and transmission has come from studying its behavior in vivo and in vitro (Tuite et al., 2015, review). [*PSI*$^+$] was originally identified by Cox (1965) as a extrachromosomal modifier of nonsense suppression mediated by mutant tRNAs. Such a phenotype can arise through a defect in recognition of the offending stop codon by the eRF1/eRF3 release factor (RF) complex (Stansfield et al., 1995). This shifts the competition for the stop codon in favor of the nonsense suppressor tRNA and hence one gets translation of the stop codon as a sense codon. Given that the PFP that gives rise to the [*PSI*$^{+}$] prion is eRF3 (more usually referred to by its historical name of Sup35), the [*PSI*$^{+}$]-linked phenotype is entirely consistent with a loss of function of eRF3 possibly interfering with its key functional interaction with eRF1.

The potential benefit of decoding a premature (i.e., mutant) stop codon as sense is evident since this would restore synthesis to some level of full-length protein. The potential benefits of decoding *authentic* stop codons at the end of each reading frame as "sense" codons are less evident. In fact such "termination readthrough" events should be detrimental to the cell since they extend the protein sequence beyond its native C-terminus. Alternatively, the ability to extend the C-terminus of a protein may give rise to a form of that protein with an altered potentially beneficial function although one example of this in *S. cerevisiae* has so far been uncovered (Namy et al., 2002). However, translation termination is rarely 100% efficient even at authentic stop codons with the efficiency being modulated by the choice of stop codon, the nucleotide context in which the stop codon is placed, and the presence of endogenous tRNAs able to decode a stop codon albeit inefficiently (von der Haar and Tuite, 2007; Dabrowski et al., 2015; reviews).

Yeast cells have therefore adapted to a low level of stop codon readthrough as is evident by the neutral impact of most form of the [*PSI*$^{+}$] prion on cell growth (Byrne et al., 2009) even though some strains of *S. cerevisiae* contain a significant number of inactivating stop codon mutations in their genome (Fitzpatrick et al., 2011). Analysis of the yeast translatome, that is, ribosome-associated mRNAs, using ribosome profiling has also revealed that over 100 proteins show a detectable level of C-terminal extension in a [*PSI*$^{+}$] strain compared to the isogenic [*psi*$^{-}$] strain (Baudin-Baillieu et al., 2014). However, if one introduces [*PSI*$^{+}$] into a strain expressing an efficient nonsense suppressor tRNA such as the tyrosine-inserting *SUP4* suppressor tRNA (Cox, 1971) or carries a mutation in the *SUP35* gene that impairs Sup35/eRF3 function (Cox, 1977), the cells die presumably because of the unviable levels of stop codon readthrough.

Analysis of a wide range of phenotypes plus the differences in the transcriptome, proteome and translatome of [*PSI*$^{+}$] strains compared to an otherwise isogenic [*psi*$^{-}$] strain paints a much more complex picture of the impact the [*PSI*$^{+}$] prion has on its host. Most strikingly, the ability of a yeast cell to successfully reduce the disadvantageous impact of a variety of physical and chemical stresses (Eaglestone et al., 1999; True and Lindquist, 2000; True et al., 2004) is evident although some detrimental phenotypes are exacerbated by the prion. Some of the phenotypic differences can be explained by the finding that [PSI^+] cells show a significant increase in a +1 frameshift event during the translation of the *OAZ1* mRNA thereby increasing the levels of the encoded protein, antizyme (Namy et al., 2008). Antizyme is a negative regulator of polyamine synthesis and the changes in levels of polyamines in the cell triggered by the action of the [PSI^+] prion can explain many—but not all—of the phenotypic differences between [PSI^+] and [PSI^-] cells first reported by True and Lindquist (True and Lindquist, 2000; True et al., 2004).

The +1 frameshift event in the *OAZ1* gene promoted by the [PSI^+] prion indicates that the effects of the prion form of eRF3 on translation might not be restricted to just the consequences of stimulating stop codon readthrough. Mutations in the *SUP35* gene can also act as suppressors of certain +1 frameshift mutations (Wilson and Culbertson, 1988) although whether [PSI^+] has a similar suppressor activity has not been reported. Comparisons of the transcriptome and translatome also revealed both effects of [PSI^+] on the transcription of subset of some 75 genes and a significant level of errors in selecting the correct reading frame in certain genes (Baudin-Baillieu et al., 2014). This raises the possibility that the phenotypic impact of [PSI^+] may be via several different mechanisms not all of which are linked to a defect in translation termination. This understanding is further complicated by a report that the aggregated prion form of Sup35 actually remains active as a translation termination factor (Pezza et al., 2014).

The most likely explanation for the phenotypes linked to [PSI^+], but not accountable for by a translation termination defect is that the Sup35 amyloid aggregates sequester other functionally unrelated proteins (Baudin-Baillieu et al., 2014). This is supported by a mass spectrometric analysis of Sup35 aggregates which revealed ~40 different proteins associated with these aggregates including translation factors, molecular chaperones, and proteins involved in the oxidative stress response (Nevzglyadova et al., 2009). There is also evidence that [PSI^+] amyloid aggregates sequester the native binding partner of Sup35, namely eRF1 and this latter interaction can account for the observation that overexpression of Sup35 in a [PSI^+] cells is toxic (Vishveshwara et al., 2009).

[$MOT3^+$]/Mot3: controlling multicellularity in response to environmental triggers

[*MOT3*$^+$] the prion form of the Mot3 protein, was uncovered through a systematic screen of proteins that had predicted prion domains (Alberti et al., 2009). [*MOT3*$^+$] shows all of the expected genetic and biochemical properties of a fungal prion but it also has several unique properties. These include: (1) the rate of de novo formation of [*MOT3*$^+$] is relatively high ($\sim 10^{-4}$), but this rate is not influenced by the [*PIN*$^+$] prion (Alberti et al., 2009) and (2) amongst the various [*MOT3*$^+$]-linked phenotypes, some can explained by a loss of function of the Mot3 protein while others are consistent with a gain-of-function (Holmes et al., 2013).

Mot3 is a complex transcription factor that has both transcriptional activator and repressor functions acting on the promoters of a large number of target genes that impact on a wide variety of cellular processes including stress responses (Martínez-Montañés et al., 2013). Among the many Mot3-regulated genes is *FLO11*, which encodes a flocculin, a GPI-anchored cell surface glycoprotein that controls cell-cell adhesion. Flocculin is a therefore a major player in the complex cell wall reprogramming that occurs when cells switch from an aerobic state to an anaerobic state. This remodeling includes the formation of multicellular structures such as cell flocculants and biofilms. [*MOT3*$^+$] directly impacts on this multicellular developmental program via its control expression of the *FLO11* gene although such [*MOT3*$^+$]-mediated reprogramming of colony morphology is only evident under conditions which naturally induce *FLO11* expression, for example, nitrogen starvation. In its native structure Mot3 represses transcription of *FLO11*, but this repression is lifted when Mot3 switches to its prion form and presumably is no longer able to bind to and hence repress transcription from the *FLO11* promoter. While the effects of [*MOT3*$^+$] on colony morphology appear to reflect a loss of Mot3 function, the fact that a deletion of the *MOT3* gene in a [*mot3*$^-$] background gives a different spectrum of effects on colony morphology implies there may be additional gain-of-function aspects to the prionization of Mot3 such as sequestration of a Mot3-interacting corepressor the identity of which remains unknown (Holmes et al., 2013).

The role of [*MOT3*$^+$] in regulating the development of multicellularity in *S. cerevisiae* is intimately linked with sugar metabolism (Holmes et al., 2013). The rate at which Mot3 switches to its [*MOT3*$^+$] prion form is significantly increased by ethanol that is derived by the fermentation of glucose. The resulting

[*MOT3*$^+$]-mediated switch in colony morphology to a multicellular state may protect the cells from ethanol stress, but as the cells begin to respire the ethanol during the diauxic phase of growth, this leads to a significant reduction in oxygen levels and consequently hypoxia. The low oxygen levels then trigger a loss of the [*MOT3*$^+$] from cells and the reemergence of a [*mot3*$^-$] population of cells. The formation and the propagation of the [*MOT3*$^+$] prion are therefore controlled by rapid changes in cellular metabolism. It remains to be seen whether or not the switch to [*MOT3*$^+$] results in wholesale changes in the transcription of the multiplicity of genes under the control of the Mot3 transcription factor.

[*SWI*$^+$]/Swi1: an impact on global transcriptional regulation

The Swi1 protein is a key component of the *S. cerevisiae* SWI/SNF complex, which remodels chromatin and thus represents one of the major global transcriptional regulators in this organism. Swi1 was recognized early on as a potential PFP through the presence of a typical Asn/Gln-rich prion-forming domain in the protein (Michelitsch and Weissman, 2000). Confirmation that it could switch to a prion form came via the discovery that it was one of a small subset of endogenous yeast proteins that when overexpressed could replace the function of the [*PIN*$^+$] prion to facilitate de novo formation of other prions (Derkatch et al., 2001). Following confirmation by a range of biochemical and genetic methods that the Swi1 protein was a PFP, the resulting prion was designated [*SWI*$^+$] (Du et al., 2008; Alberti et al., 2009) although in comparison to most other fungal prions it is relatively unstable during cell division.

A switch to a loss-of-function prion form for Swi1 in a [*SWI*$^+$] strain would be expected to have a significant effect on the host phenotype if this conformational switch impaired the function of the all-important SWI/SNF regulatory complex. This complex is responsible for regulating the transcription of ~6% of all yeast genes (Sudarsanam et al., 2000) and comparison of the transcriptomes of [*SWI*$^+$] and [*SWI*$^-$] cells demonstrated there are significant differences in the levels of transcription for many of the genes known to be regulated by the SWI/SNF complex (Malovichko et al., 2019). However, a comparison of the transcriptome of [*SWI*$^+$] cells versus *swi1*Δ gene knockout cells reveals that the transcription of only a fraction of the SWI/SNF-dependent genes were affected by the prion switch (Malovichko et al., 2019). This finding raises the question of whether or not the [*SWI*$^+$]-linked phenotypes are indicative of a loss of Swi1, and consequently SWI/SNF, function.

Among the genes controlled by the SWI/SNF complex are genes controlling carbon metabolism, DNA replication, cell adhesion, mating type switching, and sporulation. The phenotype first linked to [*SWI*$^+$] was impaired growth on a variety of carbon sources such as galactose and raffinose although there was no effect on glucose metabolism (Du et al., 2008). A range of other phenotypes has also been linked to the presence of [*SWI*$^+$] including impacts on cell adhesion, pseudohyphae formation, flocculation, and adhesion to agar surfaces (Goncharoff et al., 2018; review). Of particular importance in this context is the repression of expression of the *FLO11* gene, which encodes flocculin and that leads to a loss of multicellularity. As described earlier, a second prion, [*MOT3*$^+$] also controls expression of the *FLO11* gene but does so in a manner opposite to [*SWI*$^+$], that is, [*SWI*$^+$] switches off *FLO11* expression and hence promotes unicellularity while [*MOT3*$^+$] derepresses *FLO11* expression and hence promotes multicellularity. The switching between a unicellular and a multicellular existence in response to a changing environment is clearly subject to complex epigenetic control in *S. cerevisiae.*

What is the benefit to the host of [*SWI*$^+$] when [*SWI*$^+$]-containing cells are less likely to form multicellular structures such as biofilms that are believed to be beneficial to the organism when subjected to nutritional deprivation and other potentially harmful stresses (Brückner and Mösch, 2012; review)? As [*SWI*$^+$] cells are more likely to exist as unicellular structures this would aid cell dispersal in, for example, aqueous environments (Du et al., 2015; Newby and Lindquist, 2017). This would ensure that the [*SWI*$^+$] cells are able to readily move to new environments and seek new mating partners thereby increasing the potential for genetic diversity.

[*GAR*$^+$]/Pma1/Std1: broadening the choice of sugars

While fungal prions are considered to be heritable amyloid forms of a single protein, there is one prion that does not conform to this simple definition, namely the [*GAR*$^+$] prion. Following the discovery of mutants of *S. cerevisiae* that were resistant to glucosamine, but showed a non-Mendelian pattern of inheritance (Kunz and Ball, 1977), the nature of the underlying epigenetic determinant remained a mystery. The resistance to glucosamine phenotype indicated that in [*GAR*$^+$] cells the genes required to utilize respirative carbon sources such as ethanol or glycerol and whose expression is repressed by the presence of glucose, no longer respond to this repression mechanism. Glucosamine is a glucose mimetic that acts as a glucose signal without being metabolized. Consequently, [*GAR*$^+$] cells can be detected using a combination of glucosamine and glycerol; only [*GAR*$^+$] cells can override the repression by glucosamine to metabolize the glycerol. [*GAR*$^+$] cells can therefore be considered as "metabolic generalists" in that they can utilize a much wider range of sugars than [*GAR*$^-$] cells who are therefore "metabolic specialists" committed to one preferred carbon source, namely, glucose.

Although [*GAR*$^+$] shows the epigenetic behavior associated with other fungal prions, it has a number of properties that distinguishes it from other yeast prions. Of particular note is that its continued propagation is not dependent on the disaggregase activity Hsp104, but rather is dependent on another chaperone Hsp70 (Brown and Lindquist, 2009). Furthermore, two proteins have been linked to the [*GAR*$^+$] determinant, neither of which are naturally amyloidogenic in nature. These are the plasma membrane proton pump Pma1 and Std1, a component of the control of glucose regulated genes (Brown and Lindquist, 2009). In [*GAR*$^+$] cells, Pma1 undergoes some form of conformational switch that increases its interaction with Std1 but how this then overrides glucose repression in cells remains to be fully established.

As seen with other yeast prions, cells can switch between the [*PRION*$^+$] and [*PRION*$^-$] states to reprogram carbohydrate metabolism utilization without the need for an underlying change in the host cell genome sequence. What is remarkable about this switch with the [*GAR*$^+$] prion is that it can be triggered by a chemical messenger produced by certain bacterial species (Jarosz et al., 2014a,b), a unique form of cross-kingdom chemical communication. Although the identity of the chemical messenger has now been identified as L-lactic acid (Garcia et al., 2016), exactly how it triggers the formation of [*GAR*$^+$] has not been established.

Where this novel prion switch becomes important is in wine fermentations where lactic acid-producing bacteria are well known spoilers of fermentation. The L-lactic acid produced by the bacteria trigger the formation of [*GAR*$^+$] in the yeast in the fermentation and one outcome is that the levels of ethanol produced by the yeast are dramatically reduced (Garcia et al., 2016). This then provides a low ethanol environment that does not inhibit bacterial growth and eventually the bacteria flourish. However there has been a suggestion that the formation of [*GAR*$^+$] in sake yeast may be beneficial in a sake fermentation because the [*GAR*$^+$] cells show increased survival in the sake starter mash (Watanabe et al., 2018).

As is evident from the aforementioned examples, yeast prions have the potential to have a major impact on their host. In considering whether these impacts are of benefit to the host in the normal environment in which yeasts are found it is necessary to consider whether prion-containing cells are present in wild populations of cells. Several surveys have revealed the presence of [*PIN*$^+$], [*MOT3*$^+$], and [*PSI*$^+$], but in each case these are only present in a low percentage of the wild strains screened (Resende et al., 2003; Halfmann et al., 2012a,b; Kelly et al., 2012). It is possible though that those strains identified as prion-free may have a few cells in a population carrying one or other of these prions as a bet-hedging strategy that may allow the cells to readily adapt to rapid changes in the environment (Newby and Lindquist, 2013; review).

Conformational diversity generates phenotypic diversity

There are a number of different fungal prions each of which can generate a range of phenotypes (Table 5.1) but there maybe a number of others that have yet to be confirmed as fully satisfying the definition "prion" and/or whose phenotypic impact remains to be established (Alberti et al., 2009; Chakrabortee et al., 2016). Yet there is a further dimension to the ways by which prions can modify the phenotype of the host without fixing mutational change into the nuclear genome. Several of the yeast PFPs are able to fold into more than one transmissible, self-templating polymorph with each conformational variant giving rise to a subtly different phenotype. These are referred to as prion "variants" rather than strains; the latter term is used to refer to strains of yeast with different nuclear genotypes or different natural isolates.

Most of what we know about prion variants again comes from studies with the [*PSI*$^+$] prion primarily because the impact on the prion phenotype can be readily analyzed using a colony color-based assay and there are a number of distinct variants (Fig. 5.2). The standard assay for detecting the presence of [*PSI*$^+$] in cells uses strains carry alleles of either the *ADE1* or *ADE2* gene as reporters which contain nonsense mutations. The resulting loss of function of either reporter gene leads to the accumulation a red intermediate of the adenine biosynthetic pathway that results in red colonies on agar plates. They are also adenine auxotrophs, that is, require adenine for growth. In

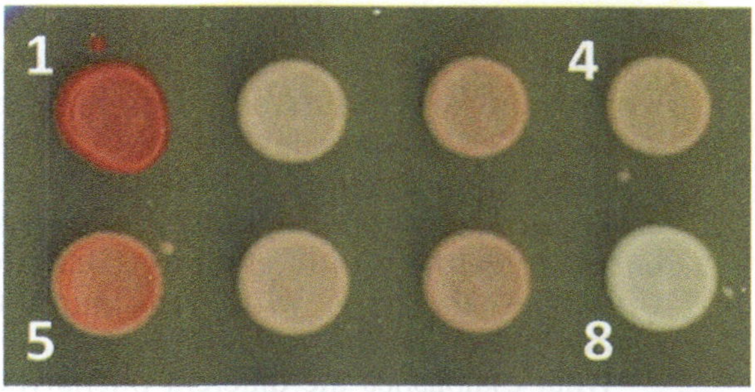

Figure 5.2 The [PSI^+] prion can give rise to a number of stable variants that can be differentiated on the basis of the efficiency of nonsense suppression. In this assay all variants have arisen in the same genetic strain carrying the suppressible *ade1–14* nonsense allele. Isolate 1 is [PSI^-] while isolates 2–8 are different [PSI^+] variants with isolate 8 being a "strong" variant.

[PSI^+] cells, the resulting termination defect leads to a loss of both the red intermediate and regaining prototrophy. Conformational variants of the [PSI^+] prion show different degrees of nonsense suppression and these are visualized by colony color with the "strong" variants giving rise to white colonies and the "weak" variants producing colonies with various shades of pink (Fig. 5.2).

Prion variants were first noticed when [PSI^+] cells were induced by overexpression of the *SUP35* gene (Derkatch et al., 1996) and different [PSI^+] variants can also be produced by transforming [PSI^-] cells with fibrils of Sup35 generated in vitro at different temperatures (Tanaka et al., 2004, 2006; King and Diaz-Avalos, 2004). A recent survey of a range of [PSI^+] strains of different origins and using various genetic and phenotypic criteria suggested that there are more than 20 different [PSI^+] variants (Huang and King, 2019), although structural studies indicate they can still be classified as either weak or strong (Dergalev et al., 2019).

Variants of other yeast prions have been described including [*URE3*] (Schlumpberger et al., 2001) and [PIN^+] (Bradley et al., 2002). In the case of the [PIN^+] variants, these dare differentiated by the frequency with which the [PSI^+] prion arises de novo. Furthermore, not all prion variants necessarily recapitulate the "normal" phenotype associated with a given prion. This is best example of this is the identification of different variants of the [PSI^+] that kill the host cell—so-called "Suicidal [PSI^+]" (McGlinchey et al., 2011).

Prion variants are structural polymorphs of the amyloidogenic PFP with no change in primary amino acid sequence. Biophysical and structural studies of prion variants are most advanced for Sup35/[*PSI*$^+$] whose variants are typically classified as weak or strong although as described above, there are a spectrum of variants based on a variety of criteria (Huang and King, 2019; Dergalev et al., 2019). Besides distinguishable nonsense suppression phenotypes (Fig. 5.2) the [*PSI*$^+$] variants also show other distinct properties; for example, different degrees of mitotic stability (Derkatch et al., 1996, 1997; Uptain et al., 2001) with weak variants being much more unstable than strong variants.

What establishes the different phenotypes in the case of [*PSI*$^+$] relates to the ability of the endogenous chaperone system driven by the disaggregase Hsp104, to fragment the amyloid fibrils to generate transmissible propagons (Toyama et al., 2007). Fibrils formed by weak variants are much less susceptible to fragmentation by Hsp104 than the strong variants. Consequently, the balance of Sup35 molecules in the cells of weak variants shifts in favor of functional nonprion and therefore fully functional forms of Sup35 because there are less seeds, that is, propagons present. The result is more efficient termination, less nonsense suppression.

Attempts to define how the amyloid fibrils differ between the different [*PSI*$^+$] variants has provided some insight into the regions of the Sup35 protein that undergo the structural rearrangements during the establishment of a specific conformation (Krishnan and Lindquist, 2005; Toyama et al., 2007; Gorkovskiy et al., 2014; Ohhashi et al., 2018; Dergalev et al., 2019). The key regions showing distinct conformational differences between the different as expected lie in the N-terminal prion domain of Sup35. No structural studies have yet been reported on the nature of the conformational forms associated with other fungal prions although there is considerable detailed knowledge of the amyloid form of the HET-S protein of *P. anserina* (Riek and Saupe, 2016; review).

Concluding remarks

The discovery of prions in fungi and in particular *S. cerevisiae* has opened a new vista on research into the control of phenotypes in fungi. With almost a dozen verified prions described (Table 5.1) plus a significant number of other proteins that show some of the properties but to which no particular phenotype has been ascribed, the contribution of epigenetic regulation of phenotypes cannot be underestimated. However, prions are not the only form of protein-based inheritance in yeast. Several examples have recently emerged of phenotypes being linked to atypical protein behavior. The proteins associated with these phenotypes share some of the properties associated with prions, for example, the formation of high molecular weight protein aggregates. In contrast to prions they are not amyloid forming but which have their own characteristic properties. These include mnemons, which are nontransmissible protein aggregates that impart a form of molecular "memory" on the host cell (Caudron and Barral, 2013), and a class of proteins which when overexpressed in a cell lead to structural changes and leading to a variety of stable phenotypes (Chakrabortee et al., 2016). In both cases these might also be considered examples of "protein-based epigenetic memory" (Tuite, 2016).

The evidence to date therefore points to prions and potentially the other protein-based epigenetic elements as being important contributors to the generation of phenotypic diversity in *S. cerevisiae.* These epigenetic elements provide the organism with the potential to rapidly adapt to the harsh environments although not all prions necessarily are beneficial. It remains to be seen whether other fungal species exploit this novel means of generating phenotypic diversity.

Acknowledgments

Where appropriate I have used review articles rather than a long list of primary research papers and therefore would like to apologize to those authors whose important papers I have not cited. I would like to gratefully acknowledge the support of the Biotechnology and Biological Sciences Research Council (BBSRC, BB/J000191/1), the Leverhulme Trust, and the University of Kent.

References

Aguzzi, A., Lakkaraju, A.K.K., Frontzek, K., 2018. Toward therapy of human prion diseases. Ann. Rev. Pharmacol. Toxicol. 8, 331–351.

Aigle, M., Lacroute, F., 1975. Genetical aspects of [*URE3*], a non-Mendelian, cytoplasmically-inherited mutation in yeast. Mol. Gen. Genet. 136, 327–335.

Alberti, S., Halfmann, R., King, O., Kapila, A., Lindquist, S., 2009. A systematic survey identifies prions and illuminates sequence features of prionogenic proteins. Cell 137, 146–158.

Ali, M., Chernova, T.A., Newnam, G.P., Yin, L., Shanks, J., Karpova, T.S., et al., 2014. Stress-dependent proteolytic processing of the actin assembly protein Lsb1 modulates a yeast prion. J. Biol. Chem. 289, 27625–27639.

An, L., Fitzpatrick, D., Harrison, P.M., 2016. Emergence and evolution of yeast prion and prion-like proteins. BMC Evol. Biol. 16, 24.

Baudin-Baillieu, A., Legendre, R., Kuchly, C., Hatin, I., Demais, S., Mestdagh, C., et al., 2014. Genome-wide translational changes induced by the prion [PSI^+]. Cell Rep. 8, 439–448.

Becker, B., Schmitt, M.J., 2017. Yeast killer toxin K28: biology and unique strategy of host cell intoxication and killing. Toxins 9, E333.

Bradley, M.E., Edskes, H.K., Hong, J.Y., Wickner, R.B., Liebman, S.W., 2002. Interactions among prions and prion "strains" in yeast. Proc. Natl. Acad. Sci. U.S.A. 99, 16392–16399.

Brown, J.C., Lindquist, S., 2009. A heritable switch in carbon source utilization driven by an unusual yeast prion. Genes. Dev. 23, 2320–2332.

Brückner, S., Mösch, H.U., 2012. Choosing the right lifestyle: adhesion and development in *Saccharomyces cerevisiae.* FEMS Microbiol. Rev. 36, 25–58.

Byrne, L.J., Cole, D.J., Cox, B.S., Ridout, M.S., Morgan, B.J., Tuite, M.F., 2009. The number and transmission of [*PSI*] prion seeds (propagons) in the yeast *Saccharomyces cerevisiae.* PLoS One 4, e4670.

Caudron, F., Barral, Y., 2013. A super-assembly of Whi3 encodes memory of deceptive encounters by single cells during yeast courtship. Cell 155, 1244–1257.

Chakrabortee, S., Byers, J.S., Jones, S., Garcia, D.M., Bhullar, B., Chang, A., et al., 2016. Intrinsically disordered proteins drive emergence and inheritance of biological traits. Cell 167, 369–381.

Chernoff, Y.O., Derkach, I.L., Inge-Vechtomov, S.G., 1993. Multicopy *SUP35* gene induces *de-novo* appearance of psi-like factors in the yeast *Saccharomyces cerevisiae.* Curr. Genet. 24, 268–270.

Chernova, T.A., Kiktev, D.A., Romanyuk, A.V., Shanks, J.R., Laur, O., Ali, M., et al., 2017. Yeast short-lived actin-associated protein forms a metastable prion in response to thermal stress. Cell Rep. 18, 751–761.

Chiti, F., Dobson, C.M., 2017. Protein Misfolding, Amyloid formation, and human disease: a summary of progress over the last decade. Ann. Rev. Biochem. 86, 27–68.

Cox, B.S., 1965. Ψ, a cytoplasmic suppressor of super-suppressors in yeast. Heredity 20, 505–521.

Cox, B.S., 1971. A recessive lethal super-suppressor mutation in yeast and other psi phenomena. Heredity 26, 211–232.

Cox, B.S., 1977. Allosuppressors in yeast. Genet. Res. 30, 187–205.

Dabrowski, M., Bukowy-Bieryllo, Z., Zietkiewicz, E., 2015. Translational readthrough potential of natural termination codons in eucaryotes–the impact of RNA sequence. RNA Biol. 12, 950–958.

DePace, A.H., Santoso, A., Hillner, P., Weissman, J.S., 1998. A critical role for amino-terminal glutamine/asparagine repeats in the formation and propagation of a yeast prion. Cell 93, 1241–1252.

Derdowski, A., Sindi, S.S., Klaips, C.L., DiSalvo, S., Serio, T.R., 2010. A size threshold limits prion transmission and establishes phenotypic diversity. Science 330, 680–683.

Dergalev, A.A., Alexandrov, A.I., Ivannikov, R.I., Ter-Avanesyan, M.D., Kushnirov, V.V., 2019. Yeast Sup35 prion structure: two types, four parts, many variants. Int. J. Mol. Sci. 20, E2633.

Derkatch, I.L., Chernoff, Y.O., Kushnirov, V.V., Inge-Vechtomov, S.G., Liebman, S. W., 1996. Genesis and variability of [*PSI*] prion factors in *Saccharomyces cerevisiae.* Genetics 144, 1375–1386.

Derkatch, I.L., Bradley, M.E., Zhou, P., Chernoff, Y.O., Liebman, S.W., 1997. Genetic and environmental factors affecting the *de novo* appearance of the [*PSI* +] prion in *Saccharomyces cerevisiae.* Genetics 147, 507–519.

Derkatch, I.L., Bradley, M.E., Hong, J.Y., Liebman, S.W., 2001. Prions affect the appearance of other prions: the story of [PIN^+]. Cell 106, 171–182.

Doronina, V.A., Staniforth, G.L., Speldewinde, S.H., Tuite, M.F., Grant, C.M., 2015. Oxidative stress conditions increase the frequency of *de novo* formation of the yeast [PSI^+] prion. Mol. Microbiol. 96, 163–174.

Du, Z., Park, K.W., Yu, H., Fan, Q., Li, L., 2008. Newly identified prion linked to the chromatin-remodeling factor Swi1 in *Saccharomyces cerevisiae*. Nat. Genet. 40, 460–465.

Du, Z., Zhang, Y., Li, L., 2015. The yeast prion [SWI^+] abolishes multicellular growth by triggering conformational changes of multiple regulators required for flocculin gene expression. Cell Rep. 13, 2865–2878.

Eaglestone, S.S., Cox, B.S., Tuite, M.F., 1999. Translation termination efficiency can be regulated in *Saccharomyces cerevisiae* by environmental stress through a prion-mediated mechanism. EMBO J. 18, 1974–1981.

Fitzpatrick, D.A., O'Brien, J., Moran, C., Hasin, N., Kenny, E., Cormican, P., et al., 2011. Assessment of inactivating stop codon mutations in forty *Saccharomyces cerevisiae* strains: implications for [*PSI*] prion-mediated phenotypes. PLoS One 6, e28684.

Ganusova, E.E., Ozolins, L.N., Bhagat, S., Newnam, G.P., Wegrzyn, R.D., Sherman, M.Y., et al., 2006. Modulation of prion formation, aggregation, and toxicity by the actin cytoskeleton in yeast. Mol. Cell Biol. 26, 617–629.

Garcia, D.M., Dietrich, D., Clardy, J., Jarosz, D.F., 2016. A common bacterial metabolite elicits prion-based bypass of glucose repression. Elife 5, e17978.

Goncharoff, D.K., Du, Z., Li, L., 2018. A brief overview of the Swi1 prion-[SWI^+]. FEMS Yeast Res. 18. Available from: https://doi.org/10.1093/femsyr/foy061.

Gorkovskiy, A., Thurber, K.R., Tycko, R., Wickner, R.B., 2014. Locating folds of the in-register parallel β-sheet of the Sup35p prion domain infectious amyloid. Proc. Natl. Acad. Sci. U.S.A 111, E4615–E4622.

Halfmann, R., Alberti, S., Lindquist, S., 2010. Prions, protein homeostasis, and phenotypic diversity. Trends Cell Biol. 20, 125–133.

Halfmann, R., Alberti, S., Krishnan, R., Lyle, N., O'Donnell, C.W., King, O.D., et al., 2011. Opposing effects of glutamine and asparagine govern prion formation by intrinsically disordered proteins. Mol. Cell. 43, 72–84.

Halfmann, R., Jarosz, D.F., Jones, S.K., Chang, A., Lancaster, A.K., Lindquist, S., 2012a. Prions are a common mechanism for phenotypic inheritance in wild yeasts. Nature 482, 363–368.

Halfmann, R., Wright, J.R., Alberti, S., Lindquist, S., Rexach, M., 2012b. Prion formation by a yeast GLFG nucleoporin. Prion 6, 391–399.

Harbi, D., Harrison, P.M., 2014. Interaction networks of prion, prionogenic and prion-like proteins in budding yeast, and their role in gene regulation. PLoS One 9, e100615.

Holmes, D.L., Lancaster, A.K., Lindquist, S., Halfmann, R., 2013. Heritable remodeling of yeast multicellularity by an environmentally responsive prion. Cell 153, 153–165.

Huang, Y.W., King, C.Y., 2019. A complete catalog of wild-type Sup35 prion variants and their protein-only propagation. Curr. Genet. 66, 97–122. Available from: https://doi.org/10.1007/s00294-019-01003-8.

Ironside, J.W., Ritchie, D.L., Head, M.W., 2017. Prion diseases. Handb. Clin. Neurol. 145, 393–403.

Jarosz, D.F., Lancaster, A.K., Brown, J.C.S., Lindquist, S., 2014a. An evolutionarily conserved prion-like element converts wild fungi from metabolic specialists to generalists. Cell 158, 1072–1082.

Jarosz, D.F., Brown, J.C.S., Walker, G.A., Datta, M.S., Ung, W.L., Lancaster, A.K., et al., 2014b. Cross-kingdom chemical communication drives a heritable, mutually beneficial prion-based transformation of metabolism. Cell 158, 1083–1093.

Kabani, M., Melki, R., 2016. More than just trash bins? Potential roles for extracellular vesicles in the vertical and horizontal transmission of yeast prions. Curr. Genet. 62, 265–270.

Kelly, A.C., Shewmaker, F.P., Kryndushkin, D., Wickner, R.B., 2012. Sex, prions, and plasmids in yeast. Proc. Natl. Acad. Sci. U.S.A. 109, E2683–E2690.

King, C.Y., Diaz-Avalos, R., 2004. Protein-only transmission of three yeast prion strains. Nature 428, 319–323.

Kotta-Loizou, I., Coutts, R.H.A., 2017. Mycoviruses in *Aspergilli:* a comprehensive review. Front. Microbiol. 8, 1699.

Krishnan, R., Lindquist, S.L., 2005. Structural insights into a yeast prion illuminate nucleation and strain diversity. Nature 435, 765–772.

Kunz, B.A., Ball, A.J., 1977. Glucosamine resistance in yeast. II. Cytoplasmic determinants conferring resistance. Mol. Gen. Genet. 153, 169–177.

Lancaster, A.K., Bardill, J.P., True, H.L., Masel, J., 2010. The spontaneous appearance rate of the yeast prion [PSI^+] and its implications for the evolution of the evolvability properties of the [PSI^+] system. Genetics 184, 393–400.

Lancaster, A.K., Nutter-Upham, A., Lindquist, S., King, O.D., 2014. PLAAC: a web and command-line application to identify proteins with prion-like amino acid composition. Bioinformatics 30, 2501–2502.

Malovichko, Y.V., Antonets, K.S., Maslova, A.R., Andreeva, E.A., Inge-Vechtomov, S.G., Nizhnikov, A.A., 2019. RNA sequencing reveals specific transcriptomic signatures distinguishing effects of the [SWI^+] prion and *SWI1* deletion in yeast *Saccharomyces cerevisiae.* Genes 10, E212.

Manogaran, A.L., Hong, J.Y., Hufana, J., Tyedmers, J., Lindquist, S., Liebman, S. W., 2011. Prion formation and polyglutamine aggregation are controlled by two classes of genes. PLoS Genet. 7, e1001386.

Marchante, R., Rowe, M., Zenthon, J., Howard, M.J., Tuite, M.F., 2013. Structural definition is important for the propagation of the yeast [PSI^+] prion. Mol. Cell 50, 675–685.

Martínez-Montañés, F., Rienzo, A., Poveda-Huertes, D., Pascual-Ahuir, A., Proft, M., 2013. Activator and repressor functions of the Mot3 transcription factor in the osmostress response of *Saccharomyces cerevisiae.* Eukaryot. Cell 12, 636–647.

McCarthy, C.G.P., Fitzpatrick, D.A., 2019. Pan-genome analyses of model fungal species. Microb. Genom. 5, e000243.

McGlinchey, R.P., Kryndushkin, D., Wickner, R.B. 2011. Suicidal [PSI^+] is a lethal yeast prion. *Proc. Natl. Acad. Sci. U.S.A.* **108**, 2011, 5337–5341.

Michelitsch, M.D., Weissman, J.S., 2000. A census of glutamine/asparagine-rich regions: implications for their conserved function and the prediction of novel prions. Proc. Natl. Acad. Sci. U.S.A. 97, 11910–11915.

Namy, O., Duchateau-Nguyen, G., Rousset, J.P., 2002. Translational readthrough of the PDE2 stop codon modulates cAMP levels in *Saccharomyces cerevisiae.* Mol. Microbiol. 43, 641–652.

Namy, O., Galopier, A., Martini, C., Matsufuji, S., Fabret, C., Rousset, J.P., 2008. Epigenetic control of polyamines by the prion [PSI^+]. Nat. Cell Biol. 10, 1069–1075.

Ness, F., Ferreira, P., Cox, B.S., Tuite, M.F., 2002. Guanidine hydrochloride inhibits the generation of prion "seeds" but not prion protein aggregation in yeast. Mol. Cell Biol. 22, 5593–5605.

Ness, F., Cox, B.S., Wongwigkarn, J., Naeimi, W.R., Tuite, M.F., 2017. Over-expression of the molecular chaperone Hsp104 in *Saccharomyces cerevisiae* results in the malpartition of [PSI^+] propagons. Mol. Microbiol. 104, 125–143.

Nevzglyadova, O.V., Artemov, A.V., Mittenberg, A.G., Solovyov, K.V., Kostyleva, E.I., Mikhailova, E.V., et al., 2009. Prion-associated proteins in yeast: comparative analysis of isogenic [*PSI*$^+$] and [*psi-*] strains. Yeast 26, 611–631.

Newby, G.A., Lindquist, S., 2013. Blessings in disguise: biological benefits of prion-like mechanisms. Trends Cell Biol. 23, 251–259.

Newby, G.A., Lindquist, S., 2017. Pioneer cells established by the [*SWI*$^+$] prion can promote dispersal and out-crossing in yeast. PLoS Biol. 15, e2003476.

Newnam, G.P., Birchmore, J.L., Chernoff, Y.O., 2011. Destabilization and recovery of a yeast prion after mild heat shock. J. Mol. Biol. 408, 432–448.

Ohhashi, Y., Yamaguchi, Y., Kurahashi, H., Kamatari, Y.O., Sugiyama, S., Uluca, B., et al., 2018. Molecular basis for diversification of yeast prion strain conformation. Proc. Natl. Acad. Sci. U.S.A. 115, 2389–2394.

Osherovich, L.Z., Weissman, J.S., 2001. Multiple Gln/Asn-rich prion domains confer susceptibility to induction of the yeast [*PSI*$^+$] prion. Cell 106, 183–194.

Patel, B.K., Gavin-Smyth, J., Liebman, S.W., 2009. The yeast global transcriptional co-repressor protein Cyc8 can propagate as a prion. Nat. Cell Biol. 11, 344–349.

Peter, J., De Chiara, M., Friedrich, A., Yue, J.X., Pflieger, D., Bergström, A., et al., 2018. Genome evolution across 1,011 *Saccharomyces cerevisiae* isolates. Nature 556, 339–344.

Pezza, J.A., Villali, J., Sindi, S.S., Serio, T.R., 2014. Amyloid-associated activity contributes to the severity and toxicity of a prion phenotype. Nat. Commun. 5, 4384.

Prusiner, S.B., 1982. Novel proteinaceous infectious particles cause scrapie. Science 216, 136–144.

Prusiner, S.B., 2013. Biology and genetics of prions causing neurodegeneration. Ann. Rev. Genet. 47, 601–623.

Resende, C.G., Outeiro, T.F., Sands, L., Lindquist, S., Tuite, M.F., 2003. Prion protein gene polymorphisms in *Saccharomyces cerevisiae*. Mol. Microbiol. 49, 1005–1017.

Riek, R., Saupe, S.J., 2016. The HET-S/s prion motif in the control of programmed cell death. Cold Spring Harb. Perspect. Biol. 8i, a023515.

Rizvi, S.M.A., Prajapati, H.K., Ghosh, S.K., 2018. The 2 micron plasmid: a selfish genetic element with an optimized survival strategy within *Saccharomyces cerevisiae*. Curr. Genet. 64, 25–42.

Rogoza, T., Goginashvili, A., Rodionova, S., Ivanov, M., Viktorovskaya, O., Rubel, A., et al., 2010. Non-Mendelian determinant [*ISP*$^+$] in yeast is a nuclear-residing prion form of the global transcriptional regulator Sfp1. Proc. Natl. Acad. Sci. U.S.A. 107, 10573–10577.

Ross, E.D., Maclea, K.S., Anderson, C., Ben-Hur, A., 2013. A bioinformatics method for identifying Q/N-rich prion-like domains in proteins. Methods Mol. Biol. 1017, 219–228.

Sato, Y., Castón, J.R., Suzuki, N., 2018. The biological attributes, genome architecture and packaging of diverse multi-component fungal viruses. Curr. Opin. Virol. 33, 55–65.

Saupe, S.J., Jarosz, D.F., True, H.L., 2016. Amyloid prions in fungi. Microbiol. Spectr. 4. Available from: https://doi.org/10.1128/microbiolspec.FUNK-0029-2016.

Schlumpberger, M., Prusiner, S.B., Herskowitz, I., 2001. Induction of distinct [*URE3*] yeast prion strains. Mol. Cell Biol. 21, 7035–7046.

Schmitt, M.J., Breinig, F., 2006. Yeast viral killer toxins: lethality and self-protection. Nat. Rev. Microbiol. 4, 212–221.

Schneider, K.L., Nyström, T., Widlund, P.O., 2018. Studying spatial protein quality control, proteopathies, and aging using different model misfolding proteins in *S. cerevisiae*. Front. Mol. Neurosci. 11, 249.

Serio, T.R., 2018. [PIN +]ing down the mechanism of prion appearance. FEMS Yeast Res. 18, foy026. Available from: https://doi.org/10.1093/femsyr/foy026.

Si, K., 2015. Prions: what are they good for? Ann. Rev. Cell Dev. Biol. 31, 149–169.

Sondheimer, N., Lindquist, S., 2000. Rnq1: an epigenetic modifier of protein function in yeast. Mol. Cell 5, 163–172.

Speldewinde, S.H., Doronina, V.A., Tuite, M.F., Grant, C.M., 2017. Disrupting the cortical actin cytoskeleton points to two distinct mechanisms of yeast [PSI^+] prion formation. PLoS Genet. 13, e1006708.

Stansfield, I., Jones, K.M., Kushnirov, V.V., Dagkesamanskaya, A.R., Poznyakovski, A.I., Paushkin, S.V., et al., 1995. The products of the *SUP45* (eRF1) and *SUP35* genes interact to mediate translation termination in *Saccharomyces cerevisiae*. EMBO J. 14, 4365–4373.

Sudarsanam, P., Iyer, V.R., Brown, P.O., Winston, F., 2000. Whole-genome expression analysis of snf/swi mutants of *Saccharomyces cerevisiae*. Proc. Natl. Acad. Sci. U.S.A. 97, 3364–3369.

Suzuki, G., Shimazu, N., Tanaka, M., 2012. A yeast prion, Mod5, promotes acquired drug resistance and cell survival under environmental stress. Science 336, 355–359.

Suzuki, G., Weissman, J.S., Tanaka, M., 2015. [*KIL-d*] protein element confers antiviral activity via catastrophic viral mutagenesis. Mol. Cell 60, 651–660.

Tanaka, M., Chien, P., Naber, N., Cooke, R., Weissman, J.S., 2004. Conformational variations in an infectious protein determine prion strain differences. Nature 428, 323–328.

Tanaka, M., Collins, S.R., Toyama, B.H., Weissman, J.S., 2006. The physical basis of how prion conformations determine strain phenotypes. Nature 442, 585–589.

Taylor, J.W., Branco, S., Gao, C., Hann-Soden, C., Montoya, L., Sylvain, I., et al., 2017. Sources of fungal genetic variation and associating it with phenotypic diversity. Microbiol. Spectr. 5. Available from: https://doi.org/10.1128/microbiolspec.FUNK-0057-2016.

Toyama, B.H., Kelly, M.J., Gross, J.D., Weissman, J.S., 2007. The structural basis of yeast prion strain variants. Nature 449, 233–237.

Tuite, M.F., 2013. The natural history of yeast prions. Adv. Appl. Microbiol. 84, 85–137.

Tuite, M.F., 2016. Remembering the past: a new form of protein-based inheritance. Cell 167, 302–303.

True, H.L., Lindquist, S.L., 2000. A yeast prion provides a mechanism for genetic variation and phenotypic diversity. Nature 407, 477–483.

True, H.L., Berlin, I., Lindquist, S.L., 2004. Epigenetic regulation of translation reveals hidden genetic variation to produce complex traits. Nature 431, 184–187.

Tuite, M.F., Staniforth, G.L., Cox, B.S., 2015. [PSI^+] turns 50. Prion. 9, 318–332.

Tyedmers, J., Madariaga, M.L., Lindquist, S., 2008. Prion switching in response to environmental stress. PLoS Biol. 6, e294.

Tyedmers, J., Treusch, S., Dong, J., McCaffery, J.M., Bevis, B., Lindquist, S., 2010. Prion induction involves an ancient system for the sequestration of aggregated proteins and heritable changes in prion fragmentation. Proc. Natl. Acad. Sci. U.S.A. 107, 8633–8638.

Uptain, S.M., Sawicki, G.J., Caughey, B., Lindquist, S., 2001. Strains of [PSI^+] are distinguished by their efficiencies of prion-mediated conformational conversion. EMBO J. 20, 6236–6245.

Vishveshwara, N., Bradley, M.E., Liebman, S.W., 2009. Sequestration of essential proteins causes prion associated toxicity in yeast. Mol. Microbiol. 73, 1101–1114.

von der Haar, T., Tuite, M.F., 2007. Regulated translational bypass of stop codons in yeast. Trends Microbiol. 15, 78–86.

Watanabe, D., Kumano, M., Sugimoto, Y., Ito, M., Ohashi, M., Sunada, K., et al., 2018. Metabolic switching of sake yeast by kimoto lactic acid bacteria through the [GAR^+] non-genetic element. J. Biosci. Bioeng. 126, 624–629.

Wickner, R.B., 1994. [*URE3*] as an altered URE2 protein: evidence for a prion analog in *Saccharomyces cerevisiae*. Science 264, 566–569.

Wickner, R.B., Fujimura, T., Esteban, R., 2013. Viruses and prions of *Saccharomyces cerevisiae*. Adv. Virus Res. 86, 1–36.

Wilson, P.G., Culbertson, M.R., 1988. SUF12 suppressor protein of yeast. A fusion protein related to the EF-1 family of elongation factors. J. Mol. Biol. 199, 559–573.

6

Bistability in virus–host interaction networks underlies the success of hepatitis C treatments

Pranesh Padmanabhan[1], Rubesh Raja[2] and Narendra M. Dixit[2,3]

[1]Clem Jones Centre for Ageing Dementia Research, Queensland Brain Institute, The University of Queensland, St. Lucia, Brisbane, Australia [2]Department of Chemical Engineering, Indian Institute of Science, Bangalore, India [3]Centre for Biosystems Science and Engineering, Indian Institute of Science, Bangalore, India

Introduction

Hepatitis C virus (HCV) infection poses a significant health burden, affecting nearly 150 million people worldwide (Lavanchy, 2011). HCV is a small, positive-strand RNA virus that primarily infects hepatocytes. In ~26% of the individuals infected, HCV is spontaneously cleared by our immune system. In the rest, HCV evades the immune defense and establishes chronic infection, which if left untreated can lead to cirrhosis, hepatocellular carcinoma, and liver-related deaths (Micallef et al., 2006; Hajarizadeh et al., 2013). There are no preventive or therapeutic vaccines for HCV infection. Until recently, the standard-of-care for HCV infection involved a combination of two drugs, pegylated interferon and ribavirin (Heim, 2013a). Interferon acts by stimulating the expression of several hundred genes with activity against HCV, whereas ribavirin is thought to potentiate the anti-HCV activity of interferon (Dixit et al., 2004; Rotman et al., 2013). This combination (PR) required intravenous drug administration for 24–48 weeks and cured ~50% of the HCV patients treated (Horner and Naggie, 2015). The treatment also resulted in strong side effects. In the last 10–15 years, a greater understanding of the structures of HCV proteins and of the HCV life cycle has paved the way for the development of drugs targeting specific stages of the HCV life

Phenotypic Switching. DOI: https://doi.org/10.1016/B978-0-12-817996-3.00024-4

cycle, termed as direct-acting antiviral agents (DAAs) (Götte and Feld, 2016). DAAs are revolutionizing the treatments of HCV infection. Addition of HCV protease inhibitors, a class of DAAs, to PR improved the cure rates to ~70% (Ghany et al., 2011). Today all-oral combinations of DAAs can cure nearly all patients treated, with reduced side effects and treatment durations of 12 weeks or shorter (WHO, 2016). The treatments are expensive, however, and >98% of hepatitis C patients worldwide are yet to receive DAA-based treatments. Significant efforts are therefore ongoing to identify optimal DAA combinations that would reduce costs, side effects and treatment durations, and improve accessibility (Padmanabhan and Dixit, 2017; Ohashi et al., 2017; Aghemo and Colombo, 2017).

Over the years, mathematical models of HCV kinetics have provided critical insights into HCV pathogenesis and treatment outcomes and guided strategies of interventions (Dixit et al., 2004; Padmanabhan et al., 2014; Raja et al., 2018; Baral et al., 2018; Neumann et al., 1998; Padmanabhan and Dixit, 2011; Perelson and Guedj, 2015; Padmanabhan and Dixit, 2015b; Padmanabhan and Dixit, 2012). They are today playing a key role in identifying optimal DAA-based interventions. In this chapter, we discuss mathematical models that have identified systems-level properties of the intracellular and the intercellular virus–host interaction networks underlying HCV infection and treatment outcomes, providing quantitative frameworks for rational treatment optimization (Padmanabhan et al., 2014; Raja et al., 2018; Baral et al., 2018; Baral et al., Submitted; Raja et al., 2019; Venugopal et al., 2018). At the heart of these models are bistability and the resulting phenotypic heterogeneity at the cellular and the infected individual levels. We begin by discussing a model of the interferon signaling network in HCV-infected cells that identified HCV-induced bistability in the network and the resulting cellular phenotypes with distinct interferon responsiveness (Padmanabhan et al., 2014). In the section titled, 'Bistability in the interferon signaling network,' we describe a framework that integrates the latter description with the standard models of HCV viral kinetics to predict outcomes of interferon-based and DAA-based treatments (Padmanabhan et al., 2014; Raja et al., 2019; Venugopal et al., 2018). In the section titled, 'Potential considerations and strategies for optimizing DAA treatments,' we discuss potential avenues to personalize DAA-based treatments by quantifying the diverse drug resistance pathways available to HCV and by exploiting the bistability arising from the essential interactions between the virus, target cells, infected cells and immune effector cells (Baral et al., 2018; Baral et al., Submitted). We end with concluding remarks.

Bistability in the interferon signaling network

Interferons and HCV infection

Interferon has been the mainstay of HCV treatments for about 25 years (Heim, 2013a). Following the onset of interferon-based therapy, the viral load in plasma typically decays in a biphasic manner, with a rapid first-phase decline lasting ~1–2 days followed by a slower second-phase decline typically lasting up to the end of treatment (Neumann et al., 1998). The slope of the second-phase decline, which is attributed to treatment effectiveness, is correlated with treatment response and outcome. The cause of variations in the treatment effectiveness across patients and, hence, treatment failure has remained poorly understood.

Interferons are a family of cytokines that are produced endogenously and secreted by cells in response to viral infections (Heim, 2013b; Thimme et al., 2012). These secreted molecules bind cell surface receptors to activate a complex series of intracellular signaling events involving the Janus kinase-signal transducer and activator of transcription (JAK-STAT) pathway, culminating in the transcription of several hundred genes known as interferon-stimulated genes (ISGs), which together induce an antiviral state in cells. HCV, like many other viruses, has evolved several mechanisms to circumvent the action of interferon. For example, HCV can block the production of endogenous interferons, interact with signaling components of the JAK-STAT pathway to inhibit ISG transcription, suppress ISG mRNA translation, and neutralize the effector functions of ISG proteins (Heim, 2013b; Thimme et al., 2012). The outcome of this multifaceted battle between HCV and interferons dictates whether HCV establishes persistent infection in the host and whether interferon-based treatments cure an HCV patient. A mechanistic model of the interferon signaling network in HCV-infected cells identified an emergent, systems-level property of the network as underlying the success of interferon-based treatments (Padmanabhan et al., 2014). We discuss this model and its key predictions next.

HCV induces bistability in the interferon signaling network

Upon cellular infection, HCV hijacks the host cellular machineries and replicates inside cells to generate multiple copies of its genomic RNA and proteins (Dahari et al., 2007a,b; Dahari et al., 2009; Martinez et al., 2011; Sardanyes et al., 2012) (Fig. 6.1A). The newly synthesized genomes and viral proteins

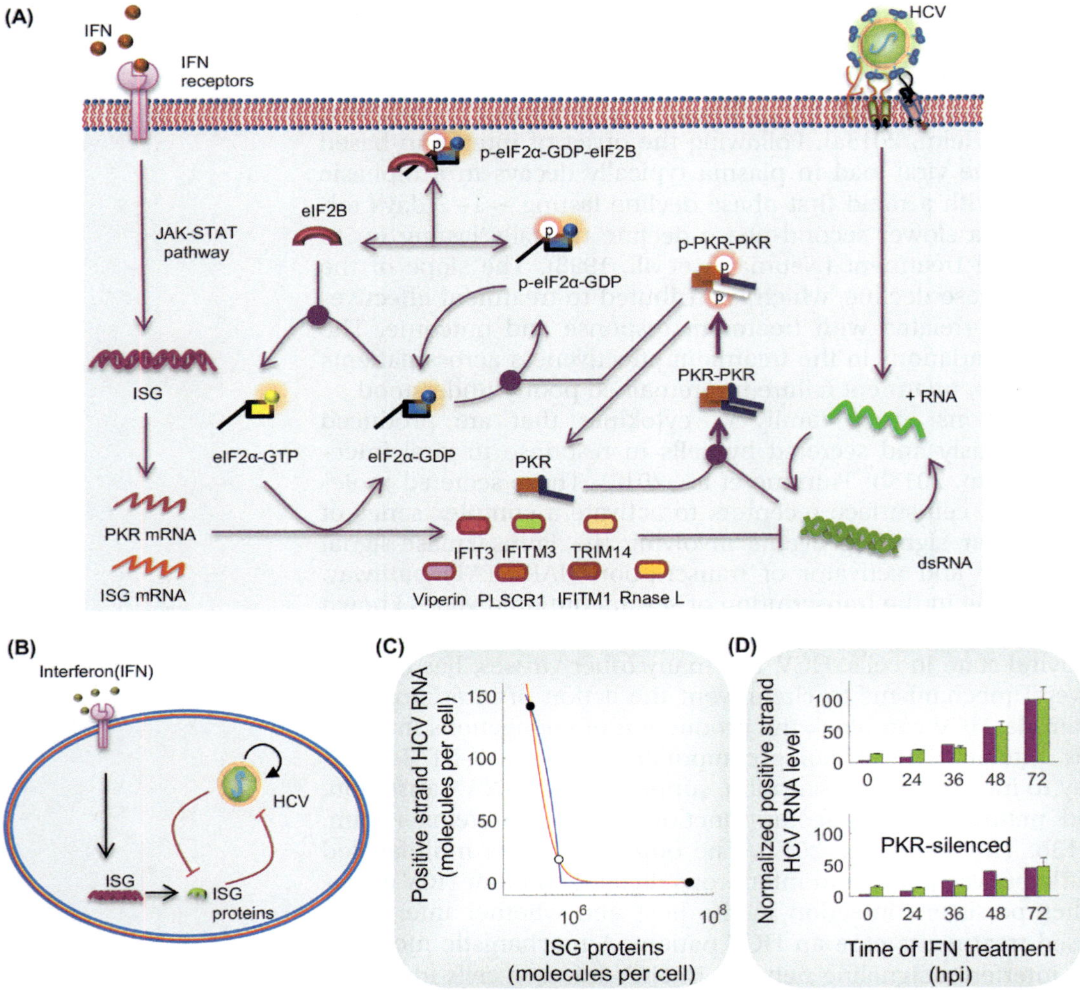

Figure 6.1 HCV induces bistability in the interferon signaling network. (A) A schematic of the interferon signaling network in an HCV-infected cell. Interferon exposure triggers the JAK-STAT pathway leading to the expression of several ISGs, which control HCV replication inside a cell. HCV catalyzes PKR dimerization and autophosphorylation, resulting in the sequestration of eIF2B, the depletion of eIF2α-GTP and the subsequent suppression of ISG translation. (B) A schematic of the double negative feedback loop in the network due to the competing interactions between interferon and HCV. (C) Model predictions of concentration-effect curves showing the steady-state levels of HCV RNA for different fixed ISG protein levels *(blue line)* and the steady-state levels of ISG proteins for different fixed HCV RNA levels *(red line)*. The points of intersections represent the steady states of the network. Closed circles are stable steady states and the open circle is an unstable steady state. (D) Experimental measurements *(green)* and model predictions *(purple)* of HCV RNA levels at 20 h postinterferon treatment at different times of interferon addition postinfection in normal *(top)* and PKR-silenced cells *(bottom)*.

are then assembled and released as progeny virions, which spread the infection to other cells. Interferon exposure triggers ISG transcription via the JAK-STAT pathway and subsequent ISG mRNA translation to ISG proteins. These proteins interfere with different stages of the HCV life cycle and inhibit HCV replication. For instance, IFITM proteins inhibit HCV entry into target cells and block HCV replication, whereas RNase L cleaves and degrades HCV RNA (Metz et al., 2013). HCV counteracts interferon action by employing several strategies, including the global suppression of cellular protein synthesis and of ISG protein expression through the phosphorylation of protein kinase R (PKR) and eukaryotic translation initiation factor 2α (eIF2α) (Garaigorta and Chisari, 2009). The latter mechanism is thought to be the dominant mode of HCV interference both in vitro (Garaigorta and Chisari, 2009) and in vivo (Wieland et al., 2014). HCV replication proceeds via a negative strand RNA intermediate. Positive and negative strand RNA associate to form a double-stranded RNA (dsRNA) complex, which catalyzes PKR dimerization and its subsequent autophosphorylation at multiple sites (Dabo and Meurs, 2012; Cole, 2007). The phosphorylated PKR, in turn, phosphorylates eIF2α bound to guanosine diphosphate (eIF2α-GDP). The guanosine triphosphate-bound form of eIF2α (eIF2α-GTP) is essential for global protein translation, and gets hydrolyzed to eIF2α-GDP during translation initiation (Proud, 2005). The guanine nucleotide exchange factor eIF2B mediates the recycling of eIF2α-GDP to eIF2α-GTP (Proud, 2005; Rowlands et al., 1988). Phosphorylated eIF2α-GDP does not undergo GDP to GTP conversion and is a competitive inhibitor of eIF2B (Rowlands et al., 1988). As a result, the concentration of eIF2α-GTP is reduced and ISG protein production is impaired. HCV, however, has developed mechanisms of translation in an eIF2α-independent manner and thrive (Garaigorta and Chisari, 2009; Koev et al., 2002; Shimoike et al., 2009).

A recent study developed a comprehensive model of the above molecular interactions (Fig. 6.1A), using a system of coupled differential equations, capturing the essence of the battle fought between HCV and interferon within individual infected cells (Padmanabhan et al., 2014). The different parts of the model were validated, and parameter values estimated, by mimicking several corresponding cell culture experiments, viz., those measuring the HCV intracellular replication kinetics in the absence of interferons (Garaigorta and Chisari, 2009; Keum et al., 2012), the expression patterns of key ISGs in uninfected cells stimulated by interferon (Garaigorta and Chisari, 2009), and the effectiveness of each of the ISG proteins in blocking intracellular HCV replication (Metz et al., 2012; Helbig et al., 2011). The model recognized that the network exhibits double negative feedback: HCV inhibits ISG protein production, whereas ISG proteins inhibit intracellular HCV replication (Fig. 6.1B). The model predicted that, consequently, HCV infection induces bistability in the interferon signaling network (Fig. 6.1C). In one stable steady state, HCV persists by suppressing ISG protein expression, thus representing the cellular state of viral persistence. In the other, ISG protein expression dominates, clearing HCV and marking the cellular state of viral clearance. The network displayed bistability over wide parameter ranges and model architectures, indicating that HCV-induced bistability is a robust systems-level feature of this network.

The two stable states of viral persistence and clearance are separated by an unstable steady state with intermediate viral and ISG levels. Following infection, viral levels rise due to replication. The model predicts that if interferon is added to cells before HCV levels cross the unstable boundary, the system settles into the cellular state where HCV is cleared. Else, persistence results. This prediction was validated by interferon time-of-addition experiments in cell cultures, presenting strong evidence of the underlying bistability in the network (Garaigorta and Chisari, 2009) (Fig. 6.1D). In the latter experiments, cells were treated with a fixed concentration of interferon either before HCV infection or at different times postinfection, and the levels of HCV RNA and ISG proteins were measured after a specific amount of time after interferon exposure. Adding interferon at early time points of HCV infection ($\leq$24 h postinfection) resulted in high ISG protein levels and low HCV RNA levels, whereas adding it at later time points ($\geq$48 h postinfection) led to a switch in the expression levels, with lower levels of ISG proteins and higher levels of HCV RNA (Garaigorta and Chisari, 2009). Using independent estimates of parameter values, the model quantitatively described these observations and attributed the observed switch-like behavior to the underlying bistability (Padmanabhan et al., 2014). The switch-like response was not observed in experiments performed in PKR-silenced cells, where HCV levels remained low irrespective of the time of interferon addition (Garaigorta and Chisari, 2009). The HCV-mediated translational block is precluded in PKR-silenced cells. These cells admit the viral clearance steady state alone, resulting in HCV clearance irrespective of the time of interferon addition, consistent with the model predictions (Padmanabhan et al., 2014).

Phenotypic heterogeneity in interferon responsiveness

Owing to the known cell-to-cell variations in the expression levels of proteins regulating the interferon signaling events or the HCV life cycle (Rand et al., 2012; Levin et al., 2011), interferon effectiveness against HCV is expected to vary across cells. According to the model, cells with high interferon effectiveness would admit the viral clearance steady state alone, whereas cells with low interferon effectiveness would realize only the viral persistence steady state. Cells with moderate interferon effectiveness would allow both states. Whether HCV thrives or gets cleared in these cells depends on the initial conditions, as illustrated with the time-of-addition aforementioned experiments. Consequently, the response to interferon-based treatment would depend on the relative abundance of these three cellular phenotypes in patients (Fig. 6.2A-B). Models of HCV kinetics that explicitly accounted for the distinct cellular responses to interferon were therefore constructed to predict treatment outcomes. We discuss these models next.

Phenotypic heterogeneity, viral kinetics, and treatment outcomes

Interferon-based treatment outcome

The standard model of HCV viral kinetics describes the viral load in plasma to be a result of the interactions between populations of uninfected hepatocytes, HCV-infected hepatocytes, and free HCV virions (Neumann et al., 1998). Uninfected cells are assumed to be produced, die, or be infected by HCV at fixed per capita rates. The resulting infected cells die at an enhanced rate and produce new virions, which are later cleared due to natural degradation or immune-mediated clearance. Without treatments, viral production and clearance rates, as well as infected cell production and death rates, are in balance, resulting in a steady viral load (also known as the baseline or set-point viral load) in chronically infected individuals. Interferon-based treatments perturb this balance, causing a biphasic decline in viral load. The standard model predicts that the rapid first-phase slope is determined by the virion clearance

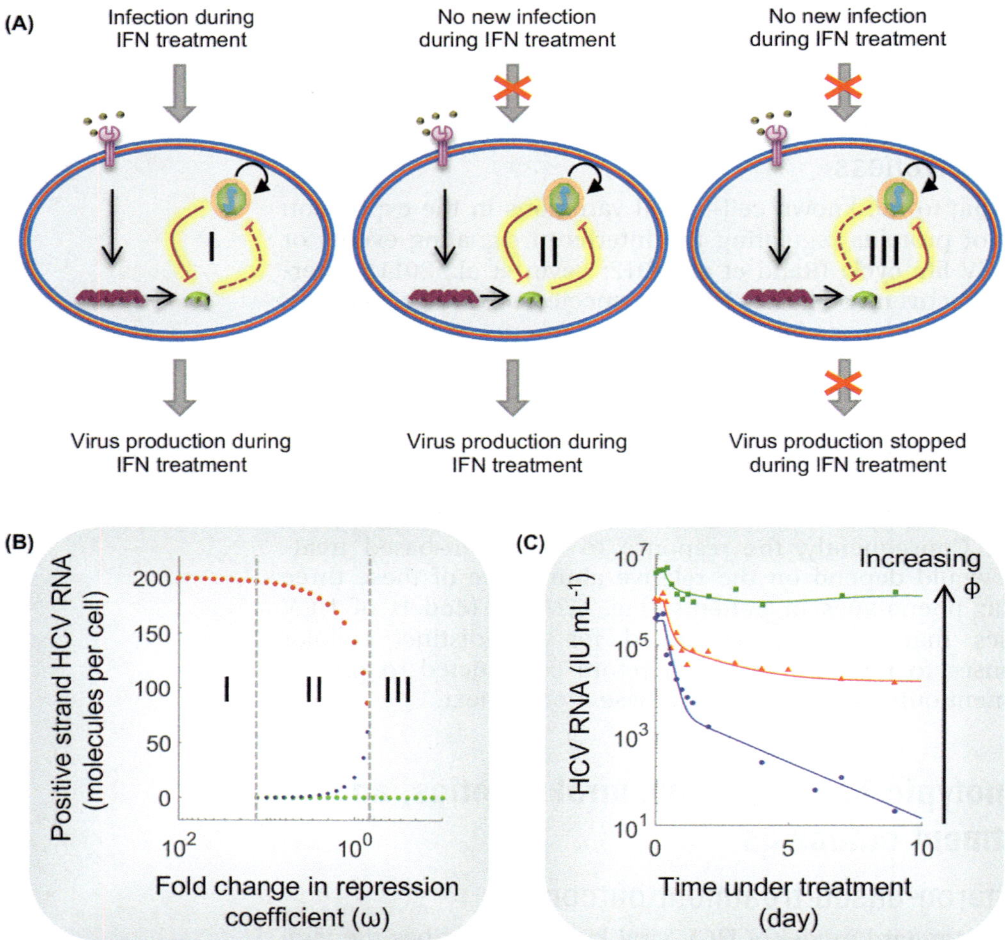

Figure 6.2 Viral kinetics with distinct cellular responses to interferon. (A) A schematic of HCV kinetics in cells with distinct interferon response phenotypes. Cells with weak *(left)*, comparable *(middle)*, and strong *(right)* interferon responses relative to the ability of HCV to subvert the interferon response represent the three distinct cellular phenotypes (see text). (B) The steady states of the system for different values of the repression coefficient, ω, for the ISG control of HCV replication. As ω decreases, ISGs exert stronger control over HCV. In A and B, I, II, and III represent interferon-refractory, bistable, and interferon-sensitive cellular phenotypes, respectively. (C) Fits of model predictions *(lines)* to viral load decline patterns measured in patients treated with interferon monotherapy *(symbols)*. The viral decay patterns mimic rapid responders *(blue)*, partial responders *(red)*, and null responders *(green)*, and correspond to increasing values of the fraction, ϕ, of cells refractory to interferon.

rate and treatment effectiveness, whereas the slow second-phase slope is defined by the infected cell death rate and treatment effectiveness. The model provided an excellent fit to the patient data, estimated the viral kinetics parameters and identified critical interferon effectiveness necessary for treatment success (Neumann et al., 1998). Extensions of the standard model yielded critical insights into the impact of drug pharmacokinetics on treatment effectiveness (Powers et al., 2003; Conway and Perelson, 2014), the mechanisms of drug action (Dixit et al., 2004; Guedj et al., 2013a,b), and synergy (Padmanabhan and Dixit, 2015a), the influence of liver homeostasis on viral kinetics (Dahari et al., 2007a,b), the minimum genetic barrier of DAAs necessary for treatment success (Rong et al., 2010), and the growth dynamics of drug-resistant strains (Venugopal et al., 2018; Rong et al., 2010; Adiwijaya et al., 2010). Multiscale models linking intracellular viral replication with extracellular HCV kinetics were then developed to describe the rapid, complex viral decline patterns observed with new DAA treatments (Guedj et al., 2013a) and to predict the diverse drug resistance pathways accessible to HCV (Raja et al., 2019). The models, however, did not factor in the cellular heterogeneity, due to the HCV-induced bistability in the interferon signaling network, often precluding accurate descriptions of treatment response rates.

A recent study advanced the standard model by dividing the hepatocytes into three subpopulations with distinct interferon responsiveness (Padmanabhan et al., 2014) (Fig. 6.2A and B). The first subpopulation of cells admits only the viral persistence steady state. In this subpopulation, even in the presence of interferon, uninfected cells continue to get infected, and infected cells produce new HCV virions. The infection dynamics of this subpopulation is not affected by PR treatment. These cells are thus refractory to interferon. Cells in the second subpopulation can realize both the viral persistence and clearance steady states. Infected cells in this subpopulation already contain high levels of intracellular HCV RNA and proteins before treatment initiation and attain the state of viral persistence following treatment. These infected cells continue to produce new virions in the presence of interferon until they die. Uninfected cells, however, get primed by interferon, reach the state of viral clearance and are resistant to new infections. The third subpopulation of cells admits the viral clearance steady state alone. In this subpopulation, uninfected cells are protected from further infection, and infected cells clear HCV upon interferon exposure. These cells are thus sensitive to interferon.

A key prediction of the model is that the fraction of cells in the first subpopulation (interferon-refractory cells), ϕ_1, determines the success of interferon-based treatment. By simply varying ϕ_1, the model recapitulated all the viral kinetics patterns observed in patients, namely rapid response, partial response and null response, following the start of PR treatment (Fig. 6.2C; also see below). The model quantitatively captured HCV viral kinetics in patients treated with interferon monotherapy and provided an alternative interpretation of the biphasic viral load decline (Fig. 6.2C). The first-phase slope is governed by the virion clearance rate and the fraction of cells in the third subpopulation (interferon-sensitive cells), ϕ_3, the latter also determining the extent of the first phase. The second-phase is dictated by the infected cell death rate and ϕ_1. As ϕ_1 increases, the slope of the second-phase decreases. Indeed, by analyzing a large cohort of patients treated with interferon monotherapy, ϕ_1 was estimated and found to be negatively correlated with the second-phase slope. Further, ϕ_1 estimated in patients with a favorable single-nucleotide polymorphism in the interferon lambda gene locus and better treatment response (Hayes et al., 2012; Ge et al., 2009) was significantly lower than in patients with an unfavorable polymorphism and weaker treatment response. Importantly, the model identified a critical value of ϕ_1 below which PR treatment would succeed in clearing HCV. The interferon activity required to lower the fraction of interferon-refractory cells below this critical value is similar to the critical interferon effectiveness defined by the standard model of HCV kinetics.

Paradoxically, studies have found that the pretreatment expression levels of ISGs in the livers of nonresponders to PR were higher than in the responders (Sarasin-Filipowicz et al., 2008; Lau et al., 2013; Feld et al., 2007). Furthermore, PR treatment induced a significant increase in ISG expression in the responders but not in the nonresponders (Sarasin-Filipowicz et al., 2008; Lau et al., 2013; Feld et al., 2007). The model provided a plausible explanation of this phenomenon. According to the model, rapid responders to PR treatment are predicted to have low endogenous interferon production and potent ISG effector function. HCV establishes persistent infection due to insufficient endogenous interferon stimulation. Exogenous interferon addition triggers ISG expression, and due to robust ISG effector function, clears HCV. In these individuals, ϕ_1 is low, and the viral load declines rapidly in a biphasic manner resulting in HCV clearance following PR treatment. Nonresponders, on the other hand, are predicted to have high endogenous interferon production and poor ISG effectiveness, and infection is established despite strong interferon stimulation and ISG expression. Interferon treatment does not induce any further ISG expression and fails to clear HCV. In such individuals, ϕ_1 is high, and there is little or no change in the viral load during PR treatment. Partial responders are those with low endogenous interferon levels and ISG effector function. Interferon addition might increase the ISG expression levels, but the therapy fails due to inadequate ISG effectiveness. In these individuals, ϕ_1 is smaller than in nonresponders but greater than the critical value, and hence, the viral load decreases but HCV is not cleared during PR treatment. Individuals with both high endogenous interferon production and ISG effectiveness are likely to spontaneously clear HCV infection. The adaptive immune response has also been implicated in the spontaneous clearance of HCV (Thimme et al., 2012). We return to discuss this phenomenon in detail in the section titled, 'Potential considerations and strategies for optimizing DAA treatments.'

Overall, the model provided a new conceptual understanding of HCV persistence and response to PR treatment. It also identified a new mechanism of synergy between DAAs and PR. DAAs target specific HCV proteins and reduce HCV replication rates, which, in turn, lowers the control of HCV on interferon action. Consequently, DAAs could potentially convert interferon-refractory cells into interferon-sensitive cells, thereby synergizing with interferons to clear HCV. Whether this mechanism underlies the synergy between DAAs and interferon observed in vitro and in vivo (Ghany et al., 2011; Padmanabhan and Dixit, 2015a; Welsch et al., 2012; McHutchison et al., 2010) and whether this mode of synergy can be exploited to improve treatments remains to be tested. Nonetheless, interferon responsiveness has been argued to have a significant influence on DAA-based treatment outcomes. We consider the latter arguments next.

Leveraging endogenous interferon responsiveness to improve DAA treatments

HCV has a high mutation rate and can rapidly develop resistance against individual DAAs (Ribeiro et al., 2012). For DAA treatments to succeed, the growth of drug-resistant strains, termed as resistance-associated amino acid variants (RAVs), during treatment must be controlled. An effective endogenous interferon response might limit the replication space available for the growth of RAVs and improve DAA treatments. Direct measurements of the interferon responsiveness in the liver and correlating it with DAA treatment outcomes have been challenging. As discussed earlier, nonresponders to PR are expected to have weaker endogenous interferon responsiveness than treatment naïve individuals. If endogenous interferons were to control the growth of RAVs and influence DAA treatment outcomes, one would expect HCV cure rates to be lower in a population of prior nonresponders to PR than in a population of treatment-naïve individuals treated with the same drug combinations. To investigate such a relationship between interferon responsiveness and DAA treatment outcomes, a recent study collated data from >50 clinical trials and found that HCV cure rates were significantly higher in treatment-naïve individuals than in previous null responders to PR (Venugopal et al., 2018) (Fig. 6.3). The difference was significant when both interferon-free and interferon-based treatments were analyzed separately and even when difficult-to-treat cirrhotic patients alone were considered. The difference vanished when potent DAA combinations that elicit ~100% cure were used (Venugopal et al., 2018). Nevertheless, these clinical studies provide strong evidence of a correlation between responsiveness to PR and DAAs.

To mechanistically describe the underlying correlation, the model of HCV kinetics with distinct cellular responses to

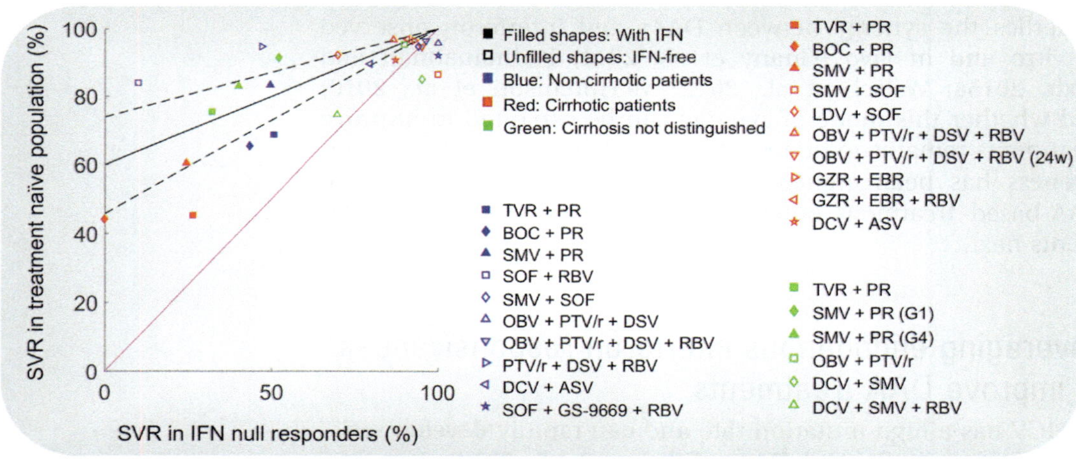

Figure 6.3 Relationship between interferon responsiveness and DAA treatment outcome. Correlation between SVR or cure rates elicited by different treatment regimens in treatment-naïve and prior null responders to interferon-based treatment. Comparison of model predictions *(black line)* and their confidence intervals *(dashed lines)* with data from clinical trials. The datasets were categorized depending on whether the treatment regimens contained interferon and whether the patients were cirrhotic or noncirrhotic. The pink line represents the x = y boundary, indicating the absence of any underlying relationship. Abbreviations for drugs: SOF, sofosbuvir; RB, ribavirin; BOC, boceprevir; PR, pegylated interferon and ribavirin; TVR, telaprevir; SMV, simeprevir; PTV/r, paritaprevir/ritonavir; DSV, dasabuvir; DCV,- daclatasvir; ASV, asunaprevir; BCV, beclabuvir; LDV, ledipasvir; OBV, ombitasvir; GZR, grazoprevir; EBR, elbasvir; GS-9669, radalbuvir.

interferon was extended to incorporate within-host viral evolution (Venugopal et al., 2018). The modified model divided the viral population into two subpopulations, one containing the wild-type viral strain and the other a mutant strain carrying an RAV. Due to the fitness penalties caused by the resistance mutation, the infection rates and/or the production rates of the mutant were assumed to be lower than the wild-type. Infected cells produced both the wild-type and mutant strains, albeit with different rates depending on the infecting strain and the mutation rate. Cells infected with the wild-type predominantly produced wild-type progeny virions, whereas cells infected with the mutant produced a majority of mutant progeny virions. DAA treatments exerted higher antiviral effectiveness against the wild-type than the mutants. The model could thus examine the dependence of the pretreatment RAV frequencies and the growth rates of RAVs during DAA treatments on endogenous interferon responsiveness.

The model predicted that the fraction of interferon-refractory cells, which depends on the endogenous levels of interferon in infected individuals, did not affect the pretreatment frequencies of RAVs. The absolute abundance of infected cells and viral load, however, increased with the interferon-refractory fraction. For fixed values of DAA effectiveness against the wild-type and mutant strains, the model identified a critical fraction of interferon-refractory cells below which DAA treatments succeeded and above which they failed. This is because larger the fraction of interferon-refractory cells, higher the viral and infected cell population and, hence, greater the chance of the emergence and growth of RAVs causing treatment failure. These predictions support the notion that stronger endogenous interferon responsiveness improves interferon-free, DAA-based treatment outcomes.

To examine whether interferon responsiveness could predict cure rates of different DAA treatments, a model of interferon response at the population level was developed (Venugopal et al., 2018). Here, the fraction of interferon-refractory cells was assumed to follow a log-normal distribution across infected individuals. By comparing model predictions with the measured distribution of baseline viral load and cure rates in a subset of clinical trials, the parameters of the distribution were estimated. The model then quantitatively described several clinical observations: (1) the percentage of individuals who spontaneously cleared HCV; (2) the percentage of HCV-infected individual displaying null response to PR; (3) the percentage of nonresponders to PR who responded to subsequent treatment with a combination of PR and a protease inhibitor; and (4) the surprising relationship between cures rates in treatment-naïve and treatment-experienced patients subjected to different treatment regimens (Fig. 6.3). The multiscale model thus presented a comprehensive picture of HCV infection and treatment response and revealed new strategies to improve treatments. Given the prediction that the minimum treatment duration depends on interferon responsiveness, one potential approach would be to estimate the interferon responsiveness from the initial viral kinetics and use the measurement for determining the optimal treatment duration.

The multiscale models described so far, though complex, considered processes that are essential to a conceptual understanding of interferon-based and DAA-based treatment outcomes. Building a quantitative framework for treatment optimization, however, would require integration of additional phenomena such as the evolutionary dynamics of RAVs, the synergy between drugs, and the potential roles of the adaptive immune response in controlling HCV infection during DAA treatments. Recent studies have attempted to integrate viral evolution and adaptive immune responses with models of HCV kinetics, presenting new avenues of treatment optimization. We briefly discuss these studies next.

Potential considerations and strategies for optimizing DAA treatments

Mutational pathways of resistance to DAA-based treatments

Preexisting RAVs are a major cause of DAA treatment failures (Horner and Naggie, 2015; Perales et al., 2015). Estimating the prevalence of RAVs is, therefore, essential for identifying optimal DAA combinations. Current resistance testing assays can detect preexisting frequencies of RAVs only up to $\sim 0.1\%$ (Mangul et al., 2014). The baseline viral load in chronically HCV-infected individual is $\sim 10^6$ copies/mL (Nainan et al., 2006). Thus, with current assays, RAVs with abundances <1000 copies/mL would remain undetected, and yet they could induce treatment failure. Mathematical models provide an alternative means by which preexisting frequencies of RAVs could be estimated (Padmanabhan and Dixit, 2015b). Modeling within-host HCV evolution and estimating the prevalence of RAVs, however, poses two main challenges. First, HCV evolution and selection happen both at the intracellular and extracellular levels, albeit over different time scales, and the integration of evolutionary processes occurring at these two levels has been difficult. Second, while the positions conferring resistance to DAAs are well known, different mutations at the same positions could lead to DAA resistance. For instance, several mutations R155K/I/G/M/T/Q/C/W/N at the position 155 on the NS3 gene could yield resistance to proteases inhibitors (Sarrazin, 2016). New modeling frameworks are therefore needed to quantify the prevalence of the entire spectrum of RAVs in infected individuals.

A recent study overcame these barriers and developed a hybrid, multiscale model that integrated stochastic intracellular HCV replication and evolution processes with deterministic extracellular HCV kinetics (Raja et al., 2019) (Fig. 6.4A). Interestingly, the stochastic simulations of the intracellular model predicted that the RNA population within infected cells was dominated by the infecting strain and not the fittest strain, indicating that founder effects dictate viral evolution at the intracellular level. A comprehensive set of simulations was performed with every possible genomic variant as the infecting strain to predict all possible "input–output relationships" for individual cells. The latter relationships were then coupled to a deterministic model of extracellular HCV kinetics. This integrated framework allowed estimation of the frequencies of RAVs that could be present in chronically infected individuals.

First, the model quantitatively captured the preexisting frequency of the RAV R155K associated with NS3/4A protease inhibitors estimated from the public HCV sequence database (Bartels et al., 2008), thus presenting a successful validation of the model. Next, the model explained the confounding clinical observation of why the prevalence of the RAV R155K was lower in individuals infected with HCV genotype 1b than those infected with genotype 1a (Horner and Naggie, 2015). Because the wild-type codon at the position 155 is different between genotypes 1a and 1b, the maps of mutational pathways accessible to these genotypes are different; genotype 1a requires a single transition whereas genotype 1b requires a transversion and then a transition to generate the R155K mutant. Consequently, the model predicted the frequency of the R155K to be $\sim$100-fold lower in HCV genotype 1b infection than in genotype 1a infection. Finally, using an experimentally measured fitness landscape (Qi et al., 2014), the model predicted the entire spectrum of mutants at the position 93 in the NS5A protein. A key prediction of the model is that 13 different RAVs encoded by 44 different codons are expected to pre-exist at levels well below

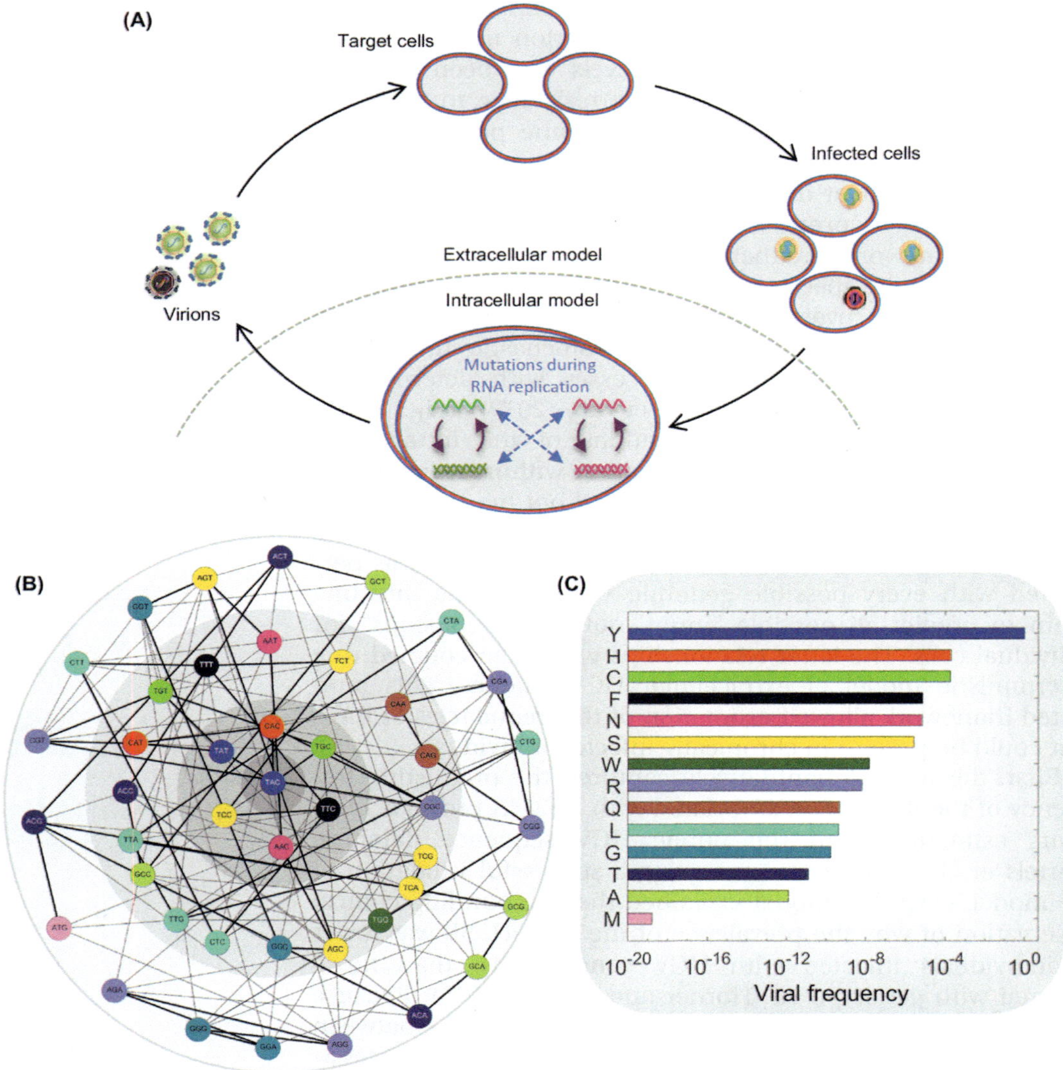

Figure 6.4 Prevalence of resistance-associated RAVs associated with NS5A inhibitors. (A) A schematic of a multiscale model where the target cells are infected with HCV virions to produce infected cells, within which HCV RNA replicates and mutates to produce new virions containing wild-type or mutant genomes. The dashed line highlights the separation of scales between intracellular and extracellular evolutionary dynamics. (B) Model prediction of the map of mutational pathways, with solid and dashed lines indicating transitions and transversions, respectively, for resistance to NS5A inhibitors at position 93 of the NS5A protein. Lighter shades of gray indicate an increasing number of mutations from the founder strain TAC. (C) Model predictions of the frequencies of the different mutants at the position 93 of the NS5A protein containing the wild-type amino acid Y.

the current detection limits (Fig. 6.4B-C), explaining why RAVs associated with NS5A inhibitors are not detected before treatment but can grow rapidly during treatment causing treatment failure. The model thus provides a framework for future studies to identify potential pathways of the growth of preexisting RAVs during treatments and design drug combinations to block these pathways and optimize DAA treatments.

Posttreatment cure

Several recent studies have made a puzzling observation, termed end of treatment (EOT^+)/sustained virological response (SVR), that in some patients subjected to DAA therapies, the viral load was detectable at the EOT, and yet HCV was cleared posttreatment without any additional intervention (Sidharthan et al., 2015). Understanding this phenomenon holds promise in identifying ways to optimize DAA treatment durations. One hypothesis, termed the virologic hypothesis, argues that DAA treatment leaves viral fitness compromised (Goyal et al., 2017; Nguyen et al., 2017). The other, perhaps equally if not more plausible, hypothesis, termed the immunologic hypothesis, argues that DAA treatments reverse HCV-induced progressive loss of the function of cytotoxic T lymphocytes (CTLs), termed CTL exhaustion, thereby facilitating immune clearance of viremia present at the EOT (Kohli et al., 2015; Sidharthan et al., 2015; Malespin et al., 2017). To test whether the latter hypothesis is consistent with the kinetics of viral load decline in patients who experienced EOT^+/SVR, a recent study developed a model of HCV kinetics incorporating CTL dynamics (Baral et al., 2018) (Fig. 6.5A).

The population dynamics of uninfected cells, HCV-infected cells, and free virions was described following the standard model of HCV kinetics (see the subsection titled 'Interferon-based treatment outcome'). Additionally, HCV-specific CTLs were assumed to be produced and die with fixed per capita rates. They were activated by infected cells, which were, in turn, killed by CTLs. Infected cells also suppressed CTLs by triggering their exhaustion. DAAs lowered viral production rates during treatment, and DAA activity ceased after the EOT. The model could thus predict the viral load decline during and after DAA therapy. Following DAA treatment, the viral load declined in a biphasic manner, as observed in patients and predicted by the standard model (Fig. 6.5B). After the EOT, viral production from infected cells was restored, resulting in a surge in viremia. During treatment, the reduction in viremia lowered the rate of CTL exhaustion and improved CTL function. Consequently,

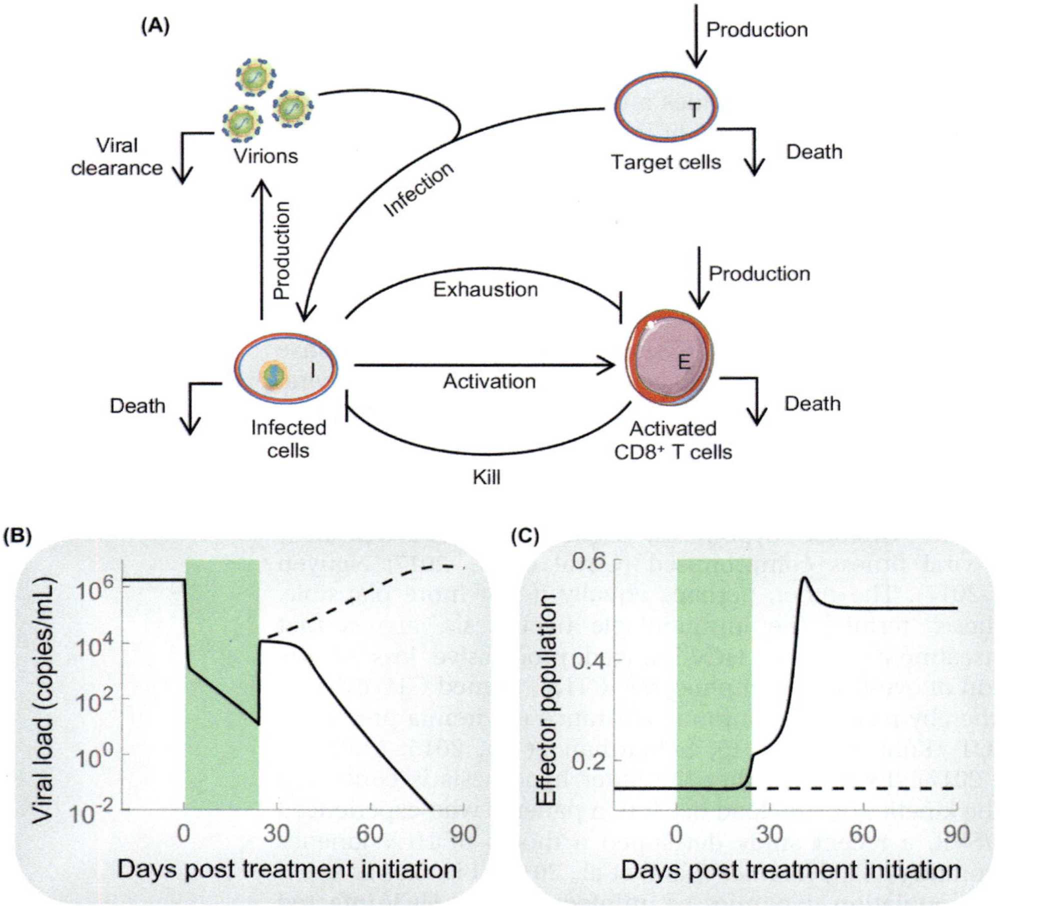

Figure 6.5 Model of viral kinetics based on the immunological hypothesis. (A) A schematic of a model of HCV viral kinetics incorporating CTL dynamics (see text). Model predictions of the time evolution of (B) the viral load and (C) the effector T cell population during and after DAA treatments with *(solid lines)* and without *(dashed lines)* the immunological hypothesis. The DAA treatment duration is shaded in *green*.

CTLs could continue to control HCV past the EOT, eventually further reducing the rate of exhaustion, and tilting the competition between infected cell-mediated CTL activation and exhaustion toward activation. The CTL population thus rose sharply, causing a continuous viral load decline and eventually clearing HCV (Fig. 6.5C). The model without the reversal of CTL exhaustion leads to DAA treatment failure, providing further support for the immunological hypothesis (Fig. 6.5B–C).

The model, importantly, identified a bistability in the underlying dynamical system as determining EOT^+/SVR. According to the model, when the effector strength of CTLs was higher in HCV-infected individuals, the system was monostable, admitting only the state where HCV is cleared. Such individuals are likely to clear HCV infection spontaneously (also see the section on the natural outcomes of HCV infection below). When the strength of CTLs was weak, the system was again monostable, now allowing only the state where HCV establishes chronic infection. These individuals are likely to experience treatment failure if viremia is detected at the EOT and would require DAA treatments until HCV is completely cured. With intermediate CTL strength, the system exhibited bistability and could realize both the stable steady states marking chronic infection and viral clearance. The two stable states are separated by an unstable steady state with intermediate HCV and CTL levels. Here, DAA treatments can achieve EOT^{+}/SVR by driving the system sufficiently past the intermediate unstable steady state such that CTL exhaustion is adequately reversed, and HCV is eventually cleared. Treatment duration can be much shorter in these individuals. The model quantitatively captured viral load decline in patients who experienced EOT^{+}/SVR, presenting a framework to estimate the minimum DAA treatment durations to cure HCV.

Natural outcomes of HCV infection

The explanation above of cure post the EOT suggests that spontaneous clearance of HCV, seen in nearly a quarter of infected individuals, may be due to a strong relative CTL response. At the same time, the studies on interferon responsiveness in the earlier sections argued that the spontaneous clearance of HCV infection may be due to strong interferon responses. Remarkably, a recent study synthesizes both these arguments. The study proposed a general dynamical motif capturing the essential interactions between antigens and CTLs to determine whether infected individuals spontaneously clear an infection or establish long-term chronic infection (Baral et al., Submitted). Antigens exert opposing effects on CTLs. On the one hand, antigen stimulation activates CTLs, which, in turn, kill infected cells. On the other hand, sustained antigen stimulation could lead to CTL exhaustion and subsequent dysfunction. The dynamical motif comprising these competing interactions could govern the different outcomes of infection. A minimal mathematical model of the dynamical motif exhibited bistability, with one state marking viral clearance and the other representing viral persistence, and recapitulated the observed distinct outcomes of viral infections.

In the context of HCV infection, the dynamical motif was expanded to include the innate immune response (Fig. 6.6A).

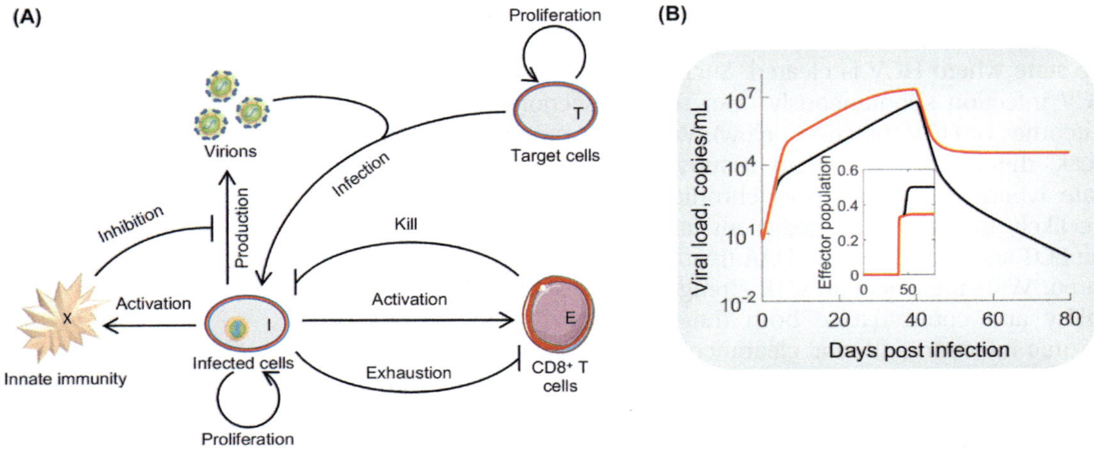

Figure 6.6 Dynamical network motif underlying the natural outcomes of HCV infection. (A) A schematic of a model of HCV viral kinetics incorporating CTL and innate immune responses (see text). (B) Model prediction of the time evolution of HCV and CTL *(inset)* levels for weak *(red)* and strong *(black)* innate immune response.

The model predicted that when the innate immune response, including the interferon response, of an individual was weak HCV established chronic infection, whereas the infection was spontaneously cleared with a stronger innate immune response. In agreement with the acute phase viral kinetics observed in chimpanzees infected with HCV, the model predicted a rapid initial rise followed by a slower increase in viremia to a peak level and then a decline in viremia during acute infection (Fig. 6.6B). The model attributed the arrest of the initial rise to the strength of innate immune response and the subsequent decline in viremia to the CTL response. A strong innate immune response curtailed the initial rise in viremia, preventing T cell exhaustion, which then allowed CTLs to effect a cure akin to the aforementioned posttreatment cure. The combined strengths of the innate immune and CTL responses therefore determined whether chronic infection or viral clearance was realized in the model (Fig. 6.6B). The dynamical motif thus presents a new conceptual understanding of the variable outcomes of HCV infection, potentially enabling tuning of the outcomes using interventions in order to achieve viral clearance.

Concluding remarks

Modeling HCV kinetics within infected individuals has over the past two decades contributed enormously to our understanding of HCV pathogenesis and treatment response. The beginning was with the standard model of HCV kinetics, which identified the origins of viral decline patterns found in patients on treatment and estimated key viral kinetic parameters (Neumann et al., 1998). Today, with a much more sophisticated understanding of the viral life cycle and the mechanisms of drug action, complex, multiscale models that bring us closer to a comprehensive framework for treatment optimization are being constructed. For instance, recent models have integrated the action of interferon at the molecular, cellular, and population level to describe interferon-based and DAA-based treatment outcomes quantitatively (Padmanabhan et al., 2014; Raja et al., 2018; Venugopal et al., 2018). These models identified a causal relationship between interferon responsiveness and DAA treatment success, thus presenting promising avenues to leverage endogenous interferon responsiveness for personalizing DAA treatments. Models have also been expanded to explore the possibility of the reversal of HCV-mediated CTL exhaustion during DAA treatments and optimize treatment durations (Baral et al., 2018). Both the latter studies have unraveled phenotypic heterogeneities at the level of intracellular and intercellular interaction networks as defining disease progression and treatment outcomes. These and other such conceptual advances present the necessary ingredients for future studies to develop reliable in silico platforms for the evaluation of different DAA combinations and the identification of optimal DAA treatment protocols. We temper this optimism, however, with a cautionary note. While DAAs indeed elicit rapid cure of HCV infection, their long-term effects need to be established. A case in point is the ongoing debate on whether DAA treatments have an impact on the development of hepatocellular carcinoma (HCC) in HCV-infected individuals who successfully clear the infection (Llovet and Villanueva, 2016; El Kassas et al., 2019). As mechanistic links between DAA treatments and progression to HCC become clearer, future studies would have to expand the above frameworks to incorporate the latter links, with the goal of developing guidelines not just for expediting cure but also for ensuring the long-term well-being of HCV patients.

Acknowledgments

This work was supported by the Welcome Trust/DBT India Alliance Senior Fellowship IA/S/14/1/501307.

References

Adiwijaya, B.S., Herrmann, E., Hare, B., Kieffer, T., Lin, C., Kwong, A.D., et al., 2010. A multi-variant, viral dynamic model of genotype 1 HCV to assess the *in vivo* evolution of protease-inhibitor resistant variants. PLoS Comput. Biol. 6 (4), e1000745.

Aghemo, A., Colombo, M., 2017. Response-guided duration of direct acting antiviral therapy for chronic hepatitis C: back to the future? Gastroenterology 152 (5), 1238–1239.

Baral, S., Antia, R., Dixit, N.M., 2019. A dynamical motif comprising the interactions between antigens and CD8 T cells may underlie the outcomes of viral infections. Proc. Natl. Acad. Sci. U. S. A. 116 (35), 17393–17398.

Baral, S., Roy, R., Dixit, N.M., 2018. Modeling how reversal of immune exhaustion elicits cure of chronic hepatitis C after the end of treatment with direct-acting antiviral agents. Immunol. Cell Biol. 96 (9), 969–980.

Bartels, D.J., Zhou, Y., Zhang, E.Z., Marcial, M., Byrn, R.A., Pfeiffer, T., et al., 2008. Natural prevalence of hepatitis C virus variants with decreased sensitivity to NS3.4A protease inhibitors in treatment-naive subjects. J. Infect. Dis. 198 (6), 800–807.

Cole, J.L., 2007. Activation of PKR: an open and shut case? Trends Biochem. Sci. 32 (2), 57–62.

Conway, J.M., Perelson, A.S., 2014. A hepatitis C virus infection model with time-varying drug effectiveness: solution and analysis. PLoS Comput. Biol. 10 (8), e1003769.

Dabo, S., Meurs, E.F., 2012. dsRNA-dependent protein kinase PKR and its role in stress, signaling and HCV infection. Viruses 4 (11), 2598–2635.

Dahari, H., Ribeiro, R.M., Perelson, A.S., 2007a. Triphasic decline of hepatitis C virus RNA during antiviral therapy. Hepatology 46 (1), 16–21.

Dahari, H., Ribeiro, R.M., Rice, C.M., Perelson, A.S., 2007b. Mathematical modeling of subgenomic hepatitis C virus replication in Huh-7 cells. J. Virol. 81 (2), 750–760.

Dahari, H., Sainz Jr., B., Perelson, A.S., Uprichard, S.L., 2009. Modeling subgenomic hepatitis C virus RNA kinetics during treatment with alpha interferon. J. Virol. 83 (13), 6383–6390.

Dixit, N.M., Layden-Almer, J.E., Layden, T.J., Perelson, A.S., 2004. Modelling how ribavirin improves interferon response rates in hepatitis C virus infection. Nature 432 (7019), 922–924.

El Kassas, M., Elbaz, T., Salaheldin, M., Abdelsalam, L., Kaseb, A., Esmat, G., 2019. Impact of treating chronic hepatitis C infection with direct-acting antivirals on the risk of hepatocellular carcinoma: the debate continues - A mini-review. J. Adv. Res. 17, 43–48.

Feld, J.J., Nanda, S., Huang, Y., Chen, W., Cam, M., Pusek, S.N., et al., 2007. Hepatic gene expression during treatment with peginterferon and ribavirin: identifying molecular pathways for treatment response. Hepatology 46 (5), 1548–1563.

Garaigorta, U., Chisari, F.V., 2009. Hepatitis C virus blocks interferon effector function by inducing protein kinase R phosphorylation. Cell Host Microbe 6 (6), 513–522.

Ge, D., Fellay, J., Thompson, A.J., Simon, J.S., Shianna, K.V., Urban, T.J., et al., 2009. Genetic variation in IL28B predicts hepatitis C treatment-induced viral clearance. Nature 461, 399–401.

Ghany, M.G., Nelson, D.R., Strader, D.B., Thomas, D.L., Seeff, L.B., 2011. An update on treatment of genotype 1 chronic hepatitis C virus infection: 2011

practice guideline by the American Association for the Study of Liver Diseases. Hepatology 54 (4), 1433–1444.
Götte, M., Feld, J.J., 2016. Direct-acting antiviral agents for hepatitis C: structural and mechanistic insights. Nat. Rev. Gastroenterol. Hepatol. 13 (6), 338.
Goyal, A., Lurie, Y., Meissner, E.G., Major, M., Sansone, N., Uprichard, S.L., et al., 2017. Modeling HCV cure after an ultra-short duration of therapy with direct acting agents. Antivir. Res. 144, 281–285.
Guedj, J., Dahari, H., Rong, L., Sansone, N.D., Nettles, R.E., Cotler, S.J., et al., 2013a. Modeling shows that the NS5A inhibitor daclatasvir has two modes of action and yields a shorter estimate of the hepatitis C virus half-life. Proc. Natl Acad. Sci. USA 110 (10), 3991–3996.
Guedj, J., Dahari, H., Shudo, E., Smith, P., Perelson, A.S., 2013b. Hepatitis C viral kinetics with the nucleoside polymerase inhibitor mericitabine (RG7128). Hepatology 55 (4), 1030–1037.
Hajarizadeh, B., Grebely, J., Dore, G.J., 2013. Epidemiology and natural history of HCV infection. Nat. Rev. Gastroenterol. Hepatol. 10 (9), 553–562.
Hayes, C.N., Imamura, M., Aikata, H., Chayama, K., 2012. Genetics of IL28B and HCV-response to infection and treatment. Nat. Rev. Gastroenterol. Hepatol. 9 (7), 406–417.
Heim, M.H., 2013a. 25 years of interferon-based treatment of chronic hepatitis C: an epoch coming to an end. Nat. Rev. Immunol. 13 (7), 535.
Heim, M.H., 2013b. Innate immunity and HCV. J. Hepatol. 58, 564–574.
Helbig, K.J., Eyre, N.S., Yip, E., Narayana, S., Li, K., Fiches, G., et al., 2011. The antiviral protein viperin inhibits hepatitis C virus replication via interaction with nonstructural protein 5A. Hepatology 54 (5), 1506–1517.
Horner, S.M., Naggie, S., 2015. Successes and challenges on the road to cure hepatitis C. PLoS Pathog. 11 (6), e1004854.
Keum, S.J., Park, S.M., Park, J.H., Jung, J.H., Shin, E.J., Jang, S.K., 2012. The specific infectivity of hepatitis C virus changes through its life cycle. Virology 433 (2), 462–470.
Koev, G., Duncan, R.F., Lai, M.M., 2002. Hepatitis C virus IRES-dependent translation is insensitive to an eIF2alpha-independent mechanism of inhibition by interferon in hepatocyte cell lines. Virology 297 (2), 195–202.
Kohli, A., Osinusi, A., Sims, Z., Nelson, A., Meissner, E.G., Barrett, L.L., et al., 2015. Virological response after 6 week triple-drug regimens for hepatitis C: a proof-of-concept phase 2A cohort study. Lancet 385 (9973), 1107–1113.
Lau, D.T., Negash, A., Chen, J., Crochet, N., Sinha, M., Zhang, Y., et al., 2013. Innate immune tolerance and the role of Kupffer cells in differential responses to interferon therapy among patients with HCV genotype 1 infection. Gastroenterology 144, 402–413.
Lavanchy, D., 2011. Evolving epidemiology of hepatitis C virus. Clin. Microbiol. Infect. 17 (2), 107–115.
Levin, D., Harari, D., Schreiber, G., 2011. Stochastic receptor expression determines cell fate upon interferon treatment. Mol. Cell Biol. 31 (16), 3252–3266.
Llovet, J.M., Villanueva, A., 2016. Liver cancer: Effect of HCV clearance with direct-acting antiviral agents on HCC. Nat. Rev. Gastroenterol. Hepatol. 13 (10), 561–562.
Malespin, M., Benyashvili, T., Uprichard, S.L., Perelson, A.S., Dahari, H., Cotler, S.J., 2017. Prevalence of end of treatment RNA-positive/sustained viral response in HCV patients treated with sofosbuvir combination therapies. Ther. Adv. Gastroenterol. 10 (1), 68–73.

Mangul, S., Wu, N.C., Mancuso, N., Zelikovsky, A., Sun, R., Eskin, E., 2014. Accurate viral population assembly from ultra-deep sequencing data. Bioinformatics 30 (12), i329–i337.

Martinez, F., Sardanyes, J., Elena, S.F., Daros, J.A., 2011. Dynamics of a plant RNA virus intracellular accumulation: stamping machine vs. geometric replication. Genetics 188 (3), 637–646.

McHutchison, J.G., Manns, M.P., Muir, A.J., Terrault, N.A., Jacobson, I.M., Afdhal, N.H., et al., 2010. Telaprevir for previously treated chronic HCV infection. N. Engl. J. Med. 362 (14), 1292–1303.

Metz, P., Dazert, E., Ruggieri, A., Mazur, J., Kaderali, L., Kaul, A., et al., 2012. Identification of type I and type II interferon-induced effectors controlling hepatitis C virus replication. Hepatology 56 (6), 2082–2093.

Metz, P., Reuter, A., Bender, S., Bartenschlager, R., 2013. Interferon-stimulated genes and their role in controlling hepatitis C virus. J. Hepatol. 59, 1331–1341.

Micallef, J.M., Kaldor, J.M., Dore, G.J., 2006. Spontaneous viral clearance following acute hepatitis C infection: a systematic review of longitudinal studies. J. Viral Hepat. 13 (1), 34–41.

Nainan, O.V., Alter, M.J., Kruszon-Moran, D., Gao, F.-X., Xia, G., McQuillan, G., et al., 2006. Hepatitis C virus genotypes and viral concentrations in participants of a general population survey in the United States. Gastroenterology 131 (2), 478–484.

Neumann, A.U., Lam, N.P., Dahari, H., Gretch, D.R., Wiley, T.E., Layden, T.J., et al., 1998. Hepatitis C viral dynamics *in vivo* and the antiviral efficacy of interferon-alpha therapy. Science 282 (5386), 103–107.

Nguyen, T.H.T., Guedj, J., Uprichard, S.L., Kohli, A., Kottilil, S., Perelson, A.S., 2017. The paradox of highly effective sofosbuvir-based combination therapy despite slow viral decline: can we still rely on viral kinetics? Sci. Rep. 7 (1), 10233.

Ohashi, H., Koizumi, Y., Fukano, K., Wakita, T., Perelson, A.S., Iwami, S., et al., 2017. Reply to Padmanabhan and Dixit: Hepatitis C virus entry inhibitors for optimally boosting direct-acting antiviral-based treatments. Proc. Natl Acad. Sci. U S A 114 (23), E4527–E4529.

Padmanabhan, P., Dixit, N.M., 2011. Mathematical model of viral kinetics *in vitro* estimates the number of E2-CD81 complexes necessary for hepatitis C virus entry. PLoS Comput. Biol. 7 (12), e1002307.

Padmanabhan, P., Dixit, N.M., 2012. Viral kinetics suggests a reconciliation of the disparate observations of the modulation of claudin-1 expression on cells exposed to hepatitis C virus. PLoS One 7 (4), e36107.

Padmanabhan, P., Dixit, N., 2015a. Modeling suggests a mechanism of synergy between hepatitis C virus entry inhibitors and drugs of other classes. CPT: Pharmacomet. Syst. Pharmacol. 4 (8), 445–453.

Padmanabhan, P., Dixit, N.M., 2015b. Models of viral population dynamics. Quasispecies: From Theory to Experimental Systems. Springer, pp. 277–302.

Padmanabhan, P., Dixit, N.M., 2017. Inhibitors of hepatitis C virus entry may be potent ingredients of optimal drug combinations. Proc. Natl Acad. Sci. U S A 114 (23), E4524–E4526.

Padmanabhan, P., Garaigorta, U., Dixit, N.M., 2014. Emergent properties of the interferon-signalling network may underlie the success of hepatitis C treatment. Nat. Commun. 5, 3872.

Perales, C., Quer, J., Gregori, J., Esteban, J., Domingo, E., 2015. Resistance of hepatitis C virus to inhibitors: complexity and clinical implications. Viruses 7 (11), 5746–5766.

Perelson, A.S., Guedj, J., 2015. Modelling hepatitis C therapy—predicting effects of treatment. Nat. Rev. Gastroenterol. Hepatol. 12 (8), 437.

Powers, K.A., Dixit, N.M., Ribeiro, R.M., Golia, P., Talal, A.H., Perelson, A.S., 2003. Modeling viral and drug kinetics: hepatitis C virus treatment with pegylated interferon alfa-2b. Semin. Liver Dis. 23 (Suppl 1), 13–18.

Proud, C.G., 2005. eIF2 and the control of cell physiology. Semin. Cell Dev. Biol. 16 (1), 3–12.

Qi, H., Olson, C.A., Wu, N.C., Ke, R., Loverdo, C., Chu, V., et al., 2014. A quantitative high-resolution genetic profile rapidly identifies sequence determinants of hepatitis C viral fitness and drug sensitivity. PLoS Pathog. 10 (4), e1004064.

Raja, R., Baral, S., Dixit, N.M., 2018. Interferon at the cellular, individual, and population level in hepatitis C virus infection: Its role in the interferon-free treatment era. Immunol. Rev. 285 (1), 55–71.

Raja, R., Pareek, A., Newar, K., Dixit, N.M., 2019. Mutational pathway maps and founder effects define the within-host spectrum of hepatitis C virus mutants resistant to drugs. PLoS Pathog. 15 (4), e1007701.

Rand, U., Rinas, M., Schwerk, J., Nohren, G., Linnes, M., Kroger, A., et al., 2012. Multi-layered stochasticity and paracrine signal propagation shape the type-I interferon response. Mol. Syst. Biol. 8, 584.

Ribeiro, R.M., Li, H., Wang, S., Stoddard, M.B., Learn, G.H., Korber, B.T., et al., 2012. Quantifying the diversification of hepatitis C virus (HCV) during primary infection: estimates of the *in vivo* mutation rate. PLoS Pathog. 8 (8), e1002881.

Rong, L., Dahari, H., Ribeiro, R.M., Perelson, A.S., 2010. Rapid emergence of protease inhibitor resistance in hepatitis C virus. Sci. Transl. Med. 2 (30), 30ra2.

Rotman, Y., Noureddin, M., Feld, J.J., Guedj, J., Witthaus, M., Han, H., et al., 2013. Effect of ribavirin on viral kinetics and liver gene expression in chronic hepatitis C. Gut 63, 161–169.

Rowlands, A.G., Panniers, R., Henshaw, E.C., 1988. The catalytic mechanism of guanine nucleotide exchange factor action and competitive inhibition by phosphorylated eukaryotic initiation factor 2. J. Biol. Chem. 263 (12), 5526–5533.

Sarasin-Filipowicz, M., Oakeley, E.J., Duong, F.H., Christen, V., Terracciano, L., Filipowicz, W., et al., 2008. Interferon signaling and treatment outcome in chronic hepatitis C. Proc. Natl Acad. Sci. USA 105 (19), 7034–7039.

Sardanyes, J., Martinez, F., Daros, J.A., Elena, S.F., 2012. Dynamics of alternative modes of RNA replication for positive-sense RNA viruses. J. R. Soc. Interface 9, 768–776.

Sarrazin, C., 2016. The importance of resistance to direct antiviral drugs in HCV infection in clinical practice. J. Hepatol. 64 (2), 486–504.

Shimoike, T., McKenna, S.A., Lindhout, D.A., Puglisi, J.D., 2009. Translational insensitivity to potent activation of PKR by HCV IRES RNA. Antivir. Res. 83 (3), 228–237.

Sidharthan, S., Kohli, A., Sims, Z., Nelson, A., Osinusi, A., Masur, H., et al., 2015. Utility of hepatitis C viral load monitoring on direct-acting antiviral therapy. Clin. Infect. Dis. 60 (12), 1743–1751.

Thimme, R., Binder, M., Bartenschlager, R., 2012. Failure of innate and adaptive immune responses in controlling hepatitis C virus infection. FEMS Microbiol. Rev. 36 (3), 663–683.

Venugopal, V., Padmanabhan, P., Raja, R., Dixit, N.M., 2018. Modelling how responsiveness to interferon improves interferon-free treatment of hepatitis C virus infection. PLoS Comput. Biol. 14 (7), e1006335.

Welsch, C., Jesudian, A., Zeuzem, S., Jacobson, I., 2012. New direct-acting antiviral agents for the treatment of hepatitis C virus infection and perspectives. Gut 61 (Suppl), i36–i46.

WHO, 2016. Guidelines for the Screening Care and Treatment of Persons with Chronic Hepatitis C Infection. World Health Organization, Geneva.

Wieland, S., Makowska, Z., Campana, B., Calabrese, D., Dill, M.T., Chung, J., et al., 2014. Simultaneous detection of hepatitis C virus and interferon stimulated gene expression in infected human liver. Hepatology 59 (6), 2121–2130.

7

Quantifying Waddington landscapes, paths, and kinetics of cell fate decision making of differentiation/development

Li Xu[1] and Jin Wang[2]
[1]*State Key Laboratory of Electroanalytical Chemistry, Changchun Institute of Applied Chemistry, Chinese Academy of Sciences, Changchun, P.R. China*
[2]*Department of Chemistry and of Physics and Astronomy, State University of New York at Stony Brook, Stony Brook, NY, United States*

Introduction

Decision making is essential for cell functions and realized through the underlying cellular processes, such as signaling, gene regulation, the development and differentiation in viruses, bacteria, yeast, metazoans, and mammals (Balázsi et al., 2011; Ferrell, 2012). Cells under regulations, environments, and fluctuations can experience fate changes. Understanding the cell decision making for differentiation and reprograming, the biological process from a multipotent stem or progenitor state to a mature cell and back is crucial not only for uncovering its underlying mechanisms (Balázsi et al., 2011; Ferrell, 2012; Hood and Flores, 2012; Roeder and Glauche, 2006; Chickarmane and Peterson, 2008; Takahashi and Yamanaka, 2006; Yu et al., 2007; Takahashi et al., 2007; Saha and Jaenisch, 2009; Zhou and Melton, 2008; Zhou et al., 2008; Ieda et al., 2010; Chang et al., 2011; Gore et al., 2011; Hussein et al., 2011; Huang and Zhou, 2011; Lister et al., 2011; Zhao et al., 2011; Chia et al., 2010) but also for the practice of tissue regeneration (Hood and Flores, 2012; Takahashi and Yamanaka, 2006; Takahashi et al., 2007).

Stem cells appear in all the multicellular organisms. Stem cells themselves can divide and differentiate to a collection of distinct specialized cell types. Stem cells can self-renew and generate furthermore stem cells (Roeder and Glauche, 2006; Chickarmane and Peterson, 2008; Takahashi and Yamanaka,

Phenotypic Switching. DOI: https://doi.org/10.1016/B978-0-12-817996-3.00002-5

2006; Yu et al., 2007; Takahashi et al., 2007; Saha and Jaenisch, 2009; Zhou and Melton, 2008; Zhou et al., 2008; Ieda et al., 2010; Chang et al., 2011; Gore et al., 2011; Hussein et al., 2011; Huang and Zhou, 2011; Lister et al., 2011; Zhao et al., 2011; Chia et al., 2010). In mammals, two major types of stem cells emerge. One type of stem cells is the embryonic stem cells, which are isolated from the inner cell mass of blastocysts. The other type of stem cells is the adult stem cells, which appeared in different tissues. In adult organisms, stem cells and progenitor cells play the role of repairing for the body and refreshing adult tissues. In the development, for example, in a developing embryo, stem cells can differentiate to the specialized cells (called pluripotent cells). Moreover, stem cells also have the functions to maintain the normal turnover of regenerative organs, such as blood, skin, or intestinal tissues (Roeder and Glauche, 2006; Chickarmane and Peterson, 2008; Takahashi and Yamanaka, 2006; Yu et al., 2007; Takahashi et al., 2007; Saha and Jaenisch, 2009; Zhou and Melton, 2008; Zhou et al., 2008; Ieda et al., 2010; Chang et al., 2011; Gore et al., 2011; Hussein et al., 2011; Huang and Zhou, 2011; Lister et al., 2011; Zhao et al., 2011; Chia et al., 2010).

It is important to realize that the organisms are all originated from a single cell by differentiation. The differentiation process leads to different types of cells, illustrated as in Fig. 7.1. It is now believed that the expressions of the tissue specific genes are regulated by the underlying gene regulatory network (Roeder and Glauche, 2006; Chickarmane and Peterson, 2008; Takahashi and Yamanaka, 2006; Yu et al., 2007; Takahashi et al., 2007; Saha and Jaenisch, 2009; Zhou and Melton, 2008; Zhou

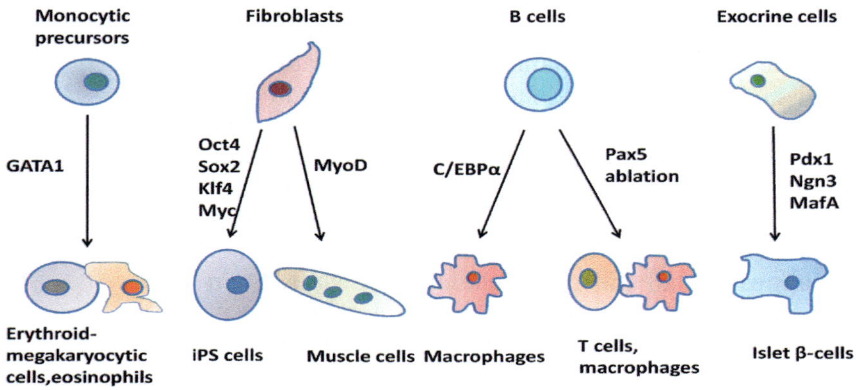

Figure 7.1 Different cells upon regulations experience fate changes.

et al., 2008; Ieda et al., 2010; Chang et al., 2011; Gore et al., 2011; Hussein et al., 2011; Huang and Zhou, 2011; Lister et al., 2011; Zhao et al., 2011; Chia et al., 2010). This shows the importance of the gene regulatory network for differentiation of the cell. The appearance of different cell types in the differentiation/development process is quite often believed to have a tree-like structure under a hierarchical organization, in analogy to the evolution branching of species (Roeder and Glauche, 2006; Chickarmane and Peterson, 2008; Takahashi and Yamanaka, 2006; Yu et al., 2007; Takahashi et al., 2007; Saha and Jaenisch, 2009; Zhou and Melton, 2008; Zhou et al., 2008; Ieda et al., 2010; Chang et al., 2011; Gore et al., 2011; Hussein et al., 2011; Huang and Zhou, 2011; Lister et al., 2011; Zhao et al., 2011; Chia et al., 2010). At the end of tree branching, the differentiated cells emerge. At the branch point, an undecided stem cell (pluripotent or multipotent) will need to decide to whether go for one way or another to differentiation. Thus, one can see that the decision-making process is essential for the stem cell differentiation. An integrated approach that relates the underlying gene regulatory network and the cell fate decision making is crucial for quantifying and understanding the cell differentiation and development.

The first attempt to understand such cell fate decision-making processes in the developmental context was made by Waddington (1957). In his seminar studies, Waddington proposed that the developmental process can be viewed as a ball rolling downhill from the top of the mountain landscape. Here the top of the mountain represents the stem cell state or fate while the states at the bottom of the mountain landscape represent the differentiated cell states or fates. Waddington landscape gives a qualitative picture of the developmental process. However, it lacks physical basis and quantitative understanding. Recently, there are efforts on quantifying the Waddington landscape for understanding the cell fate decision-making process for development based on the underlying gene regulatory networks (Balázsi et al., 2011; Ferrell, 2012; Hood and Flores, 2012; Roeder and Glauche, 2006; Chickarmane and Peterson, 2008; Takahashi and Yamanaka, 2006; Yu et al., 2007; Takahashi et al., 2007; Saha and Jaenisch, 2009; Zhou and Melton, 2008; Zhou et al., 2008, 2012; Ieda et al., 2010; Chang et al., 2011; Gore et al., 2011; Hussein et al., 2011; Huang and Zhou, 2011; Lister et al., 2011; Zhao et al., 2011; Chia et al., 2010; Wang et al., 2010b; Bhattacharya et al., 2011).

Despite significant efforts and progresses made in this field (Balázsi et al., 2011; Ferrell, 2012; Hood and Flores, 2012; Roeder and Glauche, 2006; Chickarmane and Peterson, 2008;

Takahashi and Yamanaka, 2006; Yu et al., 2007; Takahashi et al., 2007; Saha and Jaenisch, 2009; Zhou and Melton, 2008; Zhou et al., 2008; Ieda et al., 2010; Chang et al., 2011; Gore et al., 2011; Hussein et al., 2011; Huang and Zhou, 2011; Lister et al., 2011; Zhao et al., 2011; Chia et al., 2010), there are still major challenges in the cell fate decision making in differentiation and development: (1) How to define the cell states and cell fates quantitatively and how to uncover the physical mechanism and dynamics of decision making? (2) How to quantify the differentiation and reprograming paths for the cell fate decision-making process? (3) How to quantify the kinetic speed of the underlying cell fate switching or decision making in the development? (4) How to identify the key elements determining the cell fates? (5) What are the effects of epigenetics on differentiation and reprograming (Dovey et al., 2010; Singh et al., 2008; Xu et al., 2009; Judson et al., 2009; Li et al., 2011b; Subramanyam et al., 2011; Samavarchi-Tehrani et al., 2010; LipchinaI et al., 2011; Pfaff et al., 2011; Ye et al., 2012; Bao et al., 2013; Dvorak et al., 2005; Dvorak and Hampl, 2005; Ying et al., 2003; Xu et al., 2002; Nusse, 2008; Chiba, 2006)? (6) How is the study related to the reprograming and practice of tissue regeneration (Hood and Flores, 2012; Takahashi and Yamanaka, 2006; Takahashi et al., 2007)? Addressing these issues requires explicit quantifications of stem cell fate decision-making process and a solid physical theory as the foundation.

In this chapter, we will review the recent landscape and flux theory for quantifying the cell fate decision making of stem cell differentiation and reprograming. We also aim to clarify some critical issues and comment on the current understanding of the cell decision-making processes.

Potential and flux landscape theory of cell fate decision of differentiation and reprograming

Potential and flux as the driving force for stem cell differentiation and development

We will first review the landscape and flux theory of the nonequilibrium systems and networks (Wang et al., 2008, 2010a,b, 2011; Wang, 2015; Xu et al., 2014; Kim and Wang, 2007; Hornos et al., 2005; Yan and Wang, 2012; Yan et al., 2013; Hensel et al., 2012; Feng et al., 2012; Fang et al., 2018). This includes the identification of the cell states and the driving force for stem cell differentiation process from the state dynamics in terms of landscape and flux, quantifying kinetic paths for the cell differentiation process, and determining the speed of the differentiation and reprogramming processes.

The differentiation and development are tightly regulated by the underlying gene networks. To quantify the dynamics of cell differentiation and development, we first explore the dynamics of the associated gene networks. Let us identify the building blocks of the cell networks. The gene expression levels of each node can be used for the purpose. The dynamics of the underlying gene regulatory network of cell decision making can be formulated as (Lecca et al., 2013; Kruse and Elf, 2006):

$$\frac{dx}{dt} = F(x) \tag{7.1}$$

Here x or $\vec{x}$ is a vector, where $\vec{x} = \{x_1, x_2, x_3, \ldots x_n\}$ represents the expression levels of 1, 2, 3, ...nth genes and $F(x)$ or $\vec{F}(\vec{x}) = \{F_1(\vec{x}), F_2(\vec{x}), F_3(\vec{x}) \ldots F_n(\vec{x})\}$ represents the driving force characterizing the regulations among genes for the underlying gene network dynamics. Here, $\vec{x} = \{x_1, x_2, x_3, \ldots x_n\}$ represents a collection of certain expressions of each gene. $\vec{x}$ can be used to define a state of the gene network. If each gene can have q possible expression values, then there are q^N states in the whole state space of the gene network. The biological functioning states such as stem cell state, differentiated cell states, normal, disease states, etc. can then be characterized this way. The dynamics of the gene network determining the cell fate decision-making process for differentiation can be viewed as the time evolution from one state to another following the nonlinear dynamics (Lecca et al., 2013; Kruse and Elf, 2006).

In the conventional nonlinear dynamics, it is often useful to find the fixed points and explore its local stability (Jackson, 1990). This can provide some insights toward the preference of certain states or gene expression patterns. However, this local stability analysis cannot give the description of the global connection between local stable states, for example, how one can go from a stem cell state to a differentiated state (Jackson, 1990). One might think that the potential landscape approach in equilibrium statistical mechanics can help to address this issue. However, the underlying gene regulatory networks are in general not in equilibrium. This is due to the energy, material, and information exchange with the environments often in the form of the ATP hydrolysis, phosphorylation, etc. As a result, the system is not in equilibrium and the associated dynamics does not usually follow a potential landscape gradient. Therefore, the global approach using directly the equilibrium landscape ideas successfully applied to the biomolecular scale processes such as protein folding and binding does not work here for the cellular networks.

In general, the fluctuations from intrinsic and extrinsic sources are present (McAdams and Arkin, 1997; Thattai and van Oudenaarden, 2001; Ackers et al., 1982; Swain et al., 2002). The extrinsic fluctuations from the environments may play more important roles in cell fate decision making of differentiation and development in mammalian cells due to the relatively large number of molecular numbers. The intrinsic fluctuations can be considered in the form of concentration dependent fluctuations or multiplicative noise (Gillespie, 2000). The dynamics becomes stochastic when including the fluctuations: $dx/dt = F(x) + \eta$, where η is the stochastic force from the fluctuations of the intrinsic or extrinsic sources. The fluctuations can be quantified by the correlations where $< \eta(t)\ \eta(0) > = 2DD\delta(t)$. D is a scale factor quantifying the strength of the fluctuations. Following the individual stochastic evolution trajectory is not meaningful due to its unpredictability. On the other hand, the probability evolution of the underlying gene network obeys a Fokker-Planck diffusion equation, which is linear in the probability and thus predictable.

$$\frac{\partial P}{\partial t} = -\nabla \cdot J; \quad J = FP - D\nabla \cdot (DP) \tag{7.2}$$

The physical meaning is straightforward. The change of the probability P is equal to the flux J in and out. The flux J is determined by the driving force F of the gene network and diffusion from the fluctuations. In steady state, $\partial P/\partial t = 0$, divergence of flux is zero, $\nabla \cdot J = 0$.

In steady state, there are two cases worth discussion. One is the case when the flux J is zero. We can then define a potential function U where $U = -\ln P_{ss}$ and P_{ss} represents the steady-state probability. The driving force of the dynamics F can thus be represented as the gradient of the potential U, $-DD \cdot \nabla U$ and the gradient of the diffusion $D\nabla \cdot D$, which is related to the inhomogeneity of the fluctuations. The flux $J = 0$ means that there is no net in or out flux. This is the detailed balance condition where the system is in equilibrium. In an equilibrium system, the flux is zero and the system can be characterized by the potential U related to equilibrium probability $U = -\ln P_{eq}$, following the Boltzmann law. Here the P_{eq} represents the equilibrium probability. Moreover, the dynamics of a detailed balanced equilibrium system is determined by the gradient of the potential and the inhomogeneity of the fluctuations. Here, we uncovered the link between equilibrium systems and dynamical systems under the detailed balance (Wang et al., 2008, 2010a, 2011; Xu et al., 2014;

Kim and Wang, 2007; Hornos et al., 2005; Schultz et al., 2007; Kepler and Elston, 2001; Feng et al., 2011; Das et al., 2007; Singh and Weinberger, 2009; Choi et al., 2008; Sasai and Wolynes, 2003; Metzler and Wolynes, 2002).

In another case, the dynamics of the general gene networks in steady state does not necessarily satisfy the detailed balance condition due to the input or output of the energy, material, and information from or to the environment, which gives arise to $J_{ss} \neq 0$. In this case, we still can define a potential landscape U, which is related to the steady-state probability distribution as $U = -\ln P_{ss}$. Since the steady-state probability quantifies the weight of each state, the potential function can then be used to globally quantify the system. The driving force now can be decomposed into three terms, the gradient of the potential landscape U, $-DD\cdot\nabla U$; the steady-state flux related force J_{ss}/P_{ss}; and inhomogeneity of the fluctuations, $D\nabla\cdot D$. Since the flux is nonzero, the detailed balance condition is broken (Wang et al., 2008, 2010a, 2011; Xu et al., 2014; Kim and Wang, 2007; Hornos et al., 2005). The system is in nonequilibrium state even at steady state. Therefore, we can now establish a link between the general network dynamics for cell fate decision making and the degree of nonequilibriumness.

As stated, the global nature of the network can still be described by a potential landscape U. The dynamics however is not only determined by the gradient of the potential and inhomogeneity of the fluctuations as it is in the equilibrium systems. Instead, there is an additional contribution to the driving force originated from the nonzero flux which breaks the detailed balance. Because the flux is divergent free in steady state, the contribution to the driving force has a rotational nature. The rational is as follows. There can be no sources or sinks for the divergent free flux lines to go into and come out. The nonzero flux will be either constant or rotational in state space. The general network dynamics is thus closely linked to the nonequilibriumness characterized by the nonzero steady-state flux of the dynamical systems. In nonequilibrium systems with broken detailed balance, the global features can still be captured by the potential landscape in analogy to the equilibrium case, with the distinction that the potential landscape is related to the steady-state probability rather than equilibrium probability. However, distinctly different from the equilibrium dynamics, the driving force for the nonequilibrium dynamics is not only dictated by the gradient of the potential but also by the curl flux (Fig. 7.2). Therefore, the equilibrium dynamics is analogous to an electron moving in electric field while the nonequilibrium dynamics is

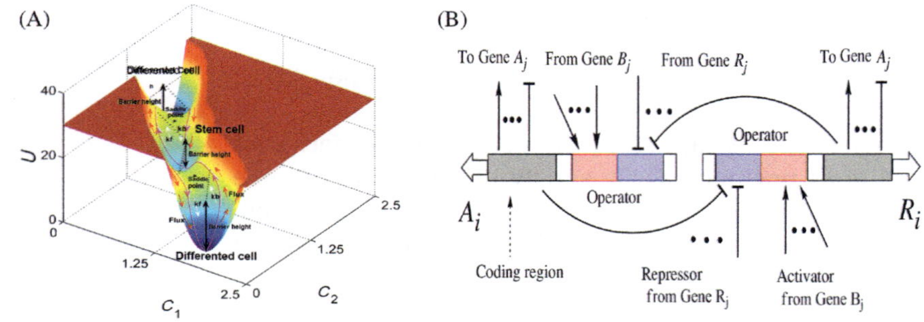

Figure 7.2 (A) Potential landscape *U* and flux *(red)* of cell fate decision as function of concentration variables. Cell fates are quantified by the basins of attractions. The cell decision paths do not go through saddle points. Barrier heights between basin and highest potential on the path determine the cell fate decision kinetic rates. (B) Gene regulations through protein regulators, the activators, and repressors (Li and Wang, 2013b).

analogous to an electron moving in an electric and magnetic field. In practice, for cell fate decision making of differentiation/reprograming, the global landscape and flux analysis through this general force decomposition requires numerically solving either the probabilistic Fokker-Planck diffusion equation, or the probabilistic master equation (including intrinsic fluctuations), or the corresponding stochastic Langevin equations, based on the underlying gene networks.

Optimal paths for quantifying the differentiation/development and reprograming processes

A cell regulated by its underlying gene networks often has huge number of states. Not every state has necessarily the same weight. Certain states can have higher chances of appearance than others. The observed functional states are often more probable and therefore with higher weights. They can then be identified as the cell attractor states. From the probability landscape topography, one can then quantify both the basin depths and the barriers between the basins of the attractors. A key question here is how one cell fate can switch from one to another. In the landscape view, this is equivalent to ask how a cell can switch from one basin of attraction to another. One can further ask the question what the optimal pathway from one cell state to another is. Understanding these issues can help us to uncover the underlying physical mechanisms and detailed process of the cell fate decision making in the differentiation/development. In order to address these, a path integral formulation (Wang et al., 2005, 2006, 2010c; Faccioli et al., 2006; Maier and Stein, 1993; Aurell and Sneppen, 2002; Roma et al., 2005; Weinan and Vanden-Eijnden, 2010) can be developed for the cell fate decision-making process.

Starting from the stochastic dynamical equation for the cell state $dx/dt = F(x) + \eta$, one can assume Gaussian probability distribution of environmental fluctuation η due to the many degrees of the freedom from the environments (large number theorem). From the relationship between $\eta = dx/dt - F(x)$ one can find out the probability of a path $x(t)$. The general path probability of starting from the initial cell state x_i (basin of attraction) at $t=0$ and end up at the final cell state of x_f at time t, from a path integral formulation, is given as (Wang et al., 2010c):

$$P(x_f, t|x_i, 0) = \int Dx\exp\left[-\int Ldt\right];\ L = \frac{1}{4D}\dot{x}\cdot\mathbf{D}^{-1}\cdot\dot{x} - \frac{1}{2D}D^{-1}\cdot F\cdot\dot{x} + V_{\text{eff}};$$
$$V_{\text{eff}} = \frac{1}{4D}F\cdot D^{-1}\cdot F + \frac{1}{2}D\cdot\nabla\cdot\left(D^{-1}\cdot F\right) \tag{7.3}$$

Each path $x(t)$ connecting initial and finite cell state has specific weight $\exp\left[-\int Ldt\right]$. L is the Lagrangian of the system, $\int Ldt$ is the action of the system, while the V_{eff} is the effective potential of the system. The path probability is then equal to the sum of the weights of all possible paths Dx. Since not every path gives the same weight, the paths with the maximum weights give the optimal contribution to the path integral and they are the dominant paths. Because the weight contribution through Lagrangian L or action is on the exponential, the contributions to the weights from the subleading paths will be exponentially smaller than those from the dominant paths. Therefore, one can quantify the dominant paths by optimizing the weights, $\exp\left[-\int Ldt\right]$. This leads to the optimal path equation: $\frac{\partial L}{\partial x} - \frac{d}{dt}\frac{\partial L}{\partial \dot{x}} = 0$.

Let us probe some general properties of the paths. If the driving force F is a gradient, $F = -D\cdot\nabla U$, then $\int dtD^{-1}\cdot F\cdot\dot{x} = \int D^{-1}\cdot F\cdot \boldsymbol{dx} = \int \nabla U\cdot dx = U_f - U_i$ is a constant, which only depends on initial and final positions and thus does not contribute to the dynamics of the paths (path independent). Therefore, in equilibrium system, the surface term (the aforementioned line integral) does not influence the optimal path equation.

However, when the driving force cannot be written as a potential gradient, as in the cases of the general nonequilibrium network dynamics of cell fate decision making, one can clearly see that $\int D^{-1}\cdot F\cdot \boldsymbol{dx}$ is not a constant and is in general path dependent. Since the general force can be decomposed by the potential gradient and curl flux as mentioned, the curl flux force contributes to the actual paths, which will be deviated from the gradient paths. In other words, the optimal paths will be distinctly different from the conventionally expected gradient path of the landscape. Furthermore, due to the rotational nature of the flux force operating in addition to the gradient paths, the forward path from x_i to x_f and backward path from x_f to x_i do not usually follow the same route. This has significant consequences on the cell fate decision-making process. It indicates that the cell fate decision-making progress is in general irreversible. Furthermore, the differentiation path and reprograming paths are in general distinctly different from each other (Fig. 7.2). To quantify the paths, one can reduce the multidimensional path integral into a one-dimensional line integral in the Hamilton-Jacobian representation.

$$P(x) = \frac{\partial L}{\partial \dot{x}}; \; H = -L + P(x)\cdot\dot{x} = E_{\text{eff}}; \; \int L dt = \int [P(x)\cdot\dot{x} - H] dt;$$
$$\int P(x)\cdot\dot{x}\, dt = \int P(x)\cdot dx = \int P_l dl \tag{7.4}$$

Therefore, optimizing the weights $\int Ldt$ becomes optimizing the line integral of the canonical momentum P (P is defined as $\partial L/\partial \dot{x}$ earlier and P_l is the projection of P to the path) with constant Hamiltonian, H (total energy E_{eff} conservation) (Wang et al., 2005, 2006, 2010c; Faccioli et al., 2006; Maier and Stein, 1993; Aurell and Sneppen, 2002; Roma et al., 2005; Weinan and Vanden-Eijnden, 2010). In this way, one can numerically search for the dominant paths through the optimization of a one-dimensional line integral in multidimensional state space, which can be performed by the Monte Carlo sampling. This can greatly simplify the computation. The method can be used for studying the large cell networks.

Kinetic rates of differentiation/development and reprograming cell fate decision-making processes

Once the dominant paths between the cell fates can be specified, one can find out the detailed process of the cell differentiation/development. A natural next question to ask is how fast the differentiation and development process is. In the landscape view, this is equivalently of asking how fast the transition is from one cell state described by the basin of attraction to another.

In equilibrium systems, the kinetic rate is determined by the barrier height between the two basins of attractions. The barrier heights are determined by the difference of the landscape depths at the basin of attraction and landscape height at the saddle point between the basins of the attraction. This is sometimes called the transition state theory or Kramer's rate theory (Eyring, 1935; Kramers, 1940). However, the gene regulatory network dynamics of the cell fate decision-making process in general does not follow the equilibrium potential gradient dynamics. An additional contribution to the driving force of the dynamics is present from the nonequilibrium curl flux due to the detailed balance breaking from the energy input. The curl flux leads to the deviations of from the conventionally expected gradient paths. The optimal paths between cell states do not necessarily pass through the saddle point between the basins of attractions (Fig. 7.2). Therefore, the conventional transition state theory/ Kramer rate theory cannot be applied directly to the cell fate decision-making processes for differentiation/development.

Clearly, a new formulation needs to be developed for quantifying the kinetic speeds of the cell fate decisions. A modified nonequilibrium transition state theory beyond the conventional one for nonequilibrium systems was proposed (Feng et al., 2014). The speed or the rate of the transition is determined by the barrier on the landscape from the basin of attraction to the highest action $S = \int Ldt$ on the optimal path (Feng et al., 2014). Instead of searching for the saddle point on the landscape, one search for the highest action point along the optimal path. The location of the highest action is in general different from the saddle point of the landscape when the curl flux is present. The barrier height using this quantification is thus different from the equilibrium case measured with respect to the landscape saddle point. The transition rate quantifies the capability of escaping from the basins of attraction and therefore the communication capability among states. In this sense, the kinetic rates for transitions can be used to quantify the global stability. Since the action can be transformed to a line integral of the canonical momentum along the optimal path, the biggest contribution comes from the difference between the starting point and the highest point of the action. This provides a quantitative foundation for the proposed nonequilibrium transition state theory: the barrier height determining the kinetics is not from the barrier between the landscape saddle point and the basin but from the barrier between the maximum action along the optimal path and the basin (Fig. 7.2). Moreover, one can study the fluctuations around the optimal paths and quantify the prefactor in addition to the exponential factor for the transition rate.

Quantifying Waddington landscape and paths for differentiation/development

Gene regulatory motif circuit determining the differentiation

A human embryonic stem cell can often be characterized by the expression pattern from several transcription factors and cell surface proteins. The underlying simplified gene circuit motifs extracted from the entire network determining the cell decisions are often composed of the two fate determining regulators mutually repressing each other through protein-protein interactions. For example, the regulator pair, Cdx2 and Oct4 (Niwa et al., 2005; Rossant, 2007; Jedrusik et al., 2008; Kalmar et al., 2009; Lutz and Bujard, 1997), controls the branching to the trophectoderm and the inner cell mass fates from totipotent (pluripotent) embryonic stem cells. Cdx2 induces the trophectoderm fate, while Oct4 induces the inner cell mass fate. Since these two regulators repress each other, a cell cannot be both at an inner cell mass state and trophectoderm cell state (Fig. 7.3). In this way, the mutual exclusion of different cell types can secure the distinct cell identity. The dynamical behavior of these two alternate cell types can be characterized by the bistability of the corresponding two basins of attractors on the landscape emerged from the underlying gene circuit motif. Effective self-activation feedback loops are often found in these circuits (Fig. 7.3). This can support the chance and the stability of yet another cell state, the pluripotent (bipotent) stem cell state

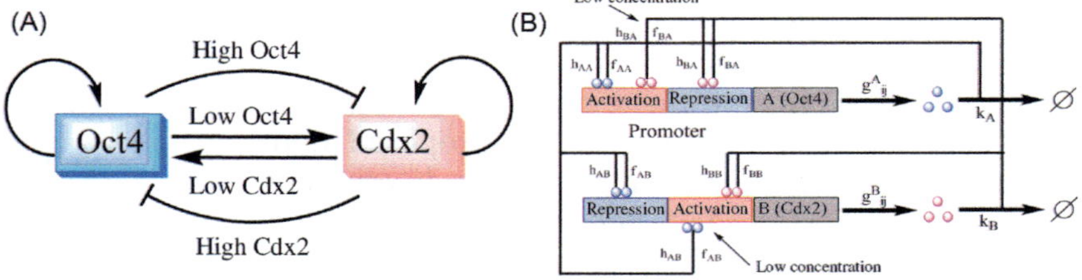

Figure 7.3 The wiring diagram of cell fate decision circuit involving Oct4 and Cdx2 controlled by mutual regulations and self-regulations (A), and controlled by slow binding/unbinding of regulators to the genes (B) (Feng and Wang, 2012).

where both expressions of the Cdx2 and Oct4 are significant (Niwa et al., 2005; Rossant, 2007; Jedrusik et al., 2008; Kalmar et al., 2009; Lutz and Bujard, 1997). Therefore, the cell fate decision making for differentiation can be studied by the process from the pluripotent stem cell state with significant expression of both Cdx2 and Oct4 to the differentiated cell states with either trophectoderm cells of dominant Cdx2 expression or inner cell mass cells of dominant Oct4 expression. The cell fate decision for cell reprograming is determined by the backward process from differentiated cell to the stem cell.

Cell fate decision for differentiation and reprograming through regulations

In cell differentiation process, the cells evolve from undifferentiated pluripotent stem state to differentiated state. The underlying gene regulatory network dictates the gene expression patterns of the cell. This forces the cell to adopt the cell type-specific phenotypes. Cells have states with the higher probability of appearance. This leads to different cell phenotypes. Different cell phenotypes should correspond to different basins of attractions on the landscape. Therefore, the differentiation and the reverse reprograming process of the cell can be thought as the evolution from one basin to another of the underlying landscape. The challenge here is to uncover quantitatively how this occurs, what the underlying mechanism is, and how to quantify the differentiation and reprograming process through the changes of the regulatory strengths of the underlying gene networks due to the genetic or environmental changes.

The landscape-flux theory was applied to study the underlying gene circuit motif dynamics of stem cell differentiation and reprograming (Wang et al., 2010b, 2011; Wang, 2015; Xu et al., 2014). The gene expression dynamics can be formulated according to the wiring diagram of Fig. 7.3 as follows:

$$\frac{dX_i}{dt} = F_i = -K[i]^*X_i + \frac{\mathrm{SA}^*(X_i)^n}{S^n + (X_i)^n} + \sum_{j=1}^{m1} \frac{A^*\left(X_j\right)^n}{S^n + \left(X_j\right)^n} + \sum_{j=1}^{m2} \frac{B^*(S^n)^n}{S^n + \left(X_j\right)^n} \tag{7.5}$$

Here X_i is the expression of the ith gene regulator. The left side of the equation represents the dynamical changes of the expression for specific gene regulator. The first term on the right side represents the degradation; the second term represents the self-activation; the third and fourth terms represent the mutual activations and repressions from the other genes to this specific gene regulator *i*. For Cdx2 and Oct4 system, one has $i = 1, 2$. The stochastic fluctuation force can be added to the right side of the equation. One can then follow the probability evolution by studying the corresponding Fokker-Planck equation. The steady-state equation can be solved and obtain the landscape *U*.

The cells can have states with the higher weight or probability giving the different cell types. Different cell types should correspond to different basins of attractions of the probability landscape.

Quantified Waddington landscape and paths for development/differentiation

One can quantify the landscapes of the development at different stages in Fig. 7.4. The vertical axis represents the potential landscape *U* (inversely related to the probability landscape $= -\ln P_{ss}$) (Wang et al., 2011). The horizontal axis represents the coordinate with the combination of expression levels of both

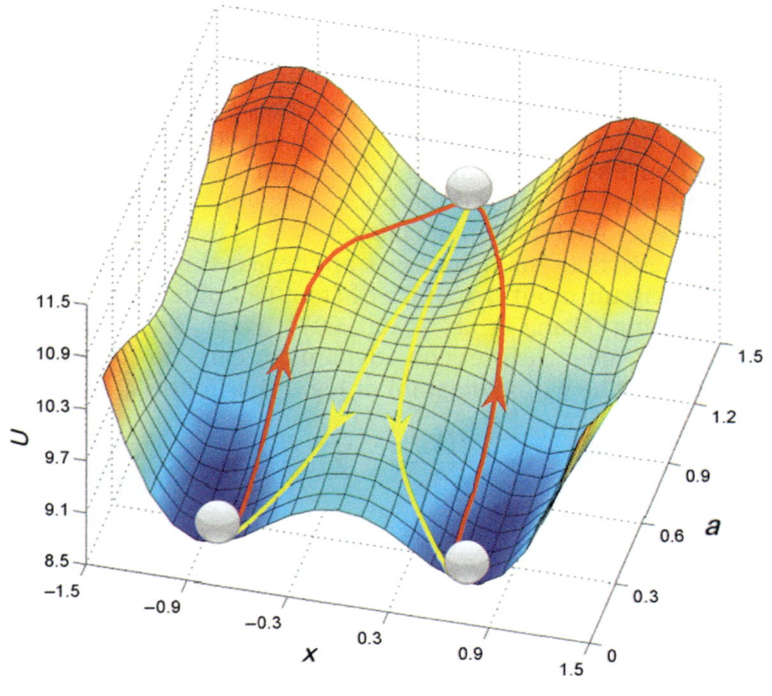

Figure 7.4 Shows the core gene network for development and its corresponding Waddington landscape. The horizontal axis *x* represents the gene expression and vertical axis *a* represents the change of gene regulation strengths in terms of the direction of development. *U* quantifies the Waddington landscape where the cell starts with stem cell state *(top)* basin and developed to differentiated state basins *(bottom)*. *Yellow lines* represent the differentiation process while the *red lines* represent the reprograming process. Transdifferentiation proceeds between the two bottom differentiated basins (Wang et al., 2011).

opposing cell fate determining transcription factors $x1$ and $x2$ ($x1 - x2$). At fixed regulations, the differentiation circuit can exhibit three basins of attractions. The center attractor C denotes the undecided multipotent stem cell state with balanced expression levels of both factors $x1$ and $x2$ (Wang et al., 2010b). The two side attractors A and B represent the differentiated states with mutually exclusive expressions of $x1$ and $x2$ (Wang et al., 2010b). The three attractor states are quantified as potential valleys. When the self-activation a becomes larger, the central basin of attraction appears deeper and the cell prefers to stay at undifferentiated state basin. When the self-activation becomes weaker, the side basins appear deeper and differentiated states are preferred. When the self-activation becomes extremely weak, the undifferentiated state appears to be unstable. Under this situation, the differentiated states are dominant. Experimental studies show clear evidences of the changes of transcription factor Klf4 in regulating the self-regulation strength during the differentiation process (Jiang et al., 2008). Therefore, the destabilization of the undifferentiated stem cell attractor as the self-activation regulation a is decreased (Fig. 7.4) (Wang et al., 2011) provides a mechanism for the development from undifferentiated stem cell C to either the differentiated cell A or B. Here the self-activation regulation strength provides a quantitative indicator for the development process.

One can now estimate the kinetic rate from the central undifferentiated state to the differentiated states under certain fluctuations along the self-activation regulation mimicking the developmental process. The kinetic rate is shown to be correlated with the landscape topography in terms of the basin depth (undifferentiated stem cell) or barrier height (between undifferentiated cell state and differentiated cell states) for differentiation (Wang et al., 2010b, 2011; Wang, 2015; Xu et al., 2014). As development proceeds (self-activation regulation a decreases), the chance characterized by the kinetic rate for differentiation becomes higher and reaches the maximum when the undifferentiated state becomes unstable. As seen, both the induction by destabilizing undifferentiated state through self-activation regulation changes and fluctuations can be in action for the developmental processes in the landscape picture.

The continuous three-dimensional quantified landscape and kinetic paths against gene expressions and development are illustrated in Fig. 7.4. As the differentiation progresses (self-activation regulation a decreases), the undifferentiated stem cell state becomes less stable, the differentiated state becomes more stable forming stable basins of attractions. We can see the quantified differentiation and the reprograming paths are irreversible, reflecting the existence of the curl flux as a driving force for the underlying nonequilibrium dynamics. We can see the obvious qualitative similarities between our quantified landscape/paths and the original Waddington's proposed picture for differentiation/development. This provides a solid physical foundation and quantification for the original Waddington proposed landscape. Therefore, the Waddington landscape is no longer a metaphor and can now be quantified from the underlying gene circuit for regulating the differentiation/development, which can give quantifiable predictions.

At fixed development/differentiation stage, the interplay between the gene regulations and fluctuations can dictate the cell fate switching from one state to another (kinetic rates and paths). During the development, the cell fate switching can be determined from the interplay among the induction through self-activation regulation changes as in this motif gene circuit, the gene regulations and the fluctuations. The cell phenotypic plasticity is therefore mainly determined by these three factors (induction, gene regulation, and fluctuations).

Epigenetics, heterogeneity, and plasticity

For the gene circuit dynamics dictating the differentiation/development, one often uses only the genes rather than both genes and protein regulators to represent the regulatory interactions among each other. The interactions among genes are mediated by the protein regulators. When the interactions among genes are very strong, the protein regulators and genes are tightly coupled together. Therefore, they can be treated as the same entity. These tight interactions act as constraints to the underlying gene regulatory network so that not all the gene expression patterns and therefore the gene states will be likely to have the equal chance of appearance. In fact, only finite number of states have high probability of appearance. They are likely the functional states. However, if the interactions among genes become weak, then there is little coupling between the genes and protein regulators. One can no longer treats the genes and protein regulators together as one entity. In fact, they have to be studied as separated entities. As a result, the constraints to the gene states in the gene network become weak. Therefore, many more gene states can have the chance of appearance due to the lift of the constraints. This results to many possible states, giving possible mechanism of heterogeneity (Wang, 2015; Feng and Wang, 2012; Li and Wang, 2013b; Zhang et al., 2013; Sasai et al., 2013; Ashwin and Sasai, 2015).

Epigenetics plays an important role in gene regulation dynamics and therefore cell decision making and phenotypic plasticity. From dynamic perspective, it involves additional time scales from histone modifications and DNA methylations. To account for this, let us consider the gene regulations and dynamics. In cells, intrinsic fluctuations are unavoidable due to the limited number of protein molecules and environments. In addition, the gene state fluctuations from the regulatory proteins binding/unbinding to the promoters can be significant for gene expression dynamics. Conventionally, it was often assumed that the binding/unbinding is significantly faster than the synthesis and degradation (adiabatic limit) (Sasai and Wolynes, 2003), leading to the strong interactions among genes through the protein regulations. The tight regulations control the emergence of finite number of functional states. This assumption may hold in some prokaryotic cells. However, one expects in eukaryotic cells, binding/unbinding can be comparable or even slower than the corresponding synthesis and degradation (nonadiabatic limit) due to the multistep processes in epigenetics, such as histone modification and DNA methylation (Hornos et al., 2005; Schultz et al., 2007; Kepler and Elston, 2001; Feng et al., 2011; Feng and Wang, 2012; Li and Wang, 2013b; Zhang et al., 2013; Sasai et al., 2013; Ashwin and Sasai, 2015).

The same core gene regulatory motif can be used to study the possible source of heterogeneity. Stem state and differentiated states can emerge as possible phenotype alternations in both adiabatic fast and nonadiabatic slow regulation regime. Slow regulations from the epigenetics of DNA methylation and histone remodification can lead to weaker coupling among genes. As a result, more states emerge. The heterogeneity is reflected from the emergence of more phenotypic states, the larger expression level fluctuations, wider kinetic distributions, and multiplicity of kinetic paths for phenotypic switching, more dissipation cost per gene switching, and weaker stability (Feng and Wang, 2012; Li and Wang, 2013b; Zhang et al., 2013; Sasai et al., 2013; Ashwin and Sasai, 2015). This shows the relationship between nonadiabatic slow regulation dynamics and epigenetic heterogeneity.

Identifying key factors of cell fate decision making in differentiation/development

Once the landscape of stem cell differentiation is quantified based on the underlying gene regulatory network, one can identify the fates of the stem cell differentiation/development by exploring the basins of attractions on the potential landscape. By applying the nonequilibrium transition state theory, one can quantify the kinetic speed of differentiation and reprograming. One can also identify dominant pathways for differentiation and reprograming. One can then quantify the barrier height on the dominant paths. By doing global sensitivity analysis on the barrier and the path, one can find out the key genes and key regulations influencing the global stability and biological paths for differentiation and reprograming. This is critical for uncovering the routes of stem cell reprogramming and design. Using this strategy, the dynamics of a human stem cell developmental gene regulatory network involving 52 genes was studied (Li and Wang, 2013a). The underlying landscape for the stem cell differentiation/development was uncovered. The landscape topography characterized by the barrier heights and switching rates quantify the global stability and speed of differentiation/development respectively (Li and Wang, 2013a). Based on the global sensitivity analysis upon landscape topography, kinetic switching and dominant paths, the key genes, and regulations for the cellular differentiation or reprogramming process were identified (Li and Wang, 2013a). These key regulations found can be used to guide the reprogramming design, consistent with some recent experiments on gene expressions and reprogramming path changes upon perturbation (Balázsi et al., 2011; Ferrell, 2012; Hood and Flores, 2012; Roeder and Glauche, 2006; Chickarmane and Peterson, 2008; Takahashi and Yamanaka, 2006; Yu et al., 2007; Takahashi et al., 2007; Saha and Jaenisch, 2009; Zhou and Melton, 2008; Zhou et al., 2008, 2012; Ieda et al., 2010; Chang et al., 2011; Gore et al., 2011; Hussein et al., 2011; Huang and Zhou, 2011; Lister et al., 2011; Zhao et al., 2011; Chia et al., 2010; Waddington, 1957; Wang et al., 2010b; Bhattacharya et al., 2011; Dovey et al., 2010; Singh et al., 2008; Xu et al., 2002, 2009; Judson et al., 2009; Li et al., 2011b; Subramanyam et al., 2011; Samavarchi-Tehrani et al., 2010; LipchinaI et al., 2011; Pfaff et al., 2011; Ye et al., 2012; Bao et al., 2013; Dvorak et al., 2005; Dvorak and Hampl, 2005; Ying et al., 2003; Nusse, 2008; Chiba, 2006; Hyslop et al., 2005; Kunarso et al., 2010).

Discussions on critical issues of cell fate decision making of differentiation and development

Cell fate decision-making dynamics of differentiation/development is not only determined by the landscape but also by the curl flux

Recently, there is significant number of studies that emphasized the role of the landscape for determining the cell fate decision-making dynamics (Wang et al., 2008, 2010b, 2011; Bhattacharya et al., 2011; Zhou et al., 2012; Wang, 2015; Xu et al., 2014; Kauffman, 1969; Huang, 2012; Moris et al., 2016). However, it is worth noticing that the landscape is not the only force determining the cell fate decision-making dynamics. In fact, a rotational curl flux force quantifying the degree of detailed balance breaking originated from the chemical energy input is also important for the dynamics. Therefore, the global dynamics of cell fate decision making is not only determined by the landscape but also by the curl flux (Wang, 2015; Wang et al., 2008).

On the historical ground, the original Waddington picture is rather qualitative and does not have a quantification of the valleys and barriers of the landscape (Waddington, 1957). The modern efforts for quantifying the original Waddington landscape starts from Kauffman and his collaborator's pioneering work (Kauffman, 1969) using dynamical system approach as a description of the underlying gene regulatory networks for the cell fate decision making. The locations of the attractors in gene expression space can be identified. However, this approach lacks the information of global quantifications the landscape, such as the weights of the states and therefore the depth of the basins or the barrier heights between the basins.

The next stage of the development began with the quantification of the landscape based on the dynamical systems of the underlying gene regulatory networks (Wang et al., 2008, 2010b, 2011; Bhattacharya et al., 2011; Zhou et al., 2012; Wang, 2015; Xu et al., 2014; Li and Wang, 2013a; Kauffman, 1969; Huang, 2012; Moris et al., 2016).

In some studies (Zhou et al., 2012; Moris et al., 2016), researchers claimed that the true potential landscape is not obtainable and only quasi potential can be utilized in an approximate way. We want to stress here that there is no problem in principle in finding the true potential landscape formally in general and the numerical simulations can be used to obtain the potential landscape for practical situations (Wang et al., 2008, 2010b, 2011; Wang, 2015; Xu et al., 2014; Li and Wang, 2013a). The nonequilibrium landscape in deterministic dynamics was identified as the Lyapunov function for the dynamical system crucial for quantifying the global stability (Wang, 2015; Zhang et al., 2012) (also the references therein).

However, for the nonequilibrium dynamical systems, the landscape alone cannot determine the whole dynamics as in the equilibrium case. Instead, the dynamics for differentiation and development as a nonequilibrium system is determined by both the landscape gradient and a rotational curl flux force quantifying the degree of the system away from equilibrium (Wang, 2015; Wang et al., 2008). The nonequilibriumness here comes from the energy input/output to the system (such as nutrition supply, ATP hydrolysis and phosphorylation, oxygen supply, etc.). This was illustrated and quantified in (Wang, 2015; Wang et al., 2008, 2010b,c, 2011; Xu et al., 2012, 2014; Feng et al., 2014; Zhang et al., 2012, 2013; Li and Wang, 2014). This is quite different from the picture of often-studied equilibrium systems where the dynamics is determined by only the gradient of the landscape. In fact, in equilibrium systems, the global dynamics is analogous to the motion of charged particles in an electric field while for nonequilibrium systems, the global dynamics is analogous to the motion of the charged particles in an electric field and magnetic field. In terms of exploring the cell fate decision-making dynamics of differentiation/development, one needs to consider both the landscape and the curl flux instead of landscape alone. This will be discussed further in the following.

The differences of the original Waddington landscape and quantified landscape for cell fate decision making in differentiation and development

When applying the landscape and flux framework (Wang, 2015; Wang et al., 2008) to cell fate decision-making circuit for differentiation and development, a quantitative mapping of the Waddington landscape (Wang et al., 2011) in the spirit of the original Waddington picture rather than merely the qualitative landscape sketched in many studies was revealed in (Wang et al., 2011) (see Fig. 7.4). Although the basic picture reflects the original Waddington's one, there are several significant differences. (1) Multipotent undifferentiated cells are stable/metastable instead of unstable at the barrier top and become unstable only when the developmental process proceeds. (2) The cell fate decision-making paths can be quantified but do not follow the gradient of the landscape as the original Waddington picture, due to the presence of the rotational curl flux force in addition to the landscape. (3) Different from Waddington's picture, the forward differentiation and backward reprogramming paths are irreversible. This irreversibility of the cell fate decision-making process is intrinsic and due to the nonequilibrium driving force from the rotational curl flux breaking the detailed balance (Wang, 2015; Wang et al., 2008, 2010c, 2011). (4) In the original Waddington picture, the developmental direction was not specified and no quantitative measure was given. Here the developmental landscape is evolving in time as the development proceeds. The developmental process and direction can be quantified by the changes in the regulation strengths in the gene regulatory networks explicitly, the self-activation strength changes in this example (Wang et al., 2011; Zhang et al., 2012). Therefore, the Waddington landscape is an evolving landscape, not a static landscape. The direction and speed for the development is determined by the regulation changes and the underlying gene network. (5) The dynamics of the differentiation/development is determined by the induction characterized by the regulation changes (self-activation, for example), the presence of stochastic fluctuations and gene network regulations. Gene network dynamics leads to the expression dynamics while the fluctuations give arise to the variations and chances of escaping from the cell fate attractors. The induction influenced by the changing environments can drive the differentiation/development process.

Origins of the bifurcations and phase transitions of cell fate decision making of differentiation/development

In addition to the nonlinear dynamics description of the bifurcation and cell fate decision-making process (Ferrell, 2012; Moris et al., 2016), the landscape and flux picture of various bifurcation mechanisms (supercritical pitchfork bifurcation, saddle-node bifurcation, and subcritical pitchfork bifurcation) have been quantified explicitly in details in reference (Xu et al., 2014). The optimal paths for cell fate decision making are determined by both the landscape and curl flux, which deviate from the naively expected ones and are irreversible. In fact, the bifurcation mechanisms for cell decision-making process are in analogy to the discrete first-order and continuous second-order phase transitions in physics (Qian et al., 2016). This allows the thermodynamic characterizations in addition to the dynamical descriptions.

Time arrow and mechanism of irreversibility originating from the curl flux breaking the detailed balance

Understanding the irreversibility and time arrow is important for uncovering the underlying mechanism of cell fate decision making. It was argued that supercritical bifurcation scheme lacks irreversibility (Ferrell, 2012; Moris et al., 2016) while the saddle-node bifurcation and subcritical bifurcation scheme would generate such irreversibility in the deterministic case (Ferrell, 2012; Moris et al., 2016). However, the fluctuations are inevitable. Then the irreversibility in a stochastic system becomes a challenge since the barriers between basins can be surpassed under sufficient fluctuations and with long times.

On the other hand, the differentiation process (Wang et al., 2010c, 2011; Xu et al., 2014; Li and Wang, 2013a), just as the self-renew (replication) process (Wang, 2015; Wang et al., 2008; Li and Wang, 2014), requires energy, and information input (Wang et al., 2010c, 2011; Xu et al., 2014; Feng et al., 2014; Li and Wang, 2013a). The net input of the energy breaks the detailed balance of the system and provides the curl flux as driving force for the dynamics (Wang, 2015; Xu et al., 2012). This explicitly breaks the time reversibility and thus provides a source of irreversibility or time arrow for both deterministic and stochastic cases (Wang, 2015; Wang et al., 2008, 2010c, 2011; Xu et al., 2014; Feng et al., 2014; Li and Wang, 2013a; Zhang et al., 2012; Li et al., 2011a; Zhang and Wang, 2018). The time arrow or irreversibility can be reflected by the differences in the forward differentiation and backward reprograming paths for cell fate decision-making processes as illustrated in Fig. 7.1 (Wang et al., 2011) and Fig. 7.2 (Feng et al., 2014). Therefore, the curl flux breaking the detailed balance originated from the net energy input provides a robust source of time arrow or irreversibility of the cell fate decision-making process of differentiation/development and reprogramming for either deterministic or stochastic systems. This mechanism of time arrow or irreversibility can be tested experimentally through single-cell measurements by comparing the differentiation and reprogramming processes.

Heterogeneity from epigenetics

Heterogeneity is widely present in differentiation and cell fate decision-making processes. A way of generating the heterogeneity was suggested from the transition states (Moris et al.,

2016) in the development due to the relatively low barrier heights between the landscape basins of attractions, which allows the communications and populations of the substates. It is worth noticing that the range of the parameters for the existence of these transition states is narrow (Xu et al., 2014) and the probabilities of these transition states can be low. The generality, the stabilities and robustness of those transition states may become an issue.

There is another alternative way of explaining the heterogeneity at the individual cell level (Feng and Wang, 2012; Li and Wang, 2013b; Zhang et al., 2013; Sasai et al., 2013; Ashwin and Sasai, 2015). The heterogeneity does not always come from the genetics since the genes themselves are typically not significantly mutated during the normal developmental process. This is contrast to the case of cancer. For differentiation and development, the heterogeneity can be from the epigenetics. In fact, histone remodifications, DNA methylations, and environmental perturbations can give arise to extra chemical steps and time scales for transcription factors to regulate genes. This can slow down the regulation processes and effectively reduces the regulation strengths or connections in the gene circuit. When a gene network has weaker links among nodes, there are lesser constraints on the expressions of individual genes. As a result, more possible states emerge. This mechanism can provide a general, stable and robust way of generating heterogeneity (Feng and Wang, 2012; Li and Wang, 2013b; Zhang et al., 2013; Sasai et al., 2013; Ashwin and Sasai, 2015). This is distinctly different from the transition state ideas (Moris et al., 2016). Even without the transient states, heterogeneity can still emerge from this mechanism. An optimal kinetics for decision making can appear (Feng and Wang, 2012; Li and Wang, 2013b). One can test this mechanism of generating heterogeneity and optimal kinetics for cell fate decision making by perturbations on the DNA methylation and histone modification levels or the environments and see how the heterogeneity and kinetics changes.

Quantifications of transition states, speed, and optimal paths of cell fate decision making of differentiation and reprograming through landscape and flux

Cell fate decision-making process of differentiation and development can be quantified by the paths. The speed of the cell fate decision-making process is important in describing the underlying kinetics but still challenging. The importance of the transition state for cell decision-making process was pointed out and an analogy to the equilibrium molecular transition state was used with the speed determined by the barrier at the transition state of the landscape to illustrate the ideas (Moris et al., 2016). The description was based on the equilibrium analogy.

However, the cell fate decision-making process is an intrinsically nonequilibrium process. The nonequilibrium dynamics is determined by both the landscape and rotational curl flux measuring the degree of the deviation from the equilibrium (Wang, 2015; Wang et al., 2008). As a result, the optimal kinetic paths between the cell fate basins of attraction do not usually pass through the saddle points (i.e., transition states) of the landscape (Wang et al., 2010c, 2011; Xu et al., 2014; Feng et al., 2014) (see Fig. 7.2). This is due to the presence of the rotational curl flux force deviating from the equilibrium. Therefore, by just exploring the landscape on the gene expression space in the single-cell experiments such as RNAseq (Weinreb et al., 2018; Wagner et al., 2018; Briggs et al., 2018; Farrell et al., 2018; Schiebinger et al., 2019; La Manno et al., 2018), in principle one cannot in general expect to be able to identify the true transition states directly for cell fate decision making in differentiation/development and reprogramming (from the landscape alone).

In fact, as shown in (Wang et al., 2010c, 2011; Feng et al., 2014), the transition states are path dependent. For example, the optimal differentiation path from multipotent undifferentiated cell fate basin A to differentiated cell basin B is in general different from the reprogramming path from B to A. Therefore, the resulting transition state from A to B is different from B to A. The transition states for cell fate decision-making process are on the optimal paths but not necessarily on the saddle point or transition state of the landscape as naively expected since the paths do not necessarily go through the saddle point or transition state of the landscape. Therefore, the speed of the decision making is determined by the path dependent transition state and the barrier at that location. The transition states for cell fate decision making can only be identified when including the effect of the rotational curl flux force correctly in the dynamics.

The identification and quantifications of the nonequilibrium transition states and the associated effective barrier have been suggested (Feng et al., 2014). The barrier on the optimal path rather than the saddle on the landscape determine the speed of the cell fate decision-making process. The position of the barrier on the optimal path locates the new nonequilibrium transition state. This gives arise to the critical bottleneck of the underlying gene regulatory network states to pass through for the cell fate decision making of differentiation/development and reprogramming to occur (Feng et al., 2014). It is important to note that the location of the transition state for cell decision making is usually not at the saddle point or the transition state of the landscape. One can test this experimentally through the single-cell real time dynamics or RNA velocity (Weinreb et al., 2018; Wagner et al., 2018; Briggs et al., 2018; Farrell et al., 2018; Schiebinger et al., 2019; La Manno et al., 2018).

Transition states or intermediate states?

The canonical definition of the transition states in chemistry and physics refers to the saddle point of the underlying energy landscape. The transition state has maximum potential in the reaction coordinate while having minima in another direction. The "transition states" defined (Moris et al., 2016) for cell fate decision are in fact the intermediate states with shallow basins of attractions in the reaction coordinates.

Unlike the localizations of the transition states in chemistry and physics, these shallow intermediate states are often dispersed or spread on the landscape, which are not necessarily very helpful for determining the kinetics and paths for the dynamics from the conventional transition state theory. Therefore, the "transition states" may not be the best name for these states. The more accurate name for this type of "transition states" seems to be the shallow intermediate states for cell fate decision making.

It was proposed that the distribution of these shallow intermediate states as the origin of the heterogeneity (Moris et al., 2016). With further considerations of more details of the full network beyond the motif level, there may still be no significant distribution of the intermediates (Li and Wang, 2013a).

For the aforementioned proposal (Moris et al., 2016), the heterogeneity can come from the intermediate state populations from the full network. We believe that even at the network motif level, the heterogeneity can emerge from the regulation changes while having the same wiring topology (Feng and Wang, 2012). In this view (Feng and Wang, 2012; Li and Wang, 2013b; Zhang et al., 2013; Sasai et al., 2013; Ashwin and Sasai, 2015), the heterogeneity can come from the effectively weaker regulations (through, e.g., epigenetics) even at the network motif level. The effectively weaker regulations can lead to lesser constraints and more freedoms for the states to appear as mentioned earlier, giving the heterogeneity.

Even when the original full underlying gene regulatory network does not have apparent intermediate populations or heterogeneity, the effectively weaker regulations can still give arise to heterogeneity (Wang et al., 2011; Xu et al., 2014; Feng and Wang, 2012; Li and Wang, 2013b). In this view (Feng and Wang, 2012; Li and Wang, 2013b; Zhang et al., 2013; Sasai et al., 2013; Ashwin and Sasai, 2015), the heterogeneity emerges not only in the intermediate cell states but also in the multipotent stem cells and differentiated cells. One should in principle be able to distinguish the heterogeneity from the intermediates (Moris et al., 2016) and heterogeneity from the weaker regulations (Feng and Wang, 2012; Li and Wang, 2013b) in the future single-cell experiments.

Discrete paths versus continuous paths

Transition or switching paths between cell fate basins of attractions can be quantified (Wang et al., 2010c, 2011; Xu et al., 2014; Li and Wang, 2013a). It is worth noticing that the "discrete jumps" between cell fate basins of attraction at the cellular decision event (Moris et al., 2016) do not mean that the continuous transition paths do not exist. In fact, the "discrete jumps" are originated from the shorter time intervals representing the actual jumps in contrast to the continuous paths with the long waiting times for the transitions. Therefore, the transition paths for the "discrete jumps" between the basins of attractions do exist and can be quantified (Wang et al., 2010c; Feng et al., 2014).

Acknowledgment

LX thanks supports by Natural Science Foundation of China nos. 11305176, 21721003 and 81874254, MOST, China, grant no. 2016YFA0203200, the Natural Science Foundation of Jilin Province, grant no. 20150520082JH. JW thanks support in part by NSF-PHY-76066 and NSF-CHE-1808474.

References

Ackers, G.K., Johnson, A.D., Shea, M.A., 1982. Quantitative model for gene regulation by lambda phage repressor. Proc. Natl. Acad. Sci. U.S.A. 79, 1129–1133.

Ashwin, S.S., Sasai, M., 2015. Effects of collective histone state dynamics on epigenetic landscape and kinetics of cell reprogramming. Sci. Rep. 5, 16746.

Aurell, E., Sneppen, K., 2002. Epigenetics as a first exit problem. Phys. Rev. Lett. 88, 048101.

Balázsi, G., van Oudenaarden, A., Collins, J.J., 2011. Cellular decision making and biological noise: from microbes to mammals. Cell 144 (6), 910–925.

Bao, X., Zhu, X., Liao, B., Benda, C., Zhuang, Q., Pei, D., et al., 2013. MicroRNAs in somatic cell reprogramming. Curr. Opin. Cell Biol. 25, 208–214.
Bhattacharya, S., Zhang, Q., Andersen, M.E., 2011. A deterministic map of Waddington's epigenetic landscape for cell fate specification. BMC Sys. Biol. 5 (1), 85.
Briggs, J.A., Weinreb, C., Wagner, D.E., Megason, S., Peshkin, L., Kirschner, M.W., et al., 2018. The dynamics of gene expression in vertebrate embryogenesis at single-cell resolution. Science 360 (6392), eaar5780.
Chang, R., Shoemaker, R., Wang, W., 2011. Systematic search for recipes to generate induced pluripotent stem cells. PLoS Comp. Biol. 7, e1002300.
Chia, N.Y., Chan, Y.S., Feng, B., Lu, X., Orlov, Y.L., et al., 2010. A genome-wide RNAi screen reveals determinants of human embryonic stem cell identity. Nature 468, 316–320.
Chiba, S., 2006. Notch signaling in stem cell systems. Stem Cell 24, 2437–2447.
Chickarmane, V., Peterson, C., 2008. A computational model for understanding stem cell, trophectoderm and endoderm lineage determination. PLoS One. 3, e3478.
Choi, P.J., Cai, L., Frieda, K., Xie, S., 2008. A stochastic single-molecule event triggers phenotype switching of a bacterial cell. Science 322, 442–446.
Das, J., Kardar, M., Chakraborty, A.K., 2007. Purely stochastic binary decisions in cell signaling models without underlying deterministic instabilities. Proc. Natl. Acad. Sci. U.S.A. 104, 18598–18963.
Dovey, O., Foster, C., Cowley, S., 2010. Histone deacetylase 1(HDAC1), but not HDAC2, controls embryonic stem cell differentiation. PNAS 107 (18), 8242–8247.
Dvorak, P., Hampl, A., 2005. Basic fibroblast growth factor and its receptors in human embryonic stem cells. Folia Histochem. Cytobiol. 43 (4), 203–208.
Dvorak, P., Dvorakova, D., Koskova, S., Vodinska, M., Najvirtova, M., Krekac, D., et al., 2005. Expression and potential role of fibroblast growth factor 2 and its receptors in human embryonic stem cells. Stem Cells 23 (8), 1200–1211.
Eyring, H., 1935. The activated complex in chemical reactions. J. Chem. Phys. 3, 107115.
Faccioli, P., Sega, M., Pederiva, F., Orland, H., 2006. Dominant pathways in protein folding. Phys. Rev. Lett. 97, 108101.
Fang, X.N., Liu, Q., Bohrer, C., Hensel, Z., Han, W., Wang, J., et al., 2018. Cell fate potentials and switching kinetics uncovered in a classic bistable genetic switch. Nat. Commun. 9, 2787.
Farrell, J.A., Wang, Y., Riesenfeld, S.J., Shekhar, K., Regev, A., Schier, A.F., 2018. Single-cell reconstruction of developmental trajectories during zebrafish embryogenesis. Science 360 (6392), eaar3131.
Feng, H.D., Wang, J., 2012. A new mechanism of stem cell differentiation through slow binding/unbinding of regulators to genes. Sci. Rep. 2, 550.
Feng, H.D., Han, B., Wang, J., 2011. Adiabatic and non-adiabatic non-equilibrium stochastic dynamics of single regulating genes. J. Phys. Chem. B 115, 1254–1261.
Feng, H., Hensel, Z., Wang, J., Xiao, J., 2012. Analytical calculation of protein production distributions in models of clustered protein expression. Phys. Rev. E 85 (3), 031904.
Feng, H.D., Zhang, K., Wang, J., 2014. Non-equilibrium transition state rate theory. Chem. Sci. 5, 3761–3769.
Ferrell Jr., J.E., 2012. Bistability, bifurcations, and Waddington's epigenetic landscape. Curr. Biol. 22, R458–R466.
Gillespie, D., 2000. The chemical Langevin equation. J. Chem. Phys. 113, 297.

Gore, A., Li, Z., Fung, H.L., Young, J.E., Agarwal, S., et al., 2011. Somatic coding mutations in human induced pluripotent stem cells. Nature 471, 63–67.

Hensel, Z., Feng, H., Han, B., Hatem, C., Wang, J., Xiao, J., 2012. Stochastic expression dynamics of transcription factor revealed by single-molecule noise analysis. Nature. Struct. Mol. Biol. 19, 797–802.

Hood, L., Flores, M., 2012. A personal view on systems medicine and the emergence of proactive P4 medicine: predictive, preventive, personalized and participatory. N. Biotechnol. 29, 613–624.

Hornos, J.E.M., Schultz, D., Innocentini, G.C.P., Wang, J., Walczak, A.M., Onuchic, J.N., et al., 2005. Self-regulating gene: an exact solution. Phys. Rev. E 72, 051907.

Huang, S., 2012. The molecular and mathematical basis of Waddington's epigenetic landscape: a framework for post-Darwinian biology? BioEssays 34 (2), 149–157.

Huang, S., Zhou, J.X., 2011. Understanding gene circuits at cell-fate branch points for rational cell reprograming. Trends Genet. 27, 55–62.

Hussein, S.M., Batada, N.N., Vuoristo, S., Ching, R.W., Autio, R., et al., 2011. Copy number variation and selection during reprogramming to pluripotency. Nature 471, 58–62.

Hyslop, L., Stojkovic, M., Armsrong, L., Walter, T., Stojkovic, P., et al., 2005. Downregulation of NANOG induces differentiation of human embryonic stem cells to extraembryonic lineages. Stem Cells 23 (8), 1035–1043.

Ieda, M., Fu, J.D., Delgado-Olguin, P., Vedantham, V., Hayashi, Y., et al., 2010. Direct reprogramming of fibroblasts into functional cardiomyocytes by defined factors. Cell 142, 375–386.

Jackson, E.A., 1990. Perspectives of Nonlinear Dynamics. Cambridge University Press, Cambridge.

Jedrusik, A., et al., 2008. Role of Cdx2 and cell polarity in cell allocation and specification of trophectoderm and inner cell mass in the mouse embryo. Genes. Dev. 22, 2692–2706.

Jiang, J., et al., 2008. A core Klf circuitry regulates self-renewal of embryonic stem cells. Nat. Cell Biol. 10, 353–360.

Judson, R.L., Babiarz, J.E., Venere, M., Blelloch, R., 2009. Embryonic stem cell-specific microRNAs promote induced pluripotency. Nat. Biotechnol. 27, 459–461.

Kalmar, T., Lim, C., Hayward, P., Munoz-Descalzo, S., Nichols, J., et al., 2009. Regulated fluctuations in Nanog expression mediate cell fate decisions in embryonic stem cells. PLoS Biol. 7, e1000149.

Kauffman, S., 1969. Homeostasis and differentiation in random genetic control network. Nature 224, 177–178.

Kepler, T.B., Elston, T.C., 2001. Stochasticity in transcriptional regulation: origins, consequences, and mathematical representations. Biophys. J. 81, 3116–3136.

Kim, K., Wang, J., 2007. Potential energy landscape and robustness of a gene regulatory network: toggle switch. PLoS Comp. Biol. 3, e60.

Kramers, H.A., 1940. Brownian motion in a field of force and the diffusion model of chemical reactions. Phys. A 7, 284–304.

Kruse, K., Elf, J., 2006. Kinetics in Spatially Extended Systems in Systems Modeling in Cellular Biology. MIT Press, Cambridge, MA.

Kunarso, G., Chia, N.Y., Jeyakani, J., Hwang, C., Lu, X., et al., 2010. Transposable elements have rewired the core regulatory network of human embryonic stem cells. Nat. Genet. 42 (7), 631–634.

La Manno, G., Soldatov, R., Zeisel, A., Braun, E., Hochgerner, H., Petukhov, V., et al., 2018. RNA velocity of single cells. Nature 560, 494.

Lecca, P., Laurenzi, I., Jordan, F., 2013. Deterministic Versus Stochastic Modelling in Biochemistry and Systems Biology. Woodhead Publishing, Oxford.
Li, C.H., Wang, J., 2013a. Quantifying cell fate decisions for differentiation and reprogramming of a human stem cell network: landscape and biological paths. PLoS Comput. Biol. 9 (8), e1003165.
Li, C.H., Wang, J., 2013b. Quantifying Waddington landscapes and paths of non-adiabatic cell fate decisions for differentiation, reprogramming and transdifferentiation. J. R. Soc. Interface 10, 20130787.
Li, C.H., Wang, J., 2014. Landscape and flux reveal a new global view and physical quantification of mammalian cell cycle. Proc. Natl. Acad. Sci. U.S.A. 111 (39), 14130–14135.
Li, C.H., Wang, E.K., Wang, J., 2011a. Landscape, flux, correlation, resonance, coherence, stability and key network wirings of stochastic circadian oscillation. Biophys. J. 101, 1335–1344.
Li, Z., Yang, C.S., Nakashima, K., Rana, T.M., 2011b. Small RNA-mediated regulation of iPS cell generation. EMBO J. 30, 823–834.
Lipchinal, I., Elkabetz, Y., Hafner, M., Sheridan, R., Mihailovic, A., Tuschl, T., et al., 2011. Genome-wide identification of microRNA targets in human ES cells reveals a role for miR-302 in modulating BMP response. Genes Dev. 25, 2173–2186.
Lister, R., Pelizzola, M., Kida, Y.S., Hawkins, R.D., Nery, J.R., et al., 2011. Hotspots of aberrant epigenomic reprogramming in human induced pluripotent stem cells. Nature 471, 68–73.
Lutz, R., Bujard, H., 1997. Independent and tight regulation of transcriptional units in Escherichia coli via the LacR/O, the TetR/O and AraC/I 1 -I 2 regulatory elements. Nucleic Acids Res. 25, 1203–1210.
Maier, R.S., Stein, D.L., 1993. Escape problem for irreversible systems. Phys. Rev. E 48, 931.
McAdams, H.H., Arkin, A., 1997. Stochastic mechanisms in gene expression. Proc. Natl. Acad. Sci. U.S.A. 94 (3), 814–819.
Metzler, R., Wolynes, P.G., 2002. Number fluctuations and the threshold model of kinetic switches. Chem. Phys. 284, 469–479.
Moris, N., Pina, C., Arias, M., 2016. Transition states and cell fates decisions in epigenetic landscapes. Nat. Rev. Gen. 17 (11), 693.
Niwa, H., et al., 2005. Interaction between Oct3/4 and Cdx2 determines trophectoderm differentiation. Cell 123, 917–929.
Nusse, R., 2008. Wnt signaling and stem cell control. Cell Res. 18, 523–527.
Pfaff, N., Fiedler, J., Holzmann, A., Schambach, A., Moritz, T., Cantz, T., et al., 2011. miRNA screening reveals a new miRNA family stimulating iPS cell generation via regulation of Meox2. EMBO Rep. 12, 1153–1159.
Qian, H., Ao, P., Tu, Y.H., Wang, J., 2016. A framework towards understanding the mesoscopic phenomena: Emergent unpredictability, symmetry breaking and dynamics across scales. Chem. Phys. Lett. 665, 153–161.
Roeder, I., Glauche, I., 2006. Towards an understanding of lineage specification in hematopoietic stem cells: a mathematical model for the interaction of transcription factors GATA-1 and PU.1. J. Theor. Biol. 241, 852–865.
Roma, D.M., O'Flanagan, R.A., Ruckenstein, A.E., Sengupta, A.M., Mukhopadhyay, R., 2005. Optimal path to epigentic switching. Phys. Rev. E 71 (011902), 1–5.
Rossant, J., 2007. Stem cells and lineage development in the mammalian blastocyst. Reprod. Fertil. Dev. 19, 111–118.
Saha, K., Jaenisch, R., 2009. Technical challenges in using human induced pluripotent stem cells to model disease. Cell Stem Cell 5, 584–595.

Samavarchi-Tehrani, P., Golipour, A., David, L., Sung, H.-K., Beyer, T.A., Datti, A., et al., 2010. Functional genomics reveals a BMP-driven mesenchymal-to-epithelial transition in the initiation of somatic cell reprogramming. Cell Stem Cell 7, 64–77.

Sasai, M., Kawabata, Y., Makishi, K., Itoh, K., Terada, T.P., 2013. Time scales in epigenetic dynamics and phenotypic heterogeneity of embryonic stem cells. PLoS Comp. Biol. 9, e1003380.

Sasai, M., Wolynes, P.G., 2003. Stochastic gene expression as a manybody problem. Proc. Natl. Acad. Sci. U.S.A. 100, 2374–2379.

Schiebinger, G., Shu, J., Tabaka, M., Cleary, B., Subramanian, V., Solomon, A., et al., 2019. Optimal-transport analysis of single-cell gene expression identifies developmental trajectories in reprogramming. Cell 176, 928–943.

Schultz, D., Onuchic, J.N., Wolynes, P.G., 2007. Understanding stochastic simulations of the smallest genetic networks. J. Chem. Phys. 126, 245102.

Singh, A., Weinberger, L.S., 2009. Stochastic gene expression as a molecular switch for viral latency. Curr. Opin. Microbiol. 12, 460–466.

Singh, S.K., Kagalwala, M.N., Thornburg, J.P., Adams, H., Majumder, S., 2008. Rest maintains self-renewal and pluripotency of embryonic stem cells. Nature 453 (7192), 223–227.

Subramanyam, D., Lamouille, S., Judson, R.L., Liu, J.Y., Bucay, N., Derynck, R., et al., 2011. Multiple targets of miR-302 and miR-372 promote reprogramming of human fibroblasts to induced pluripotent stem cells. Nat. Biotechnol. 29, 443–448.

Swain, P.S., Elowitz, M.B., Siggia, E.D., 2002. Intrinsic and extrinsic contributions to stochasticity in gene expression. Proc. Natl. Acad. Sci. U.S.A. 99, 12795–12800.

Takahashi, K., Tanabe, K., Ohnuki, M., Narita, M., Ichisaka, T., et al., 2007. Induction of pluripotent stem cells from adult human fibroblasts by defined factors. Cell 131, 861–872.

Takahashi, K., Yamanaka, S., 2006. Induction of pluripotent stem cells from mouse embryonic and adult fibroblast cultures by defined factors. Cell 126, 663–676.

Thattai, M., van Oudenaarden, A., 2001. Intrinsic noise in gene regulatory networks. Proc. Natl. Acad. Sci. U.S.A. 98 (15), 8614–8619.

Waddington, C.H., 1957. The Strategy of the Genes. Allen & Unwin, London.

Wagner, D.E., Weinreb, C., Zollins, Z.M., Briggs, J.A., Megason, S.G., Klein, A.M., 2018. Single-cell mapping of gene expression landscapes and lineage in the zebrafish embryo. Science 360 (6392), 981–987.

Wang, J., 2015. Landscape and flux theory of non-equilibrium dynamical systems with application to biology. Adv. Phys. 64 (1), 1–137.

Wang, J., Zhang, K., Lu, H.Y., Wang, E.K., 2005. Quantifying kinetic paths of protein folding. Biophys. J. 89, 1612.

Wang, J., Zhang, K., Lu, H.Y., Wang, E.K., 2006. Identifying the kinetic paths on biomolecular binding-folding energy landscape. Phys. Rev. Lett. 96, 168101.

Wang, J., Xu, L., Wang, E.K., 2008. Potential landscape and flux framework of non-equilibrium networks: robustness, dissipation and coherence of biochemical oscillations. Proc. Natl. Acad. Sci. 105, 12271–12276.

Wang, J., Li, C., Wang, E.K., 2010a. Potential and flux landscapes quantify the stability and robustness of budding yeast cell cycle network. Proc. Natl. Acad. Sci. U.S.A. 107, 8195–8200.

Wang, J., Xu, L., Wang, E.K., Huang, S., 2010b. The potential landscape of genetic circuits imposes the arrow of time in stem cell differentiation. Biophys. J. 99, 29–39.

Wang, J., Zhang, K., Wang, E.K., 2010c. Kinetic paths, time scale, and underlying landscapes: a path integral framework to study global natures of nonequilibrium systems and networks. J. Chem. Phys. 133, 125103.
Wang, J., Zhang, K., Xu, L., Wang, E.K., 2011. Quantifying the Waddington landscape and biological paths for development and differentiation. Proc. Natl. Acad. Sci. U.S.A 108 (20), 8257–8262.
Weinan, E., Vanden-Eijnden, E., 2010. Transition-oath theory and path-finding algorithms for the study of rare events. Ann. Rev. Phys. Chem. 61, 391–420.
Weinreb, C., Wolock, S., Tusi, B.K., Socolovskyb, M., Klein, A.M., 2018. Fundamental limits on dynamic inference from single-cell snapshots. Proc. Natl. Acad. Sci. 115, E2467–E2476.
Xu, R.H., Chen, X., Li, D.S., Li, R., Addicks, G.C., Glennon, C., et al., 2002. Bmp4 initiates human embryonic stem cell differentiation to trophoblast. Nat. Biotechnol. 20 (12), 1261–1264.
Xu, N., Papagiannakopoulos, T., Pan, G., Thomson, J.A., Kosik, K.S., 2009. MicroRNA-145 regulates OCT4, SOX2, and KLF4 and represses pluripotency in human embryonic stem cells. Cell 137, 647–658.
Xu, L., Zhang, K., Wang, J., 2014. Exploring the mechanisms of differentiation, dedifferentiation, reprogramming and transdifferentiation. PLoS One 9 (8), e105216.
Xu, L.F., Shi, H.L., Feng, H.D., Wang, J., 2012. The energy pump and the origin of the non-equilibrium flux of the dynamical systems and the networks. J. Chem. Phys. 136, 165102.
Yan, Z., Wang, J., 2012. Specificity quantification of biomolecular recognition and its implication for drug discovery. Sci. Rep. 2, 309.
Yan, Z., Zheng, X.L., Wang, E.K., Wang, J., 2013. Thermodynamic and kinetic specificities of ligand binding. Chem. Sci. 4, 2387–2395.
Ye, D., Wang, G., Liu, Y., Huang, W., Wu, M., Zhu, S., et al., 2012. -138 promotes induced pluripotent stem cell generation through the regulation of the p53 signaling, Stem Cells, 30.), MiR, pp. 1645–1654.
Ying, Q.L., Nichols, J., Chambers, I., Smith, A., 2003. Bmp induction of id proteins suppresses differentiation and sustains embryonic stem cell self-renewal in collaboration with stat3. Cell 115 (3), 281–292.
Yu, J., Vodyanik, M.A., Smuga-Otto, K., Antosiewicz-Bourget, J., Frane, J.L., et al., 2007. Induced pluripotent stem cell lines derived from human somatic cells. Science 318, 1917–1920.
Zhang, K., Wang, J., 2018. Exploring the underlying mechanisms of the Xenopus laevis embryonic cell cycle. J. Phys. Chem. B 122, 5487–5499.
Zhang, F., Xu, L., Zhang, K., Wang, E.K., Wang, J., 2012. The potential and flux landscape theory of evolution. J. Chem. Phys. 137, 065102.
Zhang, K., Sasai, M., Wang, J., 2013. Eddy current and coupled landscapes for nonadiabatic, nonequilibrium dynamics of complex systems. Proc. Natl. Acad. Sci. U.S.A. 110, 14930–14935.
Zhao, T., Zhang, Z.N., Rong, Z., Xu, Y., 2011. Immunogenicity of induced pluripotent stem cells. Nature 474, 212–215.
Zhou, Q., Melton, D.A., 2008. Extreme makeover: converting one cell into another. Cell Stem Cell 3, 382–388.
Zhou, Q., Brown, J., Kanarek, A., Rajagopal, J., Melton, D.A., 2008. In vivo reprogramming of adult pancreatic exocrine cells to beta-cells. Nature 455, 627–632.
Zhou, J.X., Aliyu, M.D., Aurell, E., Huang, S., 2012. Quasi-potential landscape in complex multi-stable systems. J. R. Soc. Interface 9, 3539–3553.

8

The physics of cell fate

Philip Greulich[1,2], Rosanna Smith[3] and Ben D. MacArthur[1,2,4]

[1]Mathematical Sciences, University of Southampton, Southampton, United Kingdom [2]Institute for Life Sciences, University of Southampton, Southampton, United Kingdom [3]Cancer Sciences, Faculty of Medicine, University of Southampton, Southampton, United Kingdom [4]Centre for Human Development, Stem Cells and Regeneration, Faculty of Medicine, University of Southampton, Southampton, United Kingdom

Introduction

> *You can recognize truth by its beauty and simplicity ... inexperienced students, make guesses that are very complicated, and it sort of looks as if it is alright, but I know it is not true because the truth always turns out to be simpler than you thought.*
>
> ***Richard Feynman, from the Messenger Lectures on "The Character of Physical Law" (1964)***

Biology is complex. The human body contains approximately 37 trillion cells (Bianconi et al., 2013). Almost all cells in the body carry the entire genome of 3 billion base pairs, encoding for 21,000 genes (Pertea et al., 2018), which are translated into 92,000 proteins (allowing for splice variants) (UniProt Consortium, 2014). These proteins, in turn, combine to form hundreds of thousands of different regulatory complexes (Rual et al., 2005). Extraordinarily, the human exome constitutes just 1% of genomic DNA—the remaining 99% encodes for other regulatory molecules, such as micro-RNAs, or performs functions that we do not yet understand—suggesting that the vast genetic and proteomic complexity that we are now beginning to decipher may be just a small part of an even bigger picture.

In addition to this molecular complexity, human biology is also inherently multiscale: genes and proteins instruct cell identities; cells communicate with each other and aggregate into tissues; tissues stratify to form organs; and organs reliably self-organize to form the body. At each scale there is

Phenotypic Switching. DOI: https://doi.org/10.1016/B978-0-12-817996-3.00003-7

overwhelming detail, and these details matter: even very small errors in the genetic code can have catastrophic consequences for the health of the organism [for instance, it is estimated that approximately 10,000 human diseases are monogenic (World Health Organization, 2019)] and slight misplacement of cells during development can lead to life-long disorders (Meier et al., 2000). Yet, despite this complexity, biology is also remarkably reproducible, suggesting the presence of underlying organizational principles (Kitano, 2004). Uncovering these principles, while still accounting for the inherent complexity of life, is the fundamental problem in biology.

The last 20 years have seen increasing use of mathematical and computational approaches to help understand biological complexity. Collectively, the combined use of experimental and computational methods to dissect biological systems has become known as "systems biology." Systems biology typically proceeds by constructing complex sets of equations (or other mathematical representations such as networks) that seek to encode known biology, in all its detail, as closely as possible. This approach has been enormously successful (Kitano, 2002) [e.g., in defining the "design principles" of simple biological modules, such as feedback and feedforward loops, and how they function collectively (Alon, 2006)] and systems biology approaches are now routinely incorporated into research and development pipelines.

Yet, despite this tremendous success, there are two reasons why systems biology approaches can be limited. Firstly, as computational models become more complex, they typically become harder to analyze, and ultimately become as difficult to interpret as the system they are designed to mimic, and therefore have diminished explanatory value. Secondly, by focusing on the details particular to a given system (and therefore distinct to other systems) systems biology models are inherently focused on specific contexts, and therefore of less use in the search for general *principles* [although such principles can sometimes be inferred, for example by comparing the structure of similar models in different contexts (Milo et al., 2004), or by examining the prevalence of putative regulatory "mofifs" (Milo et al., 2002; Alon, 2007)].

The systems biology process of gathering details, and then painstakingly piecing them together to obtain a model of the system as a whole, is characteristic of the biologists approach to problem solving. This approach is vitally important since without a rigorous procedure for identifying the "parts" in a complex biological system we could not begin to piece together how they function collectively. Yet, the dynamics of the whole is determined by interactions between these parts, and surprising behavior may *emerge* from these interactions in unforeseen ways (Anderson, 1972). A different way of thinking is also therefore needed if we are understand this emergence and assimilate detailed biological knowledge into a coherent understanding of biological system organization.

As many have noted, this alternative perspective may come from physics (Bialek, 2012). In contrast to the biologist's descriptive approach to experimental data, physicists tend to seek organizational principles that explain the data. These organizational principles are known as *theories* and typically consist of a set of statements that describe observed phenomena, which may, but do not need to, have a mathematical form. Importantly, the power of a physical theory is not only determined by its ability to describe observed data, but also its simplicity or elegance. In the physical sciences, a theory is more powerful if it can explain more data, and do so simply. Thus, the aforementioned Feynmans quote. To a physicist while the data may *appear* complex, this observational complexity need not be produced by complex underlying principles; and indeed should not when properly understood.

This focus on elegance has an important consequence. Because notions of beauty, elegance, or simplicity, are central to the development of physical theories, it is typical for a physical theory to become *simpler* as it develops, as the mathematical language used to phrase the theory is sharpened and deeper organizational principles are discovered [the late Murray Gell-Mann has an entertaining TED talk on this topic (Gell Mann, 2007)]. It is notable that this process of simplification is in stark contrast to the usual progression in the biological sciences, in which increased understanding typically means increased complexity, as more is learnt and more details are filled-in.

Here, we will argue that both the biologist's reductionist approach and the physicists search for principles are equally important, and can be reconciled with the use of some tools from physics. We should emphasize from the start that this perspective is not new: research at the interface between the biological and physical sciences has a long and fruitful history. Notable examples include Turing's pioneering reaction-diffusion theory of morphogenesis (Turing, 1990), Wolpert's French flag model of morphogen gradient response (Wolpert, 1969) and Waddington's interpretation of developmental processes arising from an epigenetic landscape (Waddington, 1957). These early models have been widely explored, refined, and revised in a variety of contexts and have inspired the development of new theoretical approaches to investigating cell biology (Murray, 1989). Here we will explore some of these developments, and look to how they may be developed in the future. To focus our discussion we will specifically consider some theoretical notions that are relevant to understanding phenotypic switching. In particular, we will discuss what we mean by a cell "type," how cell types relate to cell "states" and how we can use both systems biology and statistical physics to better understand cell fate transitions.

The "physical laws" of cell fate dynamics

The phenotype of a cell is determined by complicated interactions between its genome, transcriptome, proteome, and epigenome. Collectively, these interactions form an intricate molecular regulatory network (Karlebach and Shamir, 2008). Although much progress has been made in inferring the structure of these networks (Marbach et al., 2012) and making quantitative predictions of a cell's behavior based on known interactions (De Jong, 2002), we are still far from a complete knowledge of how these networks function.

Because of the intrinsic complexity of intracellular molecular regulatory networks, it is difficult to determine which aspects of a given cell fate regulatory network are specific to a particular context, and which features are more general. To address this question, there have historically been attempts to understand cellular behavior in a more abstract way, without knowledge of all the details of the molecular determinants of the dynamics. Rather than dissect the details of a particular system, these studies seek (mathematical) *laws* that describe cellular dynamics in simple way, analogous to the laws that describe physical dynamics, such as Newton's laws in mechanics or Maxwell's laws in electrodynamics. We note that these attempts do not deny the complexity of biological systems; rather their aim is to distinguish between *a set of rules,* which describe how the dynamics work in general, and how the complexity specific to a given context emerges.

To illustrate this approach we will start with a simple—although not entirely accurate, as we will discuss later—law for cellular dynamics that is well known to most developmental biologists: Waddington's epigenetic landscape (Waddington, 1957). In Waddington's view, the state of a cell can be described as a position in a complex landscape that contains multiple hills and valleys. To account for dynamics Waddington envisioned the cell "rolling" downward on the landscape, like a marble rolling in a bowl under the influence of gravity, before coming to rest at a local minimal of the landscape. The implicit rule behind these dynamics is very simple, and is indeed equivalent to the physical law governing the movement of a particle in a gravitational or electrostatic field. Because of this simplicity, we can formulate Waddington's idea mathematically with minimal technicalities.

For this purpose, let us first assume that the state of the cell can be defined by a set of numbers, x_i for $i = 1, 2, \ldots, N$, that could, for example, account for the expression levels of the all the genes in the cell (or any other molecular characteristic of interest). We can thus express the state of the cell by the vector $x = (x_1, x_2, \ldots, x_N) \in X$, which can be interpreted as the position of the cell in an N-dimensional expression space. Now let $V(x)$ denote the height of Waddington's landscape at position x. Waddington's implicit rule, that the cell rolls down the landscape $V(x)$, following the path of steepest descent, can then be expressed mathematically as:

$$\frac{dx}{dt} = -\nabla V(x) \tag{8.1}$$

where $\nabla = \left(\frac{\partial}{\partial x_1}, \frac{\partial}{\partial x_2}, \ldots\right)$ is the gradient of the landscape $V(x)$. This rule is structurally identical to the physical law governing the dynamics of a charged particle within an electrostatic potential $V(x)$ neglecting inertia (e.g., when immersed into a highly viscous liquid).

Waddington's framework provides a very simple law that describes cellular dynamics generally. Importantly, this law does not ignore biological complexity: the complexity of the biological interactions between genes is "hidden" in the shape of the landscape $V(x)$. As Waddington noted (Waddington, 1939):

> *The line followed by the process [development] is the bottom of a valley ... One might roughly say that all these genes correspond to the geological structure which moulds the form of the valley.*

Thus, in Waddington's view, the structure of the landscape is determined by complex patterns of interactions within an underpinning gene regulatory network. Due to this complexity, the form of this landscape might not be fully known (or entirely experimentally amenable), yet the governing law still remains and acts as a conceptually simply principle to guide our understanding of how the set of regulatory interactions as a whole guide changes in cell fate.

Since he introduced it in the 1930s, Waddington's model of cellular dynamics has, either implicitly or explicitly, guided much of our understanding of development. However, although conceptually appealing, Waddington's formulation is not a complete theory because it does not allow for aspects of cellular dynamics, which we know to be important. Yet, nevertheless, it is a remarkably prescient example of an attempt to find unifying principles in biology.

To see that Waddington's description is not complete, one just needs to consider the cell cycle, which can be considered as a periodic transition between different cell states, corresponding to a recurrent trajectory in X. Intuitively, in the absence of inertia, a marble rolling down a hill must eventually come to a place of rest and so cannot engage in sustained periodic trajectories. More formally, the dynamics encoded in Eq. (8.1) do not allow for recurrent dynamics and so cannot describe oscillatory phenomena such as the cell cycle. Thus, to allow for the possibility of oscillations, Waddington's law has to be adapted. More generally, the dynamics of the cell state $x(t)$ may be described by a dynamical system (Huang et al., 2005; Huang, 2012):

$$\frac{dx}{dt} = F(x) \tag{8.2}$$

where $F(x) = (F_1(x), F_2(x), \ldots)$ is a vector of functions that determines the regulatory interactions between the genes. In analogy to physics, this vector function is often called a (force) *field*.

If we interpret the entries of x as the expression states of genes, then $\boldsymbol{F}(x)$ is a direct formulation of the gene regulatory network: each row, i, of Eq. (8.2) reads $dx_i/dt = F_i(x_1, x_2, \ldots)$. The function F_i describes how the dynamics of gene i depends on all the other genes. In particular, the way in which the dynamics of gene i depends on gene j (i.e., whether, for a given value of x_j the interaction $x_j \rightarrow x_i$ is "activating" or "repressing") are given by the partial derivatives, $\partial F_i/\partial x_j$. Specifically, if $\partial F_i/\partial x_j \neq 0$, then gene i depends on gene j which means that in the genetic network a link between gene i and j exists, while the weight of this link (which can be positive or negative) is the value of $\partial F_i/\partial x_j$. Hence, the function $\boldsymbol{F}(\boldsymbol{x})$ completely defines the genetic network at state $\boldsymbol{x}$ via its Jacobian matrix $J_{ij} = \partial F_i/\partial x_j$.

Eq. (8.2) is the most general way of writing a continuous dynamical system. If no periodic trajectories exist, then $\boldsymbol{F}(\boldsymbol{x})$ is rotation-free and can be written in terms of the gradient of a scalar potential function and we recover Eq. (8.1). However, it is not generally true that a vector field can be written as the gradient of a scalar potential. In fact, this is a very strong constraint that most dynamical systems do not satisfy. In this case, Waddington's rule is not strictly valid, yet the 'force' field $\boldsymbol{F}(\boldsymbol{x})$ can nevertheless be partitioned into pieces by the so-called Hodge-Helmholtz decomposition (Bhatia et al., 2013; Wang et al., 2011; Weinreb et al., 2018):

$$\boldsymbol{F}(\boldsymbol{x}) = -\nabla V(x) + \boldsymbol{F}_r(\boldsymbol{x}), \tag{8.3}$$

where $\boldsymbol{F}_r(\boldsymbol{x})$ is a rotational component that allows for periodic trajectories, such as the cell cycle. This decomposition is useful since for arbitrary cell state dynamics, an "effective" Waddington's landscape $V(x)$ can be found that will describe the dynamics around equilibrium states of Eq. (8.2). This fact has led to increasing interest in finding the effective Waddington landscape, that is, the rotation-free component of $\boldsymbol{F}$ from experimental data (Wang et al., 2011; Weinreb et al., 2018).

More philosophically, the construction given by Eq. (8.2) is useful because it provides a way to connect cellular states to cell fates. Specifically, assuming that the cellular dynamics are dissipative (i.e., operating away from thermodynamic equilibrium: a weak condition that is satisfied whenever the cell is exchanging energy/matter with its environment) it is natural to associate *attractor* states of the dynamical system Eq. (8.2) with distinct cell "types."

While the technical definition of an attractor is complex, they are essentially preferred subsets of the state space X to which neighboring states are drawn over time (Strogatz, 2018). In the context of Waddington's formalism, the attractors are local minima of the landscape, at which cells tend to accumulate. In the dynamical systems perspective, these are known as fixed-point attractors, which are the only type of attractor that Waddington's model allows. However, more generally, attractors can have a range of different forms, including: fixed points (i.e., isolated attractor points in X), limit cycles (isolated sets of connected points that are topologically equivalent to the circle, and are associated with self-sustaining oscillations), or even topologically more complicated sets (such as limit tori, strange attractors, etc.) (Strogatz, 2018).

The correspondence between attractors and cell types was first suggested, in embryonic form, by Waddington in the 1930s and was popularized and developed by Delbrück, Szilard, Kauffman, and others (Kauffman, 1993, 1969; Delbrück, 1949) [an interesting discussion between Waddington and the French mathematician René Thom tracing the origins to this idea can be found in (Thom, 1982)]. It is a notion that has received some experimental support (Huang et al., 2005). This idea gives a simple explanation of why the same genome can give rise to a variety of different cell types: the genome of a cell determines

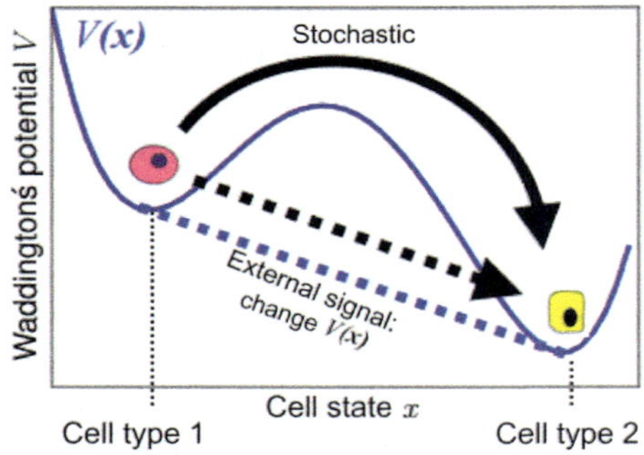

Figure 8.1 Cell types and cell differentiation. In absence of a rotational field F_n cell types can be seen as local minima of the landscape described by $V(x)$ in Eq. (8.3). Depending on the initial conditions, a cell's state will converge toward one of the two local minima. Transitions between cell types, for example, through differentiation, can then occur in two different ways. (i) Deterministically *(black dashed arrow)*: through an external signal, the landscape $V(x)$ changes its structure such that the barrier between two states disappears creating a downward path to the lower minima *(dashed blue line)*. This corresponds to a bifurcation of the corresponding dynamical system, Eq. (8.2). (ii) Stochastically *(bold arrow)*: in which the landscape $V(x)$ does not change but random fluctuations in the cell state $x(t)$ occasionally overcome the potential barrier and drive stochastic transitions between states.

the function $\boldsymbol{F}(\boldsymbol{x})$, which defines the dynamical system Eq. (8.2), that can support numerous attractors, each of which defines a distinct phenotypic cell type (Huang, 2012).

The dynamical systems description of cell states also provides an intuitive way to understand phenotypic switching. First we note that in addition to $\boldsymbol{x}$, the field $\boldsymbol{F}$ will also depend upon a set of parameters λ that may change over time (e.g., to reflect changes in environment), thus, $\boldsymbol{F} = \boldsymbol{F}(\boldsymbol{x}; \lambda(\boldsymbol{t}))$. The parameters λ are known as bifurcation parameters. Importantly, at critical points, known as bifurcation points, the field $\boldsymbol{F}$ will change its topology. Practically, this results in the loss and gain of attractor states. If a developmental process tunes a subset of bifurcation parameters such that once attractor state loses stability and another gains stability, a cell fate transition can occur. For example, in Waddington's landscape, a barrier separating two basins may disappear in response to external signals, and the cell may transit from a "higher" state (i.e., cell type) to a "lower" one. An illustration of such a transition is given in Fig. 8.1.

Statistical mechanics of cell state dynamics

The law stated in equation Eq. (8.2) does not take into account biological noise or measurement uncertainty. In fact, biological noise can play an important role in cellular dynamics, and can, for instance, trigger noise-induced phenotypic switching (Bialek, 2012; Horsthemke, 1984; Chang et al., 2008).

Noise can be incorporated in to our cell dynamics law as follows:

$$\frac{dx}{dt} = \boldsymbol{F}(x) + \boldsymbol{\sigma}(x,t)\boldsymbol{\xi}(t), \tag{8.4}$$

where $\xi(t)$ is a vector of independent Gaussian white noise terms that accounts for both biological and technical noise of magnitude $\sigma(x,t)$. Again, we find an equivalence to physical dynamics: this equation is the *Langevin equation*, which is well known from statistical physics.

This formulation is useful because it allows us to explore how stochastic mechanisms play a role in phenotypic switching. For example, consider two cell types, corresponding to local minima of the potential $V(x)$, as depicted in Fig. 8.1. If $V(x)$ does not change in structure then no deterministic transitions between the two cell types can occur. However, in the presence of stochastic fluctuations in expression, the noise term $\sigma(x,t)\xi(t)$ in Eq. (8.4) can allow for transient fluctuations in expression that are sufficiently strong to overcome the energy barrier between the states and induce spontaneous transitions between cell types. This process is sometimes called "stochastic tunneling" (Iwasa et al., 2004; Wenzel and Hamacher, 1999) and is conceptually very similar to the tunneling effect in quantum physics.

Because Eq. (8.4) contains a random term there is now no unique trajectory $x(t)$ that evolves smoothly over time. Instead, each cell will follow a different trajectory in the expression space X, in a way determined by both the deterministic and stochastic dynamics. In this case, it is not beneficial to dissect any particular trajectory in detail, but rather it is more sensible to consider the probability $p(x,t)$ that the cell will be in state x at time t. This problem has been well studied in statistical physics, and it is known that the probability $p(x,t)$ will evolve as (Cox and Miller, 1965):

$$\frac{\partial p(x,t)}{\partial t} = -\sum_{i=1}^{N}\frac{\partial}{\partial x_i}\left[F_i(x,t)p(x,t)\right] + \sum_{i=1}^{N}\sum_{j=1}^{N}\frac{\partial}{\partial x_i \partial x_j}\left[D_{ij}(x,t)p(x,t)\right], \tag{8.5}$$

where $\boldsymbol{D} = \frac{1}{2}\boldsymbol{\sigma}\boldsymbol{\sigma}^T$ is a matrix of "diffusion coefficients." This equation is known as the *Fokker-Planck equation*. Thus, again, we have found a simple, general mathematical law for the cell dynamics, inspired by a similar law in physics, in which all the system's complexity is "hidden" within the diffusion coefficients

$D(x, t)$ and the field $\boldsymbol{F}(x, t)$, both of which may depend in complex ways on the cell state x and interactions between the genes.

While the Fokker-Planck equation is typically analytically intractable, some general conclusions, valid in a wide range of situations, can be drawn. Notably, it can be shown that if the expression levels are all bounded and $\boldsymbol{D}$ and $\boldsymbol{F}$ do not explicitly depend on time, then $p(x, t)$ will converge to a (not necessarily unique) stationary distribution $p^*(x)$ as $t \to \infty$ (Zeeman, 1988). This stationary probability distribution may have numerous "peaks" (regions of high probability) and "valleys" (regions of low probability) and we may establish an alternative definition of a cell type in the stochastic regime in terms of the structure of this landscape that is analogous to Waddington's principle: cell types correspond to local maxima of the stationary probability distribution (which may be connected, for example to form a closed ridge, representing stochastic periodic trajectories). In the case that the matrix of diffusion coefficients $\boldsymbol{D}$ is constant, the local maxima of $P^*(x)$ are coincident with the attractors of Eq. (8.2), although this is not necessarily the case when the diffusion coefficients are not constant. In the case, that $\boldsymbol{D}$ is not constant, the situation is more subtle and local maxima may not directly correspond to attractors of $\boldsymbol{F}$, but rather emerge from interactions between $\boldsymbol{F}$ and $\boldsymbol{D}$. For example, if $\boldsymbol{D}$ explicitly depends on the state x then there may be regions in which noise is tightly regulated and the diffusion coefficient is small. Such regions may act as "diffusive traps" that can lead to an accumulation of probability (and therefore to peaks in $p(x)$) even in absence of any fixed points of the field $\boldsymbol{F}(x)$.

The Fokker-Planck and Langevin formalisms are equivalent and describe the stochastic dynamics of expression patterns within a single cell. However, neither formalism accounts for cell death, division, or migration and so they offer an incomplete description for collective cellular dynamics. To address this issue Weinreb et al. (2018) introduced a direct extension of the Fokker-Planck formalism to describe the dynamics of cell population density $c(x, t)$ (i.e., the expected number of cells per unit volume at x). Their extension has the form

$$\frac{\partial c(x,t)}{\partial t} = \nabla[D(x)\nabla c(x,t)] - \nabla[\boldsymbol{F}(x)c(x,t)] + r(x)c(x,t), \qquad (8.6)$$

where $r(x)$ is the rate of cell division in state x minus rate of death/migration. The first two terms are equivalent to Eq. (8.5), under the assumption that noise is constant and isotropic and can therefore be described by a scalar function $D(x)$.

The representations Eqs. (8.5) and (8.6) have some implications for the interpretation of experimental data. Modern experimental methods now allow the profiling of the entire transcriptome of thousands of individual cells in a single experiment (Grün and van Oudenaarden, 2015). To interpret this data it is common to perform clustering analysis, and associate distinct cell types with distinct clusters in the data. The presented theoretical framework reveals that this process is not entirely ad hoc: the Fokker-Planck equation and its extension Eq. (8.6) predict that regions of high probability, and thus data clusters, will emerge from interactions between the fields $\boldsymbol{F}$ and $\boldsymbol{D}$ (Risken and Frank, 1984). If the diffusion coefficient is constant then it is appropriate to assume that regions of high probability in the data will form around attractors of the deterministic dynamics. Furthermore, if the attractor is a fixed point, one would expect a globular cluster in the data; yet, from the theory we would also expect other topological structures to appear if the attractor geometry is more complex. For example, if the attractor corresponds to a periodic trajectory (a limit cycle) one would expect to find a "ring-shaped" data cloud or more complex structures for more complex attractors (Chance et al., 2014). Searching for these structures is considerably harder to perform than clustering, since it requires a means to identify nontrivial topological structures in noisy data, and we do not yet have methods to do this. We anticipate that recent advances in topological data analysis may prove powerful in this regard (Carlsson, 2009).

The Fokker-Planck formalism can also be used to investigate the dynamics of cell state changes. For example, based on their model Weinreb et al. (2018) developed a powerful inference algorithm that is able to determine Waddington's landscape $V(x)$ and, to some degree, the cell differentiation trajectories from single cell expression data. Importantly, their analysis also identifies some inherent limitations to cell state trajectory inference from single-cell expression data. These limitations arise from the invariance of Eq. (8.6) to certain transformations. In particular, since Eq. (8.6) is invariant under adding a divergence-free component $\boldsymbol{F}_r$ (i.e., with $\nabla \cdot \boldsymbol{F}_r(x) = 0$) to the field $\boldsymbol{F}(x)$, any inference algorithm based on measuring snapshots of cell distributions (e.g., from single cell data) cannot detect such components. This means that periodic trajectories cannot be inferred. Furthermore, Weinreb et al. (2018) also show that $D(x)$, $r(x)$, and $V(x)$ cannot be inferred simultaneously. Hence, in order to infer the potential $V(x)$, one needs to know $r(x)$ and $D(x)$ a priori, for instance from other experiments. Collectively these results are chastening: many features of cell fate trajectories, including reversibility of cell state transitions and oscillations, cannot be detected by dynamical inference from cell samples taken at fixed time points. To advance in this area other experimental approaches or analysis techniques based on topological features of the data are needed.

Universality in cell biology

In the theory outlined so far all the biological details are "hidden" within the functions $\boldsymbol{F}(x)$, $\boldsymbol{D}(x)$, and $\boldsymbol{r}(x)$. These functions will depend on biological context and attempts to (partially) specify these functions from biological knowledge are the aim of systems biology. However, biology is complex and so these models often tend to be complex. While biologically accurate, the high complexity and large number of free parameters in many systems biology models renders them difficult to analyse. This also poses a major problem for testing hypotheses through these models, since the statistical power of a model to describe a data set reduces with the number of free parameters. A general systems biology model may have hundreds (or thousands) of unspecified parameters, and fitting such a model to data may easily lead to overfitting. While these models can be rendered predictive by using appropriate regularization (Neumaier, 1998), there may still be large areas of parameter space for which the model fits the data, and so fitted parameters do not unambiguously represent the underlying biological truth. Hence, the straightforward approach of encoding all the known details of a biological process into a mathematical model is unlikely to succeed in finding the underlying biological principles.

However, there are some principles from physics that again may rescue the situation. In particular, it is well known that when considering the statistical properties of models of large systems, often only very few features of the model affect its predictions and sets of similar models that share these features may therefore all explain the data equally well. This principle is called *universality* in the physics literature, a phenomenon originally described in the context of statistical and quantum physics (Stanley, 1999). In physics, universality refers to the observation that, close to a phase transition, apparently very different systems will behave in very similar ways and, furthermore, there are only a small number of ways that a system can behave. Accordingly, physical systems can be categorized into *universality classes* by their common behavior near a phase transition. Importantly, often only a few salient features of a system determine the universality class, and these features tend to be general properties—such as the dimension, symmetries, or sets of conserved quantities—rather than fine details.

This notion, that apparently different processes may behave the same way, when properly viewed, can be extended beyond its original statistical physics context. For example, it is a well-known fact of statistics that for large mean values or sample sizes, statistical distributions tend to converge to a handful of limiting distributions, which under suitable rescaling of the measured outcomes can be rendered parameter-free. The most famous manifestation of this is the *central limit theorem*, which states that if a random variable of interest (x say) arises as a sum of N other independent random variables (x_i, say) which each have finite mean and variance, then the distribution of x will converge toward a normal distribution, irrespective of the details of the distributions of the x_i's as N gets large (see Fig. 8.2). Furthermore, in terms of the effective variable $z = (x-\mu)/\sigma$, where μ is the mean of x and σ its variance (i.e., the Z-score), the limiting distribution is the standard normal, that is, $z \sim \mathcal{N}(0,1)$. Since both μ and σ can usually be directly estimated from sample data, when phrased in terms of the rescaled variable z, this universality prediction is parameter-free and can be tested in direct way, without fitting. For instance, if the measured data is not simply the sum of independent random variables with finite first and second moments, then it will possess features that deviate from the standard normal. For example, if the underlying random variables do not have finite means then the limiting distribution may have a fat power-law tail characteristic of a Levy flight (Mandelbrot, 1982).

In recent years the principle of universality has also been used in biology (Bialek, 2012; Rulands and Simons, 2017). An example is the dynamics of stem cell populations observed via cell lineage tracing experiments (Kretzschmar and Watt, 2012). In these experiments, individual stem cells are labeled and their progeny are counted. From theoretical considerations it can be shown that if the cells divide stochastically—sometimes dividing symmetrically and sometimes asymmetrically in a probabilistic way—then the clone (or colony) size distribution will converge to an exponential distribution (Klein and Simons, 2011; Blanpain and Simons, 2013). Conversely, if the stem cells divide purely asymmetrically then the clone size distribution will, in the limit of large clone sizes, converge to a normal distribution (Parigini and Greulich, 2020). Alternatively

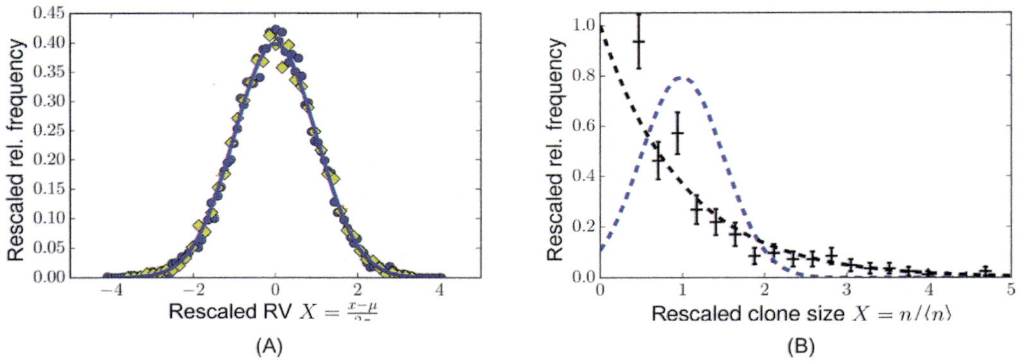

Figure 8.2 Universality in mathematics and biology. (A) The central limit theorem. The frequency distributions of outcomes of three different stochastic processes are shown: *(blue discs)* the mean value of N rolls of a fair dice; *(red crosses)* a symmetric random walk with N steps; and *(yellow diamonds)* the sum of N exponentially distributed random numbers. In all cases $N = 10{,}000$, and outcomes are given in terms of the Z-score. The *black line* is a standard normal distribution. All processes exhibit the same statistics, when rescaled, indicating universality. (B) Universality in stem cell dynamics. Model predictions of two simple models of clonal dynamics, from different universality classes. *(Black dashed line)* Prediction from a neutral competition model of stem cell proliferation in which each cell can divide symmetrically or asymmetrically in a probabilistic way. *(Blue dashed line)* Prediction from a model that allows only purely asymmetric divisions. Points with error bars show clonal data from mouse esophagus. *Source*: (A) Taken from Doupe, D.P., Alcolea, M.P., Roshan, A., Zhang, G., Klein, A.M., Simons, B.D., et al., 2012. A single progenitor population switches behavior to maintain and repair esophageal epithelium. Science 337, 1091. doi:10.1126/science.1218835. Available from: <http://www.ncbi.nlm.nih.gov/pubmed/22821983> (B) Greulich, P., 2019. Mathematical modelling of clonal stem cell dynamics incomputational stem cell biology. In: Cahan, P. (Ed.), Methods in Molecular Biology. Humana Press.

if, in addition, clones fragment and merge then a log-normal distribution of labeled cell clusters is expected (Rulands and Simons, 2017). These few possibilities represent the universality classes for stem cell dynamics: regardless of the biological mode of regulation of the divisions, in the limit of large mean clone sizes all models that give rise to stochastic division patterns that support neutral competition will give rise to an exponential clone size distribution, and all models that maintain purely asymmetric divisions will give rise to a normal clone size distribution. Thus, the principle of universality provides a means to test for broad principles without having to commit to biological details that may not be fully known (Greulich, 2019). Furthermore, since models of the same class produce the same scaling distribution, it is sufficient to analyze the data in light of the *simplest* model of that universality class, since even the simplest model captures the salient features of the biology. This is a tremendous advantage because it allows us to significantly simplify comparisons of models with data, and thereby maintain statistical power.

Critique and outlook

We have seen that, inspired by physical laws, simple laws can also be found to elegantly describe the dynamics of cell state and fate changes. However, these laws are not yet complete. For instance, Waddington's notion of a cell type as an attractor of the regulatory network dynamics is beautifully simple and useful in many contexts, yet biological considerations tell us that it is not always appropriate. Consider the cell cycle. During the cell cycle, cells transit through G1, S, G2, and M phases in a periodic way. This repetition can be seen as a limit cycle, that is, an attractor of the dynamics, and it is reasonable to view cell types independently of cell cycle phase. However, what if a cell pauses the cell cycle by entering the G0 phase? Such a paused state may emerge as a distinct attractor of the dynamics and accordingly it would be seen as a separate cell type. Yet, few biologists would view classify it as such. Here the biological context matters, and this case provides an example of how a physically motivated theory should be adapted in the light of biological detail.

Conclusions

Physicists typically seek unifying principles, while biologists usually seek detailed explanations. Here we have argued that both are important and a synergy of biological and physical approaches can help us understand cell fate dynamics, clarify what we mean by cell "types," and how cells change type.

We are still in the early stages of identifying unifying principles that underpin cellular dynamics. Any theory should satisfy the physicists desire for simplify or elegance, yet not ignore the complexity that is inherent to biology. The search for principles must therefore involve both experimental biologists and theoreticians. We anticipate the search for a theory of the physics of living matter will generate some of the most exciting future problems in both the physical and biological sciences.

References

Alon, U., 2006. An Introduction to Systems Biology: Design Principles of Biological Circuits. Chapman and Hall/CRC.

Alon, U., 2007. Simplicity in biology. Nature 446, 497.

Anderson, P.W., 1972. More is different. Science 177, 393–396.

Bhatia, H., Norgard, G., Pascucci, V., Bremer, P.T., 2013. The Helmholtz-Hodge decomposition–a survey. IEEE Trans. Vis. Comput. Graph. 19, 1386–1404. Available from: https://doi.org/10.1109/TVCG.2012.316.

Bialek, W., 2012. Biophysics: Searching for Principles. Princeton University Press.
Bianconi, E., Piovesan, A., Facchin, F., Beraudi, A., Casadei, R., Frabetti, F., et al., 2013. An estimation of the number of cells in the human body. Ann. Hum. Biol. 40, 463–471.
Blanpain, C., Simons, B.D., 2013. Unravelling stem cell dynamics by lineage tracing. Nat. Rev. Mol. Cell Biol. 14, 489–502. Available from: https://doi.org/10.1038/nrm3625.
Carlsson, G., 2009. Topology and data. Bull. Am. Math. Soc. 46, 255–308.
Chance, B., Ghosh, A.K., Pye, E.K., 2014. Biological and Biochemical Oscillators. Academic Press.
Chang, H.H., Hemberg, M., Barahona, M., Ingber, D.E., Huang, S., 2008. Transcriptome-wide noise controls lineage choice in mammalian progenitor cells. Nature 453, 544.
Cox, D.R., Miller, H.D., 1965. The Theory of Stochastic Processes. Chapman and Hall, London.
De Jong, H., 2002. Modeling and simulation of genetic regulatory systems: a literature review. J. Comput. Biol. 9, 67–103.
Delbrück, M., 1949. Unités biologiques douées de continuité génétique. Colloques internationaux du Centre national de la recherche scientifique, pp. 33–34.
Doupe, D.P., Alcolea, M.P., Roshan, A., Zhang, G., Klein, A.M., Simons, B.D., et al., 2012. A single progenitor population switches behavior to maintain and repair esophageal epithelium. Science 337, 1091. Available from: https://doi.org/10.1126/science.1218835. Available from: <http://www.ncbi.nlm.nih.gov/pubmed/22821983>.
Gell Mann, M., 2007. Beauty, truth and ... physics? <https://www.ted.com/talks/murray_gell_mann_on_beauty_and_truth_in_physics?language = en> (accessed 12.06.19.).
Greulich, P., 2019. Mathematical modelling of clonal stem cell dynamics. Computational Stem Cell Biology: Methods Protocols. Springer Science + Business Media, pp. 107–129. Available from: https://doi.org/10.1007/978-1-4939-9224-9.
Grün, D., van Oudenaarden, A., 2015. Design and analysis of single-cell sequencing experiments. Cell 163, 799–810.
Horsthemke, W., 1984. Noise Induced Transitions. Springer.
Huang, S., 2012. The molecular and mathematical basis of Waddington's epigenetic landscape: a framework for post-Darwinian biology? BioEssays 34, 149–157. Available from: https://doi.org/10.1002/bies.201100031.
Huang, S., Eichler, G., Bar-Yam, Y., Ingber, D.E., 2005. Cell fates as high-dimensional attractor states of a complex gene regulatory network. Phys. Rev. Lett. 94, 128701. Available from: https://doi.org/10.1103/PhysRevLett.94.128701.
Iwasa, Y., Michor, F., Nowak, M.A., 2004. Stochastic tunnels in evolutionary dynamics. Genetics 166, 1571–1579.
Karlebach, G., Shamir, R., 2008. Modelling and analysis of gene regulatory networks. Nat. Rev. Mol. Cell Biol. 9, 770.
Kauffman, S., 1969. Homeostasis and differentiation in random genetic control networks. Nature 224, 177.
Kauffman, S.A., 1993. The Origins of Order: Self-Organization and Selection in Evolution. OUP USA.
Kitano, H., 2002. Systems biology: a brief overview. Science 295, 1662–1664.
Kitano, H., 2004. Biological robustness. Nat. Rev. Genet. 5, 826.

Klein, A.M., Simons, B.D., 2011. Universal patterns of stem cell fate in cycling adult tissues. Development 138, 3103. Available from: https://doi.org/10.1242/dev.060103.

Kretzschmar, K., Watt, F.M., 2012. Lineage tracing. Cell 148, 33–45. Available from: https://doi.org/10.1016/j.cell.2012.01.002. Available from: <http://www.ncbi.nlm.nih.gov/pubmed/22265400>.

Mandelbrot, B., 1982. The Fractal Geometry of Nature. W. H. Freeman.

Marbach, D., Costello, J.C., Küffner, R., Vega, N.M., Prill, R.J., Camacho, D.M., et al., 2012. Wisdom of crowds for robust gene network inference. Nat. Methods 9, 796.

Meier, P., Finch, A., Evan, G., 2000. Apoptosis in development. Nature 407, 796.

Milo, R., Shen-Orr, S., Itzkovitz, S., Kashtan, N., Chklovskii, D., Alon, U., 2002. Network motifs: simple building blocks of complex networks. Science 298, 824–827.

Milo, R., Itzkovitz, S., Kashtan, N., Levitt, R., Shen-Orr, S., Ayzenshtat, I., et al., 2004. Superfamilies of evolved and designed networks. Science 303, 1538–1542.

Murray, J.D., 1989. Mathematical Biology. Springer, Berlin, Germany.

Neumaier, A., 1998. Solving ill-conditioned and singular linear systems: a tutorial on regularization. SIAM Rev. 40, 636–666. Available from: https://doi.org/10.1137/s0036144597321909.

Parigini, C., Greulich, P., 2020. Universality of clonal dynamics poses fundamental limits to identify stem cell self-renewal strategies, bioRxiv 2020.02.10.941286v1 P.

Pertea, M., Shumate, A., Pertea, G., Varabyou, A., Breitwieser, F.P., Chang, Y.C., et al., 2018. CHESS: a new human gene catalog curated from thousands of large-scale RNA sequencing experiments reveals extensive transcriptional noise. Genome Biol. 19, 208. Available from: https://doi.org/10.1186/s13059-018-1590-2.

Risken, H., Frank, T., 1984. The Fokker-Planck Equation: Methods of Solution and Applications. Springer-Verlag.

Rual, J.F., Venkatesan, K., Hao, T., Hirozane-Kishikawa, T., Dricot, A., Li, N., et al., 2005. Towards a proteome-scale map of the human protein-protein interaction network. Nature 437, 1173.

Rulands, S., Simons, B.D., 2017. Emergence and universality in the regulation of stem cell fate. Curr. Opin. Syst. Biol. 5, 57–62. Available from: https://doi.org/10.1016/j.coisb.2017.08.001.

Stanley, H.E., 1999. Scaling, universality, and renormalization: three pillars of modern critical phenomena. Rev. Mod. Phys. 71, S358–S366. Available from: https://doi.org/10.1103/revmodphys.71.s358.

Strogatz, S.H., 2018. Nonlinear Dynamics and Chaos: With Applications to Physics, Biology, Chemistry, and Engineering. CRC Press.

Thom, R., 1982. Modeles Mathematiques de la Morphogenese. Springer.

Turing, A.M., 1990. The chemical basis of morphogenesis. Bull. Math. Biol. 52, 153–197.

UniProt Consortium, 2014. UniProt: a hub for protein information. Nucleic Acids Res. 43, D204–D212.

Waddington, C., 1939. An Introduction to Modern Genetics. The Macmillan Company, New York.

Waddington, C., 1957. The Strategy of the Genes. Routledge, London.

Wang, J., Zhang, K., Xu, L., Wang, E., 2011. Quantifying the Waddington landscape and biological paths for development and differentiation. Proc. Natl Acad. Sci. 108, 8257–8262. Available from: https://doi.org/10.1073/pnas.1017017108.

Weinreb, C., Wolock, S., Tusi, B.K., Socolovsky, M., Klein, A.M., 2018. Fundamental limits on dynamic inference from single-cell snapshots. In: Proceedings of the National Academy of Sciences. Available from: <http://www.pnas.org/lookup/doi/10.1073/pnas.1714723115>, <https://doi.org/10.1073/pnas.1714723115>, arXiv:170118.

Wenzel, W., Hamacher, K., 1999. Stochastic tunneling approach for global minimization of complex potential energy landscapes. Phys. Rev. Lett. 82, 3003–3007. Available from: https://doi.org/10.1103/PhysRevLett.82.3003.

Wolpert, L., 1969. Positional information and the spatial pattern of cellular differentiation. J. Theor. Biol. 25, 1–47.

World Health Organization, 2019. Monogenic disorders. <https://www.who.int/genomics/public/geneticdiseases/en/index2.html> (accessed 20.09.19.).

Zeeman, E.C., 1988. Stability of dynamical systems. Nonelinearity 1, 115–155.

9

Disentangling the environmentally induced and stochastic developmental components of phenotypic variation

Günter Vogt
Faculty of Biosciences, University of Heidelberg, Heidelberg, Germany

Introduction

The phenotype of an organism is determined by the genome, environmental impacts, and stochastic developmental events (e.g., Gärtner, 1990, 2012; Falconer and Mackay, 1996; Lewontin, 2000; Vogt et al., 2008; Feinberg and Irizarry, 2010; Verhoeven and Preite, 2014; Skinner, 2015; Vogt, 2015, 2017a; Leung et al., 2016; Angers et al., 2020). These factors together shape the individual phenotype in all life stages leading to the establishment of phenotypic variation in populations. Phenotypic variation is manifested either as a continuum of phenotypes in a given environment, alterations of these phenotypes in different environments (phenotypic plasticity), discrete alternative phenotypes in response to environmental signals (polyphenism), and reversible switches between distinct phenotypes (phenotypic switching) (Fig. 9.1) (e.g., West-Eberhard, 1989; Pigliucci, 2001; Vogt et al., 2008; Simpson et al., 2011; Soll, 2014; Mallon et al., 2016).

The two nongenetic components of phenotypic variation are here called "environmentally induced variation" (EIV) and "stochastic developmental variation" (SDV) due to reasons explained in the next section. They correspond to phenotypic plasticity *sensu stricto* and developmental noise, respectively. Because of the difficulty to determine SDV in natural populations, EIV and SDV were rarely distinguished from each other and mostly treated together under the term "environmental variation" or "phenotypic plasticity" (Pigliucci, 2001; Nanjundiah, 2003;

Phenotypic Switching. DOI: https://doi.org/10.1016/B978-0-12-817996-3.00010-4

Figure 9.1 Different manifestations of nongenetic phenotypic variation from various kingdoms of life. (A) Collage of differently raised isogenic clutch mates of the parthenogenetic marbled crayfish, *Procambarus virginalis*, showing remarkable phenotypic plasticity with respect to growth. Shown are the largest specimens from three culture systems. (B) Iterative flowers of the orchid *Ponerorchis graminifolia* on the same stem showing marked differences in coloration due to SDV. The color pattern of the labellum varies between flowers and between right and left sides in each flower. (C) Nymphs of desert locust *Schistocerca gregaria* as an example of polyphenism showing green solitary phase and brown gregarious phase. (D) Phenotypic switching of pathogenic fungus *Candida albicans* between highly virulent spherical white form (top) and less virulent hyphal opaque form (bottom). (E) Differential gene expression in clone of pathogenic bacterium *Salmonella* ser. Typhimurium. The cells were cultured from a single cell over six generations. The green color shows the stochastic expression of green fluorescent protein under the control of the flagellar promoter flgK in a fraction of cells only. *Source*: (A) from Vogt, G., 2008. The marbled crayfish: a new model organism for research on development, epigenetics and evolutionary biology. J. Zool. 276, 1–13. (B) photo by Rogier van Vugt. (C) from Burrows, M., Rogers, S.M., Ott, S.T., 2011. Epigenetic remodelling of brain, body and behaviour during phase change in locusts. Neural Syst. & Circuits, 1, article 11. (D) from Tao, L., Du, H., Guan, G., Dai, Y., Nobile, C.J., Liang, W., et al., 2014. Discovery of a "white-gray-opaque" tristable phenotypic switching system in Candida albicans: roles of non-genetic diversity in host adaptation. PLoS Biol., 12, article e1001830. (E) from Ackermann, M., 2013. Microbial individuality in the natural environment. ISME J., 7, 465–467.

Fusco and Minelli, 2010). All triggers of phenotypic variation, the genes, the environment, and developmental stochasticity, act through the modification of developmental pathways.

This chapter discusses the differences and relationships between EIV and SDV on a theoretical and empirical basis and presents examples on their extent in animals. I will first define the use of the terms EIV and SDV, examine their occurrence in the various kingdoms of life, and discuss their differences and relationships. Thereafter, I will present laboratory experiments, field studies, and a mathematical simulation on the dissection of EIV and SDV in animals to demonstrate their extent and relative proportions in the phenotypically most diverse organisms on earth. Finally, I will discuss molecular mechanisms underlying EIV and SDV in different organisms and introduce a promising research model for investigating their molecular underpinning and ecological and evolutionary relevance. The chapter mostly deals with continuous phenotypic variation in populations and addresses polyphenism and classical phenotypic switching only briefly.

Considerations on phenotypic, genetic, environmentally induced, and stochastic developmental variations

The phenotypic variation (V_P) of a population or species is traditionally considered to be composed of genetically caused variation ("genetic variation," V_g) and environmentally caused variation ("environmental variation," V_E) (e.g., Lynch and Walsh, 1998).

$$V_P = V_G + V_E$$

Genetic variation is caused by differences in the DNA sequence among population members.

The environmental variation is often subdivided into general environmental variation (V_{Eg}), specific environmental variation (V_{Es}), and variation resulting from genotype by environment interaction (V_{GxE}).

$$V_E = V_{Eg} + V_{Es} + V_{GxE}$$

V_{Eg} is attributed to environmental conditions that are experienced by all individuals in a population, whereas V_{Es} is attributed to environmental conditions experienced only by some specimens. V_{Es} is often called residual variation, error, or noise. V_{GxE} refers to different responses of genetic lineages to variation of the general environment. It is either considered as part of V_E as shown in the above formula or as a separate component besides V_G and V_E. V_{GxE} is difficult to measure in practice and is therefore often ignored. It is experimentally avoided by working with specific breeds within a specific environment.

However, there is convincing evidence for the existence of a third source of phenotypic variation besides genetic variation and environmental variation, namely SDV, the difference in phenotypic outcomes that occurs when genotype and environment are fixed (e.g., Waddington, 1957; Bonner, 1965; Gärtner, 1990, 2012; Falconer and Mackay, 1996; Finch and Kirkwood, 2000; Lajus and Alekseev, 2004; Vogt et al., 2008; Vogt, 2015). This third source of phenotypic variation was already recognized at the beginning of the 20th century (Warren, 1902; Astauroff, 1930) but gained little attention when compared to genetic variation and environmental variation. A detailed history of research on SDV is given in Vogt (2015). Thus, the formula for phenotypic variation should be expanded as follows:

$$V_P = V_G + V_{\mathrm{EI}} + V_{\mathrm{SD}}$$

where G is genetically based, EI is environmentally induced and SD is stochastic developmental.

The genetic component of phenotypic variation can experimentally be distinguished from the nongenetic components by rearing of genetically diverse populations in the same environment and rearing of isogenic populations in different environments. Researchers on phenotypic plasticity have usually not distinguished between variations coming from the external environment and the so-called "internal environment." They treated both on a common conceptual footing (e.g., Nanjundiah, 2003). Thus, SDV was included together with V_{Es} in the noisy proportion of phenotypic variation, which often exceeded 50% of the total phenotypic variation measured (Vogt, 2015; Tikhodeyev and Shcherbakova, 2019). However, summarizing phenotypic variation resulting from individually experienced impacts of the external environment and random intrinsic processes in the same category makes conceptually little sense. V_{Es} and SDV have only in common that they are difficult to measure and remain as residual variance or "noise" if the genetic and macroenvironmental causes of phenotypic variability are determined. To avoid misunderstandings, I will here use the term EIV for the proportion of phenotypic variation that is exclusively generated by cues of the external environment. It is synonymous with environmental variation or phenotypic plasticity *sensu stricto*, excluding variation coming from stochastic internal processes, which is the realm of SDV.

Researchers working on different aspects of intrinsic stochasticity generation at different levels of biological organization coined many different terms for this phenomenon such as gene expression noise, cellular noise, developmental noise, phenotypic noise, intangible variation, range variation, third component, intrinsic chance variation, random variability, developmental instability, etc. (e.g., Waddington, 1957; Bonner, 1965; Gärtner, 1990, 2012; Falconer and Mackay, 1996; Finch and Kirkwood, 2000; Elowitz et al., 2002; Lajus and Alekseev, 2004; Vogt et al., 2008). The word "noise" is usually negatively connoted impairing laboratory experiments, pharmaceutical testing, and true-to-type breeding but SDV has also positive effects since it contributes to individuation, resistance to diseases, and ecological adaptation as discussed earlier (Vogt, 2015). It may even be a potential driver of evolution (Feinberg and Irizarry, 2010; Verhoeven and Preite, 2014; Skinner, 2015; Vogt, 2015, 2017b). The need for an umbrella term that unifies all negative and positive stochastic cellular and developmental processes that contribute to phenotypic variation prompted me to introduce the term "stochastic developmental variation" in analogy to "genetic variation" and "environmental variation" (Vogt, 2015). The term "development" does not only include embryonic development but all progressive and regressive changes of the individual from the zygote to death.

The difference between EIV and SDV is best illustrated on the example of genetically identical populations raised in simple, highly standardized laboratory environments. Under these circumstances, V_G, V_{Es}, and V_{GxE} are zero or very close to zero and cues of the external environment determine the mean or target phenotype in each environment (Fig. 9.2A). When the experimental conditions are changed, the mean phenotype is shifted to another position on the scale of genetically possible phenotypes (Fig. 9.2A), suggesting that EIV is directional. The mean phenotypes of all environments together constitute the range of EIV or the reaction norm (Schlichting and Pigliucci, 1998).

Despite of the absence of V_G, V_{Es} and V_{GxE} in our example, there are still ranges of phenotypes observed around the mean phenotypes of the various environments (Fig. 9.2A), and these ranges are produced by SDV. Because of the stochastic nature of SDV, the ranges of phenotypes around the mean can vary between replicates and subsequent generations grown in the same environment (Fig. 9.2B), but the mean phenotype remains constant in these conditions. EIV is thought to help in acclimatization and adaptation to the inhabited environment, whereas SDV is assumed to serve for risk spreading that enhances the chance to stay in the game of life when the environmental conditions change (Vogt, 2015, 2017a).

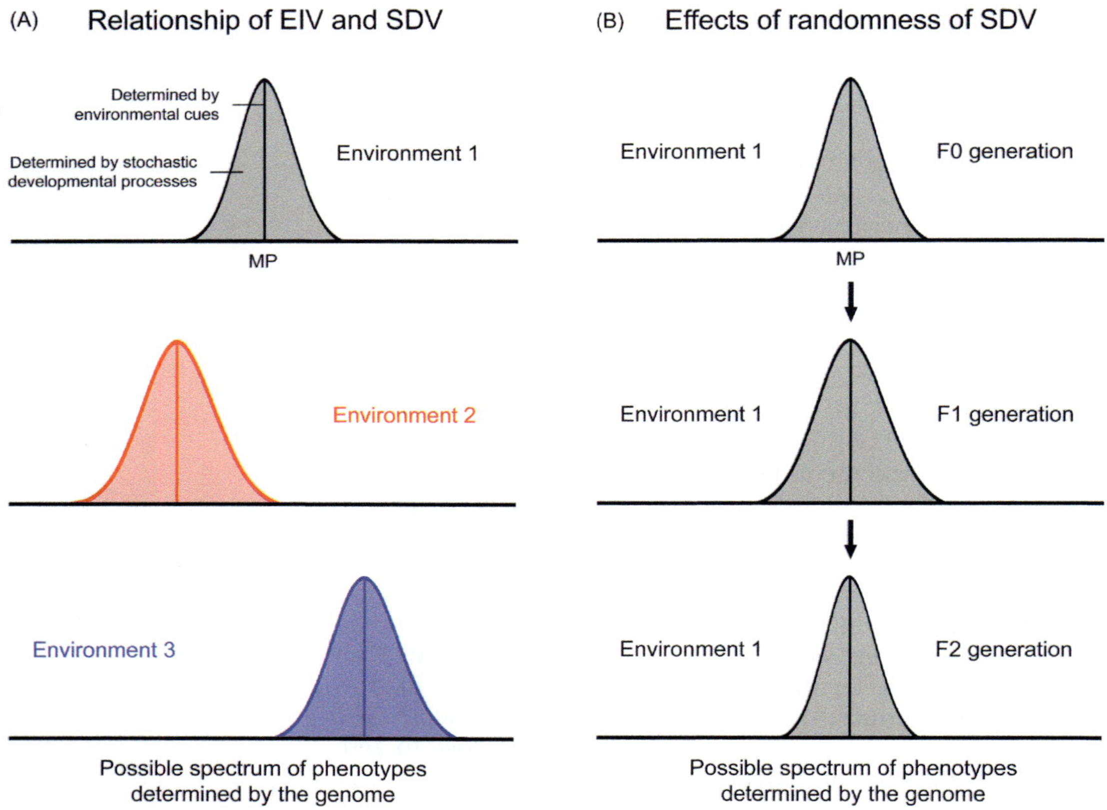

Figure 9.2 Schematic illustration of the relationship of EIV and SDV on the example of genetically identical populations reared in highly standardized laboratory environments. In this condition, genetic variation is zero. However, the genome determines the possible spectrum of phenotypes. (A) Cues of the external environment determine the position of the mean or target phenotype (MP) in a particular environment. Different environments produce different MPs, which together constitute the norm of reaction. Despite genetic identity of the population members, the MP in each environment is surrounded by a range of phenotypes that is produced by SDV. Specific environmental variation, which could also produce a range of phenotypes around the MP, is zero or close to zero under these conditions. (B) In a given environment the MP holds its position on the scale of genetically possible phenotypes throughout subsequent generations (F0-F2) but the ranges of SDV-caused phenotypes around it may vary to some degree due to the stochastic nature of SDV. *Source*: Modified after Vogt, G., 2017a. Facilitation of environmental adaptation and evolution by epigenetic phenotype variation: insights from clonal, invasive, polyploid, and domesticated animals. Environ. Epigenetics, 3, article dvx002.

Not all environmental cues are capable of producing EIV. Many environmental impacts result only in a short-term physiological response and others result in no response at all. Toxicants, food limitation, predator pressure, and harsh environmental conditions are thought to be the most potent elicitors of EIV (Skinner, 2014, 2015; Guillette et al., 2016). In contrast, all traits seem to vary due to SDV but at different degrees, depending on species and condition, as will be demonstrated in the succeeding text.

There are obviously sensitive windows of development for the production of EIV and SDV, depending on trait and organism (Burggren and Mueller, 2015; Skinner, 2015; Vogt, 2015). The environmental exposure at a critical window of development can apparently alter the epigenetic programming and subsequently change gene expression (Skinner, 2014). In the case of SDV, the critical windows seem to depend on the life history of the species considered. As a rule of thumb, embryonic development seems to be of prime importance in determinately growing mammals and insects, whereas the adult life period appears equally important in indeterminately growing crustaceans, mollusks, and fish (Vogt, 2015). For example, differences in adult body weight among identically raised litter-mates of inbred rat had their roots in stochastic events in the zygote and first cleavage stages (Gärtner, 2012), whereas in isogenic clutch mates of marbled crayfish such differences were more attributable to the adult life (Vogt et al., 2008). SDV-related differences in ageing and diseases of humans can have their origin in both early development and the later life (Aranda-Anzaldo and Dent, 2003; Martin, 2014).

EIV and SDV are nongenetic, meaning that they are not caused by variations of the DNA sequence. Instead, they are caused by epigenetic mechanisms. Molecular biologists restrict the term "epigenetics" to stable, mitotically and sometimes meiotically inheritable alterations of gene expression, for example, by changes in DNA methylation, alterations of histone acetylation and methylation and further mechanisms of gene regulation like microRNAs (e.g., Jaenisch and Bird, 2003; Lennartsson and Ekwall, 2009; Moutinho and Esteller, 2017). Unlike the genetic code, the epigenetic marks on the DNA and histones can be erased and rewritten during the lifetime of an organism, although there are examples of the marks persisting across generations (transgenerational epigenetic inheritance), e.g., Richards (2006), Jablonka and Raz (2009), and Perez and Lehner (2019). Organismal and evolutionary biologists additionally consider higher-level epigenetic processes that can lead to the variable expression of phenotypic traits via chemical and mechanical cell-to-cell interactions, self-organization of tissues and self-reinforcing circuitries involving behavior, metabolism and neuroendocrine control (e.g., Newman and Müller, 2005; Hallgrímsson and Hall, 2011; Vogt, 2015). In this paper, the term epigenetics is used in this broader sense.

Genetic variation, EIV, and SDV are sources of phenotypic variation in their own right but, of course, they are not totally independent from each other. For example, no environment is capable of inducing elephant-sized individuals from the genome of a mouse and also no stochastic developmental process will achieve this. Thus, the genome sets the frame in which EIV and SDV can vary. Moreover, in cases where EIV and SDV are mediated via epigenetic marks on the DNA and the histones, both types of nongenetic phenotypic variation are to some degree dependent on each other. For example, if a certain cytosine of the DNA is unmethylated in a given environment it can stochastically only shift into a methylated state and vice versa.

The previous discussion focused on populations consisting of a continuum of phenotypes produced by the same genotype (DNA sequence and identical copies). However, there are also cases in which the same genetic template generates distinct bistable or multistable phenotypes instead of a continuum. These manifestations of phenotypic variation are called polyphenism and phenotypic switching. Polyphenism is the phenomenon where two or more distinct and permanently stable phenotypes are produced by the same genotype. It is triggered by environmental signals and mediated by neurochemical and hormonal pathways and alterations of epigenetic marks on the chromatin (Simpson et al., 2011; Yang and Pospisilik, 2019). Phenotypic switching describes the frequent switch between two or more phenotypes. It is of considerable importance in microbial pathogens.

The developmental stages of holometabolous insects (larvae, pupa and imago) are well known examples of polyphenism. Further examples are seasonal morphs in aphids, density-dependent phenotypes in locusts and diet-mediated phenotypes in bees (Simpson et al., 2011). For example, in the desert locust, *Schistocerca gregaria*, the solitary phase is green and the gregarious phase is brown (Fig. 9.1C) (Burrows et al., 2011; Ayali, 2019). The phenotypic divide between queens and workers in honeybee, *Apia mellifera*, is triggered by differential feeding (presumptive queens are fed with royal jelly and presumptive workers with pollen). The production of discrete alternative phenotypes allows insects to partition life history stages to either feeding and growth (larvae) or reproduction and dispersal (adults), to best adapt to predictable environmental changes (seasonal morphs), to cope with temporally heterogeneous environments (dispersal morphs) and to partition labor (castes of eusocial insects) (Simpson et al., 2011).

Phenotypic switching is best known for pathogenic fungi (Fig. 9.1D) and bacteria (Fig. 9.1E) (e.g., Balaban et al., 2004; Soll, 2014). A particularly well known example is the white-opaque switching system of *Candida albicans*. This human-pathogenic fungus can switch between two or three stable forms (Soll, 2014; Tao et al., 2014). The white and opaque cells show differences in cell morphology (Fig. 9.1D), gene expression, and virulence. Phenotypic switching is apparently caused by developmental stochasticity (Balaban et al., 2004; Ackermann, 2013) but can be influenced to some degree by the external environment (Soll, 2014).

Occurrence of environmentally induced variation and stochastic developmental variation in the kingdoms of life

EIV is well known for all organisms including bacteria, protists, fungi, plants, and animals as documented by numerous papers on phenotypic plasticity (references in Schlichting, 1986; Schlichting and Pigliucci, 1998; Pigliucci, 2001; Nanjundiah, 2003; Justice et al., 2008; Kümmerli et al., 2009; Slepecky and Starmer, 2009; Mulot et al., 2017; Arnold et al., 2019; Fox et al., 2019). In this section I will examine if this also holds for the less well-investigated SDV.

A literature survey demonstrated that SDV occurs in all major evolutionary lineages of the animals (Vogt, 2015). Examples of SDV of morphological, biochemical, physiological, behavioral, and life history traits are available for the Porifera, the most basal branch of the Animalia, the diploblastic Cnidaria, the protostomian Bryozoa, Platyhelminthes, Rotifera, Nematoda, Mollusca, and Arthropoda, and the deuterostomian Ascidiacea and Vertebrata (Table 9.1).

In fungi, the closest relatives of the animals, SDV is documented for gene expression, hyphal morphology, phenotypic switching between morphs (Fig. 9.1D), and resistance to antimycotics in several species of the Ascomycota and Basidiomycota, the two main branches of the fungi (Table 9.1; Li et al., 2010).

In plants, information on SDV comes from the morphology of leaves, branches and roots, the color of flowers (Fig. 9.1B), fruits, and leaves, some life history traits, and molecular parameters such as gene expression and epigenetic profiles. Table 9.1 lists examples for the monocotyledonous orders Asparagales and Zingiberales and the dicotyledonous orders Brassicales, Lamiales, Solanales, Ericales, Fabales, Malvales, and Malpighiales.

Table 9.1 Examples of SDV from different kingdoms of life.

Group	Species	Traits investigated	Reference
Animalia			
Rodentia	*Rattus norvegicus* (I)	Broad spectrum of traits	Gärtner et al. (1976)
Artiodactyla	*Sus scrofa domestica* (C)	Broad spectrum of traits	Archer et al. (2003a,b)
Carnivora	*Felis silvestris catus* (C)	Coat coloration	Shin et al. (2002)
Cingulata	*Dasypus novemcinctus* (P)	Broad spectrum of traits	Storrs and Williams (1968)
Osteichthyes	*Kryptolebias marmoratus* (S)	Behavior, gill morphology	Turko et al. (2011)
Ascidiacea	*Botryllus schlosseri* (A)	Egg production	Stewart-Savage et al. (1999)
Decapoda	*Procambarus virginalis* (AP)	Broad spectrum of traits	Vogt et al. (2008)
Cladocera	*Daphnia magna* (AP)	Size, age, reproductive traits	Pietrzak (2011)
Branchiopoda	*Artemia parthenogenetica* (AP)	Reproductive traits, life span	Browne et al. (2002)
Diptera	*Drosophila melanogaster* (I)	Asymmetry of bristles	Indrasamy et al. (2000)
Hemiptera	*Longicaudus trirhodus* (AP)	Morphology, life history traits	Warren (1902)
Gastropoda	*Melanoides tuberculata* (AP)	Speed of development	Ben-Ami and Hodgson (2005)
Nematoda	*Caenorhabditis elegans* (S)	Life span, locomotion	Herndon et al. (2002)
Rotifera	*Brachionus calyciflorus* (AP)	Reproductive traits	Gilbert and Schröder (2007)
Platyhelminthes	*Maritrema novaezealandensis* (A)	Morphology, behavior	Koehler et al. (2011)
Bryozoa	*Electra pilosa* (A)	Morphometric traits	Hageman et al. (1999)
Cnidaria	*Hydrallmania falcata* (A)	Morphometric traits	Ponczek and Blackstone (2001)
Porifera	*Tethya wilhelma* (A)	Patterning of buds	Hammel et al. (2009)
Fungi			
Saccharomycetes	*Candida albicans* (A)	Switching between morphs	Tao et al. (2014)
Eurotiomycetes	*Aspergillus niger* (A)	Variation of mycelium	Vinck et al. (2005)
Tremellomycetes	*Cryptococcus neoformans* (A)	Resistance to antibiotics	Avery (2006)
Plantae			
Asparagales	*Ledebouria graminifolia* (T)	Morphology, life history traits	Shushu et al. (2009)
Zingiberales	*Musa acuminata* (T)	Pigmentation of leafs and fruits	Sahijram et al. (2003)
Brassicales	*Arabidopsis thaliana* (T)	Root system	Forde (2009)
Lamiales	*Nyctanthes arbor-tristis* (IO)	Morphology of leaves	Roy (1963)
Solanales	*Nicotiana tabacum* (IO)	Leaf and flower morphology	Sakai and Shimamoto (1965)
Ericales	*Rhododendron simsii* (T)	Flower color in bud sports	De Schepper et al. (2003)
Fabales	*Retama sphaerocarpa* (IO)	Branch morphology	Fungairiño et al. (2005)
Malvales	*Theobroma cacao* (T)	Epigenetic profiles	Rodríguez López et al. (2010)
Malpighiales	*Populus tremuloides* (A)	Biochemical traits	Smith et al. (2011)

(*Continued*)

Table 9.1 (Continued)

Group	Species	Traits investigated	Reference
Protista			
Choanoflagellata	*Salpingoeca rosetta* (A)	Size and morphology of cells	Dayel et al. (2011)
Mesomycetozoa	*Psorospermium haeckeli* (A)	Size and shape of spores	Vogt and Rug (1999)
Amoebozoa	*Dictyostelium discoideum* (A)	Cell fate, spore formation	Nanjundiah and Bhogle (1995)
Cryptophyta	*Chilomonas paramecium* (A)	Length and width of cells	Pearl (1906)
Ciliophora	*Tetrahymena thermophila* (I)	rDNA expression	Orias and Bradshaw (1992)
Apicomplexa	*Plasmodium falciparum* (A)	Expression of surface antigens	Avery (2006)
Euglenozoa	*Trypanosoma brucei* (A)	Expression of surface antigens	Figueiredo et al. (2009)
Chlorophyta	*Volvox sp.* (A)	Cell differentiation	Shelton et al. (2012)
Bacteria			
Enterobacteriales	*Salmonella* ser. Typhimurium (A)	Differential gene expression	Ackermann (2013)
Bacillales	*Bacillus subtilis* (A)	Gene expression, cell fate	Maamar et al. (2007)
Pseudomonadales	*Pseudomonas aeruginosa* (A)	Bistability and cytotoxicity	Smits et al. (2006)
Lactobacillales	*Streptococcus pneumoniae* (A)	Development of competence	Smits et al. (2006)
Actinomycetales	*Mycobacterium tuberculosis* (A)	Resistance to antibiotics	Avery (2006)
Archaea			
Sulfolobales	*Sulfolobus acidocaldarius* (A)	Motility	Lewus and Ford (1999)
Methanococcales	*Methanococcus jannaschii* (A)	Cell size, DNA content	Malandrin et al. (1999)
Halobacteriales	*Halobacterium halobium* (A)	Swimming behavior	Schimz and Hildebrand (1992)
Viruses			
Siphoviridae	phage-λ	Lysis and lysogeny	Singh and Weinberger (2009)
Baculoviridae	*Gilpinia hercyniae* NPV	Length of virion rods	Ackermann and Smirnoff (1983)

A, Asexual reproduction; *AP*, apomictic parthenogenesis; *C*, artificial cloning; *I*, inbreeding; *IO, iterative organs; NPV*, nuclearpolyhedrovirus; *P*, polyembryony; *S*, self-fertilization; *T*, cloning by tissue culture (modified after Vogt, 2015).

In protists, there are examples of SDV for morphometric parameters, cell fate determination, spore formation, and surface antigen expression in species from distantly related clades like Choanoflagellata, Mesomycetozoa, Amoebozoa, Cryptophyta, Ciliophora, Apicomplexa, Euglenozoa, and Chlorophyta (Table 9.1).

There is also sound evidence of SDV in bacteria from different orders concerning growth, gene expression (Fig. 9.1E), protein content, behavior, virulence, and resistance to antibiotics (Table 9.1). In the prokaryotic Archaea, the third domain of life besides Eukaryota and Bacteria (Woese et al., 1990), there are examples of SDV for cell size, DNA content, and swimming behavior in species from different orders (Table 9.1). The Archaea show the simplest cellular architecture of all independently living organisms.

There are even examples of SDV in viruses. Viruses require other organisms for replication and are interpreted as primitive precellular "organisms" or as a form at the edge of life (Koonin et al., 2006). SDV has been demonstrated in phage-λ viruses with respect to lysis–lysogeny switching and in rod-shaped baculoviruses with respect to form and length (Table 9.1).

The occurrence of SDV in all kingdoms of life and all clades investigated suggests that it is a universal biological phenomenon inherent in all living beings (Vogt, 2015).

Determination of stochastic developmental variation in laboratory experiments

In this section, I will compare the ranges of SDV of morphological, physiological, biochemical, behavioral, and life history traits in animals. The respective experiments were done with apomictic parthenogenetic, polyembryonic, inbred, and artificially cloned species and lineages in highly standardized environments.

SDV is easiest recognized with respect to body size, the shape of prominent morphological structures, and coloration. Examples are size and color differences in the offspring of apomictic pea aphid, *Acyrthosiphon pisum* (Fig. 9.3A), variation of body size, body form, and egg number in parthenogenetic water flea, *Daphnia pulex* (Fig. 9.3B), differences in patterns of skeleton shields in a quadruplet of polyembryonic nine-banded armadillo (Fig. 9.3C), body height differences in clone mates of pigs (Fig. 9.3D), and horn shape differences in clone mates of cattle (Fig. 9.3E,F). SDV is also pronounced in fingerprints, irises, and retinas of monozygotic human twins (Daugman and Downing, 2001; Jain et al., 2002; Kong et al., 2006). These traits can be used for individual authentication of monozygotic twins, which are indistinguishable by routine DNA fingerprinting.

SDV of meristic and metric morphological traits is relatively small when compared to biochemical and life history traits (Table 9.2). For example, the coefficients of variation (CV)

Figure 9.3 SDV of morphological traits. (A) Differences in coloration and size among parthenogenetically produced offspring of pea aphid, *Acyrthosiphon pisum*. (B) Variation of body shape, size, and egg number (arrows) in highly synchronized parthenogenetic water flea, *Daphnia pulex*. (C) Differences in patterns of skeleton shields in a quadruplet of polyembryonic nine-banded armadillo, *Dasypus novemcinctus*. Red circles show variation in head shields of two littermates. (D) Size differences between 27-week-old cloned Duroc pigs originating from the same nuclear donor cell line and born to the same recipient mother. The pigs were kept communally and had unlimited access to food and water. (E, F) Variation of horn pattern in cloned Hanwoo cattle from the same mother. *Source*: (A–F) from Vogt, G., 2015. Stochastic developmental variation, an epigenetic source of phenotypic diversity with far-reaching biological consequences. J. Biosci., 40, 159–204; (A) Original photo by Alex Wild; (B) original photo by Winfried Lampert; (C) original photo by Brian Bagatto; (D) original photo from Archer, G.S., Dindot, S., Friend, T.H., Walker, S., Zaunbrecher, G., Lawhorn, B., et al., 2003a. Hierarchical phenotypic and epigenetic variation in cloned swine. Biol. Reprod., 69, 430–436; (E and F) original photos from Yang, B.-C., Lee, S.-H., Hwang, S., Lee, H.-C., Im, G.-S., Kim, D.-H., et al., 2012. Phenotypic characterization of Hanwoo (native Korean cattle) cloned from somatic cells of a single adult. BMB Rep., 45, 38–43.

Table 9.2 Examples of the extent of SDV in identically reared groups of isogenic animals.

Species	Trait	Range/Mean[a]	CV (%)	Reference
Rattus norvegicus (I), n = 18	Mandible length	26.8 mm	1.49	Flamme (1977)
	Body weight	333 g	12.91	
	Heart weight	0.87 g	10.34	
	Liver weight	11.24 g	11.12	
	Serum protein	67.3 g/L	9.66	
	GOT	43.3 U/L	36.72	
Dasypus novemcinctus (P), n = 4	No. of scutes in BR	526–531	0.39	Storrs and Williams (1968)
	Body weight	52.61–60.30 kg	5.72	
	Brain weight	5.23–5.86% BW	5.52	
	Heart weight	0.45–0.64% BW	14.65	
	Spleen weight	0.13–0.24% BW	29.99	
	Glutamate in brain	12.24–20.57 RL	21.06	
	Alanine in brain	3.33–12.29 RL	55.80	
	Adrenaline in AG	0.05–1.60 μg/g	102.41	
Sus scrofa domestica (C), *n* = 5	Weight at 27 weeks	81.6–102.1 kg	9.25	Archer et al. (2003a)
	Serum protein	7.0–7.7 g/dL	3.73	
	Blood calcium	10.7–10.9 mg/dL	0.93	
	Blood albumin	3.6–4.3 g/dL	7.25	
	Blood glucose	70–88 mg/dL	9.20	
	Cortisol	3.2–6.7 μg/dL	28.98	
	Triiodothyronine	43.41–54.63 ng/dL	20.54	
	Blood urinary nitrogen	8.9–11.6 mg/dL	14.04	
Capra aegagrus hircus (C), *n* = 5	Weight at 52 weeks PW	43.8 kg	15.34	Landry et al. (2005)
	Growth hormone	3.4 ng/mL	135.29	
	Insulin-like GF I	177.9 ng/mL	44.74	
	Thyroxine	4.3 μg/dL	27.91	
	Insulin	17.7 μIU/mL	66.67	
Oncorhynchus masou macrostomus (C), *n* = 22	Standard length	8.0 cm	5.00	Iguchi et al. (2001)
	Body weight	8.2 g	12.20	
	Horizontal movement	7.64 grids/min	112.43	
	Hiding	1.08 freq/12 min	215.74	
	Benthic feeding	31.28 freq/12 min	96.23	

(*Continued*)

Table 9.2 (Continued)

Species	Trait	Range/Mean[a]	CV (%)	Reference
Procambarus virginalis (AP), $n = 8$	Total length at 152 days	3.4–4.4 cm	10.26	Vogt et al. (2008), Vogt (2010)
	Carapace length at 152 days	1.6–2.0 cm	9.55	
	Body weight at 152 days	0.99–2.40 g	30.91	
	Life span	437–910 days	21.31	
	Reproduction cycles	1–5	49.52	
	First spawning	157–531 days	52.46	
	No. of offspring at 430 days	0–219	90.68	

All groups were reared in captivity in highly standardized environments. *AG*, adrenal gland; *AP*, apomictic parthenogenesis; *BR*, banded region; *BW*, body weight; *C*, artificial cloning; *CV*, coefficient of variation; freq, frequency; *GF*, growth factor; *GOT*, glutamic-oxaloacetic transaminase; *I*, inbreeding; *IU*, international unit; *P*, polyembryony; *PW*, postweaning; *RL*, relative level.
[a]Means are given when data on ranges were not available. Modified after Vogt, 2015.

of body length in two clones of red-spotted cherry salmon *Oncorhynchus masou macrostomus* were 4.65% ($n = 24$) and 5.00% ($n = 22$) (Iguchi et al., 2001). The CVs of the number of scales in the banded region of polyembryonic armadillos were even smaller, varying only between 0.39% and 3% in 16 quadruplets (Storrs and Williams, 1968). In clutches of marbled crayfish juveniles, the CVs of carapace length, body length, and the numbers of olfactory aesthetascs and gustatory corrugated setae varied between 2.73% and 18.57% but were mostly below 10% (Vogt et al., 2008; Vogt, unpublished data).

In a given group of animals, the CVs of different morphological or anatomical traits were usually different. For example, in a newborn armadillo quadruplet, CVs of the weights of brain, heart, and spleen were 5.52%, 14.65%, and 29.99%, respectively (Table 9.2). Likewise, in a group of 12 juvenile stage-3 clutch mates of marbled crayfish CVs of carapace length, number of aesthetascs, and number of corrugated setae were 2.73%, 6.43%, and 3.72%, respectively (Table 9.2).

Coat coloration can vary widely between genetically identical animals raised in the same environment. For instance, inbred A^{vy} mice show coat colors from pure yellow to agouti (Fig. 9.4A). Much more variable than the color hue is color

Figure 9.4 SDV of coloration. (A) Inbred A^{vy} mouse litter mates showing color variation from pure yellow (left) to pseudoaguti (right). (B) Monozygotic twins of Holstein cattle derived from a split embryo together with their donor mother showing differences in spotting pattern between twins and to the mother. (C, D) Variation of color and spotting pattern between nuclear donor mother (C) and cloned offspring (D) of cat. (E–H) Marmoration pattern of lateral cephalothorax in apomictic parthenogenetic marbled crayfish showing pronounced variation between mother (E) and her genetically identical offspring (F–H) and between clutchmates (F–H). *Source*: (A–D) from Vogt, G. (2015). Stochastic developmental variation, an epigenetic source of phenotypic diversity with far-reaching biological consequences. J. Biosci., 40, 159–204; (A) original photo from Cropley, J.E., Suter, C.M., Beckman, K.B., & Martin, D.I.K., 2010. CpG methylation of a silent controlling element in the murine Avy allele is incomplete and unresponsive to methyl donor supplementation. PLoS ONE, 5, article e9055; (B) original photo from Seidel, G.E., Jr., Elsden, R.P., & Hasler, J.F., 2003. Embryo transfer in dairy cattle. Fort Atkinson: Hoards and Sons; (C and D) original photo from Shin, T., Kraemer, D., Pryor, J., Liu, L., Rugila, J., Howe, L., et al., 2002. A cat cloned by nuclear transplantation. Nature, 415, 859. (E–H) from Vogt, G., Huber, M., Thiemann, M., van den Boogaart, G., Schmitz, O.J., Schubart, C.D., 2008. Production of different phenotypes from the same genotype in the same environment by developmental variation. J. Exp. Biol., 211, 510–523.

patterning as shown for inbred guinea pigs (Chase, 1939), cloned Holstein cattle (Fig. 9.4B), cloned cats (Fig. 9.4C and D), and parthenogenetic marbled crayfish (Fig. 9.4E). The marmoration motifs on the lateral cephalothorax of marbled crayfish differed greatly between mother and offspring and among clutch mates (Fig. 9.4E). Each of the numerous genetically identical marbled crayfish raised by us in the same laboratory setting had a unique marmoration pattern that identified the individuals unambiguously (Vogt et al., 2008).

SDV of biochemical traits can either be small or high depending on trait. For example, the blood parameters serum protein, glucose, and albumin in cloned pigs had CVs below 10% (Table 9.2). Blood calcium, which is generally narrowly regulated, had a CV below 1%. Water content, fat content, and ash had CVs between 10% and 20% in inbred mice (Dawson, 1970). Higher CVs of 20%–135% were determined for the hormones cortisol, growth hormone, triiodothyronine, thyroxine, insulin, and adrenalin in polyembryonic armadillos and cloned pigs and goats (Table 9.2).

Variations of life history traits such as growth, reproduction, life span, and mortality are particularly important because they markedly influence demography and fitness of a population. Considerable variation of SDV of growth was noted when early life stages from the same clutch of marbled crayfish were individually raised in the wells of 12-well microplates, the simplest environment possible (Fig. 9.5A). In the lecithotrophic embryos and juvenile stages 1 and 2, the speed of development was rather uniform but from juvenile stage 3, the first feeding stage, the individuals developed with different speed although they were fed ad libitum with the same food. An increase of SDV of growth with age was also observed in a group of seven marbled crayfish fed ad libitum, which had CVs of body weight of 10.29%, 14.37%, 20.98%, 30.91%, and 48.05% at days 71, 101, 143, 152, and 351 after hatching, respectively (Fig. 9.5B). Marked differences of the SDV of growth were also recorded among four clutch mates of marbled crayfish that were individually raised in identical environments (Fig. 9.5C). This approach excludes the possible influence of social interactions.

An example of SDV of reproductive traits is presented in Table 9.2 for eight clutch mates of marbled crayfish reared communally from hatching to death. The time between hatching and first spawning varied from 157 to 531 days, the number of breeding cycles from 1 to 5 and the number of offspring per female at day 430 from 0 to 219, corresponding to CVs of 52.46%, 49.52%, and 90.68%. Other examples of SDV of reproductive traits come from parthenogenetic water flea, *Daphnia*

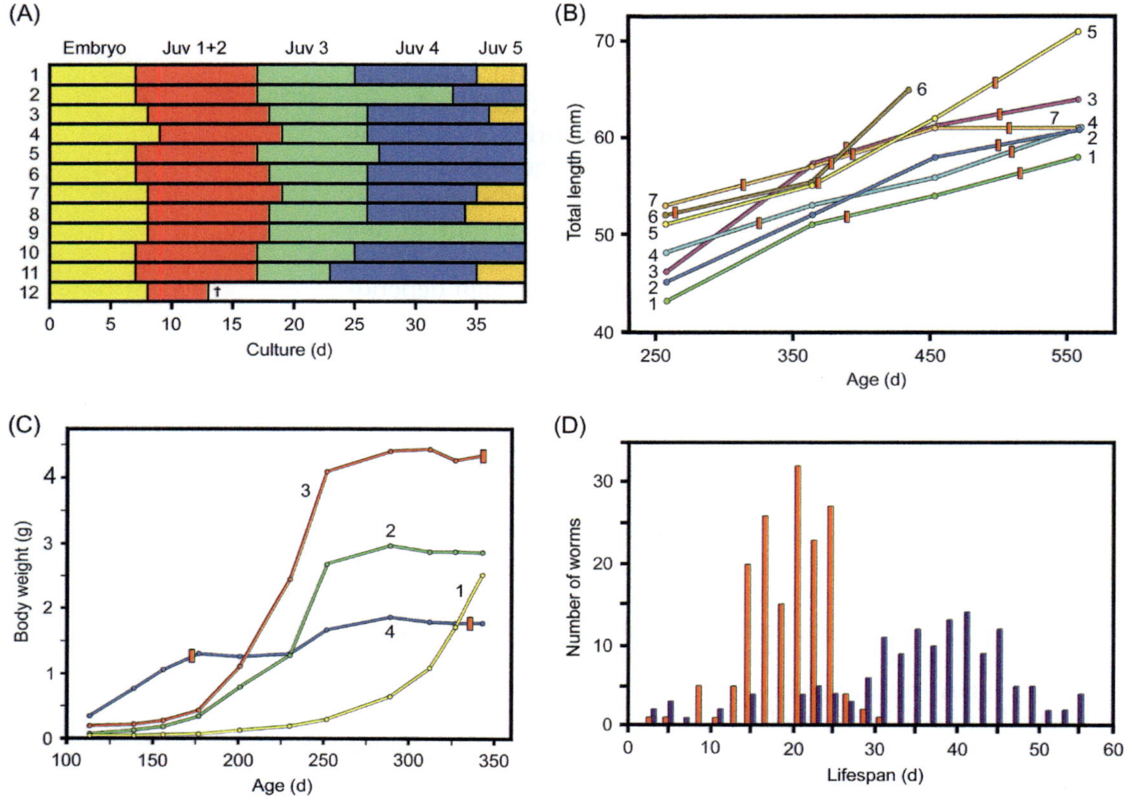

Figure 9.5 SDV of life history traits. (A) Variation of speed of development among 12 clutch mates of parthenogenetic marbled crayfish raised individually in the wells of the same microplate. Shown is a 39-day period from 80% embryonic development to juvenile stage 5. Differences in development were small in the lecithotrophic period (embryo to juvenile 2) and increased markedly from stage 3, the first feeding stage. (B) Variation of growth and reproduction between communally reared clutch mates (1–7) of marbled crayfish. Note position changes of individuals in group over time with respect to size and differences in frequency and time of oviposition (red vertical bars). (C) Variation of growth and reproduction between clutch mates of marbled crayfish (1–4) kept individually under identical environmental conditions. (D) Variation of life spans between and within wild-type *(red)* and long-lived age-1 *(blue)* strains of selfing nematode *Caenorhabditis elegans. Source*: (A–C) from Vogt, G., Huber, M., Thiemann, M., van den Boogaart, G., Schmitz, O.J., Schubart, C.D., 2008. Production of different phenotypes from the same genotype in the same environment by developmental variation. J. Exp. Biol., 211, 510–523, (D) from Vogt, G., 2015. Stochastic developmental variation, an epigenetic source of phenotypic diversity with far-reaching biological consequences. J. Biosci., 40, 159–204; redrawn and modified after Kirkwood, T.B.L., & Finch, C.E., 2002. The old worm turns more slowly. Nature, 419, 794–795.

pulex, which had different numbers of embryos in their brood chamber (Fig. 9.3B).

SDV of life span was also quite high among isogenic and identically reared animals. For instance, in a group of 14 inbred mice this trait varied from 27 to 147 days, and in 16 inbred rats from 16 to 113 days (Williams and Pelton, 1966). Life spans ranging from 3 to 30 days and 3 to 55 days were found in two identically raised but genetically different strains of selfing nematode, *Caenorhabditis elegans* (Fig. 9.5F) (Kirkwood and Finch, 2002). In addition, in a group of eight marbled crayfish that had reached the reproductive stage, longevity varied between 437 and 910 days, corresponding to a CV of 21.31% (Table 9.2).

SDV of behavior is mainly known from vertebrates. For example, Bierbach et al. (2017) measured movement behavior in clone mates of Amazon molly, *Poecilia formosa,* and recorded considerable differences between individuals (Fig. 9.6A). Archer et al. (2003b) quantified food preference, temperament, and time budget in cloned pigs and revealed that intraclonal variation could reach dimensions comparable to variation in naturally bred control groups (Fig. 9.6B). Tordoff et al. (2007a, b) investigated voluntary uptake of water, sodium, and calcium in 40 strains of inbred mice and found considerable variation between and within strains. Iguchi et al. (2001) analyzed SDV of movement, feeding, alerting, and threat behavior in clonal salmon *Oncorhynchus masou macrostomus* and revealed CVs between 96.23% and 215.74% (Table 9.2), indicating pronounced behavioral individuality.

There are also a few interesting examples of SDV of behavior in invertebrates. For example, Schuett et al. (2011) recorded different escape responses to predator attack among clone mates of the parthenogenetic pea aphid: dropping off the plant, nondropping, and inconsistent behavior. In marbled crayfish, a remarkable divergence of behavior from the same initial status occurred when stage-6 clutch mates were placed in groups of five into culture vessels with only a net as shelter (Vogt et al., 2008). In the following 34 days, social hierarchies were gradually established. At the end of the experiments, each group consisted of 1 dominant, 1–2 subdominants, and 2–3 subordinates. During establishment of the hierarchy, the dominant developed increasingly offensive behaviors, whilst the subordinates developed increasingly defensive and avoiding behaviors. Interestingly, growth of the dominant speeded up compared to the subdominant and subordinates (Fig. 9.6C) although all specimens had unlimited access to the food and fed regularly as revealed by the externally visible filling of the intestine. These differences in behavior and growth probably developed from small stochastic behavioral differences via self-reinforcing circuitries involving metabolism and neuroendocrine feedback.

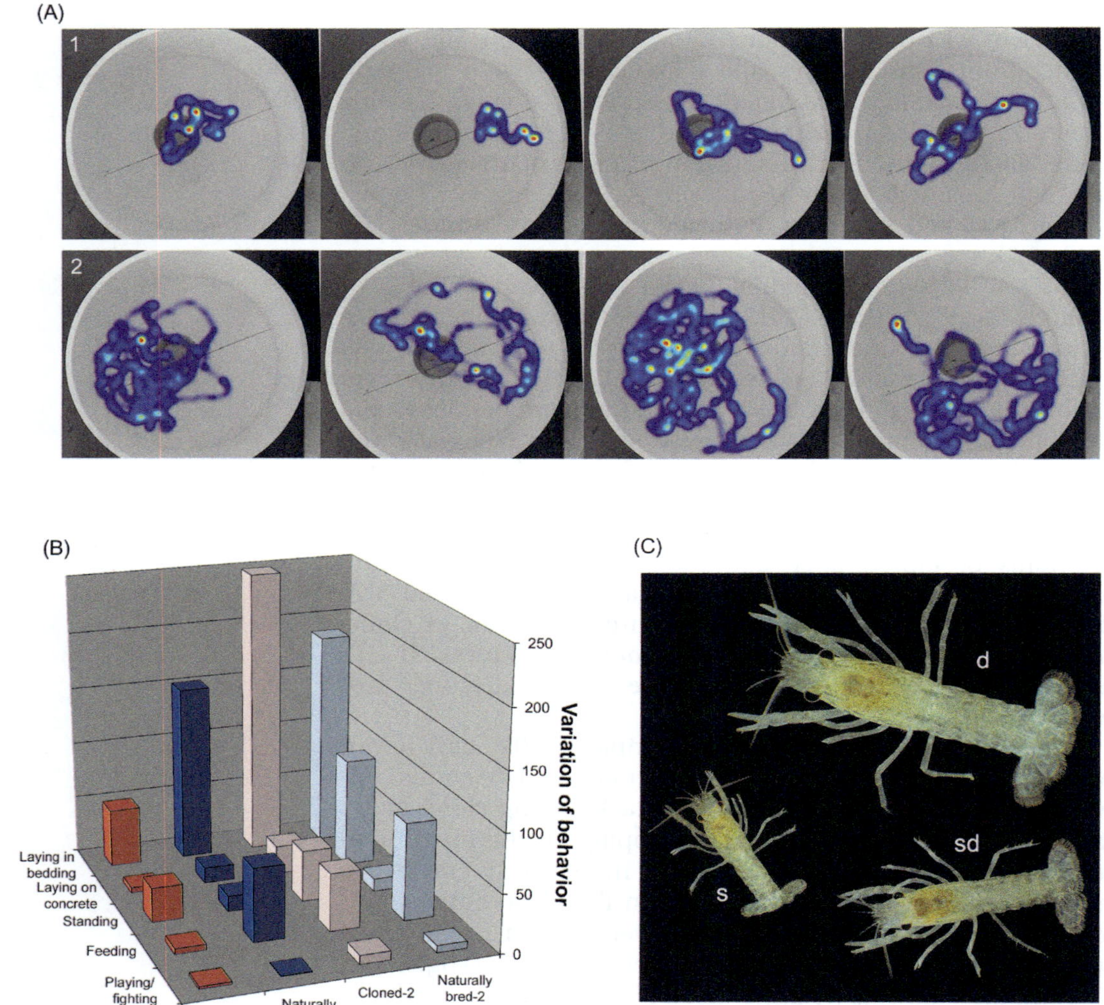

Figure 9.6 SDV of behavior. (A) Movement behavior of two clone mates of Amazon molly, *Poecilia formosa,* in an open-field test over four repeated observations. Shown are heat maps of a less active (1) and a highly active (2) individual. (B) Variation of time budget in two litters of cloned female pigs from the same fetal cell line in comparison to two naturally bred litters. (C) Variation of agonistic behavior and concomitant growth differences in clutch mates of marbled crayfish kept together for 34 days. The experiment was started with five size-matched clutch mates of indifferent agonistic behavior and ended with one dominant (d), one subdominant (sd), and three subordinates (s) of remarkably different sizes, although food was available in excess and not monopolized. *Source:* (A) from Bierbach, D., Laskowski, K.L., & Wolf, M., 2017. Behavioural individuality in clonal fish arises despite near-identical rearing conditions. Nat. Commun., 8, article 15361. (B) from Vogt, G., 2015. Stochastic developmental variation, an epigenetic source of phenotypic diversity with far-reaching biological consequences. J. Biosci., 40, 159–204; redrawn and modified after Archer, G.S., Friend, T.H., Piedrahita, J., Nevill, C.H., & Walker, S., 2003b. Behavioral variation among cloned pigs. Appl. Anim. Behav. Sci., 82, 151–161. (C) from Vogt, G., Huber, M., Thiemann, M., van den Boogaart, G., Schmitz, O.J., Schubart, C.D., 2008. Production of different phenotypes from the same genotype in the same environment by developmental variation. J. Exp. Biol., 211, 510–523.

Disentangling genetic variation, environmentally induced variation, and stochastic developmental variation in the laboratory

Most animal populations are sexually reproducing having males and females. Sexual reproduction generates genetic variation, and males and females are genetically different if they possess different sex chromosomes. In such populations, the genetic proportion of phenotypic variation can be experimentally distinguished from the nongenetic proportions by raising the animals in single or different environments and analyzing males and females separately. Examples are given in textbooks on quantitative genetics and phenotypic plasticity (e.g., Falconer and Mackay, 1996; Pigliucci, 2001). Thomson et al. (2018) have summarized recent developments to the "multiple-matrix animal model" of quantitative genetics that now allows relatively precise calculation of individual-based measures of nongenetic phenotypic variance.

Disentangling all three types of phenotypic variation, namely genetic variation, EIV, and SDV, is difficult in genetically diverse and sexually reproducing populations, even in the laboratory. Therefore, experiments on the distinction of EIV and SDV were usually done with highly inbred lineages in which genetic variation is very close to zero. An example is provided by Ashoub et al. (1958), who exposed inbred mice to cold and warm laboratory environments. Differences between groups reflected EIV and differences within groups reflected SDV. The authors also showed that the intra-group variance depended to some degree on the environment.

Gärtner and colleagues (Gärtner et al., 1976; Flamme, 1977; Gärtner, 1990) made particularly intense attempts to partition the genetic and nongenetic sources of phenotypic variation in laboratory raised animals. They tried for more than 30 years to reduce the variability of qualitative and quantitative traits in laboratory mice and rats by standardizing genotypes and environmental conditions. Fig. 9.7A shows an example on the phenotypic variation of the kidney weights of 1160 adult rats kept in 58 groups. These groups differed in genotype (inbred, outbred, or hybrid), state of health (pathogen-free or infected by *Mycoplasma pulmonis*) and rearing conditions. The rats lived either in cages in highly standardized animal rooms (one or four specimens per cage) or freely in a wild, fenced-in area.

The authors distinguished between two different kinds of phenotypic variability: fixed effects and random variability (Fig. 9.7A). Fixed effects referred to the distribution of different animal groups in

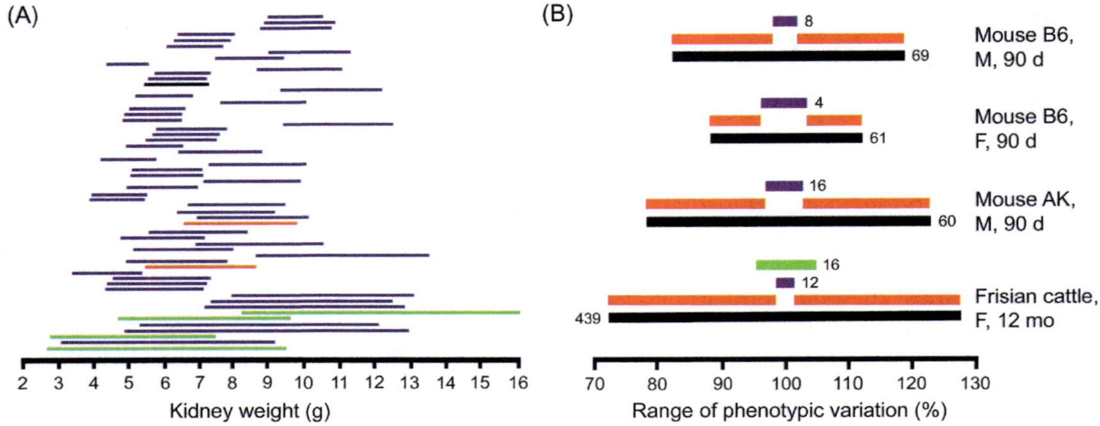

Figure 9.7 Experiments on the dissection of the genetic, environmental, and stochastic developmental components of phenotypic variation in sexually reproducing mammals. (A) Ranges of kidney weights in 58 groups of rats. Included are animals from different inbred, outbred, and hybrid strains and different age, sex, and health status. Each group consisted of 20 animals, caged single or in groups of four. The groups indicated by *blue bars* lived in the same animal house under highly standardized environmental condition and were disease free. *(Green bars)* infected with *Mycobacterium pulmonis*; *(red bars)* living in the wild. Two different kinds of variability were distinguished: fixed effects and random variability. Fixed effects refer to the polar distribution of different groups in the spectrum of kidney weights. They are mainly determined by differences in genetics, age, and health status of the animals. Random variability refers to the range within groups, which is not explained by these factors. (B) Determination of the environmental and stochastic proportions of the random variability of body weight by transplanting split eight-cell twin embryos of inbred mouse strains (B6 and AK) and Frisian cattle into the same *(blue bars)* and different *(green bar)* uterine foster mothers. *Black bars* indicate ranges of total random variability (4 × coefficient of variation). *Red bars* indicate the stochastic developmental component before fertilization, *blue bars* indicate the stochastic developmental component post fertilization, and the difference between *green bar* and *blue bar* indicates the environmental proportion. F, females; M, males. Numbers at bars give specimens investigated.
Source: Redrawn and modified after Gärtner, K., 1990. A third component causing random variability beside environment and genotype. A reason for the limited success of a 30 year long effort to standardize laboratory animals? Laboratory Anim., 24, 71–77.

the weight spectrum. They were due to genetic differences, age, sex, and environment. Random variability referred to the range within the groups and consisted of SDV plus V_{Es}. Standardization reduced to some degree the fixed effects but the random component of phenotypic variability could not markedly be reduced by inbreeding and standardization of the living conditions. Neither did an increase of the environmental variability such as living in natural settings markedly increase the random variability. Only diseases increased random variability markedly (Fig. 9.7A). Gärtner and colleagues (Gärtner et al., 1976; Flamme, 1977; Gärtner, 1990) investigated 25 further traits in the same manner and revealed similarly large but trait-specific ranges of the random variability.

Gärtner (1990) tried to determine the environmental effects on random variability by implanting split eight-cell embryos of mice and calves into either the same or different foster mothers. They revealed that only 3%–30% of the total random variability measured in naturally born mice and cattle were caused by the different environments experienced post fertilization (Fig. 9.7B), reflecting V_{Es}. The rest is attributable to a third component, which is SDV. Gärtner (1990, 2012) assumed that the observed Gaussian distribution of body weights in inbred and identically raised animals that results from the third component is an arrangement supporting natural selection.

Lajus and Alekseev (2004) used a polyphenic trait of *Daphnia pulicaria* to disentangle EIV and SDV in the laboratory. These water fleas can either produce normal eggs or resting eggs with different morphology. The authors revealed that phenotypic plasticity (EIV) explains only 1%–2% of the phenotypic variation of this trait whereas developmental instability (SDV) explains 98%–99%.

Fluctuating asymmetry (FA), the deviation of morphological structures from perfect symmetry in bilaterally symmetric animals (Palmer and Strobeck, 1986) or bilaterally symmetric structures of plants (Fig. 9.1B) was occasionally propagated as an indicator of developmental stochasticity. FA can easily be determined in the laboratory and the wild, including sexually reproducing and genetically diverse populations. However, FA can be caused not only by SDV but also by genetic disturbances and environmental stress (Parsons, 1992; Graham et al., 2010). Surprisingly, in a given group of marbled crayfish siblings, FA of a trait was usually considerably smaller than SDV of the same trait, which was the result of repeated corrections of the asymmetry toward symmetry during lifetime (Vogt et al., 2008). Therefore, measurement of FA does not fully reflect SDV.

Disentangling genetic variation, environmentally induced variation, and stochastic developmental variation in field studies

Partitioning of phenotypic variation in wild populations is much more difficult than in the laboratory because natural populations are usually genetically diverse and the environments are highly complex. Genetic variability can be reduced by choosing asexually reproducing lineages for studying.

In order to dissect genetic and nongenetic variation Thorson et al. (2017) examined the morphological and epigenetic variation in asexual populations of the freshwater snail, *Potamopyrgus antipodarum,* from different habitats in Oregon and Washington

(USA). *Potamopyrgus antipodarum* is native to New Zealand and is characterized by the frequent coexistence of diploid sexual and polyploid parthenogenetic lineages. A single clone was introduced into the western United States fewer than 30 years ago. Dybdahl and Drown (2011) used genetic marker studies to investigate these Western snail populations and demonstrated the near absence of genotypic variation within and among these populations. Despite genetic uniformity, these snail populations have evolved habitat specific differences in shell shape, and these differences were consistent with adaptation to water current speed. Thorson et al. (2017) found significant genome wide DNA methylation differences between lakes and rivers, suggesting that environmentally induced epigenetic diversity (probably supplemented by SDV-induced epigenetic diversity) caused adaptive phenotypic changes in less than 100 generations. Recently, Thorson et al. (2019) compared genetically identical snail populations in rural and polluted urban lakes and identified numerous differentially methylated DNA regions (DMRs). They associated these DMRs to different gene categories like metabolism, development, etc. The presence of site-specific differences in DMRs between these lake populations suggest an epigenetic response to varied environmental factors.

Leung et al. (2016) hypothesized that if environmentally induced and stochastic epimutations were associated with phenotypic plasticity and bet-hedging, respectively, they should be distinguishable because both strategies are differentially selected according to environmental uncertainty. Epimutations are reversible changes in the chemical structure of the DNA and chromatin (e.g., the addition of methyl groups to cytosines of the DNA and methyl and acetyl groups to the histones) that, unlike genetic mutations, do not change the DNA coding sequence. In the model plant *Arabidopsis thaliana*, spontaneous epimutations of CpG-dinucleotides are about five orders of magnitude more frequent than genetic mutations (10^{-4} compared with 10^{-9}) (Van der Graaf et al., 2015). Johannes and Schmitz (2019) summarized the present knowledge on spontaneous epimutations in plants and presented examples with heritable phenotypic effects.

To test their hypothesis on the ecological and evolutionary relevance of environmentally induced and stochastic epimutations, Leung et al. (2016) assessed the sources of epigenetic changes in clonal fish, *Chrosomus eos-neogaeus* from predictable and unpredictable environments. *Chrosomus eos-neogaeus* is a gynogenetic all-female fish, which occurs in North America in 14 isogenic lineages originating from different hybridization events between the redbelly dace *Chrosomus eos* and the fine-scale dace *Chrosomus neogaeus* (Angers and Schlosser, 2007). Each hybrid lineage apparently originated from a single hybrid zygote and is genetically rather uniform. The analysis of DNA methylation in *Chrosomus eos-neogaeus* lineages from predictable (lakes) and unpredictable (intermittent headwater streams) environments in southern Quebec, Canada identified the relative contributions of EIV and SDV to phenotype variation. Environmentally induced variation was inferred from epigenetic differences between sampling sites while stochastic changes were inferred from the residual epigenetic variation, explained neither by genetic variation nor sampling site. Comparison of clones revealed that directional EIV is predominant in predictable environments whereas risk-spreading SDV prevails in unpredictable environments (Fig. 9.8), as expected. Differences in environmental effects on epigenetic variation between sympatric but genetically different lineages (Fig. 9.8) showed that the epigenetic response to environmental signals is influenced by the genotype. Common garden experiments revealed that the proportion of pure environmental effects can considerably change when clone members are transplanted into a new environment (Fig. 9.8).

The example of *Chrosomus eos-neogaeus* demonstrates that EIV and SDV always occur together but have different weighting in different environments. In a recent paper, Angers et al. (2020) discussed the relevance of assessing the different sources of epigenetic variation in natural populations and their applications in ecology and conservation.

Disentangling genetic variation plus environmentally induced variation from stochastic developmental variation by mathematical modeling

Kiskowski et al. (2019) used patterning of leopard gecko, *Eublepharis macularius*, head color spots to distinguish phenotypic variation due to genotype plus environment from phenotypic variation due to developmental noise (SDV) (Fig. 9.9A–C). They applied computational models of developmental reaction-diffusion mechanisms (e.g., Turing mechanism) to generate phenotypes. In their mathematical simulation, fixed (predetermined) parameters correspond to genetic and environmental factors while intrinsic stochastic variation within simulations corresponds to SDV.

The authors modified the parameters of simulations corresponding to genetic and environmental variation to generate the full range of phenotypic variation in color patterns seen on the heads of eight leopard geckos. They observed that over the range of these

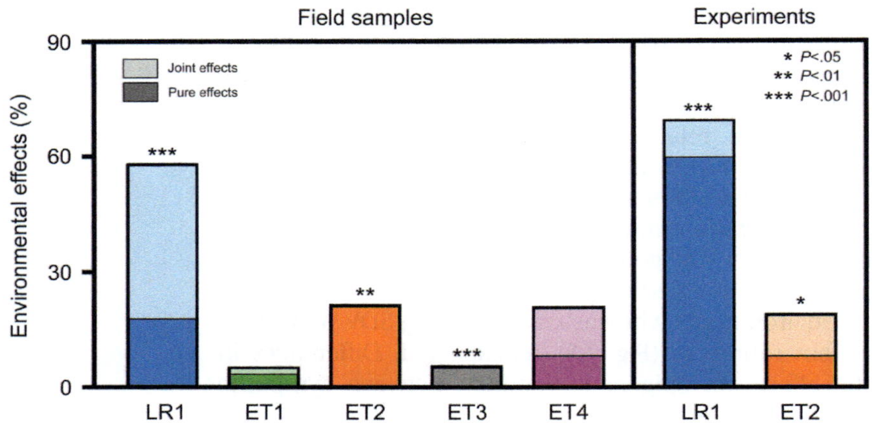

Figure 9.8 Determination of the environmental and stochastic developmental effects on epigenetic variation in asexual fish *Chrosomus eos-neogaeus* under natural and experimental conditions. The pure environmental effect *(darker colors in columns)* is separated from the environmental plus genetic joint effect *(lighter colors in columns).* The remaining amount to 100% is due to developmental stochasticity. Epigenetic differences were determined by MSAP (methylation sensitive amplification polymorphism) and genetic differences by microsatellite analysis. LR1 are from environmentally stable lakes and ET1–ET4 are from environmentally unstable streams. The animals for the laboratory experiments were sampled as larvae from the indicated sites and raised for 5 months until adults. *P*-values refer to pure site effects. The graph shows differences of site effects on epigenetic variation between predictable (LR1) and unpredictable environments (ET1–ET4) and among lineages even if they occur in sympatry (ET3 and ET4). Site effects were highest in predictable environments and stochastic developmental effects were highest in unpredictable environments. Comparison of LR1 and ET2 between the field and laboratory, respectively, suggests rapid epigenetic response to environmental change. *Source*: Redrawn and modified after Leung, C., Breton, S., & Angers, B., 2016. Facing environmental predictability with different sources of epigenetic variation. Ecol. Evolution, 6, 5234–5245.

parameters, the component of variation due to genotype and environment exceeds that due to SDV, but the effect of SDV on patterning was also substantial (Fig. 9.9C). Kiskowski and colleagues concluded that this approach can be applied to any morphological trait that results from self-organized processes, including patterning of striped and spotted color coats, patterning of bones, and body segmentation in animals.

Identification of the molecular mechanisms underlying environmentally induced variation and stochastic developmental variation

For a long time it was unclear how environmentally induced and stochastic developmental phenotypic variation might be produced. The enormous progress in epigenetics in the last two

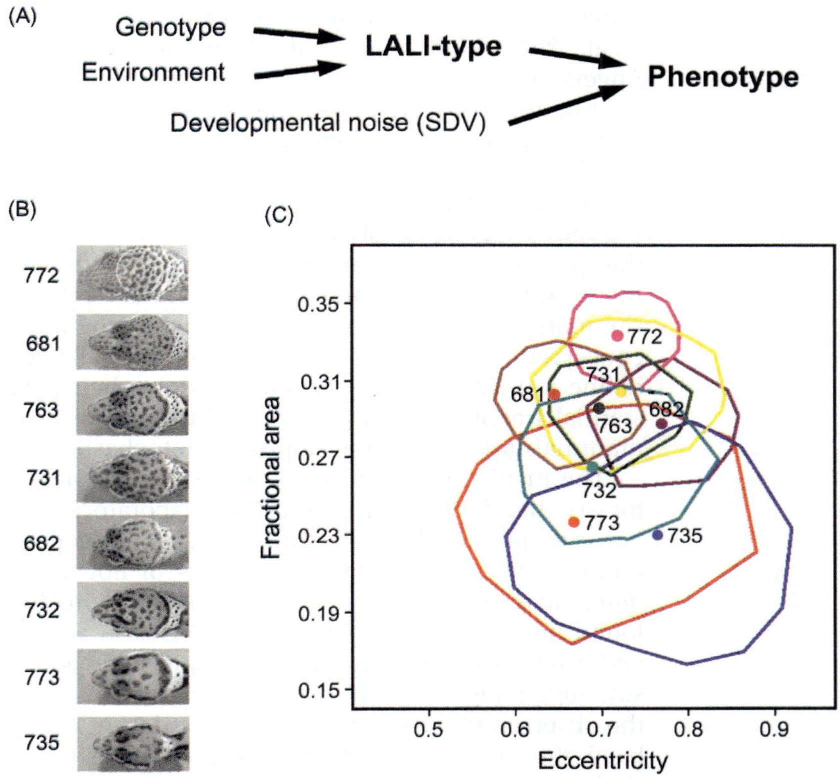

Figure 9.9 Determination of SDV by mathematical simulation. (A) Conceptual model of the local activation long-range inhibition (LALI)-type to phenotype map. The LALI-type summarizes the genetic and environmental factors of a LALI pattern. The phenotype is a product of the LALI-type and developmental noise (SDV). (B) Images of eight leopard gecko heads used for the simulation. Disk-shaped postorbital regions *(white circle)* were selected for pattern analysis. (C) Intragroup variation of simulations of head spot patterns shown in B. The fractional area is the total number of dark image pixels divided by the total number of pixels in the disk-shaped region of the head spots. The eccentricity of a spot is a value between 0 (perfect circle) and 1 (perfect line). The closed curves show the outer contour of the 95% phenotype cloud of 1000 simulations for eight LALI-types. The potential of one LALI-type to produce more than one of the live gecko phenotypes is shown by the inclusion of more than one phenotype in a phenotype cloud. Although the phenotype clouds overlap, even the largest phenotype cloud does not contain all of the phenotype variations of the group, indicating that the random variation (reflecting SDV) is substantial but not large enough on its own to account for all of the variations. *Source*: Redrawn and modified after Kiskowski, M., Glimm, T., Moreno, N., Gamble, T., & Chiari, Y., 2019. Isolating and quantifying the role of developmental noise in generating phenotypic variation. PLoS Computational Biol., 15, article e1006943.

decades put forward that directional and random alterations of the epigenetic marks on the DNA and histones are potent generators of nongenetic phenotypic variation. Experiments revealed that the DNA methylation and histone acetylation and methylation signatures can be altered by environmental impacts (Jaenisch and Bird, 2003; Dai and Wang, 2014; Schrey et al., 2016; Lindeman et al., 2019; Norouzitallab et al., 2019; Angers et al., 2020) but they can also change stochastically (Feinberg and Irizarry, 2010; Becker et al., 2011; Feinberg, 2014; Angers et al., 2020). Such epigenetic changes can obviously modify the expression of genes and change phenotypic traits. Ballouz et al. (2019) analyzed the transcriptomes of wild monozygotic quadruplets of the nine-banded armadillo and found that persistent transcriptional signatures occur early in development and uniquely characterize individuals relative to siblings. Comparing these results to human twins, the authors detected conserved co-expression of transcriptional signatures that define individuals. They concluded from their data that a substantial fraction of phenotypic and disease discordance within mammals arises from developmental stochasticity.

The mediating role of DNA methylation is better investigated for EIV than for SDV and comes from laboratory and field studies. For example, Liebl et al. (2013) investigated genome-wide DNA methylation in expanding populations of house sparrow, *Passer domesticus,* which was introduced to Kenya in the 1950s. They found high levels of variation in methylation across the genome and a negative correlation between epigenetic and genetic diversity, suggesting that DNA methylation contributes significantly to the observed phenotypic plasticity. Bossdorf et al. (2010) manipulated the DNA methylation pattern in thale cress, *Arabidopsis thaliana,* and revealed dramatic impacts on ecologically important traits and their variability. And Liu et al. (2019) measured differential DNA methylation between populations of social spider, *Stegodyphus dumicola,* and revealed significant correlation with differential gene expression. These results are consistent with a possible role of DNA methylation in environmental adaptation.

Sommer et al. (2017) investigated the molecular mechanism underlying feeding plasticity in the nematode *Pristionchus pacificus.* This nematode lives preferentially on scarab beetles. On living beetles, it exists as an arrested dauer stage and after the beetle's death, this dauer stage develops into adults to exploit the cadaver. *Pristionchus pacificus* adopt during larval development in an irreversible manner either a eurystomatous or a stenostomatous mouth form, which allows predatory or bacterivorous feeding, respectively. Mouth-form plasticity is regulated by conditional factors such as starvation and crowding but also contains stochastic elements of regulation. The authors showed that feeding plasticity in this nematode is controlled by developmental switch genes that are themselves under epigenetic control (histone modifications).

Anastasiadi and Piferrer (2019) investigated domestication in farmed European seabass (*Dicentrarchus labrax*) and revealed that epimutations in developmental genes underlie the onset of domestication, which leads to pronounced phenotypic alterations when compared to the wild type. About one fifth of epimutations that persisted into adulthood were established by the time of gastrulation and affected genes involved in developmental processes. Some of these genes were differentially expressed in seabass with lower jaw malformations, a key feature of the domestication syndrome. The authors assumed that these epimutations might be fixed as genetic variants explaining Darwin's domestication syndrome.

Whole genome mapping approaches revealed that DNA methylation is involved in animal polyphenism (Lyko et al., 2010; Mallon et al., 2016). In honeybee, morphologically, behaviorally, and reproductively different queens and workers are produced from the same genotype by differential feeding, as explained earlier. Lyko et al. (2010) reported that the DNA of the brain of reproducing queens and sterile workers differed in methylation of more than 550 genes, including genes involved in metabolism, RNA synthesis, nucleic acids binding, signal transduction, brain development, and neural functions. In a similar study, Herb et al. (2012) could not verify these results but instead found substantial differences in DNA methylation between nurses and forager subcastes. Reverting foragers back to nurses reestablished methylation signatures for a majority of genes. Wojciechowski et al. (2018) produced the first genome-wide maps of chromatin structure in honeybee at a key larval stage in which developmental canalization into queen or worker was virtually irreversible. They found extensive genome-wide differences in histone modifications (H3K4me3, H3K27ac, and H3K36me3), many of which correlate with caste-specific transcription. These results demonstrate a key role for DNA and chromatin modifications in inducing caste-specific phenotypes in honeybee upon environmental signals.

Differences in DNA methylation between castes were also observed in social ants and wasps. However, in the primitively eusocial wasp *Polistes dominula* genome-wide DNA methylation was close to zero and only seven genes were methylated (Standage et al., 2016), calling into question the general importance of DNA methylation in caste differentiation and social behavior. Interestingly, morphologically and behaviorally distinct worker castes of the carpenter ant, *Camponotus floridanus,* were shown to be determined by histone acetylation rather than by DNA methylation (Simola et al., 2016). These examples confirm that epigenetic modifications of the chromatin are involved in the expression of polyphenism but there is apparently no simple universal relationship.

The marbled crayfish as a promising model for investigating the nongenetic components of phenotypic variation

In-depth investigation of EIV and SDV requires appropriate research models. Some of them have already been introduced in the previous sections. In this section, I would like to introduce the marbled crayfish as a particularly promising model for investigating EIV and SDV in the laboratory and the field. I will also sketch how this crayfish might help to identify the underlying molecular mechanisms and estimate the ecological and evolutionary relevance of EIV and SDV.

The marbled crayfish or Marmorkrebs is an apomictic parthenogenetic all-female crayfish that was detected in 1995 in the German aquarium trade (Scholtz et al., 2003; Vogt, 2018a, 2020). It has a maximum length (without chelae) of ca. 12 cm and is an auto triploid descendant of the sexually reproducing slough crayfish, *Procambarus fallax,* that occurs in Florida and southern Georgia (Martin et al., 2010). Marbled crayfish is now considered as a separate asexual species named *Procambarus virginalis* (Vogt et al., 2015; Lyko, 2017; Vogt, 2020). Wild populations of marbled crayfish have not been found in the native range of its parent species giving rise to the hypothesis that it might have originated in captivity (Vogt et al., 2015; Vogt 2019, 2020). Marbled crayfish is kept by aquarists and research laboratories worldwide. It has already established numerous wild populations in tropical to cold-temperate biomes in Europe, (15 countries), Africa (Madagascar), and Asia (Israel and Japan) resulting from releases (Vogt, 2020). In Europe, it was found in individual lakes and rivers from France to the Ukraine and from Estonia to Malta. In Madagascar, it has spread from an initial introduction near the capital Antananarivo in about 2003 over more than 100.000 km^2, inhabiting different biomes and ecosystems (Jones et al., 2009; Kawai et al., 2009; Gutekunst et al., 2018; Andriantsoa et al., 2019, 2020).

With almost 300 publications, the marbled crayfish is now one of the best-studied crayfish species (references in Vogt, 2018a, b, 2020). All marbled crayfish known today are genetically identical suggesting that they originated from a single individual (Vogt et al., 2008, 2015; Gutekunst et al., 2018). The genome of marbled crayfish is fully sequenced and a reference genome generated from a 3-year-old individual from my *Petshop* laboratory lineage is published (Gutekunst et al., 2018). Having a sequenced and annotated genome is of outstanding importance for research on genotype-phenotype relationships. The haploid genome size is ~3.5 GB, which is a bit larger than the human genome. Marbled crayfish produces clutches of ~50–650 offspring, depending on female size, providing an extraordinary source of genetically identical animals for experimentation. They can be raised throughout life in very simple laboratory settings, the early life stages even in microplates. All life stages can be fed with a single pellet food (e.g., Tetra WaferMix) (Vogt, 2008, 2020). A higher degree of genetic and experimental standardization is hardly conceivable for animals.

In 2008, I proposed to use the marbled crayfish as a model for investigating the nongenetic sources of phenotypic variation and their underlying molecular mechanisms (Vogt, 2008), and in the following years, I tried to push this topic forward. Meanwhile, the extent of SDV has been investigated under laboratory conditions for numerous traits (Vogt et al., 2008, 2009; Vogt, 2015) as exemplified in the previous sections of this paper. The extent of EIV has not been studied so systematically but some examples are available. For instance, specimens raised under stringent laboratory conditions reproduced well and grew old but reached maximum total lengths (tip of rostrum to end of telson) of only 9 cm whereas their relatives in the wild grew to 12 cm. My laboratory specimens also had significantly longer pleons and broader carapaces when compared to equally sized specimens from Lake Moosweiher, Germany (Vogt et al., 2018). Moreover, wild specimens had prominent sharp spines on their carapace and chelipeds but laboratory-raised specimens lacked these spines or had only small blunt spines (Fig. 9.10C) (Vogt et al., 2018). Andriantsoa et al. (2019) analyzed populations from ponds, lakes, rivers, and rice fields in Madagascar and revealed marked differences in size-frequency distribution among populations (Fig. 9.11), despite genetic identity. Linzmaier (2019) revealed that marbled crayfish is highly plastic with respect to trophic position and niche breath, depending on habitat, the availability of food and shelter, and the presence of competitors and predators.

The phenotypic differences between the various laboratory-raised and wild populations cannot be attributed to genetic variation, because all marbled crayfish investigated so far are genetically identical with the exception of a few random mutations (Gutekunst et al., 2018). Comparison of whole genome sequences of specimens from different laboratory lineages and

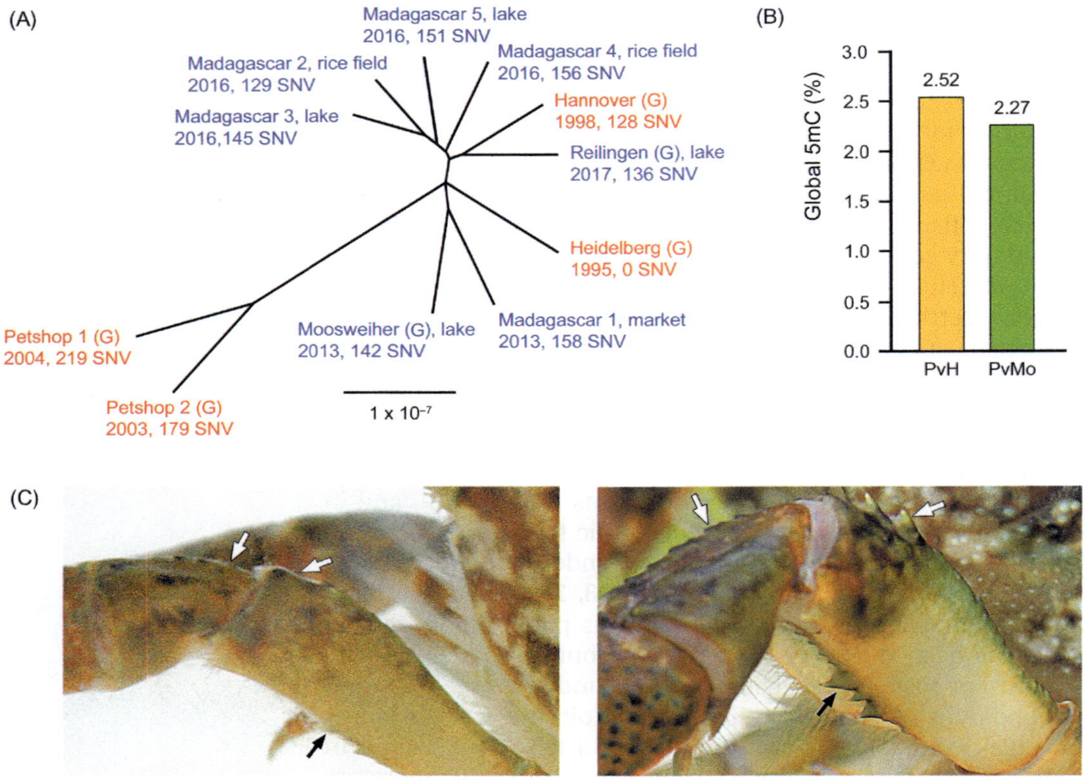

Figure 9.10 Genetic uniformity and epigenetic and morphological diversity in parthenogenetic marbled crayfish, *Procambarus virginalis*, from different ecosystems. (A) Phylogenetic tree of 11 marbled crayfish from diverse laboratory *(red)* and field sources *(blue)* in Germany (G) and Madagascar, based on the comparison of whole genome sequences. The maximum difference between specimens was only 219 single nucleotide variants (SNV). (B) Global DNA methylation values for specimens from my Heidelberg laboratory lineage (PvH) and Lake Moosweiher (PvMo). The methylation value of the wild sample is smaller than that of the laboratory-raised specimen. (C) Differences of spination between laboratory-raised and wild specimens. Wild specimens from Lake Moosweiher *(right)* had bigger and sharper spines *(arrows)* on their chelipeds than specimens from my Heidelberg laboratory lineage *(left)*. *Source*: (A) redrawn and modified after Gutekunst, J., Andriantsoa, R., Falckenhayn, C., Hanna, K., Stein, W., Rasamy, J.R., et al., 2018. Clonal genome evolution and rapid invasive spread of the marbled crayfish. Nat. Ecol. & Evolution, 2, 567–573. (B) redrawn and modified after Vogt, G., 2018a. Investigating the genetic and epigenetic basis of big biological questions with the parthenogenetic marbled crayfish: a review and perspectives. J. Biosci., 43, 189–223. (C) from Vogt, G., Lukhaup, C., Williams, B.W., Pfeiffer, M., Dorn, N.J., Schulz, R., et al., 2018. Morphological characterization and genotyping of the marbled crayfish and new evidence on its origin. Zootaxa, 4524, 329–350.

wild populations of known age from Germany and Madagascar revealed only very small differences of 129–219 single nucleotide variants (SNV) (Fig. 9.10A). The vast majority of these SNV were silent mutations. The maximum number of nonsynonymous SNV

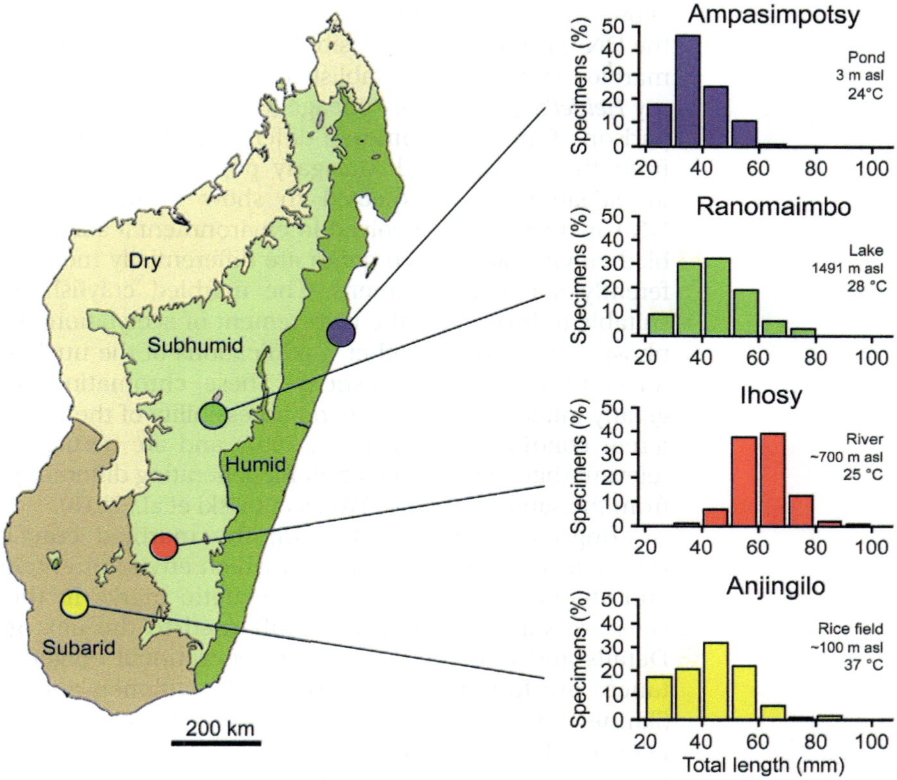

Figure 9.11 Size-frequency distribution of differently adapted, isogenic populations of marbled crayfish in Madagascar. The analyzed populations inhabited different water bodies in different bio-climatic regions. Values on right side of graphs give altitude above sea level and water temperature measured at time of sampling (8–10 a.m.) in 10 cm water depth. The particularly high water temperature in Anjingilo was caused by thermal water. Marbled crayfish were significantly larger ($P < 0.05$, Kruskal–Wallis one-way analysis of variance) in the Ihosy River than in the other sites. *Source*: Redrawn and modified after Andriantsoa, R., Tönges, S., Panteleit, J., Theissinger, K., Coutinho Carneiro, V., Rasamy, J., et al., 2019. Ecological plasticity and commercial impact of invasive marbled crayfish populations in Madagascar. BMC Ecol., 19, article 8.

that change the amino acid sequence of proteins was only four between animals (Gutekunst et al., 2018).

If the recorded phenotypic differences between the differently adapted marbled crayfish were not caused by genetic variation, then they must depend on epigenetic variation, for example, on differences in DNA methylation or histone acetylation and methylation. First measurements of global DNA methylation revealed higher values in a specimen from my laboratory (2.52%) than in a wild specimen from Lake Moosweiher (2.27%), indeed (Fig. 9.10B) (Vogt, 2018a). Comparison of genome-wide methylomes, which allows the identification of methylation differences at each base of the DNA, is meanwhile possible because a reference methylome of marbled crayfish was established for a 2-year-old individual from my *Heidelberg* laboratory lineage (Gatzmann et al., 2018). The comparison of the methylomes of differently adapted marbled crayfish from the German and Malagasy populations is currently under investigation. It is expected to show in more detail whether DNA methylation is involved in environmental adaptation of marbled crayfish and which genes are differentially methylated in differently adapted specimens. The marbled crayfish is also well suitable to investigate the involvement of acetylation, methylation, phosphorylation, and other modifications of the nucleosomal histones in phenotypic plasticity. These chromatin modifications greatly influence packaging and accessibility of the DNA under different conditions (Kouzarides, 2007) and are perhaps even more relevant than DNA methylation for generating different phenotypes from the same genotype (Wojciechowski et al., 2018).

Long-term experiments with the marbled crayfish should reveal, if there are conditions in which environmentally induced and stochastically produced epigenetic marks in the germline cells are stable enough to be inherited to the next generation. Details and references on "transgenerational epigenetic inheritance" are found in Richards (2006), Jablonka and Raz (2009), Skinner (2015) and Perez and Lehner (2019). In mammals, the DNA methylation marks are largely erased and reprogrammed in the primordial germ cells, the zygote, and early cleavage stages (Seisenberger et al., 2012). In zebrafish, the methylomes of the gametes are less intensely demethylated and there are significant differences between males and females (Jiang et al., 2013). The paternal methylome is largely maintained throughout early embryogenesis, whereas the maternal methylome is maintained only until the 16-cell stage and then progressively reprogrammed by losses and gains of methylation markers (Jiang et al., 2013). For marbled crayfish and other parthenogenetically reproducing invertebrates such data is not yet available.

Long-term experiments with the marbled crayfish should also reveal if phenotypically relevant epimutations can end up in stabilizing genetic mutations, for example, by conversion of 5-methylcytosine to thymine by deamination. The cytosine-to-thymine transition is by far the most common single nucleotide mutation and this mutation is apparently promoted by methylation of the cytosine. In bacteria and mammals, many sites of cytosine methylation in the DNA were identified as hot spots for C to T mutations. 5-methylcytosines at CpG-dinucleotides exhibit mutation rates higher than an order of magnitude above that of unmodified cytosines (Lutsenko and Bhagwat, 1999). Such epimutation-to-genetic mutation shifts might help to permanently fix phenotypes that have initially been induced by EIV and SDV. It might be an alternative scenario to "genetic assimilation," which assumes that a phenotype originally produced in response to an environmental condition later becomes genetically encoded via random mutation and natural selection (Waddington, 1953; Ehrenreich and Pfennig, 2016).

The demonstration of the conversion of reversible epimutations with phenotypic effects to irreversible, phenotype-fixing genetic mutations would significantly contribute to the "plasticity and epigenetics-first" alternative of evolution, which was propagated in different variants by Waddington (1953), Jablonka and Lamb (1995), West-Eberhard (2003), Skinner (2015), Vogt (2017a, b), Levis et al. (2018), and others. This concept posits that novel phenotypes cannot only arise from random genetic mutations but also from preexisting, epigenetically mediated phenotypic variants that later become genetically fixed. Such a mechanism with a Lamarckian touch would help to explain why evolution is often much faster than expected from the mere combination of random genetic mutation and natural selection.

Conclusions

The phenotypic variation in populations is the result of genetic variation and two nongenetic sources of variation, namely environmentally induced variation (EIV) and stochastic developmental variation (SDV). EIV and SDV occur in all kingdoms of life and are thus considered general biological principles. They have in common that they are mediated by epigenetic mechanisms but they differ in quality and function. EIV is directional and contributes to phenotype optimization in the inhabited environment, whereas SDV is random and contributes to evolutionary bet-hedging. They always act together but in different proportions, depending on species, trait, and environment. Together, EIV and SDV seem to facilitate environmental acclimatization and adaptation, which becomes increasingly important in the times of habitat destruction, climate change, and increased worldwide introduction of nonindigenous species. EIV and SDV probably also contribute to the evolution of phenotypic traits and species.

In sexually reproducing wild populations EIV and SDV are difficult to distinguish but differently adapted clonal lineages have proven useful to dissect both components to some degree. In the laboratory, EIV and SDV can be determined more precisely by exposing genetically identical populations to either single or different environments. Computer simulations of developmental reaction-diffusion mechanisms are also suitable to estimate the phenotypic variation produced by SDV. There is increasing evidence that alterations of the methylation marks on the DNA and acetylation and methylation marks on the histones are among the molecular mechanisms that produce different phenotypes from the same genome, either stochastically or by environmental induction.

Future experiments are expected to identify in more detail the molecular mechanisms underlying EIV and SDV. They are also hoped to elucidate in more detail the role of EIV and SDV in environmental adaptation and evolution. It should be tested if and under what conditions reversible epimutations with phenotypic effects, caused either by EIV or SDV, are inherited across generations and transformed into genetic mutations that irreversibly fix the corresponding phenotype. The experimental verification of such a mechanism would revolutionize our understanding of evolution. Suitable model organisms with known genomes, transcriptomes and methylomes, which have colonized diverse habitats in different biomes despite genetic identity, are at hand (e.g., the monoclonal marbled crayfish).

Acknowledgments

Many thanks to Guanghua Huang (Beijing, China) for Fig. 1D and Rogier van Vugt (Leiden, Netherlands) for Fig. 1B. I am particularly grateful to Vidyanand Nanjundiah (Bangalore, India) for suggestions that helped to improve the manuscript.

References

Ackermann, H.-W., Smirnoff, W.A., 1983. A morphological investigation of 23 baculoviruses. J. Invertebrate Pathol. 41, 269–280.

Ackermann, M., 2013. Microbial individuality in the natural environment. ISME J. 7, 465–467.

Anastasiadi, D., Piferrer, F., 2019. Epimutations in developmental genes underlie the onset of domestication in farmed European Sea Bass. Mol. Biol. Evolution 36, 2252–2264.

Andriantsoa, R., Tönges, S., Panteleit, J., Theissinger, K., Coutinho Carneiro, V., Rasamy, J., et al., 2019. Ecological plasticity and commercial impact of invasive marbled crayfish populations in Madagascar. BMC Ecol. 19, article 8.

Andriantsoa, R., Jones, J.P.G., Achimescu, V., Randrianarison, H., Raselimanana, M., Andriatsitohaina, M., et al., 2020. Perceived socio-economic impacts of the marbled crayfish invasion in Madagascar. PLoS One 15, e0231773.

Angers, B., Perez, M., Menicucci, T., Leung, C., 2020. Sources of epigenetic variation and their applications in natural populations. Evol. Appl. in press. https://doi.org/10.1111/eva.12946.

Angers, B., Schlosser, I.J., 2007. The origin of *Phoxinus eos-neogaeus* unisexual hybrids. Mol. Ecol. 16, 4562–4571.

Aranda-Anzaldo, A., Dent, M.A.R., 2003. Developmental noise, ageing and cancer. Mechanisms Ageing Dev. 124, 711–720.

Archer, G.S., Dindot, S., Friend, T.H., Walker, S., Zaunbrecher, G., Lawhorn, B., et al., 2003a. Hierarchical phenotypic and epigenetic variation in cloned swine. Biol. Reprod. 69, 430–436.

Archer, G.S., Friend, T.H., Piedrahita, J., Nevill, C.H., Walker, S., 2003b. Behavioral variation among cloned pigs. Appl. Anim. Behav. Sci. 82, 151–161.

Arnold, P.A., Kruuk, L.E.B., Nicotra, A.B., 2019. How to analyse plant phenotypic plasticity in response to a changing climate. N. Phytologist 222, 1235–1241.

Ashoub, M.R., Biggers, J.D., McLaren, A., Michie, D., 1958. The effect of the environment on phenotypic variability. Proc. R. Soc. Lond. B 149, 192–203.

Astauroff, B.L., 1930. Analyse der erblichen Störungsfälle der bilateralen Symmetrie im Zusammenhang mit der selbstständigen Variabilität ähnlicher Strukturen. Z. für Indukt. Abstammungs- und Vererbungslehre 55 (3), 183–262.

Avery, S.V., 2006. Microbial cell individuality and the underlying sources of heterogeneity. Nat. Rev. Microbiology 4, 577–587.

Ayali, A., 2019. The puzzle of locust density-dependent phase polyphenism. Curr. Opin. Insect Sci. 35, 41–47.

Balaban, N.Q., Merrin, J., Chait, R., Kowalik, L., Leibler, S., 2004. Bacterial persistence as a phenotypic switch. Science 305, 1622–1625.

Ballouz, S., Pena, M.T., Knight, F.M., Adams, L.B., Gillis, J.A., 2019. The transcriptional legacy of developmental stochasticity. bioRxiv preprint, https://doi.org/10.1101/2019.12.11.873265.

Becker, C., Hagmann, J., Müller, J., Koenig, D., Stegle, O., Borgwardt, K., et al., 2011. Spontaneous epigenetic variation in the *Arabidopsis thaliana* methylome. Nature 480, 245–249.

Ben-Ami, F., Hodgson, A.N., 2005. Ovoviviparity and the structure of the brood pouch in *Melanoides tuberculata* (Gastropoda: Prosobranchia: Thiaridae). J. Morphology 263, 322–329.

Bierbach, D., Laskowski, K.L., Wolf, M., 2017. Behavioural individuality in clonal fish arises despite near-identical rearing conditions. Nat. Commun. 8, article 15361.

Bonner, J.T., 1965. Size and Cycle: An Essay on the Structure of Biology. Princeton University Press, Princeton.

Bossdorf, O., Arcuri, D., Richards, C.L., Pigliucci, M., 2010. Experimental alteration of DNA methylation affects the phenotypic plasticity of ecologically relevant traits in *Arabidopsis thaliana*. Evolut. Ecol. 24, 541–553.

Browne, R.A., Moller, V., Forbes, V.E., Depledge, M.H., 2002. Estimating genetic and environmental components of variance using sexual and clonal *Artemia*. J. Exp. Mar. Biol. Ecol. 267, 107–119.

Burggren, W.W., Mueller, C.A., 2015. Developmental critical windows and sensitive periods as three-dimensional constructs in time and space. Physiological Biochemical Zool. 88, 91–102.

Burrows, M., Rogers, S.M., Ott, S.T., 2011. Epigenetic remodelling of brain, body and behaviour during phase change in locusts. Neural Syst. & Circuits 1, article 11.

Chase, H.B., 1939. Studies on the tricolor pattern of the Guinea pig. I. The relations between different areas of the coat in respect to the presence of color. Genetics 24, 610–621.

Cropley, J.E., Suter, C.M., Beckman, K.B., Martin, D.I.K., 2010. CpG methylation of a silent controlling element in the murine A^{vy} allele is incomplete and unresponsive to methyl donor supplementation. PLoS ONE 5, article e9055.
Dai, H., Wang, Z., 2014. Histone modification patterns and their responses to environment. Curr. Environ. Health Rep. 1, 11–21.
Daugman, J., Downing, C., 2001. Epigenetic randomness, complexity and singularity of human iris patterns. Proc. R. Soc. B 268, 1737–1740.
Dawson, N.J., 1970. Body composition of inbred mice (*Mus musculus*). Comp. Biochem. Physiol. 37, 589–593.
Dayel, M.J., Alegado, R.A., Fairclough, S.R., Levin, T.C., Nichols, S.A., McDonald, K., et al., 2011. Cell differentiation and morphogenesis in the colony-forming choanoflagellate *Salpingoeca rosetta*. Developmental Biol. 357, 73–82.
De Schepper, S., Debergh, P., van Bockstaele, E., de Loose, M., Gerats, A., Depicker, A., 2003. Genetic and epigenetic aspects of somaclonal variation: flower colour bud sports in azalea, a case study. South. Afr. J. Botany 69, 117–128.
Dybdahl, M.F., Drown, D.M., 2011. The absence of genotypic diversity in a successful parthenogenetic invader. Biol. Invasions 13, 1663–1672.
Ehrenreich, I.M., Pfennig, D.W., 2016. Genetic assimilation: a review of its potential proximate causes and evolutionary consequences. Ann. Botany 117, 769–779.
Elowitz, M.B., Levine, A.J., Siggia, E.D., Swain, P.S., 2002. Stochastic gene expression in a single cell. Science 297, 1183–1186.
Falconer, D.S., Mackay, T.F.C., 1996. Introduction to Quantitative Genetics, 4th ed. Longman, Harlow.
Feinberg, A.P., 2014. Epigenetic stochasticity, nuclear structure and cancer: the implications for medicine. J. Intern. Med. 276, 5–11.
Feinberg, A.P., Irizarry, R.A., 2010. Stochastic epigenetic variation as a driving force of development, evolutionary adaptation, and disease. Proc. Natl Acad. Sci. U.S.A. 107 (Supplement 1), 1757–1764.
Figueiredo, L.M., Cross, G.A.M., Janzen, C.J., 2009. Epigenetic regulation in African trypanosomes: a new kid on the block. Nat. Rev. Microbiology 7, 504–513.
Finch, C.E., Kirkwood, T.B.L., 2000. Chance, Development, and Aging. Oxford University Press, New York.
Flamme, A., 1977. Untersuchungen über die Ursachen der phänotypischen Varianz quantitativer Merkmale bei Laboratoriumsratten. Doktorarbeit, Technische Universität Hannover, Hannover.
Forde, B.G., 2009. Is it good noise? The role of developmental instability in the shaping of a root system. J. Exp. Botany 60, 3989–4002.
Fox, R.J., Donelson, J.M., Schunter, C., Ravasi, T., Gaitán-Espitia, J.D., 2019. Beyond buying time: the role of plasticity in phenotypic adaptation to rapid environmental change. Philos. Trans. R. Soc. B 374, article 20180174.
Fungairiño, S.G., Fernández, C., Serrano, J.M., López, F., Acosta, F.J., 2005. Developmental instability and plant potential fitness in a Mediterranean perennial plant, *Retama sphaerocarpa* (L.) Boiss. Acta Oecologia 27, 43–48.
Fusco, G., Minelli, A., 2010. Phenotypic plasticity in development and evolution: facts and concepts. Philos. Trans. R. Soc. B 365, 547–556.
Gärtner, K., 1990. A third component causing random variability beside environment and genotype. A reason for the limited success of a 30 year long effort to standardize laboratory animals? Laboratory Anim. 24, 71–77.
Gärtner, K., 2012. Commentary: random variability of quantitative characteristics, an intangible epigenomic product, supporting adaptation. Int. J. Epidemiol. 41, 342–346.

Gärtner, K., Bube, P., Flamme, A., Peters, K., Pfaff, J., 1976. Komponenten biologischer Variabilität und die Grenzen ihrer Manipulierbarkeit. Z. für Versuchstierkunde 18, 146–158.

Gatzmann, F., Falckenhayn, C., Gutekunst, J., Hanna, K., Raddatz, G., Coutinho Carneiro, V., et al., 2018. The methylome of the marbled crayfish links gene body methylation to stable expression of poorly accessible genes. Epigenetics & Chromatin 11, article 57.

Gilbert, J.J., Schröder, T., 2007. Intraclonal variation in propensity for mixis in several rotifers: variation among females and with maternal age. Hydrobiologia 593, 121–128.

Graham, J.H., Raz, S., Hel-Or, H., Nevo, E., 2010. Fluctuating asymmetry: methods, theory, and applications. Symmetry 2, 466–540.

Guillette Jr., L.J., Parrott, B.B., Nilsson, E., Haque, M.M., Skinner, M.K., 2016. Epigenetic programming alterations in alligators from environmentally contaminated lakes. Gen. Comp. Endocrinol. 238, 4–12.

Gutekunst, J., Andriantsoa, R., Falckenhayn, C., Hanna, K., Stein, W., Rasamy, J.R., et al., 2018. Clonal genome evolution and rapid invasive spread of the marbled crayfish. Nat. Ecol. & Evolution 2, 567–573.

Hageman, S.J., Bayer, M.M., Todd, C.D., 1999. Partitioning phenotypic variation: genotypic, environmental and residual components from bryozoan skeletal morphology. J. Nat. History 33, 1713–1735.

Hallgrímsson, B., Hall, B.K., 2011. Epigenetics: the context of development. In: Hallgrímsson, B., Hall, B.K. (Eds.), Epigenetics: Linking Genotype and Phenotype in Development and Evolution. University of California Press, Berkeley, pp. 424–438.

Hammel, J.U., Herzen, J., Beckmann, F., Nickel, M., 2009. Sponge budding is a spatiotemporal morphological patterning process: insights from synchrotron radiation-based X-ray microtomography into the asexual reproduction of *Tethya wilhelma*. Front. Zool. 6, article 19.

Herb, B.R., Wolschin, F., Hansen, K.D., Arye, M.J., Langmead, B., Irizarry, R., et al., 2012. Reversible switching between epigenetic states in honeybee behavioral subcastes. Nature Neurosci. 15 (2012), 1371–1373.

Herndon, L.A., Schmeissner, P.J., Dudaronek, J.M., Brown, P.A., Listner, K.M., Sakano, Y., et al., 2002. Stochastic and genetic factors influence tissue-specific decline in ageing *C. elegans*. Nature 419, 808–814.

Iguchi, K., Matsubara, N., Hakoyama, H., 2001. Behavioural individuality assessed from two strains of cloned fish. Anim. Behav. 61, 351–356.

Indrasamy, H., Woods, R.E., McKenzie, J.A., Batterham, P., 2000. Fluctuating asymmetry for specific bristle characters in *Notch* mutants of *Drosophila melanogaster*. Genetica 109, 151–159.

Jablonka, E., Lamb, M.J., 1995. Epigenetic Inheritance and Evolution: The Lamarckian Dimension. Oxford University Press, Oxford.

Jablonka, E., Raz, G., 2009. Transgenerational epigenetic inheritance: prevalence, mechanisms, and implications for the study of heredity and evolution. Q. Rev. Biol. 84, 131–176.

Jaenisch, R., Bird, A., 2003. Epigenetic regulation of gene expression: how the genome integrates intrinsic and environmental signals. Nat. Genet. 33, 245–254.

Jain, A.K., Prabhakar, S., Pankanti, S., 2002. On the similarity of identical twin fingerprints. Pattern Recognit. 35, 2653–2663.

Jiang, L., Zhang, J., Wang, J.-J., Wang, L., Zhang, L., Li, G., et al., 2013. Sperm, but not oocyte, DNA methylome is inherited by zebrafish early embryos. Cell 153, 773–784.

Johannes, F., Schmitz, R.J., 2019. Spontaneous epimutations in plants. N. Phytologist 221, 1253–1259.

Jones, J.P.G., Rasamy, J.R., Harvey, A., Toon, A., Oidtmann, B., Randrianarison, M.H., et al., 2009. The perfect invader: a parthenogenic crayfish poses a new threat to Madagascar's freshwater biodiversity. Biol. Invasions 11, 1475–1482.

Justice, S.S., Hunstad, D.A., Cegelski, L., Hultgren, S.J., 2008. Morphological plasticity as a bacterial survival strategy. Nat. Rev. Microbiology 6, 162–168.

Kawai, T., Scholtz, G., Morioka, S., Ramanamandimby, F., Lukhaup, C., Hanamura, Y., 2009. Parthenogenetic alien crayfish (Decapoda: Cambaridae) spreading in Madagascar. J. Crustacean Biol. 29, 562–567.

Kirkwood, T.B.L., Finch, C.E., 2002. The old worm turns more slowly. Nature 419, 794–795.

Kiskowski, M., Glimm, T., Moreno, N., Gamble, T., Chiari, Y., 2019. Isolating and quantifying the role of developmental noise in generating phenotypic variation. PLoS Computational Biol. 15, article e1006943.

Koehler, A.V., Springer, Y.P., Keeney, D.B., Poulin, R., 2011. Intra- and interclonal phenotypic and genetic variability of the trematode *Maritrema novaezealandensis*. Biol. J. Linn. Soc. 103, 106–116.

Kong, A.W.-K., Zhang, D., Lu, G., 2006. A study of identical twins' palmprints for personal verification. Pattern Recognit. 39, 2149–2156.

Koonin, E.V., Senkevich, T.G., Dolja, V.V., 2006. The ancient Virus World and evolution of cells. Biol. Direct 1, article 29.

Kouzarides, T., 2007. Chromatin modifications and their function. Cell 128, 693–705.

Kümmerli, R., Jiricny, N., Clarke, L.S., West, S.A., Griffin, A.S., 2009. Phenotypic plasticity of a cooperative behaviour in bacteria. J. Evolut. Biol. 22, 589–598.

Lajus, D.L., Alekseev, V.R., 2004. Phenotypic variation and developmental instability of life-history traits: a theory and a case study on within-population variation of resting eggs formation in *Daphnia*. J. Limnology 63 (Supplement 1), 37–44.

Landry, A.M., Landry, D.J., Gentry, L.R., Green, H.L., Reggio, B., Koonce, K.L., et al., 2005. Endocrine profiles and growth patterns of cloned goats. Cloning Stem Cell 7, 214–225.

Lennartsson, A., Ekwall, K., 2009. Histone modification patterns and epigenetic codes. Biochimica et Biophysica Acta 1790, 863–868.

Leung, C., Breton, S., Angers, B., 2016. Facing environmental predictability with different sources of epigenetic variation. Ecol. Evolution 6, 5234–5245.

Levis, N.A., Isdaner, A.J., Pfennig, D.W., 2018. Morphological novelty emerges from pre-existing phenotypic plasticity. Nat. Ecol. & Evolution 2, 1289–1297.

Lewontin, R., 2000. The Triple Helix: Gene, Organism and Environment. Harvard University Press, Cambridge.

Lewus, P., Ford, R.M., 1999. Temperature-sensitive motility of *Sulfolobus acidocaldarius* influences population distribution in extreme environments. J. Bacteriol. 181, 4020–4025.

Li, J., Min, R., Vizeacoumar, F.J., Jin, K., Xin, X., Zhang, Z., 2010. Exploiting the determinants of stochastic gene expression in *Saccharomyces cerevisiae* for genome-wide prediction of expression noise. Proc. Natl Acad. Sci. U.S.A. 107, 10472–10477.

Liebl, A.L., Schrey, A.W., Richards, C.L., Martin, L.B., 2013. Patterns of DNA methylation throughout a range expansion of an introduced songbird. Integr. Comp. Biol. 53, 351–358.

Lindeman, L.C., Thaulow, J., Song, Y., Kamstra, J.H., Xie, L., Asselman, J., et al., 2019. Epigenetic, transcriptional and phenotypic responses in two generations of *Daphnia magna* exposed to the DNA methylation inhibitor 5-azacytidine. Environ. Epigenetics 5, article dvz016.

Linzmaier S.M., Ecology and Evolution in Novel Communities: The Marbled Crayfish and its Interaction Partners, 2019. PhD-Thesis. Free University of Berlin, Berlin.

Liu, S., Aagaard, A., Bechsgaard, J., Bilde, T., 2019. DNA methylation patterns in the social spider, *Stegodyphus dumicola*. Genes 10, article 137.

Lutsenko, E., Bhagwat, A.S., 1999. Principal causes of hot spots for cytosine to thymine mutations at sites of cytosine methylation in growing cells. A model, its experimental support and implications. Mutat. Res. 437, 11–20.

Lyko, F., 2017. The marbled crayfish (Decapoda: Cambaridae) represents an independent new species. Zootaxa 4363, 544–552.

Lyko, F., Foret, S., Kucharski, R., Wolf, S., Falckenhayn, C., Maleszka, R., 2010. The honey bee epigenomes: differential methylation of brain DNA in queens and workers. PLoS Biol. 8, article e1000506.

Lynch, M., Walsh, B., 1998. Genetics and Analysis of Quantitative Traits. Sinauer Associates, Sunderland.

Maamar, H., Raj, A., Dubnau, D., 2007. Noise in gene expression determines cell fate in *Bacillus subtilis*. Science 317, 526–529.

Malandrin, L., Huber, H., Bernander, R., 1999. Nucleoid structure and partition in *Methanococcus jannaschii*: an archaeon with multiple copies of the chromosome. Genetics 152, 1315–1323.

Mallon, E.B., Amarasinghe, H.E., Ott, S.R., 2016. Acute and chronic gregarisation are associated with distinct DNA methylation fingerprints in desert locusts. Sci. Rep. 6, article 35608.

Martin, G.M., 2014. Nature, nurture, chance: their roles in interspecific and intraspecific modulations of aging. Annu. Rev. Gerontology Geriatrics 34, 267–284.

Martin, P., Dorn, N.J., Kawai, T., van der Heiden, C., Scholtz, G., 2010. The enigmatic Marmorkrebs (marbled crayfish) is the parthenogenetic form of *Procambarus fallax* (Hagen, 1870). Contributions Zool. 79, 107–118.

Moutinho, C., Esteller, M., 2017. MicroRNAs and epigenetics. Adv. Cancer Res. 135, 189–220.

Mulot, M., Marcisz, K., Grandgirard, L., Lara, E., Kosakyan, A., Robroek, B.J.M., et al., 2017. Genetic determinism vs. phenotypic plasticity in protist morphology. J. Eukaryot. Microbiology 64, 729–739.

Nanjundiah, V., 2003. Phenotypic plasticity and evolution by genetic assimilation. In: Müller, G.B., Newman, S.A. (Eds.), Origination of Organismal Form: Beyond the Gene in Development and Evolutionary Biology. MIT Press, Cambridge, pp. 245–264.

Nanjundiah, V., Bhogle, A.S., 1995. The precision of regulation in *Dictyostelium discoideum*: implications for cell-type proportioning in the absence of spatial pattern. Indian. J. Biochem. Biophysics 32, 404–416.

Newman, S.A., Müller, G.B., 2005. Origination and innovation in the vertebrate limb skeleton: an epigenetic perspective. J. Exp. Zool. B 304, 593–609.

Norouzitallab, P., Baruah, K., Vanrompay, D., Bossier, P., 2019. Can epigenetics translate environmental cues into phenotypes? Science of the Total Environment 647, 1281–1293.

Orias, E., Bradshaw, A.D., 1992. Stochastic developmental variation in the ratio of allelic rDNAs among newly differentiated, heterozygous macronuclei of *Tetrahymena thermophila*. Developmental Genet. 13, 87–93.

Palmer, A.R., Strobeck, C., 1986. Fluctuating asymmetry: measurement, analysis, patterns. Annu. Rev. Ecol. Syst. 17, 391–421.

Parsons, P.A., 1992. Fluctuating asymmetry: a biological monitor of environmental and genomic stress. Heredity 68, 361–364.

Pearl, R., 1906. Variation in *Chilomonas* under favourable and unfavourable conditions. Biometrika 5, 53–72.

Perez, M.F., Lehner, B., 2019. Intergenerational and transgenerational epigenetic inheritance in animals. Nat. Cell Biol. 21, 143–151.

Pietrzak, B., 2011. Interclonal differences in age-specific performance in *Daphnia magna*. J. Limnology 70, 345–352.

Pigliucci, M., 2001. Phenotypic Plasticity: Beyond Nature and Nurture. Johns Hopkins University Press, Baltimore.

Ponczek, L.M., Blackstone, N.W., 2001. Effect of cloning rate on fitness-related traits in two marine hydroids. Biol. Bull. 201, 76–83.

Richards, E.J., 2006. Inherited epigenetic variation — revisiting soft inheritance. Nat. Rev. Genet. 7, 395–401.

Rodríguez López, C.M., Wetten, A.C., Wilkinson, M.J., 2010. Progressive erosion of genetic and epigenetic variation in callus-derived cocoa (*Theobroma cacao*) plants. N. Phytologist 186, 856–868.

Roy, S.K., 1963. The variation of organs of individual plants. J. Genet. 58, 147–176.

Sahijram, L., Soneji, J.R., Bollamma, K.T., 2003. Analyzing somaclonal variation in micropropagated bananas (*Musa* spp.). Vitro Cell Developmental Biol. Plant. 39, 551–556.

Sakai, K.-I., Shimamoto, Y., 1965. Developmental instability in leaves and flowers of *Nicotiana tabacum*. Genetics 51, 801–813.

Schimz, A., Hildebrand, E., 1992. Nonrandom structures in the locomotor behavior of *Halobacterium*: a bifurcation route to chaos? Proc. Natl Acad. Sci. U.S.A 89, 457–460.

Schlichting, C.D., 1986. The evolution of phenotypic plasticity in plants. Annu. Rev. Ecol. Syst. 17, 667–693.

Schlichting, C.D., Pigliucci, M., 1998. Phenotypic Evolution: A Reaction Norm Perspective. Sinauer Associates, Sunderland.

Scholtz, G., Braband, A., Tolley, L., Reimann, A., Mittmann, B., Lukhaup, C., et al., 2003. Parthenogenesis in an outsider crayfish. Nature 421, 806.

Schrey, A.W., Robins, T.R., Lee, J., Dukes Jr., D.W., Ragsdale, A.K., Thawley, C.J., et al., 2016. Epigenetic response to environmental change: DNA methylation varies with invasion status. Environ. Epigenetics 2, article dvw008.

Schuett, W., Dall, S.R.X., Baeumer, J., Kloesener, M.H., Nakagawa, S., Beinlich, F., et al., 2011. "Personality" variation in a clonal insect: the pea aphid, *Acyrthosiphon pisum*. Developmental Psychobiology 53, 631–640.

Seidel Jr., G.E., Elsden, R.P., Hasler, J.F., 2003. Embryo Transfer in Dairy Cattle. Hoards and Sons, Fort Atkinson.

Seisenberger, S., Peat, J.R., Hore, T.A., Santos, F., Dean, W., Reik, W., 2012. Reprogramming DNA methylation in the mammalian life cycle: building and breaking epigenetic barriers. Philos. Trans. R. Soc. B 368, 20110330.

Shelton, D.E., Desnitskiy, A.G., Michod, R.E., 2012. Distributions of reproductive and somatic cell numbers in diverse *Volvox* (Chlorophyta) species. Evolut. Ecol. Res. 14, 707–727.

Shin, T., Kraemer, D., Pryor, J., Liu, L., Rugila, J., Howe, L., et al., 2002. A cat cloned by nuclear transplantation. Nature 415, 859.

Shushu, D.D., Comar, J.M., Abegaz, B.M., 2009. Somaclonal variation in *in vitro* regenerated *Ledebouria graminifolia* (Hyacinthaceae), an indigenous bulb in Botswana and its potential exploitation as an ornamental plant. J. Biol. Sci. 9, 152–158.

Simola, D.F., Graham, R.J., Brady, C.M., Enzmann, B.L., Desplan, C., Ray, A., et al., 2016. Epigenetic (re)programming of caste-specific behavior in the ant *Camponotus floridanus*. Science 351, article acc6633.
Simpson, S.J., Sword, G.A., Lo, N., 2011. Polyphenism in insects. Curr. Biol. 21, R738–749.
Singh, A., Weinberger, L.S., 2009. Stochastic gene expression as a molecular switch for viral latency. Curr. Opin. Microbiol. 12, 460–466.
Skinner, M.K., 2014. Endocrine disruptor induction of epigenetic transgenerational inheritance of disease. Mol. Cell Endocrinol. 398, 4–12.
Skinner, M.K., 2015. Environmental epigenetics and a unified theory of the molecular aspects of evolution: a neo-Lamarckian concept that facilitates neo-Darwinian evolution. Genome Biol. Evolution 7, 1296–1302.
Slepecky, R.A., Starmer, W.T., 2009. Phenotypic plasticity in fungi: a review with observations on *Aureobasidium pullulans*. Mycologia 101, 823–832.
Smith, E.A., Collette, S.B., Boynton, T.A., Lillrose, T., Stevens, M.R., Bekker, M.F., et al., 2011. Developmental contributions to phenotypic variation in functional leaf traits within quaking aspen clones. Tree Physiol. 31, 68–77.
Smits, W.K., Kuipers, O.P., Veening, J.-W., 2006. Phenotypic variation in bacteria: the role of feedback regulation. Nat. Rev. Microbiology 4, 259–271.
Soll, D.R., 2014. The role of phenotypic switching in the basic biology and pathogenesis of *Candida albicans*. J. Oral. Microbiology 6, article 22993.
Sommer, R.J., Dardiry, M., Lenuzzi, M., Namdeo, S., Renahan, T., Sieriebriennikov, B., et al., 2017. The genetics of phenotypic plasticity in nematode feeding structures. Open. Biol. 7, article 160332.
Standage, D.S., Berens, A.J., Glastad, K.M., Severin, A.J., Brendel, V.P., Toth, A.L., 2016. Genome, transcriptome and methylome sequencing of a primitively eusocial wasp reveal a greatly reduced DNA methylation system in a social insect. Mol. Ecol. 25, 1769–1784.
Stewart-Savage, J., Wagstaff, B.J., Yund, P.O., 1999. Developmental basis of phenotypic variation in egg production in a colonial ascidian: primary oocyte production versus oocyte development. Biol. Bull. 196, 63–69.
Storrs, E.E., Williams, R.J., 1968. A study of monozygous quadruplet armadillos in relation to mammalian inheritance. Proc. Natl Acad. Sci. U.S.A. 60, 910–914.
Tao, L., Du, H., Guan, G., Dai, Y., Nobile, C.J., Liang, W., et al., 2014. Discovery of a "white-gray-opaque" tristable phenotypic switching system in *Candida albicans*: roles of non-genetic diversity in host adaptation. PLoS Biol. 12, article e1001830.
Thomson, C.E., Winney, I.S., Salles, O.C., Pujol, B., 2018. A guide to using a multiple-matrix animal model to disentangle genetic and nongenetic causes of phenotypic variance. PLoS ONE 13, article e0197720.
Thorson, J.L.M., Smithson, M., Beck, D., Sadler-Riggleman, I., Nilsson, E., Dybdahl, M., et al., 2017. Epigenetics and adaptive phenotypic variation between habitats in an asexual snail. Sci. Rep. 7, article 14139.
Thorson, J.L.M., Smithson, M., Sadler-Riggleman, I., Beck, D., Dybdahl, M., Skinner, M.K., 2019. Regional epigenetic variation in asexual snail populations among urban and rural lakes. Environ. Epigenetics 2019, 1–12.
Tikhodeyev, O.N., Shcherbakova, O.V., 2019. The problem of non-shared environment in behavioral genetics. Behav. Genet. 49, 259–269.
Tordoff, M.G., Bachmanov, A.A., Reed, D.R., 2007a. Forty mouse strain survey of water and sodium intake. Physiol. & Behav. 91, 620–631.
Tordoff, M.G., Bachmanov, A.A., Reed, D.R., 2007b. Forty mouse strain survey of voluntary calcium intake, blood calcium, and bone mineral content. Physiol. & Behav. 91, 632–643.

Turko, A.J., Earley, R.L., Wright, P.A., 2011. Behaviour drives morphology: voluntary emersion patterns shape gill structure in genetically identical mangrove rivulus. Anim. Behav. 82, 39–47.

Van der Graaf, A., Wardenaar, R., Neumann, D.A., Taudt, A., Shaw, R.G., Jansen, R.C., et al., 2015. Rate, spectrum, and evolutionary dynamics of spontaneous epimutations. Proc. Natl Acad. Sci. U.S.A. 112, 6676–6681.

Verhoeven, K.J.F., Preite, V., 2014. Epigenetic variation in asexually reproducing organisms. Evolution 68, 644–655.

Vinck, A., Terlou, M., Pestman, W.R., Martens, E.P., Ram, A.F., van den Hondel, C. A.M.J.J., et al., 2005. Hyphal differentiation in the exploring mycelium of *Aspergillus niger*. Mol. Microbiology 58, 693–699.

Vogt, G., 2008. The marbled crayfish: a new model organism for research on development, epigenetics and evolutionary biology. J. Zool. 276, 1–13.

Vogt, G., 2010. Suitability of the clonal marbled crayfish for biogerontological research: a review and perspective, with remarks on some further crustaceans. Biogerontology 11, 643–669.

Vogt, G., 2015. Stochastic developmental variation, an epigenetic source of phenotypic diversity with far-reaching biological consequences. J. Biosci. 40, 159–204.

Vogt, G., 2017a. Facilitation of environmental adaptation and evolution by epigenetic phenotype variation: insights from clonal, invasive, polyploid, and domesticated animals. Environ. Epigenetics 3, article dvx002.

Vogt, G., 2017b. Evolution of epigenetic mechanisms in animals and their role in speciation. In: Tollefsbol, T.O. (Ed.), Handbook of Epigenetics: The New Molecular and Medical Genetics, 2nd edition Academic Press, New York, pp. 409–426.

Vogt, G., 2018a. Investigating the genetic and epigenetic basis of big biological questions with the parthenogenetic marbled crayfish: a review and perspectives. J. Biosci. 43, 189–223.

Vogt, G., 2018b. Annotated bibliography of the parthenogenetic marbled crayfish, *Procambarus virginalis*, a new research model, potent invader and popular pet. Zootaxa 4418, 301–352.

Vogt, G., 2019. Estimating the young evolutionary age of marbled crayfish from museum samples. J. Nat. Hist. 53, 2353–2363.

Vogt, G., 2020. Biology, ecology, evolution, systematics and utilization of the parthenogenetic marbled crayfish, *Procambarus virginalis*. In: Ribeiro, F.B. (Ed.), Crayfish: Evolution, Habitat and Conservation Strategies. Nova Science Publishers, New York, pp. 137–227.

Vogt, G., Rug, M., 1999. Life stages and tentative life cycle of *Psorospermium haeckeli*, a species of the novel DRIPs clade from the animal-fungal dichotomy. J. Exp. Zool. 283, 31–42.

Vogt, G., Falckenhayn, C., Schrimpf, A., Schmid, K., Hanna, K., Panteleit, J., et al., 2015. The marbled crayfish as a paradigm for saltational speciation by autopolyploidy and parthenogenesis in animals. Biol. Open. 4, 1583–1594.

Vogt, G., Huber, M., Thiemann, M., van den Boogaart, G., Schmitz, O.J., Schubart, C.D., 2008. Production of different phenotypes from the same genotype in the same environment by developmental variation. J. Exp. Biol. 211, 510–523.

Vogt, G., Lukhaup, C., Williams, B.W., Pfeiffer, M., Dorn, N.J., Schulz, R., et al., 2018. Morphological characterization and genotyping of the marbled crayfish and new evidence on its origin. Zootaxa 4524, 329–350.

Vogt, G., Wirkner, C.S., Richter, S., 2009. Symmetry variation in the heart-descending artery system of the parthenogenetic marbled crayfish. J. Morphology 270, 221–226.

Waddington, C.H., 1953. Genetic assimilation of an acquired character. Evolution 7, 118–126.
Waddington, C.H., 1957. The Strategy of the Genes. MacMillan, New York.
Warren, E., 1902. Variation and inheritance in the parthenogenetic generations of the aphis *Hyalopterus trirhodus* (Walker). Biometrika 1, 129–154.
West-Eberhard, M.J., 1989. Phenotypic plasticity and the origins of diversity. Annu. Rev. Ecol. Syst. 20, 249–278.
West-Eberhard, M.J., 2003. Developmental Plasticity and Evolution. Oxford University Press, New York.
Williams, R.J., Pelton, R.B., 1966. Individuality in nutrition: effects of vitamin A-deficient and other deficient diets on experimental animals. Proc. Natl Acad. Sci. U.S.A 55, 126–134.
Woese, C.R., Kandler, O., Wheelis, M.L., 1990. Towards a natural system of organisms: proposal for the domains Archaea, Bacteria, and Eucarya. Proc. Natl Acad. Sci. U.S.A. 87, 4576–4579.
Wojciechowski, M., Lowe, R., Maleszka, J., Conn, D., Maleszka, M., Hurd, P.J., 2018. Phenotypically distinct female castes in honey bees are defined by alternative chromatin states during larval development. Genome Res. 28, 1–11.
Yang, B.-C., Lee, S.-H., Hwang, S., Lee, H.-C., Im, G.-S., Kim, D.-H., et al., 2012. Phenotypic characterization of Hanwoo (native Korean cattle) cloned from somatic cells of a single adult. BMB Rep. 45, 38–43.
Yang, C.-H., Pospisilik, J.A., 2019. Polyphenism – a window into gene-environment interactions and phenotypic plasticity. Front. Genet. 10, article 132.

10

The evolution of cell differentiation in animals: biomolecular condensates as amplification hubs of inherent cell functions

Stuart A. Newman
Department of Cell Biology and Anatomy, New York Medical College, Valhalla, NY, United States

Introduction

Development of multicellular organisms involves *morphogenesis*, that is, generation of structural motifs—layers, segments, branches, and fruiting bodies, depending on the species, *differentiation* of distinct cell types, for example, muscle, bone, root, spores, and *pattern formation*, that is, reliable arrangement of cell types within the generated forms [reviewed in (Newman, 2019a)]. Each of these processes may employ nonlinear effects and therefore can potentially mediate phenotypic switching. But only in the case of cell differentiation is switching between discrete states a defining attribute.

In the case of animal development, cloning experiments with somatic cell nuclei by John Gurdon and others beginning in the 1960s suggested that terminally differentiated cells retained all the genes necessary for full-term development from a single cell [reviewed in Blau (2014)]. This quickly led to models of differentiation based on differential gene activity rather than the conceivable alternatives of progressive gene loss or gene innovation. This a paradigm that continues to prevail, although exceptions for some developmental lineages such as the immune system are acknowledged and uncontroversial.

Genes interact with one another via their RNA or protein products. A 1969 paper by Roy Britten and Eric Davidson "Gene regulation for higher cells: a theory," presented a network representation

Phenotypic Switching. DOI: https://doi.org/10.1016/B978-0-12-817996-3.00011-6

of animal and plant genomes that featured reciprocal control by regulatory RNAs or transcription factors (TFs) of "batteries" (linked sets) of genes characteristic of distinct cell types (Britten and Davidson, 1969). The model described what came to be called the "regulatory genome" (Davidson, 2006), since the functions of cell-specific products at the bottom of the hierarchy were not part of the logical structure. The means by which such gene regulatory networks (GRNs) could be modeled mathematically and studied in silico was not part of the initial formulation although this is a requirement for testing the predictive capabilities of systems of such complexity.

The Britten-Davidson model was inspired by studies of bacterial gene regulation via operons that were well underway at the time [reviewed in Loison and Morange (2017)]. The logical structure of the "higher cells" version was a concatenation of bacteria-like modules, with each cell type programmed independently and hierarchically. Even in a recent version of the model cell types are described as the outcomes of developmentally programmed, "directed, oriented networks in which information flows in only one direction" (Peter and Davidson, 2015, p. 43). I will refer to this account of cell differentiation as the "modular gene regulatory network" (MGRN) model.

By the late 1960s, it was evident that the structure of the DNA-protein complex—chromatin—in the eukaryotic nucleus made the allocation of genes to transcriptionally active and inactive states a different kind of mechanism from that of the bacterial operons that provided the inspiration for the MGRN's logical structure. The 1969 paper acknowledged that switching between alternative states of multigene activity would eventually need to take this level of organization into account (Britten and Davidson, 1969).

Another paper from 1969, Stuart Kauffman's "Metabolic stability and epigenesis in randomly constructed genetic nets," provided a theory of the regulatory genome, which like the MGRN model was inspired by bacterial gene regulatory systems. In contrast to Britten and Davidson's model, however, this model was intrinsically nonhierarchical (gene interactions are bidirectional) and came with a whole-system mathematical representation (Kauffman, 1969). Components representing genes (a fixed N for a given network) were wired up so that two inputs regulated each component via a Boolean function (e.g., AND and OR) selected at random from among all of the 16 possible ones with this topology. The network state was updated in synchronous time-steps and the "modes of behavior" (discrete subsets of states that cyclically recur once the system finds itself in one of them, also referred to as "dynamical attractors") were computed for nets of different component numbers. Depending on the initial conditions, the same network would arrive at one of *k* [less than N; (Paul et al., 2006)] gene activity patterns, which Kauffman identified with cell types.

Different versions of this model with more realistic assumptions have been advanced since then (Raeymaekers, 2002; Mojtahedi et al., 2016), but the main conceptual innovation was that the cell types of an organism were the global attractors of a unitary dynamical system identified with the regulatory genome. This was carried over to instances in which gene control was implemented by continuous rather than discrete functions, and the GRNs represented as systems of ordinary differential equations [ODEs; e.g., (Glass and Kauffman, 1973; Keller, 1995)], or by networks in which the Boolean functions were given a simplifying mean-field interpretation (Andrecut and Kauffman, 2006). As stated by the latter authors "it is almost an inevitable hypothesis that the distinct cell types of an organism correspond to the distinct attractors of the network." I term this theoretical framework (in contrast to the MGRN approach described earlier) the "global genome regulatory dynamics" (GGRD) model of cell type diversification.

Modeling gene switches by Boolean functions is computationally more tractable than using continuous functions. Studies within the GGRD framework have therefore largely continued to employ Boolean networks. Work explicitly following the Britten-Davidson MGRN program also came to adopt Boolean representations (Davidson, 2011; Cui et al., 2017; Peter and Davidson, 2011). There is an important difference between the perspectives, however. The MGRN school conceives of the "regulatory genome" as machine-like, such that many thousands of *cis*-regulatory control modules are wired together in large networks—"genomic computers"—which control major processes-like development and regeneration (Istrail et al., 2007). Connections between modules, and thus new cell types and their functions, are built stepwise by evolution (Royo et al., 2011). For the GGRD school, in contrast, the organization of the system is not modular. While links among components are, in principle, established and refined by evolution, cell types are emergent modes of activity of the entire regulatory system. Each successive change involves a de novo specification of cell types. Thus, it is under the requirement to preserve the identity and stability of many existing modes.

Because of the indifference of MGRN models to global network properties, the assumptions behind them have not been obviously challenged by the increasingly recognized complexity of chromatin-level gene activating and silencing processes. As mentioned, this level of control was anticipated in the earliest presentation of Britten-Davidson model, and the role of histone modifications in effecting persistence of alternative gene regulatory states (one finding, but not the most novel, from new work in this field) is noted in an update of the framework five decades later (Peter and Davidson, 2015). Modular networks are relatively tolerant to such complications and can exploit them for stability.

In contrast, the GGRD theoretical program has never contended with the features of genome organization, including chromatin architecture, that distinguish eukaryotic gene regulation from bacterial promoter-based mechanisms. The question of whether gene on-off switches are realistically rendered as Boolean functions is more critical for models in which cell types are identified with dynamical attractors of a global system. This is in part due to effects of biochemical noise on the robustness of the dynamical attractors of such systems (Villani et al., 2011). Also, most developmentally critical TFs (at least 90% of them in animal systems) contain intrinsically disordered protein regions (IDRs), which introduce uncharacterized conditionalities into their binding of DNA sequences and their partnering with cofactors (Liu et al., 2006). This further undermines the plausibility of global logic-based switching models of either category, both of which depend on stable network topologies (Niklas et al., 2015).

The two classes of model also encounter difficulties concerning their evolutionary origination and the nature (e.g., function and internal structure) of the cell types they purport to specify. We know much more about the emergence and early evolution of Metazoa than we did 50 years ago, and the order of appearance of cell types in successively divergent animals is well established (Newman, 2020). The original cell type novelty, found in all animals, from the early branching Porifera (sponges) and Placozoa to the most elaborate triploblasts (e.g., arthropods and vertebrates) is the epithelial cell. The protective barriers formed by layers of these cells (epithelia) and the liquid-like behaviors of epithelioid masses are the distinguishing features of the animals and the fundamental basis of their morphogenetic properties [reviewed in (Newman, 2019a; Arnellos and Moreno, 2016)]. The epithelial cell only exists because animals contain so-called "classical" cadherins, cell-cell adhesion proteins with a cytoskeleton-engaging cytoplasmic domain not found in the unicellular and colonial organisms ancestral to the animals [reviewed in Newman (2016b)]. [The ctenophores, a sister- or early-branching clade of the animals use a different unique cadherin domain to similar effect (Belahbib et al., 2018)].

To be realistic accounts of the evolutionary development of animals, the MGRN and GGRD models would have to posit an original specialized cell type (i.e., the epithelial cell). With subsequent increase of phyletic complexity, leading to progressive acquisition of mesenchyme (loosely organized cells in which cadherin function is downregulated), muscle, neurons, skeletal cells, gland cells, and so forth, the earlier-appearing types would have to be retained in the new clades' developmental repertoire. Furthermore, developmental hierarchies among and between such functionalities (e.g., ectoderm→neuronal-glial stem cell→neuron; mesoderm→chondrocyte-connective tissue stem cell→fibroblast), would need to continue to characterize all newly appearing animal types.

For GGRD-type models based on the attractor sets of dynamical systems, this scenario is highly implausible. From an evolutionary point of view, such models would need to postulate that a minimal number of mutually compatible cell types—attractors of the regulatory genome of an ancestral metazoan—emerged early in evolution. As the small number of cell types of the earliest diverging animals, for example, epithelial and mesenchymal cells, were complemented in later-appearing forms with additional cell types, the regulatory genomes of the later-diverging forms would have been required to exhibit attractors representing the old and new cell types, as well as conserved developmental lineages. Studies of the attractor organization of dynamical systems do not point to such scenarios being likely (Craciun et al., 2013; Conradi and Pantea, 2019).

A different type of GGRD model, "isologous diversification," is inherently multicellular in contrast to Kauffman-type models, in that it postulates molecular communication between replicates of a single dynamical system (Kaneko and Yomo, 1999). In isologous diversification, mutual gene regulatory interactions are represented as coupled ODEs rather than Boolean networks. These systems have a number of novel and interesting properties not seen in any of the nonmulticellular GGRDs. Hierarchies of nested cell types naturally emerge in the simulated "cell societies," as well as attractor states that are not characteristic of the unicell prototype (Kaneko and Yomo, 1999; Furusawa and Kaneko, 2002). However, these models still have the same functionality problem of the original GGRDs: unless most of the mathematically determined attractors of a global genome correspond to functional cell types, the biological relevance of genomic attractors and transitions between them is just a formal speculation. Indeed, the metazoan gene regulatory apparatus may not even a category of system that exhibits dynamical multistability (Conradi and Pantea, 2019; Craciun et al., 2013).

The modular, logical circuit-based MGRN models, in which there are no system-level constraints, superficially avoid these formal problems associated with GGRD models. They are presumed to emerge in an evolutionary scenario in which subnetworks can be sequentially acquired by natural selection, employing genes and gene products organized around new functional demands. Hierarchically organized lineages would follow, in analogy to nested subroutines in computer programs (Peter and Davidson, 2011). For this reason, MGRN-type approaches have become the default for computational analyses of cell type switching during development regardless of the original theoretical commitment of the investigators (Cui et al., 2017; Huang et al., 2009; Corson and Siggia, 2017). In all cases, function is sidelined by focusing on transitions between cell types adjacent to one another in developmental lineages in experimentally characterized systems. These include emergence of endoderm, mesoderm and skeletogenic micromeres from blastomeres in the sea urchin gastrula (Peter et al., 2012); differentiation of blood progenitor cells into cells of the erythroid or myeloid lineages (Huang et al., 2009); determination of alternative epidermal Pn.p-cell fates in the nematode vulva (Corson et al., 2017). In these models, the cell types in question are locally accessible attractor states of narrowly defined dynamical networks in which there are marginally different TF complexions; in no cases are the cell types claimed to be attractors of the full regulatory genome.

New findings, however, discussed in the remainder of this chapter, indicate that the mechanism of metazoan cell differentiation is unique to this clade and differs dramatically from the bacterial operon systems that inspired the conceptual framework for both MGRN- and GGRD-type models [see also (Newman, 2020)]. I propose that the functional "content" of animal cell types (i.e., the specialized function-related genes they express), a concept that presents theoretical difficulties for both MGRN and GGRD models, actually represents exaggerated versions of functional modes intrinsic to ancestral unicellular organisms, and that these functionalities have been partitioned into distinct cellular phenotypes over evolution. I show that this ability to amplify and compartmentalize functions is a novel capability of the metazoan gene regulatory apparatus.

Rather than being mediated by hierarchical logical circuits as in MGRN models, or reflecting the attractor states of nonlinear dynamical systems, as in GGRD models, switching between differentiated states occurs by a previously unanticipated set of liquid-protein phase transformations, a very different category of physical process. This phenomenon is finding increasing relevance throughout cell biology, including in activities outside the nucleus (Nakamura et al., 2019; Holehouse and Pappu, 2018); see also Nicholson (2019).

Finally, I consider the intertwined roles of the ancestral dual-function cell-cell adhesion/transcriptional modulator β-catenin and the Grainyhead-like transcription factor in the origination of Metazoa. These connections suggest that the novel means by which animals partition ancestral functions into distinct cell types may be intrinsically tied to multicellularity, but of a type only found in animals (Newman, 2019b; Niklas and Newman, 2016).

Metazoan-specific modes of transcriptional regulation

The animals arose within a clade of unicellular and occasionally colonial unicellular organisms termed Holozoa. In addition to Metazoa, this group contains subclades such as Choanoflagellatea, Filasterea, and Ichthyosporea, whose extant members serve as the basis of inferences about the genetics and gene regulatory mechanisms of the last common holozoan ancestor of the animals (King et al., 2008; Ruiz-Trillo et al., 2007). Holozoa is contained within a broader clade, Opisthokonta, which also includes Fungi (reviewed in (Newman, 2016a)). Nonholozoan opisthokonts such as yeasts and molds can therefore serve as outgroups by which conserved and novel aspects genomic evolution of the animals can be traced (Ruiz-Trillo, 2016). While multicellular organisms outside the opisthokonts in the domain Eukarya (all organisms with nucleated cells), most prominently the vascular plants, generate a variety of specialized cell types, only metazoans have this capacity among the holozoans. In this section I will discuss the specific features of animal gene regulatory systems that distinguish them from those of their sister clades and enable a cell differentiation capability that is the most prolific among all known life forms.

All eukaryotes have their nuclear DNA organized by chromatin-associated proteins, most prominently histones. These organizing proteins are capable of being chemically modified, typically by acetylation or methylation, in a way that affects DNA accessibility. With these in place, transcriptional programs can be remembered after they have been activated, thus becoming independent of the conditions that first brought them about (Prohaska et al., 2010). Archaea, and in rare cases Bacteria (Dilweg and Dame, 2018), both exhibit this feature, but eukaryotes appear to be unique among cellular forms of life in having enzymes that can erase and remodel the chemical marks recorded on histone-type proteins. Such organisms thus have "write-read-rewrite" genomic machinery capable of storing and processing, as a form of experiential information, what in simpler cells is just a succession of condition-dependent biochemical states (Prohaska et al., 2010); see also (Shapiro, 2011).

Nuclei of all eukaryotic cells also contain the large multisubunit protein adaptor complex Mediator [mediator of RNA polymerase II transcription (Verger et al., 2019)]. Mediator consists of a core of 15 different subunits, with up to 15 additional ones in metazoans, many of the latter having appeared concomitantly with increasing organismal complexity. The proteins of the complex are rich in IDRs that potentially interact with thousands of different eukaryotic TFs which themselves contain disordered domains (Niklas et al., 2015). Mediator transduces signals from transcriptional activators bound to enhancers (in the case of metazoans), or *cis*-acting upstream activating sequences (in nonmetazoan eukaryotes), to the transcriptional machinery, which is assembled at promoters that are linearly closer to the target genes.

While the write-read-rewrite machinery for recording gene activity and the mediator transcription-regulating adaptor are conserved throughout the eukaryotes, the evolution of animals required additional steps that set first the holozoans, and then the metazoans, apart from their progenitors. The multidomain protein complex p300/CBP initiates transcription by acetylating histones and thereby relaxing the structure of nucleosomes at promoters of all or most expressed genes. This facilitates the recruitment of RNA polymerase II and TFs to those sites (Chan and La Thangue, 2001). While p300/CBP is best characterized in animal systems and is absent in fungi [the closest counterpart in yeast is an unrelated protein (Dahlin et al., 2015)], extant unicellular holozoans such as *Capsaspora owczarzaki* contain homologs of this complex, and their promoter nucleosomes exhibit p300/CBP-specific histone acetylation marks. This strongly suggests that the associated gene control mechanism predated metazoan evolution (Grau-Bové et al., 2017).

Mediator, p300/CBP and TFs congregate in topologically associating domains (TADs) within the nucleus (Furlong and Levine, 2018; Plys and Kingston, 2018; Galupa and Heard, 2017), which form by phase-separation (like oil and water) from the surrounding nuclear sap as proteinaceous "biomolecular condensates" (Alberti et al., 2019; Shin et al., 2019). In these TADs, p300/CBP serves as an interaction hub in which a multitude of TFs and other coregulators such as nuclear receptors (NRs) mediate tissue responses to developmental and physiological cues (Dyson and Wright, 2016). A given gene is expressed when p300/CBP binds to TFs and NRs targeting the gene's promoter, apparently followed by TF-dependent phase separation of Mediator into gene-activating droplets (Boija et al., 2018).

Metazoans also have "silencing" mechanisms that suppress gene expression during development and in terminally differentiated cells. These first appeared in the form of the histone methylating enzymes Polycomb Group II (PcGII) proteins and SUV39H (Jih et al., 2017) in a common ancestor of holozoans and fungi (Kingston and Tamkun, 2014; Steffen and Ringrose, 2014). In metazoans, however, a new class of silencing proteins, Polycomb Group I (PcGI) appeared (Grossniklaus and Paro, 2014; Steffen and Ringrose, 2014). In contrast to the earlier-evolving ones, these operate at the level of biomolecular condensates by a phase-separation process similar to that of the transcription hubs (Tatavosian et al., 2019). Silenced sequences are converted into stably inactive heterochromatin (Chan et al., 2018).

While genes of all cells are regulated by upstream *cis*-acting promoters, metazoans [with the apparent exception of Placozoa (Sebé-Pedrós et al., 2016)] also employ *enhancers*, promoter-like sequences which can be located upstream or downstream of their target genes, in introns, or even on different chromosomes. Enhancers are apparently unique to Metazoa and its immediate antecedents, being absent in extant unicellular holozoans (Sebé-Pedrós et al., 2016). They are essential and integral to the development and maintenance of differentiated cell types. In metazoans, chromatin loops containing cell-type regulated DNA sequences enter or exit TAD condensates within the interphase nucleus along with other loops containing enhancers and Mediator (Furlong and Levine, 2018; Plys and Kingston, 2018; Galupa and Heard, 2017). Up to a thousand or more enhancers can be recruited to a given gene by lineage-determining transcription factors (LDTFs) (Heinz and Glass, 2012; Link et al., 2015) such as the muscle and bone "master regulators," respectively MyoD (Blum and Dynlacht, 2013) and Runx2 (Vimalraj et al., 2015). This occurs in a highly cooperative and synergistic fashion, with the number of enhancers recruited being responsive to developmental signals.

While the TF-binding motifs of enhancers are similar to those of promoters (Arenas-Mena, 2017) they function differently. Due to their responsiveness to patterning cues, and numerousness [as many as 50,000 in mammalian genomes, for example (Heinz et al., 2015)], enhancers are highly suited to mediating the high levels of expression of the characteristic genes of terminally differentiated cells, as well as the precise and quantitatively calibrated spatiotemporal gene control of the patterning of cells and tissues during development (Lenhard et al., 2012; Zabidi and Stark, 2016). As discussed below (see *Inherent cell functions in the origin of differentiation*), enhancers that regulate genes involved in stage- and cell type-specific "developmental" and "terminal" differentiation function differently, and via partly different sets of TFs, from enhancers regulating ubiquitously expressed "housekeeping" genes (Zabidi et al., 2015). Enhancers are used in sponges and all eumetazoans, with an indication of their presence in ctenophores, although not the placozoan *Trichoplax adhaerens*, which appears to rely solely on promoters to generate its few cell types (Sebé-Pedrós et al., 2018).

Enhancers show evidence of conservation, but also divergent usage over evolution and between species (Chen et al., 2018). Wong and coworkers have shown that enhancers from the sponge *Amphimedon queenslandica* can drive cell type-specific reporter gene expression in zebrafish and mouse, despite sponge and vertebrate lineages having diverged over 700 million years ago (Wong et al., 2019). Although sponge enhancers have no significant sequence identity with vertebrate genomic sequences, the type and frequency of TF binding motifs in these enhancers allow for the identification of homologous enhancers in bilaterians. Enhancers identified in human and mouse *Scaper* genes drive reporter expression patterns in zebrafish that are almost identical to those of the sponge *Islet* enhancer, although the Scaper and Islet proteins are unrelated and apparently do not have similar cellular functions. This suggests animal development is controlled in part by TF-enhancer DNA interactions that were present in some of the first multicellular animals. These have an evolutionarily conserved dynamics that is independent of changes in enhancer sequence, TF function, and cell type identity, but is reflected in aspects of spatiotemporal regulation (Wong et al., 2019).

In summary, gene regulatory mechanisms based on activating p300/CBP-containing biomolecular condensates and PcGII- and SUV-type methylation-based gene silencing effects were conveyed to metazoans from ancestral unicellular holozoans. Enhancers and PcGI biomolecular condensate-based gene sequestration mechanisms, in contrast, are unknown outside the animals. The ancestral proteins β-catenin and Grainyhead-like (the roles of which are described in the following section) rounded out the constitution of Metazoa as a new type of organism and established a deep connection between multicellularity and cell differentiation. While not all metazoan groups have equally copious complements of cell types, cell differentiation in animal species were built on these foundations.

Beta-catenin, Grainyhead-like and the role of multicellularity in the evolution of differentiation

Whether holozoan multicellularity preceded or followed the appearance of metazoan-specific modes of gene regulation in the ancestors of animals is unknown. However, the transition between cells that are free-living or transiently colonial, and cells that can switch on surface properties that enable their long-term association and cohesion, was the founding differentiation event of the metazoan lineage. It is also a precondition for the existence of stable organisms whose cells can diversify further into functionally specialized types.

As mentioned in the *Introduction*, cell-cell attachment and its modulation during animal development is mediated by classical cadherins containing a cytoplasmic domain with no counterpart in any other sequenced organisms, including non-metazoan holozoans. The role of the classical cadherins in cell-cell attachment depends strictly on the binding of this domain to β-catenin, which is part of the complex that links these transmembrane adhesion proteins to the actin cytoskeleton. Although there is a homolog of β-catenin in *Dictyostelium discoideum* (a member of a group of eukaryotes more distantly related to metazoans than the fungi) (Dickinson et al., 2011), no related proteins have been identified in nonmetazoan opisthokonts. This protein plays a key role not only in the evolution and development of the metazoan multicellular state, but also in the diversification of cell types in this group. This differentiation function is tied to β-catenin's relationship to the pan-metazoan morphogen Wnt.

Wnt, a secreted protein, is unique to the metazoan clade [reviewed in (Newman and Bhat, 2009)]. One of its major effects is causing the reorganization (via cell surface receptors) of the cytoskeleton by inducing modifications to the PAR complex of adapters and protein kinases that evolved prior to the metazoans (Lang and Munro, 2017). The cytoskeletal effects occur by two pathways (Karner et al., 2006a,b): one, which changes the shape or spatial orientation of a cell (the planar cell polarity pathway), is independent of β-catenin. The other, which depends on β-catenin (the "canonical" pathway), influences the apicobasal polarity of a cell, that is, the arrangement of cadherins and other proteins on its surface. Apart from its role in cell polarity, which is a key mechanism of tissue morphogenesis (Newman and Bhat, 2009), the canonical Wnt pathway is a major determinant of cell differentiation. Its main role in this process is to inactivate the enzyme complex that would otherwise degrade cytoplasmic β-catenin (when released, for example, from its association with cadherins on the inner cell membrane), permitting it to enter the nucleus. There β-catenin activates the expression of cell type-specific genes.

Nuclear β-catenin performs this function by associating with TFs of the TCF/Lef family. In the absence of the Wnt-initiated signal, these factors act as transcriptional repressors. The binding of β-catenin converts TCF/Lef proteins into transcriptional activators that induce the transcription of specific target genes (Najdi et al., 2011; Archbold et al., 2012). In variations on this general pattern, β-catenin can in some cases function independently of TCF/LEF (Doumpas et al., 2018), and can also be induced independently of Wnt. This may result from external physical stimuli (Zhang et al., 2016; Popova et al., 2012), as well as during early development of some unusual species (Pang et al., 2010). Finally, both β-catenin and TCF/LEF proteins are highly promiscuous with respect to the TFs they partner with, due to the IDRs at the N- and C-terminal regions of β-catenin and disordered domains in developmental TFs (Zhao and Xue, 2016). This makes β-catenin an enormously versatile effector of cell differentiation, particularly in its synergy with other conditionally expressed TFs and transcriptional cofactors such as STATs (Wang et al., 2013), SMADs (Zhang et al., 2013), GATAs (Afouda et al., 2008), and GLIs (Wang et al., 2018).

There is thus an antagonistic relationship between β-catenin's dual roles in both preserving the integrity of a multicellular metazoan embryo (or a protometazoan ancestral form) as an essential component of cadherin-based adhesion, and in promoting cell differentiation. However, this is just one side of the "multicellularity-differentiation" dyad. Cell division is commonly followed by cell separation, and this must be suppressed for organismal identity to be preserved and determinate. The regulation of this function is deeply embedded in the opisthokonts. In the fungus *Neurospora*, for example, the GRHL (Grainyhead-like) TF is involved in spore dispersal by regulating the extracellular material between these cells (Pare et al., 2012). Significantly, members of the Grainyhead-like (Grhl) family of metazoan TFs have a DNA-binding specificity similar to the fungal one, as well as a related function in modulating cell associations. Specifically, they positively regulate the classical E-cadherin and the junctional protein claudin4 (Werth et al., 2010; Riesgo et al., 2014).

While Grhl activates the expression of adhesion-related genes, it simultaneously decreases the expression of other genes whose products, metalloproteinases, degrade cadherins, and thereby undermine cell detachment. Like other "pioneer" TFs (Iwafuchi-Doi and Zaret, 2016), it acts at upper levels of enhancer hierarchies to prime target regions for the binding of activating or inhibitory cofactors (Jacobs et al., 2018; Mayran and Drouin, 2018; Gehrke et al., 2019). Grainyhead-like is unusual among such factors in that it directly inhibits the histone acetylase domain of p300 (Pifer et al., 2016). Grainyhead-like TFs are therefore multifunctional gene regulators with reciprocal effects on maintaining cohesion of cell clusters. Further, since these functions are mediated by control of cadherin levels, Grhl titrates the availability of free β-catenin, described above, the most ubiquitous and versatile of metazoan cofactors for differential gene expression.

Grainyhead-like's role is not confined to the establishment of epithelial tissue in early development but continues to be a central factor in later organotypic cell differentiation. In an unbiased analysis in vivo of GRNs involved in the control of Drosophila eye development, Grhl emerged as the central cofactor in the network. Based on combined assays of functional TF-gene interactions and physical TF-DNA interactions Grhl was determined to be bound to thousands of nucleosome-free regions associated with the genes regulated during this developmental process (Potier et al., 2014). Grainyhead-like is also involved in epithelial organ development and epithelial barrier repair after tissue damage (Wang and Samakovlis, 2012).

With such evidence we can speculate that the repurposing of Grhl from its role in opisthokont and holozoan ancestors [the filasterean *C. owczarzaki* has a Grainyhead homolog, for example (Sebé-Pedrós et al., 2011)] was an originating step in both metazoan multicellularity and metazoan developmental gene regulation. Along with other components inherited from unicellular ancestors, most importantly β-catenin, p300/CBP transactivating hubs and PcGII proteins, and additional ones emerging at the unicellular holozoan-metazoan interface: enhancers and PcGI silencing factors, a new type of inherently multicellular metazoan gene regulatory system was brought into existence. The patterns of gene activity afforded by this system appear to be based on the temporal and spatial arrangements of phase separating and coalescing proteinaceous condensates. Such entities are directly observed in living cells and have been analyzed in vitro, where they are found to undergo abrupt state transformations based on different principles from those of the chemical networks that inspired the MGRN and GGRD models (Bracha et al., 2018; Nakashima et al., 2019; Wan et al., 2018).

Inherent cell functions in the origin of differentiation

While we do not know the order of appearance, or the pace at which the unique elements of the metazoan gene regulatory apparatus evolved, once it was in place differential expression of selected genes was readily achievable. Based on the mechanisms described earlier, this can theoretically occur one gene at a time, or in coordinated sets. Concerted gene expression is facilitated by the action of the mentioned pioneer TFs (e.g., Oct4, FoxA, Sox2, and GATA4) (Iwafuchi-Doi and Zaret, 2016), which open "closed" chromatin for transcription, in conjunction with LDTFs mentioned earlier. The latter initiate (in pioneer TF-capacitated chromatin, and in conjunction with accessory factors) expression of suites of genes that ultimately result in one or more terminally differentiated cell types (Obier and Bonifer, 2016). Examples of LDTF-cell type correspondences include MyoD: skeletal muscle, Nkx3-1: cardiac muscle, neurogenin: neurons, Sox9: cartilage, and PPAR-γ: white fat [reviewed in (Newman et al., 2009)]. Grainyhead-like serves as both a pioneer TF and an LDTF for epithelium. Some LDTFs act in a mutually antagonistic fashion in concert with silencing effectors like PcGI proteins, to ensure that mixed-identity cell types are not produced (Sunadome et al., 2014).

When pioneer and lineage-defining TFs bind to the unusual enhancers of cell identity-associated or -specific genes, the LDTFs (in the early differentiation of skeletal muscle and fat, for example, and perhaps more generally) recruit the bromodomain-containing protein Brd4 (Lee et al., 2017). This complex in turn recruits mediator and RNA polymerase II to condensed, p300-containing phase separated liquid protein droplets, into which hundreds or thousands of enhancers from distant chromosomal sites may be drawn. These enhancer-rich loci are the structurally characterized TADs described earlier, but to emphasize their functional role in cell differentiation, I will refer to them as "function-amplifying centers," or FACs.

How are the sets of expressed genes characteristic of a given cell type (the "content" mentioned in the *Introduction*) determined? Understanding the role of the system of FACs by which metazoan cell type-related genes are differentially transcribed allows us to address this question, ignored by both the MGRN and GGRD models, in an entirely new fashion. In the remainder of this section, I explore the hypothesis that cell types are based on inherent functionalities already found in unicellular holozoan ancestors. Those that were based on the concerted activity of suites of genes in progenitor cells would have been susceptible to coordinated amplification by the unique gene regulatory apparatus that emerged with the metazoans [see also Newman (2019a)].

Premetazoan opisthokonts and holozoans, like their present-day counterparts, almost certainly exhibited alternative cell phenotypes and states. While some of these may have been antecedents of differentiated cell types in metazoans as suggested by (Brunet and King, 2017), the more than 250 cell types that characterize complex animals such as the vertebrates could not have arisen from ancestral ones. Furthermore, while GRNs underlying specific cell types (e.g., neurons, skeletal myoblasts) exhibit a fair degree of conservation across animal phyla (Arendt et al., 2016), the "same" cell types and their homologous genes can also be regulated differently, across taxa and even in individuals within a species (Halfon, 2017; Nowotschin et al., 2019). What seems to be a constant of animal cell types, however, is their reflection of preexisting (and thus inherent) unicellular physiological functions, reflected in gene "coexpression networks" (Singh et al., 2018). Examples include motility/myoblast, extracellular matrix production/chondrocyte, osteocyte, detoxification/hepatocyte, lipid storage/adipocyte, light responsivity/retinal rods, and cones, oxygen capture/erythrocyte. In contrast, inherent functions of unicellular ancestors of other multicellular lineages, such as the capture of light energy and the storage of starch in plants, or the polarisome vesicular system of fungi, appear in differentiated cells of those organisms, but not the animals. Under the assumption that they are recruited from preexisting functions, the cell types in any group of multicellular organisms could be multifarious, but not unlimited.

If one or more of the cells in a multicellular protometazoan came to perform any of these inherent functions in a partitioned, intensified fashion, the novel differentiated form could have been at a selective advantage over organisms with no such division of labor. This kind of variability may have been facilitated by the appearance, in the earliest metazoans, of enhancers that were specialized for allocating different gene expression profiles between spatially and temporally distinct cells of the developing embryo, as described earlier (Wong et al., 2019). Regulatory elements of this type would have been components of an engine of cell type innovation.

The question remains of how some inherent unicellular functions were selected over others in determining the actual content of the new cell types. In a proposed scenario for the origin of developmental promoters and enhancers, Arenas-Mena

(2017) has suggested that these regulatory elements evolved from an interaction of two distinct promoter architectures in ancestral cells. Unicellular holozoans have inducible promoters that respond to external cues, as well as promoters that regulate constitutive genes, that is, which are transcribed continually. Inducible promoters, but not constitutive ones, have nearby TATA boxes. TFs in nonmetazoan holozoans have the constitutive-type promoter architecture, but Arenas-Mena hypothesizes that some of these TF promoters became responsive to developmental signals by acquiring inducible-type architectures. A few small changes (described in the paper) then converted them into developmental enhancers.

Inducible TFs did not become LDTFs however, until they came to regulate sets of constitutive genes with shared sets of *cis*-regulatory modules, that is, DNA sequences with binding sites for the signal-induced TFs. Expression of suites of such commonly regulated genes, if confined to a limited set of cells in the developing embryo, would have constituted a cell type. In metazoans, interactions of enhancers with gene-proximal promoters across distant sites in chromatin mediate this mode of regulation. Developmental systems with the logic of the MGRN model of Britten and Davidson (1969) can be produced by this evolutionary scenario, but so could less hierarchically organized ones.

Remarkably, the long-range chromatin associations of functionally related genes seen in animal cells (Cao and Cheng, 2015; Laarman et al., 2019; Stodola et al., 2018) may have predated the evolution of metazoan differentiation, since similar arrangements are also found in fungi and thus appear to be an opisthokont synaptomorphy (Tanizawa et al., 2010; Diament and Tuller, 2017). This raises the possibility that the cells that gave rise to the animals contained "pre-loaded" sets of associated genes poised to engage with the evolutionarily novel metazoan gene regulatory apparatus to generate specialized cell types based on accentuation of ancestral functions.

The concept of metazoan cell type origination presented here can be distinguished from another recent proposal in which novel cell types are suggested to have resulted from the response of established animal species to external stress. In this view [similar to one proposed earlier by Nanjundiah (2003)], protective gene-expression responses to these stimuli, a form of phenotypic plasticity, would lead to variation in the complement of cell types. Natural selection could then lead to assimilation of the responses into the respective lineage's developmental repertoire (Wagner et al., 2019). In terms of the categorization of models described in the *Introduction*, the outcome of this process is an MGRN-type mechanism. As noted, the MGRN research program (when evolution has been considered) has posited a conventional stepwise variation and selection scenario for the emergence of adaptive modules [e.g., (Peter and Davidson, 2011; Royo et al., 2011)]. The recent proposal by Wagner and coworkers incorporates phenotypic plasticity into this picture.

In contrast, the hypothesis for the origin of cell type functionalities outlined in this chapter emphasizes how the metazoan innovation of a biomolecular condensate-based gene regulatory apparatus (the FAC system) was poised to allocate ancestral cell functions to separate compartments (cell types) once multicellularity had emerged. "Stress" is part of this picture, but only tangentially, in that [following (Arenas-Mena, 2017)] I have suggested that the promoters of developmental regulatory TFs arose from the promoter structure of inducible genes. However, this is an evolutionary precondition of a developmental regulatory apparatus already present in the first metazoans. Furthermore, the model I have described contains no implication that those TFs [most of which were present in holozoan ancestors (Sebé-Pedrós et al., 2011)] originally regulated protective responses to external insults. Rather, the suite of genes regulated by an LDTF [either carried over from ancestral nonmetazoan holozoans or newly appearing in Metazoa (Grau-Bové et al., 2017)] were related to inherent functions already involved in maintenance and propagation of cellular life. These functions, conveyed from the unicellular world to the multicellular one, may therefore not have required multiple cycles of natural selection of incremental variations to provide the basis for new cell types.

Conclusion: Prolific cell differentiation as a metazoan evolutionary innovation

In this chapter, I have presented a perspective on metazoan cell differentiation that is based on regulatory mechanisms only found in this group. The FAC system, the unique developmental gene regulatory apparatus of animal cells, operates on the general eukaryotic platform of write-read-rewrite transcription control but makes use of novelties like enhancers and chromatin-structural silencing mechanisms to accentuate and partition life-sustaining functions that were first established in premetazoan cells. This view draws on comparative phylogenomics of animals and their direct ancestors that has only emerged over the past decade and a half, as well as new discoveries in the physical biochemistry of protein structure and condensed phases in nuclear organization that are even more recent. Apart from leading to a picture that is at odds with the mechanistic concepts of cell differentiation that have dominated the field for 50 years, this proposal challenges certain underlying evolutionary assumptions of previous models (Newman, 2020).

The MGRN and GGRD models both bear the stamp of their origins in mid-20th century scientific thought. For the modular framework, the main influence was cybernetics, a machine-based theory of regulated systems and their coding that accompanied the invention of the programmable computer. The physicist Erwin Schrödinger famously wrote "The chromosome structures are...instrumental in bringing about the development they foreshadow. They are law-code and executive power – or, to use another simile, they are architect's plan and builder's craft – in one" (Schrödinger, 1945). A similar gene-centrism of the MGRN paradigm is reflected in such assertions as "Information necessary and sufficient to specify any spatial pattern of gene expression in development is resident in genomic DNA sequence... [T]he naked, inherited DNA sequence does indeed contain information for specific control of gene expression in development" (Peter and Davidson, 2015, p. 10). However, organismal DNA never does anything outside the complex environment of the nucleus, which is equally inherited across generations, and in the context of multilevel developmental systems and life cycles. Further, attempting to conceive of how "naked DNA" acquired such information over evolution without the participation of other forms of cellular information and organization shows the inadequacy of what seems to be more a metaphysical than a scientific position.

The GGRD-type models exemplify a different, frankly insurgent movement of the same period, a theoretical biology that hoped to see living processes yield to explanations in terms of mathematical dynamical systems theory and the new physics of self-organization, a marked departure from cybernetics. An attractive selling point of the Kauffman model of cell differentiation and allied models was that they would bootstrap evolution, potentially providing "order for free" via self-organization (Kauffman, 1995). It is clear regarding the GGRD differentiation model, however, that even if the different combinations of TFs characterizing the (mathematically defined) attractors of an evolutionary early-arising, self-organizing network happened to specify a coherent set of biological functionalities, these would be exceptionally unlikely to maintain their identities and coherence in organisms with additional TFs, and new mathematically defined sets of attractors.

The linear computer logic of the MGRN framework and the transitions among dynamical attractors in state space of the GGRD framework both continue survive in heuristic models for narrowly proscribed developmental lineages described earlier. These models base the "content" of the simulated cell types on experimental data, making no attempt to generate them from ancestral organisms as any truly evolutionary model would need to. The two categories of models also share the assumptions of another mid-20th century conceptual framework that was part of intellectual landscape when they were first proposed—the "modern evolutionary synthesis" (Huxley, 2010/1942). This theory contends that evolution occurs in an opportunistic fashion, and that phenotypic outcomes reflect adaptations to external challenges rather than any internal organismal templates. Newer, development-informed concepts of evolution ["EvoDevo" (Müller and Newman, 2003; Müller, 2007)] have expanded the synthesis to include the role in evolution of morphological or functional properties that are inherent to multicellular forms and their constituent cells (Newman, 2018, 2019a). The model of cell differentiation proposed here reflects this change in theoretical orientation.

The metazoan differentiation system of condensed, liquid-like cell-function-amplifying centers, or FACs, evolved in stages, with components and processes present in a narrowing set of ancestral forms (the "FAC Pyramid"; Fig. 10.1) but is qualitatively different from anything that came before. The main components are (1) the pan-eukaryotic Mediator-scaffolded, write-read-rewrite transcriptional state-encoding system, (2) the holozoan p300/CBP complex, suited to interfacing with the pioneer and other TFs that became progressively richer in IDRs in the ancestral cell populations most closely related to the animals, and finally, (3) liquid-phase silencing complexes, and enhancers capable of looping over great linear distances across chromatin and congregating in large numbers in the amplifying hubs. Under these conditions, the exaggeration of amplification of some cellular functions is virtually assured, and potentially very rapid.

The functions amplified (from the evidence of the specialized activities of cell types across the animal phyla) are ones that were already inherent in the unicellular and transiently colonial holozoan cell populations from which they arose. The manner in which these ancestral cell functions were harnessed into distinct cell types is obscure, but from the thin evidence that currently exists it depended on (1) preformed associations of functionally related genes which appear to be mediated in

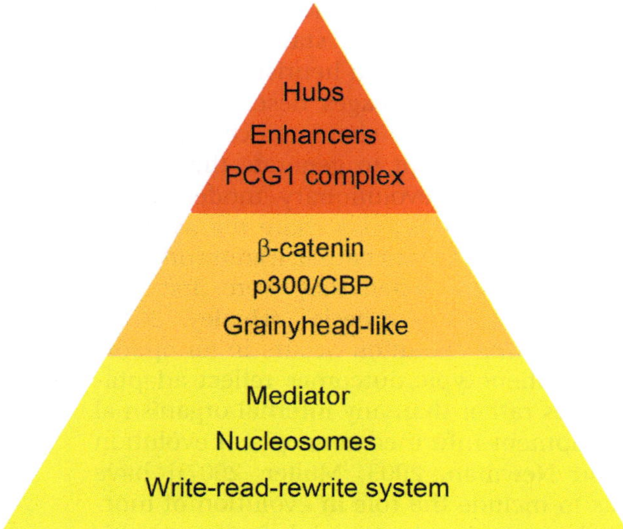

Figure 10.1 The FAC pyramid. Schematic representation of the relationship of the metazoan cell differentiation-enabling gene regulatory system, consisting of function-amplifying centers (FACs) to the various key components that appeared, in part, in earlier-evolving forms. The yellow sector represents all eukaryotic cells, the orange (with one exception) holozoans, and the red, metazoans. The features of the wider sectors are included in the narrower ones above them. The components listed (which are described in the main text) are the most important ones for gene regulation in their respective groups. The pan-eukaryotic components of the yellow sector distinguish this group from other domains of cellular life. The dual-function protein β-catenin, essential to metazoan cell differentiation, has not yet been identified in any nonmetazoan holozoan or opisthokont, but is present in the more distantly related Amoebozoa.

yeast by distal promoter-promoter interactions and shared TFs, many of which were converted into LDTFs in metazoans, (2) the multicellularity-enforcing role of the pioneer TF Grainyhead-like, (3) the dual role of β-catenin in gene expression and maintenance of the multicellular state, and (4) the spatiotemporal regulation of β-catenin's activity in promoting differential gene activity by the morphogen Wnt, possibly in conjunction with spatiotemporally acting enhancers that appeared early in metazoan evolution.

As noted, the standard evolutionary narrative affords no way to anticipate novelties such as new cell types, since something cannot be selected until it already exists. For the model proposed here, however, the functions amplified in eventual metazoan cell types were inherent in ancestral cells. Considering this, it would be surprising if the highly elaborate FAC system had not coevolved with a chromatin framework enabling multiple functional gene regulatory associations (i.e., proto-cell types) in common opisthokont ancestors of metazoans and fungi, rather than appearing full-blown in the Metazoa with no connections to its later role in cell differentiation.

Abbreviations

FAC Function-amplifying center
GGRD Global gene regulatory dynamics
LDTF Lineage-determining transcription factor
MGRN Modular gene regulatory networks
TAD Topologically associating domain
TF Transcription factor

References

Afouda, B.A., Martin, J., Liu, F., Ciau-Uitz, A., Patient, R., Hoppler, S., 2008. GATA transcription factors integrate Wnt signalling during heart development. Development 135 (19), 3185–3190.

Alberti, S., Gladfelter, A., Mittag, T., 2019. Considerations and challenges in studying liquid-liquid phase separation and biomolecular condensates. Cell 176 (3), 419–434.

Andrecut, M., Kauffman, S.A., 2006. Mean-field model of genetic regulatory networks. N. J. Phys. 8, 148.

Archbold, H.C., Yang, Y.X., Chen, L., Cadigan, K.M., 2012. How do they do Wnt they do?: regulation of transcription by the Wnt/beta-catenin pathway. Acta Physiol. (Oxf.) 204 (1), 74–109.

Arenas-Mena, C., 2017. The origins of developmental gene regulation. Evol. Dev. 19 (2), 96–107.

Arendt, D., Musser, J.M., Baker, C.V.H., Bergman, A., Cepko, C., Erwin, D.H., et al., 2016. The origin and evolution of cell types. Nat. Rev. Genet. 17 (12), 744–757.

Arnellos, A., Moreno, A., 2016. Integrating constitution and interaction in the transition from unicellular to multicellular organisms. In: Niklas, K.J., Newman, S.A. (Eds.), Multicellularity: Origins and Evolution Vienna Series in Theoretical Biology. The MIT Press, Cambridge, MA, pp. 249–275.

Belahbib, H., Renard, E., Santini, S., Jourda, C., Claverie, J.M., Borchiellini, C., et al., 2018. New genomic data and analyses challenge the traditional vision of animal epithelium evolution. BMC Genomics 19 (1), 393.

Blau, H.M., 2014. Sir John Gurdon: father of nuclear reprogramming. Differentiation 88 (1), 10–12.

Blum, R., Dynlacht, B.D., 2013. The role of MyoD1 and histone modifications in the activation of muscle enhancers. Epigenetics 8 (8), 778–784.

Boija, A., Klein, I.A., Sabari, B.R., Dall'Agnese, A., Coffey, E.L., Zamudio, A.V., et al., 2018. Transcription factors activate genes through the phase-separation capacity of their activation domains. Cell 175 (7), 1842–1855.e16.

Bracha, D., Walls, M.T., Wei, M.T., Zhu, L., Kurian, M., Avalos, J.L., et al., 2018. Mapping local and global liquid phase behavior in living cells using photo-oligomerizable seeds. Cell 175 (6), 1467–1480.e13.

Britten, R.J., Davidson, E.H., 1969. Gene regulation for higher cells: a theory. Science 165 (3891), 349–357.

Brunet, T., King, N., 2017. The origin of animal multicellularity and cell differentiation. Dev. Cell 43 (2), 124–140.

Cao, R., Cheng, J., 2015. Deciphering the association between gene function and spatial gene-gene interactions in 3D human genome conformation. BMC Genomics 16, 880.

Chan, H.L., Beckedorff, F., Zhang, Y., Garcia-Huidobro, J., Jiang, H., Colaprico, A., et al., 2018. Polycomb complexes associate with enhancers and promote oncogenic transcriptional programs in cancer through multiple mechanisms. Nat. Commun. 9 (1), 3377.

Chan, H.M., La Thangue, N.B., 2001. p300/CBP proteins: HATs for transcriptional bridges and scaffolds. J. Cell Sci. 114 (Pt 13), 2363–2373.

Chen, L., Fish, A.E., Capra, J.A., 2018. Prediction of gene regulatory enhancers across species reveals evolutionarily conserved sequence properties. PLoS Comput. Biol. 14 (10), e1006484.

Conradi, C., Pantea, C., 2019. Multistationarity in biochemical networks: results, analysis, and examples. In: Robeva, R., Macauley, M. (Eds.), Algebraic and Combinatorial Computational Biology. Academic Press, pp. 279–317.

Corson, F., Couturier, L., Rouault, H., Mazouni, K., Schweisguth, F., 2017. Self-organized Notch dynamics generate stereotyped sensory organ patterns in Drosophila. Science 356 (6337).

Corson, F., Siggia, E.D., 2017. Gene-free methodology for cell fate dynamics during development. Elife 6, e30743.

Craciun, G.F., Nazarov, F., Pantea, C., 2013. Persistence and permanence of mass-action and power-law dynamical systems. SIAM J. Appl. Math. 73, 305–329.

Cui, M., Vielmas, E., Davidson, E.H., Peter, I.S., 2017. Sequential response to multiple developmental network circuits encoded in an intronic cis-regulatory module of sea urchin hox11/13b. Cell Rep. 19 (2), 364–374.

Dahlin, J.L., Chen, X., Walters, M.A., Zhang, Z., 2015. Histone-modifying enzymes, histone modifications and histone chaperones in nucleosome assembly: lessons learned from Rtt109 histone acetyltransferases. Crit. Rev. Biochem. Mol. Biol. 50 (1), 31–53.

Davidson, E.H., 2006. The Regulatory Genome: Gene Regulatory Networks in Development and Evolution. Elsevier/Academic Press, Amsterdam; London.

Davidson, E.H., 2011. Evolutionary bioscience as regulatory systems biology. Dev. Biol. 357 (1), 35–40.

Diament, A., Tuller, T., 2017. Tracking the evolution of 3D gene organization demonstrates its connection to phenotypic divergence. Nucleic Acids Res. 45 (8), 4330–4343.

Dickinson, D.J., Nelson, W.J., Weis, W.I., 2011. A polarized epithelium organized by beta- and alpha-catenin predates cadherin and metazoan origins. Science 331 (6022), 1336–1339.

Dilweg, I.W., Dame, R.T., 2018. Post-translational modification of nucleoid-associated proteins: an extra layer of functional modulation in bacteria? Biochemical Soc. Trans. 46, 1381–1392.

Doumpas, N., Lampart, F., Robinson, M.D., Lentini, A., Nestor, C.E., Cantu, C., et al., 2018. TCF/LEF dependent and independent transcriptional regulation of Wnt/beta-catenin target genes. EMBO J 38 (2), pii: e98873.

Dyson, H.J., Wright, P.E., 2016. Role of intrinsic protein disorder in the function and interactions of the transcriptional coactivators CREB-binding protein (CBP) and p300. J. Biol. Chem. 291 (13), 6714–6722.

Furlong, E.E.M., Levine, M., 2018. Developmental enhancers and chromosome topology. Science 361 (6409), 1341–1345.

Furusawa, C., Kaneko, K., 2002. Origin of multicellular organisms as an inevitable consequence of dynamical systems. Anat. Rec. 268 (3), 327–342.

Galupa, R., Heard, E., 2017. Topologically associating domains in chromosome architecture and gene regulatory landscapes during development, disease, and evolution. Cold Spring Harb. Symp. Quant. Biol. 82, 267–278.

Gehrke, A.R., Neverett, E., Luo, Y.J., Brandt, A., Ricci, L., Hulett, R.E., et al., 2019. Acoel genome reveals the regulatory landscape of whole-body regeneration. Science 363 (6432), eaau6173.

Glass, L., Kauffman, S.A., 1973. The logical analysis of continuous, non-linear biochemical control networks. J. Theor. Biol. 39 (1), 103–129.

Grau-Bové, X., Torruella, G., Donachie, S., Suga, H., Leonard, G., Richards, T.A., et al., 2017. Dynamics of genomic innovation in the unicellular ancestry of animals. Elife 6, pii: e26036.

Grossniklaus, U., Paro, R., 2014. Transcriptional silencing by polycomb-group proteins. Cold Spring Harb. Perspect. Biol. 6 (11), a019331.

Halfon, M.S., 2017. Perspectives on gene regulatory network evolution. Trends Genet. 33 (7), 436–447.

Heinz, S., et al., 2015. The selection and function of cell type-specific enhancers. Nat Rev Mol Cell Biol 16 (3), 144–154.

Heinz, S., Glass, C.K., 2012. Roles of lineage-determining transcription factors in establishing open chromatin: lessons from high-throughput studies. Curr. Top. Microbiol. Immunol. 356, 1–15.

Holehouse, A.S., Pappu, R.V., 2018. Functional implications of intracellular phase transitions. Biochemistry 57 (17), 2415–2423.

Huang, A.C., Hu, L., Kauffman, S.A., Zhang, W., Shmulevich, I., 2009. Using cell fate attractors to uncover transcriptional regulation of HL60 neutrophil differentiation. BMC Syst. Biol. 3, 20.

Huxley, J., 2010/1942. Evolution: The Modern Synthesis, Definitive ed. MIT Press, Cambridge, MA.

Istrail, S., De-Leon, S.B., Davidson, E.H., 2007. The regulatory genome and the computer. Dev. Biol. 310 (2), 187–195.

Iwafuchi-Doi, M., Zaret, K.S., 2016. Cell fate control by pioneer transcription factors. Development 143 (11), 1833–1837.

Jacobs, J., Atkins, M., Davie, K., Imrichova, H., Romanelli, L., Christiaens, V., et al., 2018. The transcription factor Grainy head primes epithelial enhancers for spatiotemporal activation by displacing nucleosomes. Nat. Genet. 50 (7), 1011–1020.

Jih, G., Iglesias, N., Currie, M.A., Bhanu, N.V., Paulo, J.A., Gygi, S.P., et al., 2017. Unique roles for histone H3K9me states in RNAi and heritable silencing of transcription. Nature 547 (7664), 463–467.

Kaneko, K., Yomo, T., 1999. Isologous diversification for robust development of cell society. J. Theor. Biol. 199 (3), 243–256.

Karner, C., Wharton, K.A., Carroll, T.J., 2006a. Apical-basal polarity, Wnt signaling and vertebrate organogenesis. Semin. Cell Dev. Biol. 17 (2), 214–222.

Karner, C., Wharton Jr., K.A., Carroll, T.J., 2006b. Planar cell polarity and vertebrate organogenesis. Semin. Cell Dev. Biol. 17 (2), 194–203.

Kauffman, S.A., 1969. Metabolic stability and epigenesis in randomly constructed genetic nets. J. Theor. Biol. 22 (3), 437–467.

Kauffman, S.A., 1995. At Home in the Universe: The Search for Laws of Self-Organization and Complexity. Oxford University Press, New York.

Keller, A.D., 1995. Model genetic circuits encoding autoregulatory transcription factors. J. Theor. Biol. 172 (2), 169–185.

King, N., Westbrook, M.J., Young, S.L., Kuo, A., Abedin, M., Chapman, J., et al., 2008. The genome of the choanoflagellate Monosiga brevicollis and the origin of metazoans. Nature 451 (7180), 783–788.

Kingston, R.E., Tamkun, J.W., 2014. Transcriptional regulation by trithorax-group proteins. Cold Spring Harb. Perspect. Biol. 6 (10), a019349.

Laarman, M.D., Geeven, G., Barnett, P., Netherlands Brain, B., Rinkel, G.J.E., de Laat, W., et al., 2019. Chromatin conformation links putative enhancers in intracranial aneurysm-associated regions to potential candidate genes. J. Am. Heart Assoc. 8 (9), e011201.

Lang, C.F., Munro, E., 2017. The PAR proteins: from molecular circuits to dynamic self-stabilizing cell polarity. Development 144 (19), 3405–3416.

Lee, J.E., et al., 2017. Brd4 binds to active enhancers to control cell identity gene induction in adipogenesis and myogenesis. Nat Commun 8 (1), 2217.

Lenhard, B., Sandelin, A., Carninci, P., 2012. Metazoan promoters: emerging characteristics and insights into transcriptional regulation. Nat. Rev. Genet. 13 (4), 233–245.

Link, V.M., Gosselin, D., Glass, C.K., 2015. Mechanisms underlying the selection and function of macrophage-specific enhancers. Cold Spring Harb. Symp. Quant. Biol. 80, 213–221.

Liu, J., Perumal, N.B., Oldfield, C.J., Su, E.W., Uversky, V.N., Dunker, A.K., 2006. Intrinsic disorder in transcription factors. Biochemistry 45 (22), 6873–6888.

Loison, L., Morange, M., 2017. L'invention de la régulation génétique: les Nobel 1965 (Jacob, Lwoff, Monod) et le modèle de l'opéron dans l'histoire de la biologie. Les Rencontres de Normale Sup'. Éditions Rue d'Ulm, Paris.

Mayran, A., Drouin, J., 2018. Pioneer transcription factors shape the epigenetic landscape. J. Biol. Chem. 293 (36), 13795–13804.

Mojtahedi, M., Skupin, A., Zhou, J., Castano, I.G., Leong-Quong, R.Y., Chang, H., et al., 2016. Cell fate decision as high-dimensional critical state transition. PLoS Biol. 14 (12), e2000640.

Müller, G.B., 2007. Evo-devo: extending the evolutionary synthesis. Nat. Rev. Genet. 8 (12), 943–949.

Müller, G.B., Newman, S.A. (Eds.), 2003. Origination of Organismal Form: Beyond the Gene in Developmental and Evolutionary Biology. MIT Press, Cambridge, MA.

Najdi, R., Holcombe, R.F., Waterman, M.L., 2011. Wnt signaling and colon carcinogenesis: beyond APC. J. Carcinog. 10, 5.

Nakamura, H., DeRose, R., Inoue, T., 2019. Harnessing biomolecular condensates in living cells. J. Biochem. 166 (1), 13–27.

Nakashima, K.K., Vibhute, M.A., Spruijt, E., 2019. Biomolecular chemistry in liquid phase separated compartments. Front. Mol. Biosci. 6, 21.

Nanjundiah, V., 2003. Phenotypic plasticity and evolution by genetic assimilation. In: Müller, G.B., Newman, S.A. (Eds.), Origination of Organismal Form: Beyond the Gene in Developmental and Evolutionnary Biology. MIT Press, Cambridge, MA, pp. 245–263.

Newman, S.A., 2016a. Origination, variation, and conservation of animal body plan development. Rev. Cell Biol. Mol. Med. 2 (3), 130–162.

Newman, S.A., 2016b. Origination, variation, and conservation of animal body plan development. Rev. Cell Biol. Mol. Med. 2, 130–162.

Newman, S.A., 2018. Inherency. In: Nuno de la Rosa, L., Müller, G.B. (Eds.), Evolutionary Developmental Biology. Springer, Cham, Switzerland.

Newman, S.A., 2019a. Inherency of form and function in animal development and evolution. Front. Physiol. 10 (702).

Newman, S.A., 2019b. Inherent forms and the evolution of evolution. J. Exp. Zool. B Mol. Dev. Evol. 332 (8), 331–338.

Newman, S.A., 2020. Cell differentiation: what have we learned in 50 years? J. Theor. Biol. 485, 110031.

Newman, S.A., Bhat, R., 2009. Dynamical patterning modules: a "pattern language" for development and evolution of multicellular form. Int. J. Dev. Biol. 53 (5–6), 693–705.

Newman, S.A., Bhat, R., Mezentseva, N.V., 2009. Cell state switching factors and dynamical patterning modules: complementary mediators of plasticity in development and evolution. J. Biosci. 34 (4), 553–572.

Nicholson, D.J., 2019. Is the cell really a machine? J. Theor. Biol. 477, 108–126.

Niklas, K.J., Bondos, S.E., Dunker, A.K., Newman, S.A., 2015. Rethinking gene regulatory networks in light of alternative splicing, intrinsically disordered protein domains, and post-translational modifications. Front. Cell Dev. Biol. 3, 8.

Niklas, K.J., Newman, S.A. (Eds.), 2016. Multicellularity: Origins and Evolution. The MIT Press, Cambridge, MA.

Nowotschin, S., Hadjantonakis, A.K., Campbell, K., 2019. The endoderm: a divergent cell lineage with many commonalities. Development 146 (11).

Obier, N., Bonifer, C., 2016. Chromatin programming by developmentally regulated transcription factors: lessons from the study of haematopoietic stem cell specification and differentiation. FEBS Lett. 590 (22), 4105–4115.

Pang, K., Ryan, J.F., Program, N.C.S., Mullikin, J.C., Baxevanis, A.D., Martindale, M.Q., 2010. Genomic insights into Wnt signaling in an early diverging metazoan, the ctenophore Mnemiopsis leidyi. Evodevo 1 (1), 10.

Pare, A., Kim, M., Juarez, M.T., Brody, S., McGinnis, W., 2012. The functions of grainy head-like proteins in animals and fungi and the evolution of apical extracellular barriers. PLoS One 7 (5), e36254.

Paul, U., Kaufman, V., Drossel, B., 2006. Properties of attractors of canalyzing random Boolean networks. Phys. Rev. E Stat. Nonlin Soft Matter Phys 73 (2 Pt 2), 026118.

Peter, I.S., Davidson, E.H., 2011. Evolution of gene regulatory networks controlling body plan development. Cell 144 (6), 970–985.

Peter, I.S., Davidson, E.H., 2015. Genomic Control Process: Development and Evolution. Academic Press, London; San Diego, CA, is an imprint of Elsevier.

Peter, I.S., Faure, E., Davidson, E.H., 2012. Predictive computation of genomic logic processing functions in embryonic development. Proc Natl Acad Sci U S A 109 (41), 16434–16442.

Pifer, P.M., Farris, J.C., Thomas, A.L., Stoilov, P., Denvir, J., Smith, D.M., et al., 2016. Grainyhead-like 2 inhibits the coactivator p300, suppressing tubulogenesis and the epithelial-mesenchymal transition. Mol. Biol. Cell 27 (15), 2479–2492.

Plys, A.J., Kingston, R.E., 2018. Dynamic condensates activate transcription. Science 361 (6400), 329–330.

Popova, A.P., Bentley, J.K., Anyanwu, A.C., Richardson, M.N., Linn, M.J., Lei, J., et al., 2012. Glycogen synthase kinase-3beta/beta-catenin signaling regulates neonatal lung mesenchymal stromal cell myofibroblastic differentiation. Am. J. Physiol. Lung Cell Mol. Physiol 303 (5), L439–L448.

Potier, D., Davie, K., Hulselmans, G., Naval Sanchez, M., Haagen, L., Huynh-Thu, V.A., et al., 2014. Mapping gene regulatory networks in Drosophila eye development by large-scale transcriptome perturbations and motif inference. Cell Rep. 9 (6), 2290–2303.

Prohaska, S.J., Stadler, P.F., Krakauer, D.C., 2010. Innovation in gene regulation: the case of chromatin computation. J. Theor. Biol. 265 (1), 27–44.

Raeymaekers, L., 2002. Dynamics of Boolean networks controlled by biologically meaningful functions. J. Theor. Biol. 218 (3), 331–341.

Riesgo, A., Farrar, N., Windsor, P.J., Giribet, G., Leys, S.P., 2014. The analysis of eight transcriptomes from all poriferan classes reveals surprising genetic complexity in sponges. Mol. Biol. Evol. 31 (5), 1102–1120.

Royo, J.L., Maeso, I., Irimia, M., Gao, F., Peter, I.S., Lopes, C.S., et al., 2011. Transphyletic conservation of developmental regulatory state in animal evolution. Proc. Natl Acad. Sci. U. S. A. 108 (34), 14186–14191.

Ruiz-Trillo, I., 2016. What are the genomes of premetazoan lineages telling us about the origin of Metazoa? In: Niklas, K.J., Newman, S.A. (Eds.), Multicellularity: Origins and Evolution Vienna Series in Theoretical Biology. The MIT Press, Cambridge, MA, pp. 171–184.

Ruiz-Trillo, I., Burger, G., Holland, P.W., King, N., Lang, B.F., Roger, A.J., et al., 2007. The origins of multicellularity: a multi-taxon genome initiative. Trends Genet. 23 (3), 113–118.

Schrödinger, E., 1945. What Is Life? The Physical Aspect of the Living Cell. The University Press; The Macmillan Company, Cambridge, New York.

Sebé-Pedrós, A., Ballare, C., Parra-Acero, H., Chiva, C., Tena, J.J., Sabido, E., et al., 2016. The dynamic regulatory genome of capsaspora and the origin of animal multicellularity. Cell 165 (5), 1224–1237.

Sebé-Pedrós, A., Chomsky, E., Pang, K., Lara-Astiaso, D., Gaiti, F., Mukamel, Z., et al., 2018. Early metazoan cell type diversity and the evolution of multicellular gene regulation. Nat. Ecol. Evol. 2 (7), 1176–1188.

Sebé-Pedrós, A., de Mendoza, A., Lang, B.F., Degnan, B.M., Ruiz-Trillo, I., 2011. Unexpected repertoire of metazoan transcription factors in the unicellular holozoan Capsaspora owczarzaki. Mol. Biol. Evol. 28 (3), 1241–1254.

Shapiro, J.A., 2011. Evolution: A View From the 21st Century. FT Press Science, Upper Saddle River, NJ.

Shin, Y., Chang, Y.C., Lee, D.S.W., Berry, J., Sanders, D.W., Ronceray, P., et al., 2019. Liquid nuclear condensates mechanically sense and restructure the genome. Cell 176 (6), 1518.

Singh, A.J., Ramsey, S.A., Filtz, T.M., Kioussi, C., 2018. Differential gene regulatory networks in development and disease. Cell Mol. Life Sci. 75 (6), 1013–1025.

Steffen, P.A., Ringrose, L., 2014. What are memories made of? How Polycomb and Trithorax proteins mediate epigenetic memory. Nat. Rev. Mol. Cell Biol. 15, 340.

Stodola, T.J., Liu, P., Liu, Y., Vallejos, A.K., Geurts, A.M., Greene, A.S., et al., 2018. Genome-wide map of proximity linkage to renin proximal promoter in rat. Physiol. Genomics 50 (5), 323–331.

Sunadome, K., Suzuki, T., Usui, M., Ashida, Y., Nishida, E., 2014. Antagonism between the master regulators of differentiation ensures the discreteness and robustness of cell fates. Mol. Cell 54 (3), 526–535.

Tanizawa, H., Iwasaki, O., Tanaka, A., Capizzi, J.R., Wickramasinghe, P., Lee, M., et al., 2010. Mapping of long-range associations throughout the fission yeast genome reveals global genome organization linked to transcriptional regulation. Nucleic Acids Res. 38 (22), 8164–8177.

Tatavosian, R., Kent, S., Brown, K., Yao, T., Duc, H.N., Huynh, T.N., et al., 2019. Nuclear condensates of the polycomb protein chromobox 2 (CBX2) assemble through phase separation. J. Biol. Chem. 294 (5), 1451–1463.

Verger, A., Monte, D., Villeret, V., 2019. Twenty years of mediator complex structural studies. Biochem. Soc. Trans. 47 (1), 399–410.

Villani, M., Barbieri, A., Serra, R., 2011. A dynamical model of genetic networks for cell differentiation. PLoS One 6 (3), e17703.

Vimalraj, S., Arumugam, B., Miranda, P.J., Selvamurugan, N., 2015. Runx2: structure, function, and phosphorylation in osteoblast differentiation. Int. J. Biol. Macromol. 78, 202–208.

Wagner, G.P., Erkenbrack, E.M., Love, A.C., 2019. Stress-induced evolutionary innovation: a mechanism for the origin of cell types. Bioessays 41 (4), e1800188.

Wan, G., Fields, B.D., Spracklin, G., Shukla, A., Phillips, C.M., Kennedy, S., 2018. Spatiotemporal regulation of liquid-like condensates in epigenetic inheritance. Nature 557 (7707), 679–683.

Wang, J., Sun, Z., Zhang, Z., Saadi, I., Wang, J., Li, X., et al., 2013. Protein inhibitors of activated STAT (Pias1 and Piasy) differentially regulate pituitary homeobox 2 (PITX2) transcriptional activity. J. Biol. Chem. 288 (18), 12580–12595.

Wang, S., Samakovlis, C., 2012. Grainy head and its target genes in epithelial morphogenesis and wound healing. Curr. Top. Dev. Biol. 98, 35–63.

Wang, Y., Lin, P., Wang, Q., Zheng, M., Pang, L., 2018. Wnt3a-regulated TCF4/ beta-catenin complex directly activates the key Hedgehog signalling genes Smo and Gli1. Exp. Ther. Med. 16 (3), 2101–2107.

Werth, M., Walentin, K., Aue, A., Schonheit, J., Wuebken, A., Pode-Shakked, N., et al., 2010. The transcription factor grainyhead-like 2 regulates the molecular composition of the epithelial apical junctional complex. Development 137 (22), 3835–3845.

Wong, E.S., Tan, S.Z., Garside, V., Vanwalleghem, G., Gaiti, F., Scott, E., et al., 2019. Early origin and deep conservation of enhancers in animals. bioRxiv 633651.

Zabidi, M.A., Arnold, C.D., Schernhuber, K., Pagani, M., Rath, M., Frank, O., et al., 2015. Enhancer-core-promoter specificity separates developmental and housekeeping gene regulation. Nature 518 (7540), 556–559.

Zabidi, M.A., Stark, A., 2016. Regulatory enhancer-core-promoter communication via transcription factors and cofactors. Trends Genet. 32 (12), 801–814.

Zhang, L., Liu, W., Zhao, J., Ma, X., Shen, L., Zhang, Y., et al., 2016. Mechanical stress regulates osteogenic differentiation and RANKL/OPG ratio in periodontal ligament stem cells by the Wnt/beta-catenin pathway. Biochim. Biophys. Acta 1860 (10), 2211–2219.

Zhang, R., Oyajobi, B.O., Harris, S.E., Chen, D., Tsao, C., Deng, H.W., et al., 2013. Wnt/beta-catenin signaling activates bone morphogenetic protein 2 expression in osteoblasts. Bone 52 (1), 145–156.

Zhao, B., Xue, B., 2016. Self-regulation of functional pathways by motifs inside the disordered tails of beta-catenin. BMC Genomics 17 (Suppl 5), 484.

11

Phenotypic switching and its evolutionary consequences

Dragan Stajic and Claudia Bank
Gulbenkian Science Institute, Oeiras, Portugal

Biological systems are characterized by stunning diversity. This diversity is reflected in variation of observable characters (i.e., phenotypes) both between individuals within the same and between different species (Darwin, 1895). Studying how this variation among individuals in the same population is generated and maintained, and how it translates into the differences that we can observe within and between species is the essence of evolutionary biology. The main source of observable phenotypic diversity is genetic changes that are stably transmitted to subsequent generations (Fig. 11.1). Natural selection acts on these inherited phenotypes by preserving the beneficial and erasing the deleterious ones, causing a corresponding change in frequency of the underlying genetic determinants in the population. However, phenotypes can also be determined and maintained by molecular mechanisms that are independent from the underlying functional DNA sequence. These epigenetic

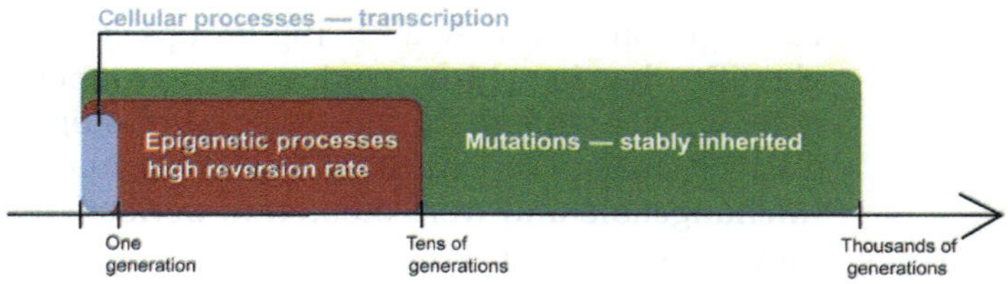

Figure 11.1 The timescale of inheritance. The boxes represent the time period for which particular forms of inheritance (differently colored boxes) persist. Mutations are stably inherited and are a substrate for natural selection. The timescale over which the inherited mutation is maintained in the population depends on its consequences on the phenotype. Certain cellular processes are maintained for only one generation and are of no direct importance for evolution. Epigenetic states of gene expression can be maintained for tens of generations. *Source*: The figure was designed and made by Inês Amaro.

Phenotypic Switching. DOI: https://doi.org/10.1016/B978-0-12-817996-3.00019-0

mechanisms act through the regulation of gene expression causing different developmental outcomes.

Epigenetically induced phenotypic states are not as stable as genetically determined ones, and are characterized by high reversion rates to the original phenotype. Due to the unstable nature of epigenetic inheritance, its contribution to the process of evolution is questioned. In this chapter, we cover the current knowledge of molecular mechanisms of epigenetic inheritance and its evolutionary consequences, specifically in the context of discrete phenotypic switching. To better understand why the role of epigenetic mechanisms in the process of evolution is contentious, it is important to first introduce the historical development of evolutionary thought and the importance of inheritance in this process.

Evolution and the principle of inheritance

During the 17th century, the vast diversity of the natural world became apparent to the scientific community, and the first attempts to catalogue and document phenotypic variation began. One of the scientific pioneers in this field was John Ray (1627–1705). He classified groups depending on their similarities, and termed as species any group that shares characteristics, which distinguish them from another group. For species to exist, he realized that these characteristics must be stably transmitted from parents to their offspring, making him one of the first to recognize the connection between inheritance and interspecies variation.

A similar stance was adopted by Carl Linnaeus (1707–78), who embarked on one of the most comprehensive efforts to catalogue and classify all living groups. He devised a system of classification in which species were grouped into higher order associations depending on their morphological similarities. For Carl Linnaeus, and most of his contemporaries, species were unchangeable and were created to perfectly fit the environment they inhabit (Mayr, 1982).

The view of immutability of species was challenged by, among others, Jean-Baptiste Lamarck (1744–1829), who provided the first theoretical framework for the evolution of species. In his work *Zoological Philosophy-An exposition with Regard to the Natural History of Animals,* Lamarck postulated that through constant use or disuse of organs an individual would further develop the organ or degenerate it. Moreover, this change in the morphology of the organ would then be passed on to the next generation (Lamarck, 1809). Whether an organ would be used or not depends on the environmental conditions. In Lamarck's view, environmental cues act as inducers of a morphological change, creating in such a way variation within a population that is beneficial in the given environment. Inheritance in Lamarck's theory plays a significant role, as it enables the transfer of environmental information between generations.

The crucial importance of inheritance was also acknowledged in the fundamental formulation of evolutionary theory by Charles Darwin (1809–82), and independently, by Alfred Wallace (1823–1913). Darwin conducted research on understanding the variation between individuals, as well as the geographical distribution of species. The results from his studies led him to the formulation of the theory of evolution through natural selection that he presented in his work *On the Origin of Species*. For Darwin, variation between individuals is the basis of the evolutionary process. On the other hand, the mere existence of variation is not enough for evolution to proceed. Inheritance of variation is of paramount importance, or how Darwin formulated it—*Any Variation that is not Inherited is Unimportant for Us* (Darwin, 1895). In Darwin's view, evolution stems from the Malthusian principle by which populations with infinite resources will geometrically increase in number. However, since resources are always limited, competition arises between individuals. Those phenotypic characteristics that enable better survival in a given environment will be preserved. On the other hand, those that are not beneficial will be purged, over time, from the population. In contrast with Lamarck's theory of inheritance of acquired characteristics, where the environment acts as both an inductive and selective force, in the theory of natural selection the environment has the sole role to select for beneficial variation. Variation is created in the parents' germline and transmitted to the offspring. Darwin's theory holds as long as there is heritable variation within a population, no matter the underlying mechanism of inheritance and the generation of the variation. In Darwin's words "owing to this struggle of life, any variation, however slight and from whatever cause proceeding, if it be in any degree profitable to an individual of any species..., if useful, is preserved."

As we have seen, inheritance is an inseparable part of the evolutionary process. At the time Darwin published *On the Origin of Species* the nature of inheritance was a complete mystery. Discovery of its underlying mechanism only happened during the end of the 19th and the beginning of the 20th century. Today, we know that phenotypic information is transferred through the genetic code. The discovery of genes as the carriers of the inheritance of phenotypic characters proved to be paramount for explaining evolutionary processes. However, observations of phenotypes that can be transmitted by mechanisms that are on first sight independent of the genetic code, i.e. epigenetic inheritance, in the 20th century brought the question of inheritance back into evolutionary theory. In the following paragraphs we will cover the relevant discoveries of the genetic basis of inheritance and its importance for the evolutionary theory.

Unknown to Darwin, and in parallel to his work, Gregor Mendel (1822–84) was conducting his research on the nature of inheritance. In his studies on the phenotypic variation in plants, he crossed pea plants with different phenotypic variants (seed shape, flower color, and plant length) and examined the phenotype of their offspring. He concluded that the subsequent generation (F_1 generation) will usually inherit the characteristic form of one parent, which he termed as dominant, and the other character as recessive. However, after another round of crossing (F_2 generation) the recessive character would reappear always in a 1:3 ratio within the offspring, compared to the dominant character. As Mendel pointed out, this particular pattern of inheritance is possible only if these characters are determined by independent particles that are segregated in the parents' reproductive organs and assorted independently in the offspring (Mendel, 1865). This was the first evidence of the particulate basis of the inheritance and a first hint of its molecular mechanism. At first thought incompatible, Darwin's theory and Mendel's discovery complemented each other and proved crucial for the further development of evolutionary theory.

The nature of Mendel's factors remained a mystery until the end of the 19th century. It was Wiliam Johannsen (1857–1927) who coined the term "genes" for these factors (Churchill, 1974), as well as the terms "phenotype" for the observable characters of an individual and "genotype" for the underlying genetic determinants. Around the same time in his work Hugo de Vries (1848–1935) referred to "mutations" as any changes in the observable characters that deviate from those found in other individuals of the same species (Lenay, 2000). With the work of Thomas Morgan (1866–1945) and coworkers it became evident that genes are located on chromosomes, cellular structures that were suggested previously to be carriers of phenotypic information (Gilbert, 1998). Moreover, Morgan was also the first to suggest that mutations are alterations in these cellular structures.

However, since chromosomes were shown to be complex structures that are composed of proteins and deoxyribonucleic acid (DNA), the molecular basis of inheritance was elusive. Two schools of thoughts competed, one claiming that the proteins are the carriers of information and the other pointing out the importance of DNA in the inheritance process. The debate was put to rest by the experiments conducted by Oswald Avery (1877–1955) and coworkers. In his studies, he examined the transformation of a non-virulent bacteria (rough colony, R strain) into a virulent phenotype (smooth colony, S strain) when exposed to heat inactivated, virulent S strain. He noted that if the solution with mortalized virulent bacteria was treated with proteases, enzymes that degrade proteins, before the transformation procedure, the R strain would acquire the virulence capacity. However, if the solution was treated with enzymes that specifically degrade DNA, the R strain would not become virulent. These results clearly showed that the phenotypic information, in this case virulent phenotype, is encoded within the DNA (Avery et al., 1944). The discovery of its chemical composition and of the secondary structure showed that the information is contained in the sequence of the nucleotides within the DNA chains and the mutations are the result of the alteration of this sequence. These discoveries shed light on evolutionary processes as well, and completed the theory of natural selection.

The final unification of the particulate theory of inheritance and evolutionary theory happened in the works of Ronald Fisher (1890–1962), Sewall Wright (1889–1988), Sergei Chetverikov (1880–1959), and John Haldane (1892–1964). Fisher in his work *Genetical Theory of Natural Selection* showed that fitness increase depends on the additive genetic variation in the population (Fisher, 1930). Here, fitness is considered as the long-term reproductive success of an individual. Mutations appear randomly in genes and are the source of phenotypic variation. Some mutations are neutral and have no observable effect, whereas others decrease (deleterious mutations) or increase (beneficial mutations) fitness.

All of the four mentioned scientists showed independently how the frequency of alleles, that is, different genetic variants, in the population would change when we take into account natural selection (Provine, 1971). In an infinitely large, randomly mating, population that is not under selection, and in which different alleles exist, an equilibrium frequency of these alleles will be established (Hardy, 1908; Weinberg, 1908). If selection acts on such a population, the equilibrium will be changed. Here, the change in frequency of an allele depends on its fitness effect. Moreover, the new equilibrium frequency of alleles in the population under selection depends on the relative difference between their corresponding fitness effects (Provine, 1971).

Independently, Fisher, Haldane, and Wright also devised ways of representing the evolutionary process by means of a fitness landscape (Fig. 11.2; Fisher, 1930; Wright, 1932; Haldane, 1932). The concept of the fitness landscape aims at constructing a map of the connection between phenotype/genotype and fitness. In a two-dimensional fitness landscape, two axes represent phenotypic values or genotypes and a third one fitness. This results in a three-dimensional landscape in which valleys represent genotypes with low fitness effects and peaks represent genotypes with great reproductive success. During evolution, mutations change the genotype, which causes populations to move on the landscape until they reach a fitness peak. According to this simplistic theoretical concept, reaching a peak means that no single step mutations are available that would further increase fitness.

Although the concept of the fitness landscape can be misleading because of the gigantic complexity of the true genotype space, various evolutionary hypotheses have been derived from fitness landscapes models. For example, if the population is well adapted and thus at a fitness peak, most new mutations are deleterious or neutral. On the other hand, if the population is further away from the peak, for example due to a change in the environment, there is a greater proportion of mutations that are beneficial (Fisher, 1930; Fragata et al., 2019; Orr, 2006).

Following these works, the Modern Synthesis, the final unification of Mendelian laws of inheritance and Darwin's theory of natural

Figure 11.2 Representation of a fitness landscape. The scheme represents a two-dimensional fitness landscape with genotypes as two axes and fitness as the third. The landscape consists of two fitness peaks separated by a fitness valley. A population (here represented as a figure of a cat) evolves by acquiring mutations (represented by arrows) that move it toward the fitness optimum (represented by a figure of a lion) at one of the fitness peaks. *Source*: The figure was designed and made by Inês Amaro.

selection was born (Huxley, 1942). In summary, heritable variation is created by random changes in the DNA sequence. These mutations can have a significant impact on reproductive success. Depending on their fitness effect and the population size and structure, evolution happens when the frequency of genotypes changes over time.

However, the phenotype is not only the result of a nucleotide sequence. It is an outcome of the complex interactions that occur between the organism and the environment. Since the beginning of the 20th century, it is known that the same genotype can produce different phenotypes depending on the environment that the organism is experiencing during the development, a phenomenon known as phenotypic plasticity (Bradshaw, 1965). Moreover, in some cases the phenotype of an organism can switch between generations at a certain rate, a phenomenon known as phenotypic switching. This can happen due to the stochastic nature of the underlying developmental mechanisms or it can be induced by an environmental cue.

A beautiful example of an environmentally induced phenotype is seen in the butterfly *Byciclus anynana*. This butterfly exists in two distinct morphological forms depending on the season. During the rainy season, the butterfly develops colorful wings with ring patterns, whereas in the dry season it produces wings that are more conspicuous and resemble the color of its surroundings (Shapiro, 1984). It was shown that this phenomenon is adaptive, since the color patterns help the butterfly to avoid predators (Lyytinen et al., 2003). On first sight, this case resembles the principles of Lamarck's view of evolutionary mechanisms. However, the phenomenon is quite distinct. In Lamarck's view, an environmental cue acts as an inducer of phenotypic change but at the same time as the selective force. In this example of environmentally induced phenotypic switching, the inductive cue (availability of water) is quite different from the selective force (predators).

The first attempt to explain phenotypic plasticity and its possible contribution to evolution was made by Conrad Waddington (1905–75). To this end, he introduced the concept of the epigenetic landscape. According to this concept, genes that determine a phenotype are functionally interacting with each other resulting in a network that interacts with complex environmental cues. Out of this interaction, an epigenetic landscape arises that dictates how an organism will develop. The environmental cues can alter the landscape through epigenetic changes, resulting in the development of different phenotypes (Waddington, 1957). Today, the concept of epigenetic inheritance goes beyond Waddington's model. In modern terms, epigenetic inheritance is comprised of many different mechanisms (some of which can be inherited seemingly independently of the underlying DNA sequence) that modulate gene expression and cause production of different phenotypes. In the next part, we will introduce the molecular mechanisms that can create epigenetically determined phenotypic switching and highlight examples of epigenetic mechanisms of gene expression and its effect on phenotypes and their stability.

Molecular basis and examples of epigenetically determined phenotypic switching

Epigenetic inheritance represents the transfer of information from one generation to the other independently of the underlying DNA sequence, both in the case of mitotic and meiotic transmissions. Epigenetic inheritance is enabled by many interconnected molecular mechanisms that mainly involve the regulation of gene expression: DNA methylation, histone modification, positive protein feedback loops, and transfer of regulatory RNAs (Jablonka and Raz, 2009). As explained below, through these mechanisms different phenotypic states can be achieved in the same population. As a result of environmental input or due to an intrinsic stochastic nature, phenotypes induced by epigenetic means can switch between generations (i.e., phenotypic switching). Here, we mainly focus on the molecular basis and phenotypic effects of stochastic epigenetically determined phenotypic switching. Before we explain the molecular mechanisms in detail, it is important to note that even though an epigenetic state is inherited independently of the underlying DNA sequence, the ability to produce epigenetic changes is encoded genetically. Particular epigenetic states are determined by the underlying sequence and are strongly dependent on this sequence. What makes the described mechanisms epigenetic is that for the same genotype different gene expression states, for example, active or silent, can be observed and sometimes propagated through cell divisions during the development of an individual, or from one generation to the other (Moazed, 2011).

Firstly, we will briefly describe each of the four known epigenetic mechanisms that are involved in gene expression regulation and their mode of inheritance. DNA methylation is a process that consists of the addition of methyl group residues to the cytosine ring of the DNA chain. Methylation does not happen randomly along the genome, but on sequence specific stretches of DNA that are recognized by specific enzymes, called methyl-transferases. The sequences that are methylated are usually palindromic and are methylated on both strands. After cell division, the next generation inherits DNA in hemi-methylated state, where only one strand contains a methyl group. In the subsequent step, this methyl group acts as an anchor for the molecular machinery that adds a methyl group to the complementary strand, which restores the methylation pattern (Heard and Martienssen, 2014).

Transfer of information through modifications of histones is similar, in principle, to DNA methylation. Here, the modification happens on the protein level. Histones are proteins that form multiprotein complexes around which DNA is wrapped. Some of these histones (H3 and H4) have extensive N-terminal arms protruding from the complex that can be modified by the addition of different chemical groups such as phosphoryl, acetyl, methyl, or ubiquitin. The effect of modifications on gene expression depends on the chemical group that is added, the genomic context, and also on the position and amino acid to which the chemical group is attached. For example, the addition of a methyl group on the lysine residue of histone H3 at positions 9 or 27 leads to gene repression. However, the same modification on lysine at position 4 leads to gene activation (Berger, 2007). The principle of inheritance of histone modifications is similar to DNA methylation. The modified histones are distributed randomly between cells during the division and act as anchors for histone modifying protein complexes that, in turn, modify neighboring histones, which reestablishes the local chromatin state (Fig. 11.3).

Regulatory RNAs (such as siRNAs and miRNAs) are short 20–25 nucleotide-long chains that bind to the growing mRNA during transcription through complementary binding, which forms a short double strand stretch that represents a recruiting platform for histone modifiers that modify the local chromatin, usually causing its compaction and gene inactivation. During meiosis, regulatory RNAs are maternally inherited via the cytoplasm of the oocyte where they are usually stored during oogenesis.

Transcription regulating protein feedback loops represent a system of gene regulation in which a protein is acting as an indirect (e.g., via cross-feedback) or direct transcriptional activator of the gene that encodes its own synthesis. The protein itself is usually inherited through the cytoplasm as is the case

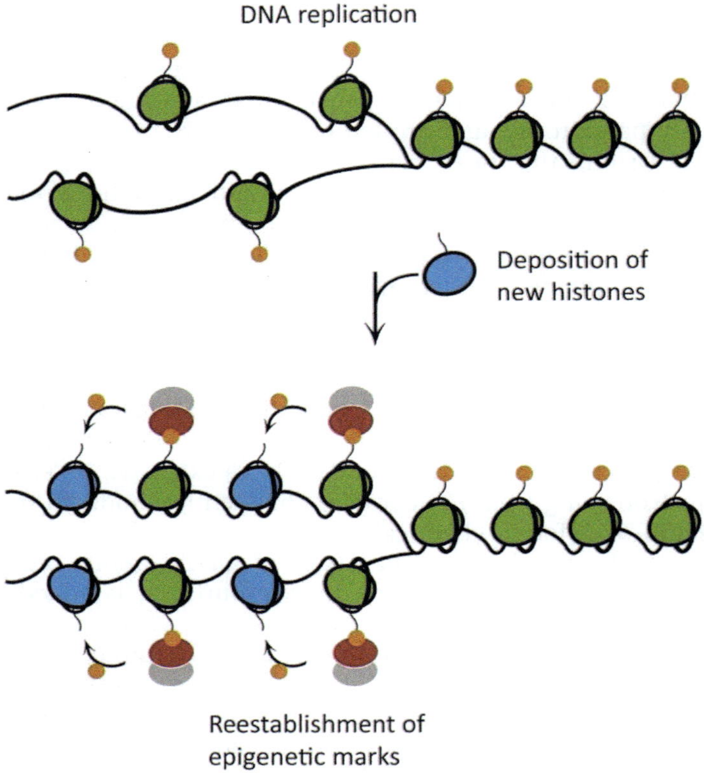

Figure 11.3 Inheritance of histone modifications. Histones are proteins that form multiprotein complexes around which DNA is wrapped. During DNA replication parental histones (*green*) with particular epigenetic modifications (*orange*) are randomly distributed between the two newly synthetized DNA strands. Subsequently, newly produced histones (*blue*) are incorporated. The chromatin remodeling complex, consisting of a subunit that recognizes the parental histone modification (*red*) and an enzyme that modifies in the same pattern the neighboring histone (*gray*), reestablishes the pattern of paternal epigenetic marks in the daughter cells. *Source*: The figure was designed and made by Inês Amaro.

with regulatory RNAs. The transcriptional state of the gene in this case is determined by the protein concentration thresholds in the cytoplasm.

All these mechanisms can have a profound effect on the phenotype (as we will discuss via examples below) and create phenotypic variation within a population, which, as we saw, is the first necessary prerequisite for evolution. However, they are all characterized by low stability and low fidelity of inheritance compared to genetic changes. Since inheritance is paramount for evolution, the effect of such a system of generation of phenotypic variance on adaptation is questionable. However, there is increasing evidence that they provide short-term adaptive value, which we will cover in the rest of this chapter.

Phenotypic switching and epigenetic inheritance are present in all life forms, from bacteria to mammals and plants. One of the oldest examples of phenotypic switching was described by Carl Linnaeus. As a strong advocate of the immutability of species, he thought that each group of beings was created to fit perfectly to its natural surroundings. However, this view was challenged in 1742 by the finding of student Magnus Zioberg, who discovered, on an island near Stockholm, a particular variety of a common toad-flax (*Linaria vulgaris,* an outcrossing plant). This variety exhibited a completely different flower shape and morphology from the more common plant form. Regular flowers of *Linaria* have five petals that are united to form a corolla tube, which is characterized by clear dorsoventral asymmetry (Fig. 11.4A). The variety that was found by Zioberg contained flowers with petals that formed five spurs with a distinct radial symmetry (Fig. 11.4B) (Gustafsson, 1979).

However, the peculiarity of the specimen was not only in the flower morphology, but was reflected also in the fact that from one generation to the other the "aberrant" plants would produce offspring with a regular flower shape, more typical to the common toad-flax. Carl Linnaeus noted what is probably one of the oldest observations of phenotypic switching: "Nothing can, however, be more fantastic than that which has occurred, namely, that a malformed offspring of a plant which has previously always produced irregular flowers now has produced regular ones. As a result of this, it does not only deviate from its mother genus but also completely from the entire class and thus is example of something that is unparalleled in botany so

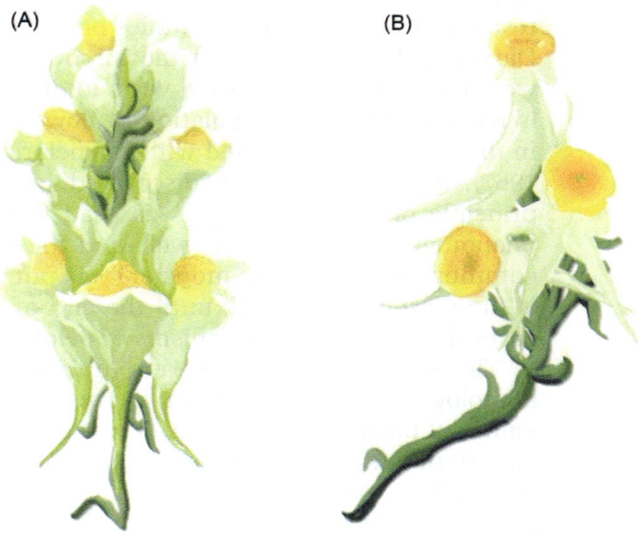

Figure 11.4 Pelorism in *Linaria vulgaris.* (A) Common form of flower shape in *Linaria.* The petals form a corolla tube with a distinct dorso-ventral asymmetry. (B) Peloric flower shape in *Linaria.* Petals form five spurs with a distinct radial symmetry. *Source*: The figure was designed and made by Inês Amaro.

owing to the difference in the flowers no one can recognize the plant anymore." This plant eventually led Linnaeus to revise his position on immutability of species and adopt a view in which species could change through the process of hybridization, the view that he would also use to explain the appearance of the variety of toad-flax with malformed flowers. However, as a consequence of this, the "reputation" of the plant would suffer as Linnaeus infamously named it a "monster" (Peloria in Greek) (Gustafsson, 1979).

The phenomenon of peloric flower morphology was, soon after, observed in other plant species as well, and was taken up by the first pioneers of genetic theory of inheritance as an example of the effect of random mutational changes in the genetic code (de Vries, 1901). In his studies of the mutations, De Vries estimated the mutation rate for the peloric variant of toad-flax to be around 1%, which is, as we now know, an unusually high mutation rate for a genetic change. Further observations of the nature of pelorism, made not in toad-flax, but in a related species, *Antirrhinum*, showed that the peloric phenotype is inherited in Mendelian fashion and results in a 1:3 ratio of phenotypes in the F_2 generation, indicating that a genetic change is responsible for the phenomenon. This observation was first made by Charles Darwin in experiments in which he crossed the mutant specimens with the regular plant variety (Darwin, 1868). Even though he essentially obtained the same result as Mendel did in his studies on the nature of inheritance, Darwin did not understand the importance of the findings. It would take almost a hundred years to discover the *cyc* gene that is responsible for the regulation of flower shape. The mutation in this gene was associated with the peloric phenotype (Stubbe, 1967). Although this was discovered in another species, the principle of genetic change was used to explain the pelorism phenomenon in *Linaria* as well, and for most of the scientific community the story of "monster" was put to rest. Probably this is the reason why it took until the end of 20th century to discover the molecular mechanisms of pelorism in toad-flax.

The symmetry of the flowers in *Linaria* is controlled by a homolog of the *cyc* gene of *Antirrhinum, Lcyc*. In the study of the nature of pelorism in toad-flax, it was shown that the expression pattern of this gene in the peloric form of *Linaria* corresponds to the pattern observed in mutant *Antirrhinum*. However, no genetic change was associated with the peloric form of toad-flax. The phenotypic information and its inheritance in this case was determined to be dependent on one of the molecular mechanisms of epigenetic inheritance. The mechanism in question, in the case of *Linaria*, is DNA methylation (Cubas et al., 1999). This modification leads to a change in gene expression. In the peloric form of toad-flax, the *Lcyc* gene was shown to be highly methylated in contrast to the form with regular flowers. Two hundred years after the discovery of pelorism, this study clearly showed that the phenomenon was independent of genetic change. In one of the finest examples of scientific irony, what was considered as the first example of the effect of genetic mutations on the phenotype became the oldest example of the effect of the epigenetic inheritance on selectable traits, and the first measurement of mutation rate turned out to be the oldest analysis of the rate of phenotypic switching.

Examples of phenotypic switching are not exclusive to plants. However, due to the nonexistence of a clear distinction between somatic and germ cell lines, the evidence of epigenetic inheritance is more abundant in plants than in animals. In animals, where the germline-soma distinction is established very early on during the postfertilization development, epigenetic reprogramming erases most epigenetic marks (Monk et al., 1987; Platz et al., 1975). However, some of these marks can still be transmitted and affect the phenotype of subsequent generations (Brykczynska et al., 2010; Hammoud et al., 2009; Messerschmidt, 2012). One such case was observed in *agouti* mice. These mice carry an insertion of a transposable element just upstream of the *agouti* gene that determines, among other phenotypic traits, the color of the fur. The transposon itself can be differentially methylated and, consequently, change the transcription level of the neighboring gene (Morgan et al., 1999). This results in mice with variegating yellow color of the fur and altered susceptibility to obesity. The methylation state of the locus is maternally transmitted and can also be influenced by the environment. Food with high levels of methyl donors can increase the methylation state and increase the frequency of yellow fur color in the offspring (Daxinger and Whitelaw, 2012). The phenotypic trait in this case can persist for up to two generations before switching to the initial methylation state (Morgan et al., 1999).

Phenotypic switching is also present in unicellular organisms. The bacterium *Photorhabdus luminescens*, a mutualistic partner of the nematode *Heterorhabditis bacteriophora*, can exist in two distinct phenotypic forms, long-shaped rods, and shorter rods. Additionally, the form with long rods produces enzymes and metabolites such as proteases and cell-clumping factors that the shorter form cannot produce (Akhurst, 1980). This affects the capability of the bacteria to establish a mutualistic relationship with the nematode. Shorter rods are no longer capable of supporting nematode growth and development. The switch from the long form to the shorter one occurs upon infection of an insect larva by the nematode that contains the bacteria. Once the nematode is inside the insect, the bacteria are released into the hemolymph of the larva, causing an infection and subsequent death of the infected victim that then serves as food for the nematode (Eckstein and Heermann, 2019). The phenotypic switch is induced by stress that bacteria experience upon their release into the hemolymph of the insect (Joyce et al., 2006). The molecular basis of this phenotypic change is a positive protein feedback loop. The responsible protein, HexA, acts as an autoactivator of its production and was shown to be responsible for the increased pathogenicity of the shorter form (Joyce and Clarke, 2003; Langer et al., 2017). It is produced as part of cellular stress response mechanism and its levels are maintained by self-perpetuating mechanisms that consequently preserve a certain proportion of the shorter form of bacteria in the population.

These examples clearly show that epigenetic phenotypic switching can have a profound effect on phenotype and on the generation of phenotypic variation. In order to determine its effect on adaptation, it is of paramount importance to understand the stability of inheritance and the rate of switching. This rate is dependent on the organism, the mechanism of switching, and the environment. However, in all identified examples, it is consistently higher than mutation rate.

In plants, particular epigenetic marks can be maintained for several generations. Studies that used *Arabidopsis* epigenetic recombinant inbred lines (epiRILs), that were created through several rounds of inbreeding of epigenetically diverse population with varying DNA methylation patterns (Johannes et al., 2009; Reinders et al., 2009), showed that certain unmethylated loci can be inherited for up to 7 generations (Johannes et al., 2009). Moreover, particular epigenetic marks explained up to 90% of the broad-sense heritability of flowering time, and >50% of the observed phenotypic variance (Cortijo et al., 2014).

Epigenetic inheritance can be maintained in animals to a similar extent. In the nematode *Caenorabditis elegans,* high temperature was used as an inducing cue that caused a depression of the heat-shock responsive HSP90 gene array. This change in expression was maintained for up to 14 generations and was clearly linked to the changes in the methylation levels of lysine at position 9 of histone H3 (Klosin et al., 2017).

The direct measurement of the rate of switching is an experimental challenge. Firstly, this is due to the difficulty in choosing the phenotypic readout to measure in the process, and secondly, because the switching rate tends to be much higher than the genetic mutation rate. Therefore, the traditional protocols for the measurement of mutation rate, such as the fluctuation test, cannot be used. Nevertheless, in a recent study using budding yeast, a more direct experimental estimation of the rate was made (Dodson and Rine, 2015). The yeast strain was labeled using a GFP marker (conferring green fluorescent color to the cells) flanked by 3′and 5′ parts of RFP gene (conferring red fluorescent color to the cells). In a constitutively epigenetically silenced region, a gene encoding Cre recombinase was introduced. Upon the activation of the gene due to epigenetic switching, the produced Cre recombinase cut out the GFP gene and reconstituted the RFP gene, causing a switch in the color of the cells from green to red. The frequency of the switch in color allowed for the measurement of switching rate from one epigenetic state to the other. It was determined that the epigenetic switching rate was on the order of 10^{-3}–10^{-4} (Dodson and Rine, 2015), which is much higher than the mutation rate in yeast, which is estimated to be on the order of 10^{-7}–10^{-8} (Lang and Murray, 2008).

In summary, phenotypic variation can be maintained through mechanisms independently of the DNA sequence and result in epigenetically determined phenotypic switching. This epigenetically generated variation can be maintained for several generations, though not as stably as mutations (Fig. 11.1). In the rest of the chapter, we will present possible mechanisms by which phenotypic switching might shape evolutionary outcomes.

Evolutionary consequences of phenotypic switching

As new examples of epigenetic phenotypic switching were discovered, the question about its role in evolution became more pertinent. Conrad Waddington, who was the first to introduce and define the epigenetic landscape, also made the first attempt to define the role of phenotypic switching in evolution. As we described previously, a single genotype can produce several phenotypes randomly or through the interaction with its environment. These phenotypes could serve as a transition step for new or preexisting genetic variation, which, when subsequently fixed in the population, make the respective phenotype permanent. According to this view, the phenotype appears first in the population (Waddington, 1959; Behera and Nanjundiah,

2004). This process is known as genetic assimilation and represents the general theoretical framework that underlies modern concepts of evolutionary consequences of phenotypic switching.

From a theoretical point of view, the presence of stochastic phenotypic switching can have converse effects. In the simplest scenario of a stable, homogeneous environment, phenotypic variation produced in the same genotype should result in a fitness disadvantage, because not all individuals express the optimal phenotype. On the other hand, in a fluctuating or heterogeneous environment or upon environmental change, phenotypic switching can theoretically provide a fitness advantage for two reasons. Firstly, an existing phenotypic switch can create the crucial phenotypic variation to deal with a new challenge more rapidly and at higher proportions than (even preexisting) genetic mutation. Here, the epigenetic switching provides a bet-hedging strategy (Cohen, 1966), where the fitness disadvantage in the current environment is compensated later by the potential advantage upon the environmental change (Carja et al., 2014; Furrow and Feldman, 2014). Secondly, this absolute fitness advantage that phenotypic switching confers in a new environment enables populations to persist at higher numbers, which results in a higher probability of subsequent genetic mutations (Bonduriansky and Day, 2009; Klironomos et al., 2013). Thus, phenotypic switching might greatly promote the survival of populations in new environments, especially in scenarios of evolutionary rescue, in which a population is exposed to a deleterious environment where it would face extinction if it remained at its current phenotype.

Furthermore, theoretical studies have proposed that epigenetic switching can facilitate the transition of a population across a fitness valley (Fig. 11.5). For example, if both a genetic mutation and an epigenetic switch confer the same fitness effect, that is, a knock-out mutation versus the transcriptional silencing of a gene, then the epigenetic switch could alter the fitness landscape and the distribution of fitness effects of genetic mutations (Klironomos et al., 2013). That is because the silencing of the gene would render all genetic mutations in the respective gene neutral. Genetic mutations that are deleterious in the active transcription form of the gene would have no effect on the phenotype once the gene is silenced. This could allow a population to maintain cryptic genetic diversity and to explore a greater part of the fitness landscape. Ultimately, this can change the evolutionary fate of the population (Klironomos et al., 2013). Here, the buffering mechanism of epigenetic switching attenuates deleterious fitness effects of mutations, which makes the valley appear shallower (Tadrowski et al., 2018).

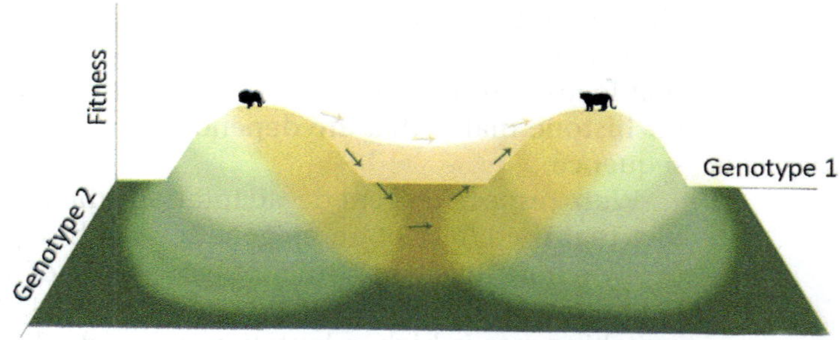

Figure 11.5 Representation of the possible effect of an epigenetically induced phenotypic switching on fitness landscape. A two-dimensional fitness landscape with two fitness peaks with a valley between them (see Fig. 11.2) is presented. A population in one fitness peak (here represented with a figure of a lion) could move to another fitness peak (represented with a figure of a panther) by acquiring mutations that would alter its fitness (darker arrows). These genetic mutations require the population to cross a fitness valley, which might result in the possible extinction of the population and would ultimately inhibit the transition. Here, epigenetic switching could buffer the deleterious effects of the mutations, resulting in the alteration of the fitness landscape (yellow transparent cloud connecting the peaks) by making the fitness valley shallower. This would, consequently, facilitate the transition between the peaks. *Source*: The figure was designed and made by Inês Amaro.

Epigenetic switching can not only affect evolution by changing the fitness effect of genetic mutations, but also by modifying the local mutation rate. It is known, for example, that methylated cytosine tends to convert to thymidine at a higher rate than the nonmethylated form, usually through the process of spontaneous deamination (Ehrlich et al., 1986). As a result, methylated DNA has a higher propensity of acquiring mutations. On the other hand, methylation of the DNA can inactivate transposable elements and prevent their jumping across the genome. Indeed, it was shown in plants that lack of the methylation machinery causes an increase in the rate of transposable element insertions (Kakutani et al., 1995; Mlura et al., 2001). Histone modifications can also affect the mutation rate, because highly compacted regions of the genome are thought to protect the DNA from possible mutagenic factors. For example, it was shown that in human cancer cell lines, variation in a particular histone mark, the methylation of lysine on histone H3, can account for up to 40% of the variation in mutation rate (Schuster-Böckler and Lehner, 2012). Finally, in addition to altering mutation rates, chromatin structure can also alter the frequency of recombination, as it was shown in yeast where the absence of epigenetic silencing increases recombination events (Batté et al., 2017). On the other hand, a genetic mutation can cause a change in the epigenetic switching rate, or its presence altogether. For example, mutations of cytosine in CpG islands (a type of DNA methylation) remove the epigenetic marks from that locus and thus change the epigenetic state. This is also true for histone marks that are dependent on the underlying genetic sequence.

With respect to genetic adaptation, theoretical models predict that the effect of an epigenetic switch depends on the rate of switching itself and the fitness effect of epigenetic states (Kronholm and Collins, 2016; Tadrowski et al., 2018). When the switching rate is high, adaptation proceeds slowly and is mostly driven by genetic mutations, regardless of the fitness effect of the epigenetic change, due to the cost of switching to the suboptimal phenotype. Lower switching rates tend to speed up adaptation, but the final fitness value reached is usually smaller than in the case of high reversion rates.

A second determinant of the contribution of an epigenetic switch to genetic adaptation is the relative fitness effect of the switched phenotype compared to the effect of the respective genetic mutation. If those fitness effects have the same value, genetic adaptation will be slow and genetic mutations might never fix in the populations. On the other hand, if the effect of an epigenetic switch is too small there would be no benefit of switching at all. Populations reap the largest benefit if the epigenetic switch has a fitness effect that is close to the effect of mutations, but not the same. Thus, theory proposes that there exist intermediate optimal switching rates and fitness effects of epigenetic changes that are most beneficial for adaptation (Kronholm and Collins, 2016; Tadrowski et al., 2018).

In spite of extensive theoretical work, experimental evidence for the contribution of epigenetically induced phenotypic switching to evolution is still scarce (Charlesworth et al., 2017). The reason for this, as exemplified above, is due to the intertwined connection between the epigenetic and genetic systems of inheritance and the difficulty of disentangling the phenotypic effect of one system from the other. However, certain attempts were made in this direction recently.

One experimental way of understanding the contribution of epigenetic phenotypic switching to evolution is to observe adaptation in an organism in which the known epigenetic machinery was shut off. In such a study (Kronholm et al., 2017), conducted in *Chlamydomonas,* histone acetylation and DNA methylation were manipulated both by knocking-out the responsible gene and by using an inhibitor of the enzymes likely to be involved in epigenetic regulation. The resulting bacterial strains that had an impairment in the epigenetic machinery were exposed to different stressful environments (salt stress, nitrogen starvation, or high concentration of carbon dioxide). The results showed that the impaired strains did not adapt to these stressors as easily as the epigenetically capable wild type. The genomic and epigenomic data showed differences in methylation pattern in evolved clones, indicating that new epigenetic patterns arose or became active during the experiment.

One route to quantifying the adaptive value of epigenetic phenotypic switching is the use of controllable model systems where epigenetic and genetic effects can be more easily distinguished. Yeast offers an experimental system with great possibilities for genetic manipulations. This advantage was used in a study that examined the impact of protein feedback loops (i.e., dynamic regulatory interactions between genes) on adaptation (Bódi et al., 2017; see also Braun, 2015 and David et al., 2013). In the study of Bódi and coworkers, two yeast strains were constructed, one in which a positive feedback loop controlling the expression of a multidrug transporter gene was engineered (i.e., gene expression was variable and self-reinforcing, resulting in potential phenotypic switching), and another strain in which the expression of the transporter gene was enforced and constant. The expression of the transporter gene in the strains that contained the positive feedback loop showed bimodal expression patterns (implying phenotypic switching) and a higher variance in expression compared to the strain with constant expression. During a subsequent adaptation experiment, the strain with the ability to switch phenotypes evolved to higher levels of drug resistance. Moreover, the effect of the beneficial mutations that accumulated during the adaptation experiment depended on the presence of the feedback loop. This was shown by experimentally shutting off the feedback loop in the evolved strains, which resulted in lower fitness.

Another recent study conducted in budding yeast provided direct evidence that epigenetic phenotypic switching can aid genetic adaptation (Stajic et al., 2019), thereby confirming theoretical predictions (Behera and Nanjundiah, 2004; Lande, 2009). *Saccharomyces cerevisiae* has the particular characteristic that heterochromatin-like epigenetic silencing of genes occurs at three distinct locations in the genome. In one of these locations, the subtelomeric region, a *URA3* reporter gene was inserted at different distances from the telomere (the end of the chromosome). Because distance to the telomere correlates with epigenetic silencing, this resulted in several yeast strains with different epigenetic silencing/switching rates. The reporter gene *URA3* is an essential gene that is responsible for the production of uracil, an essential nucleotide component of RNAs. On the other hand, in the presence of a particular drug (5-FOA), the activity of Ura3 protein is deleterious, since it converts the drug into a toxin that eventually kills the cell (Boeke et al., 1987). In an evolution experiment, the drug was used to select against the activity of the gene. Moreover, by analyzing the proportion of cells that were resistant to the drug but at the same time able to produce uracil, the experimental system allowed to easily distinguish phenotypically switching clones from carriers of genetic mutations that accumulated during the evolution. The authors observed that as the rate of gene silencing increases (up to 10^{-2}) the populations survived better in the drug environment. Consequently, the proportion of populations that escaped extinction was lowest in cases when the rate of silencing was low (10^{-6}). However, too high rates of switching delayed the spread of beneficial genetic mutations, probably because the efficiency of the switching system reduced the effective selective advantage of adaptive genetic changes. This supported the aforementioned hypothesis that there exists an optimal switching rate that enables the fast appearance and spread of beneficial genetic mutations. Moreover, Stajic et al. showed that the spectrum of genetic mutations depended on the rate of epigenetic phenotypic switching. At low rates of silencing, most beneficial mutations were found in the uracil biosynthesis pathway. At higher rates of switching, mutations were increasingly observed in genes known to control epigenetic silencing. Interestingly, these mutations were shown to directly change the epigenetic switching rate by making an epigenetic state more stable. Thus, not only the presence but also the rate of phenotypic switching affected the speed and mechanism of evolution.

The empirical and theoretical examples highlighted in this chapter show that phenotypic switching might indeed be a significant factor for evolution. An important step is to quantify how much of the observed phenotypic variation in nature is due to epigenetic variation. This is a challenge due to the intricate connection between the epigenetic and genetic systems, as highlighted above. Moreover, our experimental examples show the potential power of epigenetic switching for adaptation. However, the aforementioned studies were conducted in organisms that reproduce asexually. Sexual reproduction adds another level of complexity, because epigenetic marks tend to be erased during the process of gametogenesis. To demonstrate the importance of epigenetic switching for adaptation across the tree of life, it will be important to perform similar studies in more complex organisms in the future.

For a better assessment of the contribution of epigenetic phenotypic switching to evolution, it is crucial to understand the fitness effects of epigenetically determined phenotypes across environments. It is likely that the effect size of functional epigenetic changes tends to be smaller than that of functional genetic mutations. That is because genetic mutations can completely alter the role and activity of a protein, whereas epigenetic changes only alter the gene expression patterns. Nevertheless, this can have profound effects on the interacting genes in biological pathways and the resulting phenotype. Considering the difficulties of detecting genetic mutations of small effect, it will be a challenge for the field of evolutionary biology to develop approaches to detecting epigenetic mechanisms of small effect in natural data.

In this chapter we have argued that any variation in a phenotype that is inherited can be the basis for evolutionary processes. Specifically, epigenetically induced phenotypic switching can cause considerable changes to the phenotype and influence how organisms adapt to their environment. Considering epigenetic switching in evolutionary studies can be important, especially with respect to the response of populations to rapid environmental change or environmental fluctuations.

Acknowledgments

We thank Vidyanand Nanjundiah, Laasya Samhita, Irene Adrian-Kalchhauser, and the members of the Evolutionary Dynamics lab for their careful reading and insightful comments on the manuscript. This work was supported by Fundação Calouste Gulbenkian. D.S. was supported by a postdoctoral fellowship from the FCT (Fundação para a Ciência e a Tecnologia) within the project JPIAMR/0001/2016. C.B. is grateful for support by EMBO Installation Grant IG4152 and by ERC Starting Grant 804569 - FIT2GO.

References

Akhurst, R.J., 1980. Morphological and functional dimorphism in Xenorhabdus spp., bacteria symbiotically associated with the insect pathogenic nematodes neoaplectana and heterorhabditis. J. Gen. Microbiol. 79 (121), 303–309.

Avery, O.T., MacLeod, C.M., McCarty, M., 1944. Studies on the chemical nature of the substance inducing transformation of pneumococcal types: induction of transformation by a deoxyribonucleic acid fraction isolated from Pneumococcus Type III. J. Exp. Med. 79 (2), 137–158.

Batté, A., Brocas, C., Bordelet, H., Hocher, A., Ruault, M., Adjiri, A., et al., 2017. Recombination at subtelomeres is regulated by physical distance, double-strand break resection and chromatin status. EMBO J. 36, 2609–2625.

Behera, N., Nanjundiah, V., 2004. Phenotypic plasticity can potentiate rapid evolutionary change. J. Theor. Biol. 226 (2), 177–184.

Berger, S.L., 2007. The complex language of chromatin regulation during transcription. Nature 447, 407–412.

Bódi, Z., Farkas, Z., Nevozhay, D., Kalapis, D., Lázár, V., Csörgő, B., et al., 2017. Phenotypic heterogeneity promotes adaptive evolution. PLoS Biol. 15, e2000644.

Boeke, J.D., Trueheart, J., Natsoulis, G., Fink, G.R., 1987. 5-Fluoroorotic acid as a selective agent in yeast molecular genetics. Meth. Enzymol. 154, 164–175.

Bonduriansky, R., Day, T., 2009. Nongenetic inheritance and its evolutionary implications. Annu. Rev. Ecol. Evol. Syst. 40, 103–125.

Bradshaw, A.D., 1965. Evolutionary significance of phenotypic plasticity in plants. Adv. Genet. 13, 115–155.

Braun, E., 2015. The unforeseen challenge: from genotype-to-phenotype in cell populations. Rep. Prog. Phys. 78, 036602.

Brykczynska, U., Hisano, M., Erkek, S., Ramos, L., Oakeley, E.J., Roloff, T.C., et al., 2010. Repressive and active histone methylation mark distinct promoters in human and mouse spermatozoa. Nat. Struct. Mol. Biol. 17, 679–687.

Carja, O., Liberman, U., Feldman, M.W., 2014. The evolution of phenotypic switching in subdivided populations. Genetics 196, 1185–1197.

Charlesworth, D., Barton, N.H., Charlesworth, B., 2017. The sources of adaptive variation. Proc. R. Soc. B-Biological Sci. 284, 20162864.

Churchill, F.B., 1974. William Johannsen and the genotype concept. J. Hist. Biol. 7 (1), 5–30.

Cohen, D., 1966. Optimizing reproduction in a randomly varying environment. J. Theor. Biol. 12, 119–129.

Cortijo, S., Wardenaar, R., Colomé-Tatché, M., Gilly, A., Etcheverry, M., Labadie, K., et al., 2014. Mapping the epigenetic basis of complex traits. Science 343, 1145–1148.

Cubas, P., Vincent, C., Coen, E., 1999. An epigenetic mutation responsible for natural variation in floral symmetry. Nature 401, 157–161.

Darwin, C., 1895. On the Origin of the Species, John Murray, London.

Darwin, C., 1868. The Variation of Animals and Plants under Domestication, John Murray. London.

David, L., Ben-Harosh, Y., Stolovicki, E., Moore, L.S., Nguyen, M., Tamse, R., et al., 2013. Multiple genomic changes associated with reorganization of gene regulation and adaptation in yeast. Mol. Biol. Evol. 30, 1514–1526.

Daxinger, L., Whitelaw, E., 2012. Understanding transgenerational epigenetic inheritance via the gametes in mammals. Nat. Rev. Genet. 13, 153–162.

de Vries, H., 1901. Die mutationstheorie. Versuche und beobachtungen über die entstehung von arten im pflanzenreich.

Dodson, A.E., Rine, J., 2015. Heritable capture of heterochromatin dynamics in Saccharomyces cerevisiae. Elife 4, e05007.

Eckstein, S., Heermann, R., 2019. Regulation of phenotypic switching and heterogeneity in Photorhabdus luminescens cell populations. J. Mol. Biol 431, 4559–4568.

Ehrlich, M., Norris, K.F., Wang, R.Y., Kuo, K.C., Gehrke, C.W., 1986. DNA cytosine methylation and heat-induced deamination. Biosci. Rep 6, 387–393.

Fisher, R.F., 1930. The Genetical Theory of Natural Selection. Oxford Press.

Fragata, I., Blanckaert, A., Dias Louro, M.A., Liberles, D.A., Bank, C., 2019. Evolution in the light of fitness landscape theory. Trends Ecol. Evol 34 (1), 69–82.

Furrow, R.E., Feldman, M.W., 2014. Genetic variation and the evolution of epigenetic regulation. Evolution 68 (3), 673–683.

Gilbert, S.F., 1998. Bearing crosses: a historiography of genetics and embryology. Am. J. Med. Genet. 76, 168–182.

Gustafsson, Å., 1979. Linnaeus' peloria: the history of a monster. Theor. Appl. Genet 54, 241–248.

Haldane, J.B., 1932. The Causes of Evolution. Princeton University Press.

Hammoud, S.S., Nix, D.A., Zhang, H., Purwar, J., Carrell, D.T., Cairns, B.R., 2009. Distinctive chromatin in human sperm packages genes for embryo development. Nature 460, 473–478.

Hardy, G.H., 1908. Mendelian proportions in a mixed population. Science. 28 (706), 49–50.

Heard, E., Martienssen, R.A., 2014. Transgenerational epigenetic inheritance: Myths and mechanisms. Cell 157, 95–109.

Huxley, J., 1942. Evolution: The Modern Synthesis. Allen & Unwin, London.

Jablonka, E., Raz, G., 2009. Transgenerational epigenetic inheritance: prevalence, mechanisms, and implications for the study of heredity and evolution. Q. Rev. Biol. 84, 131–176.

Johannes, F., Porcher, E., Teixeira, F.K., Saliba-Colombani, V., Simon, M., Agier, N., et al., 2009. Assessing the impact of transgenerational epigenetic variation on complex traits. PLoS Genet. 5, e1000530.

Joyce, S.A., Clarke, D.J., 2003. A hexA homologue from Photorhabdus regulates pathogenicity, symbiosis and phenotypic variation. Mol. Microbiol 47 (5), 1445–1457.

Joyce, S.A., Watson, R.J., Clarke, D.J., 2006. The regulation of pathogenicity and mutualism in Photorhabdus. Curr. Opin. Microbiol 9 (2), 127–132.

Kakutani, T., Jeddeloh, J.A., Richards, E.J., 1995. Characterization of an Arabidopsis thaliana DNA hypomethylation mutant. Nucleic Acids Res. 23, 130–137.

Klironomos, F.D., Berg, J., Collins, S., 2013. How epigenetic mutations can affect genetic evolution: model and mechanism. BioEssays 35, 571–578.

Klosin, A., Casas, E., Hidalgo-Carcedo, C., Vavouri, T., Lehner, B., 2017. Transgenerational transmission of environmental information in C. elegans. Science 356, 320–323.

Kronholm, I., Collins, S., 2016. Epigenetic mutations can both help and hinder adaptive evolution. Mol. Ecol. 25, 1856–1868.

Kronholm, I., Bassett, A., Baulcombe, D., Collins, S., 2017. Epigenetic and genetic contributions to adaptation in chlamydomonas. Mol. Biol. Evol. 34, 2285–2306.

Lamarck, 1809. Zoological Philosophy: An Exposition with Regard to the Natural History of Animals, National Museum of Natural History, Paris.

Lande, R., 2009. Adaptation to an extraordinary environment by evolution of phenotypic plasticity and genetic assimilation. J. Evol. Biol. 22 (7), 1435–1446.

Lang, G.I., Murray, A.W., 2008. Estimating the per-base-pair mutation rate in the yeast Saccharomyces cerevisiae. Genetics 178, 67–82.

Langer, A., Moldovan, A., Harmath, C., Joyce, S.A., Clarke, D.J., Heermann, R., 2017. HexA is a versatile regulator involved in the control of phenotypic heterogeneity of Photorhabdus luminescens. PLoS One 12 (4), e0176535.

Lenay, C., 2000. Hugo De Vries: From the theory of intracellular pangenesis to the rediscovery of Mendel. Comptes Rendus l'Academie Des. Sci. - Ser. III 323, 1053–1060.

Lyytinen, A., Brakefieid, P.M., Mappes, J., 2003. Significance of butterfly eyespots as an anti-predator device in ground-based and aerial attacks. Oikos 100 (2), 373–379.

Mayr, E., 1982. The Growth of Biological Thought: Diversity, Evolution, and Inheritance. Belknap Press, pp. 616–617.

Mendel, G., 1865. Experiments in Plant Hybridization (1865). Sch. Publ., pp. 3–47.

Messerschmidt, D.M., 2012. Should I stay or should I go, protection and maintenance of DNA methylation at imprinted genes. Epigenetics 7, 969–975.

Mlura, A., Yonebayashi, S., Watanabe, K., Toyama, T., Shimada, H., Kakutani, T., 2001. Mobilization of transposons by a mutation abolishing full DNA methylation in Arabidopsis. Nature 411, 212–214.

Moazed, D., 2011. Mechanisms for the inheritance of chromatin states. Cell 146, 510–518.

Monk, M., Boubelik, M., Lehnert, S., 1987. Temporal and regional changes in DNA methylation in the embryonic, extraembryonic and germ cell lineages during mouse embryo development. Development 99, 371–382.

Morgan, H.D., Sutherland, H.G.E., Martin, D.I.K., Whitelaw, E., 1999. Epigenetic inheritance at the agouti locus in the mouse. Nat. Genet. 23, 314–318.

Orr, H.A., 2006. The distribution of fitness effects among beneficial mutations in Fisher's geometric model of adaptation. J. Theor. Biol 238 (2), 279–285.

Platz, R.D., Grimes, S.R., Meistrich, M.L., Hnilica, L.S., 1975. Changes in nuclear proteins of rat testis cells separated by velocity sedimentation. J. Biol. Chem. 250, 5791–5800.

Provine, W.B., 1971. The Origins of Theoretical Population Genetics. University of Chicago Press.

Reinders, J., Wulff, B.B.H., Mirouze, M., Marí-Ordóñez, A., Dapp, M., Rozhon, W., et al., 2009. Compromised stability of DNA methylation and transposon immobilization in mosaic Arabidopsis epigenomes. Genes. Dev. 23, 939–950.

Schuster-Böckler, B., Lehner, B., 2012. Chromatin organization is a major influence on regional mutation rates in human cancer cells. Nature 488, 504–507.

Stubbe, H., 1966. Genetik Und Zytologie Von Antirrhinum L Sect Antirrhinum, Jena: G. Fischer.

Stajic, D., Perfeito, L., Jansen, L.E.T., 2019. Epigenetic gene silencing alters the mechanisms and rate of evolutionary adaptation. Nat. Ecol. Evol. 3, 491–498.

Tadrowski, A.C., Evans, M.R., Waclaw, B., 2018. Phenotypic switching can speed up microbial evolution. Sci. Rep. 8, 8941.

Waddington, C.H., 1957. The Strategy of the Genes. A Discussion of Some Aspects of Theoretical Biology. Allen & Unwin, London.

Waddington, C.H., 1957. The Strategy of the Genes. A Discussion of Some Aspects of Theoretical Biology. Allen & Unwin, London.

Waddington, C.H., 1959. Canalization of development and genetic assimilation of acquired characters. Nature 183, 1654–1655.

Weinberg, W., 1908. Über den nachweis der vererbung beim menschen. Jahreshefte des. Ver. für vaterländische Naturkunde Württemberg 64, 368–382.

Wright, S., 1932. The roles of mutation, inbreeding, crossbreeding and selection in evolution. In: Proceedings of the Sixth International Congress of Genetics, pp. 356–366.

12

Cell-state organization by exploratory sloppy dynamics

Erez Braun
Department of Physics, Technion-Israel Institute of Technology, Haifa, Israel

Introduction

Living cells are prime examples for the emergence of order from molecular disorder in natural complex, far from equilibrium, systems. At specific conditions, cells exhibit well-defined traits, manifested in their morphology, metabolism, and functionality, with long-term stability even under fluctuating environments. We coin this property, *cell-state organization*. On the other hand, cells also show remarkable capacity to adapt and develop novel properties (Gerhart and Kirschner, 1997). This duality of robustness and flexibility is one of the hallmarks of biological systems at different levels of organization; a property enabling them to evolve and create new forms (Gilbert and Epel, 2009; West-Eberhard, 2003; Wilkins, 2002). What class of dynamical systems supports such a duality? To gain insight on this issue, we need to better understand the physics underlying the emergence of order, an organized state of the living cell, which is the subject of this essay.

From the biological point of view, cell-state organization is a manifestation of *genotype-to-phenotype* associations. Genotype and phenotype represent two types of realm in the living cell, connected by the dynamics of regulatory processes. The genotype, manifested by the genome—the DNA sequence, is a central source of information in the cell, integrating its evolutionary history and coding for the potential set of expressed proteins. The phenotype, generally encompasses any observable property of the cell, is manifested in its metabolism, form and function. From the physics perspective, the cell phenotype results from the many-body interactions of its constituent particles, with no privileged causal mechanism assigned to the genome (Moss, 2003; Noble, 2006). The main reason for the inability to assign a privileged role to any of the cellular constituents is the lack of a clear scale separation

Phenotypic Switching. DOI: https://doi.org/10.1016/B978-0-12-817996-3.00007-4

between the different physiological processes determining the cell phenotype. For instance, in many cases intracellular responses extend over time scales longer than a cell generation time, which is not well separated from the time scales of protein production, degradation, or metabolic processes. These processes, in turn, depend on the population's history through genetic and nongenetic inheritance (Jablonka and Raz, 2009). On evolutionary extended time scales, the genome may indeed play a special role in maintaining stable information, but these extremely long time scales are not our concern here. Moreover, while it is true that changes in the genome are usually found to occur on long (evolutionary) time scales, DNA itself is a passive object; its participation in cellular processes is only through a complex set of interactions with proteins which are largely determined by epigenetic and structural processes, by themselves part of the phenotype (Nijhout, 1990). In other words, not only that the phenotype is not solely determined by the genotype, the latter is also not entirely isolated from the former, nor is the sole source of inheritance. There is ample evidence for genomic modifications occurring on fast (physiological) time scales and reflecting physiological responses to environmental and internal cues. There is also recent evidence for long-term nongenomic inheritance across generations (for at least multiple generations and in our yeast experiments for hundreds of generations—see the succeeding text); another clue for the nonprivileged role of the genome as a memory "device" (Jablonka and Raz, 2009; Jablonka and Lamb, 2014; Sabine and Joseph, 2015; Posner et al., 2019).

From the physics perspective taken here, the genotype provides constraints on the phenotypic dynamics, the organization of a cell state, much like boundary conditions in physical dynamical systems. The expressed protein content of the cell, which plays a central role in determining its phenotype, is regulated by a complex dynamical system, operating across levels and on a wide range of time scales (Alberts et al., 2008). Different cell states can represent changes in the spectrum of expressed proteins, but there is *no one-to-one mapping* between protein expression levels and other phenotypic manifestations of the cell such as metabolism, morphology and function (Braun, 2015; Braun and Marom, 2015). The main reason is the deep degeneracy in protein expression levels; different patterns of expression can lead to similar phenotypes. The opposite is also true; a similar pattern of expression can support different phenotypes, due to environmental issues, the history of the system and the context. Accordingly, gene expression by itself is not a phenotype, but rather an auxiliary process supporting the phenotype. Although the protein content of a cell in general determines its type, stabilization of a cell state is a dynamical process determined by convoluted correlations of protein interactions over time. Because of the lack of direct mapping between gene-expression and the emerging phenotypes, a snapshot of the molecular content of the cell does not capture its actual state; the latter can only be understood by studying the dynamics. Therefore, inquiring into the emergence of order from the molecular disorder in cell-state organization requires a shift in focus from *structure to dynamics*, from the molecular *stuff* of the cell to its temporal *organization*.

We can recapitulate the contrast between the approach taken here and the common one in analyzing the organization of cell states as follows. The prevailing approach in biology regards the living cell as operating by codes and programs, which are manifested in underlying structured networks of interactions (including protein-DNA and protein–protein interactions) shaped in evolution. In this approach, the cell is a kind of an "engineered" control system, which is amenable to the process of *reverse engineering*. The latter is a relatively well-defined procedure in the context of engineering and technology, denoting the process of detailed examination of a functional system in the face of limited a priori knowledge of its design principles. In its broader definition, reverse engineering is an approach to figure out the mechanisms underlying an observed object. In many studies of a biological phenomenon, this is usually taken to the limit of maximal reductionism, disintegrating the cell and attempting to understand its "design principles" from molecular mechanisms. However, the living cell is a natural phenomenon rather than an engineered system and as discussed later, in our view the cell indeed is not amenable to reverse engineering (Braun and Marom, 2015; Marom et al., 2009; Muller et al., 2003).

A crucial difference between natural and engineered control systems is that in the latter one can assign a well-defined target function and measure the distance from it during its operation; this distance is translated into a measured error, which can be used as feedback, allowing convergence and stabilization of the functional system. In living systems, on the other hand, there is no a priori target function and no machinery for measuring the distance from such a putative target. Natural functional systems must rely on some form of *self-organization* rather than predesign controllers. The alternative concept of self-organization, for the emergence of phenotypic order in the living cell proposed here, relies on *exploratory dynamics*, which again shifts the focus from the structure of underlying putative networks into dynamical organization principles (Gerhart and Kirschner, 1997; Edelman, 1987).

Self-organization is a commonly used term to explain different inanimate and living systems. It generally means systems with self-referential dynamical capabilities leading to ordered states. Living systems, the cell included, present a unique challenge among systems exhibiting self-organization dynamics, since they are historical objects; not only because they went through evolutionary selection processes, but also because any given realization of a developmental trajectory goes through a particular history. More specifically, the challenge for a self-organization framework of the living cell is the lack of methodology to study historical systems, as well as systems with no clear scale separation (Marom, 2015; Vygotsky, 1978). As we further discuss below, these features present a significant barrier to our ability to identify the *relevant variables*, or degrees of freedom, underlying the observed dynamics in cell organization. The meaning of relevant variables, as defined here, is a set of degrees of freedom of the system that cause significant functional changes when modulated; as explained below, these degrees of freedom do not coincide with the set of variables that are directly measured in the system.

The complexity of the cell also manifested in the heterogeneity of its molecular building blocks, in particular its protein content and the unfathomable potential of protein expression patterns, is also a major issue. The combinatorial high-dimensional space of protein expression levels stands in sharp contrast to the discrete finite number of stable cell states observed in an adult animal. A rough estimation is that the body-plan of a human embryo is constructed by a few hundred types of cells (e.g., skin cells, neurons, etc.) (Vickaryous and Hall, 2006). Even if the exact counting of cell types is challenging, since the cell state depends on the dynamics of expressed proteins, the discrete small number of states means an enormous *dimensionality reduction* from the potential phase-space of possible expression patterns to any specific realization. This is roughly analogous and an extension of the well-known protein folding problem; for an unbiased folding process, the exponential increase in search time and potential folded states, with the length of the amino-acid chain due to its combinatorial complexity (Karplus, 1997; Gianni and Jemth, 2016; Zwanzig et al., 1992). The conventional picture in biology relates the observed realization of specific cell states (types), say in the emergence of a body-plan during embryonic development, to selection processes in evolution. From the physics point of view, however, this explanation is unsatisfactory since the level of control required to eliminate deviations from the selected phenotypes, should be of the same level of complexity as manifested in protein expressions. In other words, the constraints set in evolution should eliminate a huge number of possible expression patterns, which potentially could emerge in the high-dimensional phase-space of protein expressions. The experiments described below, were designed to demonstrate an alternative framework to this evolutionary-selected programmatic view. These experiments demonstrate that the naïve reductionist view relying on selection of the complex intracellular interactions is not compatible with the level of plasticity and adaptation capability of the cells.

Toward a physical picture of cell-state organization, we took a phenomenological approach and studied the dynamics of cell populations under an *unforeseen challenge*—a novel perturbation the cells had not encountered before in their evolutionary history (Braun, 2015). The assumption behind this approach is that biological systems most often show two types of responses; fast, *pre-organized* responses to familiar challenges based on mechanisms instilled in the system during its evolutionary history, and *improvised* responses based on a more general capability of the system to learn and adapt (Soen et al., 2015). The improvised adaptation could be followed by a longer-term evolutionary change that stabilizes it. These two facets of biology appear in many systems and at many levels, such as protein interactions, sensory systems in a behaving animal and processes in developmental systems (Stern et al., 2012; Elgart et al., 2015). Our experiments demonstrate that the living cell indeed exhibits these two types of responses. Adaptation of the cell to an unforeseen challenge exposed a complex response that was not based on pre-organized processes but rather on exploratory dynamics; wandering of expression patterns till convergence. Thus, our experiments uncovered the hidden potential of the cell, allowing it to respond in an adaptive way to challenges not encountered before in its evolutionary history; in other words, enabling the emergence of cellular *novelty*. The experimental results indeed painted a picture of the living cell that in many ways was not compatible with the prevailing picture. In particular, they demonstrated that inherited adaptation could result from physiological processes; there was a strong cross-talk between levels of organization so the population level played a role in shaping the cell dynamics; the environment played a dynamic role in adaptation rather than being simply a passive selection filter; and finally, there was a metabolic cost for adaptation which therefore relied on active reorganization processes.

This picture places the process of cell-state organization with other physics problems at the forefront of complexity science, seeking to understand more broadly the convergence dynamics in high-dimensional systems far from thermal equilibrium. We come back to this important issue later.

The organization of this essay is as follows: first, we discuss our experiments on yeast populations, studying the phenomenology of cell adaptation to an unforeseen challenge. Second, we discuss the implications of the experimental results on our understanding of cell-state organization. Next, we discuss a picture of the cell as an active exploratory dynamical system. This analysis leads to our hypothesis that the cell is a "sloppy" dynamical system and we further discuss the implications of this hypothesis. We finally summarize the main points of this essay.

Experimental approach: cell adaptation to an unforeseen challenge

To study the response of cells to an unforeseen challenge, we "rewired" the genome in the budding yeast, exclusively linking an essential metabolic gene to a foreign regulatory system. Under frustrating conditions, in which the regulatory system was highly repressed while the rewired gene was essential, the cells faced a severe challenge. This synthetic modification of the genome, in fact mimicked a natural evolutionary process responsible for numerous innovative changes in developmental and other systems (Wilkins, 2002). In our experiment, the gene *HIS3*, an essential enzyme in the histidine biosynthesis system, was detached from its natural regulatory system and was placed exclusively under a promoter of the GAL system, responsible for galactose utilization (Fig. 12.1) (Stolovicki et al., 2006). The rewiring of *HIS3* under GAL was utilized in different ways, either on plasmids or integrated into the genome. The main experimental results were found to be insensitive to the specific setup.

The major and most notable challenge to the cells was evoked by placing these rewired cells in a frustrating environment of glucose and lack of histidine. The GAL system behaves as a strong switch, highly induced in galactose and strongly repressed in glucose (Carlson, 1999; Johnston et al., 1994). In the absence of histidine in the medium, *HIS3* was absolutely essential and its repression together with the GAL genes placed the cells under a severe challenge (Stolovicki et al., 2006). Although there were multiple other challenges due to the rewiring event, for example, overexpression of *HIS3* in galactose or insensitivity of it to processes affecting the amino acid metabolic systems, we concentrate here

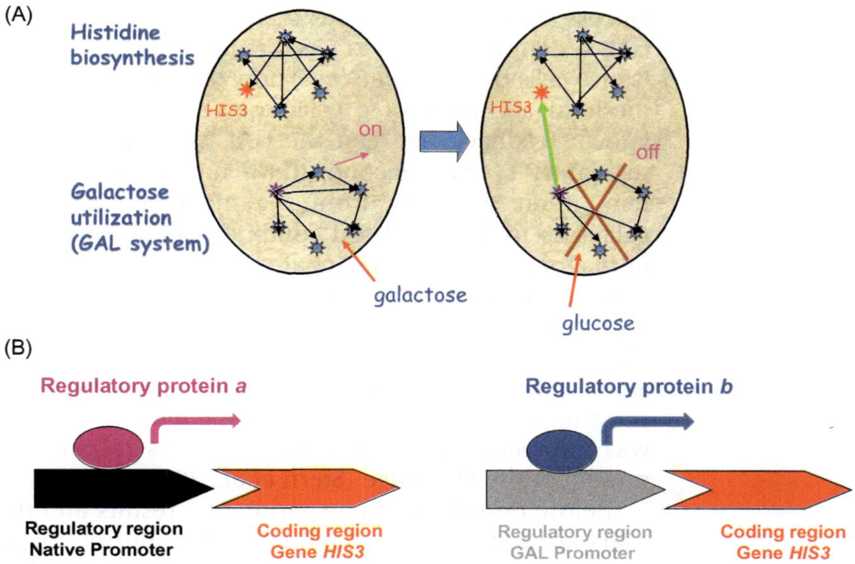

Figure 12.1 Genome rewiring in the budding yeast. (A) *(Left)* A schematic representation of the two modules involved in our genome rewiring experimental setup, the histidine biosynthesis pathway and the GAL system responsible for galactose utilization. *(Right)* In our rewiring setup, the essential gene *HIS3* from the histidine pathway is detached from its original regulatory linkage and placed exclusively under the promoter of the GAL system. The GAL system is highly induced when the sole sugar in the environment is galactose and is highly repressed when it is glucose. (B) In practice, the native regulatory region in front of the *HIS3* coding region is replaced with the GAL promoter (the promoter in front of the gene *GAL1*). In our experiments, the insertion was done by standard genetic engineering approach in different ways, by a constructed plasmid as well as integration into the genome. The basic phenomena discussed in this article are insensitive to the mode of construction of the rewired strain. *Source*: Reproduced from Braun, E., 2015. The unforeseen challenge: from genotype-to-phenotype in cell populations. Rep. Prog. Phys. 78, 036602, with permission.

on the challenge due to the GAL repression in glucose. This serves as a prime example for the dynamic cell-state organization we wish to discuss. In fact, it was not clear a priori that the cells can survive this type of a severe stress. Note that cells defective in *HIS3* enzymatic function, or missing this gene, could not survive in a medium lacking histidine. The experiments, however, demonstrated that cells had the ability to adapt to this challenge by multiple alternative trajectories, reflecting the richness of the underlying potential of cells to self-organize under novel conditions.

The dynamics of *HIS3*-GAL rewired cells under glucose reflected a novel response showing that indeed they faced an *unforeseen challenge*. This was evident from the following: first, as far as we know, yeast cells have not experienced the control of the amino-acid histidine biosynthesis system by a carbon source (galactose to glucose transition), nor does there exist an example in which the GAL system was shown to react to amino-acid metabolic demands. Native cells use the same regulatory system for amino-acid biosynthesis in a glucose or a galactose environment and there is no known feedback between the GAL system and the amino-acid regulatory system. Second, the dynamic range of the GAL system, upon switching from the highly induced conditions in galactose to the repressed conditions in glucose, is three orders of magnitude, while *HIS3* changes only within a limited range of twofold to threefold under its native control (Hinnebusch, 1992; Johnston and Carlson, 1992; Struhl and Davis, 1981). This large change in dynamic range required significant cellular reorganization, which was observed in our experiments. Finally and most importantly, the richness of the dynamics described below was unprecedented in any known stress response, proving the novelty of the challenge (Stern et al., 2007).

Before describing the experimental results and their implications, we note that the main characteristics of adaptation to an unforeseen challenge were found by us to be similar in systems other than rewiring an essential gene to the GAL system. We also rewired *HIS3* to a set of promoters regulating the cell cycle at the level of transcription (Katzir et al., 2012). Harnessing cell-cycle regulators to directly control an essential metabolic process increased the load on the cell-cycle's regulatory network at a specific phase of the cycle and demanded re-distribution of its resources (Katzir et al., 2013). As in the case of the GAL system, such a perturbation introduced a complex challenge to the cells by requiring the cell-cycle regulators to operate outside their natural context and in concert with arbitrarily chosen metabolic demands. An important demonstration of the ability of living systems to reorganize when exposed to unforeseen challenges was also provided in fly developmental processes, showing that the capability for novel responses is not limited to single-cell organisms but is extended to multicellular ones (Stern et al., 2012). In these experiments, the food was supplemented with the antibiotic G418 at concentrations that are lethal to the wild-type larvae and a resistance gene was placed under the regulation of different, spatiotemporally restricted, developmental promoters. The larvae could develop under these conditions by operating the developmental promoters outside of their natural domains.

Coming back to the case of *HIS3* under the GAL system, the ability of rewired cells to accommodate the challenge had been studied by different culture techniques; batch culture, plating on agar and most notably in chemostats (Braun, 2015).

All culturing methods demonstrated similar overall qualitative behavior. However, only the chemostat allowed us to study the dynamic response of cells to the imposed challenge, by continuous growth of large populations over extended time scales in a noninvasive way (Novick and Szilard, 1950; Smith and Waltman, 1995; Monod, 1950). This was achieved by continuous dilution of the population at a constant rate while injecting fresh medium to keep the total volume of the culture in the reactor fixed. Under steady conditions, one of the nutrients serving as a limiting factor sets the average growth rate of the population equals to the dilution rate of the chemostat. The population density, however, is a dynamic variable and its measurement reports on the population metabolic state and growth conditions (Smith and Waltman, 1995).

Let us now briefly summarize the main experimental observations relevant to our discussion of cell-state organization. More details on the experimental results and their biological implications can be found in (Braun, 2015) and references therein.

Our basic experimental observation is that the population of rewired cells, upon switching medium from galactose to glucose, exhibited a relatively fast adaptation process following the severe challenge. Chemostat experiments showed that the adaptation process occurred on time scales of 10–30 generations (measured by chemostat dilution rates); an extremely short adaptation time compared to those observed for microorganisms in evolutionary lab experiments (David et al., 2010). As discussed later, this adaptation occurs also in isolated cells and do not involve existing genomic variations in the population nor selection processes, so it is not of the type discovered by Waddington (genetic assimilation) (Waddington, 1942; Waddington, 1953), nor it was based on genetic exchange in the yeast population. This adaptation was manifested by the ability of cells to resume growth and cell division, after residing in a latent phase of nongrowth for some duration following the medium switch to glucose. This transition from a latent phase to normal growth reflected a genuine adaptation process, since the adapted phenotype was faithfully inherited across generations for hundreds of generations (Fig. 12.2) (Stolovicki et al., 2006; David et al., 2010). Our experiments showed that the inherited adaptation was not due to selection of a rare advantageous subpopulation; every cell in the population had, in principle, the potential ability to adapt. This was also manifested in single cell experiments, showing the capacity of individuals to switch their phenotypes from a latent state of nongrowers into a dividing state followed by normal growth (Woronoff et al., 2019). The fact that multiple individual cells can adapt, the high rate of adaptation, and the emerging of adapted states without the involvement of genetic changes, make the conventional neo-Darwinian adaptation mechanism of mutation and selection improbable.

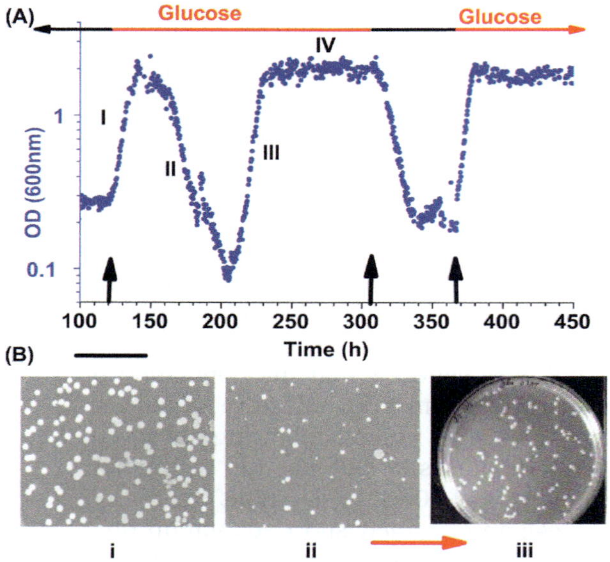

Figure 12.2 Fast inherited adaptation. (A) Typical population adaptation dynamics in the chemostat. The optical density (OD at 600 nm) measured in-line with the chemostat as a function of time. The OD is proportional to the population cell density. The population was first stabilized in a galactose medium lacking histidine. Next, the medium was switched from galactose to glucose at the left arrow, leaving all other nutrients the same. The medium was switched back from glucose to galactose at the middle arrow, and to glucose again at the right arrow. Four phases of the population dynamics are marked I-IV; fast exponential increase (I), followed by an exponential decline (II). The adapted population exhibits an exponential increase in density at (III) and the stabilization of a new steady state in (IV). At the second medium switch to glucose, the population density was immediately increased (exponentially growing at maximal rate) to the steady-state level of phase (IV), indicating that the population adapted state is inherited. The chemostat dilution time is 7 h. Bar, 10 cell generations (generation time equals chemostat dilution time $\times$ ln $2 \sim 5$ h). Note the y-axis logarithmic scale. Repeated experiments show similar phases of the dynamics with variations in the population adaptation time. (B) The chemostat experiment in (A) was repeated on agar plates lacking histidine. Cells from a batch culture propagated in galactose were placed on an agar plate with (1) galactose (first visible colonies after 2–3 days, saturation after 4 days) and (2) glucose (first visible colonies after $\sim$6 days, image taken at day 14, showing $\sim$40% of the colonies on galactose). Each colony was grown from a single cell on the plate. Both plates were initiated with the same number of cells. Note the uniform colonies in (i) and the variations in colony-size in (ii), indicating variation in adaptation times of the different lineages. A visible colony contains about 10^6 cells. Single cells from a colony from (ii) were redistributed on a glucose agar plate in (iii), showing uniform colonies after 2 days of growth, indicating that adaptation is inherited. *Source*: Reproduced from Stolovicki, E., et al., 2006. Synthetic gene recruitment reveals adaptive reprogramming of gene regulation in yeast. Genetics. 173, 75-85; Braun, E., 2015. The unforeseen challenge: from genotype-to-phenotype in cell populations. Rep. Prog. Phys., 78, 036602, with permission.

Although the actual mechanisms leading to the inherited adaptation remain elusive, we nevertheless can extract a few leading principles from these processes, serving as guidelines to the discussion of the physics of cell-state organization. We first note that our experiments demonstrated that the challenged populations exhibited multiple alternative trajectories to adaptation. While in some of the cases mutations emerging during this process could be identified, there were clear cases in which inherited adaptation was established without the involvement of any mutations (David et al., 2010; David et al., 2013). Moreover, in some of the cases in which mutations were detected, they had been shown by us to emerge late in the process, after cells resumed growth, so they were in fact induced mutations, arising due to the adaptation dynamics rather than causing it (Moore et al., 2014). This was shown by whole genome sequencing of cell lineages and by direct mapping techniques. At any rate, we have shown that there was *no global mutagenesis* involved in the adaptation process so when mutations emerged, they were part of the underlying dynamics. In a set of experiments, we could also map specific genomic loci associated with the inheritance of the adapted state (David et al., 2013). Importantly, while some of these loci did not show mutations (so they involved epigenetic processes of some sort), in some others the emerging mutations have been demonstrated *not to be sufficient* for resolving the unforeseen challenge; they therefore had to be complemented by other physiological processes. For example, one potentially interesting mutation that emerged in some of the cells was in the gene *GAL80*, a major participant in glucose repression in yeast. However, a mutation in *GAL80* did not alleviate the *HIS3* repression in glucose, so adaptation had to proceed with the utilization of other processes (David et al., 2013). In summary, based on detailed experiments, we concluded that *genetic changes could not account for the spectrum of adaptation trajectories* observed. This means that cells have mechanisms to faithfully transmit their phenotypic state across generations, without the involvement of the genome. The multiple alternative trajectories of adaptation to an unforeseen challenge suggest that this memory was not simply supported and maintained by a familiar epigenetic molecular process. Such an epigenetic process will probably lead to adaptation along more restricted and well-defined trajectories, for example, as manifested in patterns of gene expression. Rather, we proposed that one should think of the global dynamics as involving stable attractors, supporting the memorized cell states (Kaneko, 2006).

In more recent experiments, we studied the metabolic response of single cells during the adaptation process (Woronoff et al., 2019). Toward this end, we utilized a droplet-based microfluidic technology allowing us to measure, in a noninvasive way, the sugar uptake of individual yeast cells simultaneously with their growth and division. Yeast cells having *HIS3* rewired to the GAL system were first grown in a batch culture in galactose and then switched to glucose, after which they stopped their growth and division. These latent cells were then encapsulated in droplets, containing the same glucose medium and embedded in oil for measurements. It was demonstrated before that consumption of nutrients (glucose) in a cell-containing droplet, created an osmotic imbalance, resulting in water flux and inducing the shrinkage of the droplet and the swelling of neighboring cell-free droplets. The rate of the droplet shrinkage therefore reported on the rate of metabolism, allowing us to monitor simultaneously the rate of sugar uptake and cell growth and division (Boitard et al., 2012).

Surprisingly, the large majority of the cells show diverse rates of sugar uptake (droplet shrinkage rates) while not dividing or growing (Woronoff et al., 2019). These significant amounts of energy consumption, as measured in our drop experiments, were utilized by the cells for *reorganization* rather than for increasing their biomass. Over the course of the experiment, a sizeable fraction of these latent cells switched their metabolic state, either by accelerating the rate of sugar uptake or arresting it. Remarkably, a large fraction of the cells accelerating their metabolism also resumed division, on the average at the same time. Details of the experiments and discussion of the results can be found in (Woronoff et al., 2019). The results of the single cell experiments were consistent with those of the population experiments discussed at the beginning of this section, but they added the important information on the role of metabolism in adaptation, never measured before. First, as recognized also in previous experiments, adaptation occurred simultaneously in numerous individual cells on physiological time scales, so it was indeed *not the result of a selection process.* Second, adaptation is an *active dynamical process,* requiring the consumption of significant amounts of energy. Third, the single cell experiments verified that the adaptation rate ($\sim 10^{-3}\,h^{-1}$) was orders of magnitude higher than expected from conventional evolutionary processes relying on genetic mutations (10^{-5} to $10^{-9}\,h^{-1}$) (Hall, 1992). Finally, the droplet experiments demonstrated the tight coupling between the metabolic and the regulatory processes in adaptation (Varahan et al., 2019).

Going back to the chemostat experiments, an interesting glimpse of the underlying processes supporting the adaptation process was achieved by studying the gene expression response of the challenged rewired cells (Stern et al., 2007; Stolovicki and Braun, 2011). Our experiments demonstrated a few notable characteristics of this response (Fig. 12.3): (1) Adaptation involved *global reorganization* of gene regulation, manifested in a genome-wide transcriptional response observed in the population dynamics (Stern et al., 2007). Moreover, attempts to identify a "*biological logic*" in the global expression response, based on known metabolic pathways, failed. The emerging correlations measured for the expression patterns, did not coincide with expectations based on known metabolic pathways; even genes within the same pathways did not show a coherent response. Similarly, genes from different functional groups showed coherent dynamics. The emerging picture was of a multimode pattern of expression, involving groups of genes that reorganized under the challenge, but did not reflect prior known functional relations. (2) The experiments demonstrated the emergence of a novel feedback between the histidine pathway and the GAL genes, not recognized in wild-type cells (Stolovicki et al., 2006; Stolovicki and Braun, 2011). This emerging feedback allowed the GAL system to control the expression levels according to the *constraints of the histidine metabolic system*, although the GAL metabolic system itself was basically nonfunctional in glucose. Thus, feedback loops enabling the stabilization of the cell state, emerged by the dynamics rather than being parts of a "predesigned" control system. (3) The global gene expression response was found to be irreproducible between repeated experiments (Stern et al., 2007; Stolovicki and Braun, 2011). More significantly, twin populations constructed from a single population divided into two large subpopulations upon the switch of the medium to glucose, diverged in their gene expression response quickly after their separation. Thus, their common history was not enough to support a similar pattern of gene expression. By contrast, the pattern of gene expression within each subpopulation exhibited stability and long-term correlations for many generations (Stolovicki and Braun, 2011). This indicated that each population developed its own coherent gene expression response and the divergence between populations was not noise (there are no finite-size effects, as these populations were very large, containing $\sim 10^{10}$ cells each). The coherent response of cells in a population, measured both at the level of mRNA and proteins, was unlikely to be due to specific intercellular signals; this was demonstrated by mixing marked populations from different chemostats and observing that the subpopulations specific dynamics were maintained for long durations over more than 20 generations (unpublished data). This demonstrated a tight connection between levels of organization, a hallmark of biological systems in general. In this case, the coupling was between the intracellular

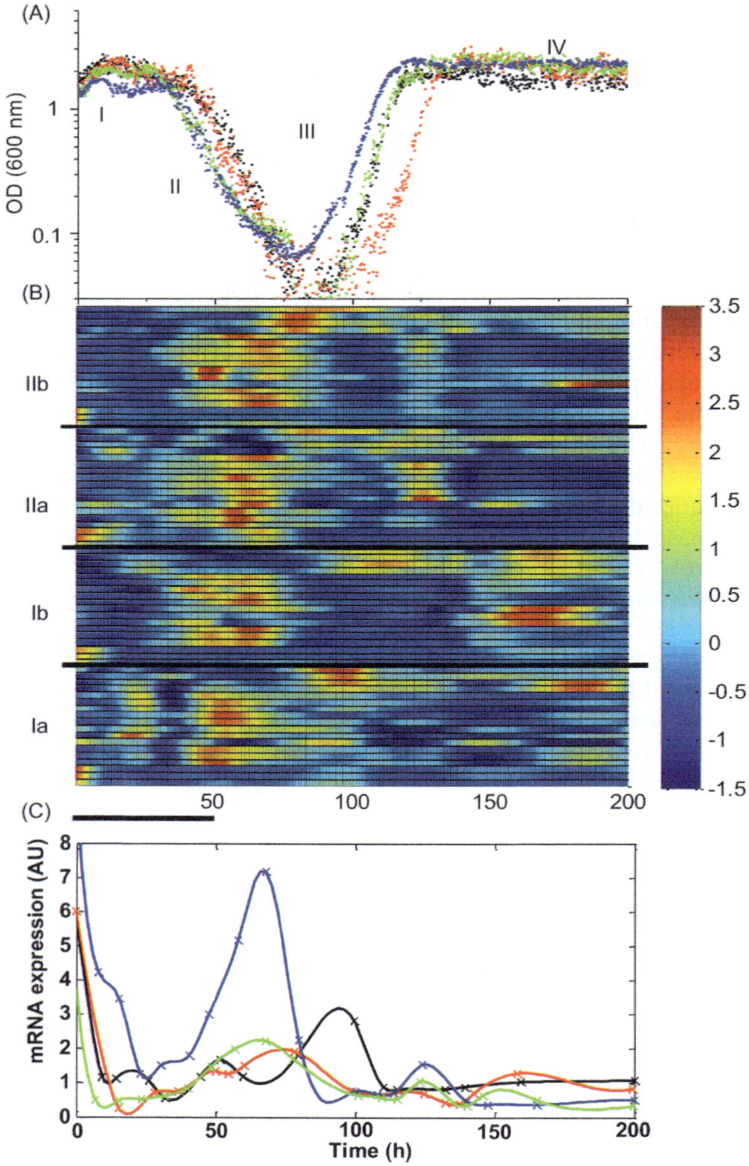

Figure 12.3 Phenotypes and gene expression profiles for "twin" populations. (A) Cell density (OD at 600 nm) as a function of time for two pairs of "twin" chemostats with populations of rewired cells (Ia-black and Ib-red are "twin" populations and so are IIa-blue and IIb-green). The histidine lacking medium was switched from galactose to glucose as a sole carbon source at $t = 0$, leaving all other nutrients the same. A steady-state typical of galactose metabolism was first established for each pair of twin chemostats, mixed at rates faster than their dilution rate, therefore behaving as a single population. The two chemostats were decoupled prior to the medium

(*Continued*)

processes responsible for gene expression, and the population level, due to collective cell dynamics. As discussed above, the chemostat experiments of mixed populations show that collective cell dynamics was more likely to arise from indirect cell-cell interactions via their common environment rather than specific signaling. It demonstrated the outstanding role of the environment, which was not merely a passive agent but an active one, coordinating cells within the dynamic population. (4) Finally, the intrapopulation expression patterns demonstrated coherent modes, relaxing on time scales of ~ 10 generations. This implied stable transgenerational inheritance of expression levels and synchronization among large groups of cells (Stolovicki and Braun, 2011). The emergence of slow dynamics (over many generations) is a signature of high-dimensional systems, as recognized in physical complex systems far from thermal equilibrium involving many-body interactions (Kurchan and Laloux, 1996). We come back to this important point in the succeeding text.

To summarize the lessons from the gene expression response during the adaptation process: (1) Gene expression was not mapped to metabolism in a straightforward way and the mapping was certainly *not one-to-one.* (2) The population developed *collective* modes, exhibiting *slow relaxations* over extended time-scales and memory over generations. (3) Each population developed its own unique pattern of expression and these patterns were not reproducible in repeated experiments.

In summary of the experimental findings:

- Rewiring an essential gene to a foreign regulatory system was a perturbation of regulatory modes. Under challenging environmental conditions, cells resolved the regulatory frustration by *global* reorganizing of gene regulation, involving large sets of genes and types of interactions, rather than by local specific processes.

◀ switch into glucose. Thereafter, each population of the twin was grown in a separate chemostat (but fed from the same medium). Note the y-axis logarithmic scale. Different phases of the dynamics are marked I–IV as in Fig. 12.2. (B) Color-coded raster plot of the mRNA expression profiles: Ia–Ib and IIa–IIb mark the same twin populations as in (A). The expression level of a gene as a function of time is marked by the color variation across the horizontal line. Each line is a measurement of a different gene. The plot shows the measurements of 18 genes (measured by real-time PCR) belonging to different metabolic functional modules, repeated for each population. The measured expression levels were normalized for each gene to zero mean and unit standard deviation across its entire time profile. Bar-10 chemostat generations. (C) mRNA expression levels of *HIS3*, the rewired gene, for the four populations shown in (A) and (B). The data are normalized as in (B). *Source*: Data from Stolovicki, E., Braun, E., 2011. Collective dynamics of gene expression in cell populations. PLoS One. 6, e20530; Braun, E., 2015. The unforeseen challenge: from genotype-to-phenotype in cell populations. Rep. Prog. Phys. 78, 036602, with permission.

- The reorganization dynamics was driven by an *active* process, requiring a significant amount of energy. The fact that this process was global, irreproducible between populations and not directed, suggests that this response reflected an unbiased *exploratory* process (Schreier et al., 2017).
- Adaptation was inherited across generations but nevertheless could be realized by multiple alternative trajectories not necessarily involving genomic mutations. The existence of multiple trajectories to adaptation was another manifestation of the exploratory nature of the process and revealed the degeneracy of regulatory modes and the flexibility of the coupling between metabolic pathways and gene regulatory modes.

Implications of the yeast adaptation experiments

The lessons learned from our yeast experiments open a wide vista on some fundamental issues in cell biology, most notably on the emergence of stable cell states. Of particular importance here is our finding that an adapted cell state is essentially an *active exploratory process* (more on this issue below). This observation has significant consequences for our understanding of the physical basis of cell-state organization as we now discuss.

What is the connection between yeast adaptation to an unforeseen challenge and the general issue of cell-state organization? To get a deeper insight into this question, let us first describe a concrete (but major) example; the emergence of cell types in developmental processes. In mature animals, it is possible to distinguish between different types of cells, for example, heart cells, skin cells, neurons, etc. However, it could well be that there exists a more-or-less continuous spectrum of variants spanning the entire range of morphologies, biochemical composition and functionalities between the extreme cases we identify as a discrete set of differentiated cell types (Soen et al., 2006). Nevertheless, the classification into a finite discrete set of cell types is possible; in spite of the large phenotypic variability, member cells of the same type are more similar to each other in many characteristics than to nonmember cells. However, correlating types with gene expression profiles remains problematic (Trapnell, 2015). In humans, recent work identified $\sim$400 cell types, and the number required for the construction of the body plan in a developing embryo is more limited (Vickaryous and Hall, 2006). The fascinating process of cell differentiation in multicellular organisms, the stabilization of a functionally specific type during embryonic development is far from being understood. In particular, it raises the following issues: what are the relevant degrees of freedom that determine the identity and stabilization of a cell type? What type of dynamics ensures the robust convergence and then stabilization of a cell type on physiological time scales? Finally, what are the relations between gene regulation and protein expression levels and cell types, and how do they integrate with metabolism to ensure homeostasis, given the huge degeneracy of the metabolic/regulatory phase-space?

We can recognize an immediate connection between these questions and the issues raised following the experiments on adaptation of yeast cells to an unforeseen challenge; the underlying physical concepts turned out to be closely related.

Although the stabilization of a cell type in developmental systems is usually regarded as a programmed process, preshaped in evolution to ensure convergence and stabilization of the adult organism, recent advance shed new light on this concept (Newman, 2020). In particular, it has been known for long time that cell differentiation is reversible. Fully differentiated cells, can be triggered to trans-differentiate into another type and remarkably enough, to dedifferentiate back into stem cells with various degrees of plasticity—a process coined *reprogramming* (Yamanaka, 2012; Yamanaka and Blau, 2010; Graf, 2011). These processes seriously challenge our understanding of the stability of the differentiated state. Transdifferentiation and reprogramming raise a question about the dual characteristics discussed before; *how is the integrity of mature cells stably maintained in the face of perturbations while simultaneously enabling plasticity, trans-differentiation and reprogramming?* A similar issue is raised in the case of cancer. The emergence of cancer, from the cellular perspective, reflects de-differentiation and gain of plasticity. In fact, cancer can be regarded as adaptation of a cell population in a tissue; the emergence of novel phenotypes under a challenge (Braun, 2015; Huang et al., 2009). From the organism viewpoint, it is hard to accept the common justification, assigning advantage to any observed biological process via evolutionary selection, when the ability to dedifferentiate seems like a dangerous potential. It reflects flexibility in regulatory organization that can hardly be explained as a program-like process. In fact, regarded as adaptation processes, trans/dedifferentiation and cancer processes are closely related and are based on similar physics as the process of cell-state organization discussed in this essay.

To gain insight on the physics of this problem, two issues need to be addressed: First, the high dimensionality of gene-activity phase-space dictates dynamic characteristics that are far from the type of attractors, utilized to explain the stability of cell states; these attractors are objects of low-dimensional systems and their relevance for the a priori high-dimensional biological system is questionable (Newman, 2020). Most of the examples analyzed in the literature thus far, adhere to the low-dimensional type of systems and therefore their results may not be applicable to the realistic high-dimensional cases presented by gene regulation systems in the living cell (Kaneko, 2006; Kauffman, 1993; Bornholdt and Kauffman, 2019; Jia et al., 2017; Mojtahedi et al., 2016; Huang et al., 2005; Huang, 2011). Second, what class of dynamical systems allows efficient convergence under the complexity of the living cell? We next discuss our hypothesis that these processes reflect exploratory self-organization dynamics.

The exploratory dynamics of cell-state organization

The capability of biological systems to respond to familiar as well as to novel perturbations is well recognized outside the domain of cell biology; the realization of novel responses characterizes a broad class of "*exploratory*" systems, the prototypes of which are the nervous and immune systems (Edelman, 1987). The exploratory dynamics of these systems are constrained by their interactions with the environment. Although much less recognized as such, our work on yeast highlights the fact that the *living cell belongs to this class.* The exploratory nature of adaptation reflects the complexity of gene activity phase-space. The characteristics of the adaptation dynamics and the remarkable efficiency in which cells resolve the severe unforeseen challenge, indicate that although a priori the dimensionality of this phase-space is extremely large, eventually gene regulation is reorganized on a relatively low-dimensional manifold, ensuring convergence of the dynamics. A physical framework for cell-state organization however, cannot simply assume this low-dimensional dynamics but rather should seek to understand this nontrivial dimensionality reduction. Indeed, *dimensionality reduction by self-organization* highlights cell-state organization as an exciting frontier in the physics of complex systems.

Current research gives ample examples of exploratory dynamics at the subcellular level. A concrete and important example for a mechanism supporting exploratory dynamics is given by the class of proteins identified as *disordered.* There are indications from experiments over decades, showing evidence that many proteins lack fixed structures or are disordered under physiological conditions (Oldfield and Dunker, 2014). Namely, many proteins are unfolded and yet play important functional roles. More recent studies show that *intrinsically disordered proteins* (IDPs) or proteins having disordered domains span all life forms from bacteria to human (Dunker et al., 2015). The functions of many proteins in fact depend on the unstructured rather than structured state, and these functions also span all domains; catalysts, binding proteins, and most importantly for our discussion here, many of the transcription factors are disordered. This picture of complex proteins' structure/function relations stands in sharp contrast to the older views such as the "lock and key" hypothesis, in which protein binding to its target assumed to be highly specific due to its unique three-dimensional folding structure, mainly dictated by its amino-acid sequence. The more modern view hypothesized that a substantial portion of the IDPs undergoes disorder-to-order transitions upon binding to their cognate partners (Dunker et al., 2008). This structural flexibility and binding plasticity enable IDPs to interact with a broad class of partners. Binding partners in the form of small-molecule ligands, macromolecules, or any posttranslational modification can induce order, namely inducing IDPs or protein disordered regions to become structured (Oldfield and Dunker, 2014). This disorder-to-order transition is due to the active role of the environment of these IDPs and highlights the fact, that cellular interactions and organization is highly context dependent. These characteristics that are essential features of an exploratory system, open a new dimension of complexity; dynamical systems in which their underlying "elementary particles" themselves exhibit plasticity and exploratory characteristics present unprecedented challenge for modeling and theory.

The general picture of *promiscuous* protein interactions serves as an example here for the underlying exploratory nature of biological systems at all levels of organization, molecular, cellular, tissue, and whole organism, and plays important role in their ability to support novel organizations and evolvability (Gerhart and Kirschner, 1997; Tokuriki and Tawfik, 2009).

The next issue for a physical framework of cell-state organization is the identification of the relevant variables, or observables, underlying the convergence adaptive dynamics in adaptation. This might sound strange, since it is clear that the protein content of a cell determines its state. However, two types of observations from our experiments show that the level of expression of each gene by itself cannot be the relevant observable determining the dynamics. First, the protein level in a cell population, even for essential functional proteins (such as *HIS3*), is highly variable among cells, exhibiting a *broad non-Gaussian* distribution (Salman et al., 2012; Moore et al., 2013). The shape of the protein-level distribution, for the majority of proteins (excluding perhaps low-expression level proteins and others showing gradual and multimodal distributions) was found to be universal, similar for bacteria and yeast and independent of the biological context and the specificity of central processes (e.g., metabolism, transcription, and copy number) affecting gene expression (Fig. 12.4) (Salman et al., 2012). This distribution of protein expression levels reflects fast intracellular processes and the flexibility of the cell in utilizing different patterns of expressed proteins to support organized functional responses (Brenner et al., 2015). Second, the irreproducible collective multimode expression response discussed before, reflects slowly varying dynamics that are specific to the underlying biological context, but nevertheless highly degenerate; different gene expression responses can lead to similar phenotypes (Stolovicki and Braun, 2011). There is a high degree of flexibility also here, as different populations exhibit alternative patterns of expression although all converging in their functional outcome. This shows again the flexibility in mapping protein

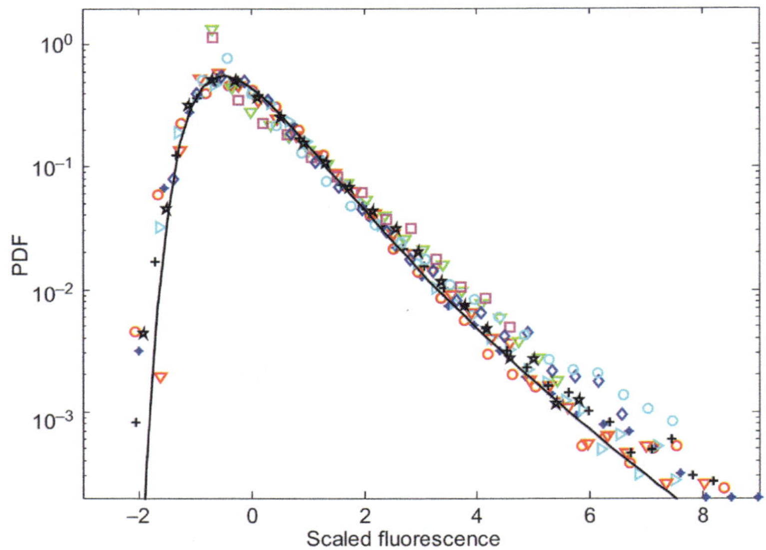

Figure 12.4 Universal protein distribution in cell populations. The distributions of expression levels (probability distribution function, PDF) for different populations of bacteria and yeast under different conditions. The distributions were estimated by the GFP fluorescence measured by flow cytometry. GFP was either placed under different promoters or fused with functional genes. See reference (Salman et al., 2012) for details. All distributions, either for essential genes integrated into the genome or for marker genes on plasmids collapse to a universal distribution after scaling, by subtracting the mean and dividing by the standard deviation. The symbols represent different experiments. The black line is the Frechet distribution best fit to the data. Note that the distribution is skewed and non-Gaussian, and the y-axis logarithmic scale. *Source*: Reproduced from Salman, H., et al., 2012. Universal protein fluctuations in populations of microorganisms. Phys. Rev. Lett., 108(23): p. 238105, with permission.

expression levels to metabolism and function; the fact that coexpression and cofunctionality represent two types of dynamic correlations, not necessarily any causal relations. Taken together, the fast process leading to broad universal protein expression distributions, and the slow processes leading to irreproducible protein expression patterns between populations, imply that the protein content of single genes are not the *relevant degrees of freedom* determining the cell phenotype. The profiles of protein expression in a cell, by themselves are not a unique determinant of the cell state (metabolism and other phenotypes). In many cell types, there are of course single genes, or groups of genes, that can serve as well-defined markers of the type (e.g., certain ion channels in a neuron type or a heart cell), but the rest of the expression pattern is usually not unique to the type and depends on the environment and other factors (Briggs et al., 2018; Crocker et al., 2016; Efroni et al., 2015). We are interested here not in identifying specific markers, but rather to understand how a global protein expression pattern supports the robustness and stability of a cell state. The set of expressed proteins (i.e., protein concentrations) and their rates of change, contain the relevant information affecting the phenotype. However, they determine the phenotype through some yet unknown *coarse-grained* processes, reflecting the overall correlations underlying the multigene expression patterns. We currently do not have a logical framework relating the global protein expression levels to the phenotype. In our attempt to construct a physics-based framework of cell-state organization, the inability to identify the relevant observables is a major barrier.

Next, we remark on dynamics in high-dimensional phase-spaces. The relevant phase-space here is spanned by all the microscopic (molecular) degrees of freedom determining gene expression and metabolism. This space is of a gigantic dimension. Consider the space spanned by the protein expression levels. In each cell of the human body, there are around 10,000 different types of proteins expressed at any given moment. Different cells, even within the same group of cell types (e.g., skin cells), may exhibit different patterns of protein expressions. Considering the level of expression of each protein as a degree of freedom, requires to describe the combinatorial dynamics in a phase-space with at least the dimension determined by the potential expression of all proteins, namely one that is measured in tens of thousands. This is truly an *unfathomable space*.

The implications of the high dimensionality of phase-space on the dynamics are profound. For example, in a high-dimensional phase-space, in contrast to a low-dimensional one, the system may not directly flow into a nearby attractor since there are an exponentially large number of directions away from it. This situation is familiar in physical complex systems (e.g., glassy systems) and to some views, might be the origin of slow relaxations and aging in these systems, due to their wandering on saddle trajectories rather than relaxing towards attractors (Kurchan and Laloux, 1996; Stein and Newman, 2012; Cugliandolo et al., 1997). High-dimensional systems are nonintuitive and unfortunately currently, there is no general theoretical foundations for the dynamics in such spaces. In functional systems like the biological cell, it is not enough to realize the type of dynamical modes possible; we seek to understand how the natural system maintains its robust functionality under such dynamics. In fact, complex functional systems like the living cell must converge on physiologically relevant time scales in order to maintain their viable functional state. Therefore, it suggests that in the case of cells, the actual dynamics are converging quickly to effective low-dimensional manifolds. These manifolds are determined by the effective coarse-grained degrees of freedom. Adopting this approach, the problem from the physics viewpoint is to develop a methodology allowing to identify the relevant coarse-grained variables determining the dynamic convergence to functional organization state. This is akin to a renormalization procedure familiar in physical systems near a critical phase transition. In some cases, the effective low dimensionality of the system is apparent and the search for a procedure for reducing phase-space dimensionality is successful (Jordana et al., 2013; Ori et al., 2018). However, without the benefit of equilibrium and given the complexity of the living cell, the approach toward such a methodology remains elusive. This is the essence of the problem; a physics theory of cell-state organization must rely on such methodology.

The current approach to analyze cell dynamics concentrates on prestructured networks of interactions (see an extension in (Schreier et al., 2017)). Our conceptual approach presented here, proposes to go beyond preorganized networks and pathways. Again, we do not in any way dismiss the analysis of networks as an effective description for some purposes. Indeed, as explained before, living systems show dual types of dynamics: fast relaxation to familiar situations and slow exploratory dynamics to novel ones. While network analysis might be useful for the former, we believe it does not provide a proper framework for the latter. The essence of systems epigenetics (in the broad sense, including all physiological interactions) underlying adaptation to an unforeseen challenge, is that gene interactions are highly *fluidic*, exhibiting a high degree of plasticity and depending on the environment, context and the history of the system [see discussion in (Nanjundiah,

2003)]. The lesson from our yeast experiments suggests that the reorganization processes toward adaptation should be regarded as *triggered* by the perturbation rather than induced by it. The eventual protein interactions leading to converging dynamics reflect an ad hoc reorganization process rather than relaxation of pre-existing networks of interactions.

How does a new cell-state organization emerge? We propose that cells utilize exploratory dynamics to converge and stabilize an organized cell state along the following steps:

- The stress due to *mismatch* between the environmental constrains and the internal cell state—its metabolic fluxes and other intracellular processes—serves as a *global driving* force for a *nonspecific* response. This stress drive is analogous to heat in a physical system. The emerging dynamics are based on the labile protein-DNA and protein–protein interactions due to their weak (compared to k_BT) intermolecular forces.
- The effect of this *global driving* enables new functional states, for example, by changing the spectrum of metabolic fluxes. These fluxes are *nonspecific* toward resolving the challenge.
- The emergence of a new spectrum of internal states, for example, a new set of metabolic fluxes, is adaptive if increasing the match between the external constraints and the internal state. The cell eventually converges to a "solution" supporting its proper functionality (its viability) under the constraints (e.g., resumption of normal growth). The requirement from the global nonspecific response is to be broad enough to allow convergence. Note that the large degeneracy of the system allows multiple possible solutions, it does not need to search for a particular configuration.
- The actual metabolism and more generally, other cellular functionalities, affect the cellular state, for example, protein expression levels, by a *feedback* causing relaxation and alleviation of the stress.
- The adapted state is stabilized through the process of *drive reduction*. The drive in the case of a cell is stress and drive reduction means alleviating the stress. This important concept, familiar from neural systems, does not necessarily mean that the drive is turned off, but rather that the state of the cell becomes immune to perturbations and the system maintains stability under the given internal and external constraints (Shahaf and Marom, 2001). As long as there is no drive reduction, the stress continues to drive the system, causing it to wander among new states.

This exploratory picture of adaptation, the emergence of a spectrum of metabolic and expression modes followed by physiological selection and drive reduction is again compatible with our understanding of a broad class of biological systems, most notably the neural and the immune systems. During non-specific exploration, the internal states of the cells could become deleterious, leading to cell death. Indeed, adaptation under exploration-exploitation processes represents a successful *tug-of-war* between viable states and deleterious ones. This was observed in our yeast experiments when the metabolic response of single cells is measured; a significant number of cells, initially metabolizing and exploring, eventually encounter a state of metabolic arrest and do not resume cell growth and division.

What types of dynamical processes allow convergence into an adapted state for those cells that can survive the tug-of-war? We propose that the living cell belongs to a class of systems identified as "*sloppy*." We turn now to briefly discuss the cell as a sloppy dynamical system.

The living cell as a sloppy dynamical system

We reached the conclusion that the protein content of each gene by itself is not a relevant variable determining cell-state organization. Although proteins indeed determine the cell functionality, they do so by collective modes involving large groups of interacting particles. The groups themselves are dynamic entities; the same protein can participate in different groups and the group assignment can change over time. Thus, the relevant variables are the collective dynamical modes of these groups and their interrelations. A similar approach to multicomponent organization, involving dynamic group organization was proposed long ago for neuronal systems (Edelman, 1987). Overall, theses group dynamics are highly degenerate and the macroscopic dynamics are highly insensitive to the details of the underlying interactions.

This property, the insensitivity of the macroscopic global dynamics to the precise values of the microscopic degrees of freedom, characterizes a class of systems identified as "*sloppy*." Sethna and coworkers, analyzed a class of models (not necessarily biological ones—but almost all biological models belong to this class) with several fitting parameters which have this property (Brown and Sethna, 2003; Daniels et al., 2008; Gutenkunst et al., 2007; Mark et al., 2015). They suggested that models belonging to this class, present a challenge for the identification of the relevant variables, mostly discussing this class from the perspective of the difficulty in fitting parameters to the models. We would like to take a step further and propose that the *living cell itself is sloppy*, not just models describing it. In other words, we propose that sloppiness is an essential property, allowing adaptive cell-state organization. The microscopic degrees of freedom in the case of the cell are its intracellular interactions. There are certain important consequences of this proposal on the required methodology for analyzing and understanding the living cell and the picture it paints of cell-state organization.

In a sloppy system, changes of each microscopic variable by itself, causes little modification of the system's dynamics. Most of the directions in phase-space are therefore not constraining for the dynamics, justifying the sloppiness characterization. Only a few directions in phase-space, not necessarily along any of the original variables-axes spanning this space, are "stiff," in the sense that perturbations along them have large effects on the dynamics (Daniels et al., 2008). For models, this property is highly challenging if one wishes to fit each variable of the model to the experimental findings. For a dynamical system, like the living cell, this property allows flexibility in its internal kinetics and freedom when facing unforeseen challenges, which cannot be accommodated by pre-designed kinetic parameters nor rigid network structures. The consequence of this line of thinking is that cells accommodate novel challenges by exploratory dynamics, which are largely possible due to the sloppy characteristics of the system. Rather than operating prewired networks of interactions with well-defined kinetics, the overall dynamics are determined by converging on a low-dimensional manifold determined by the stiff directions, a small number of combinations of the microscopic variables (e.g., the protein content of each gene).

We note that indeed, as discussed before, many of the protein interactions themselves are dynamically determined according to the biological context. One example discussed above is that of the IDPs. Another known example is provided by the *Hox* genes, which play critical roles in determining the body-plan during development. These are transcription factors that lack specificity in their DNA binding in vitro, but specificity is achieved in vivo by the joint effect of other proteins during development.

How does the cell converge to this low-dimensional manifold determined by the stiff combination of variables? This is a major open issue. For a theoretical framework of cell-state organization, this picture provides a highly coarse-grained description of the complex system. Indeed, the schematic picture of the cell as a sloppy dynamical system is appealing. Unfortunately, currently there is no theoretical foundation or methodology allowing us to identify the stiff directions in sloppy systems, outside of special cases (Mark et al., 2015).

There are some other important consequences of identification the cell as a sloppy system. Reverse engineering of sloppy systems is impossible, since the inverse problem in sloppy models is ill posed. Thus, it is impossible to deduce the underlying microscopic kinetics from measurements of the global coarse-grained dynamics (Marom et al., 2009). This conclusion is in sharp contrast to the widely spread reductionist approach in biology, relying on breaking the cell into its molecular components and reconstructing the cellular dynamic response from the parts. The hierarchy of interactions in the cell reflects different levels of organization. At each level of organization, processes emerge from the collective dynamics of the underlying processes, reshaped by the upper-level processes. Sloppiness in fact arises at all these different levels of organization, whenever an observable emerges due to the collective interactions of the underlying correlated microscopic degrees of freedom. We conclude then that cell-state organization is an *emergent phenomenon.*

Summary and open issues

The aim of this essay is to characterize the physics underling cell-state organization, its metabolism, morphology, and function. This characterization is based on our experiments, studying the adaptation of yeast cells to an unforeseen challenge. Although the mechanisms of adaptation remain elusive, the experimental results contain ample evidence allowing us to paint a general physical framework for the emergence of order from the molecular disorder in the living cell. We propose that the ordered cell-state reflects the process of *self-organization based on exploratory dynamics.* We further propose that the apparent high dimensionality of variable space reflects our inability to identify the relevant observables. The rapid adaption of challenged cells, suggests the adaptation dynamics converge to a low-dimensional manifold determined by the effective relevant variables. We finally hypothesized that the cell is a sloppy dynamical system, in which the outcome of the dynamics is largely immune to changes of most variables and is affected mainly by a small number of stiff directions in variable space, reflecting collective effects of some the original system's variables. This property enables efficient exploration and convergence of the dynamics, which are eventually stabilized by the process of drive reduction. The barriers toward a physical framework of cell-state organization are significant at the forefront of the physics of complex systems. The main open issues are the lack of methodology to identify the relevant observables, and the construction of a theoretical basis for exploratory sloppy dynamics under the duality of robustness and flexibility characterizing the living cell.

This essay presents more open issues than answers. I hope that the discussion of the physics of cell-state organization will stimulate further discussions and a theoretical work along the lines presented here.

Acknowledgments

Over the years I benefited from discussions of the experiments and ideas painted here with several friends. I am grateful to Paul Chaikin, Shimon Marom, Naama Brenner, Yoav Soen, Yitzhak Rabin, Kinneret Keren, and Hanna Salman, for their encouragement and fruitful discussions. I would like to thank Jerome Bibette and Andrew Griffiths and their groups at ESPCI Paris, for a fruitful collaboration and discussions over the years.

References

Alberts, B., et al., 2008. Molecular Biology of the Cell, 5th ed. Garland Science.

Boitard, L., et al., 2012. Monitoring single-cell bioenergetics via the coarsening of emulsion droplets. Proc. Natl Acad. Sci. 109 (19), 7181–7186.

Bornholdt, S., Kauffman, S., 2019. *Ensembles, dynamics, and cell types: revisiting the statistical mechanics perspective on cellular regulation*. J. Theor. Biol. 467, 15–22.

Braun, E., Marom, S., 2015. *Universality, complexity and the praxis of biology: two case studies*. Stud. History Philosophy Sci. Part. C: Stud. History Philosophy Biol. Biomed. Sci. 53, 68–72.

Braun, E., 2015. The unforeseen challenge: from genotype-to-phenotype in cell populations. Rep. Prog. Phys. 78, 036602.

Brenner, N., et al., 2015. Single-cell protein dynamics reproduce universal fluctuations in cell populations. Eur. Phys. J. E 38 (9), 102.

Briggs, J.A., et al., 2018. The dynamics of gene expression in vertebrate embryogenesis at single-cell resolution. Science 360 (6392), eaar5780.

Brown, K.S., Sethna, J.P., 2003. Statistical mechanical approaches to models with many poorly known parameters. Phys. Rev. E 68.

Carlson, M., 1999. Glucose repression in yeast. Curr. Opin. Microbiol. 2, 202.

Crocker, A., et al., 2016. Cell-type-specific transcriptome analysis in the Drosophila mushroom body reveals memory-related changes in gene expression. Cell Rep. 15, 1580–1596.

Cugliandolo, L.F., et al., 1997. Glassy behaviour in disordered systems with nonrelaxational dynamics. Phys. Rev. Lett. 78 (2), 350–353.

Daniels, B.C., et al., 2008. *Sloppiness, robustness, and evolvability in systems biology*. Curr. Opin. Biotechnol. 19, 389–395.

David, L., et al., 2010. Inherited adaptation of genome-rewired cells in response to a challenging environment. HFSP J. 4, 131–141.

David, L., et al., 2013. Multiple genomic changes associated with reorganization of gene regulation and adaptation in yeast. Mol. Biol. Evol. 30, 1514–1526.

Dunker, A.K., et al., 2015. Intrinsically disordered proteins and multicellular organisms. Semin. Cell & Developmental Biol. 37, 44–55.

Dunker, A.K., et al., 2008. The unfoldomics decade: an update on intrinsically disordered proteins. BMC Genomics 9 (2), S1.

Edelman, G.M., 1987. Neural Darwinism: The Theory of Neuronal Group Selection. BasicBooks.

Efroni, I., et al., 2015. Quantification of cell identity from single-cell gene expression profiles. Genome Biol. 16 (1), 9.

Gerhart, J., Kirschner, M., 1997. *Cells, embryos, and evolution*. Blackwell Science Inc.

Gianni, S., Jemth, P., 2016. Protein folding: vexing debates on a fundamental problem. Biophysical Chem. 212, 17–21.

Gilbert, S.F., Epel, D., 2009. Ecological Developmental Biology. Sinauer Associates, Inc, Sunderland, Massachusetts U.S.A.

Graf, T., 2011. Historical origins of transdifferentiation and reprogramming. Cell Stem Cell 9 (6), 504–516.

Gutenkunst, R.N., et al., 2007. Universally sloppy parameter sensitivities in systems biology models. PLoS Computational Biol. 3, 1871–1878.

Hall, B.G., 1992. Selection-induced mutations occur in yeast. Proc. Natl Acad. Sci. 89 (10), 4300–4303.

Hinnebusch, A.G., 1992. General and pathway-specific regulatory mechanisms controlling the synthesis of amino acid biosynthesis enzymes in Saccharomyces cerevisiae. In: Jones, E.W., Pringle, J.R., Broach, J.R. (Eds.), The Molecular and Cellular Biology of the Yeast Saccharomyces Gene Expression. Cold Spring Harbor Laboratory Press.

Huang, S., et al., 2005. Cell fates as high-dimensional attractor states of a complex gene regulatory network. Phys. Rev. Lett. 94, 128701–128704.

Huang, S., Ernberg, I., Kauffman, S., 2009. Cancer attractors: a systems view of tumors from a gene network dynamics and developmental perspective. Semin. Cell & Developmental Biol. 20, 869–876.

Huang, S., 2011. The molecular and mathematical basis of Waddington's epigenetic landscape: a framework for post-Darwinian biology? BioEssays 34 (2), 149–157.

Jablonka, E., Raz, G., 2009. *Transgenerational epigenetic inheritance: prevalence, mechanisms and implications for the study of heredity and evolution*. Q. Rev. Biol. 84, 131–176.

Jablonka, E., Lamb, M.J., 2014. *Evolution in Four Dimensions: Genetic, Epigenetic, Behavioral, and Symbolic Variation in the History of Life*. MIT Press, Cambridge, Mass.

Jia, D., et al., 2017. Phenotypic plasticity and cell fate decisions in cancer: insights from dynamical systems theory. Cancers 9, 70.

Johnston, M., Carlson, M., 1992. Regulation of carbon and phosphate utilization. In: Jones, E.W., Pringle, J.R., Broach, J.R. (Eds.), The Molecular and Cellular Biology of the Yeast Saccaromyces: Gene Expression. Cold Spring Harbor Laboratory Press, p. 193.

Johnston, M., Flick, J.S., Pexton, T., 1994. Multiple mechanisms provide rapid and stringent glucose repression of GAL gene expression Saccharomyces cerevisiae. Mol. Cell. Biol. 14, 3834.

Jordana, D., et al., 2013. Behavioral diversity in microbes and low-dimensional phenotypic spaces. Proc. Natl. Acad. Sci. USA 110, 14018–14023.

Kaneko, K., 2006. *An Introduction to Complex Systems Biology* Understanding Complex Systems. Springer-Verlag, Berlin.

Karplus, M., 1997. The Levinthal paradox: yesterday and today. Sci. Direct 2 (Supplement 1), S69–S75.

Katzir, Y., et al., 2012. Cellular plasticity enables adaptation to unforeseen cell-cycle rewiring challenges. PLoS One 7, e45184.

Katzir, Y., et al., 2013. Dynamics of the cell-cycle network under genome-rewiring perturbations. Phys. Biol. 10, 066001.

Kauffman, S.A., 1993. The Origins of Order; Self-organization and Selection in Evolution. Oxford University Press, New York.

Kurchan, J., Laloux, L., 1996. Phase space geometry and slow dynamics. J. Phys. A: Math. Gen. 29, 1929–1948.

Elgart, M., Snir, O., Soen, Y., 2015. Stress-mediated tuning of developmental robustness and plasticity in flie. Biochimica et. Biophysica Acta (BBA) - Gene Regulatory Mechanisms 1849 (4), 462–466.

Mark, K.T., et al., 2015. *Perspective: sloppiness and emergent theories in physics, biology, and beyond.* J. Chem. Phys. 143 (1), 010901.

Marom, S., et al., 2009. On the precarious path of reverse neuro-engineering. Front. Computational Neurosci. 3 (5).

Marom, S., 2015. *Science, Psychoanalysis, and the Brain: Space for Dialogue.* Cambridge University Press, New York.

Mojtahedi, M., et al., 2016. Cell fate decision as high-dimensional critical state transition. PLoS Biol. 14, e2000640.

Monod, J., 1950. The technique of continuous culture. Ann. Inst. Pasteur 79, 390–410.

Moore, L.S., Stolovicki, E., Braun, E., 2013. Population dynamics of metastable growth-rate phenotypes. PLoS ONE 8, e81671.

Moore, L.S., et al., 2014. Induced mutations in yeast cell populations adapting to an unforeseen challenge. PLoS ONE 8, e111133.

Moss, L., 2003. *What Genes Can't Do.* Basic Bioethics Series. MIT Press, Cambridge, Massachusetts, p. 228.

Origination of organismal form: beyond the gene in developmental and evolutionary biology. In: Muller, G.B., Newman, S.A., Muller, G.B., Wagner, G. P., Callebaut, W. (Eds.), Vienna Series in Theoretical Biology. The MIT Press, Cambridge, MA.

Nanjundiah, V., 2003. In: Muller, G.B., Newman, S.A. (Eds.), *Phenotypic plasticity and evolution by genetic assimilation,* in *Origination of Organismal Form: Beyond the Gene in Developmental and Evolutionary Biology.* The MIT Press, Cambridge MA, pp. 245–263.

Newman, S.A., 2020. Cell differentiation: what have we learned in 50 years? J. Theor. Biol. 485, 110031.

Nijhout, H.F., 1990. Problems and paradigms: metaphors and the role of genes in development. BioEssays 12 (9), 441–446.

Noble, D., 2006. The Music of Life: Biology Beyond Genes. Oxford University Press, Oxford.

Novick, A., Szilard, L., 1950. Experiments with the chemostat on spontaneous mutations of bacteria. Proc. Nat. Acad. Sci. USA 36, 708.

Oldfield, C.J., Dunker, A.K., 2014. Intrinsically disordered proteins and intrinsically disordered protein regions. Annu. Rev. Biochem. 83 (1), 553–584.

Ori, H., Marder, E., Marom, S., 2018. Cellular function given parametric variation in the Hodgkin and Huxley model of excitability. Proc. Natl Acad. Sci. 115 (35)), E8211–E8218.

Posner, R., et al., 2019. Neuronal small RNAs control behavior transgenerationally. Cell 177 (7), 1827–1841.e12.

Sabine, S., Joseph, H.N., 2015. *The genetics of epigenetic inheritance: modes, molecules, and mechanisms.* Q. Rev. Biol. 90 (4), 381–415.

Salman, H., et al., 2012. Universal protein fluctuations in populations of microorganisms. Phys. Rev. Lett. 108 (23), 238105.

Schreier, H.I., Soen, Y., Brenner, N., 2017. Exploratory adaptation in large random networks. Nat. Commun. 8, 14826.

Shahaf, G., Marom, S., 2001. Learning in networks of cortical neurons. J. Neurosci. 21, 8782–8788.

Smith, H.L., Waltman, P., 1995. The Theory of the Chemostat Dynamics of Microbial Competition. Cambridge University Press, New York.

Soen, Y., et al., 2006. Exploring the regulation of human neural precursor cell differentiation using arrays of signaling microenvironments. Mol. Syst. Biol. 2 (1), 37.

Soen, Y., Knafo, M., Elgart, M., 2015. A principle of organization which facilitates broad Lamarckian-like adaptations by improvisation. Biol. Direct 10 (1), 68.

Stein, D.L., Newman, C.M., 2012. Rugged landscapes and timescale distributions in complex systems. Emergence, Complex. Computation 5, 65–79.

Stern, S., et al., 2012. Epigenetically heritable alteration of fly development in response to toxic challenge. Cell Rep. 1, 528–542.

Stern, S., et al., 2007. Genome-wide transcriptional plasticity underlies cellular adaptation to novel challenge. Mol. Syst. Biol. 3, 106.

Stolovicki, E., Braun, E., 2011. Collective dynamics of gene expression in cell populations. PLoS One 6, e20530.

Stolovicki, E., et al., 2006. Synthetic gene recruitment reveals adaptive reprogramming of gene regulation in yeast. Genetics 173, 75–85.

Struhl, K., Davis, R.W., 1981. Transcription of the his3 gene region in Saccharomyces cerevisiae. J. Mol. Biol. 152, 535–552.

Tokuriki, N., Tawfik, D.S., 2009. Protein dynamism and evolvability. Science 324, 203–207.

Trapnell, C., 2015. Defining cell types and states with single-cell genomics. Genome Res. 25, 1491–1498.

Varahan, S., et al., 2019. Metabolic constraints drive self-organization of specialized cell groups. eLife 8, e46735.

Vickaryous, M.K., Hall, B.K., 2006. *Human cell type diversity, evolution, development, and classification with special reference to cells derived from the neural crest.* Biol. Rev. 81, 425–455.

Vygotsky, L.S., 1978. In: Cole, M., et al., (Eds.), *Problems of Method,* in *Mind in Society; the Development of Higher Psychological Processes.* Harvard University Press, Cambridge Mass, pp. 58–79.

Waddington, C.H., 1942. Canalization of development and the inheritance of acquired characters. Nature 150, 563–565.

Waddington, C.H., 1953. Genetic assimilation of an acquired character. Evolution 7, 118–126.

West-Eberhard, M.J., 2003. Developmental Plasticity and Evolution. Oxford University Press, Oxford.

Wilkins, A.S., 2002. The Evolution of Developmental Pathways. Sinauer Associates, Inc, Sunderland.

Woronoff, G., et al., 2019. Metabolic cost of rapid adaptation of single yeast cells. bioRxiv, p. 564419; accepted to the *Proc. Natl Acad. Sci.* (2020).

Yamanaka, S., Blau, H.M., 2010. Nuclear reprogramming to a pluripotent state by three approaches. Nature 465, 704–712.

Yamanaka, S., 2012. *Induced pluripotent stem cells: past, present, and future.* Cell Stem Cell 10, 678–684.

Zwanzig, R., Szabo, A., Bagchi, B., 1992. Levinthal's paradox. Proc. Natl Acad. Sci. 89 (1), 20–22.

13

Emergence of metabolic heterogeneity in cell populations: lessons from budding yeast

Sandeep Krishna[1] and Sunil Laxman[2]
[1]*Simons Centre for the Study of Living Machines, National Centre for Biological Sciences, Tata Institute of Fundamental Research, Bangalore, India*
[2]*Institute for Stem Cell Science and Regenerative Medicine (inStem), Bangalore, India*

Introduction: cell state heterogeneity in isogenic microbial populations

Phenotypic heterogeneity in genetically identical cell populations is ubiquitous in nature (Niklas, 2014). This is true for most unicellular microbes, which typically never live alone. Instead, they live as communities of varying complexity, which can either be single species ("isogenic"), or multi/mixed species groups (Ackermann, 2015). A major area of interest is in understanding how genetic differences enable microbial communities to thrive, and in the consequences, benefits, and disadvantages of phenotypic heterogeneity. In this chapter, we focus on a simpler, but widely prevalent form of heterogeneity, of differences in cellular phenotypes within genetically identical (isogenic) populations of cells raised in the same environment. In particular, we use examples from a prototypical microbe, budding yeast, to describe how certain types of constraints, especially those arising from metabolism, can determine the emergence of reversible, phenotypically heterogeneous cell states in groups of isogenic cells. We examine experimental data (coming from specific contexts) that describe the phenomenon as well as suggest mechanisms for this, and provide theoretical and mathematical explanations for how such phenomena can arise.

The observations that isogenic microbial cells growing in (ostensibly) the same environment can be phenotypically heterogeneous

Phenotypic Switching. DOI: https://doi.org/10.1016/B978-0-12-817996-3.00009-8

are quite old, and come from studies of very different types of microbes (bacteria, eukaryotes like budding yeast, and protists). For spatially organized phenotypic heterogeneity in genetically identical (isogenic) cells raised in the same environment, perhaps the best examples come from Dictyostelid social ameba. These go from individual protists to forming aggregates that become motile multicellular "slugs" and eventually fruiting bodies. Within these slugs and fruiting bodies, cells with tremendous phenotypic heterogeneity are organized in complex spatial patterns (Bonner, 1957; Du et al., 2015; Kaiser, 1986; Saran et al., 1994). Nearly all microbes show some such spatial cellular heterogeneity, as seen in clonal bacterial biofilms, in bacterial swarms, in the individuality within *Escherichia coli*, or very heterogeneous, organized communities in budding yeasts (Ackermann, 2015; Stewart and Franklin, 2008). Phenotypic heterogeneity is also routinely observed even within very well mixed clonal microbial populations, as seen within the lab (Thattai and van Oudenaarden, 2004; Wolf et al., 2005; Tu et al., 2005; Stolovicki and Bran, 2011). This is typically observed in liquid medium batch cultures, when nutrients start depleting (e.g., when cells near stationary phase), but can even be observed in cells growing in the "log" phase.

The phenomenon of phenotypic heterogeneity in isogenic populations has been a subject of considerable interest, in different contexts. One context has been in studies of emergent multicellularity—when isogenic cells must grow and differentiate into phenotypically distinct groups of cells that have to coexist, and spatially be present with distinct shapes and patterns (Niklas, 2014). Another context has been to study microbial populations in fluctuating environments, where heterogeneity provides adaptive benefits to the single-species community (Stewart and Franklin, 2008; Thattai and van Oudenaarden, 2004). A third context is that of long-term adaptation (and evolution experiments), where the emphasis has been on identifying genetic changes that allow heterogeneous populations to emerge and persist or coexist.

Most studies attempt to address the reasons for the existence of heterogeneity, and therefore focus on genetic and epigenetic changes that can maintain heterogeneity within the population (Stewart and Franklin, 2008; Niklas, 2014). A particularly powerful idea is one of stochastic gene expression (Paldi, 2012; Laforge et al., 2005). If regulators present in small numbers control the expression of a large number of genes in a cell, it is inevitable that there will be variations in the expression levels of these different genes. This will therefore enable phenotypic heterogeneity within a large enough population. This idea works for both well-mixed populations of cells, and spatially restricted populations. Another complementary idea comes from the spontaneous origins of stress-induced heterogeneity, observed for example using budding yeast as a model cellular system (Stolovicki and Braun, 2011; David et al., 2013). A second area of emphasis is to understand how cells stay together: here cells must produce adhesion molecules that hold them together, or else some codependencies (such as commensal or mutual dependencies on shared resources) must exist within the populations. These provide another perspective of how phenotypically heterogeneous cells can stay together, but not on how such heterogeneity emerges.

In most of these examples of phenotypic heterogeneity in clonal cell populations (irrespective of whether they are well-mixed or spatially restrained systems), there is a consistent underlying theme. The theme is that heterogeneity increases when there is some kind of environmental stress, most typically nutrient depletion/metabolic stress, in the system. All these studies therefore suggest that constraints, coming from metabolism, must play a key role in the emergence and maintenance of phenotypic heterogeneity within clonal populations. In this chapter, we put together some general principles that illustrate how phenotypic heterogeneity can emerge and sustain in clonal cell populations based on metabolic constraints, using extensive experimental and theoretical studies from budding yeast. The principles point to sufficient conditions, which may not be always necessary (there will be exceptions). However, by formulating them, our intention is to conceptualize one way by which phenotypic heterogeneity can emerge and sustain.

Metabolic heterogeneity and spatial organization within yeast colonies

As such, our ability to investigate different microbial communities using integrative approaches that combine high-end microscopy, genome and gene expression analysis, proteomics, and also analyze metabolites have now revealed incredible phenotypic heterogeneity and organization within microbial colonies and communities. These include studies with gram positive and gram negative bacteria, a range of "unicellular" fungi, and salt and freshwater algae (Ackermann, 2015). Images using electron or light microscopes all show remarkable architectures of cells within these colonies. Using engineered fluorescent reporters of gene expression, and visually observing cells within a colony using fluorescence microscopy, we can clearly observe that different cells express different amounts of these genes, and in several examples, groups of cells within such colonies have entirely opposite trends of gene expression. Strikingly, beyond just differences in gene expression, there also appears to be clear spatial organization of cells within a community. This has led to many investigations of possible division of labor within cell groups, specialization of function, and the advantages of heterogeneity (Hamant et al., 2019; Aktipis et al., 2015; Cerulus et al., 2016; Wolf et al., 2005; Ackermann, 2015).

Three broad ideas seem to drive such organized phenotypic heterogeneity within clonal microbial colonies. These are: stochastic gene expression and internal oscillatory transcription systems (Thattai and van Oudenaarden, 2004; Stewart and Franklin, 2008; Raj and van Oudenaarden, 2008), cellular age and differential partitioning of cellular material during cytokinesis (Ackermann, 2015; Huh and Paulsson, 2011), and the production and sharing of resource or signaling molecules (Ackermann, 2015; D'Souza et al., 2018). A classic example of the effect of stochastic gene expression has been transcriptional noise within and between cells involving the lactose promoter of *E. coli*. In the simplest experiments, a cyan fluorescent protein and a yellow fluorescent protein were fused to the lactose promoter, and both integrated into the chromosome of *E. coli*. The simple assumption now is that all cells will express the same amounts of both yellow and cyan fluorescent protein. Remarkably, tremendous heterogeneity was observed, both at the level of individual cells expressing yellow or cyan, and across the population (Elowitz et al., 2002). This study powerfully demonstrates how amounts of controlling molecules (in this case transcription factors regulating the activity of the promoter) could dramatically alter amounts of gene expression. As the numbers of cells increase, and correspondingly the heterogeneity of gene expression within the population also increases, groups of cells start to stochastically exhibit similar profiles of expression, resulting in patterns or oscillations of expression across the colony. The second idea of heterogeneity comes from what happens to many microbial cells when they divide. Asymmetric division is very common amongst many cells. Here, cells can easily be grouped into mother cells and daughter cells, and daughter cells tend to inherit the newest, most active proteins (Levy et al., 2012; Bergmiller and Ackermann, 2011; Lindner et al., 2008). This means that eventually, the functional properties of cells will differ (depending on the age of the mother and daughter cells).

The idea of a produced resource that acts as a signaling molecule to control cell states (and therefore heterogeneity) has been well studied in the context of bacterial quorum sensing. Quorum sensing (as studied in bacteria) is a way by which bacterial cells determine cell density. Here, cells produce and secrete a signaling molecule, which will bind to receptors in cells, and this will induce a signaling cascade leading to differential gene expression. In such systems, for this signaling molecule to be sensed, much relies on diffusion (and the ability to concentrate the molecule). Second, differential expression levels of the receptor that induces the transcriptional response will automatically result in a range of responses in the cells. Collectively, this can lead to phenotypic heterogeneity, depending on the location of the cell, the relative amount of the signaling molecule, and the relative expression of the receptor/inducer (Miller and Bassler, 2001).

However, an even simpler, powerful variation of this idea can control cellular state outcomes. If cells that are growing in a particular nutrient environment end up producing certain metabolites, and if these metabolites can be differentially used by some other cells, this can lead to new metabolic (and therefore phenotypic) states. These cells may even go on to spatially organize within a colony into different states or groups. Essentially, the idea reduces to that of an existing metabolic state (expected in cells present in a given nutrient environment) allowing the creation of new environments by producing a usable resource/metabolite, which then allows new phenotypic states to emerge. This means that *threshold amounts of a controlling resource*, when reached, can enable cells to switch their phenotypic states. A striking, illustrative example of this comes from again from budding yeast (Varahan et al., 2019), and we will use this example to fully explain this idea.

The idea of a threshold, controlling resource

Budding yeast (or baker's yeast) love to grow on high glucose. However, as long as there are other nutrients (plenty of amino acids, or ethanol, or glycerol and so on), they are perfectly capable of growing in the absence of glucose. In typical lab experiments, they are grown in high glucose medium, and in these conditions they form clonal colonies that look smooth, and largely uniform. However, when they are grown on solid medium without high amounts of glucose, these same yeast form beautifully structured colonies (Fig. 13.1). This of course is only the external appearance. However, what happens inside is even more remarkable. In low glucose conditions, all cells typically must carry out a metabolic process called gluconeogenesis. This is a process where other carbon sources (from amino acids, or ethanol etc.) are channeled to make glucose and storage carbohydrates (like trehalose). In yeast cells seeded to form a colony in low glucose, as expected most cells initially carried out high rates

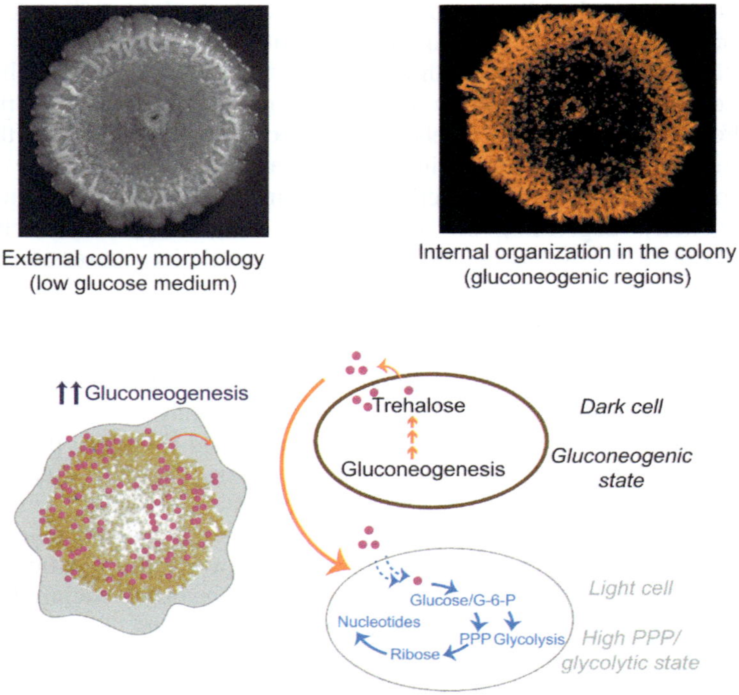

Figure 13.1 Spatially organized metabolically heterogeneous cells in a yeast colony [as shown by (Varahan et al., 2019)].

of gluconeogenesis. Remarkably, after the colony develops for several days, suddenly new groups of cells emerge that have almost no detectable gluconeogenesis. Instead, their metabolic state resembles those of cells growing in very high glucose (with high flux through the glycolysis and the pentose phosphate pathways). This is instantly a paradox: how can cells growing in environments where there is low glucose show these properties? This can only happen if, suddenly, glucose becomes available for these cells. Indeed, something similar happens. As yeast cells in the colony do more and more gluconeogenesis, large amounts of a gluconeogenic end-product, the disaccharide trehalose, is produced and secreted into the medium. As amounts of trehalose increase above a threshold, some cells spontaneously (stochastically) start taking up this trehalose, and break it down into two molecules of glucose. This means now they switch to this new metabolic and phenotypic state. Of course, when they do this, the amounts of trehalose outside start to decrease (below the threshold), so the remaining cells remain trapped in the original gluconeogenic state, while these cells can continue in the new phenotypic state. This creates a self-organizing system of cells in different states, arranged spatially with distinct patterns (Varahan et al., 2019). This idea is illustrated in Fig. 13.1. Collectively, we see here how producing a resource in a given condition can enable *new* states to emerge. In the next section, we will explore theoretical considerations with which we can model and predict the manifestation of multiple states in a population of genetically identical cells.

1. On the left is a photograph of a yeast colony, with millions of cells growing under low glucose conditions on solid medium. Note the complex architecture of the colony. In high glucose, similar colonies are round and smooth, with no apparent ridges. On the right is a similar colony in low glucose, imaged using a fluorescence microscope. This colony is expressing a genetically encoded fluorescent reporter for gluconeogenic activity. As is apparent, gluconeogenic activity is spatially restricted within regions of the colony, while cells in other regions, especially the external circumference, have no detectable gluconeogenic activity.
2. A simple model, illustrating how threshold amounts of a resource can control phenotypic heterogeneity. In a colony, initially all cells are gluconeogenic, and produce trehalose as an end product. As trehalose accumulates in the medium to high enough amounts, some cells take up the trehalose and metabolize it (as illustrated in the schematic on the right), switching their metabolic (and phenotypic) state. The system self-organizes into this patterned architecture, with regions of cells exhibiting contrary phenotypes.

Lesson: *when certain metabolites accumulate to threshold amounts, they can be utilized differently by some cells, and therefore will enable these cells to switch phenotypes.*

General considerations on the manifestation of multiple cellular states at the population level

We have seen above that budding yeast cells are capable of being in two different metabolic states and, in spatially restricted colonies growing (on solid medium) in the absence of glucose, both kinds of cells coexist and indeed interact with each other. Some of the questions we would like to answer in more generality are:

1. If single cells can be in multiple states, when will the population be in a "mixed" state exhibiting subpopulations in

different states, and when will it be in a "pure" state consisting of cells of only one type?

2. Is it possible for there to exist many different "mixed" states for the same parameter values, which differ in the proportion of cells of different types?

To shed light on such questions we turn to some simple theoretical considerations. Imagine a well-mixed population of isogenic cells, each of which can be in one of two phenotypic states A and B. The composition of the population in this case is described entirely by one number, n, the fraction of cells in the A state (we will ignore the total size of the population here—imagine the system is in a well-stirred chemostat at constant volume). Cells of each type can divide, at rates γ_A and γ_B, respectively, and cells can switch from one state to another, at rates ν_{AB} and ν_{BA}. All of these rates may depend in complex ways on the composition, through a variety of mechanisms. For instance, cells of one or both types might produce a quorum-sensing signal that enhances or inhibits any of these rates. Alternatively, if cells in different states consume some resource at different rates then the growth rate could depend on the concentration of this resource, which would in turn depend on the composition. Thus, in general, all of these rates could depend on n, and in turn the dynamics of n over time will depend on these rates.

Fig. 13.2(A) shows schematically how one can write a simple population dynamics equation to describe how cell division and cell state-switching affect the composition of such a well-mixed population of cells over time. In the Appendix we describe in more detail the derivation of this equation and the meaning of each of the terms. The equation looks simple, but has hidden complexity in the way the rates depend on n, as described earlier. However, even if the rates depend on n in complex ways it is easy to understand geometrically how many and what kind of stable states we may find the population in. Fig. 13.2(B) shows how to do this by drawing the right-hand side of Eq. (1) in Fig. 13.2(A) as a function of n. Wherever this curve touches or crosses the horizontal axis, that value of the composition corresponds to a steady state for the population. However, not all of these states are stable, that is, even a small perturbation may cause the composition to "run away" to a very different value. The arrows on the horizontal axis show the directions in which the composition will move under this dynamics for various starting values of n. Stable steady states are composition values where the curve crosses zero from *above* – at these points all the arrows

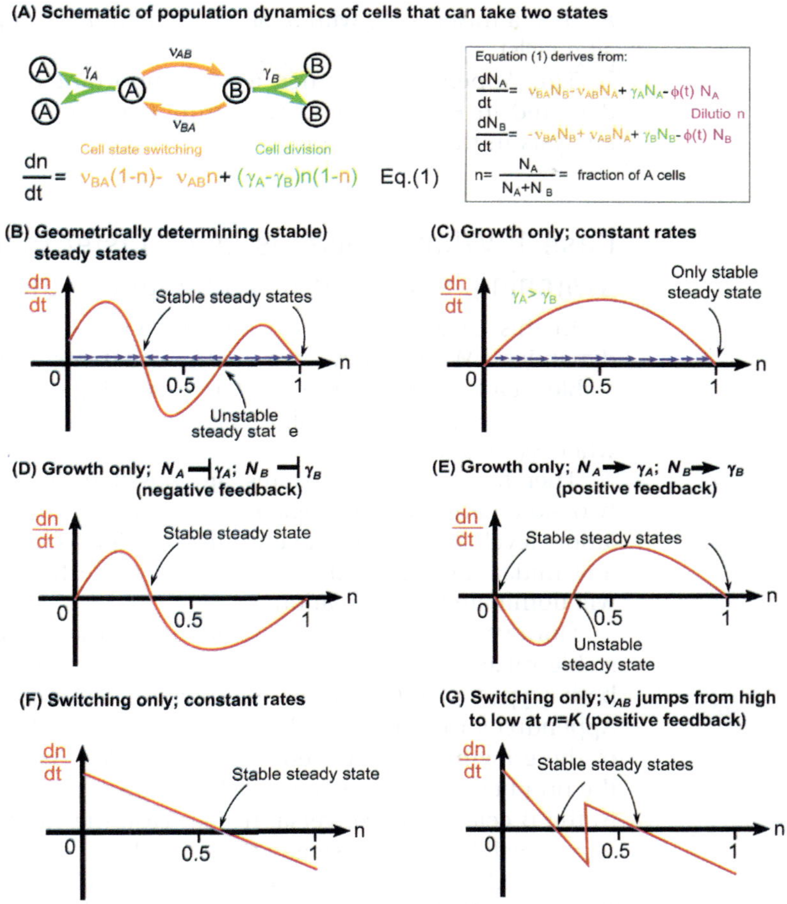

Figure 13.2 (A) The interplay between growth and state switching in a well-mixed population of cells, each of which can be in two distinct states A and B. The equation shows how the dynamics of the population composition, n, the fraction of A cells, depends on the switching and growth rates. Each of these rates may depend in complex ways on n. (B) How to determine the stable steady states for the population composition. Steady states occur at the values of n where the curve depicting the right-hand side of the equation touches or crosses the horizontal axis, that is, where $dn/dt = 0$. Of these, the ones where the curve crosses the axis from above are stable. This can be seen from the arrows on the horizontal axis, which show the direction in which n will move under this dynamics. Arrows point to the right where $dn/dt > 0$, and to the left where $dn/dt < 0$. The length of an arrow (roughly) indicates the magnitude of dn/dt at that point. Thus, if the arrows point toward a steady state from all directions then it is stable to small perturbations. (C–G) An exploration of different cases (see text, and Appendix) where growth or switching are dominant, and where there exist negative or positive feedback loops connecting the growth and switching rates to the composition of the population.

point toward the steady state, which means that the population will come back to this steady state even if perturbed by a small amount. The particular curve shown has stable steady states at

$n = 0.3$ (a mixed state with 30% A cells) and $n = 1$ (a pure state with all A cells).

The Appendix examines a few important cases, which are depicted in Fig. 13.2, and we summarize the lessons from this analysis in the following text:

Case 1: When growth is much faster than switching, such that one can ignore the latter

In this case only the two growth rates, γ_A and γ_B, affect the dynamics. When they are both constants, then the only stable steady state is the pure state consisting of all B cells ($n = 0$) when $\gamma_A < \gamma_B$, and $n = 1$ (all A cells) is the only stable steady state when $\gamma_A > \gamma_B$ (see Fig. 13.2C).

Therefore, only pure states are possible and the underlying two-state nature of the single cells does *not* manifest at the population level. This is intuitively quite obvious—in the absence of any mitigating effects the cell type with the highest growth rate will dominate the population.

However, if γ_A and γ_B depend on n, then it is possible for stable mixed states to exist where the population exhibits cells of both types. Under some reasonable assumptions (see the Appendix) a mixed state will exist and be stable if the dependence of the growth rates on n encodes a *negative feedback loop*: when the number of A cells increases, their growth rate *decreases* relative to the B cells, and vice versa. In that case the value of n at which $\gamma_A = \gamma_B$ will be a stable steady state (see Fig. 13.2D). This is similar to the idea of "frequency-dependent selection" in ecology (Lewontin, 1958), wherein the net growth rate of an allele or species (here γ_A-γ_B) depends on the frequency of occurrence of that allele or species (here n).

If the frequency dependence was the opposite, a *positive feedback loop*—where the relative growth rate of A cells *increases* as the number of A cells increases, and vice versa, then only the pure states $n = 0$ and $n = 1$ will be stable (see Fig. 13.2E).

A particularly interesting subcase in the cellular context is where type A cells are proliferating, while type B cells are quiescent (i.e., $\gamma_B = 0$). In this case, it is *not* possible for there to be a stable mixed state (see Appendix).

Lessons:

Growth differences alone, in the absence of switching between states, only allow the two-state nature of cells to manifest at the population level when the composition exerts negative feedback on the growth rates.

Case 2: When switching is much faster than growth, such that one can ignore the latter

When the switching rates are constant, in the absence of growth, there will always be one stable steady state at an intermediate value of $n = \nu_{BA}/(\nu_{AB} + \nu_{BA})$ (see Fig. 13.2F). Any individual cell in this state will keep switching between types stochastically. Note that this requires there to be switching both ways. If either state is irreversible (i.e., either ν_{AB} or $\nu_{BA} = 0$) then a stable mixed population is *not* possible.

If the switching rates depend on n, for example due to some quorum-sensing signal, then more complex behavior is possible. If the dependence on *n* results in a negative feedback loop (e.g., if ν_{BA} is a constant and ν_{AB} increases with *n*), then typically there will be a single stable mixed state at the population level. For certain types of negative feedback, the value of n at this steady state will be determined by aspects of the negative feedback rather than the switching rates themselves. If this is the case, then, unlike the mixed state obtained when rates are constant, the stable mixed state observed can be very robust to changes in switching rates. The Appendix explores this scenario for the case where the switching rates are sharp, switch-like functions of *n*, inspired by the yeast case, which suggested a metabolic threshold that triggered state switching.

When the switching rate dependence on *n* creates a positive feedback loop (e.g., if ν_{BA} is a constant and ν_{AB} decreases with *n*), one can obtain a situation as in Fig. 13.2G, where there are *two* stable mixed states possible—a bistability at the population level. Here there are two different bistabilities, one at the single cell level, which allows individual cells to be in two distinct states, and an additional one at the population level, which allows the population to exist in two distinct mixed states. This can in fact form the basis for even more complex behavior as we will discuss in the next section.

Lesson: *Switching between states at a constant rate allows the two states to co exist at the population level. If the switching rate depends on* n, *that is, there is some communication between cells, then one can get more complex coexistence (e.g., an additional bistability at the population level, beyond the bistability at the single cell level). This can in turn lead to more complex behaviors such as the yeast metabolic cycle discussed in the Section "Cell-state heterogeneity of yeast in a well-mixed chemostat". Positive feedback is necessary for this possibility.*

Case 3: Both switching and growth have to be taken into account

This is a complex case, where many things are possible. However, an interesting subcase is where one of the states (say, B) is irreversible, that is, $\nu_{BA} = 0$. In this case, even if growth rates are much smaller than switching rates, it is incorrect to ignore them in the equation. Here, a stable mixed state is possible where cells switch irreversibly from A to B, and this is balanced by the faster growth of A cells.

Lesson: *Even if growth is much slower than switching, it may nevertheless play an important role in determining the population state, when some cell states are irreversible.*

Here, we have only discussed some simple cases where individual cells are already capable of being in two distinct stable states, and the question is whether this would manifest at the population level. It is also possible for two states to be seen at a population level even when individual cells are incapable of exhibiting two stable states on their own. This is only possible, of course, if cells in a population interact. Then individual cells can stabilize into two (or more) distinct states purely because of their interactions with other cells, and not because of an individual ability to exhibit two states. This phenomenon has been theoretically suggested for some chemical systems (Kamimura et al., 2019; Yamagishi et al., 2016), but it is unclear how common such a phenomenon is in biology. In many biological scenarios, such as the yeast colonies discussed above, or cell differentiation in multicellular organisms, the cells are individually capable of being in two, or multiple, stable states and this is the class of scenarios we discuss in this paper.

Cell-state heterogeneity of yeast in a well-mixed chemostat

In the previous section, we saw how a positive feedback loop could result in a bistability at the population level. Note that this is distinct from the bistability at the single-cell level, which allows individual cells to be in two states A and B. A population level bistability means that the *entire* population can stably exist in two states with different composition. Lest readers think this is a peculiarity of only academic interest, we will describe in this section how such a population level bistability allows complex behavior that is in fact observed in yeast growing in well-mixed, continuous cultures maintained in chemostats.

The complex behavior we will refer to extensively in this chapter is the well-known yeast metabolic cycle, or metabolic oscillations, where, under conditions of low glucose, yeast cells exhibit incredibly robust oscillations between two distinct cell states over days and weeks (Murray et al., 2007; Laxman and Tu, 2010; Tu et al., 2005). The first observations of oscillatory behavior in yeast related to changes in glycolysis or oxygen consumption (Chance et al., 1964; Hommes, 1964; Hess and Boiteux, 1971). Many continuing studies of these oscillations suggest that they might involve some tradeoff between robustness and efficiency (Chandra et al., 2011). In a well-mixed environment, ostensibly all cells have identical opportunities to use any resources available. At an overly simple level of understanding, all cells must therefore behave identically. However, we know this is not true, and indeed in any well mixed system of microbial cells, we always observe some heterogeneity in the ability of cells to grow and divide, or counter different environmental stresses, and so on. So is the presence of cells in phenotypically heterogeneous states purely a stochastic event, with no identifiable cause, or can there be a more systematic investigation of this? In addition, what can we learn about this from the aforementioned yeast metabolic cycles? What, indeed, are they? This leads us to a more gentle introduction to the phenomenon of the yeast metabolic cycles, and an outline of the apparently irreconcilable data observed in these cycles.

The observations that large, well-mixed populations of yeast exhibit robust oscillatory behavior were made years ago, soon after chemostats were invented for industrial scale fermentation (Chance et al., 1964; Hommes, 1964; Hess and Boiteux, 1971). In chemostats, as high density populations of yeast cells were provided continuous but restricted nutrients (usually glucose or ethanol), the entire population apparently synchronized and went into oscillations of oxygen consumption. Many studies left these oscillations as just that, and mathematically modeled just the oscillations in oxygen consumption. But the more recent studies of these oscillations examined actual cell states during the course of these cycles, and also used global gene expression or metabolite analysis of bulk populations of these cells, isolated from different points in these cycles (Murray et al., 2007; Tu et al., 2005, 2007). Two observations stood out. First, it appeared that in these oscillations, there was some predictable, collective behavior of cells. For long periods of each cycle, cells would be in a nondividing (or apparently quiescent) state. However, for brief periods in each cycle, a fraction of cells (which was fairly constant in each cycle, depending on the amount of nutrients fed) would enter into growth and cell division (i.e., complete the cell division cycle). So this was a clear case of cells in a two-state system (growth and nongrowth/quiescence), and an oscillation between the two states. Second, even with unsorted, bulk populations of cells that were isolated from different time points in these cycles, there was a striking pattern of gene expression. A majority of the yeast genome showed periodic gene expression, with the same groups of genes peaking (or having highest expression) at different time points in the oscillations. When organized by function, they sorted into groups that were related to growth and cell division, exactly when a fraction of cells switched to growth, and also for quiescence and survival, during the long phase of nondivision (Tu et al., 2005). This was also reflected when dozens of central metabolites were similarly analyzed from samples across the cycle, and the oscillations could be grouped into distinct metabolic states (Tu et al., 2007). This is shown in Fig. 13.3.

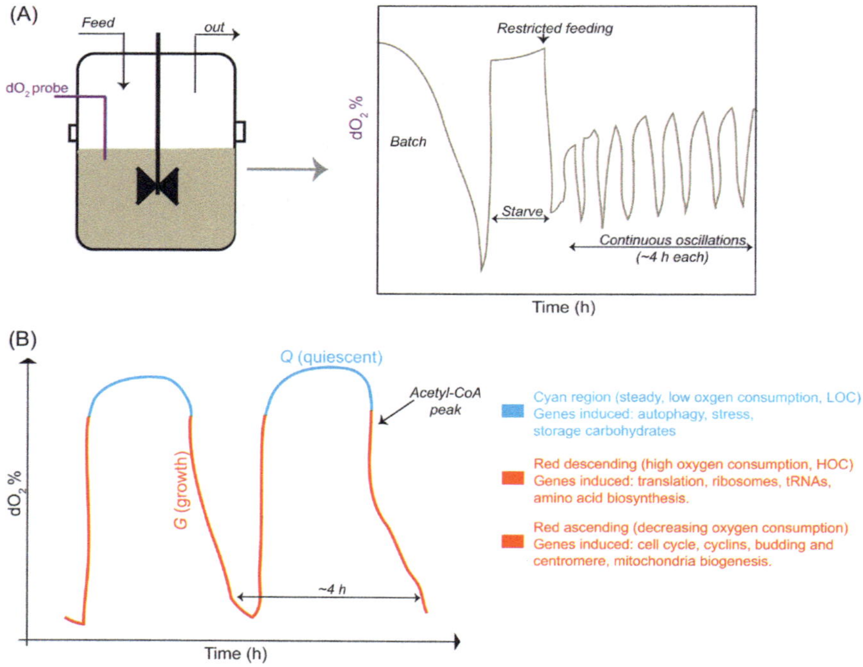

Figure 13.3 (A) Shows an example of this yeast metabolic cycle—the oscillations in dissolved oxygen indicate cyclic rounds of growth and division every 4 h, interspersed by periods of quiescence. (B) Analysis of gene expression and imaging data over the course of these oscillations suggests that in the low oxygen consumption (LOC) phase of the oscillation almost all the cells are in a quiescent state, while in the high oxygen consumption (HOC) phase about 30%–40% of cells switch to a proliferating state.

As a well-mixed group of isogenic cells, this provides a remarkable, experimentally tractable system with which very basic ideas of phenotypic heterogeneity can be probed. In particular, we can try to build a theoretical model that can explain this oscillatory behavior of cells between two very different states (a quiescent or nondividing state, and a growing/dividing state) (Krishna and Laxman, 2018). Using the framework of the previous section, we will think of the population as consisting of cells that can take two states (quiescent, *Q*, or growing/dividing, *G*). The composition of the population is described by the fraction of quiescent cells, q. In this case, quiescent cells do not grow, so the equation describing the dynamics is a variant of Eq. (1) from the previous section:

$$dq/dt = \nu_{GQ}(1-n) - \nu_{QG}n - \gamma n(1-n), \qquad (13.2)$$

where γ is the division rate of the proliferating cells and ν_{GQ} and ν_{QG} are the switching rates from the quiescent (*Q*) state to the growing (*G*) state, and vice versa. With just this equation, it is impossible to get oscillations, even if the switching rates and the growth rate are functions of q. However, we have shown that it is possible to obtain oscillations if this system has a population bistability (as in Case 2 in the previous Section), and then additional processes cause destabilization of those two stable states. Box 13.1 describes how this theoretical hypothesis of obtaining oscillations using such "frustrated bistability" works. In the yeast chemostat this works as follows: Let us assume that the switching rates and growth rate depend on q in such a way as to allow a population level bistability as in Case 2 in the previous Section, where one mixed state consists of almost all quiescent cells, while the other mixed state has say 30%–40% proliferating cells. Now we assume that in addition to dependence on q, the rates would also depend on the concentration of some metabolic resource that is necessary for cell proliferation and is consumed by cells when they divide. When (almost) all cells are in the quiescent *Q* state in the LOC phase, this resource that is supplied into the chemostat at a constant rate, will build up. If this accumulation increases the rate at which cells switch to the *G* state, then this population mixed state is eventually destabilized causing the population to jump to the other mixed state with the different composition consisting of many more proliferating cells. However, this second mixed state is also eventually unstable in the long run, because the proliferating cells will consume the resource rapidly thereby reducing the switching of cells from the *Q* to *G* state. The population will then jump back to the state where almost all cells are quiescent and the cycle starts again.

Box 13.1 Frustrated bistability.

It is relatively easy (theoretically) to convert a bistable system into an oscillator by adding a few judiciously chosen interactions (Krishna et al., 2009). Typically, a bistable system arises from a positive feedback loop (Thomas, 1998). When a positive feedback loop is present, for example, in gene regulatory or metabolic networks, any perturbation in a node of the network within that loop affects a series of other nodes and then returns to *reinforce* the perturbation in the first node. In contrast, in a negative feedback loop, the perturbation would return with the opposite sign to negate the original perturbation. By amplifying perturbations in gene expression or chemical concentration, positive feedback loops can push systems into two distinct states (Sneppen et al., 2010). Such systems are also usually bistable only for a certain range of a controlling parameter. For example, a transcription factor that enhances its own production will exhibit the following behavior: for small values of the promoter strength, the protein will have only one stable steady state where it has a low concentration; as one increases the promoter strength, this low concentration increases slowly [see the lower branch in Fig. 13.4(A)], but at some critical value the positive feedback loop is triggered and the concentration jumps to a high value [see the upper branch in Fig. 13.4(A)]. Interestingly, when the promoter strength is now decreased, no jump is seen when passing this critical value—the concentration of the transcription factor decreases smoothly along the upper branch of Fig. 13.4(A). However, there is a second lower critical value of the promoter strength at which a sharp jump does occur to the lower branch. This phenomenon is called hysteresis, and indicates that for promoter strengths in a certain range (between the two critical values), the system is bistable, that is, it can exist in two distinct stable steady states. Which steady state the system reaches depends on where it starts.

The way to obtain oscillations from a system like this is to add a negative feedback loop that makes the controlling parameter decrease when the system is on the upper branch and increase when it is on the lower branch. For the self-activating protein we have been describing, imagine that the promoter is also controlled by a second transcription factor that inhibits the promoter. If this second transcription factor is produced from a gene that is activated by the original self-activating protein, then we have such a negative feedback loop: when A increases, B's production is enhanced, but this in turn decreases production of A (see Fig. 13.4B). Let us assume that the dynamics of this second protein proceeds *much slower* than the dynamics of the original self-activating protein. Then it is easy to see how oscillation may emerge. Imagine the system starts with high A and low B, with A sitting somewhere on the upper branch. As A is high, B will start to increase, and the system will slowly creep along the upper branch toward larger values of B. Eventually, it will "fall off" the edge and A will jump to a low value on the lower branch. At this point, since A is low, B will start to decrease. This will cause the system to slowly creep up the lower branch until it again falls off the edge and the cycle is established. We call this "frustrated bistability" (Krishna et al., 2009) because the negative feedback loop essentially destabilizes both stable states making the system endlessly jump from one to the other.

In the yeast context we take this idea from a single self-activating protein to the population level. Here, the bistability is not in the concentration of some protein in a single cell, but in the composition of the entire population (i.e., the fraction of quiescent cells, q). Just as the self-activating protein showed two stable states for a certain range of a controlling parameter, let us assume that the yeast population can exhibit two stable composition states for a certain range of a controlling parameter. If the controlling parameter is some metabolic resource that is

(*Continued*)

Box 13.1 (Continued)

supplied at a constant rate but consumed rapidly by dividing cells, then it is easy to see that a negative feedback loop is established: When q is high, cells are quiescent and the resource accumulates, but when cells start dividing, the resource is consumed and rapidly decreases. If we further assume that cell-state switching occurs on a faster timescale than resource consumption and accumulation then the frustrated bistability scenario described above can be achieved. The oscillation will proceed as follows: the population starts with mostly quiescent cells and a low level of the resource; since there is little consumption of the resource it increases and the population creeps along the upper branch of the underlying bistable system where q values are high; at a critical level of resource cells start switching rapidly to the dividing state and q drops sharply; now that q is much lower, the resource starts decreasing because it is being consumed and the population creeps along the lower branch; finally once the resource passes the second low critical level, most cells switch to being quiescent again and we are back to the beginning of the oscillation. Notice, how such an oscillation naturally produces a waveform that has two relatively flat regions separated by sharp transitions, just like the observed yeast metabolic cycle.

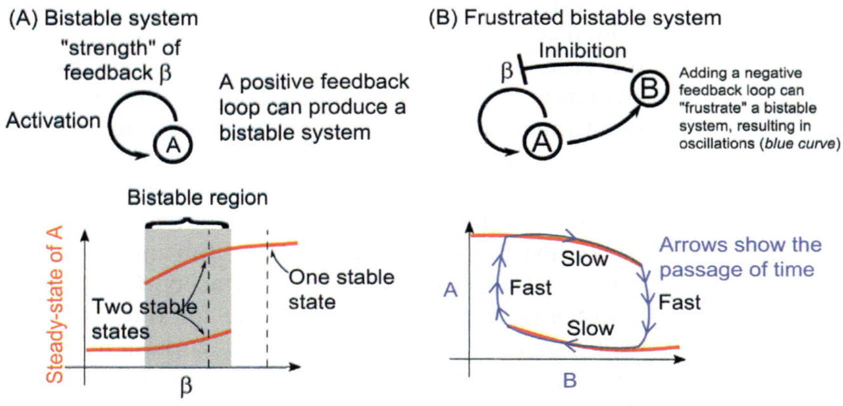

Figure 13.4 An illustration of "frustrated bistability" that can be used to generate oscillations by modifying an underlying bistable system. (A) Characteristics of a bistable system consisting of a simple positive feedback loop. (B) Such a system can be made into an oscillator by adding a negative feedback loop.

Several strands of circumstantial evidence support such a picture of the yeast metabolic cycle. First, the shape of the oscillation is characteristic of such frustrated bistable oscillations—there are two periods of relatively slow change (the LOC and HOC phases) punctuated by very rapid transitions between them. This exactly corresponds to the relatively slow changes in the model oscillations during the periods of time when the resource is accumulating or being consumed, punctuated by rapid changes in the composition of the population when it jumps from one bistable branch to the other. Second, within this overall pattern of slow changes and fast transitions, different strains of yeast exhibit slightly different shapes of the oscillation waveform—the model is able to explain these different shapes by differences in the shapes of the bistable branches caused by variations in the switching rates. Third, as the supply rate of the resource is increased the time period of the oscillation increases (more precisely, the total time period increases due to more time spent in the LOC phase, while the time spent in the HOC phase decreases a little), which is also what the model predicts. Fourth, adding glucose or ethanol during the LOC phase resets the oscillation but adding them during the LOC phase does not, which is also explained by the model. Finally, there exist metabolites, such as acetyl-coA, whose dynamics across the yeast metabolic cycle exactly match the predicted dynamics of the resource in the model, in particular that the resource peaks at or just before the transition to the HOC phase.

Thus far, we have shown how the interplay of growth and cell state switching can produce a number of interesting and complex phenomena at the population level. What kind of biologically useful insights can one derive from such considerations? In the case of the yeast, an important inference is that cell state switching is ultimately controlled by a metabolic trigger. For the yeast metabolic cycle, comparing the dynamics of various metabolites with the expected dynamics of the resource in our model, we find that Acetyl-CoA is the most likely trigger (Krishna and Laxman, 2018). This makes sense not only because acetyl-CoA peaks just before the HOC phase, but also because of the very special place in metabolism that acetyl-CoA occupies, and the compelling experimental evidence for these roles. The lines of experimental evidence that are consistent with acetyl-CoA amounts within a cell as the trigger resource to switch to growth starts with the following observation: for reentry of quiescent cells into the growth state, many studies show that storage sugars (primarily trehalose) have to be rapidly consumed to produce glucose, or a supplement of acetate, acetaldehyde, or ethanol can do something similar. Notably, all these carbon sources will be metabolized to produce acetyl-CoA. Once cells have enough acetyl-CoA, they can do several things. First, acetyl-CoA is the primary donor for acetyl groups, which can modify (acetylate) histones. Several studies now show that at high enough concentrations of acetyl-CoA, histones (placed around "growth related" genes) are acetylated, allowing genes required for growth to be expressed. Next, acetyl-CoA can fuel the many needs of a growing cell. It is essential for the energy (or ATP) producing tri-carboxylic acid cycle (TCA cycle), which will also fuel DNA replication (required for cell division). This ATP will also drive the ribosome to allow sufficient protein synthesis, required for a growing cell. Acetyl-CoA is also converted to the many fatty acids required to build new cell membranes. So, starting from predictions made by this mathematical model, we can (with some degree of confidence) determine what a controlling metabolic resource might be (Krishna and Laxman, 2018). Note again, how a threshold amount of a resource can control cell states, and therefore phenotypic heterogeneity in a large cell population.

Lesson: *Acetyl-CoA is a good example of a critical "threshold" metabolite that can control different phenotypic outcomes in a cell, depending upon how much of it accumulates.*

Discussion

We used yeast as a particular example and inspiration, and we have discussed the ways in cell division and cell state switching interplay in the manifestation (or not!) of multiple cell states at the population level. However, what we have described are fairly universal rules at play, and many of these rules will be conserved regardless of the cellular system of interest. In Fig. 13.1 and related text, we had discussed how such rules play out in a growing colony, where cells organize to form distinct structures across space.

In all of this, one might ask, while there are metabolic triggers for cell-state switching, is metabolism the top-level of control? This will require deeper investigation, but *prima facie* evidence suggests that this might be true, as explained earlier with the specific yeast examples. This is interesting to speculate about, since metabolism is one of the most primordial factors driving cell state. Extrapolating from this point, it is quite plausible that the emergence of metabolic heterogeneity is a first step in one path to multicellularity. In the example from yeast colonies discussed, there is a clear mutualism (or at least commensalism), division of labor and functional specialization, all of which are thought to be prerequisites for emergent multicellularity. In addition, for such a scenario, where metabolic heterogeneity emerges as a first step on the path to multicellularity, such heterogeneity must confer some advantage to the population. We find in fact that wild-type colonies of yeast expand much more over extended time frames (many days) than mutants which cannot break-down or take up trehalose, and therefore do not exhibit two metabolic states (Varahan et al., 2019). However, there must be some cost associated with converting the provided amino acids to oxaloacetate and then trehalose, secreting this out, for the light cells to take in and then process back down the metabolic pathway to convert trehalose to useable metabolites (instead of directly making these metabolites). Therefore, the important comparison would be between the wild-type colonies and a hypothetical single-state yeast that directly utilizes amino acids for growth.

Why would a two-state population that incurs the metabolic cost described grow better than a hypothetical single-state yeast that avoids this cost? We can speculate that cells in one state might form a transport network, which effectively allows food that is present deep inside the colony, and is unutilizable there, to be transported to the periphery of the colony where it can be utilized by the light cells. In other words, perhaps the benefit of redistributing food from locations where growth is no longer possible to locations where it is possible may overcome the metabolic cost. This idea is not very different from the well-known idea of reaction-diffusion patterns, first described by Turing (Turing, 1952), and subsequently expanded to include parameters such as cell/tissue growth (Gierer and Meinhardt, 1972; Gierer, 1981). Clearly, this idea of efficiencies could impact the development of even more complex systems, as a first step or prerequisite for multicellularity. This would also tie in neatly with the fact that such emergence of two metabolic states occurs only when the colony is growing on a low quality food source—in the presence of plentiful glucose the colony remains uniform. However, there is a speculative component that requires further experimental study, but it does suggest some interesting directions for further investigation. A collective take home for the reader is that addressing this problem of metabolic heterogeneity in cells requires a multipronged approach. This will have to include experiments that characterize the cellular phenotypes, as well as theoretical approaches (similar to those described here).

The astute reader may perceive some asymmetry here—while the experimental analysis of the microbial (especially yeast) populations is sophisticated and full of biochemical details, our theoretical considerations are simple. This is deliberate. The role of theory here is not to make detailed predictions but rather to provide a framework for thinking about the processes involved, for which starting simple is essential. The simplest frameworks laid out in the Section "General considerations on the manifestation of multiple cellular states at the population level" can then be progressively expanded (by adding a resource, as we did in the Section "Cell-state heterogeneity of yeast in a well-mixed chemostat" and ref. (Krishna and Laxman, 2018), or by including space and diffusion, as in ref. (Varahan et al., 2019), depending on how many elements or how much detail one wants to put in. This allows one to infer not only the minimum ingredients that are sufficient to produce some behavior, but also how these behaviors are modified as one adds more details. For instance, when switching back and forth between cell states occurs, the compositional state of the population can attain a long-lived mixed state even though each individual cell can maintain its own state only for a much shorter time. A few simple take home messages emerge. First, the idea of a resource that is produced by a group of cells, which can control different outcomes (depending on how it is sensed, taken up, and utilized) is a powerful idea through which heterogeneous cell groups can emerge. Second, there is an underlying idea of bistable systems (and how metabolism might create a bistable system). Third, there clearly is a population (and cell density) dependent phenomenon, in the emergence of heterogeneous states. Finally, we suggest how, in general, metabolism and metabolic outputs provide a set of constraints as a crucible to allow clonal cells to attain heterogeneous states.

Finally, we do note that in this chapter we have largely addressed only a "two phenotypic state" problem. We know however that in colonies, cells may be in more than two states. We may speculate that such multistate cells can arise by combining bistable switches in various hierarchical control structures—two independent switches can give four states, whereas a switch that depends on another switch can give three states. The framework described in the Section "General considerations on the manifestation of multiple cellular states at the population level" can easily be extended to such situations, and similar questions about when a mixed population state is possible and what determines the composition of these states may be posed, although the analysis will become correspondingly more complex. This kind of situation has not been extensively modeled, nor has this been extensively experimentally dissected, and so will be an obvious area for future enquiry.

Acknowledgments

We thank Vidyanand Nanjundiah for discussions and comments. SK is funded by the Simons Foundation, and the Department of Atomic Energy, Government of India, under project no. 12-R&D-TFR-5.04-0800.

Appendix

We consider a system consisting of N_A cells of type A and N_B cells of type B in a well-mixed chemostat. Individual cells can switch from one state to the other (at rates ν_{AB} and ν_{BA}) and also grow and divide (at rates γ_A and γ_B). The growth and division is balanced by dilution at rate ϕ (which could vary in time). Such a system can be described by the following differential equations:

$$dN_A/dt = \gamma_A N_A + \nu_{BA} N_B - \nu_{AB} N_A - \phi N_A \tag{13.3}$$

$$dN_B/dt = \gamma_B N_B - \nu_{BA} N_B + \nu_{AB} N_A - \phi N_B \tag{13.4}$$

Since we are only interested in what we call the "composition" of the system, which is the fraction of cells in state A, $n = N_A/(N_A + N_B)$, we can reduce this to one equation:

$$\begin{aligned} dn/dt &= \left(dN_A/dt\right)/(N_A + N_B) - N_A\left(dN_A/dt + dN_B/dt\right)/(N_A + N_B)^2 \\ &=> dn/dt = \nu_{BA}(1-n) - \nu_{AB} n + (\gamma_A - \gamma_B) n(1-n) \end{aligned} \tag{13.5}$$

Notice that in this description the dilution rate drops out. Any value of n that leaves the right hand side of Eq. (13.5) equal to zero (i.e., for which $dn/dt = 0$) is a steady-state composition. However, some steady-states can be unstable in the sense that an arbitrarily small perturbation to the composition might result in the system moving away from the steady state. The condition for a steady-state $n = n_{SS}$ to be (linearly) stable is that:

$$\partial\left(dn/dt\right)/\partial n < 0 \text{ when } n = n_{SS} \tag{13.6}$$

This corresponds to the more intuitive notion of the dn/dt vs n curve cutting the x-axis from above as described in Fig. 13.2(A).

We now systematically examine a few cases.

Case 1: When growth is much faster than switching, such that one can ignore the latter

Here, the equation effectively becomes:

$$dn/dt = (\gamma_A - \gamma_B) n(1-n) \tag{13.7}$$

Thus, no matter how γ_A and γ_B depend on n, there must be steady-states at $n = 0$ (all cells in state B) and $n = 1$ (all cells in state A). If γ_A and γ_B are constants, then these are the only steady states possible, and it is also easy to see which is stable: $n = 0$ (all B cells) is stable and $n = 1$ (all A cells) is unstable if $\gamma_A < \gamma_B$, and vice versa (see Fig. 13.2C). So, only pure states are possible and the underlying two-state nature of the single cells does *not* manifest at the population level.

Now we examine what is possible if γ_A and γ_B depend on n. In general, it seems reasonable to assume that $\gamma_A - \gamma_B$ is a monotonic function of n, i.e., it either increases with n or decreases with n. If that is the case, then it is possible for there to exist a stable mixed state if and only if $\gamma_A - \gamma_B > 0$ for $n = 0$, then decreases as n increases, and becomes <0 for $n = 1$. In that case the value of n at which $\gamma_A = \gamma_B$ will be a stable steady-state (See Fig. 13.2D).

Above, we also mentioned the sub-case where type A cells are proliferating, while type B cells are quiescent (i.e., $\gamma_B = 0$). In this case, the equation for the composition becomes

$$dn/dt = \gamma_A n(1-n) \tag{13.8}$$

In this case, even if γ_A is a monotonic function of n, it is *not* possible for there to be a stable mixed state.

Case 2: When switching is much faster than growth, such that one can ignore the latter

Here the equation becomes

$$dn/dt = \nu_{BA}(1 - n) - \nu_{AB}n \tag{13.9}$$

Let us first assume the switching rates are constant. In that case, there will always be one stable steady state at an intermediate value of $n = \nu_{BA} / (\nu_{AB} + \nu_{BA})$. Note that if either state is irreversible (i.e., either ν_{AB} or $\nu_{BA} = 0$) then a stable mixed population is *not* possible.

Now, we consider the case where the switching rates depend on n. In general, there are infinite ways in which switching rates could depend on n, so we will have to choose some simple classes of functions to examine. Taking inspiration from the yeast case, which suggested a metabolic threshold that triggered state switching, we will take each switching rate to be either a constant or a switch-like function. As an example of the latter, ν_{AB} may take one value ν_{AB0} when $n < K$ and a different (higher or lower) value ν_{AB1} when $n > K$. Let us suppose that ν_{AB} takes this form, while ν_{BA} is a constant.

If ν_{AB} is a switch that *increases* with n, this results in a negative feedback loop and a single stable mixed state at the population level. This is because in this case dn/dt takes the positive value ν_{BA} at $n = 0$, and monotonically decreases to the negative value $-\nu_{AB1}$ at $n = 1$. Therefore, it must take the value zero at some intermediate n, which will then be a stable mixed steady state.

When $\nu_{AB1} > \nu_{BA}(1 - 1/K) > \nu_{AB0}$, this steady state value will be $n = K$, and small changes in the switching rates will leave the steady-state composition unchanged. Thus, in this case, the composition of the mixed state is determined by K rather than the switching rates, and, unlike the mixed state obtained when rates are constant, will be robust to changes in switching rates.

More interesting is the case where ν_{AB} decreases with n, i.e., it has a large value when $n < K$ and switches to a low value when $n > K$ (i.e., $\nu_{AB0} > \nu_{AB1}$). In this case, the situation is as in Fig. 13.2G, and *two* stable mixed states may coexist – *a bistability at the population level.* The condition for this to occur is when the rates satisfy: $\nu_{AB1} > \nu_{BA}(1 - 1/K) > \nu_{AB0}$. Generally, bistabilities are associated with positive feedback loops (Sneppen, Krishna, and Semsey, 2010), and in this case the positive feedback arises because when n increases beyond K, ν_{AB} decreases, resulting in a positive effect on n.

Case 3: Both switching and growth have to be taken into account

In the main text, we mentioned the scenario where one of the states (say, B) is irreversible, i.e., $\nu_{BA} = 0$. In this case, even if growth rates are much smaller than switching rates, it is incorrect to ignore them in the equation:

$$dn/dt = -\nu_{AB}n + (\gamma_A - \gamma_B)n(1 - n) \tag{13.10}$$

Here, the steady-states are $n = 0$ (all B cells) and, if all growth and switching rates are constant, $n = 1 - \nu_{AB} / (\gamma_A - \gamma_B)$ (we would have missed this mixed state if we had ignored growth here). The former is stable if $\gamma_B > \gamma_A$ (i.e., B grows faster than A), and the latter is stable when $\gamma_A > \gamma_B$ (A grows faster than B).

References

Ackermann, M., 2015. A functional perspective on phenotypic heterogeneity in microorganisms. Nat. Rev. Microbiol. 13 (8), 497–508. Available from: https://doi.org/10.1038/nrmicro3491.

Aktipis, C.A., Boddy, A.M., Jansen, G., Hibner, U., Hochberg, M.E., Maley, C.C., et al., 2015. Cancer across the tree of life: cooperation and cheating in multicellularity. Philos. Trans. R. Soc. Lond. Ser. B, Biol. Sci. 370 (1673). Available from: https://doi.org/10.1098/rstb.2014.0219pii: 20140219.

Bergmiller, T., Ackermann, M., 2011. Pole age affects cell size and the timing of cell division in methylobacterium extorquens AM1. J. Bacteriol. 193 (19), 5216–5221. Available from: https://doi.org/10.1128/JB.00329-11.

Bonner, J.T., 1957. A theory of the control of differentiation in the cellular slime molds. Q. Rev. Biol. 32 (3), 232–246. Available from: http://www.jstor.org/stable/2815258.

Cerulus, B., New, A.M., Pougach, K., Verstrepen, K.J., 2016. Noise and epigenetic inheritance of single-cell division times influence population fitness. Curr. Biol. 26 (9), 1138–1147. Available from: https://doi.org/10.1016/j.cub.2016.03.010.

Chance, B., Estabrook, R.W., Ghosh, A., 1964. Damped sinusoidal oscillations of cytoplasmic reduced pyridine nucleotide in yeast cells. Proc. Natl Acad. Sci. U.S.A. 51 (6), 1244–1251.

Chandra, F.A., Buzi, G., Doyle, J.C., 2011. Glycolytic oscillations and limits on robust efficiency. Science 333 (6039), 187–192. Available from: https://doi.org/10.1126/science.1200705.

D'Souza, G., Shitut, S., Preussger, D., Yousif, G., Waschina, S., Kost, C., 2018. Ecology and evolution of metabolic cross-feeding interactions in bacteria. Nat. Product. Rep. 35 (5), 455–488. Available from: https://doi.org/10.1039/c8np00009c.

David, L., Ben-Harosh, Y., Stolovicki, E., Moore, L.S., Nguyen, M., Tamse, R., et al., 2013. Multiple genomic changes associated with reorganization of gene regulation and adaptation in yeast. Mol. Biol. Evolution 30 (7), 1514–1526. Available from: https://doi.org/10.1093/molbev/mst071.

Du, Q., Kawabe, Y., Schilde, C., Chen, Z.-H., Schaap, P., 2015. The evolution of aggregative multicellularity and cell-cell communication in the dictyostelia. J. Mol. Biol. 427 (23), 3722–3733. Available from: https://doi.org/10.1016/j.jmb.2015.08.008.

Elowitz, M.B., Levine, A.J., Siggia, E.D., Swain, P.S., 2002. Stochastic gene expression in a single cell. Science 297 (5584), 1183–1186. Available from: https://doi.org/10.1126/science.1070919.

Gierer, A., 1981. Generation of biological patterns and form: some physical, mathematical, and logical aspects. Prog. Biophysics Mol. Biol. 37 (1), 1–47.

Gierer, A., Meinhardt, H., 1972. A theory of biological pattern formation. Kybernetik 12 (1), 30–39. Available from: https://doi.org/10.1007/BF00289234.

Hamant, O., Bhat, R., Nanjundiah, V., Newman, S.A., 2019. Does resource availability help determine the evolutionary route to multicellularity? Evolution & Dev. 21 (3), 115–119. Available from: https://doi.org/10.1111/ede.12287.

Hess, B., Boiteux, A., 1971. Oscillatory phenomena in biochemistry. Annu. Rev. Biochem. 40, 237–258.

Hommes, F.A., 1964. Oscillatory reductions of pyridine nucleotides during anaerobic glycolysis in brewers' yeast. Arch. Biochem. Biophysics 108, 36–46.

Huh, D., Paulsson, J., 2011. Non-genetic heterogeneity from stochastic partitioning at cell division. Nat. Genet. 43 (2), 95–100. Available from: https://doi.org/10.1038/ng.729.

Kaiser, D., 1986. Control of multicellular development: dictyostelium and myxococcus. Annu. Rev. Genet. 20, 539–566. Available from: https://doi.org/10.1146/annurev.ge.20.120186.002543.

Kamimura, A., Matsubara, Y.J., Kaneko, K., Takeuchi, N., 2019. Horizontal transfer between loose compartments stabilizes replication of fragmented ribozymes. PLoS Comput. Biol. 15 (6), e1007094. Available from: https://doi.org/10.1371/journal.pcbi.1007094.

Krishna, S., Laxman, S., 2018. A minimal 'push–pull' bistability model explains oscillations between quiescent and proliferative cell statesEdited by Daniel J LewMol. Biol. Cell 29 (19), 2243–2258. Available from: https://doi.org/10.1091/mbc.E18-01-0017.

Krishna, S., Semsey, S., Jensen, M.H., 2009. Frustrated bistability as a means to engineer oscillations in biological systems. Phys. Biol. 6 (3), 036009.

Laforge, B., Guez, D., Martinez, M., Kupiec, J.-J., 2005. Modeling embryogenesis and cancer: an approach based on an equilibrium between the autostabilization of stochastic gene expression and the interdependence of cells for proliferation. Prog. Biophysics Mol. Biol. 89 (1), 93–120. Available from: https://doi.org/10.1016/j.pbiomolbio.2004.11.004.

Laxman, S., Tu, B.P., 2010. Systems approaches for the study of metabolic cycles in yeast. Curr. Opin. Genet. & Dev. 20 (6), 599–604. Available from: https://doi.org/10.1016/j.gde.2010.09.008.

Levy, S.F., Ziv, N., Siegal, M.L., 2012. Bet hedging in yeast by heterogeneous, age-correlated expression of a stress protectant. PLoS Biol. 10 (5), e1001325. Available from: https://doi.org/10.1371/journal.pbio.1001325.

Lewontin, R.C., 1958. A general method for investigating the equilibrium of gene frequency in a population. Genetics 43 (3), 419–434. Available from: https://www.ncbi.nlm.nih.gov/pubmed/17247767.

Lindner, A.B., Madden, R., Demarez, A., Stewart, E.J., Taddei, F., 2008. Asymmetric segregation of protein aggregates is associated with cellular aging and rejuvenation. Proc. Natl Acad. Sci. U.S.A. 105 (8), 3076–3081. Available from: https://doi.org/10.1073/pnas.0708931105.

Miller, M.B., Bassler, B.L., 2001. Quorum sensing in bacteria. Annu. Rev. Microbiol. 55, 165–199.

Murray, D.B., Beckmann, M., Kitano, H., 2007. Regulation of yeast oscillatory dynamics. Proc. Natl Acad. Sci. U.S.A. 194 (7), 2241–2246.

Niklas, K.J., 2014. The evolutionary-developmental origins of multicellularity. Am. J. Botany 101 (1), 6–25. Available from: https://doi.org/10.3732/ajb.1300314.

Paldi, A., 2012. What makes the cell differentiate? Prog. Biophysics Mol. Biol. 110, 41–43.

Raj, A., van Oudenaarden, A., 2008. Nature, nurture, or chance: stochastic gene expression and its consequences. Cell 135 (2), 216–226. Available from: https://doi.org/10.1016/j.cell.2008.09.050.

Saran, S., Azhar, M., Manogaran, P.S., Pande, G., Nanjundiah, V., 1994. The level of sequestered calcium in vegetative amoebae of dictyostelium discoideum can predict post-aggregative cell fate. Differentiation 57 (3), 163–169.

Sneppen, K., Krishna, S., Semsey, S., 2010. Simplified models of biological networks. Annu. Rev. Biophys. 39 (1), 43–59. Available from: https://doi.org/10.1146/annurev.biophys.093008.131241.

Stewart, P.S., Franklin, M.J., 2008. Physiological heterogeneity in biofilms. Nat. Rev. Microbiol. 6 (3), 199–210. Available from: https://doi.org/10.1038/nrmicro1838.

Stolovicki, E., Braun, E., 2011. Collective dynamics of gene expression in cell populations. PLoS One 6 (6), e20530.

Thattai, M., van Oudenaarden, A., 2004. Stochastic gene expression in fluctuating environments. Genetics 167 (1), 523–530. Available from: https://www.ncbi.nlm.nih.gov/pubmed/15166174.

Thomas, R., 1998. Laws for the dynamics of regulatory networks. Int. J. Dev. Biol. 42, 479–485.

Tu, B.P., Tu, B.P., Kudlicki, A., Rowicka, M., Mcknight, S.L., 2005. Logic of the yeast metabolic cycle : of cellular processes. Science 310 (5751), 1152–1158. Available from: https://doi.org/10.1126/science.1120499.

Tu, B.P., Mohler, R.E., Liu, J.C., Dombek, K.M., Young, E.T., Synovec, R.E., et al., 2007. Cyclic changes in metabolic state during the life of a yeast cell. Proc. Natl Acad. Sci. U.S.A. 104 (43), 16886–16891. Available from: https://doi.org/10.1073/pnas.0708365104.

Turing, A.M., 1952. The chemical basis of morphogenesis. Philos. Trans. R. Soc. B 237 (641), 37–72.

Varahan, S., Walvekar, A., Sinha, V., Krishna, S., 2019. Metabolic constraints drive self-organization of specialized cell groups. ELife 8, e46735.

Wolf, D.M., Vazirani, V.V., Arkin, A.P., 2005. Diversity in times of adversity: probabilistic strategies in microbial survival games. J. Theor. Biol. 234 (2), 227–253. Available from: https://doi.org/10.1016/j.jtbi.2004.11.020.

Yamagishi, J.F., Saito, N., Kaneko, K., 2016. Symbiotic cell differentiation and cooperative growth in multicellular aggregates. PLoS Computational Biol. 12 (10), e1005042. Available from: https://doi.org/10.1371/journal.pcbi.1005042.

14

Stochastic phenotypic switching in endothelial cell heterogeneity

Erzsébet Ravasz Regan

Biochemistry and Molecular Biology, The College of Wooster, Wooster, OH, United States

Introduction

The importance of randomness in molecular binding interactions and diffusion underlying the biochemistry of life is well established in chemistry and statistical physics. Until recently, this randomness was thought of as background, inconsequential to our understanding of biological function. The idea that it could cause more than trivial variation among genetically identical cells was slow to take root (Balazsi et al., 2011; Raj and van Oudenaarden, 2008; Eldar and Elowitz, 2010; Symmons and Raj, 2016). Indeed, the field of "biological noise" matured alongside single-cell live imaging technologies capable of probing protein dynamics in individual cells (Elowitz et al., 2002; Ozbudak et al., 2002; Blake et al., 2003). Quantifying protein expression in single cells showed that moment-to-moment variation in the rate of transcription and translation can lead to drastic cell-to-cell heterogeneity across isogenic populations (Elowitz et al., 2002; Ozbudak et al., 2002; Blake et al., 2003; Raser and O'Shea, 2004; Yu et al., 2006; Cai et al., 2006; Raj et al., 2006). Among the most convincing early experiments to quantify biological noise was a demonstration of large fluctuations in the expression of two proteins driven by identical promoters coexpressed in single cells (Elowitz et al., 2002). These two reporters not only fluctuated between near-zero to high expression over time, but also showed strong changing expression differences within the same cell. This stochastically generated variation—biological noise—was initially thought to be part of the messy background of processes which, while undoubtedly present, are unlikely to be relevant to cellular or organismal function. A direct

Phenotypic Switching. DOI: https://doi.org/10.1016/B978-0-12-817996-3.00004-9

demonstration that such variation could lead to the reliable survival of a random subpopulation among genetically identical cells exposed to an antibiotic was the first in a series of experiments to challenge this misconception (Blake et al., 2006). A large body of subsequent work showed that biological noise can fundamentally alter a cell's phenotype, relying on but not limited to its ability to generate fluctuations in protein expression (reviewed in Balazsi et al., 2011; Raj and van Oudenaarden, 2008; Eldar and Elowitz, 2010; Symmons and Raj, 2016). We now know that biological noise can influence differentiation (Huang, 2009; Magklara and Lomvardas, 2013; Chang et al., 2008; Wernet et al., 2006; Macfarlan et al., 2012), tune irreversible cell fate decisions (Chen et al., 2012; Spencer et al., 2009; Spencer et al., 2013), play an evolutionary role in fluctuating environments (Blake et al., 2006; Acar et al., 2008; Beaumont et al., 2009; Veening et al., 2008), and alter cellular function in ways that can impact tissue function (Yuan et al., 2016).

Key to understanding the way random processes can alter a cell's phenotype is the way fluctuations in gene expression interact with *phenotypic plasticity* (Kelly et al., 2012). The birth of this term was motivated by mounting evidence that organisms with identical genomes can display more than one phenotype (Bradshaw, 1965; Scheiner, 1993). Thus, phenotypic plasticity was first used to describe irreversible morphological changes in response to variation in an organism's developmental environment (Giuseppe and Alessandro, 2010). As genetics wrestled with the context-dependent relationship between genes and phenotypes, its definition expanded to cover environment-induced changes throughout the life of an organism, such as acclimatization (Murren et al., 2015; DeWitt et al., 1998). Adaptive responses to environmental variation, however, require phenotypic plasticity at the cellular level, including a developmental ability to generate and maintain distinct cell lineages sharing the same genome. It is the balance of these lineages that shifts when developmental processes are irreversibly altered by their environment (Kanji et al., 2011; Santini et al., 2016; Domyan and Sun, 2011; Lin et al., 2013; Rothenberg and Dionne, 2002; Posfai et al., 2017). Later in an organism's life, it is usually a specific subset of mature cell lineages that respond to changing environments by altering their function, though their response is not always reversible by an environmental reset (Ordovas-Montanes et al., 2018; Kudo Fabio et al., 2007). The bewildering complexity of observed gene—environment interactions fostered an early appreciation for the question: how do organisms sharing the same genome generate distinct cell types?

Answering this requires understanding the mechanisms that bring about phenotypic plasticity. As we detail in the text below, these insights also explain how biological noise can exerts profound effects on cell fate and function.

From genes to phenotypic plasticity

The question of how a single genome can "code for" multiple cell types inspired a highly influential metaphor, published in 1957 by the developmental biologist Conrad Waddington (Waddington, 2019). According to his metaphor, the irreversible commitment to increasingly restrictive cell lineages during differentiation is akin to balls rolling down a hillside with multiple ridges (Fig. 14.1A). Stem cells on Waddington's *"epigenetic landscape"* are represented by balls high up on the hillside. Unstable and poised to roll down, stem cells eventually commit to distinct irreversible fates—the bottoms of district ravines. Thus,

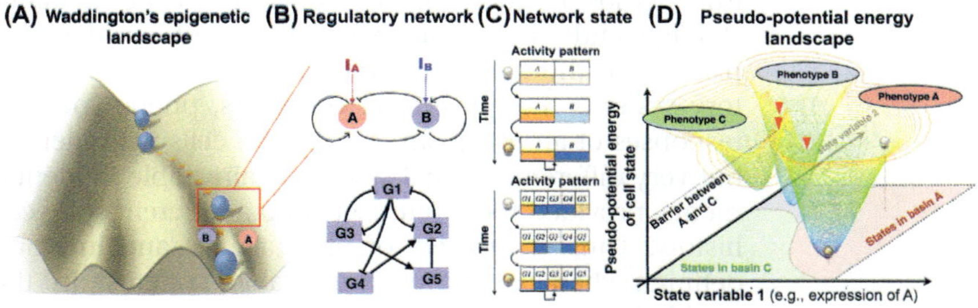

Figure 14.1 A dynamical systems framework for understanding the genetic roots of phenotypic plasticity. (A) Waddington's epigenetic landscape (figure by Kevin J. Mitchell). Reproduced from https://figshare.com/articles/_Waddington_s_8220_Epigenetic_Landscape_8221_/620879 (Waddington, 2019); *balls & dotted line*: trajectory of a differentiating stem cell; *ravines*: discrete differentiation pathways; *valley bottoms*: distinct differentiated cell types; *A/B* master regulators characteristic of two specific lineages. (B) *Top*: Architecture of a transcriptional toggle switch where transcription factors A and B mutually inhibit each other, guaranteeing that states in which both factors are highly expressed are dynamically unstable. Most cell differentiation circuits also contain positive autoregulatory feedback by the two factors, further stabilizing the two differentiated states. *Bottom*: small abstract regulatory network with five genes. (C) Example state sequences from unstable expression patterns to stable network stables enforced by the internal links of the regulatory switch (*top*)/small network (*bottom*); *orange/blue*: ON/OFF; *silver/gold balls*: dynamically unstable/stable state. (D) Pseudo-potential energy landscape illustrating key dynamical behaviors of the example networks; *x,y axes*: internal state variables (expression of *A* and *B* in case of the switch; expressions of two marker genes or distinct combinations of markers in case of the 5-gene network); *z axis*: relative stability of each network state on the (*x,y*) plane, dictated by the regulatory links between genes (akin to a pseudo-potential energy); *valley bottoms and gold ball*: stable expression patterns corresponding to distinct cell phenotypes; *valley walls/basins and silver ball*: unstable expression patterns that converge to a specific valley/phenotype; *saddle points (gray arrows)*: transitional network states marking energetically favorable paths between distinct phenotypes; *landscape color scale*: green/low energy/stable to red/high energy/unstable.

phenotypic plasticity during development could be understood as the environment subtly altering the trajectory of these balls, leading to irreversible and often drastic fate changes (Huang, 2009; Huang, 2012). Waddington metaphor was published the same year Francis Crick articulated his ideas on how genes encode for proteins and thus formulated with little understanding about how DNA encodes for heritable traits (Crick, 1958; Cobb, 2017). Yet, it came remarkably close to describing a quantifiable property of interacting gene systems responsible for development—the energy landscapes that governs their dynamics (Huang, 2012; Wang et al., 2011; Huang, 1999; Zhou et al., 2012). We have since recognized that positive feedback regulation between two or more genes can give rise to molecular switches that exhibit bistable or multistable dynamics (Balazsi et al., 2011; Huang, 2009; Huang et al., 2007; Heinäniemi et al., 2013; Andrecut et al., 2011; Foster et al., 2009). Multistablility is a general property shared by a large class of dynamical systems. It is defined as having multiple *stable* internal states in the same environment, maintained against small perturbations that push the system away from its local equilibria. The similarity in the definition of multistability and phenotypic plasticity is not accidental; phenotypic plasticity is the consequence of the multistability of genetic regulatory systems that govern all cell types, phenotypes, and adaptive responses of an organism. As a result, cells with identical genomes but very different internal states remain stable in identical environments, provided they were "set" into a different state earlier in their history. Thus, a *memory* of prior signals and internal states is a defining characteristic of multistable systems. In the nonliving world, familiar bistable systems include mechanical light-switches or ballpoint pens. Ferromagnetic materials magnetized in a particular direction are also multistable, in that they resist the reversal of the direction of their magnetization—a property we exploit to store information on magnetic hard drives. Common to all these systems is an internal structure with several low-energy states; locally stable states that echo Waddington's ravines. These states are separated by *energy barriers* (Waddington's ridges). In the absence of large perturbations, these barriers block the system from spontaneously transitioning between its states.

The first bistable biological circuit characterized using a dynamical systems approach is the toggle switch in control of lysis versus lysogen in the λ bacteriophage (Arkin et al., 1998). This switch is composed of two mutually inhibitory transcription factors (e.g., Fig. 14.1B). Their double-negative feedback guarantees that simultaneous high expression of *both* factors is unstable, as if a ball was balanced at the tip of a ridge. Thus, the smallest variation in expression can drive this system into either a lytic state, where the virus multiplies and bursts its host to spill its progeny (lytic factor is high), or a lysogenic state where the viral DNA integrates into its host and remains dormant (lysogenic factor is high). This switch can flip from lysogenic (integrated and dormant) state back to lysis when exposed to specific environmental signals such as starvation or stress in the host, setting off a burst of viral reproduction and subsequent lysis (Howard-Varona et al., 2017). This toggle-switch architecture inspired the first synthetic toggle switch engineered into bacteria (Gardner et al., 2000). Much like a light-switch, this synthetic circuit could be flipped on demand with transient signals and maintained its new state through dozens of divisions. We have since learned that the lion's share of differentiation switches in control of diverging lineages have the same basic architecture: mutual inhibition between master regulations of the two lineages (Fig. 14.1B, *top*) (Huang, 2009). The bistability introduced by this feedback is usually stabilized by positive auto-regulation by each transcription factors (Huang, 2009). Thus, at each ridge-top on Waddington's landscape, a small switch in the differentiating stem cell locks into one of two states, creating an energy barrier to trans-differentiation. The master lineage regulators in these switches often control the recruitment of DNA and histone-modifying enzymes to their target promoters, further raising the barrier between lineages by maintaining robust epigenetic silencing of other lineage drivers (Iwafuchi-Doi and Zaret, 2016; Mayran and Drouin, 2018; Iwafuchi-Doi and Zaret, 2014; Zaret and Carroll, 2011).

The simple toggle-switch circuits in charge of differentiation are not isolated from the dense web of interactions that form a cell's regulatory network. The same way the toggle switch in Fig. 14.1B is unstable if both factors are expressed (Fig. 14.1C, top), the majority of gene expression patterns in larger interaction networks are also unstable. If forced into one of these states and left alone, interactions within these networks trigger a cascade of changes until the network reaches one of its stable states (Fig. 14.1C, bottom). Dynamical systems theory calls these stable regions of the state space dynamical *attractors* (last state of time courses in Fig. 14.1C). In biological regulatory networks, attractors correspond to stable phenotypes; cell states (*fixed-point attractors*, single "points" in the system's state space), or rhythmic behaviors (*limit cycle attractors*, robust sequences of states the system oscillates through without settling into a stable state) (Huang, 1999; Huang and Ingber, 2000; Albert and Othmer, 2003; Albert and Wang, 2009). It is rare for large networks to have a single attractor (Socolar and Kauffman, 2003). Indeed, in the 1960s Stuart Kauffman showed that randomly wired networks of genes turning each other ON/OFF based on arbitrary logic rules are generally multistable (Kauffman, 1969).

Each attractor is surrounded by an *attractor basin*, an unstable domain in state space from which all trajectories converge to the attractor (red/green/blue domains on the *x-y* plane of Fig. 14.1D). It is easy to appreciate that the more distant a state is from an attractor the more unstable it is; the more energy it takes to push the network into such a state (e.g., silver ball on Fig. 14.1D). This idea can be quantified using statistical mechanics, allowing us to calculate a potential energy-like quantity for each network state (*z* axis on Fig. 14.1D) (for continuous/discrete regulatory models see Wang et al., 2011; Zhou et al., 2012 and Zhang et al., 2006; Deritei et al., 2016; Steinway et al., 2014, respectively). The resulting pseudo-potential energy landscape echoes Waddington's idea of the epigenetic landscape and adeptly showcases the energy barriers between distinct stable states of a biological system. It drives home the idea that a small change in a cell's internal state is generally not sufficient to trigger a phenotype change, rendering it robust against a broad range of perturbations. At the same time, regulatory networks are only "locally" homeostatic (to perturbations *within* a basin); they balance this with phenotypic plasticity by responding to specific signals that push the cell's state into new phenotype basins (Regan and Aird, 2012).

Due to the dense regulatory coupling of our genes (Zhang and Zhang, 2009), our genome is responsible for defining the *entire* energy landscape of possible cell states, rather than directly specifying a single phenotype as a simple genotype-phenotype map would imply. As a result, the principal difference between cell lineages and distinct phenotypes of a differentiated cell is the *height* of the energy barrier that separates these states. That said, regulatory networks are equipped with receptors and signaling pathways that grant certain environments the ability to push the system past particular barriers (Huang, 2012) (e.g., commitment to cell cycle (Yao et al., 2008; Overton et al., 2014) or apoptosis (Spencer et al., 2009)). In contrast, trans-differentiation between mature cell types is rarely built-in; achieving it usually requires an artificial change in cell state via transient expression of key transition factors (reprogramming) (Takahashi and Yamanaka, 2016; Plath and Lowry, 2011). The distinction between distinct cell types and phenotypes within a single lineage is especially blurry for endothelial cells, which form the inner lining of all our blood and lymphatic vessels (Aird, 2007; Aird, 2007). These endothelial monolayers act as barriers between our tissues and blood, but also perform a wide array of physiological and pathological roles. These cells are readily partitioned by the literature into blood versus lymphatic endothelial cells, then again into venous versus arterial versus capillary blood endothelial cells, with good cause (Kriehuber et al., 2001; dela Paz and D'Amore, 2009). The above subtypes show remarkable stability under a wide range of conditions in vivo as well as in vitro (Regan and Aird, 2012). Yet, certain environments can induce transdifferentiation from venous toward arterial phenotype (Kudo Fabio et al., 2007), a reversible shift from blood to a lymph-like endothelial cell state (Cooley et al., 2010), or an irreversible one into a smooth muscle cell (Coll-Bonfill et al., 2015). Categorizing capillary cells as a single group is even more fraught, as capillary endothelial cells of different organs are as distant from each other in expression space as venous versus arterial cells (Regan and Aird, 2012; Chi et al., 2003). Moreover, some of these differences all but disappear when capillary cells are extracted from their native environment and cultured in identical conditions (e.g., lung vs skin capillary ECs), while others remain distinct (e.g., myocardial vs intestinal ECs) (Chi et al., 2003).

Understanding the molecular mechanisms that give rise to the energy landscape of cell states is key to our ability to characterize and alter how cells and tissues function. As a result, these mechanisms are of intense interest to biomedicine, from cancer to cardiovascular research. The endothelium is an especially promising target for therapeutically intervention due to its remarkable heterogeneity in structure, function, and gene expression in small vessels across our tissues (Regan and Aird, 2012; Aird, 2007; Aird, 2007). This heterogeneity translates to differential involvement in disease, as well as differential response to therapy. At the molecular level, the unique vascular bed-specific protein expression profile (or vascular "zip code") of capillary endothelial cells opens the possibility of targeting a pharmacological intervention to individual tissues (Ruoslahti and Rajotte, 2000; Ruoslahti, 2004). This opportunity is exploited best when the phenotypic repertoire and plasticity of the endothelium is well understood, including the potential role of biological noise in generating endothelial heterogeneity.

Biological noise meets phenotypic plasticity

Biological noise is due to random fluctuations in the rate of regulatory processes such as transcription, signal transduction or localization (Balazsi et al., 2011; Raj and van Oudenaarden, 2008; Eldar and Elowitz, 2010; Symmons and Raj, 2016). As a result, it manifests as variability in the activity or expression of cellular regulatory network components, continuously altering the system's internal state (red trajectory on Fig. 14.2A, bottom). As both external signals and intrinsic noise act on the level and/or activity of proteins—that is, the cell's internal state, their effect on phenotypic plasticity parallel each other. These parallels play out in developmental processes, in response to external signals in somatic cells, and in the way bistability of the regulatory system leads to heterogeneity among isogenic cells (Fig. 14.2).

Biological noise in developmental plasticity

During development, differences in the extracellular environment of differentiating stem cells can result in a heterogeneous mix of two cell types (Fig. 14.2A, *top*). For example, the physical surroundings of individual blastomere cells, the first set of undifferentiated cells in a dividing zygote, dictate their fate (Marikawa and Alarcón, 2009). Cells in the interior differentiate into inner cell mast, while the outer layer assumes a trophoblast fate. Alternatively, developing cells can interact via short-range patter-forming signals to break the homogeneity of their microenvironment and autogenerate the signaling differences leading to separate cell fates (Briscoe and Small, 2015). This occurs during arteriovenous specification of endothelial cells during embryonic development (Fish and Wythe, 2015). Endothelial progenitors in the lateral plate mesoderm exposed to extracellular *VEGF* start a signaling cascade that promotes *ephrin-B2* expression, which not only potentiates *VEGF* signaling in *ephrin-B2* expressing cells, but also activates *EphB4* receptors on neighboring endothelial cells to shut down their ability to produce *ephrin-B2*. This lateral inhibition-type interaction results in a mix of *ephrin-B2* positive and negative cells locked into arterial and venous fates, respectively. Thus, *ephrin-B2* is only expressed on the arterial and arteriole side of the vasculature; it disappears abruptly along individual capillary vessels as they connect arterioles to venules (Gale et al., 2001).

In addition to external signals, random fluctuations in the internal state of stem cells can also alter their trajectory, even in cells exposed to identical external environments (Fig. 14.2A, *bottom*; red arrows to cell type *B* indicate expected fate). Thus, biological noise overlaid on a differentiation signal can generate mixtures of two differentiated cell types, albeit with a higher percentage of cells following the trajectory driven by the

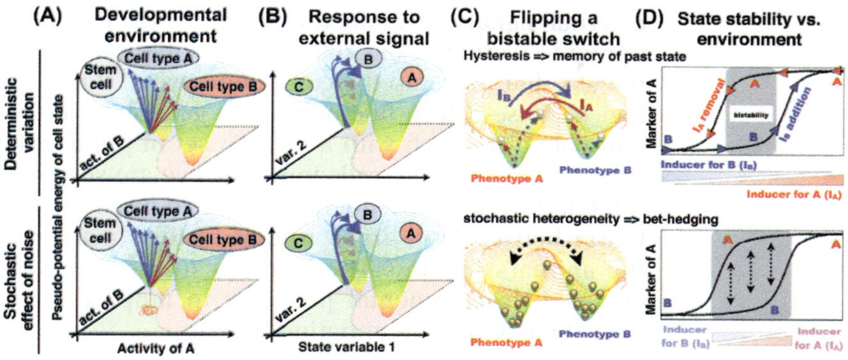

Figure 14.2 Parallels between plasticity in response to environmental signals and biological noise. (A) *Top*: Variation in the microenvironments of differentiating cells can lead to internal state changes that impose slightly different differentiation trajectories (*blue/red arrows*), leading to distinct environment-dependent differentiated cells. *Diverging blue/red arrows*: A uniform drive to differentiate combined with varying environmental signals resulting in a mix of cell types *A* and *B*. *Bottom*: Biological noise-driven fluctuations in the internal state of the cell (*red trajectories projected onto x-y*) can alter its differentiation trajectory; while the majority of cells follow the external signal driving commitment to lineage *B*, a minority commit to lineage *A* instead. *Diverging blue/red trajectories*: external drive to differentiate to lineage *B* combined with biological noise-driven state changes, resulting in heterogeneous differentiation trajectories and commitment to either B or A. (B) *Top*: Cells respond differently to slight differences in external signal strength (*blue/gray arrows*). *Blue/gray trajectories*: External signals leading to increased expression of state variable 2 leads to a stable phenotype change if it crosses the barrier from *C* to *B*, but only a short-lived reversible response otherwise. *Bottom*: Biological noise-driven fluctuations in the internal state of the cell can alter their response to a uniform environmental signal (*thick blue trajectory*), resulting in a stable response in a random subset of cells (*blue trajectories*), and a short-lived reversible pulse in others (*gray trajectories*). (C) *Top*: Environmental signals can force transitions between two stable states of a multistable regulatory network (often in either direction), but their removal or even mild reversal does not necessarily reverse the phenotype. *Arrows*: State changes brought about by external signals that exceed (*solid arrow*)/do not exceed (*dashed arrow*) the threshold for flipping the switch; *silver balls*: reversible, short-lived state-changes in the regulatory system in response to subthreshold signals. *Bottom*: Biological noise can lead to transitions between two stable states of a multistable regulatory network with a low barrier. *Arrow*: State changes due to the internal stochasticity of gene and protein expression occasionally push the cell past the barrier between distinct phenotypic states; *gold balls*: noise-driven heterogeneity in the system's internal state, with higher occupancy at low-energy states representing distinct phenotypes. (D) *Top*: Due to the abrupt change in behavior at a signal-threshold which flips a biological switch, all multistable systems show hysteresis; a direction-dependent response to the increase versus decrease of stimuli that push them past their barrier. A bistable switch going from a high concentration of the B phenotype inducer I_B toward the loss of this inducer can nevertheless maintain its *B* state. A strong inducer of *A* (I_A) is required to switch it to phenotype *A (blue arrows)*. Once flipped, withdrawal of I_A is not sufficient to flip the cell back into state *B*; a new strong pulse of I_B is required. As a result, the cell can stably maintain both phenotypes at low levels of both I_A and I_B, showing bistability in these environments. The characteristic direction-dependent difference in the cell's response to small changes in its environment (i.e., the balance of I_A and I_B) is called hysteresis. *Bottom*: Whenever the barrier between two stable states *A* and *B* is low enough to be susceptible to biological noise-driven transitions, cells in the window of bistability stochastically toggle between two states. Any population of such cells generates and maintains a robust mosaic mixture of the two states. *Landscape color scale*: Green/low energy/stable to red/high energy/unstable.

external signal. This is actually exploited by fruit fly eye development (Wernet et al., 2006). Differentiation of the ommatidia, or cells that make up the compound eye of the fly, relies on biological noise-driven fluctuations to generate a mosaic of yellow and pale cells, each with a distinct set of photoreceptors. While the fate of individual cells in the fly eye are stochastically determined, they are subsequently locked in by bistable mutual inhibition between two transcription factors (*warts* ⇆ *melted)* for the animal's lifetime. Thus, noise can interact with genetically encoded developmental plasticity to amplify fluctuations and alter the fate of differentiating cells.

Biological noise in response to environmental change

Variation in extracellular signals can lead to drastic differences in how cells respond, especially when the signal strength is near their response threshold (Fig. 14.2B, *top*). An early demonstration of such ultrasensitivity in *MAPK* signaling showed that this signaling cascade can generate drastically different activation profiles within a narrow range of inputs (Huang and Ferrell, 1996). Mammalian G-protein coupled receptors show similar ultrasensitivity to both the concentration and duration of extracellular signals (Lipshtat et al., 2010). In endothelial cells, expression of *Robo4* is induced by the loss of laminar shear flow (Mura et al., 2012; Zhuang et al., 2011), resulting in highly localized expression at the junction of intercostal arteries and the aorta but no other region of the aortic wall (Regan and Aird, 2012; Okada et al., 2008). Due to inevitable biological noise acting on the same pathways these sensitive signals engage, drastic differences in cellular response are not only observed at slight changes in external signal, but also among individual cells at specific near-threshold signals (Paliwal et al., 2007). In other words, ultrasensitive systems often respond to a signal in a digital all-or-none fashion where a fraction of cells respond in full, while the rest all but ignore the signal (Fig. 14.2B, *bottom*). This noisy digital response is used to generate tunable mixes of two mutually exclusive phenotypic outcomes across the tree of life (Balazsi et al., 2011; Raj and van Oudenaarden, 2008; Eldar and Elowitz, 2010; Symmons and Raj, 2016). These mixes are random at the level of individual cells, but responsive to the environment via the relative abundance of outcomes, an abundance that changes with signal strength. Examples include mixtures of lytic/lysogenic fates assumed by bacteriophages and HIV upon infecting their host (tuned by host size, starvation or the site of HIV integration) (Ptashne, 2004; Weinberger et al., 2005;

Weinberger et al., 2008), the switch to competence in bacteria (tuned by environmental starvation signals) (Süel et al., 2006, 2007), budding/cell cycle arrest/"shmooing" (mating) phenotypes in yeast (tuned by pheromones) (Paliwal et al., 2007), or quiescence/cell cycle entry/apoptosis in mammalian cells (tuned by growth factors and death signals) (Spencer et al., 2009; Spencer et al., 2013; Overton et al., 2014).

Stochastic phenotype switching and cellular memory

Be it a developmental switch or a circuit driving the behavior of a mature cell, multistable systems all have a feature in common. In the absence of (strong) biological noise, they are all capable of a phenomenon called *hysteresis* (Fig. 14.2C). Hysteresis occurs when the response of a system to *increasing* levels of extracellular signal does not match its response to *decreasing* levels of the same signal (Fig. 14.2D). This is due to the energy barrier between two stable states, which allow the system to "remember" its previous state—at least in extracellular environments that are not forceful enough to lock it into a single state (*gray zone* on Fig. 14.2D). Experiments have shown hysteresis in cells poised to commit to mitosis (Solomon, 2003; Sha et al., 2003), in response to entry and exit from the cell cycle (Yao et al., 2008; Tyson and Novak, 2001; Bai et al., 2003), and in lymphocyte activation (Das et al., 2009). In the presence of biological noise that is relatively strong compared to the energy barrier of the switch, noise alone can occasionally push the cell from one state to another (Fig. 14.2D, *bottom*). As a result, cell populations display dynamic mixtures of two distinct phenotypes, mixtures they reliably regenerate from homogeneous initial populations (Balazsi et al., 2011; Raj and van Oudenaarden, 2008; Eldar and Elowitz, 2010; Symmons and Raj, 2016). Stochastic phenotype switching has been documented in a wide range of organisms, including mammalian stem cells (Chang et al., 2008; Macfarlan et al., 2012). In the endothelium, the endothelial-restricted von Willebrand Factor (*vWF*) protein displays a stochastically generated ON/OFF expression pattern in culture (Yuan et al., 2016). As the case study in this chapter reveals, this stochastic phenotype switching is not an artifact of cell culture. It also occurs in vivo, in the heart capillaries of healthy animals.

Is biological noise-mediated heterogeneity adaptive?

Studies in prokaryotes and single-cell eukaryotes have shown that stochastic phenotype switching can play an adaptive evolutionary role (Acar et al., 2008; Beaumont et al., 2009; Veening et al., 2008). It allows a population of cells to solve task sharing and task allocation problems by performing them one at a time, without the cost of loosing the ability to perform alternate tasks in the future (as differentiated cells do) (Wahl, 2002; Wolf et al., 2005). Moreover, in fluctuating environments where the highest-fitness phenotype can fluctuate faster than cells are equipped to respond, populations can "hedge their bets" by dynamically generating a mix of phenotypes (Acar et al., 2008). As no individual cell is locked into a fate, cells in bet hedging populations continually roll the die on how well equipped they are at the next turn of the tide. Computational models have shown that bet hedging confers a fitness advantage in fluctuating environments (Acar et al., 2008), a finding subsequently validated by *de novo* evolution of bet hedging under fluctuating bacterial culture conditions (Beaumont et al., 2009).

Several mammalian cells studies have hinted at the possibility that a bet hedging strategy may be relevant to tissue renewal and/or function, but the evidence is circumstantial. First, stochastic responses to growth (Spencer et al., 2013; Yao et al., 2008; Overton et al., 2014) or death signals (Spencer et al., 2009) make sense in a tissue context. While most cells are exposed to similar levels of extracellular signal, ideally only a fraction of them should respond by dividing or undergoing apoptosis. Second, stochastic state fluctuations in stem cell populations were shown to alter the likelihood of individual cells differentiating into specific lineages (Chang et al., 2008; Macfarlan et al., 2012; Chen et al., 2012), hinting at the possibility that the stem cell population dynamically maintains distinct subsets prone to differentiate into alternate lineages. Until our study detailed in the succeeding text (Yuan et al., 2016), however, direct evidence for stochastic phenotype switching in mammals was only available in vitro. Its potential role in adult tissue homeostasis was beyond our reach.

Noise-mediated endothelial cell heterogeneity in vivo

Dynamic heterogeneity of vWF expression in vivo

vWF is a large glycoprotein secreted into the bloodstream by endothelial cells and megakaryocytes (Ruggeri, 2007). Inside our vessels, *vWF* binds and stabilizes *Factor VIII*, a key coagulation cascade protein primed for activation by *vWF*. In addition, *vWF* forms a bridge between collagen exposed under injured endothelium and platelets, aiding blood clot formation under high shear flow (Peyvandi et al., 2011). Expression of *vWF in vivo*, however, does not correlate well with its known sites of action (Yamamoto et al., 1998; Aird et al., 1997). For example, *vWF* is more abundant in veins and venules than in arteries and arterioles, even though its coagulative role is more critical in arteries that carry high shear flow. It is also expressed by capillaries in the heart, skeletal muscle, brain, and lung, but not those of the kidney or liver (Yuan et al., 2016). The function of capillary *vWF* remains unclear. It is possible that adequate *vWF* secretion into the bloodstream requires a larger endothelial cell area than the large vessels alone can supply. In line with this, capillary *vWF* expression in lung tissue is increased by hypoxia, likely boosting *vWF* plasma levels across the body (Anahita et al., 2013). Alternatively, capillary *vWF* may act locally to promote clothing in damaged microvessels (Peyvandi et al., 2011), modulate barrier function (Suidan et al., 2013), or regulate angiogenesis (Starke et al., 2011; Randi et al., 2013).

One of the most intriguing facets of *vWF* expression in the capillary vasculature is its mosaic ON/OFF pattern (Fig. 14.3) (Yuan et al., 2016). Variations in local tissue architecture could in theory create the signal differences responsible, but the mosaic shows no apparent spatial structure. Alternatively, the history of the vessel may explain the pattern, as subsets of cells could have experienced different signals in the past that locked them into distinct states. Finally, the mosaic may be dynamically generated by the stochastic toggling of *vWF* between ON and OFF transcriptional states, or by the local secretion of *vWF* protein from Weibel-Palade bodies. In order to investigate the origins of this mosaic, we generated a pair of mouse reporter lines (Yuan et al., 2016). First, we replaced a *vWF* allele with a *LacZ* reporter gene driven by the endogenous *vWF* promoter (Fig. 14.3A). This $vWF^{LacZ/+}$ line allowed us to rule out differential protein expression or release as the source of the ON/OFF mosaic, as transcription of the reporter *LacZ* gene followed the same pattern as that of the protein (Figs. 14.3C-i and 14.3C-v). Second, we generated a fate-mapping mouse model that allowed us to track cumulative *vWF* expression in individual cells (Fig. 14.3B). Here we replaced a *vWF* allele with *Cre* recombinase and crossed the $vWF^{Cre/+}$ mice with the ROSA26R reporter line. This reporter has a *LoxP*-flanked STOP cassette blocking the expression of *LacZ*, inserted at the constitutively active Rosa26 locus. In double transgenic offspring (*vWF*–*Cre*–ROSA26R mice), transient activation of the *vWF* promoter driving *Cre* expression leads to the removal of the STOP codon in front of *LacZ*, resulting in its

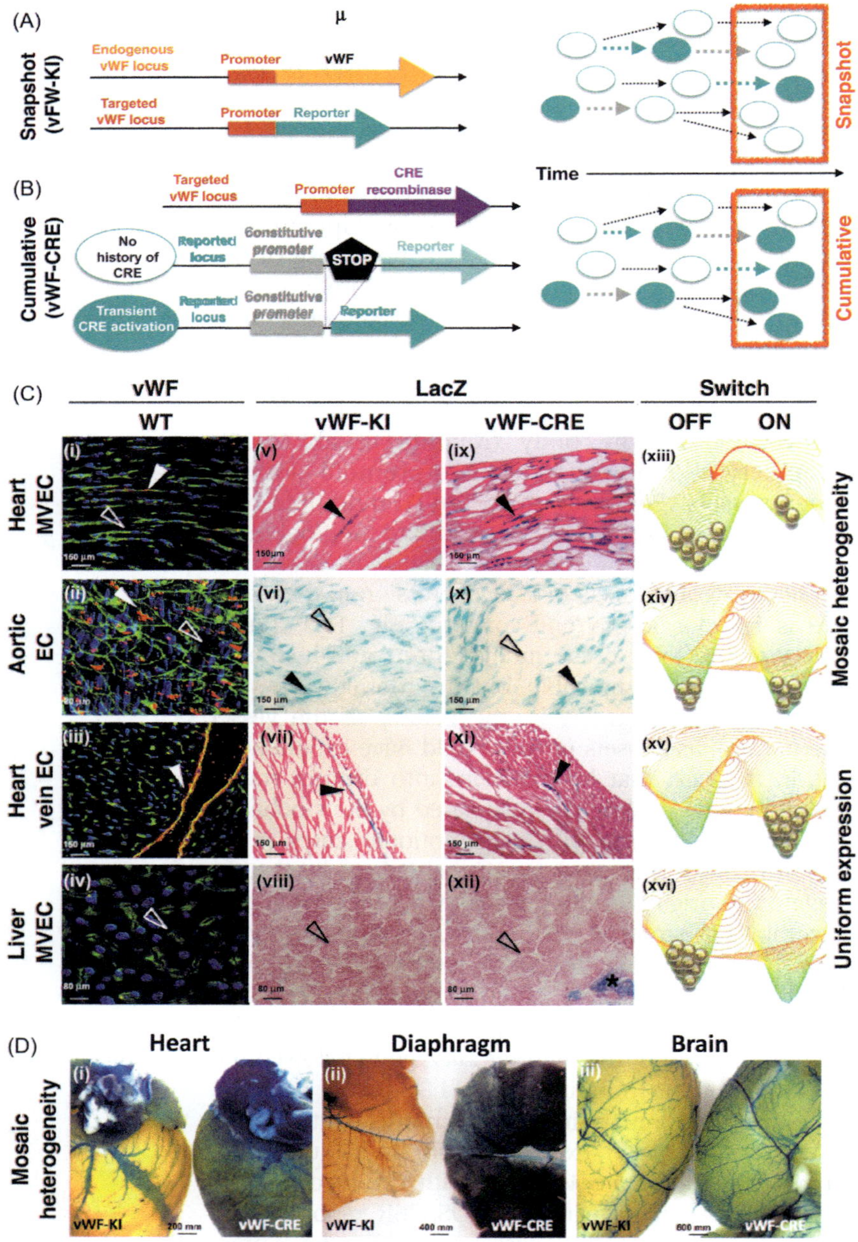

Figure 14.3 In vivo *vWF* expression displays organ-specific mosaic heterogeneity. (A) *Left*: schematics of gene targeting strategy for creating a mouse a model to assay *vWF* gene expression in a snapshot in time ($vWF^{LacZ/+}$ or *vWF*-KI mouse). The *LacZ* reporter gene is targeted to the endogenous *vWF* locus; *LacZ* expression reflects the

(*Continued*)

permanent and heritable *LacZ* expression. Thus, *LacZ* activity in our *vWF*–*Cre*–ROSA26R mice reflects cumulative expression over the lifetime of the animal. This mouse line allowed us to probe the temporal dynamics of *vWF*, and test whether its mosaic expression in the capillaries changed over time.

Comparison of capillary *LacZ* expression between $vWF^{LacZ/+}$ and *vWF*–*Cre*–ROSA26R mice points to a highly dynamic *vWF* promoter. In contrast to mosaic *LacZ* expression in our snapshot $vWF^{LacZ/+}$ line (Fig. 14.3C-v), capillary endothelial cells in the heart, lung, skeletal muscle, and brain of adult *vWF*–*Cre*–ROSA26R mice were ~100% *LacZ* positive (Figs. 14.3C-ix and 14.4D). Thus, virtually every capillary endothelial cell had expressed *LacZ* (hence *vWF*) at some point during the lifetime of the animal. As only a minority of these cells are positive at any given time, the data is consistent with stochastic ON/OFF transitions of the *vWF* promoter driving a dynamically changing mosaic.

Interestingly, mosaic *vWF* expression is also present in the aorta of adult mice (Fig. 14.3C-ii, vi). In contrast to its dynamic nature in capillaries, the aortic mosaic is frozen in time. This is evident from the patchy expression of *LacZ* in the cumulative

◀ current transcriptional state of the endogenous *vWF* promoter. *Right*: *LacZ* expression in individual endothelial cells over time; *filled/open circles*: ON/OFF; *black/blue/gray arrows*: no change in promoter activity/OFF→ON transition/ON→OFF transition. (B) *Left*: schematics of gene targeting strategy for a fate-mapping mouse model that allows for assessment of cumulative expression over time (*vWF*–*Cre*–ROSA26R mouse). *Cre* recombinase is targeted to the endogenous *vWF* locus, and the resulting $vWF^{Cre/+}$ mice are bred to the ROSA26R reporter line. In ROSA26R mice *LacZ* has been targeted to the ubiquitously expressing ROSA26 locus, which contains the functional equivalent of a stop codon flanked by *loxP* sites. In double transgenic offspring, cells that express from the *vWF* promoter result in *Cre*-mediated excision of the stop codon, and permanent expression of *LacZ* in that cell and all of its progeny. Thus, *LacZ* expression reflects the cumulative (present and past) transcriptional state of the endogenous promoter. *Right*: *LacZ* expression in individual endothelial cells over time; *filled/open circles*: ON/OFF; *black/blue/gray arrows*: no change in promoter activity/OFF-ON transition/ON-OFF transition. (C) Double immunofluorescence for *vWF* (red) and *CD31* (green) of tissue sections and en face aortae from adult male wild-type (WT) mice (i–iv), and *LacZ* staining of tissue sections and en face aortae from adult male *vWF*–KI mice (v–viii) and *vWF*–*Cre* mice (xi–xii). (i, v, ix) Heart sections with capillaries cut longitudinally. (ii, vi, x) The inner lining of an aorta that has been opened and laid flat under a coverslip. (iii, vii, xi) Heart sections with capillaries and veins cut in cross section. (iv, viii, xii) Liver sections with hepatocytes surrounded by sinusoids (arrows). (i–xii) *solid /empty arrows*: *vWF*- or *LacZ*-expressing/non-expressing cells; *: *LacZ*-expressing hepatocytes. (xiii–xvi) bistable switch regulating *vWF* expression, dynamically toggling between ON and OFF in heart capillaries, frozen into one of two states in the aorta, or locked into a single ON or OFF state in heart veins and liver sinusoidal endothelial cells, respectively. (D) Whole-mount *LacZ* staining of heart, muscle (diaphragm), and brain from adult male *vWF*–KI and *vWF*–*Cre* mice. *Scale bars*: (a–i) 100 mm; (q–s) 3 mm. $n = 3$ with three replicates. *Source*: Adapted from Yuan, L., Chan, G.C., Beeler, D., Janes, L., Spokes, K.C., Dharaneeswaran, H., et al., 2016. A role of stochastic phenotype switching in generating mosaic endothelial cell heterogeneity. Nat. Commun. 7, 10160 (https://www.nature.com/articles/ncomms10160).

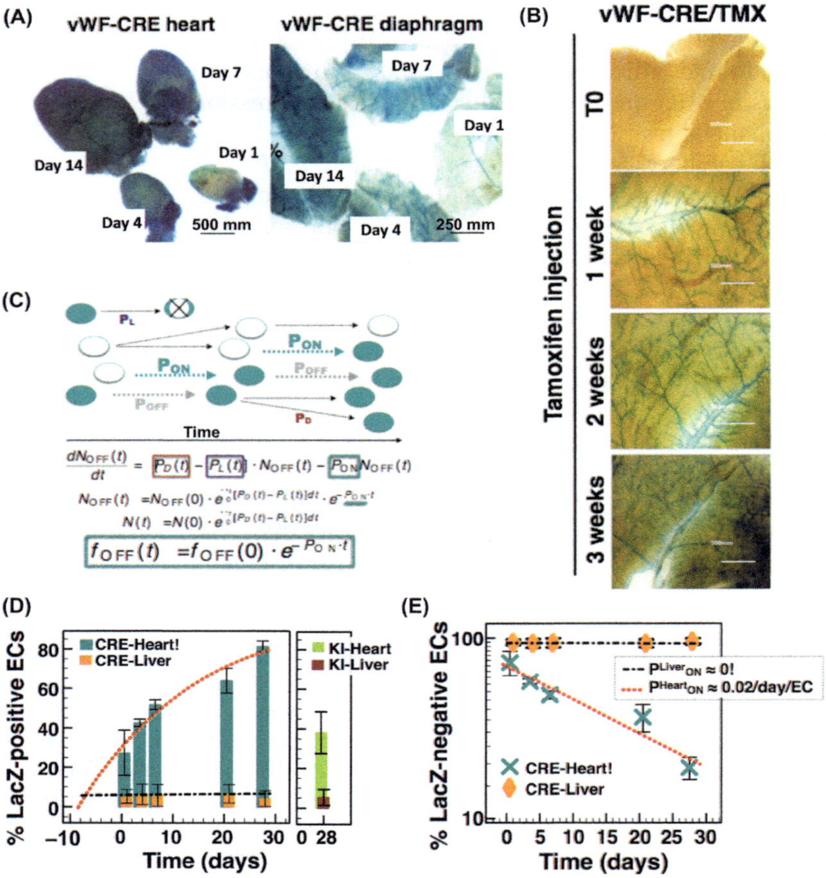

Figure 14.4 *vWF* toggles ON/OFF in the capillary endothelium of some, but not all adult organs. (A) Whole-mount *LacZ* staining of *vWF–Cre*–ROSA26R (*vWF–CRE*) hearts and diaphragms collected at indicated time points following birth (stained together and imaged side by side). *Scale bar*: 5 mm. (B) Time course of whole-mount diaphragm *LacZ* staining from tamoxifen-treated *vWF–CreERT2*–ROSA26R (*vWF–CRE/TMX*) mice. Time 0 (T_0) indicates no exposure to tamoxifen. The results reveal time-dependent accumulation of *LacZ* in the capillaries, which are arranged in linear arrays parallel to the plain of the tissue. *Scale bar*: 2.5 mm. (C) Exponential decrease of *LacZ*-negative endothelial cells (ECs) in *vWF–Cre*–ROSA26R mice modeled by assuming random OFF–ON transitions with constant, division-independent rate P_{ON}, time-dependent cell division with rate $P_D(t)$ and cell loss with rate $P_L(t)$. *Teal circle/white circle/circle with X*: *LacZ*-positive/*LacZ*-negative/dead cell; *black/blue/gray arrows*: no promoter transition/OFF-ON transition/ON-OFF transition of the promoter, masked by locked-in *LacZ* expression. (D) *Left*: time-dependent *LacZ* accumulation in *vWF–Cre*–ROSA26R mouse heart and liver capillaries (*LacZ/CD31* double stained sections); *right*: *LacZ*-positive EC percentage in heart and liver capillaries of 4-week-old $vWF^{LacZ/+}$ mice (*vWF*–KI). (E) Log-linear plot of the *LacZ*-negative EC fraction, fitted by an exponential with rate $P_{ON} = 0.02$/day/EC for heart, and $P_{ON} = 0$/day/EC for liver. $n = 3$ with three replicates. *Source*: Adapted from Yuan, L., Chan, G.C., Beeler, D., Janes, L., Spokes, K.C., Dharaneeswaran, H., et al., 2016. A role of stochastic phenotype switching in generating mosaic endothelial cell heterogeneity. Nat. Commun. 7, 10160 (https://www.nature.com/articles/ncomms10160).

vWF–*Cre*–ROSA26R mice (Fig. 14.4C-x), a pattern that is remarkably stable in time. Indeed, we have found *vWF*-negative cell clusters that have *never* expressed *vWF* in the aortic wall of ~2-year-old mice (near the end of their natural life span; see Yuan et al., 2016). Unlike the aorta, both vein and venule endothelial cells were consistently *LacZ* (*vWF*) positive in both mouse lines (*LacZ*, Fig. 14.3C-iii, vii, xi), while capillary cells in the kidney glomerular and liver sinusoidal vasculature were permanently *LacZ* (*vWF*)-negative (Fig. 14.3C-iv, viii, xii shows liver). Taken together, *vWF* expression across the mouse vasculature is consistent with the existence of a bistable switch driving *vWF* promoter activity. This switch toggles between *vWF* ON/OFF states in a subset of capillary beds (Fig. 14.3C-xiii), creates a robust barrier between stable ON/OFF states in the aorta (Fig. 14.3C-xiv), it is locked ON in veins (Fig. 14.3C-xv), and locked OFF in liver/kidney capillaries (Fig. 14.3C-xvi).

In order to prove that the capillary mosaic is indeed dynamic in adult tissue—rather than a relic of an early developmental window where the ancestors of all heart, lung, muscle, and brain capillary cells were briefly *vWF* positive, we first documented the increase in capillary *vWF* expression in *vWF*–*Cre*–ROSA26R mice following birth (Fig. 14.4A). Whole-mount staining of newborn *vWF*–*Cre*–ROSA26R mouse hearts and diaphragms shows faint staining outside of large veins. As the mice grow, there is a gradual increase in *LacZ* across the tissue, indicating that the number of capillaries carrying the memory of a *vWF*-positive past grows over time. To rule out extensive cell proliferation characteristic of young tissue as a potential cause of temporary *vWF* expression after birth, and to explicitly demonstrate that *vWF* expression flickers in healthy adult capillaries, we generated an inducible version of our *vWF*–*Cre*–ROSA26R fate mapping line. Here we used a tamoxifen-inducible version of *Cre* recombinase in the procedure shown on Fig. 14.3B, which allowed us to choose the start time after which cells start "remembering" *vWF* promoter activity (*vWF*–*CreERT2*–ROSA26R line). As Fig. 14.4B shows, tracking the cumulative expression of *vWF* in adult *vWF*–*CreERT2*–ROSA26R mice reveals a gradual filling in of the capillary vasculature on a similar timescale to that of newborn animals. These data clearly demonstrate that (i) at any given time, the majority capillary cells of the healthy heart, diaphragm, lung, and brain are *vWF*-negative in adult animals, but (ii) all capillary cells go through brief *vWF*-positive periods with the approximate frequency of 3–4 weeks.

The remarkable plasticity of *vWF* expression in adult capillaries led us to investigate whether the spontaneous toggling between *vWF* ON and OFF states was due to biological noise. If *vWF* ON/OFF states are driven by a bistable switch, the barrier of this switch may be sufficiently low to allow for stochastic transitions in the capillaries of heart, skeletal muscle, lung, and brain, yet high enough to preclude such transitions in the aorta (Fig. 14.3C). Furthermore, if dynamic *VWF* expression is the consequence of a constant background of biological noise toggling this switch with a fixed probability, then we expect the number of *LacZ*-negative cells in *vWF–Cre–*ROSA26R heart capillaries to decrease exponentially over time (Fig. 14.4C). Thus, we quantified the time-dependent accumulation of *LacZ*-positive cells in *vWF–Cre–*ROSA26R heart and liver capillaries, coimmunostained for *LacZ* and pan-endothelial *CD31* (Fig. 14.4D). Indeed, the time-dependent percentage of *LacZ*-negative endothelial cells in heart capillaries fell along a log-linear fit (turquoise bars/crosses and red curve/line on Fig. 14.4E), with a slope indicating a *vWF* OFF —> ON transition rate of $P_{ON} = 0.02 \pm 0.003$/day/endothelial cell. Assuming that the *LacZ*-positive fraction of endothelial cells in adult hearts, $f_{ON} = 38\%$, is the result of a dynamic equilibrium between ON and OFF transitions, we calculated $P_{OFF} = P_{ON}(1 - f_{ON})/f_{ON} = 0.03$/day/cell. In contrast, liver capillary endothelial cells were locked into a *vWF*-negative state (94.5%, constant over time; orange/dark-red bars; orange diamonds on Fig. 14.4D and E). Together, these findings are consistent with random, noise-driven *vWF* heterogeneity.

Dynamic vWF mosaicism in vitro is driven by biological noise

Our in vivo studies cannot rule out the possibility that dynamic *vWF* heterogeneity is a largely deterministic cellular response to a variable, fluctuating microenvironment near individual capillary endothelial cells. To test whether cells growth in a uniform environment can nevertheless stochastically generate a *vWF* mosaic, we turned to cell culture. Using fluorescence in situ hybridization (FISH) and immunofluorescence assays, we probed single-cell *vWF* expression in primary endothelial cell monolayers cultured from different organs and vessel types. In contrast to the prevailing view that *vWF* is a universal endothelial cell marker (especially in cell culture), our in vitro results echoed those in vivo. Namely, we have found that *vWF* expression follows a mosaic pattern in all tested endothelial cultures, from mouse or human tissue alike. Specifically, *vWF* showed a mosaic ON/OFF expression pattern at the mRNA as well as protein level in human umbilical vein, coronary artery, and pulmonary artery endothelial cells (Fig. 14.5A shows human coronary artery; also see Fig. 14.3 in Yuan et al., 2016). While endothelial-specific *VE-cadherin* is expressed at a similar level in all cells, distributions of *vWF* mRNA normalized to *VE-cadherin* were bimodal; as were *vWF* protein distributions assayed with flow cytometry. Moreover, clonally expanded endothelial cell populations obtained via limiting dilution recreated the same mix of *vWF* low- and high-expressing cells as their parent populations (Fig. 14.5B shows human aortic endothelial cells). These findings further support the idea that *vWF* expression is controlled by a bistable switch, the in vitro barrier of which is low enough to allow the dynamic ON/OFF toggle required to generate stochastic phenotypic heterogeneity. Indeed, under in vitro conditions even human aortic endothelial cells generate dynamic mosaics, even though they are robustly locked into a static mosaic in the mouse aorta.

Epigenetic control of vWF expression and heterogeneity

The slow rate of ON/OFF transitions measured in vivo point to a surprising stability of both the ON and OFF *vWF* transcriptional state. As both are stable on the timescale of weeks (switching rate of ~2–3 weeks/cell), we next asked whether

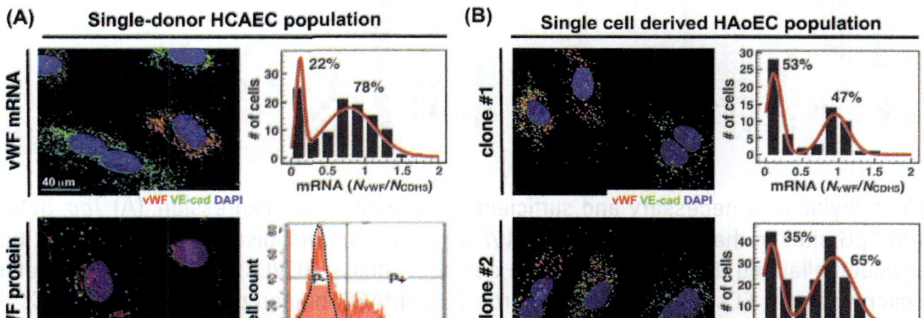

Figure 14.5 *vWF* is expressed as a dynamic mosaic in cultured endothelial cells. (A) *vWF* and *VE-cadherin* mRNA (top) and protein (*bottom*) distribution in human coronary artery endothelial cells (HCAEC). *Left*: *Top*—fluorescence in situ hybridization (FISH) for *vWF/VE-cadherin* (*red/green*); *bottom*—immunostaining for *vWF/VE-cadherin* (*green/red*); *blue*: nuclear DAPI. *Right*: *Top*—*vWF* mRNA distributions in single cells, normalized to *VE-cadherin*; *bottom*—flow cytometry analysis of IgG control (*dotted line*) and *vWF* protein expression (*red*); P −/P +: *vWF* negative/positive. (B) Clonal populations derived from a single HAoEC. *Left*: FISH for *vWF/VE-cadherin* (*red/green*); *blue*: nuclear DAPI. *Right*: *vWF* mRNA distribution normalized to *VE-cadherin*. *Scale bar*: 20 mm; $n = 6$ with three replicates. *Source*: Adapted from Yuan, L., Chan, G.C., Beeler, D., Janes, L., Spokes, K.C., Dharaneeswaran, H., et al., 2016. A role of stochastic phenotype switching in generating mosaic endothelial cell heterogeneity. Nat. Commun. 7, 10160 (https://www.nature.com/articles/ncomms10160).

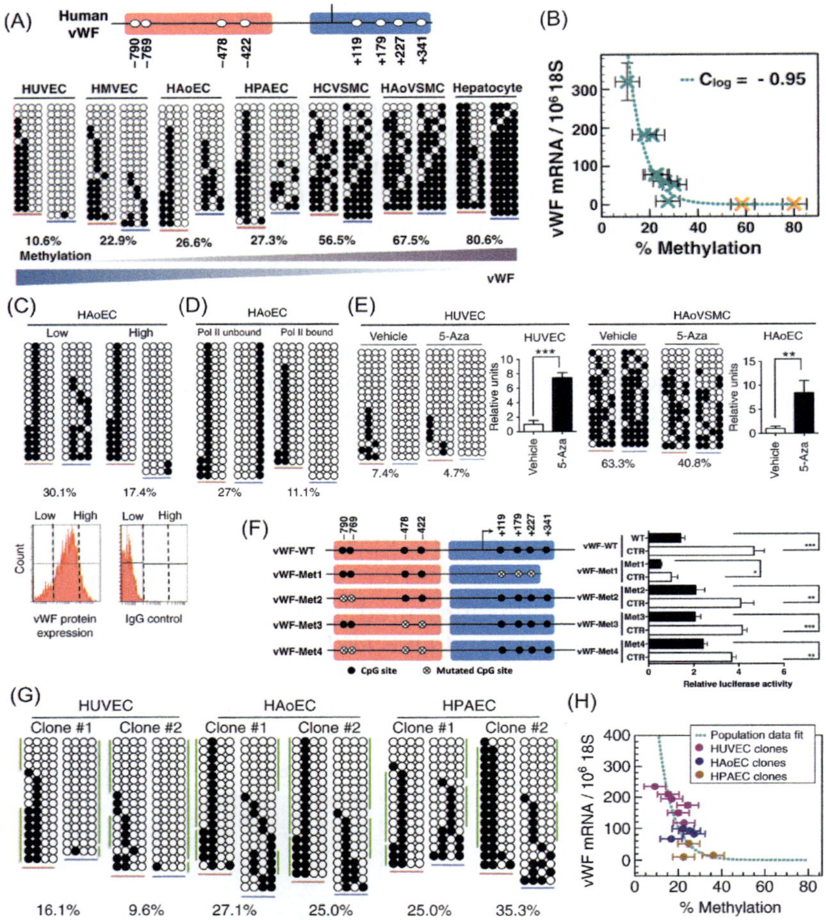

Figure 14.6 DNA methylation is necessary and sufficient for silencing *vWF* expression. (A) *Top*: Schematic showing the eight CpG sites in the proximal human *vWF* promoter; *bottom*: bisulfite sequencing of the human *vWF* promoter in four endothelial and three nonendothelial cell populations; HUVEC: human umbilical veins ECs; HMVEC: human cardiac microvascular ECs; HAoEC: human aortic ECs; HPAEC: human pulmonary artery ECs; HCVSMC: human coronary artery vascular smooth muscle cells; HAoVSMC: human aortic vascular smooth muscle cells. Each continuous row of circles represents the upstream/core region of a single *vWF* allele; ●/○: methylated/unmethylated CpG; *pink/blue underline*: upstream/core promoter regions; *percentages*: methylated/total CpG for all eight CpG sites. (B) *vWF* promoter DNA methylation versus *vWF* mRNA levels (relative to 10^6 18S copies) in endothelial (*teal X*) and nonendothelial (*orange X*) cell populations; *teal dashed line*: log-normal fit (exponential decrease of mRNA expression with DNA methylation); *vertical/horizontal error bars*: standard deviation of mRNA expression/estimated DNA methylation sampling error. (C) Top, Bisulfite sequencing of the human VWF promoter in human aortic endothelial cells that were FACS-sorted into subpopulations that expressed low/high levels of VWF. Bottom, FACS profiles of VWF expression (left) and IgG control (right). Vertical lines indicate gating thresholds used for sorting. (D) Pol II ChIP assay followed by bisulfite sequencing of Pol II-bound and -unbound chromatin in HAoEC. (E) VWF promoter methylation (left) and mRNA expression (right) of VWF following 5-azacytidine (5-Aza)

(*Continued*)

there may be an epigenetic mechanism at play in stabilizing the two distinct outcomes (Jones, 2012; Jeltsch and Jurkowska, 2014; Singer et al., 2014). DNA methylation of CpG dinucleotides near a gene's promoter can robustly silence transcription (Jones, 2012), though this mechanism is generally associated with the lifelong silencing of unused transcriptional programs in differentiated cells (e.g., stem cell and alternate lineage genes). In contrast to the dense islands of CpG dinucleotides often involved in permanent silencing events (Deaton and Bird, 2011), the human *vWF* promoter has only eight CpG sites between nucleotide positions -900 and +400 around its transcriptional start site (Fig. 14.6A). To test whether DNA methylation at these CpG sites impacts *vWF* expression, we carried out sodium bisulfite genomic sequencing in several endothelial and nonendothelial cell types. Sodium bisulfite sequencing reveals the methylation state of the eight CpG sites on individual alleles from a pool of cellular DNA; each sampled allele is represented by a single line on Fig. 14.7A. As the figure shows, human endothelial cells have low levels of DNA methylation at their *vWF* promoter (11–28% of the eight CpG sites). In contrast, *vWF*-negative lineages such as vascular smooth muscle cells and hepatocytes show high CpG methylation (59–81% of the eight CpG sites). Moreover, *vWF* expression is inversely correlated with the average methylation of CpGs near the promoter; a pattern that holds true in mouse cell lines as well (Fig. 14.7B shows human cells). As DNA methylation is generally accompanied by heterochromatin-forming histone modifications, we have

treatment in HUVEC and HAoVSMC. VWF mRNA expression levels are shown relative to untreated vehicle controls. (F) Luciferase assay for methylated versus unmethylated VWF promoter/reporter constructs. *Left:* generation of 50 VWF promoter/reporter constructs with mutations (CG-TG) at indicated CpG sites (VWF promoter/reporter constructs were methylated in vitro using SssI methylase). *Right:* Luciferase activity in HUVEC transiently transfected with mock-methylated (white) or methylated (black) VWF constructs (normalized to control vector) in HAoVSMC. Error bars indicate s.d.; $n = 3$, with three replicates; $^{*}p = 0.05$; $^{**}p = 0.01$; $^{***}p = 0.001$ (two-sided T-test). (G) Bisulfite sequencing of the human VWF promoter in clonally expanded endothelial cells (ECs). Each continuous row of circles represents the upstream/core region of a single VWF allele. The upstream/core promoter regions are marked by pink/blue underlines, respectively. Vertical green lines indicate the two dominant DNA methylation patterns, which represent the most likely methylation patterns of the two VWF alleles in the original single cell. Percentages refer to methylated/total CpG across all eight CpG sites. $n = 5$, with three replicates. (H) VWF promoter DNA methylation versus VWF mRNA expression in clonally expanded ECs. Multiple independently derived clonal populations were bisulfite sequenced for each EC type. Teal dashed line shows the log-normal dependency of VWF mRNA expression on average DNA methylation, derived from the *parent* ECs and non-EC populations shown in (B). *Source*: Adapted from Yuan, L., Chan, G.C., Beeler, D., Janes, L., Spokes, K.C., Dharaneeswaran, H., et al., 2016. A role of stochastic phenotype switching in generating mosaic endothelial cell heterogeneity. Nat. Commun. 7, 10160 (https://www.nature.com/articles/ncomms10160).

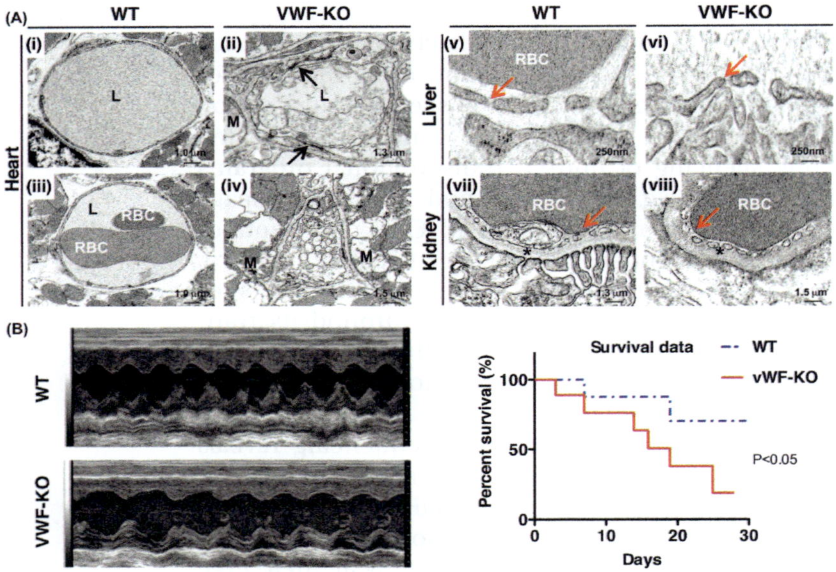

Figure 14.7 Absence of *vWF* is associated with abnormal cardiac phenotype. (A) Electron microscope images of wild type (*left*) and *vWF*$^{-/-}$ mouse (*right*) heart (i–iv), liver (v, vi), and kidney (vii, viii) capillary walls. *L*: lumen; *RBC*: red blood cells; *M*: injured mitochondria in myocytes near the capillary vessel (ii, iv); *black arrows*: normal electron-dense lateral borders/junctions (ii, iv); *red arrows*: normal fenestra showing functional liver and kidney vessel architecture; asterisk: basal lamina; *Scale bar*: 1 μm; $n = 3$, with three replicates. (B) Echocardiography in WT and KO mice. (C) Survival of WT versus *vWF* KO mice subjected to aortic banding; $n = 8$/group. *Source*: Adapted from Yuan, L., Chan, G.C., Beeler, D., Janes, L., Spokes, K.C., Dharaneeswaran, H., et al., 2016. A role of stochastic phenotype switching in generating mosaic endothelial cell heterogeneity. Nat. Commun. 7, 10160 (https://www.nature.com/articles/ncomms10160).

shown that high *vWF* -expressing endothelial cell lines such as those from human umbilical vein have lower levels of the repressive di- and trimethylated H3K9 marks near their promoter, as well as higher densities of permissive acetylated histone H3 than endothelial cell lines with lower average *vWF* expression (e.g., pulmonary arterial cells; see Yuan et al., 2016). According to this data, population-level DNA methylation and histone modifications indicative of chromatin packaging correlate well with *vWF* promoter activity, across endothelial as well as non-endothelial cell lines.

In order to show that DNA methylation is lower in *individual* vWF-positive cells within an isogenic *vWF* mosaic, we used flow-activated cell sorting to isolate single-donor human umbilical vein endothelial cells with the highest (top 10%) and lowest (bottom 10%) levels of *vWF* protein. We then compared their DNA methylation using bisulfite genomic sequencing. As expected, the *vWF*-negative subpopulation had significantly higher levels of CpG methylation at their *vWF* promoter than the *vWF*-high fraction (Fig. 14.7C, also see Yuan et al., 2016 for aortic endothelial cell). Immunoprecipitating transcriptionally active chromatin fragments with antibodies against RNA polymerase II allowed us to directly compare CpG methylation in Pol II-bound and -unbound chromatin. This assay showed that Pol II-bound *vWF* promoter regions have low- to no CpG methylation, especially at the region immediately following the transcription start site (Fig. 14.6D). The presence of repressive H3K9-trimethylated histones correlated with DNA methylation on Pol II-bound versus unbound DNA (Cedar and Bergman, 2009; Vaissière et al., 2008). In order to prove that DNA methylation is *necessary* for silencing *vWF* in individual cells, we measured *vWF* expression in various cells in the absence or presence of the DNA methyltransferase inhibitor 5-azacytidine (5-Aza). As 5-Aza partially demethylated the *vWF* promoter, a seven- to eightfold increase in *vWF* mRNA expression was observed in human umbilical vein endothelial cells. Moreover, we detected de novo expression in otherwise *vWF*-negative vascular smooth muscle cells (Fig. 14.6E). In a complementary experiment designed to show that DNA methylation is *sufficient* for silencing *vWF*, we coupled the human *vWF* promoter to a *LacZ* reporter and transfected this construct into endothelial cells on an episomal plasmid. Artificial DNA methylation of the CpG sites in these plasmids showed that the methylated promoter is significantly weaker than its unmethylated counterpart, even in the absence of the broader chromatin environment of the endogenous promoter (Fig. 14.6F). Taken together, these results indicate that DNA methylation of the *vWF* promoter is the proximal cause of the *vWF* OFF state, likely responsible for its stability on the scale of weeks.

Our epigenetics data raise the intriguing possibility that biological noise generates dynamic changes in DNA methylation, which result in transitions between ON and OFF promoter states. To test this, we performed bisulfite sequencing on clonal populations derived from single endothelial cells. These experiments reveal mixtures of CpG methylation patterns that cannot be attributed to the faithful inheritance of only two alleles present in the parental clone (Fig. 14.6G; see Yuan et al., 2016 for a likelihood estimate that the additional patterns are due to experimental error). Moreover, average *vWF* methylation in independently derived clones is inversely correlated with their *vWF* expression, obeying similar scaling to single-donor parental populations (Fig. 14.6H). A mark of active, ongoing DNA methylation/demethylation is the presence of hemimethylated CpG dyads caught in transition between symmetrically methylated and demethylated states (Pfeifer et al., 1990). Hairpin bisulfite sequencing can reveal these by linking the top and bottom strands of double-stranded DNA segments with a short hairpin, allowing us to probe methylation on both sides of individual CpG dyads (Laird et al., 2004). Using this technique, we have found that the methylation status of the three CpG sites near the *vWF* start site was relatively stable in cells at the highest and lowest ends of *vWF* expression (Pfeifer et al., 1990). In both cases, most CpG dyads were symmetrically unmethylated (*vWF*-rich human umbilical vein endothelial cells) or symmetrically methylated (*vWF*-negative human coronary artery vascular smooth muscle cells). In contrast, we observed a high incidence of hemimethylated dyads in human cardiac microvascular and pulmonary artery endothelial cells, both of which express medium levels of *vWF.* Together, our findings strongly suggest that both *vWF* expression and CpG methylation flickers ON and OFF in individual isogenic cells.

Loss of vWF-positive cells from mosaic capillary beds correlates with profound impairment of tissue function

Loss of vWF mosaic expression in heart capillaries

The slow, stochastic toggle of *vWF* expression in the healthy capillaries of the mouse heart, muscle, lung, and brain are intriguing because they raise the possibility that this changing pattern carries a biological function. However, a direct test of the requirement of this blinking mosaic for healthy tissue function is an experimental challenge. We approached it by hypothesizing that if mosaic *vWF* expression in the heart has an adaptive role, then loss of *vWF* expression from the minority of normally *vWF*-positive cells might lead to a cardiac phenotype in *vWF*-null mice. This mouse line has been well characterized prior to our study and found to display the expected bleeding phenotype due to the loss of procoagulative, circulating *vWF* (Denis et al., 1998). Our focus on their heart, however, revealed profound cell-level abnormalities as well as impaired function. Electron microscopy revealed swollen and electron-lucent capillary endothelial cells both ventricles (Fig. 14.7A, left). Excessive blebbing at the inner capillary surface left the lumen packed with membrane-bound vesicles. In spite of extensive damage, the lateral borders of endothelial cells were intact, the nuclei appeared normal, and the cytoplasm contained many polyribosomes, arguing against a preparation artifact. In neighboring cardiomyocytes, mitochondria were often swollen and electron-lucent. By contrast, the ultrastructure of the aorta, which also lost *vWF* expression from a large fraction of its cells, the normally *vWF*-negative kidney, and the liver all remained undamaged (Fig. 14.7A, right shows liver and kidney).

To determine whether the cell-level abnormalities observed in *vWF*-null mice impact cardiac function, we performed echocardiography (Fig. 14.7B) and pressure-volume loop experiments. These studies were suggestive of systolic cardiac dysfunction. To test whether *vWF*-null hearts were sensitive to stress, we performed transverse aortic constriction; a mouse model of chronic pressure overload, left ventricle dysfunction, and heart failure (deAlmeida et al., 2010). Even though *vWF* knockout mice do not show increased mortality in the absence of this procedure, their 30-day survival post transverse aortic constriction was significantly reduced (Fig. 14.7C). In contrast, functional assays for liver and kidney showed no change (Yuan et al., 2016). Together, these findings indicate that *vWF* expression is necessary for cardiac health.

The above observations do not reveal the mechanism by which *vWF* contributes to cardiac health and stress resistance. Its presence does not appear to be procoagulative, as we did not observe any sign of plasma leak or red blood cell extravasation in the capillary bed of *vWF*-null hearts. In addition to a role in hemostasis, *vWF* also controls the storage and release of endothelial *Angiopoietin 2* (*Ang2*) (Rondaij et al., 2006). Secreted *Ang2* can interfere with quiescence-promoting *Ang1-Tie2* signaling in nearby capillary endothelium, promoting an inflammatory, and proangiogenic transcriptional response (Fiedler and Augustin, 2006). Endothelial cells are known to secrete more *Ang2* in the absence of *vWF* (Starke et al., 2011). They also respond more potently to proangiogenic signaling mediated by *VEGFR2* (Starke et al., 2011; Randi et al., 2013). To test whether altered *Ang2* levels may be responsible for the damaged capillaries of *vWF*-null mice, we measured *Ang2* protein levels in blood serum, heart, and liver of *vWF*-null versus wild-type mice. Interestingly, in spite of the global loss of *vWF* in out knockout animals, circulating (serum) *Ang2* levels were unaltered. In contrast, *Ang2* protein expression was significantly increased in the heart, but not the liver. This localized increase in *Ang2* was due to a heart specific, 50% upregulation of *Ang2* mRNA expression; a strong effect considering that only ~38% of heart endothelial cells express *vWF* in wild-type mice at any given time. These data raise the intriguing possibility that in the healthy heart the changing mosaic of *vWF*-positive cells create a mixture of vessel segments that balance two mutually exclusive functions of the endothelium. On one hand, *vWF*-positive patches are well suited to establish stable and mature capillary vessels with good pericyte coverage, a process that requires low *Ang2* secretion (Gaengel et al., 2009). On the other hand, *vWF*-negative sections are primed to respond to proangiogenic and proinflammatory signals (Starke et al., 2011), potentially aiding rapid vascularization in response to damage. The dynamic nature of this mosaic may help guarantee that none of the heart tissue is exposed to *Ang2*'s inflammation-promoting effects over extended periods of time. By balancing *Ang2* secretion in both space and time, the dynamic *vWF* mosaic may help vessels remain mature and quiescent across the capillary bed.

The idea that dynamic mosaic *vWF* expression functions as a tissue-level "bet hedging" strategy, balancing two mutually exclusive endothelial phenotypes with their respective pros and cons, requires further evidence. First, a heart microvessel specific *vWF* knockout animal could be needed to confirm that the *vWF* required for healthy cardiac function must be present in the heart capillaries, rather than in circulation. Next, an animal model with uniform *vWF* expression across all heart capillary endothelial cells could help test whether continuous availability of *vWF* poses different dangers. For example, Weibel-Palade bodies in *vWF*-expressing cells may lead to the storage of as significant pool of *Ang2* and *P-selectin,* normally inert inside these organelles (Rondaij et al., 2006). In tissues where all capillary endothelial cells store *vWF, Ang2,* and *P-selectin,* rapid increase of thrombin or histamine in the blood could set off a sudden release of all three proteins from a large number of cells. This could overwhelm the circulation by pushing their levels too far up. In contrast, a mosaic vasculature can balance ongoing, steady secretion from *vWF*-negative cells with sudden release from the *vWF*-positive subpopulation. To prove the need for a *dynamic* mosaic to achieve this balance, we would ideally engineer a transgenic mouse in which we "freeze" the capillary mosaic pattern in an adult animal, then observe the morphology of its heart capillaries and cardiac health over time. This is a technical challenge, as we would not only need to turn on a *vWF*-expressing vector permanently in cells that are *vWF*-positive at a given moment, but at the same time permanently eliminate *vWF* from all negative cells. If created, however, this mouse line would allow us to probe for localized capillary damage over time and test the importance of shifting the two phenotypes/tasks from one spot to another.

Loss of vWF mosaic expression in brain capillaries

In addition to heart and skeletal muscle, our studies revealed a slowly changing dynamic *vWF* mosaic in brain capillaries. As with the heart, the role of capillary *vWF* expression in brain microvessels is not well understood. There is evidence that *vWF* plasma levels are elevated in patients with stroke (Folsom et al., 1999), traumatic brain injury (Villalba et al., 2017), and cerebral venous sinus thrombosis (Burns et al., 1995), but neither the cause or effect of elevated *vWF* has been established. A few years before our mosaic study, Suidan et al. used *vWF* loss in the brain microvasculature to probe its contribution to pathophysiology (Suidan et al., 2013). They showed that brain-specific local secretion of *vWF* is required for loosening the blood–brain barrier during brain injury, and that this loosening is protective rather than detrimental. Breakdown of the blood–brain barrier is a hallmark of pathological conditions within the central nervous system (Kaur and Ling, 2008). It occurs during prolonged hypoxia (Schoch et al., 2002), epilepsy (Marchi et al., 2009), and ischemic stroke (Kaur and Ling, 2008). In *vWF*-null mice, however, hypoxia followed by reoxygenation failed to increase permeability in brain capillary vessels, a result echoed in a mouse model of epilepsy (Suidan et al., 2013). As leaky brain microvessels with weak blood–brain barriers were thought to worsen tissue damage, it was surprising to see that *vWF*-null mice fared significantly worse under both disease models. Suidan et al. went on to characterize the mechanism by which *vWF* loosens the blood–brain barrier. They showed that the subendothelial release of *vWF* onto the basement membrane of brain capillary endothelial cells locally downregulates *Claudin 5*, depleting a key tight junction component (Suidan et al., 2013). While they did not directly show that circulating *vWF* does not alter *Claudin 5* levels, they did show that only locally released *vWF* could protect mice from seizure-induced brain damage and death. Circulating protein had no such effect.

In their discussion, Suidan et al. raise the possibility that a leakier blood–brain barrier may be protective because it allows for potentially toxic substances produced by brain cells during seizure to leave the brain parenchyma (Suidan et al., 2013). If trapped, these substances (e.g., glutamate) could promote further neuronal excitations and exacerbate seizures (Smith, 2000). That said, a blood–brain barrier that is loose under baseline conditions is equally problematic (Keep et al., 2008). Taken together with our finding that *vWF* expression in brain capillaries is a slowly blinking mosaic, this study points to another potential example of tissue-level bet hedging. A close inspection of Fig. 14.3D shows that *LacZ* accumulation is significantly slower in the brain of *vWF–Cre*–ROSA26R mice than in the heart, indicating that the steady-state percentage of *vWF*-positive cells is lower, and the timescale of their ON/OFF transitions is slower than in heart and muscle. Yet, the loss of these sporadically present *vWF*-positive loci was devastating to the mouse brain, especially under conditions that normally loosen the blood–brain barrier. It is possible that a key role of a dynamic *vWF* mosaic in the brain is to maintain a random sprinkling of positions across the vascular bed that can act as small safety valves. According to this hypothesis, during hypoxia or seizure *vWF* is released into the subendothelial space from a small, randomly positioned subset of endothelial cells. As it binds the capillary basement membrane, it signals to repress *Claudin 5* expression and create local openings in the blood–brain barrier. This system of "release valves" would help toxin uptake from the brain, and potentially buffer blood pressure fluctuations without the need to weaken the barrier across the board. As in the heart, a randomly changing subset of cells may be an optimal solution for balancing a generally tight blood–brain barrier with the need to occasionally open parts of it without weakening the capillary architecture. A similar experimental strategy to the one we proposed for the heart could also test this intriguing hypothesis.

Discussion—implications of biological noise-driven bed hedging as an adaptive trait

In humans, both deficiency and overabundance of *vWF* have been linked to disease. On one hand, lack of sufficient *vWF* production results in a bleeding disorder caller von Willebrand disease (Lillicrap, 2013). On the other hand, dysfunction of the *vWF*-clearing enzyme *ADAMTS13* and the resulting increase in circulating *vWF* is characteristic of the clothing disorder thrombotic thrombocytopenic purpura (Sadler, 2008). Paired with recent reports on *vWF*'s ability to modulate the blood–brain barrier (Suidan et al., 2013), our study cautions that a complete lack of functional *vWF* characteristic of type 3 von Willebrand disease may come with additional, unrecognized risks. These could include a weakened heart tissue and increased sensitivity to seizures and/or hypoxia followed by reoxygenation. While there is some evidence that excess *vWF* production is a risk factor for cardiovascular disease (Frankel et al., 2008), to our knowledge the risk of cardiovascular events in patients with the rare type 3 variant of von Willebrand disease is unknown. If human tissue has a similar response to *vWF* to that observed in mice, it is unclear if the injection of purified human *vWF*—the current standard treatment aimed at reducing bleeding risk—will have any impact on the damage we observed in the absence of local dynamic mosaics in heart and brain.

In spite of our near-exclusive focus on *vWF*, this gene is by no means unique in displaying a dynamic mosaic of ON/OFF expression. Fluorescence in situ hybridization analysis on four additional endothelial-restricted genes in parent and clonal endothelial populations revealed that two of the four mimicked the ON/OFF mosaic displayed by *vWF* (Fig. 14.8). These include a key mediator of sprouting angiogenesis—endothelial-specific

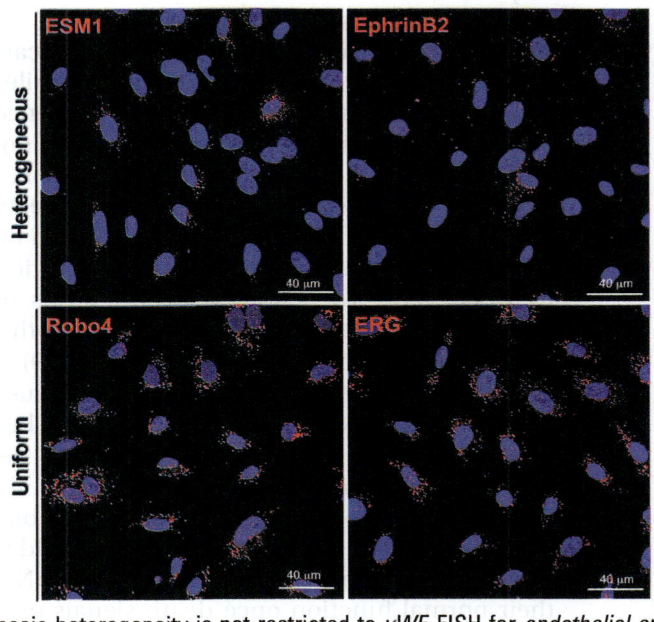

Figure 14.8 Dynamic mosaic heterogeneity is not restricted to *vWF*. FISH for *endothelial-specific molecule 1* (*ESM1*), *Ephrin B2*, *Roundabout 4* (*Robo4*), and *ETS-related gene* (*ERG*) in clonal human umbilical vein endothelial cells expanded from a single endothelial cell (EC). *ESM1* and *Ephrin B2* display mosaic mRNA heterogeneity, while *Robo4* and *ERG* are uniformly expressed in all ECs. *Red*: *ESM1/EFNB2/Robo4/ERG*; *blue*: DAPI. $n = 5$, with three replicates. *Scale bar*: 40 mm. *Source*: Adapted from Yuan, L., Chan, G.C., Beeler, D., Janes, L., Spokes, K.C., Dharaneeswaran, H., et al., 2016. A role of stochastic phenotype switching in generating mosaic endothelial cell heterogeneity. Nat. Commun. 7, 10160 (https://www.nature.com/articles/ncomms10160).

molecule 1 (ESM1) (Rocha et al., 2014), as well as the vascular patterning gene *ephrin-B2* (Gale et al., 2001). In contrast, another vascular patterning gene involved in angiogenesis, *ROBO4* (Yadav and Narayan), showed highly uniform and unimodal expression; as did the endothelial-specific transcription factor ERG (Shah et al., 2016). In vivo, *angiotensin II* protein expression in the rat femoral mesenteric artery displays a mosaic pattern of cytoplasmic versus perinuclear localization, interspersed with cells that do not express the protein at all (Tomlinson et al., 1991). Similarly, a key blood–brain barrier component, endothelial barrier antigen shows a stark mosaic expression in small vessels of the rat brain (Saubaméa et al., 2012). This growing list of examples suggests that mosaic heterogeneity is common and may represent a widespread adaptive behavior of the endothelium.

In hindsight it is perhaps not surprising that noise-driven heterogeneity among cells in identical tissue microenvironments is common. As Fig. 14.2 illustrates, biological noise can alter the response of cells whenever they cross an energy barrier between two phenotypic states (Yuan et al., 2016). Moreover, these barriers can be low enough that noise alone can generate dynamic mixtures of two or more phenotypes. While the resulting mixtures may be an unavoidable consequence of the physical/chemical processes underlying biological noise, their existence can be exploited to improve tissue function. Any challenge that demands a mix of task from a cell collective that cannot be simultaneously fulfilled by a single cell can be met by a stochastically generated mixture with tunable ratios (Wahl, 2002). For example, tissues responding to low levels of apoptotic signals leverage strong cell-to-cell variability in both the probability and timing of apoptosis (Spencer et al., 2009). In the absence of noise-driven heterogeneity, a homogeneous population would be forced to make an all-or-none choice between ignoring the signal and undergoing apoptosis. Instead, death ligands induce a dose-dependent, noise-sensitive, and thus variable-length period of apoptotic signaling that slowly builds to a threshold. Cells that do not make it past this threshold stop their apoptotic signaling with no damage to protein or DNA, and thus resume their normal function once death signals disappear. In contrast, cells pushed past the barrier by a random kick from noise commit to apoptosis in an irreversible, switch-like manner (Albeck et al., 2008). The fraction of cells with these two outcomes can be precisely tuned by the dose and exposure time to death ligands. A similar fractional response at the tissue level, paired with an all-or-none single-cell response, has been observed in response to proliferative (Spencer et al., 2013; Yao et al., 2008; Overton et al., 2014) and inflammatory signals (Kellogg and Tay, 2015).

One of the most convincing in vitro examples of bet hedging by mammalian cells was reported in neural progenitors, shown to respond to nerve growth factor by generating a reliable mix of differentiating neurons and proliferating progenitors (Chen et al., 2012). This strategy can balance two tasks that are mutually exclusive at the level of individual cells; namely maintenance of the stem cell pool and production of mature neurons. Evidence of stochastically generated phenotypic heterogeneity in other stem cell lines such as hematopoietic progenitors (Chang et al., 2008) and embryonic stem cells (Singer et al., 2014) suggests that distinct subpopulations are primed to respond differently to differentiation signals. For example, hematopoietic progenitors with high *Sca1* protein levels undergo myeloid differentiation at twice the rate of their *Sca1*-low counterparts, while the *Sca1*-low population has a distinct advantage when nudged towards the erythroid fate (Chang et al., 2008). That said, none of the progenitors are *committed.* Individual stem cells spontaneously toggle between the two phenotypes, such that a population settles into the same stable mix from any initial state. Similarly, mouse embryonic stem cells isolated from the inner cell mass of blastocysts randomly cycle in and out of a distinct cell state reminiscent of a totipotent 2-cell stage (Macfarlan et al., 2012). In contrast to "standard" embryonic stem cells which can generate all embryonic tissues, 2-cell stage cells are also capable of differentiating into extraembryonic tissues such as the yolk sack. Indeed, a sorted subpopulation of 2-cell-like stem cells injected into early embryos was shown to contribute to both trophectoderm and inner cell mast (Macfarlan et al., 2012). Taken together, the growing literature on stochastic phenotypic heterogeneity points to biological noise as an often-leveraged mechanism by which isogenic cell populations can dynamically share tasks and distribute the benefits and risks associated with each task across a tissue. Such task and risk distribution is robust and reliable, as it requires no central control (McCarron et al., 2017). At the same time, the population's behavior is highly tunable by signals that alter the barriers between regulatory stitches.

The ability of the cellular microenvironment to tune the barrier of a bistable switch is elegantly showcased by *vWF.* Indeed, the vasculature of the mouse has a use for every possible control option of the bistable switch that drives *vWF* (Fig. 14.9). In the kidney and liver, the switch is locked into a stable OFF state,

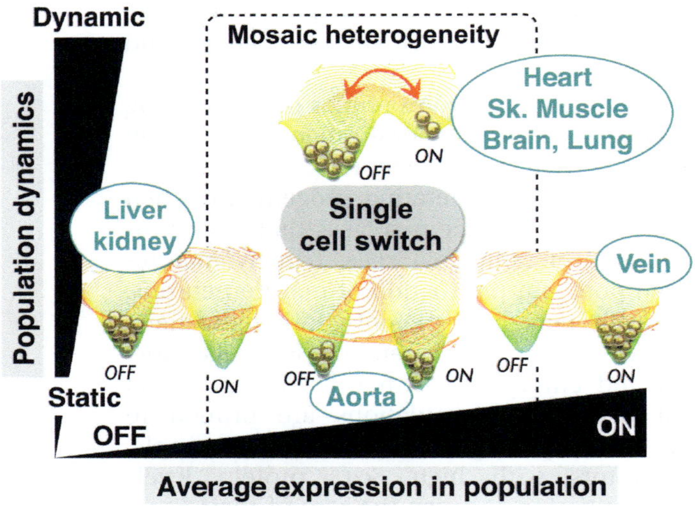

Figure 14.9 Two dimensions of bistable *vWF* regulation. Intra- and extracellular environments can fine-tune bistable regulatory switches by setting the percentage of cells in each state (*x* axis), and/or the transition rate between states (and thus the significance of biological noise; *y* axis). As any bistable regulatory switch can be placed into an environment that allows for mosaic heterogeneity, organ-specific microenvironments can tune the importance of noise relative to the barrier height of a biological switch, above and beyond setting the percentage of cells in each state. *Source*: Reproduced from Yuan, L., Chan, G.C., Beeler, D., Janes, L., Spokes, K.C., Dharaneeswaran, H., et al., 2016. A role of stochastic phenotype switching in generating mosaic endothelial cell heterogeneity. Nat. Commun. 7, 10160 (https://www.nature.com/articles/ncomms10160).

while in the venous endothelium it is kept robustly ON. In the aorta, endothelial cells carry the long-term memory of some early event that set them into *vWF*-expressing or *vWF*-negative states, but the barrier between the two states is too high for noise to breach. We suspect that this barrier height depends on the microenvironment of the aorta, as human aortic endothelial cells in vitro show dynamic mosaic heterogeneity. Similarly, in vitro human umbilical vein endothelial cells gain the ability to turn *vWF* OFF, even if they produce a lower percentage of *vWF*-negative cells than other endothelial cell types. Finally, capillary vessels of the heart, lung, skeletal muscle, and brain have a low barrier between *vWF* ON and OFF, and thus produce a dynamic mosaic.

Interestingly, we have found that the two copies of *vWF* are highly correlated in individual mouse heart endothelial cells (see Fig. 14.8 in Yuan et al., 2016), indicating that biological noise must be acting on a bistable switch upstream *vWF*, which then controls both *vWF* alleles. As it is likely that the transcription factor responsible for this regulates more than one target, *vWF* expression could be a marker of two cell states that are functionally distinct in more ways than one. Thus, it will be important to understand the broader cellular context in which this switch operates, and how other aspects of an endothelial cell's phenotype influence its ON/OFF balance as well as barrier (Fig. 14.10). Evaluating the full functional impact of dynamic mosaic heterogeneity in a tissue will require mapping not only the switch toggled by noise, but also the connections of this switch to other regulatory network modules that can modulate its barrier and/or lock it into specific states (Deritei et al., 2016). For example, finding other regulatory switches coupled to the one driving *vWF* could uncover the mechanisms that lock *vWF* tightly ON in veins in vivo, and OFF in the liver or kidney (Fig. 14.10B). Furthermore, connecting the *vWF* switch and its neighbors to the networks in charge of endothelial cell-type specification could help us delineate differences in *vWF* expression in endothelial cells versus megakaryocytes. In general, understanding

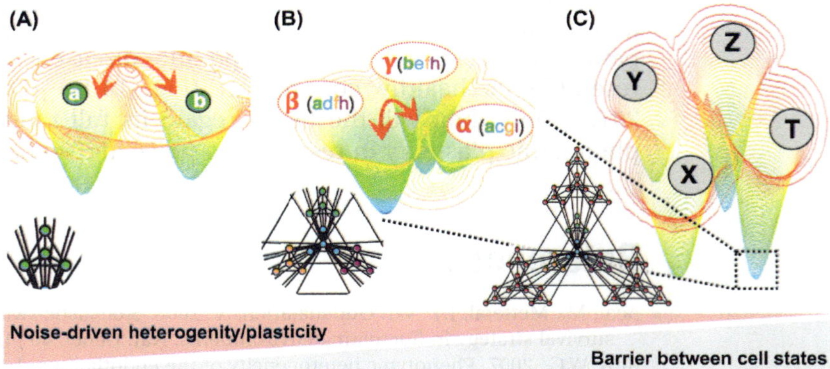

Figure 14.10 A hierarchy of Interlocking regulatory switches create epigenetic barriers at multiple scales and create context-dependent roles for biological noise. (A) Small bistable regulatory switch (*green network*) with its pseudo-energy landscape and a barrier between phenotypic states *a* and *b* (e.g., *vWF* ON and off that can be breached by biological noise (*red arrow*). (B) Landscape of several coupled multistable switches (*colored network modules*) linked to the green network (A) showing three global network states (α, β, and γ) that are combinations of phenotypic outcomes driven by individual switches (e.g., in the β cell state the green switch is in *a*, the blue switch is in state *d*, etc.). Whenever the purple switch is in state *h*, biological noise can toggle between β and γ (flipping the green and blue switches; *red arrow*). In contrast, when the purple switch is in state *i*, the green switch is robustly locked into its state *a* (e.g., *vWF* switch locked OFF in the kidney). (C) The full cell-wide regulatory network of the cell (*colored hierarchically modular network*) creates high epigenetic barriers between cell lineage states but allows for environmentally generated as well as stochastic heterogeneity at the level of individual cell lines (box: *structure at the bottom of cell-lineage valleys*). *Landscape color scale*: green/low energy/stable to red/high energy/unstable). *Source*: Hierarchical network image is reproduced from Ravasz, E., Somera, A.L., Mongru, D.A., Oltvai, Z.N., Barabási, A.L., 2002. Hierarchical organization of modularity in metabolic networks. Science 297 (5586),1551–1555. Reprinted with permission from AAAS.

the mechanisms by which multistable circuits assemble into large hierarchical networks (Fig. 14.10C) could help us map the interface at which phenotypic plasticity and biological noise interact, and uncover adaptive roles noise plays in differentiation, signal response, and bet hedging.

We do not yet have definite proof that a dynamic mosaic and bet hedging is required for tissue heath. Though technically challenging in vivo, such proof may soon force us to rethink our traditional approach to fixing a diseased tissue. Current drug-based approaches implicitly assume that inhibiting or turning on the tissue-specific activity of certain proteins can “fix” the behavior of diseased cells. In organs with adaptive bet hedging, however, a strategy that locks all cells into a single state may be ill advised. An optimal treatment should aim to restore their *flexibility* to generate mosaic heterogeneity and homeostatic bet hedging. Such a strategy cannot be pursued with one or two molecular targets in mind; it requires predictive computational models that allow us to map the landscape of individual cell functions in a wide range of physiological conditions in heath and disease. These models would allow us to search for therapeutic interventions that reshape the diseased landscape to resemble its healthy counterpart (Fig. 14.10) (Deritei et al., 2016). Only then will the tissue regain its full range of healthy phenotypic plasticity in response to noise and signals alike.

References

Acar, M., Mettetal, J.T., van Oudenaarden, A., 2008. Stochastic switching as a survival strategy in fluctuating environments. Nat. Genet. 40 (4), 471–475.

Aird, W.C., 2007. Phenotypic heterogeneity of the endothelium: I. Structure, function, and mechanisms. Circ. Res. 100 (2), 158–173.

Aird, W.C., 2007. Phenotypic heterogeneity of the endothelium: II. Representative vascular beds. Circ. Res. 100 (2), 174–190.

Aird, W.C., Edelberg, J.M., Weiler-Guettler, H., Simmons, W.W., Smith, T.W., Rosenberg, R.D., 1997. Vascular bed-specific expression of an endothelial cell gene is programmed by the tissue microenvironment. J. Cell Biol. 138 (5), 1117–1124.

Albeck, J.G., Burke, J.M., Spencer, S.L., Lauffenburger, D.A., Sorger, P.K., 2008. Modeling a snap-action, variable-delay switch controlling extrinsic cell death. PLoS Biol. 6 (12), 2831–2852.

Albert, R., Othmer, H.G., 2003. The topology of the regulatory interactions predicts the expression pattern of the segment polarity genes in Drosophila melanogaster. J. Theor. Biol. 223 (1), 1–18.

Albert, R., Wang, R.-S., 2009. Discrete dynamic modeling of cellular signaling networks. Methods Enzymol. 467, 281–306.

Anahita, M., Maryam, N.-N., Wei-Lee, P., Stephen, K., Aneta, R.-B., Paul, J., et al., 2013. Hypoxia results in upregulation and de novo activation of von

Willebrand factor expression in lung endothelial cells. Arterioscler. Thromb. Vasc. Biol. 33 (6), 1329–1338.

Andrecut, M., Halley, J.D., Winkler, D.A., Huang, S., 2011. A general model for binary cell fate decision gene circuits with degeneracy: indeterminacy and switch behavior in the absence of cooperativity. PLoS One 6 (5), e19358.

Arkin, A., Ross, J., McAdams, H.H., 1998. Stochastic kinetic analysis of developmental pathway bifurcation in phage lambda-infected Escherichia coli cells. Genetics. 149 (4), 1633–1648.

Bai, S., Goodrich, D., Thron, C.D., Tecarro, E., Obeyesekere, M., 2003. Theoretical and experimental evidence for hysteresis in cell proliferation. Cell Cycle Georget. Tex. 2 (1), 46–52.

Balazsi, G., van Oudenaarden, A., Collins, J.J., 2011. Cellular decision making and biological noise: from microbes to mammals. Cell. 144 (6), 910–925.

Beaumont, H.J.E., Gallie, J., Kost, C., Ferguson, G.C., Rainey, P.B., 2009. Experimental evolution of bet hedging. Nature. 462 (7269), 90–93.

Blake, W.J., Balázsi, G., Kohanski, M.A., Isaacs, F.J., Murphy, K.F., Kuang, Y., et al., 2006. Phenotypic consequences of promoter-mediated transcriptional noise. Mol. Cell. 24 (6), 853–865.

Blake, W.J., KAErn, M., Cantor, C.R., Collins, J.J., 2003. Noise in eukaryotic gene expression. Nature. 422 (6932), 633–637.

Bradshaw, A.D., 1965. Evolutionary significance of phenotypic plasticity in plants. In: Caspari, E.W., Thoday, J.M. (Eds.), Advances in Genetics. Academic Press [cited 2019 Jul 5]. pp. 115–155. Available from: <http://www.sciencedirect.com/science/article/pii/S0065266008600486>

Briscoe, J., Small, S., 2015. Morphogen rules: design principles of gradient-mediated embryo patterning. Development 142 (23), 3996–4009.

Burns, A., Wilson, E., Harber, M., Brunton, C., Sweny, P., 1995. Cerebral venous sinus thrombosis in minimal change nephrotic syndrome. Nephrol. Dial. Transpl. Off. Publ. Eur. Dial Transpl Assoc. - Eur. Ren. Assoc. 10 (1), 30–34.

Cai, L., Friedman, N., Xie, X.S., 2006. Stochastic protein expression in individual cells at the single molecule level. Nature. 440 (7082), 358–362.

Cedar, H., Bergman, Y., 2009. Linking DNA methylation and histone modification: patterns and paradigms. Nat. Rev. Genet. 10 (5), 295–304.

Chang, H.H., Hemberg, M., Barahona, M., Ingber, D.E., Huang, S., 2008. Transcriptome-wide noise controls lineage choice in mammalian progenitor cells. Nature. 453 (7194), 544–547.

Chen, J.-Y., Lin, J.-R., Cimprich, K.A., Meyer, T., 2012. A two-dimensional ERK-AKT signaling code for an NGF-triggered cell-fate decision. Mol. Cell. 45 (2), 196–209.

Chi, J.-T., Chang, H.Y., Haraldsen, G., Jahnsen, F.L., Troyanskaya, O.G., Chang, D.S., et al., 2003. Endothelial cell diversity revealed by global expression profiling. Proc. Natl. Acad. Sci. U S A 100 (19), 10623–108.

Cobb, M., 2017. 60 years ago, Francis Crick changed the logic of biology. PLoS Biol. 15 (9), e2003243.

Coll-Bonfill, N., Musri, M.M., Ivo, V., Barberà, J.A., Tura-Ceide, O., 2015. Transdifferentiation of endothelial cells to smooth muscle cells play an important role in vascular remodelling. Am. J. Stem Cell 4 (1), 13–21.

Cooley, L.S., Handsley, M.M., Zhou, Z., Lafleur, M.A., Pennington, C.J., Thompson, E.W., et al., 2010. Reversible transdifferentiation of blood vascular endothelial cells to a lymphatic-like phenotype in vitro. J. Cell Sci. 123 (21), 3808–3816.

Crick, F.H., 1958. On protein synthesis. Symp. Soc. Exp. Biol. 12, 138–163.

Das, J., Ho, M., Zikherman, J., Govern, C., Yang, M., Weiss, A., et al., 2009. Digital signaling and hysteresis characterize Ras activation in lymphoid cells. Cell. 136 (2), 337–351.
deAlmeida, A.C., van Oort, R.J., Wehrens, X.H.T., 2010. Transverse aortic constriction in mice. J. Vis. Exp. JoVE. (38), .
Deaton, A.M., Bird, A., 2011. CpG islands and the regulation of transcription. Genes. Dev. 25 (10), 1010–1022.
dela Paz, N.G., D'Amore, P.A., 2009. Arterial versus venous endothelial cells. Cell Tissue Res. 335 (1), 5–16.
Denis, C., Methia, N., Frenette, P.S., Rayburn, H., Ullman-Culleré, M., Hynes, R.O., et al., 1998. A mouse model of severe von Willebrand disease: Defects in hemostasis and thrombosis. Proc. Natl. Acad. Sci. U S A 95 (16), 9524–9529.
Deritei, D., Aird, W.C., Ercsey-Ravasz, M., Regan, E.R., 2016. Principles of dynamical modularity in biological regulatory networks. Sci. Rep. 6, 21957.
DeWitt, T.J., Sih, A., Wilson, D.S., 1998. Costs and limits of phenotypic plasticity. Trends Ecol. Evol. 13 (2), 77–81.
Domyan, E.T., Sun, X., 2011. Patterning and plasticity in development of the respiratory lineage. Dev. Dyn. Off. Publ. Am. Assoc. Anat. 240 (3), 477–485.
Eldar, A., Elowitz, M.B., 2010. Functional roles for noise in genetic circuits. Nature. 467 (7312), 167–173.
Elowitz, M.B., Levine, A.J., Siggia, E.D., Swain, P.S., 2002. Stochastic gene expression in a single cell. Science 297 (5584), 1183–1186.
Giuseppe, F., Alessandro, M., 2010. Phenotypic plasticity in development and evolution: facts and concepts. Philos. Trans. R. Soc. B Biol. Sci. 365 (1540), 547–556.
Fiedler, U., Augustin, H.G., 2006. Angiopoietins: a link between angiogenesis and inflammation. Trends Immunol. 27 (12), 552–558.
Fish, J.E., Wythe, J.D., 2015. The molecular regulation of arteriovenous specification and maintenance. Dev. Dyn. 244 (3), 391–409.
Folsom, A.R., Rosamond, W.D., Shahar, E., Cooper, L.S., Aleksic, N., Nieto, F.J., et al., 1999. Prospective study of markers of hemostatic function with risk of ischemic stroke. The Atherosclerosis Risk in Communities (ARIC) Study Investigators. Circulation. 100 (7), 736–742.
Foster, D.V., Foster, J.G., Huang, S., Kauffman, S.A., 2009. A model of sequential branching in hierarchical cell fate determination. J. Theor. Biol. 260, 589–597.
Frankel, D.S., Meigs, J.B., Massaro, J.M., Wilson, P.W.F., O'Donnell, C.J., D'Agostino, R.B., et al., 2008. Von Willebrand factor, type 2 diabetes and risk of cardiovascular disease: the Framingham Offspring Study. Circulation. 118 (24), 2533–2539.
Gaengel, K., Genové, G., Armulik, A., Betsholtz, C., 2009. Endothelial-mural cell signaling in vascular development and angiogenesis. Arterioscler. Thromb. Vasc. Biol. 29 (5), 630–638.
Gale, N.W., Baluk, P., Pan, L., Kwan, M., Holash, J., DeChiara, T.M., et al., 2001. Ephrin-B2 selectively marks arterial vessels and neovascularization sites in the adult, with expression in both endothelial and smooth-muscle cells. Dev. Biol. 230 (2), 151–160.
Gardner, T.S., Cantor, C.R., Collins, J.J., 2000. Construction of a genetic toggle switch in Escherichia coli. Nature. 403 (6767), 339–342.
Heinäniemi, M., Nykter, M., Kramer, R., Wienecke-Baldacchino, A., Sinkkonen, L., Zhou, J.X., et al., 2013. Gene-pair expression signatures reveal lineage control. Nat. Methods. 10 (6), 577–583.

Howard-Varona, C., Hargreaves, K.R., Abedon, S.T., Sullivan, M.B., 2017. Lysogeny in nature: mechanisms, impact and ecology of temperate phages. ISME J. 11 (7), 1511–1520.
Huang, C.Y., Ferrell, J.E., 1996. Ultrasensitivity in the mitogen-activated protein kinase cascade. Proc. Natl. Acad. Sci. 93 (19), 10078–10083.
Huang, S., 1999. Gene expression profiling, genetic networks, and cellular states: an integrating concept for tumorigenesis and drug discovery. J. Mol. Med. 77 (6), 469–480.
Huang, S., 2009. Non-genetic heterogeneity of cells in development: more than just noise. Dev. Camb. Engl. 136 (23), 3853–3862.
Huang, S., 2009. Reprogramming cell fates: reconciling rarity with robustness. BioEssays N. Rev. Mol. Cell Dev. Biol. 31 (5), 546–560.
Huang, S., 2012. The molecular and mathematical basis of Waddington's epigenetic landscape: a framework for post-Darwinian biology? BioEssays N. Rev. Mol. Cell Dev. Biol. 34 (2), 149–157.
Huang, S., Guo, Y.-P., May, G., Enver, T., 2007. Bifurcation dynamics in lineage-commitment in bipotent progenitor cells. Dev. Biol. 305 (2), 695–713.
Huang, S., Ingber, D.E., 2000. Shape-dependent control of cell growth, differentiation, and apoptosis: switching between attractors in cell regulatory networks. Exp. Cell Res. 261 (1), 91–103.
Iwafuchi-Doi, M., Zaret, K.S., 2016. Cell fate control by pioneer transcription factors. Development 143 (11), 1833–1837.
Iwafuchi-Doi, M., Zaret, K.S., 2014. Pioneer transcription factors in cell reprogramming. Genes. Dev. 28 (24), 2679–2692.
Jeltsch, A., Jurkowska, R.Z., 2014. New concepts in DNA methylation. Trends Biochem. Sci. 39 (7), 310–318.
Jones, P.A., 2012. Functions of DNA methylation: islands, start sites, gene bodies and beyond. Nat. Rev. Genet. 13 (7), 484–492.
Kanji, S., Pompili, V.J., Das, H., 2011. Plasticity and maintenance of hematopoietic stem cells during development. Recent. Pat. Biotechnol. 5 (1), 40–53.
Kauffman, S., 1969. Homeostasis and differentiation in random genetic control networks. Nature. 224 (5215), 177–178.
Kaur, C., Ling, E.A., 2008. Blood brain barrier in hypoxic-ischemic conditions. Curr. Neurovasc. Res. 5 (1), 71–81.
Keep, R.F., Xiang, J., Ennis, S.R., Andjelkovic, A., Hua, Y., Xi, G., et al., 2008. Blood-brain barrier function in intracerebral hemorrhage. Acta Neurochir. Suppl. 105, 73–77.
Kellogg, R.A., Tay, S., 2015. Noise facilitates transcriptional control under dynamic inputs. Cell. 160 (3), 381–392.
Kelly, S.A., Panhuis, T.M., Stoehr, A.M., 2012. Phenotypic plasticity: molecular mechanisms and adaptive significance. Compr. Physiol. 2 (2), 1417–1439.
Kriehuber, E., Breiteneder-Geleff, S., Groeger, M., Soleiman, A., Schoppmann, S.F., Stingl, G., et al., 2001. Isolation and characterization of dermal lymphatic and blood endothelial cells reveal stable and functionally specialized cell lineages. J. Exp. Med. 194 (6), 797–808.
Kudo Fabio, A., Muto, A., Maloney Stephen, P., Pimiento Jose, M., Bergaya, S., Fitzgerald Tamara, N., et al., 2007. Venous identity is lost but arterial identity is not gained during vein graft adaptation. Arterioscler. Thromb. Vasc. Biol. 27 (7), 1562–1571.
Laird, C.D., Pleasant, N.D., Clark, A.D., Sneeden, J.L., Hassan, K.M.A., Manley, N.C., et al., 2004. Hairpin-bisulfite PCR: assessing epigenetic methylation patterns on

complementary strands of individual DNA molecules. Proc. Natl. Acad. Sci. U S A 101 (1), 204–209.

Lillicrap, D., 2013. von Willebrand disease: advances in pathogenetic understanding, diagnosis, and therapy. Blood 122 (23), 3735–3740.

Lin, S., Marin, E.C., Yang, C., Kao, C.-F., Apenteng, B.A., Huang, Y., et al., 2013. Extremes of lineage plasticity in the drosophila brain. Curr. Biol. 23, 1908–1913.

Lipshtat, A., Jayaraman, G., He, J.C., Iyengar, R., 2010. Design of versatile biochemical switches that respond to amplitude, duration, and spatial cues. Proc. Natl. Acad. Sci. U S A 107 (3), 1247–1252.

Macfarlan, T.S., Gifford, W.D., Driscoll, S., Lettieri, K., Rowe, H.M., Bonanomi, D., et al., 2012. Embryonic stem cell potency fluctuates with endogenous retrovirus activity. Nature. 487 (7405), 57–63.

Magklara, A., Lomvardas, S., 2013. Stochastic gene expression in mammals: lessons from olfaction. Trends Cell Biol. 23 (9), 449–456.

Marchi, N., Fan, Q., Ghosh, C., Fazio, V., Bertolini, F., Betto, G., et al., 2009. Antagonism of peripheral inflammation reduces the severity of status epilepticus. Neurobiol. Dis. 33 (2), 171–181.

Marikawa, Y., Alarcón, V.B., 2009. Establishment of trophectoderm and inner cell mass lineages in the mouse embryo. Mol. Reprod. Dev. 76 (11), 1019–1032.

Mayran, A., Drouin, J., 2018. Pioneer transcription factors shape the epigenetic landscape. J. Biol. Chem. 293 (36), 13795–13804.

McCarron, J.G., Lee, M.D., Wilson, C., 2017. The endothelium solves problems that endothelial cells do not know exist. Trends Pharmacol. Sci. 38 (4), 322–338.

Mura, M., Swain, R.K., Zhuang, X., Vorschmitt, H., Reynolds, G., Durant, S., et al., 2012. Identification and angiogenic role of the novel tumor endothelial marker CLEC14A. Oncogene. 31 (3), 293–305.

Murren, C.J., Auld, J.R., Callahan, H., Ghalambor, C.K., Handelsman, C.A., Heskel, M.A., et al., 2015. Constraints on the evolution of phenotypic plasticity: limits and costs of phenotype and plasticity. Heredity. 115 (4), 293–301.

Okada, Y., Jin, E., Nikolova-Krstevski, V., Yano, K., Liu, J., Beeler, D., et al., 2008. A GABP-binding element in the Robo4 promoter is necessary for endothelial expression in vivo. Blood 112 (6), 2336–2339.

Ordovas-Montanes, J., Dwyer, D.F., Nyquist, S.K., Buchheit, K.M., Vukovic, M., Deb, C., et al., 2018. Allergic inflammatory memory in human respiratory epithelial progenitor cells. Nature. 560 (7720), 649–654.

Overton, K.W., Spencer, S.L., Noderer, W.L., Meyer, T., Wang, C.L., 2014. Basal p21 controls population heterogeneity in cycling and quiescent cell cycle states. Proc. Natl. Acad. Sci. U S A 111 (41), E4386–E4393.

Ozbudak, E.M., Thattai, M., Kurtser, I., Grossman, A.D., van Oudenaarden, A., 2002. Regulation of noise in the expression of a single gene. Nat. Genet. 31 (1), 69–73.

Paliwal, S., Iglesias, P.A., Campbell, K., Hilioti, Z., Groisman, A., Levchenko, A., 2007. MAPK-mediated bimodal gene expression and adaptive gradient sensing in yeast. Nature. 446 (7131), 46–51.

Peyvandi, F., Garagiola, I., Baronciani, L., 2011. Role of von Willebrand factor in the haemostasis. Blood Transfus. 9 (Suppl 2), s3–s8.

Pfeifer, G.P., Steigerwald, S.D., Hansen, R.S., Gartler, S.M., Riggs, A.D., 1990. Polymerase chain reaction-aided genomic sequencing of an X chromosome-linked CpG island: methylation patterns suggest clonal inheritance, CpG site autonomy, and an explanation of activity state stability. Proc. Natl. Acad. Sci. U S A 87 (21), 8252–8256.

Plath, K., Lowry, W.E., 2011. Progress in understanding reprogramming to the induced pluripotent state. Nat. Rev. Genet. 12 (4), 253–265.

Posfai, E., Petropoulos, S., de Barros, F.R.O., Schell, J.P., Jurisica, I., Sandberg, R., et al., 2017. Position- and Hippo signaling-dependent plasticity during lineage segregation in the early mouse embryo. eLife. 2017. Available from: <https://elifesciences.org/articles/22906> [cited 2019 Jul 5].

Ptashne, M., 2004. A genetic switch, Phage Lambda Revisited, third ed. Cold Spring Harbor Laboratory Press, 172 p.

Raj, A., Peskin, C.S., Tranchina, D., Vargas, D.Y., Tyagi, S., 2006. Stochastic mRNA synthesis in mammalian cells. PLoS Biol. 4 (10), e309.

Raj, A., van Oudenaarden, A., 2008. Nature, nurture, or chance: stochastic gene expression and its consequences. Cell. 135 (2), 216–226.

Randi, A.M., Laffan, M.A., Starke, R.D., 2013. Von Willebrand factor, angiodysplasia and angiogenesis. Mediterr. J. Hematol. Infect. Dis. 5 (1), e2013060.

Raser, J.M., O'Shea, E.K., 2004. Control of stochasticity in eukaryotic gene expression. Science 304 (5678), 1811–1814.

Ravasz, E., Somera, A.L., Mongru, D.A., Oltvai, Z.N., Barabási, A.L., 2002. Hierarchical organization of modularity in metabolic networks. Science 297 (5586), 1551–1555.

Regan, E.R., Aird, W.C., 2012. Dynamical systems approach to endothelial heterogeneity. Circ. Res. 111 (1), 110–130.

Rocha, S.F., Schiller, M., Jing, D., Li, H., Butz, S., Vestweber, D., et al., 2014. Esm1 modulates endothelial tip cell behavior and vascular permeability by enhancing VEGF bioavailability. Circ. Res. 115 (6), 581–590.

Rondaij, M.G., Bierings, R., Kragt, A., van Mourik, J.A., Voorberg, J., 2006. Dynamics and plasticity of Weibel-Palade bodies in endothelial cells. Arterioscler. Thromb. Vasc. Biol. 26 (5), 1002–1007.

Rothenberg, E.V., Dionne, C.J., 2002. Lineage plasticity and commitment in T-cell development. Immunol. Rev. 187 (1), 96–115.

Ruggeri, Z.M., 2007. Von Willebrand factor: looking back and looking forward. Thromb. Haemost. 98 (1), 55–62.

Ruoslahti, E., 2004. Vascular zip codes in angiogenesis and metastasis. Biochem. Soc. Trans. 32 (Pt3), 397–402.

Ruoslahti, E., Rajotte, D., 2000. An address system in the vasculature of normal tissues and tumors. Annu. Rev. Immunol. 18, 813–827.

Sadler, J.E., 2008. Von Willebrand factor, ADAMTS13, and thrombotic thrombocytopenic purpura. Blood 112 (1), 11–18.

Santini, M.P., Forte, E., Harvey, R.P., Kovacic, J.C., 2016. Developmental origin and lineage plasticity of endogenous cardiac stem cells. Development 143 (8), 1242–1258.

Saubaméa, B., Cochois-Guégan, V., Cisternino, S., Scherrmann, J.-M., 2012. Heterogeneity in the rat brain vasculature revealed by quantitative confocal analysis of endothelial barrier antigen and P-glycoprotein expression. J Cereb Blood Flow Metab Off J Int Soc Cereb Blood Flow Metab 32 (1), 81–92.

Scheiner, S.M., 1993. Genetics and evolution of phenotypic plasticity. Annu. Rev. Ecol. Syst. 24 (1), 35–68.

Schoch, H.J., Fischer, S., Marti, H.H., 2002. Hypoxia-induced vascular endothelial growth factor expression causes vascular leakage in the brain. Brain J. Neurol. 125 (Pt 11), 2549–2557.

Sha, W., Moore, J., Chen, K., Lassaletta, A.D., Yi, C.-S., Tyson, J.J., et al., 2003. Hysteresis drives cell-cycle transitions in Xenopus laevis egg extracts. Proc. Natl. Acad. Sci. U S A 100 (3), 975–980.

Shah, A.V., Birdsey, G.M., Randi, A.M., 2016. Regulation of endothelial homeostasis, vascular development and angiogenesis by the transcription factor ERG. Vasc. Pharmacol. 86–3–13.

Singer, Z.S., Yong, J., Tischler, J., Hackett, J.A., Altinok, A., Surani, M.A., et al., 2014. Dynamic heterogeneity and DNA methylation in embryonic stem cells. Mol. Cell. 55 (2), 319–331.

Smith, Q.R., 2000. Transport of glutamate and other amino acids at the blood-brain barrier. J. Nutr. 130 (4S Suppl), 1016S–1022SS.

Socolar, J.E.S., Kauffman, S.A., 2003. Scaling in ordered and critical random Boolean networks. Phys. Rev. Lett. 90 (6), 068702.

Solomon, M.J., 2003. Hysteresis meets the cell cycle. Proc. Natl. Acad. Sci. U S A 100 (3), 771–772.

Spencer, S.L., Cappell, S.D., Tsai, F.-C., Overton, K.W., Wang, C.L., Meyer, T., 2013. The proliferation-quiescence decision is controlled by a bifurcation in CDK2 activity at mitotic exit. Cell. 155 (2), 369–383.

Spencer, S.L., Gaudet, S., Albeck, J.G., Burke, J.M., Sorger, P.K., 2009. Non-genetic origins of cell-to-cell variability in TRAIL-induced apoptosis. Nature. 459 (7245), 428–432.

Starke, R.D., Ferraro, F., Paschalaki, K.E., Dryden, N.H., McKinnon, T.A.J., Sutton, R.E., et al., 2011. Endothelial von Willebrand factor regulates angiogenesis. Blood 117 (3), 1071–1080.

Steinway, S.N., Zañudo, J.G.T., Ding, W., Rountree, C.B., Feith, D.J., Loughran, T. P., et al., 2014. Network modeling of TGFβ signaling in hepatocellular carcinoma epithelial-to-mesenchymal transition reveals joint sonic hedgehog and Wnt pathway activation. Cancer Res. 74 (21), 5963–5977.

Süel, G.M., Garcia-Ojalvo, J., Liberman, L.M., Elowitz, M.B., 2006. An excitable gene regulatory circuit induces transient cellular differentiation. Nature. 440 (7083), 545.

Süel, G.M., Kulkarni, R.P., Dworkin, J., Garcia-Ojalvo, J., Elowitz, M.B., 2007. Tunability and noise dependence in differentiation dynamics. Science 315 (5819), 1716–1719.

Suidan, G.L., Brill, A., De Meyer, S.F., Voorhees, J.R., Cifuni, S.M., Cabral, J.E., et al., 2013. Endothelial Von Willebrand factor promotes blood-brain barrier flexibility and provides protection from hypoxia and seizures in mice. Arterioscler. Thromb. Vasc. Biol. 33 (9), 2112–2120.

Symmons, O., Raj, A., 2016. What's luck got to do with it: single cells, multiple fates, and biological nondeterminism. Mol. Cell. 62 (5), 788–802.

Takahashi, K., Yamanaka, S., 2016. A decade of transcription factor-mediated reprogramming to pluripotency. Nat. Rev. Mol. Cell Biol. 17 (3), 183–193.

Tomlinson, A., Van Vlijmen, H., Loesch, A., Burnstock, G., 1991. An immunohistochemical study of endothelial cell heterogeneity in the rat: observations in "en face" Häutchen preparations. Cell Tissue Res. 263 (1), 173–181.

Tyson, J.J., Novak, B., 2001. Regulation of the eukaryotic cell cycle: molecular antagonism, hysteresis, and irreversible transitions. J. Theor. Biol. 210 (2), 249–263.

Vaissière, T., Sawan, C., Herceg, Z., 2008. Epigenetic interplay between histone modifications and DNA methylation in gene silencing. Mutat. Res. 659 (1–2), 40–48.

Veening, J.-W., Stewart, E.J., Berngruber, T.W., Taddei, F., Kuipers, O.P., Hamoen, L.W., 2008. Bet-hedging and epigenetic inheritance in bacterial cell development. Proc. Natl. Acad. Sci. U S A 105 (11), 4393–4398.

Villalba, N., Sackheim, A.M., Nunez, I.A., Hill-Eubanks, D.C., Nelson, M.T., Wellman, G.C., et al., 2017. Traumatic brain injury causes endothelial dysfunction in the systemic microcirculation through arginase-1–dependent uncoupling of endothelial nitric oxide synthase. J. Neurotrauma 34 (1), 192–203.

Waddington, C.H., 2014. The Strategy of the Genes. Routledge [cited 2019 Jul 5]. Available from: <https://www.taylorfrancis.com/books/9781315765471>

Wahl, L.M., 2002. Evolving the division of labour: generalists, specialists and task allocation. J. Theor. Biol. 219 (3), 371–388.

Wang, J., Zhang, K., Xu, L., Wang, E., 2011. Quantifying the Waddington landscape and biological paths for development and differentiation. Proc. Natl. Acad. Sci. 108 (20), 8257–8262.

Weinberger, L.S., Burnett, J.C., Toettcher, J.E., Arkin, A.P., Schaffer, D.V., 2005. Stochastic gene expression in a lentiviral positive-feedback loop: HIV-1 Tat fluctuations drive phenotypic diversity. Cell. 122 (2), 169–182.

Weinberger, L.S., Dar, R.D., Simpson, M.L., 2008. Transient-mediated fate determination in a transcriptional circuit of HIV. Nat. Genet. 40 (4), 466–470.

Wernet, M.F., Mazzoni, E.O., Çelik, A., Duncan, D.M., Duncan, I., Desplan, C., 2006. Stochastic spineless expression creates the retinal mosaic for colour vision. Nature. 440 (7081), 174–180.

Wolf, D.M., Vazirani, V.V., Arkin, A.P., 2005. Diversity in times of adversity: probabilistic strategies in microbial survival games. J. Theor. Biol. 234 (2), 227–253.

Yadav, S.S., Narayan, G., 2014. Role of ROBO4 Signalling in Developmental and Pathological Angiogenesis. BioMed Research International. Available from: <https://www.hindawi.com/journals/bmri/2014/683025/> [cited 2019 Jul 9].

Yamamoto, K., de Waard, V., Fearns, C., Loskutoff, D.J., 1998. Tissue distribution and regulation of murine von Willebrand factor gene expression in vivo. Blood 92 (8), 2791–2801.

Yao, G., Lee, T.J., Mori, S., Nevins, J.R., You, L., 2008. A bistable Rb-E2F switch underlies the restriction point. Nat. Cell Biol. 10 (4), 476–482.

Yu, J., Xiao, J., Ren, X., Lao, K., Xie, X.S., 2006. Probing gene expression in live cells, one protein molecule at a time. Science 311 (5767), 1600–1603.

Yuan, L., Chan, G.C., Beeler, D., Janes, L., Spokes, K.C., Dharaneeswaran, H., et al., 2016. A role of stochastic phenotype switching in generating mosaic endothelial cell heterogeneity. Nat. Commun. 7, 10160.

Zaret, K.S., Carroll, J.S., 2011. Pioneer transcription factors: establishing competence for gene expression. Genes. Dev. 25 (21), 2227–2241.

Zhang, Y., Qian, M., Ouyang, Q., Deng, M., Li, F., Tang, C., 2006. Stochastic model of yeast cell-cycle network. Phys. Nonlinear Phenom. 219 (1), 35–39.

Zhang, Z., Zhang, J., 2009. A big world inside small-world networks. PLoS One 4 (5), e5686.

Zhou, J.X., Aliyu, M.D.S., Aurell, E., Huang, S., 2012. Quasi-potential landscape in complex multi-stable systems. J. R. Soc. Interface 9 (77), 3539–3553.

Zhuang, X., Cross, D., Heath, V.L., Bicknell, R., 2011. Shear stress, tip cells and regulators of endothelial migration. Biochem. Soc. Trans. 39 (6), 1571–1575.

15

Regulation of phenotypic plasticity from the perspective of evolutionary developmental biology

Yuichiro Suzuki[1,*], Kenneth Z. McKenna[2,*] and H. Frederik Nijhout[2,*]

[1]Department of Biological Sciences, Wellesley College, Wellesley, MA, United States [2]Department of Biology, Duke University, Durham, NC, United States

Introduction to phenotypic plasticity

Phenotypic plasticity is the ability of an organism to alter its phenotype in response to environmental changes. Virtually any trait has the potential to exhibit some phenotypic plasticity, but the degree to which plasticity manifests is shaped by natural selection. Phenotypic plasticity has been shown to play important roles during adaptive evolution. In particular, developmental plasticity, the ability of an organism to switch developmental trajectories under distinct environmental conditions, may play major roles during the evolution of novel phenotypes (West-Eberhard, 2003). Thus, understanding the mechanisms that regulate developmental plasticity provides insights into how organisms evolve.

Most traits exhibit some degree of phenotypic plasticity. At one end, phenotypic plasticity is limited or absent, even in the face of significant environmental or genetic perturbations. Such traits are said to be robust and are likely to be adaptations to stable environments. At the opposite end are traits where different environmental conditions give rise to discrete alternative phenotypes that are adapted to each environment. These traits are called polyphenisms. The developmental mechanisms that underlie polyphenisms are well studied and offer insights into how natural selection molds phenotypic plasticity to generate

*All three authors contributed equally to this paper.

Phenotypic Switching. DOI: https://doi.org/10.1016/B978-0-12-817996-3.00012-8

novel phenotypes. In this chapter, we explore how plasticity in development provides a mechanism by which novel, adaptive phenotypes might evolve.

What are polyphenisms?

Here we define polyphenisms as adaptive phenotypic plasticity wherein an identical genome can produce two or more different phenotypes in response to specific environmental cues. In some cases, organisms can develop into adults with different phenotypes because they experience a particular environmental condition. In other cases, organisms can alter their phenotype in a cyclical manner such that every season, distinct phenotypes, and behaviors are produced. Yet another type of polyphenisms involves sequential polyphenisms wherein the developing organism assumes a sequence of distinctive morphologies and behaviors during its life cycle. Metamorphosis and sequential hermaphroditism are particularly dramatic examples of sequential polyphenisms. Polyphenisms are common amongst multicellular organisms, but the developmental processes that lead to the production of alternative phenotypes are often complex and operate at multiple levels of regulation. In this chapter, we argue that homeostatic and robustness mechanisms facilitate the evolution of novel traits and life history strategies, and that understanding the regulation of polyphenic traits may offer us clues on how novel phenotype arise and how they evolve over time.

Regulation of developmental plasticity

Understanding polyphenic traits requires us to decipher how the organism senses environmental cues (sensors), integrates these inputs (integrators), and modulates effectors that alter developmental trajectories (effectors). Sensors gather information from the environment and send cues that stimulate the release of integrators. The integrators then travel through the body to target tissues, and cause effectors to direct development toward one developmental trajectory or another.

Sensors

Sensors must accumulate information from the environment during a specific period of time called the *environment-sensitive period*. Although the mechanism of environment-sensitive periods remains unclear, the environmental signals encountered during the sensitive period play a critical role in determining the subsequent developmental trajectory of the organism. In most cases, the alternative phenotypes develop long after the inductive environmental signals are sensed, sometimes weeks or months later. Therefore, the alternative phenotypes are not adaptations to the environment that is sensed, but to a future environment, for which the inductive signal is a good predictor. Shortening daylength, for instance, is a reliable predictor of winter to come. A wide range of environmental cues has been shown to influence polyphenisms, and even closely related species can use different mechanisms to acquire information about the environment. Here we identify two broad classes of sensors: neural sensors and epigenetic sensors.

Neural sensors

In many cases, the environmental signals that cue polyphenisms are sensed by the nervous system. Whether it is the detection of seasons through photoperiod, or detection of crowding, or predators, or pheromones, the nervous system is the system in the body that receives the environmental stimuli. In many seasonal polyphenisms, the organism senses day length and temperature to assess the season. For example, in many polyphenic butterflies, both daylength and temperature, or humidity, are assessed by the growing larvae. Sense organs detect these cues and send signals to the brain, which subsequently integrates these cues into meaningful physiological signals.

Many butterflies exhibit a green versus brown pupal coloration that depends on color or texture of the substrate perceived by the larva. Pupal coloration of the butterfly, *Graphium sarpedon nipponum*, can be manipulated by simply altering the substrate coloration (Hiraga, 2005). In this species, the color of the substrate is sensed optically, through the larval stemmata (Hiraga, 2005). In contrast, pupal coloration of *Papilio xuthus* depends primarily on tactile sensation, likely received from the setae on limbs (Hiraga, 2005).

Migratory locusts can develop into either dark-colored swarm-forming, gregarious phase (or form) or a green, cryptic, and more sedentary solitary phase, depending on larval crowding and to a lesser degree sight and smell of other individuals (Faure, 1932; Hagele and Simpson, 2000). Stimulation of sensory hairs on appendages is sufficient to rapidly trigger gregarious behaviors, including increased aggregation tendencies and migratory activity. Depending on the species, different appendages serve as sensors of crowding, indicating that the sensors may have evolved independently in different species (Cullen et al., 2010). For example, the femur of hindlegs act as tactile sensors in the desert locust *Schistocerca gregaria* (Simpson et al., 2001) whereas in the Australian plague locust, *Chortoicetes terminifera*, the antennae sense the presence of other individuals to elicit gregarious behavior (Cullen et al., 2010).

Multiple neurotransmitters and neurochemicals are modulated as a locust undergoes phase shifts (Rogers et al., 2004). In *S. gregaria*, serotonin is elevated in the thoracic ganglion, where nerves carrying signals from the hind-leg sensor project (Anstey et al., 2009). Serotonin has been shown be both necessary and sufficient to trigger gregarious behaviors (Anstey et al., 2009; Rogers et al., 2003). In *Locusta migratoria*, phase changes during nymphal development are associated with changes in dopamine and serotonin production (Ma et al., 2011).

In many polyphenisms, chemical signals play critical roles in inducing alternative forms. As noted earlier, odors of other locusts can influence phase shifts. Pheromones also play essential roles in inducing the development of alternative forms. For example, in termites, a volatile inhibitory pheromone comprised of n-butyl-n-butyrate and 2-methyl-1-butanol is produced by secondary queens to prevent differentiation of new queens (Matsuura et al., 2010). In the case of hymenopterans, long-chain linear and methyl-branched saturated hydrocarbons emitted by the queen suppress worker reproduction (Van Oystaeyen et al., 2014). These pheromones are detected by chemosensory receptors (Mitaka et al., 2016; Pask et al., 2017; Zhou et al., 2012). Species with sophisticated caste systems, such as ants, have expanded repertoire of chemosensory receptors that allow them to distinguish between different compounds (Zhou et al., 2012). Chemical cues, called kairomones, emitted by predators are also important in inducing defensive morphologies. The water flea *Daphnia* can develop a variety of defensive morphs in response to the presence of predators. For example, *Daphnia pulex* develops neck-teeth in the presence of *Chaoborus* midge larvae (Tollrian, 1993), whereas *Daphnia longicephala* develops a helmet-like crest on its head in the presence of hemipteran backswimmers (Barry, 2000). In *D. pulex*, glutaminergic and cholinergic neuronal pathways have been implicated in neck-teeth formation, indicating that these pathways may receive or mediate kairomone signals (An et al., 2018; Miyakawa et al., 2015; Weiss et al., 2012). When the first antennae of *D. longicephala* were impaired, they failed to form crests even when exposed to the predator's kairomone, suggesting that kairomone chemoreceptors are likely located in the first antennae (Weiss et al., 2015).

Epigenetic regulation of polyphenisms

In addition to neuronal sensors, recent studies have begun to demonstrate the importance of environmental effects on the epigenome. In particular, several polyphenic traits have been shown to be modulated by DNA methylation changes. DNA methylation involves conversion of cytosine to 5-methylcytosine by DNA cytosine-5-methyltransferases (DNMTs). These methylation changes are accompanied by alteration in the chromatin structure and lead to changes in gene expression. While the details of how DNA methylation states are altered by the environment remain unclear, in some examples of polyphenisms, DNA methylation is one of the first cellular changes observed.

Among the social insects, several studies have shown links between DNA methylation and caste determination. In honeybees, the decision to become a worker or a queen bee depends on the quality of the food a larva received during development: when royal jelly is given, the larva develops into a queen. When the expression of one of the DNMT-coding gene, *Dnmt3*, was silenced during the critical "decision-making" period using RNA interference, about ¾ of the larvae developed into queen-like adults with ovaries (Kucharski et al., 2008). In contrast, ¾ of the control larvae developed into workers with rudimentary ovaries. *Dnmt3* RNAi leads to reduced DNA methylation of the *dynactin p62* gene. When larvae are fed royal jelly, the *dynactin p62* also becomes demethylated (Kucharski et al., 2008). Thus, diet appears to reprogram the larval transcriptome to influence the caste determination process. More recent studies have found a correlation between DNA methylation sites and alternative splicing, suggesting that DNA methylation might regulate alternative splicing (Lyko et al., 2010). In support of this view, *Dnmt3* RNAi in honeybees has been shown to cause extensive alternative splicing in the fat body (Li-Byarlay et al., 2013). Similarly, DNA methylation states have been associated with caste determination in the termite *Zootermopsis nevadensis* (Glastad et al., 2016). In this species, intragenic DNA methylation states have also been found to be associated with alternative splicing. DNA methylation may not underlie caste determination in all social insects, and methodological concerns have been raised (Libbrecht et al., 2016). Thus, more studies are needed to definitively demonstrate the role of epigenetic changes in caste determination.

Additional evidence for epigenetic changes in polyphenic regulation comes from studies with *Locusta migratoria*, where methylation states differ significantly between the brain of solitary and gregarious locusts in over 90 genes (Wang et al., 2014). However, whether any of these changes are causes or consequences of phase polyphenism remains unknown.

Outside of insects, DNA methylation has been linked to temperature-sensitive sex determination in several vertebrate species. In the red-eared slider turtle *Trachemys scripta*, higher temperatures lead to female gonadal sex determination during embryogenesis. Female gonadal development is linked to upregulation of estradiol expression due to activation of the enzyme aromatase, which converts testosterone into estradiol (Matsumoto et al., 2013; Ramsey et al., 2007). The expression of aromatase-coding gene, *cyp19a*, has been shown to be epigenetically controlled by changes in DNA methylation (Matsumoto et al., 2013). When eggs are incubated at the female-producing temperature, the CpG sites in the ovary specific enhancer and the TATA box in the promoter of the *cyp19a* gene become demethylated, leading to transcriptional activation of *cyp19a*. Similarly, in the alligator, the promoter region of the *cyp19a* gene becomes demethylated under female-producing temperatures, whereas the male gonad development gene, *Sox9*, becomes hypermethylated (Parrott et al., 2014). In European sea bass *Dicentrarchus labrax*, sex determination is influenced by both genetic and environmental factors. At high temperatures, the promoter of *cyp19a* becomes hypermethylated relative to those raised at low temperatures regardless of genetic sex (Navarro-Martin et al., 2011). This leads to the reduction in *cyp19a* expression and the consequent masculinization of female gonads. Thus, temperature-dependent sex determination in many species can be traced back to changes in DNA methylation states of the *cyp19a* gene.

Although the mechanisms of environmental sensors remain unknown for many polyphenic traits, an emerging picture is that the neural sensors are diverse and sometimes species-specific. Therefore, neural sensors likely evolved independently during the evolution of polyphenisms and likely can be modified relatively easily over evolutionary time. Such evolutionary lability and relative lack of constraint would allow sensors to evolve in a way that maximizes precision and minimizes error in the assessment of environmental conditions. In contrast, the shared process of *cyp19a* methylation during temperature-dependent sex determination in vertebrates may indicate some kind of developmental constraint associated with this process.

Integrators

The central nervous system and hormones

Adaptive phenotypic plasticity is seldom if ever restricted to a single local trait. It invariably involves changes in many traits distributed throughout the body, and therefore requires developmental and physiological coordination among various tissues and organs. Systemic coordination and integration is done via the nervous and endocrine systems. Plastic seasonal adaptations, such a sexual and winter plumage in birds, antlers in deer, winter coats in mammals, and the distinctive seasonal forms of many insects, are all regulated by differential patterns of hormone signaling. Crowding-induced large-winged migratory forms in locusts are under endocrine control, as are the alternative sexual/parthenogenetic and winged/wingless phenotypes of aphids. Sex switching, such as protandry and protogyny, and repeated or serial bidirectional sex switching in fishes, are also under endocrine control.

Hormone secretion is controlled by the brain and central nervous system, either via direct nervous stimulation of endocrine glands, or, more commonly, via tropic neurosecretory hormones. In mammals, the hypothalamus-pituitary axis and in insects the brain-corpus cardiacum-corpus allatum axis, are the principal central controllers of hormones that regulate a broad diversity of developmental and physiological processes throughout the body. The brain can receive and integrate environmental factors such as temperature, photoperiod and nutrition, as well as tactile, visual, and chemical communication signals. The brain communicates this to the body via a particular temporal pattern of secretion of a variety of hormones which, in turn, control the various physiological and developmental processes that lead to the final phenotype.

The brain, through the direct or indirect control of developmental hormones, can therefore be thought of as the grand integrator of postembryonic development. Hormones are the mediators between the brain and the peripheral tissues. It is important to recognize that the target tissues are not passive responders to a hormone, but play an active role in deciding what will happen in response to a hormone. The response to a hormone is not a property of the hormone, but of the intracellular response pathways that are set up in the target tissue. The same hormone, acting on different tissues, usually provokes different responses, and any ligand that activates a hormone receptor provokes the same cellular response as the hormone itself. By expressing (or not) receptors for the hormone, a target tissue can control the timing of its response to a hormone. In addition, by altering the pathway downstream of the hormone receptor, the target cell controls which intracellular enzymes are activated and which genes are transcribed, and thus, its physiological and developmental response to the hormonal signal.

Control of developmental hormones by the brain-neuroendocrine axis enables perception and integration of complex environmental signals, which are then translated into switches in the hormonal and molecular pathways that control alternative phenotypes.

Endocrine integrators of phenotypic plasticity in vertebrates

Sex change in fish involves changes in the gonad, as well as profound changes in morphology and behavior (Todd et al., 2016). These changes are controlled by a changing balance between estradiol and testosterone that may, in turn, be controlled by changes in the expression of aromatase, the enzyme that converts testosterone to estradiol. The level of estradiol is also controlled by the gonadotropin-releasing hormone, from the hypothalamus. Cortisol, which responds to environmental stress and social cues, is also known to be involved in sex change. The environmental signals that trigger sex change are diverse. Larger body size and older age trigger male to female transition in many fish (e.g., black porgy, *Acanthopagrus schlegeli*, and barramundi, *Lates calcarifer*), or from female to male in California sheephead (*Semicossyphus pulcher*). Aggressive dominance by males the bluehead wrasse (*Thalassoma bifasciatum*), and by females in anemonefish (*Amphiprion* and *Premnas*), suppress sex switching in their mates, and when the dominant member dies or is removed, new dominance competition induces sex change.

The males of many temperate zone birds have a distinctive breeding plumage, presumably used in courtship. The color pattern of such birds is controlled by testosterone (Lindsay et al., 2016; Lindsay et al., 2011), and daylength is the environmental signal that controls the switch to mating plumage. Winter and summer fur in mammals is also controlled by daylength: Short days induce increased secretion of melatonin from the pineal gland, and reduced secretion of prolactin from the pituitary (Zimova et al., 2018). High levels of melatonin coupled with low levels of prolactin induce development of winter pelage, and the reverse is true for development of summer pelage. Seasonal growth of antlers in male deer is controlled by testosterone, whose secretion is indirectly controlled by melatonin from the pineal gland. Under short-day conditions, melatonin rises and induces production of luteinizing hormone, which, in turn, suppresses testosterone production. When days lengthen melatonin levels decline and inhibition of testosterone secretion is relieved, leading to the seasonal regrowth of antlers (Bubenik, 2006).

Endocrine integrators of phenotypic plasticity in insects

Polyphenisms are the best-studied examples of phenotypic plasticity in insects. As noted earlier, in polyphenic insects a larva can develop into one of two or more stable alternative adult phenotypes, depending entirely on signals received from the environment. The environmental signals can be as diverse as temperature, photoperiod, nutrition, and pheromones. The differences between alternative phenotypes can be modest, but can also be so profound that some were originally described as different species.

The developmental switches that lead to the alternative phenotypes are controlled by the endocrine system. Three hormones, juvenile hormone (JH), ecdysone and insulin-like peptides are the principal controllers. The secretion of JH and ecdysone is controlled by tropic hormones from the brain. The insulin-like peptides are themselves neurosecretory products. In some cases, additional neurosecretory hormones, like corazonin (Tanaka, 2000a,b), also play a role in controlling the development of alternative phenotypes. Hormone-controlled switching of developmental pathways to produce alternative phenotypes occurs during discrete critical periods (Nijhout, 1999, 2003). During these critical hormone-sensitive periods, the presence of the hormone causes a different pattern of gene expression than happens in the absence of the hormone, and this results in an altered developmental pathway.

Seasonal polyphenism in butterflies is controlled by ecdysone during a critical period that occurs shortly after pupation. In *Junonia coenia and Araschnia levana,* the summer form is induced by the presence of ecdysone above a threshold during this period, as is the wet season form in *Bicyclus anynana* (Koch and Bückmann, 1987; Kooi and Brakefield, 1999; Rountree and Nijhout, 1995; Smith, 1991). If ecdysone is absent during this period, the alternative autumn and dry season forms develop. In aphids, injection of ecdysone causes wingless forms to develop while winged forms are induced by ecdysone antagonists (Vellichirammal et al., 2017), suggesting that the winged/wingless polyphenism in aphids is also controlled by ecdysone. The temporal pattern of ecdysone secretion is controlled by photoperiod and temperature, with short daylengths and low temperatures inducing the autumn forms. In the butterfly, *Araschnia levana*, by contrast, short daylengths induce the summer form, not the autumn form, because short days also induce a winter diapause and the adults will not emerge until the following summer.

Caste determination in social insects is controlled by JH. Application of JH during a critical period in the last larval stage of the ant, *Pheidole bicarinata*, causes the larva to develop into a soldier (Wheeler and Nijhout, 1981, 1983). The soldier caste is larger than the worker (Bortolotti et al., 2001; Cnaani et al., 2000), and has a disproportionally large head. Elevated JH levels during the critical period shift the critical weight (at which the endocrine cascade that stops growth and induces metamorphosis begins) to a larger body size, and also reprograms growth of the imaginal disks of the head, resulting a larger-bodied individual with a disproportionally large head (Wheeler and Nijhout, 1983). The level of JH is controlled by the quality of nutrition, which is thought to depend on the level of protein in the food, with better nourished individual having higher levels of JH during the critical period. The sensitivity to JH during the critical period is, in turn, controlled by a pheromone produced by soldiers. The soldier-inhibiting pheromone raises the threshold of sensitivity to JH (Wheeler, 1991; Wheeler and Nijhout, 1984), so that when there are many soldiers in a colony no further soldiers are induced, even at high levels of nutrition. In honeybees (*Apis mellifera*), queens are induced by higher levels of JH during a critical period in the last larval instar (Dietz et al., 1979; Wirtz and Beetsma, 1972), and in this case higher levels of JH are, in turn, induced by the improved nutrition that queen-destined larvae receive in the form of royal jelly (Buttstedt et al., 2014; Wilson, 1971). The production of royal jelly by workers is, in turn, inhibited by the resident queen via an inhibitory pheromone, the queen substance (Butler et al., 1962). Depending on whether or not JH is present during the critical period there are two alternative massive changes in gene expression (Evans and Wheeler, 1999, 2000). Queen determination in bumblebees (*Bombus terrestris*), is likewise controlled by JH. In both bumblebees and paper wasps (*Polistes spp.*), the "queen" is established by aggressive dominance (Bloch and Hefetz, 1999; Cnaani and Hefetz, 2001; Doorn, 1989; Giray et al., 2005; Röseler et al., 1985; Tibbetts and Huang, 2010). When a reproductive "worker" is treated with JH she becomes more aggressive and her ovaries begin to develop suggesting that JH also mediates this behavioral worker/queen polyphenism.

The males of dung beetles (*Onthophagus spp.*) and rhinoceros beetles (Dynastinae) can have large horns on head and thorax, depending on their body size (Emlen, 1997; McCullough et al., 2015). Horn size scales allometrically with body size but in a strongly sigmoid fashion, so that there appears to be a threshold body size below which no (or only very small) horns are produced, and above which there is a strong positive allometry so that at larger body sizes horns are disproportionally large. In *Onthophagus*, body size is entirely environmentally determined by the size of the dung ball a larva is provisioned with (Moczek and Emlen, 2000; Shafiei et al., 2001). There is a sensitive period during the last larval instar, when body size is assessed, and male larvae that are above a given size threshold will progress to develop large horns. During the prepupal stage, the males have a JH-sensitive period during which JH suppresses horn growth. JH levels decline during the prepupal phase, and larvae that were above the body size threshold have a delayed JH-sensitive period, at which time JH levels are too low to suppress horn development (Emlen and Nijhout, 1999).

Effectors

How do alternative phenotypes arise given variation in integrators? One simple hypothesis is that integrators determine the onset of character development. When integrators pass a threshold, the developmental process is initiated. In such a scenario, intraspecific variation in the timing of integrators crossing a threshold will induce variation in the onset of developmental processes. Temporal differences in the onset, rate, and duration of developmental processes in different species are most commonly referred to as heterochrony. Adaptive polyphenisms usually show heterochrony in hormone concentrations during specific developmental windows. For instance, the seasonal color polyphenism in the buckeye butterfly, *Junonia coenia*, occurs as a result of variation in ecdysone concentration in the first 48 hours after pupation (Rountree and Nijhout, 1995). Plastic phenotypes that arise from differential growth of body parts seem to be affected by the rate and duration of growth during the final larval instar in holometabolous insects, and the final nymphal instar in hemimetabolous insects (Nijhout and McKenna, 2018). The fact that there are specific developmental contexts in which hormonal signals can alter phenotypic outcomes suggests that the morphogenetic response of tissues may be confined to specific periods in ontogeny, because those are the only times those tissues are capable of responding to a hormone. This means we need to understand how integrators induce developmental processes, that is, what are the effectors in different cases of hormonally induced polyphenisms?

What are effectors?

There are two types of effectors: those that regulate growth and those that regulate patterning at the cellular and tissue level. In the following sections, we will discuss the mechanisms and relative contribution of each of these, and attempt to synthesize.

Effectors of plastic growth

Plasticity in growth is most commonly caused by nutritional variation. Nutrition primarily alters the concentration of circulating insulin-like peptides and amino acids. The nutrition dependent release of insulins and amino acids is assessed by developing tissues via the insulin/TOR pathway (Fig. 15.1). Insulins bind to the insulin receptor (InR) on the cell membrane, initiating a phosphorylation cascade that activates phosphoinositide 3-kinase (PI3K), which thereupon phosphorylates and activates protein kinase B (PKB/AKT). Active PKB/Akt phosphorylates the growth inhibitory transcription factor FOXO, initiating its translocation from the nucleus into the cytoplasm (Aoyama et al., 2006). FOXO activity fluctuates inversely with the strength of the insulin-like signal. Under conditions of optimal nutrition and high insulin, FOXO levels are low. When nutrition is limited and transduction through the insulin pathway is low, FOXO levels are elevated. Elevated FOXO leads to the inhibition of the target of rapamycin (TOR) (Southgate et al., 2007). Counterintuitively, FOXO also upregulates the transcription, but not the activity, of the InR (Puig and Tijan, 2005), priming the tissues for increased insulin if optimal nutrition is restored (Snell-Rood and Moczek, 2012).

TOR activity is positively regulated by PKB/Akt. TOR activity is also regulated by the concentration of amino acids that come into the cell via a cationic amino acid transporter (Colombani et al., 2003). The activation of TOR by PKB/Akt and amino acids leads to cell growth via two distinct targets: activation of S6K which leads to ribosome biogenesis and inhibition of 4E-BP, which is an inhibitor of protein translation. Therefore, the ultimate outcome of the insulin/TOR cascades is protein synthesis and cell growth.

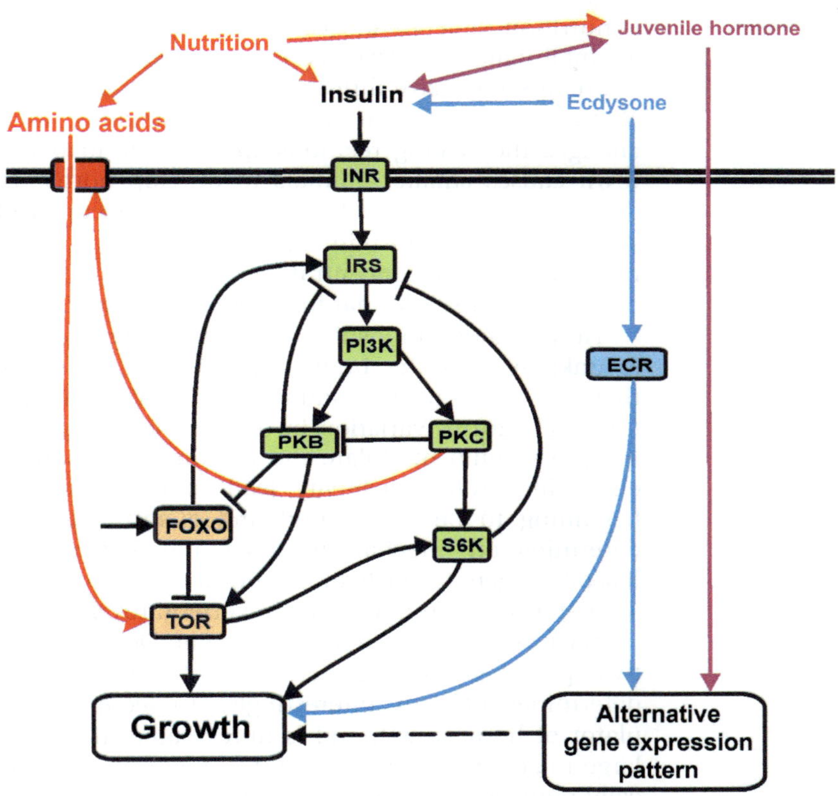

Figure 15.1 In insect development, endocrine signals (insulin, ecdysone, and juvenile hormone) control growth and patterns of gene expression. The nutrition dependent release of insulin stimulates increased uptake of amino acids in developing tissues and controls protein synthesis and growth via the insulin/TOR pathway. Insulin signaling interacts with ecdysone and juvenile hormone signals to control growth and switches in gene expression. *Source*: After Nijhout, H.F., McKenna, K.Z., 2018. The distinct roles of insulin signaling in polyphenic development. Curr. Opin. Insect Sci. 25, 58–64.

The regulation of this cascade and its sensitivity to insulin concentrations determines the plasticity in growth of different body parts. How differential utilization of the pathway produces part-specific patterns of plasticity is virtually unknown. A change in the expression of any one of the components in the pathway can alter the growth response of a part to varying concentrations of insulin and amino acids.

Horned scarabeid beetles are an exceptional example of how evolution can take advantage of tissue specific regulation of the insulin cascade to produce discrete developmental trajectories for specific parts in a sex-limited manner. In the dung beetle, *Onthophagus taurus*, horn development is limited to males and is nutrition dependent. There is virtually no horn development at small male body sizes and very enhanced horn development upon crossing a size threshold (Moczek et al., 2002; Moczek and Nijhout, 2003). Perturbing insulin signaling in this species changes the scaling relationship of horns to body size: FOXO RNAi causes smaller animals to develop horns. This suggests that evolution of the polyphenism could occur by changing the sensitivity of the insulin signaling cascade in the horns to a range of insulin concentrations that are normally below the nutritional threshold that initiates horn growth.

How variation in insulin sensitivity is achieved in this species is unknown. In the brown planthopper, *Nilaparvata lugens*, there are two kinds of insulin receptors with antagonistic effects (Xu et al., 2015). Variation in the relative expression of the two receptors could modulate the response to insulin. How differences in insulin sensitivity lead to differences in growth is beginning to be understood and is intricately linked to sex determination mechanisms. Somatic sex determination in insects is genetic instead of the hormonally induced somatic sex determination that exists in vertebrates. In most insects, the primary determinant of sex is the gene *Doublesex* (*Dsx*). *Dsx* has two splice variants, one determines female sex while the other determines male sex. Interestingly, *Dsx* seems to be a major regulator of horn growth and is upregulated in the head horns of large males (Kijimoto et al., 2012). RNAi of *Dsx* in males drastically reduces head horn size close to the size of naturally small males (Kijimoto et al., 2012). As might be expected then, *Dsx* seems to be regulated by the insulin signaling pathway. RNAi of FOXO leads to upregulation of *Dsx* whereas RNAi of the insulin receptor downregulates *Dsx* (Casasa and Moczek, 2018).

Similar mechanisms have been described in the sex-limited nutrition-dependent development of stag beetle mandibles and of drosophilid fly tibial bristles that make up the sex comb. In stag beetles, the mandibles are exaggerated in males but not in females (Fig. 15.2). When *Dsx* is knocked down, the exaggerated growth in male mandibles is drastically reduced while the normal growth in females is enhanced (Gotoh et al., 2014). Similarly, in drosophilids males develop exaggerated tibial bristles on prothoracic legs in a nutrition dependent manner (Kopp et al., 2000). In every species where the bristles are exaggerated into a sex comb, *Dsx* is highly expressed in the cells destined to become part of the sex comb. In *D. melanogaster*, *Dsx* RNAi in males reduces the sex comb considerably and expression of the male *Dsx* in females exaggerates female tibial bristle growth

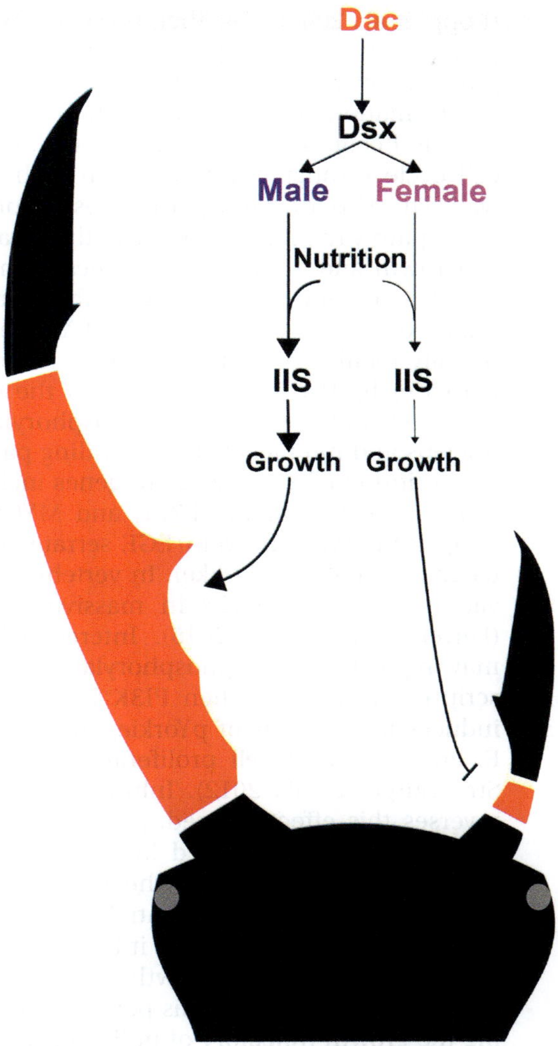

Figure 15.2 The interaction between gene products involved in cell fate, sex determination, and insulin signaling in the control of mandible polyphenism in a stag beetle. The exaggerated growth of the male mandible in stag beetles is an example of how cell identity can differentially regulate IIS signaling. The dachshund domain (Dac) makes up the medial portion of the mandible. Dac interacts with male and female *Dsx* to regulate nutritional sensitivity. When Dac interacts with male *Dsx*, the IIS pathway is hypersensitive to nutrition, allowing exaggerated growth of this region. When Dac interacts with female *Dsx*, the IIS pathway is not sensitive and inhibits growth of the medial region. *Source*: Based on data from Gotoh, H., Miyakawa, H., Ishikawa, A., Ishikawa, Y., Sugime, Y., Emlen, D.J., et al., 2014. Developmental link between sex and nutrition; doublesex regulates sex-specific mandible growth via juvenile hormone signaling in stag beetles. PLoS Genet. 10, e1004098; Gotoh, H., Zinna, R.A., Ishikawa, Y., Miyakawa, H., Ishikawa, A., Sugime, Y., et al., 2017. The function of appendage patterning genes in mandible development of the sexually dimorphic stag beetle. Dev. Biol. 422, 24–32.

(Kopp et al., 2000). Together, these studies suggest that *Dsx* may play a direct role in regulating gene products related to cellular growth, linking the insulin cascade to protein synthesis, sex determination, and cell proliferation.

It is becoming clear that the insulin cascade regulates the cell cycle through the interaction with the Hippo/Warts pathway. The Hippo/Warts pathway is a conserved growth regulatory pathway that regulates the transcription of several progrowth and antiapoptosis genes (Pan, 2010). This transcriptional regulation is achieved by the transcription factor Yorkie (Pan, 2010). Interestingly, most of Yorkie activity is posttranslationally regulated. In its inactive state, Yorkie is phosphorylated (pYorkie) by Warts and remains in the cytosol. When Warts is inhibited and Yorkie is not phosphorylated, Yorkie enters the nucleus and interacts with its binding partner Scalloped to drive transcription of a vast array of genes including cell cycle regulators (cyclin E, bantam, E2F1, and MYC) and tissue organizers (Dpp/Bmp, Wg/Wnt, vein/EGF, serrate, and heparin sulfate proteoglycans dally/dally-like). In vertebrates and insects, overactivation of Yorkie results in massive overgrowth of the tissue (Ferreira and Milán, 2015). Interestingly, the insulin cascade may regulate Yorkie phosphorylation, and therefore, its transcriptional activity. When PI3K is upregulated on the wing, it induces lower levels of pYorkie, elevated transcription of cyclin E, and increased cell proliferation (Ferreira and Milán, 2015; Straßburger et al., 2012). Inhibition of PI3K with wortmannin reverses this effect. Indeed, parts of the Hippo/Warts pathway are differentially regulated in beetle horns (Hust et al., 2018). Thus, the components of the insulin cascade are themselves effectors of growth. Differential regulation of each component in each appendage, whether it be transcriptional or posttranslational, determines the growth potential in response to varying insulin concentrations. This plays a significant role in determining the growth trajectory of individuals and the phenotypic variation produced when nutrition is variable.

Effectors involved in plastic patterning

The detailed growth and differentiation of tissues are organized by patterning networks of morphogenetic signaling molecules and transcriptional regulators. In insects, this patterning network includes the WntA homolog Wingless (Wg), the morphogen Hedgehog (Hh), and the BMP type morphogen Decapentaplegic (Dpp) (Dekanty and Milán, 2011). Each tissue utilizes the network in a unique way to set up anterior-posterior and proximo-distal axes. Within these axes, subdomains are differentiated that are defined by different transcriptional regulatory networks. This ultimately allows differential regulation of several processes including cell growth, adhesion, and cell-cycle regulation. How hormones interact with patterning networks and how temporal differences in hormone appearances influence the morphogenetic effects is almost entirely unknown. That said, recent evidence from a few systems is beginning to shed light on how morphogens can act as effectors in plastic developmental processes.

Of the domains patterned by morphogens, the proximo-distal domains have shown the greatest proclivity for plastic development. The proximo-distal axis is setup by Wg (and in some imaginal disks Wg + Dpp) and differentiates three domains: a proximal domain that expresses the homeodomain transcription factor Homothorax, a medial domain characterized by the expression of the transcriptional regulator Dachshund, and a distal domain that expresses the homeodomain transcription factor Distal-less (Angelini and Kaufman, 2005).

Interestingly, one or more of these homeobox genes plays a significant role in many insect polyphenisms. In *O. taurus*, Distal-less has been implicated in head and pronotal horn development (Moczek and Rose, 2009). In the case of stag beetles, Dachshund is required for the exaggerated growth of the mandibles (Gotoh et al., 2017). Conversely in flies, Dachshund is not expressed in the sex comb, and ectopic expression of Dachshund in the sex comb reduces the size of the sex comb considerably (Atallah et al., 2014).

The connection between intraorgan domains, insulin signaling, and *Dsx* is a common feature of each of the nutrition-dependent polyphenisms we have addressed thus far. Since the expression of *Dsx* directly leads to growth, and is regulated by insulin signaling, it may be the case that the different developmental environments, specified by the patterning genes, may differentially regulate the insulin cascade within the appendage. The potential influence of homeodomain transcription factors on the biochemical physiology of the insulin cascade helps explain how different tissues and different regions within a tissue can respond in a unique and sex-specific way to nutritional variation. In this way, patterning genes play a significant role in the evolution and deployment of developmental plasticity.

Similarly, seasonal polyphenism in butterfly color patterns may be directly attributable to variation in the expression of these patterning genes. The best example comes from the development of eyespots in Nymphalid butterflies. Early experimentation showed that eyespot development is regulated by a diffusion-threshold mechanism with a morphogen diffusing from a focus. The rings around the focus are subsequently determined by the range of the concentration gradient (Nijhout, 1980). Recent evidence suggests the morphogen diffusing from the focus is Wg (Özsu et al., 2017). This has led to the co-option of the proximo-distal patterning network as Wg induces the expression of Distal-less at the center of the eyespot (Brunetti et al., 2001). The determination of the eyespot in the larval wing imaginal disk is followed by the development of the rings of the eyespot in the first days of the pupal stage.

Some butterfly species have developed seasonal polyphenisms in which the eyespots change in size. In *Bicyclus anynana*, eyespots on the hindwing grow quite large in the wet season, but are reduced in size in the dry season (Monteiro et al., 2015). In this species, the ecdysone receptor (EcR) is highly upregulated in the focus of the eyespot. Perturbation of ecdysone signaling directly influences the size the eyespots attain (Monteiro et al., 2015). The evolution of this unique polyphenism in *Bicyclus*, but not in other nymphalids, is likely due to the co-option of EcR in the focus during the prepupal and pupal periods. Variation in the signal strength of ecdysone may affect the expression and possibly the diffusion of Wg. The net effect of this would be to vary the size of the ring domains that differentiate in a concentration-dependent manner.

Metamorphosis

Metamorphosis is a sequential polyphenism in which different life stages develop distinct morphological and physiological adaptations. Larvae are typically specialized for growth, and in marine invertebrates for dispersal, and adults are specialized for reproduction (and in amphibians and insects, for dispersal). There has been a progressive evolutionary divergence of larval and adult morphologies in most lineages that use metamorphosis as a life history strategy. Not only do tadpoles not look like frogs, or caterpillars like moths, or a brachiolaria like a starfish, a naive taxonomist, knowing only these larval forms, would be at a loss to even place them in their correct taxonomic class. Yet larval and adult forms are robust alternative phenotypes produced by the same genomes; it is hard to think of more extreme and common examples of phenotypic plasticity.

The effectors that determine life history identity have been best studied in insect metamorphosis. In the larval stages of holometabolous insects, JH maintains a larval state, hence the name "juvenile" hormone. In order for the tissues to reach pupal identity, JH must be removed from the system, which enables a developmental switch that is induced by ecdysone in the absence of JH.

The general sequential steps in the hormonal control of metamorphosis go as follows. 20-hydroxyecdysone (20E) in the presence of JH induces a molt but maintains developmental identity at current stage, and 20E in the absence of JH induces a molt that transitions to next developmental stage (Nijhout, 1994). In the larval stage, secretion of JH induces the expression of *Krüppel homolog 1* (*Kr-h1*), which represses the pupal specifier gene *broad* (Minakuchi et al., 2008; Zhou and Riddiford, 2002). Once JH is removed from the system in the final larval instar, tissues begin to express *broad* in response to low levels of 20E. This causes all tissues to take on pupal commitment. During the prepupal stage, JH levels rise again and *Kr-h1* maintains *broad* expression to maintain pupal commitment. If *broad* is inhibited during the larval pupal transition via RNAi, larvae metamorphose to larval/adult mosaics (Suzuki et al., 2008). If *Kr-h1* is inhibited at this time, animals skip the pupal stage altogether (Minakuchi et al., 2008). Thus, *broad* is required to progress pattern development to a pupal state and inhibit progression of adult patterning.

During the pupal stage, JH is no longer expressed and *Kr-h1* and *broad* expression levels decline to allow 20E to drive adult patterning. This is achieved via the expression of the ecdysone-response gene *E93* (Ureña et al., 2014). E93 inhibits *Kr-h1* and *broad* expression in pupae. If E93 is inhibited, pupa fail to progress toward adult development and both *Kr-h1* and *broad* are upregulated. Similarly, if a juvenile hormone analog (JHA) is applied during the pupal stage, a second pupal cuticle develops instead of adult cuticle, suggesting JHA drives the patterning of pupal development a second time in the presence of 20E (Ureña et al., 2016).

Let us assume that life history morphogenesis is defined by two processes then: growth and patterning. Growth proceeds with the input of insulin and amino acids, but how does patterning proceed? In every case studied in insects, progression of the appendage patterning networks that establish regional differences in the appendage requires the input of hormones (Mirth et al., 2009; Oliveira et al., 2014). We can then think of JH and 20E as having opposing roles: JH maintains life history stage identity and 20E is required for any switch in developmental state. This can be thought of as JH inhibiting the progression of a pattern. Indeed, JH application primarily affects differentiation and has even been shown to inhibit Hedgehog (Hh) expression in developing limbs (Villarreal et al., 2015). The role of 20E then is to stimulate pattern progression. In *Drosophila*, if ecdysone biosynthesis is slowed down, pattern progression on the wing slows, and if it is sped up, the rate of patterning speeds up (Oliveira et al., 2014). This effect is accomplished because 20E may regulate the expression of Wg, Hh, and Dpp (Dye et al., 2017; Mirth et al., 2009). Conceived in this way, the link between endocrine regulators and morphogenetic processes is a fundamental mechanism to control life history progression, as well as the regulation of growth, that clearly deserves more attention.

Nonadaptive plasticity and the evolution of robustness

Most traits are plastic to various degrees: for instance, undernutrition stunts growth and temperature variation alters growth and development rates in almost every biological system and can result in more or less dramatic variation in size and shape. Much of phenotypic plasticity is nonadaptive, meaning that the intermediates in the range of phenotypes produced by a variable environment are not especially more suited to the particular intermediate value of the environment that induced it. Sometimes the environmentally induced phenotypes are considered pathological, as in the case of teratologies induced by toxins, vitamin deficiencies, or viral infections.

Biological systems have evolved a diversity of mechanisms that minimize phenotypic variation even in widely varying environments. These robustness or homeostatic mechanisms are best known from physiology, but also occur widely in developmental and metabolic systems (Nijhout et al., 2019; Nijhout et al., 2017). It is conventional to distinguish between environmental robustness, which buffers against variation in the environment, and genetic robustness, which stabilizes the phenotype against genetic and mutational variation. It is likely that in many cases these two mechanisms are one and the same, for the simple reason that genes have their effect on phenotypes through same physiological and developmental mechanisms that are sensitive to environmental variation. Protection against variation in the latter would naturally confer protection against variation in the former.

Yet, in spite of a diversity of robustness and homeostatic mechanisms, many organisms vary, to some degree, with variation in external factors. The natural range of phenotypes that is produced by a gradient in an environmental factor, in organisms of the same genotype, is called a *reaction norm*. Reaction norms to temperature, light conditions, nutrition, geographic latitude, population density, and other environmental variables have been extensively documented in plants and animals (Roff, 1992; Schlichting and Pigliucci, 1998; Sultan, 2015). In most cases, reaction norms are only defined as the average phenotypes that develop in two alternative environments, and the two phenotypic values are connected by a (straight) line implying a continuity across intermediate environments. When the effect of intermediate environments is studied explicitly, the reaction norm is often found to be nonlinear (Carter et al., 2017; Reichling and German, 2000; Schlichting and Pigliucci, 1998). The temperature optima of many biological traits, for instance, can be thought of as nonlinear hump-shaped reaction norms. Unless it can be shown that each phenotype has higher fitness than alternative phenotypes in reciprocal common garden experiments, it should be assumed that the reaction norm is not an adaptation, but merely the mechanical consequence of the way in which the environmental variable affects biochemical, developmental and physiological kinetics.

Plasticity as an adaptation

Phenotypic plasticity is often thought to be an adaptation to a variable environment. This implies an assumption that the different phenotypes are adapted to, and have higher fitness in, the environments that induce them. However, if plasticity is simply due to environment-sensitive molecular and developmental processes, as argued above, then they are not adaptations, but may even be maladaptive. This is why robustness and homeostatic mechanisms evolve: they stabilize an optimal phenotype against environmental variation.

Rather than thinking of the plasticity as an adaptation to a variable environment, it is better to think of the plasticity as providing a foundation on which selection could act. Reaction norms have genetic components that can vary and that therefore can be selected on and shape the phenotype to better match a given environment (Scheiner, 2002). So how can one tell whether a reaction norm is an accident of chemistry and physics, or whether it is an adaptation? The obvious way is to test the fitness of the different phenotypes in their respective environments, and compare that to the fitness of the other phenotypes of the plastic response in that same environment. This can be done in cases where the phenotypes are threshold traits: basically two different traits in two environments. However, if there is a continuum of phenotypes across a gradient of environments, such test would be cumbersome and impractical [although see (Carter et al., 2017)]. A simple way of finding preliminary evidence for adapted phenotypes is from the shape of the reaction norm itself. If a particular phenotype is an adaptation, then it is likely to be canalized and robust to small variation in the environmental variable. Thus, nonlinear reaction norms, either sigmoidal or chair-shaped, with one or two nearly horizontal regions where variation in the environmental variable has little or no effect on the phenotype, would provide good, albeit circumstantial, evidence for adaptation. For instance, the sigmodal reaction norm of horn length in *Onthophagus* beetles is associated with increased ability to fight for access to females for the upper plateau, and increased ability to sneak for the lower one (Moczek and Emlen, 2000).

Reaction norms and allometries

Allometries describe the changes in the relative sizes of body parts with respect to variation in overall body size. If variation in body size is entirely due to an environmental factor (or factors), then the allometry is also a reaction norm that describes the change in size and shape of an organism due to environment. Intraspecific variation in body size is due to both environmental and genetic factors. Quantitative genetic analysis allows us to determine what fraction of phenotypic variance is due to additive genetic variance, and allows us to measure how gene x environment (GxE) interactions contribute to the phenotypic variance. However, this kind of statistical analysis cannot determine which specific environmental factors contributed to the phenotypic variation observed in a natural population. There are a few cases in which body size variation is entirely due to environment. In the dung beetle *O. taurus*, body size is determined entirely by how much dung a larva is supplied with. Heritability of body size is zero (Emlen, 1994, 1997). Thus, body size is determined by the amount of nutrition as the environmental variable. Similar environmental determination of body size could occur in other insects that provision their larvae with a fixed amount of food.

Are polyphenisms reaction norms?

Polyphenisms, as discussed earlier, are generally thought of as threshold traits (Roff, 1992), and exhibit two discrete alternative morphologies, induced by two different environments. It is therefore reasonable to ask whether a polyphenism is discrete, or threshold-like, because of discrete switches in the underlying developmental mechanism, or because the inducing environments are discrete and unambiguous? When polyphenic species are reared in the laboratory under conditions intermediate to those found in nature, they often produce intermediate forms, showing that the two forms found in nature are not due to some kind of binary switch-like developmental process. For instance, although horn length in *Onthophagus* beetles is strongly bimodal, in a large population in the laboratory there are always a number of individual with intermediate horn lengths (Emlen, 1994). In the butterfly, *Araschnia levana,* the summer and autumn forms are so different as to have originally been described as different species. Museum collections have a few specimens that are intermediate between the canonical forms, but those are exceptionally rare and generally classified as aberrations. In the laboratory, it is easy to produce a continuous range of intermediates, by careful timing of ecdysone injections during the sensitive period (Nijhout, 2003), or to produce an extreme morph, one that is absent in nature, e.g. by thyroid hormone-induced metamorphosis in urodeles. In this case there are no intermediate environments, because the two forms come from pupae that either diapaused or not. However, diapausing pupae can be "woken up" with an ecdysone injection, thus producing the intermediates. Not all polyphenisms can produce intermediates. The worker/soldier polyphenism of the ant, *Pheidole bicarinata,* comes about by a developmental switch that produces a discontinuous allometry of head size to body size (Fig. 15.3). Even small-bodied soldiers, within the range of worker sizes, have disproportionally large heads. In addition, very large workers, within the range of soldier sizes, have small worker-like heads (Wheeler and Nijhout, 1983).

Evolution via phenotypic plasticity

Polyphenisms and robustness are alternative ways of stabilizing phenotypic plasticity (Fig. 15.4). Selection on a plastic trait can result in canalization of an intermediate phenotype, or can lead to the stabilization of alternative phenotypes by a process known as genetic accommodation (West-Eberhard, 2003). A unique aspect of genetic accommodation is that it allows new developmental trajectories to appear in populations even without the appearance of a novel mutation. That is, standing

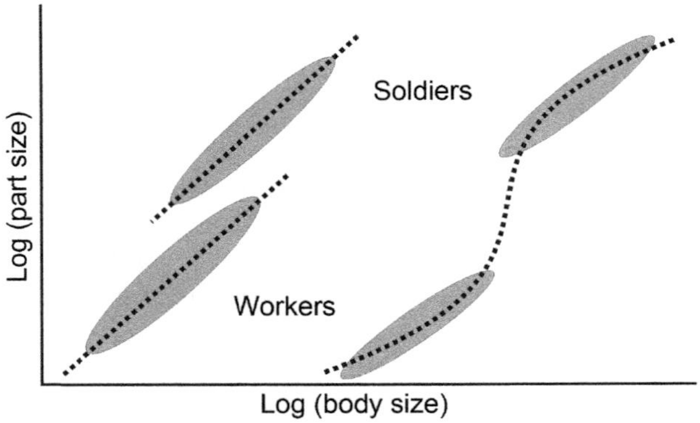

Figure 15.3 Two types of nonlinear allometries in insects. Left graph, the worker/soldier polyphenism of the ant, *Pheidole bicarinata*, comes about by a developmental switch that produces a discontinuous allometry of head size to body size, with no intermediates between the two types. Right graph, in *Onthophagus* dung beetles, a continuous sigmoid allometry produces two classes of individuals, with small or large horns, with some intermediates between the two types.

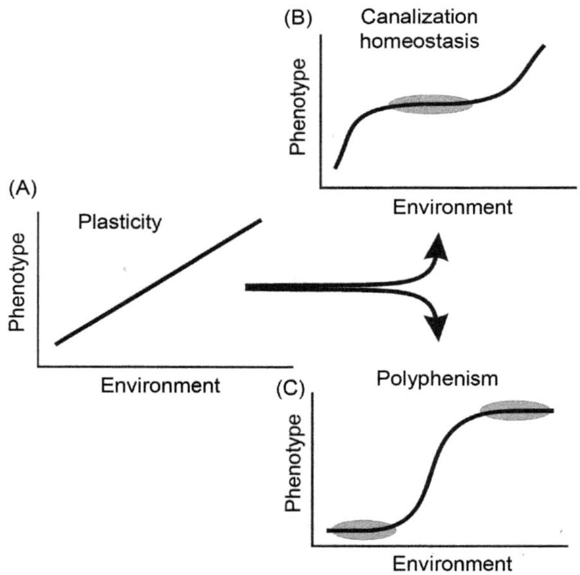

Figure 15.4 Polyphenisms and robustness are alternative ways of stabilizing phenotypic plasticity. Starting with a nonadaptive plastic trait, robustness can evolve by canalizing a particular phenotype, leading to a homeostatic chair-shaped relationship between genotype and phenotype *(top)* in which the phenotype is constant over a certain range of environments. Genetic accommodation can lead to a polyphenism *(bottom)* in which two alternative phenotypes are robustly expressed in two different environments.

genetic variation alone can generate novel developmental trajectories simply by shifts in the environment. Thus, a key tenet of genetic accommodation is that genes follow, rather than lead phenotypic changes in populations. Instead, phenotypic plasticity leads the way to novel developmental trajectories.

As a consequence of genetic accommodation, novel traits can either become fixed through a process known as genetic assimilation (Waddington, 1953, 1956), or they can become programed to appear only under specific predictable environmental conditions and become polyphenic (Suzuki and Nijhout, 2006). In both cases, extreme environments induce a range of aberrant phenotypes, and repeated cycles of environmental induction and selection on a specific phenotype, can cause that phenotype to become fixed. Robust traits may also evolve from polyphenic traits if one developmental trajectory becomes favored over the other.

Robustness is an evolved trait that promotes predictable phenotypic outcomes. Because most traits have evolved robustness, small changes in the underlying developmental processes are unlikely to manifest as major phenotypic alterations: only once a particular threshold is crossed will a phenotypic change ensue. Therefore, to generate novel traits from seemingly robust traits, the underlying developmental processes must be altered beyond the limits of the evolved mechanisms that confer robustness. Sensitizing mutations and large environmental changes facilitate this process by disrupting normal buffering mechanisms that confer robustness and hence reveal variability in development.

The role of cryptic genetic variation in genetic accommodation

One consequence to mutational robustness is that it masks the effect of mutations and therefore allows mutations to accumulate. The effect of those mutations is masked, or buffered, or counteracted, by a variety of mechanisms that stabilize the phenotype. These mechanisms are known as evolutionary capacitors because they allow the accumulation of mutations that persist in populations as crypto-genes (Landauer, 1957), or cryptic genetic variation (CGV) (Gibson and Dworkin, 2004; Rutherford, 2000), which may be important sources of variation for evolution (Gibson and Dworkin, 2004; Rice, 1998; Rutherford, 2000). Mutations that alter the activity of their gene product persist because other parts of the pathway in which they function compensate for the defect. If such a phenotype-stabilizing mechanism is disrupted, either by an extreme environmental event such as a heat shock, or by a mutation in a critical part of the stabilizing network, then the effect of the cryptic genes is no longer masked, resulting in a great increase in phenotypic variation. Repeated selection on one of those phenotypes can fix the combination of genes that produce it, and the environmental stimulus is no longer required. Once CGV is revealed, a trait can exhibit different degrees of phenotypic plasticity. A population then represents a collection of reaction norms that can be molded by natural selection. Exposure of CGVs has been shown to increase additive genetic variance in natural populations (McGuigan et al., 2011; Rohner et al., 2013). Thus, once CGVs are revealed, selection can act to stabilize the novel reaction norm through quantitative genetic changes.

As expected from selection on a reaction norm, repeated selection can remove sensitivity to the environment and stabilize a new phenotype through genetic assimilation and canalization (Fig. 15.4). If selection acts strongly to cement the novel trait, the trait becomes canalized over many generations (Scharloo, 1991; Waddington, 1942). In contrast, if, in the same lineage, one phenotype is selected after one destabilizing environmental stimulus, and a different phenotype after a different stimulus, then genetic accommodation will fix each of the phenotypes in its corresponding environment (Fig. 15.4). A plastic polyphenic response will have evolved, with each alternative phenotype canalized (Waddington, 1942). Presumably this kind of divergent selection can be done with any phenotype that exhibits a reaction norm.

Although there are many mechanisms that act as evolutionary capacitors, they can be classified under three different broad categories: (1) protein chaperones, (2) robustness that emerges from gene regulatory networks, and (3) homeostastic regulators. Using *Drosophila melanogaster*, Rutherford and Lindquist (1998) demonstrated that when the function of the chaperone Heat shock protein 90 (Hsp90) is impaired or inadequate, variation in protein folding can be exposed. As a consequence, various mutant phenotypes can be revealed, that can serve as potential target of natural selection and ultimately lead to the stabilization of novel phenotypes (Rutherford and Lindquist, 1998). In natural populations, Hsp90 has been shown to buffer against deleterious mutations (Chen and Wagner, 2012), as well as mutations that might be adaptive in novel environments (Rohner et al., 2013). Thus, Hsp90, with its extensive interactions with multiple cellular processes, can hide mutations by masking the effects of mutations that lead to improper protein folding (Zabinsky et al., 2019).

Virtually all genes in complex gene regulatory networks have pleiotropic effects and have the potential to act as an evolutionary capacitor in the sense that null or nearly null mutations can reveal genetic variation (Bergman and Siegal, 2003). Robust gene regulatory networks have multiple connections that confer redundancy and thus produce highly stable phenotypic outputs (Kitami and Nadeau, 2002; Macneil and Walhout, 2011; Siegal and Bergman, 2002; Wagner, 2005). Mutations that disrupt the connectivity and redundancy of these networks can reveal accumulated CGVs.

Finally, the mechanisms that are responsible for physiological homeostasis can act as an evolutionary capacitor. In physiology, the target phenotype is the set-point. A diversity of mechanisms maintains the system at the setpoint in a variable environment. The regulation of a constant body temperature in homeotherms, in the face of great variation in external temperature and internal energy expenditure, is the canonical example of a homeostatic mechanism. Similar homeostatic and robustness mechanisms that maintain form and function abound in physiology, biochemistry, and development. Sensitizing events that affect sensors, integrator regulation and/or effectors can all reveal CGVs. Studies in both vertebrates and invertebrates have demonstrated that changes in temporal pattern of hormone secretion can generate alternative phenotypes exhibited by closely related species (Kulkarni et al., 2017; Ledon-Rettig et al., 2010; Suzuki and Nijhout, 2006). Homeostatic mechanisms are likely to be important players during genetic accommodation because they respond readily to environmental shifts and can alter developmental trajectories without causing major disruptions in the underlying gene regulatory networks.

The role of robustness and homeostasis in speciation

Genetic accommodation has been proposed to promote speciation events (Pfennig et al., 2010; Schneider and Meyer, 2017; West-Eberhard, 2005). According to the "flexible-stem" hypothesis, phenotypic plasticity has been suggested to enable adaptive radiations (West-Eberhard, 2003) and many instances of phenotypically plastic stem lineages that give rise to a range of robust adaptive traits have been documented (Palmer, 2004; Wund et al., 2008). Hidden phenotypic plasticity exposed under laboratory conditions often phenocopies the range of phenotypes seen in fossil record or in extant species (Ledon-Rettig et al., 2010; McGuigan et al., 2011; Muschick et al., 2011; Rohner et al., 2013; Standen et al., 2014). Phenotypic plasticity in ancestral species can facilitate rapid diversification of traits because a phenotypically plastic population can generate new phenotypes that are potentially adapted to a new environment (Schneider and Meyer, 2017; West-Eberhard, 2005). Thus, relative to a phenotypically robust population, a phenotypically plastic population would be expected to colonize a broader array of habitats. A new species could then arise through sympatric or allopatric speciation. Colonization of new habitats may be accompanied by large changes in the environment, leading to the disruption of physiological homeostasis and the unleashing of cryptic mutations that allow populations to adapt. Thus, physiological homeostasis—and its disruption—may be an important first step during speciation.

Model of genetic accommodation

As mentioned above, selection on phenotypic plasticity can generate monophenic canalized traits or robust but polyphenic traits. Monophenic canalization and polyphenisms are at the opposite ends of a continuous spectrum of phenotypic variation, and genetic accommodation allows movement along this continuum.

Graphically, we can depict this relationship using a theoretical three-dimensional surface colloquially called a monkey saddle. In this surface plot, the x-axis represents different genotypes, the y-axis represents the environmental condition and can also be thought of as the strength of the integrator (e.g., hormone level), that could cause the development of alternative phenotypes. The z-axis represents phenotypic instability, with higher values representing a more unstable phenotype (Fig. 15.5). For any given genotype, a ball on the surface will tend to move downhill to a location with greater developmental stability. The shape of the landscape is determined by the interaction of genotype and environment.

We can use this surface to zoom in on the developmental choices during a sensitive period. A monophenic canalized trait corresponds to the region on the left side of Fig. 15.5B and C, where there is a single dip in the surface. If an organism has this genotype, regardless of the environmental inputs, the ball would roll down to a single phenotypically stable state. Small changes in the genotype will still lead to the ball rolling to the

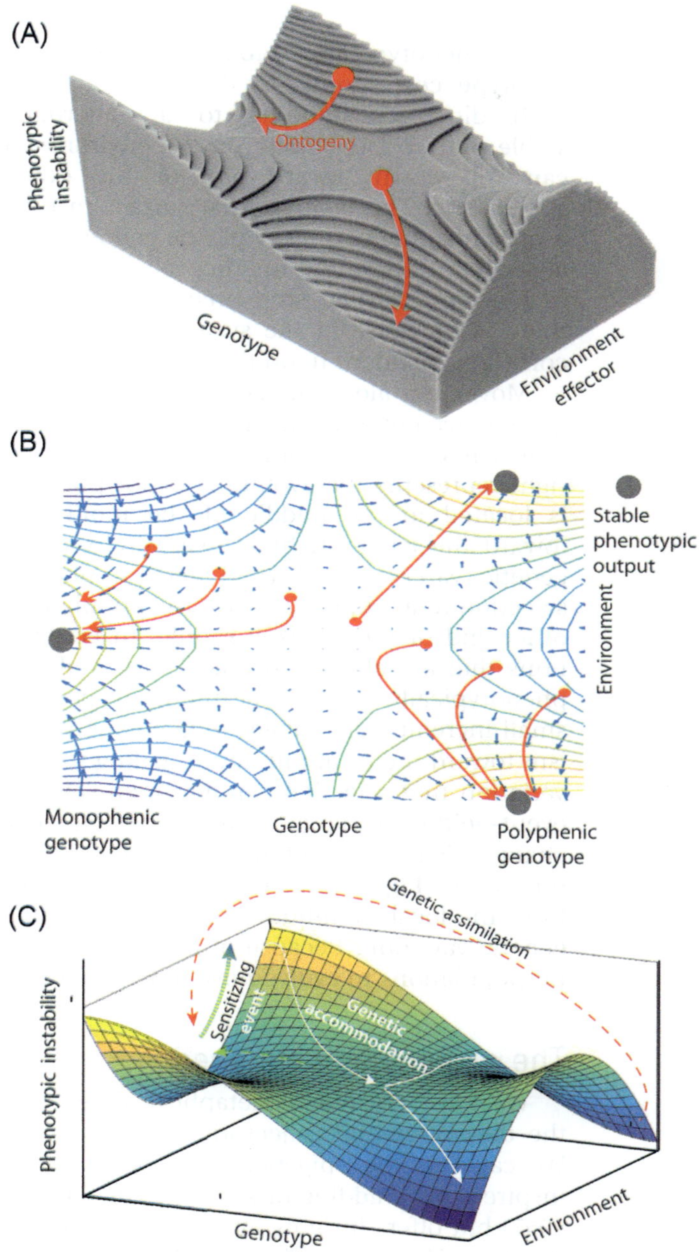

Figure 15.5 Monkey saddle model for the evolution of canalization and polyphenism. (A) Lateral perspective of the surface illustrating the meaning of the axes. The y-axis can be interpreted as either environment or the level of an effector (induced by the environment). Depending on the environment, two alternative stable phenotypes can develop. The z-axis is phenotypic "instability" meaning that lower levels are more stable than higher levels. Two ontogenetic trajectories are shown. (B) Top view of the same surface showing contours and direction arrows indicating the steepest gradient. Several hypothetical developmental trajectories are shown. (C) A monkey saddle surface showing projected trajectories for the evolution via genetic assimilation and genetic accommodation after a sensitizing event.

same phenotypically stable spot (Fig. 15.5B). A polyphenic genotype corresponds to the right-hand side of the surface, each dip corresponding to an alternative phenotypically stable output. Depending on the environmental input, the ball can roll down toward one or the other developmentally stable state. The relatively flat surface in the center (Fig. 15.5A) represents phenotypically plastic genotypes, where genetic variation has little effect on the value of the phenotype. In this region, the ball can stop practically anywhere, and, assuming there is no noise in the system, the final developmental outcome is dependent upon the environment.

Movement along the genotype axis is accomplished through genetic accommodation (Fig. 15.5C). Movement toward the left represents genetic assimilation, whereas movement toward the right represents the evolution of polyphenisms. If a population is located in the middle of the monkey saddle, movement toward the left or right is relatively easy as small changes in genetic variation will yield enough phenotypic variation for selection to act upon to move the population toward one or the other end of the monkey saddle. In contrast, moving populations that are already located in the valley at the left, with its phenotypically stable local minimum point, is difficult because small incremental changes in genotype will still pull the organism toward the local minimum, failing to reveal any phenotypic variation that selection can act upon (Fig. 15.5B). A sensitizing event—either a large mutation or a major environmental change—is needed to move the population away from the local minimum (Fig. 15.5C). If this happens, the ball can roll away from the local minimum, toward the flatter surface and reveal genetic variation. Because of this, sensitizing events are critical for populations to evolve novel phenotypes.

The epigenetic watershed

We have created a metaphorical representation to explain the developmental trajectories underlying the monkey saddle. We call this the epigenetic watershed model (Fig. 15.6). It is inspired by Waddington's epigenetic landscape (Waddington, 1957) but offers insights into sensitizing events and their consequences. The epigenetic watershed model involves a watershed with a river flowing down a valley. In this model, we define the height of the landscape as the amount of phenotypic instability: the lower the altitude, the more phenotypically stable a trait is. The length of the watershed represents size, and can also be thought of as developmental time. The horizontal axis—the

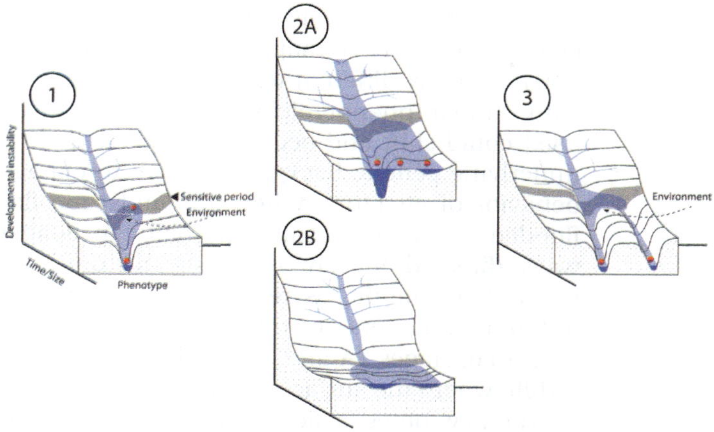

Figure 15.6 The epigenetic watershed model of genetic accommodation. In this metaphor, a ball floating down the watershed represents ontogeny of a particular trait. The water represents physiological homeostasis. The time/size axis represents both developmental time and growth. The horizontal axis represents the trait value of the phenotype. The vertical axis represents developmental instability. As development proceeds, the traits tend toward more developmental stability, depicted by a deepening of the water course. During development, there is an environment sensitive period *(gray band)*, during which the ball can be pushed toward different developmental trajectories. (Stage 1) A phenotypically robust trait always develops into the same phenotype regardless of the environmental inputs at the sensitive period. (Stage 2) Sensitizing events, such as dramatic alteration in homeostatic regulator/integrator (Stage 2A) or a change in the effectors for a trait (Stage 2B), cause water flow to broaden into previously inaccessible developmental trajectories. These novel trajectories are enabled by cryptic genetic variation, depicted here as a flattening of the landscape, which allows the stream easier access to alternative channels. (Stage 3) Upon selection, two alternative developmental trajectories may stabilize, leading to the evolution of a polyphenism, in which an environmental signal during the sensitive period pushes the stream into one of two alternative channels.

cross section of the landscape—represents the phenotype. The depth of the channel represents the relative stability of the phenotype. A ball floating down the river represents ontogeny of the trait. When the ball reaches the sensitive period, it comes upon a region where the trajectory can change.

A developmentally robust trait has one single deep valley (Fig. 15.6, Stage 1). In this case, there is only one possible outcome for the floating ball. This type of landscape underlies the tail end of the monkey saddle (Fig. 15.5). A rare sensitizing event would allow the water to spill over the valley and allow the ball to access previously inaccessible parts of the landscape (Fig. 15.6, Stage 2). Such events could be caused by either too much water flowing down the river (Stage 2A) or a flattening of the landscape (Stage 2B). The former represents a change in physiological homeostasis and the latter represents a change in effectors. A drastic change in hormone synthesis is an example of an event that would cause excess water to flow down the channel, causing a major flood that allows the ball to float down in many different ways, all ending in different phenotypes with similar phenotypic stability. Alternatively, the landscape itself could shift and become flatter as might be the case if the underlying genetic background were to change, influencing the behavior of effectors. When the landscape flattens such that a flooded watershed results, the ball can float in many different ways, directed entirely by the signal from the integrator. Again, phenotypic stability is relatively high and similar amongst the different ontogenetic outcomes. Both of these landscapes (Fig. 15.6, stages 2A and 2B) underlie the middle of the monkey saddle where the surface is relatively flat.

The new ontogenetic outcomes are analogous to the release of CGV, which is unmasked when physiological homeostasis or genetic background is disrupted. Over long periods of time, new valleys will deepen through erosion and become canalized if specific new outcomes are favored, much like the way novel developmental trajectories become stabilized through genetic accommodation. In polyphenic traits, two distinct valleys form on the landscape after the sensitizing period (Fig. 15.6, Stage 3). In highly evolved polyphenisms, the valleys deepen, forcing the ball to float down one branch or the other. The sensors regulate what valley is chosen by modifying the strength of the integrator. If only one valley is favored, that valley will deepen over time and become robust. When this happens, the trait becomes genetically assimilated.

Together, the monkey saddle and the epigenetic landscape offer complementary ways of visualizing genetic accommodation. The monkey saddle provides us with an opportunity to both visualize and model genetic accommodation. The epigenetic landscape offers a more in-depth way to visualize underlying developmental trajectories of the monkey saddle and highlights sensitizing events by illustrating how sensors experience the environment to regulate the release of integrators that in turn interact with the landscape to determine the degree to which the environment affects the developmental outcome. Together, these tools offer a novel, integrative way to conceptually model genetic accommodation.

Conclusions

We have discussed the developmental and evolutionary origin of adaptive phenotypic plasticity in the light of our current understanding of polyphenisms. Polyphenisms are bistable, or multistable forms of phenotypic plasticity in which two or more alternative phenotypes develop, triggered by specific environmental factors such as nutrition, photoperiod, temperature, and pheromones.

A key to understanding how this phenotypic plasticity develops comes from understanding how endocrine processes control postembryonic development. In polyphenisms, the environmental signals are received, processed, and integrated by the central nervous system, which then controls developmental processes throughout the body via neuroendocrine and endocrine factors that stimulate signaling pathways, which are the effectors of diverse patterns of gene expression, growth, and morphogenesis in target tissues.

Developmental plasticity provides a link between the environment and evolutionary changes in the genome. Changes in the environment are readily sensed by the organism, leading to changes in developmental trajectories. Natural selection can then alter gene frequencies at a population level leading to evolutionary changes.

Developmental homeostasis provides mechanisms that confer robustness in both monophenic and polyphenic phenotypic outcomes. These mechanisms mask the effects of mutations and this allows for accumulation of CGV that, under the right circumstances, promotes phenotypic diversification and possibly also speciation. Thus, understanding the development and evolution of adaptive phenotypic plasticity requires also a deeper understanding of the mechanisms that underly phenotypic robustness. It would be helpful, in this regard, to determine how homeostatic processes diverge in different populations and how that might facilitate or counteract different forms of adaptive plasticity.

References

An, H., Do, T.D., Jung, G., Karagozlu, M.Z., Kim, C.B., 2018. Comparative transcriptome analysis for understanding predator-induced polyphenism in the water flea Daphnia pulex. Int. J. Mol. Sci. 19.

Angelini, D.R., Kaufman, T.C., 2005. Insect appendages and comparative ontogenetics. Dev. Biol. 286, 57–77.

Anstey, M.L., Rogers, S.M., Ott, S.R., Burrows, M., Simpson, S.J., 2009. Serotonin mediates behavioral gregarization underlying swarm formation in desert locusts. Science 323, 627–630.

Aoyama, H., Daitoku, H., Fukamizu, A., 2006. Nutrient control of phosphorylation and translocation of Foxo1 in C57BL/6 and db/db mice. Int. J. Mol. Med. 18, 433–439.

Atallah, J., Vurens, G., Mavong, S., Mutti, A., Hoang, D., Kopp, A., 2014. Sex-specific repression of dachshund is required for *Drosophila* sex comb development. Dev. Biol. 386, 440–447.
Barry, M.J., 2000. Inducible defences in Daphnia: responses to two closely related predator species. Oecologia 124, 396–401.
Bergman, A., Siegal, M.L., 2003. Evolutionary capacitance as a general feature of complex gene networks. Nature 424, 549–552.
Bloch, G., Hefetz, A., 1999. Regulation of reproduction by dominant workers in bumblebee (*Bombus terrestris*) queenright colonies. Behav. Ecol. Sociobiol. 45, 125–135.
Bortolotti, L., Duchateau, M.J., Sbrenna, G., 2001. Effect of juvenile hormone on caste determination and colony processes in the bumblebee *Bombus terrestris*. Entomol. Exp. Appl. 101, 143–158.
Brunetti, C.R., Selegue, J.E., Monteiro, A., French, V., Brakefield, P.M., Carroll, S.B., 2001. The generation and diversification of butterfly eyespot color patterns. Curr. Biol. 11, 1578–1585.
Bubenik, G.A., 2006. Seasonal regulation of deer reproduction as related to the antler cycle – a review. Veterinarski Arhiv. 76 (Suppl.), S275–S285.
Butler, C.G., Callow, R.K., Johnston, N.C., 1962. The isolation and synthesis of queen substance, 9-oxodec-trans-2-enoic acid, a honeybee pheromone. Proc. R. Soc. Lond. B 155, 417–432.
Buttstedt, A., Moritz, R.F.A., Erler, S., 2014. Origin and function of the major royal jelly proteins of the honeybee (*Apis mellifera*) as members of the yellow gene family. Biol. Rev. 89, 255–269.
Carter, M.J., Lind, M.I., Dennis, S.R., Hentley, W., Beckerman, A.P., 2017. Evolution of a predator-induced, nonlinear reaction norm. Proc. Biol. Sci. 284, 20170859.
Casasa, S., Moczek, A.P., 2018. Insulin signalling's role in mediating tissue-specific nutritional plasticity and robustness in the horn-polyphenic beetle *Onthophagus taurus*. Proc. R. Soc. B: Biol. Sci. 285, 20181631.
Chen, B., Wagner, A., 2012. Hsp90 is important for fecundity, longevity, and buffering of cryptic deleterious variation in wild fly populations. BMC Evol. Biol. 12, 25.
Cnaani, J., Hefetz, A., 2001. Are queen *Bombus terrestris* giant workers or are workers dwarf queens? Solving the 'chicken and egg' problem in a bumblebee species. Naturwissenschaften 88, 85–87.
Cnaani, J., Robinson, G., Hefetz, A., 2000. The critical period for caste determination in *Bombus terrestris*, and its juvenile hormone correlates. J. Comp. Physiol. A: Neuroethology, Sensory, Neural, Behav. Physiol. 186, 1089–1094.
Colombani, J., Raisin, S., Pantalacci, S., Radimerski, T., Montagne, J., Leopold, P., 2003. A nutrient sensor mechanism controls *Drosophila* growth. Cell 114, 739–749.
Cullen, D.A., Sword, G.A., Dodgson, T., Simpson, S.J., 2010. Behavioural phase change in the Australian plague locust, Chortoicetes terminifera, is triggered by tactile stimulation of the antennae. J. Insect Physiol. 56, 937–942.
Dekanty, A., Milán, M., 2011. The interplay between morphogens and tissue growth. EMBO Rep. 12, 1003–1010.
Dietz, A., Hermann, H.R., Blum, M.S., 1979. The role of exogenous JH I, JH III and Anti-JH (precocene II) on queen induction of 4.5-day-old worker honey bee larvae. J. Insect Physiol. 25, 503–512.
Doorn, A.V., 1989. Factors influencing dominance behaviour in queenless bumblebee workers (*Bombus terrestris*). Physiol. Entomol. 14, 211–221.

Dye, N.A., Popović, M., Spannl, S., Etournay, R., Kainmüller, D., Ghosh, S., et al., 2017. Cell dynamics underlying oriented growth of the *Drosophila* wing imaginal disc. Development 144, 4406–4421.
Emlen, D.J., 1994. Environmental control of horn length dimorphism in the beetle *Onthophagus acuminatus* (Coleoptera: Scarabaeidae). Proc. R. Soc. B: Biol. Sci. 256, 131–136.
Emlen, D.J., 1997. Diet alters male horn allometry in the beetle Onthophagus acuminatus (Coleoptera: Scarabaeidae). Proc. R. Soc. B: Biol. Sci. 264, 567–574.
Emlen, D.J., Nijhout, H.F., 1999. Hormonal control of male horn length dimorphism in the dung beetle Onthophagus taurus (Coleoptera: Scarabaeidae). J. Insect Physiol. 45, 45–53.
Evans, J., Wheeler, D., 1999. Differential gene expression between developing queens and workers in the honey bee, *Apis mellifera*. Proc. Natl Acad. Sci. 96, 5575–5580.
Evans, J., Wheeler, D., 2000. Expression profiles during honeybee caste determination. Genome Biol. 2, 0001–0006. research0001.
Faure, J.C., 1932. The phases of locusts in South Africa. Bull. Entomol. Res. 23, 293–405.
Ferreira, A., Milán, M., 2015. Dally proteoglycan mediates the autonomous and nonautonomous effects on tissue growth caused by activation of the PI3K and TOR pathways. PLoS Biol. 13, e1002239.
Gibson, G., Dworkin, I., 2004. Uncovering cryptic genetic variation. Nat. Rev. Genet. 5, 681–U611.
Giray, T., Giovanetti, M., West-Eberhard, M.J., 2005. Juvenile hormone, reproduction, and worker behavior in the neotropical social wasp *Polistes canadensis*. Proc. Natl. Acad. Sci. U. S. A. 102, 3330–3335.
Glastad, K.M., Gokhale, K., Liebig, J., Goodisman, M.A.D., 2016. The caste- and sex-specific DNA methylome of the termite Zootermopsis nevadensis. Sci. Rep-UK 6, 37110.
Gotoh, H., Miyakawa, H., Ishikawa, A., Ishikawa, Y., Sugime, Y., Emlen, D.J., et al., 2014. Developmental link between sex and nutrition; doublesex regulates sex-specific mandible growth via juvenile hormone signaling in stag beetles. PLoS Genet. 10, e1004098.
Gotoh, H., Zinna, R.A., Ishikawa, Y., Miyakawa, H., Ishikawa, A., Sugime, Y., et al., 2017. The function of appendage patterning genes in mandible development of the sexually dimorphic stag beetle. Dev. Biol. 422, 24–32.
Hagele, B.F., Simpson, S.J., 2000. The influence of mechanical, visual and contact chemical stimulation on the behavioural phase state of solitarious desert locusts (Schistocerca gregaria). J. Insect Physiol. 46, 1295–1301.
Hiraga, S., 2005. Two different sensory mechanisms for the control of pupal protective coloration in butterflies. J. Insect Physiol. 51, 1033–1040.
Hust, J., Lavine, M.D., Worthington, A.M., Zinna, R., Gotoh, H., Niimi, T., et al., 2018. The Fat-Dachsous signaling pathway regulates growth of horns in *Trypoxylus dichotomus*, but does not affect horn allometry. J. Insect Physiol. 105, 85–94.
Kijimoto, T., Moczek, A.P., Andrews, J., 2012. Diversification of doublesex function underlies morph-, sex-, and species-specific development of beetle horns. Proc. Natl. Acad. Sci. U. S. A. 109, 20526–20531.
Kitami, T., Nadeau, J.H., 2002. Biochemical networking contributes more to genetic buffering in human and mouse metabolic pathways than does gene duplication. Nat. Genet. 32, 191–194.
Koch, P., Bückmann, D., 1987. Hormonal control of seasonal morphs by the timing of ecdysteroid release in *Araschnia levana* L. (Nymphalidae: Lepidoptera). J. Insect Physiol. 33, 823–829.

Kooi, R.E., Brakefield, P.M., 1999. The critical period for wing pattern induction in the polyphenic tropical butterfly Bicyclus anynana (Satyrinae). J. Insect Physiol. 45, 201–212.
Kopp, A., Duncan, I., Carroll, S.B., 2000. Genetic control and evolution of sexually dimorphic characters in Drosophila. Nature 408, 553–559.
Kucharski, R., Maleszka, J., Foret, S., Maleszka, R., 2008. Nutritional control of reproductive status in honeybees via DNA methylation. Science 319, 1827–1830.
Kulkarni, S.S., Denver, R.J., Gomez-Mestre, I., Buchholz, D.R., 2017. Genetic accommodation via modified endocrine signalling explains phenotypic divergence among spadefoot toad species. Nat. Commun. 8.
Landauer, W., 1957. Phenocopies and genotype, with special reference to sporadically-occurring developmental variants. Am. Nat. 91, 79–90.
Ledon-Rettig, C.C., Pfennig, D.W., Crespi, E.J., 2010. Diet and hormonal manipulation reveal cryptic genetic variation: implications for the evolution of novel feeding strategies. Proc. Biol. Sci. 277, 3569–3578.
Libbrecht, R., Oxley, P.R., Keller, L., Kronauer, D.J.C., 2016. Robust DNA Methylation in the Clonal Raider Ant Brain. Curr. Biol. 26, 391–395.
Li-Byarlay, H., Li, Y., Stroud, H., Feng, S., Newman, T.C., Kaneda, M., et al., 2013. RNA interference knockdown of DNA methyl-transferase 3 affects gene alternative splicing in the honey bee. Proc. Natl Acad. Sci. U. S. A. 110, 12750–12755.
Lindsay, W.R., Barron, D.G., Webster, M.S., Schwabl, H., 2016. Testosterone activates sexual dimorphism including male-typical carotenoid but not melanin plumage pigmentation in a female bird. J. Exp. Biol. 219, 3091–3099.
Lindsay, W.R., Webster, M.S., Schwabl, H., 2011. Sexually selected male plumage color is testosterone dependent in a tropical passerine bird, the red-backed fairy-wren (*Malurus melanocephalus*). PLoS One 6, e26067.
Lyko, F., Foret, S., Kucharski, R., Wolf, S., Falckenhayn, C., Maleszka, R., 2010. The honey bee epigenomes: differential methylation of brain DNA in queens and workers. PLoS Biol. 8 (11), e1000506.
Ma, Z., Guo, W., Guo, X., Wang, X., Kang, L., 2011. Modulation of behavioral phase changes of the migratory locust by the catecholamine metabolic pathway. Proc. Natl Acad. Sci. U. S. A. 108, 3882–3887.
Macneil, L.T., Walhout, A.J., 2011. Gene regulatory networks and the role of robustness and stochasticity in the control of gene expression. Genome Res. 21, 645–657.
Matsumoto, Y., Yatsu, R., Taylor, C., Crews, D., 2013. Changes in gonadal gene network by exogenous ligands in temperature-dependent sex determination. J. Mol. Endocrinol. 50, 389–400.
Matsuura, K., Himuro, C., Yokoi, T., Yamamoto, Y., Vargo, E.L., Keller, L., 2010. Identification of a pheromone regulating caste differentiation in termites. Proc. Natl Acad. Sci. U. S. A. 107, 12963–12968.
McCullough, E.L., Ledger, K.J., O'Brien, D.M., Emlen, D.J., 2015. Variation in the allometry of exaggerated rhinoceros beetle horns. Anim. Behav. 109, 133–140.
McGuigan, K., Nishimura, N., Currey, M., Hurwit, D., Cresko, W.A., 2011. Cryptic genetic variation and body size evolution in threespine stickleback. Evolution 65, 1203–1211.
Minakuchi, C., Zhou, X., Riddiford, L.M., 2008. Krüppel homolog 1 (Kr-h1) mediates juvenile hormone action during metamorphosis of *Drosophila melanogaster*. Mechanisms Dev. 125, 91–105.
Mirth, C.K., Truman, J.W., Riddiford, L.M., 2009. The ecdysone receptor controls the post-critical weight switch to nutrition-independent differentiation in *Drosophila* wing imaginal discs. Development 136, 2345–2353.

Mitaka, Y., Kobayashi, K., Mikheyev, A., Tin, M.M., Watanabe, Y., Matsuura, K., 2016. Caste-specific and sex-specific expression of chemoreceptor genes in a termite. PLoS One 11, e0146125.

Miyakawa, H., Sato, M., Colbourne, J.K., Iguchi, T., 2015. Ionotropic glutamate receptors mediate inducible defense in the water flea Daphnia pulex. PLoS One 10, e0121324.

Moczek, A.P., Emlen, D.J., 2000. Male horn dimorphism in the scarab beetle, Onthophagus taurus: do alternative reproductive tactics favour alternative phenotypes? Anim. Behav. 59, 459–466.

Moczek, A.P., Hunt, J., Emlen, D.J., Simmons, L.W., 2002. Threshold evolution in exotic populations of a polyphenic beetle. Evol. Ecol. Res. 4, 587–601.

Moczek, A.P., Nijhout, H.F., 2003. Rapid evolution of a polyphenic threshold. Evol. Dev. 5, 259–268.

Moczek, A.P., Rose, D.J., 2009. Differential recruitment of limb patterning genes during development and diversification of beetle horns. Proc. Natl Acad. Sci. 106, 8992–8997.

Monteiro, A., Tong, X., Bear, A., Liew, S.F., Bhardwaj, S., Wasik, B.R., et al., 2015. Differential expression of ecdysone receptor leads to variation in phenotypic plasticity across serial homologs. PLoS Genet. 11, e1005529.

Muschick, M., Barluenga, M., Salzburger, W., Meyer, A., 2011. Adaptive phenotypic plasticity in the Midas cichlid fish pharyngeal jaw and its relevance in adaptive radiation. BMC Evol. Biol. 11, 116.

Navarro-Martin, L., Vinas, J., Ribas, L., Diaz, N., Gutierrez, A., Di Croce, L., et al., 2011. DNA methylation of the gonadal aromatase (cyp19a) promoter is involved in temperature-dependent sex ratio shifts in the European sea bass. PLoS Genet. 7, e1002447.

Nijhout, H.F., 1980. Pattern formation on lepidopteran wings: determination of an eyespot. Dev. Biol. 80, 267–274.

Nijhout, H.F., 1994. Insect Hormones. Princeton University Press, Princeton, NJ.

Nijhout, H.F., 1999. Control mechanisms of polyphenic development in insects. Bioscience 49, 181–192.

Nijhout, H.F., 2003. Development and evolution of adaptive polyphenisms. Evol. Dev. 5, 9–18.

Nijhout, H.F., Best, J.A., Reed, M.C., 2019. Systems biology of robustness and homeostatic mechanisms. Wiley Interdiscip. Reviews: Syst. Biol. Med. 11, e1440.

Nijhout, H.F., McKenna, K.Z., 2018. The distinct roles of insulin signaling in polyphenic development. Curr. Opin. Insect Sci. 25, 58–64.

Nijhout, H.F., Sadre-Marandi, F., Best, J., Reed, M.C., 2017. Systems biology of phenotypic robustness and plasticity. Integr. Comp. Biol. 57, 171–184.

Oliveira, M.M., Shingleton, A.W., Mirth, C.K., 2014. Coordination of wing and whole-body development at developmental milestones ensures robustness against environmental and physiological perturbations. PLoS Genet. 10, e1004408.

Özsu, N., Chan, Q.Y., Chen, B., Gupta, M.D., Monteiro, A., 2017. Wingless is a positive regulator of eyespot color patterns in *Bicyclus anynana* butterflies. Dev. Biol. 429, 177–185.

Palmer, A.R., 2004. Symmetry breaking and the evolution of development. Science 306, 828–833.

Pan, D., 2010. The hippo signaling pathway in development and cancer. Dev. Cell 19, 491–505.

Parrott, B.B., Kohno, S., Cloy-McCoy, J.A., Guillette Jr., L.J., 2014. Differential incubation temperatures result in dimorphic DNA methylation patterning of the SOX9 and aromatase promoters in gonads of alligator (Alligator mississippiensis) embryos. Biol. Reprod. 90, 2.

Pask, G.M., Slone, J.D., Millar, J.G., Das, P., Moreira, J.A., Zhou, X., et al., 2017. Specialized odorant receptors in social insects that detect cuticular hydrocarbon cues and candidate pheromones. Nat. Commun. 8, 297.
Pfennig, D.W., Wund, M.A., Snell-Rood, E.C., Cruickshank, T., Schlichting, C.D., Moczek, A.P., 2010. Phenotypic plasticity's impacts on diversification and speciation. Trends Ecol. Evol. 25, 459–467.
Puig, O., Tijan, R., 2005. Transcriptional feedback control of insulin recepetor by dFOXO/FOXO1. Genes. Dev. 19, 2435–2446.
Ramsey, M., Shoemaker, C., Crews, D., 2007. Gonadal expression of Sf1 and aromatase during sex determination in the red-eared slider turtle (Trachemys scripta), a reptile with temperature-dependent sex determination. Differentiation 75, 978–991.
Reichling, T.D., German, R.Z., 2000. Bones, muscles and visceral organs of protein-malnourished rats (*Rattus norvegicus*) grow more slowly but for longer durations to reach normal final size. J. Nutr. 130, 2326–2332.
Rice, S.H., 1998. The evolution of canalization and the breaking of von Baer's laws: Modeling the evolution of development with epistasis. Evolution 52, 647–656.
Roff, D.A., 1992. The Evolution of Life Histories. Chapman and Hall, New York.
Rogers, S.M., Matheson, T., Despland, E., Dodgson, T., Burrows, M., Simpson, S.J., 2003. Mechanosensory-induced behavioural gregarization in the desert locust Schistocerca gregaria. J. Exp. Biol. 206, 3991–4002.
Rogers, S.M., Matheson, T., Sasaki, K., Kendrick, K., Simpson, S.J., Burrows, M., 2004. Substantial changes in central nervous system neurotransmitters and neuromodulators accompany phase change in the locust. J. Exp. Biol. 207, 3603–3617.
Rohner, N., Jarosz, D.F., Kowalko, J.E., Yoshizawa, M., Jeffery, W.R., Borowsky, R.L., et al., 2013. Cryptic variation in morphological evolution: HSP90 as a capacitor for loss of eyes in cavefish. Science 342, 1372–1375.
Röseler, P.-F., Röseler, I., Strambi, A., 1985. Role of ovaries and ecdysteroids in dominance hierarchy establishment among foundresses of the primitively social wasp, *Polistes gallicus*. Behav. Ecol. Sociobiol. 18, 9–13.
Rountree, D.B., Nijhout, H.F., 1995. Hormonal control of a seasonal polyphenism in *Precis coenia* (Lepidoptera: Nymphalidae). J. Insect Physiol. 41, 987–992.
Rutherford, S.L., 2000. From genotype to phenotype: buffering mechanisms and the storage of genetic information. Bioessays 22, 1095–1105.
Rutherford, S.L., Lindquist, S., 1998. Hsp90 as a capacitor for morphological evolution. Nature 396, 336–342.
Scharloo, W., 1991. Canalization - genetic and developmental aspects. Annu. Rev. of Ecol. And. Syst. 22, 65–93.
Scheiner, S.M., 2002. Selection experiments and the study of phenotypic plasticity. J. Evol. Biol. 15, 889–898.
Schlichting, C.D., Pigliucci, M., 1998. Phenotypic Evolution: A Reaction Norm Perspective. Sinauer Inc, Sunderland, MA.
Schneider, R.F., Meyer, A., 2017. How plasticity, genetic assimilation and cryptic genetic variation may contribute to adaptive radiations. Mol. Ecol. 26, 330–350.
Shafiei, M., Moczek, A.P., Nijhout, H.F., 2001. Food availability controls the onset of metamorphosis in the dung beetle *Onthophagus taurus* (Coleoptera: Scarabaeidae). Physiol. Entomol. 26, 173–180.
Siegal, M.L., Bergman, A., 2002. Waddington's canalization revisited: developmental stability and evolution. Proc. Natl Acad. Sci. U. S. A. 99, 10528–10532.

Simpson, S.J., Despland, E., Hagele, B.F., Dodgson, T., 2001. Gregarious behavior in desert locusts is evoked by touching their back legs. P Natl Acad. Sci. U. S. A. 98, 3895–3897.
Smith, K., 1991. The effects of temperature and daylength on the rosa polyphenism in the buckeye butterfly, *Precis coenia* (Lepidoptera: Nymphalidae). J. Res. Lepid. 30, 237–244.
Snell-Rood, E.C., Moczek, A.P., 2012. Insulin signaling as a mechanism underlying developmental plasticity: the role of FOXO in a nutritional polyphenism. PLoS One 7, e34857.
Southgate, R.J., Neill, B., Prelovsek, O., El-Osta, A., Kamei, Y., Miura, S., et al., 2007. FOXO1 regulates the expression of 4E-BP and inhibits mTOR signaling in mammalina skeletal muscle. J. Biol. Chem. 282, 21176–21186.
Standen, E.M., Du, T.Y., Larsson, H.C., 2014. Developmental plasticity and the origin of tetrapods. Nature 513, 54–58.
Straßburger, K., Tiebe, M., Pinna, F., Breuhahn, K., Teleman, A.A., 2012. Insulin/IGF signaling drives cell proliferation in part via Yorkie/YAP. Dev. Biol. 367, 187–196.
Sultan, S.E., 2015. Organism & Environment. Oxford University Press, Oxford.
Suzuki, Y., Nijhout, H.F., 2006. Evolution of a polyphenism by genetic accommodation. Science 311, 650–652.
Suzuki, Y., Truman, J.W., Riddiford, L.M., 2008. The role of *Broad* in the development of *Tribolium castaneum*: implications for the evolution of the holometabolous insect pupa. Development 135, 569–577.
Tanaka, S., 2000a. Hormonal control of body-color polymorphism in Locusta migratoria: interaction between [His7]-corazonin and juvenile hormone. J. Insect Physiol. 46, 1535–1544.
Tanaka, S., 2000b. Induction of darkening by corazonins in several species of Orthoptera and their possible presence in ten insect orders. Appl. Entomol. Zool. 35, 509–517.
Tibbetts, E.A., Huang, Z.Y., 2010. The challenge hypothesis in an insect: juvenile hormone increases during reproductive conflict following queen loss in *Polistes* wasps. Am. Nat. 176, 123–130.
Todd, E.V., Liu, H., Muncaster, S., Gemmell, N.J., 2016. Bending genders: the biology of natural sex change in fish. Sex Dev. 10, 223–241.
Tollrian, R., 1993. Neckteeth formation in Daphnia pulex as an example of continuous phenotypic plasticity: morphological effects of Chaoborus kairomone concentration and their quantification. J. Plankton Res. 15, 1309–1318.
Ureña, E., Chafino, S., Manjón, C., Franch-Marro, X., Martín, D., 2016. The occurrence of the holometabolous pupal stage requires the interaction between E93, Krüppel-Homolog 1 and Broad-Complex. PLoS Genet. 12, e1006020.
Ureña, E., Manjón, C., Franch-Marro, X., Martín, D., 2014. Transcription factor E93 specifies adult metamorphosis in hemimetabolous and holometabolous insects. Proc. Natl Acad. Sci. 111, 7024–7029.
Van Oystaeyen, A., Oliveira, R.C., Holman, L., van Zweden, J.S., Romero, C., Oi, C.A., et al., 2014. Conserved class of queen pheromones stops social insect workers from reproducing. Science 343, 287–290.
Vellichirammal, N.N., Gupta, P., Hall, T.A., Brisson, J.A., 2017. Ecdysone signaling underlies the pea aphid transgenerational wing polyphenism. Proc. Natl Acad. Sci. 114, 1419–1423.
Villarreal, C.M., Darakananda, K., Wang, V.R., Jayaprakash, P.M., Suzuki, Y., 2015. Hedgehog signaling regulates imaginal cell differentiation in a basally branching holometabolous insect. Dev. Biol. 404, 125–135.
Waddington, C.H., 1942. Canalization of development and the inheritance of acquired characters. Nature 150, 563–565.

Waddington, C.H., 1953. Genetic assimilation of an acquired character. Evolution 7, 118–126.

Waddington, C.H., 1956. Genetic assimilation of the bithorax phenotype. Evolution 10, 1–13.

Waddington, C.H., 1957. The Strategy of the Genes A Discussion of Some Aspects of Theoretical Biology. George Allen and Unwin, London.

Wagner, A., 2005. Robustness and Evolvability in Living Systems. Princeton University Press, Princeton, NJ.

Wang, X.H., Fang, X.D., Yang, P.C., Jiang, X.T., Jiang, F., Zhao, D.J., et al., 2014. The locust genome provides insight into swarm formation and long-distance flight. Nat. Commun. 5, 1–9.

Weiss, L.C., Kruppert, S., Laforsch, C., Tollrian, R., 2012. Chaoborus and gasterosteus anti-predator responses in Daphnia pulex are mediated by independent cholinergic and gabaergic neuronal signals. PLoS One 7, e36879.

Weiss, L.C., Leimann, J., Tollrian, R., 2015. Predator-induced defences in Daphnia longicephala: location of kairomone receptors and timeline of sensitive phases to trait formation. J. Exp. Biol. 218, 2918–2926.

West-Eberhard, M.J., 2003. Developmental Plasticity and Evolution. Oxford University Press, New York.

West-Eberhard, M.J., 2005. Developmental plasticity and the origin of species differences. Proc. Natl Acad. Sci. U. S. A. 102 (Suppl. 1), 6543–6549.

Wheeler, D.E., 1991. The developmental basis of worker caste polymorphism in ants. Am. Nat. 138, 1218–1238.

Wheeler, D.E., Nijhout, H.F., 1981. Soldier determination in ants: new role for juvenile hormone. Science 213, 361–363.

Wheeler, D.E., Nijhout, H.F., 1983. Soldier determination in *Pheidole bicarinata*: Effect of methoprene on caste and size within castes. J. Insect Physiol. 29, 847–854.

Wheeler, D.E., Nijhout, H.F., 1984. Soldier determination in *Pheidole bicarinata*: Inhibition by adult soldiers. J. Insect Physiol. 30, 127–135.

Wilson, E., 1971. The Insect Societies. Harvard University Press, Cambridge.

Wirtz, P., Beetsma, J., 1972. Induction of caste differentiation in the honeybee (*Apis mellifera*) by juvenile hormone. Entomol. Exp. Appl. 15, 517–520.

Wund, M.A., Baker, J.A., Clancy, B., Golub, J.L., Fosterk, S.A., 2008. A test of the "Flexible stem" model of evolution: Ancestral plasticity, genetic accommodation, and morphological divergence in the threespine stickleback radiation. Am. Nat. 172, 449–462.

Xu, H.-J., Xue, J., Lu, B., Zhang, X.-C., Zhuo, J.-C., He, S.-F., et al., 2015. Two insulin receptors determine alternative wing morphs in planthoppers. Nature 519, 464–467.

Zabinsky, R.A., Mason, G.A., Queitsch, C., Jarosz, D.F., 2019. It's not magic - Hsp90 and its effects on genetic and epigenetic variation. Semin. Cell Dev. Biol. 88, 21–35.

Zhou, X., Riddiford, L.M., 2002. Broad specifies pupal development and mediates the 'status quo' action of juvenile hormone on the pupal-adult transformation in *Drosophila* and *Manduca*. Development 129, 2259–2269.

Zhou, X., Slone, J.D., Rokas, A., Berger, S.L., Liebig, J., Ray, A., et al., 2012. Phylogenetic and transcriptomic analysis of chemosensory receptors in a pair of divergent ant species reveals sex-specific signatures of odor coding. PLoS Genet. 8, e1002930.

Zimova, M., Hackländer, K., Good, J.M., Melo-Ferreira, J., Alves, P.C., Mills, L.S., 2018. Function and underlying mechanisms of seasonal colour moulting in mammals and birds: what keeps them changing in a warming world? Biol. Rev. 93, 1478–1498.

16

Phenotypic plasticity and the origins of novelty

Nicholas A. Levis and David W. Pfennig
Department of Biology, University of North Carolina, Chapel Hill, NC, United States

Introduction

A long-standing goal in biology is to explain the origins of novel, complex phenotypes (Wagner and Lynch, 2010; Moczek, 2008). Historically, new phenotypes have often been assumed to arise exclusively from genetic changes (Carroll, 2008). Indeed, many distinctive features have now been traced to single gene mutations (e.g., Colosimo et al., 2004; Hoekstra et al., 2006), duplications of large regions of the genome (e.g., Conant and Wolfe, 2008; Flagel and Wendel, 2009), alterations in regulatory sequence (e.g., Chan et al., 2010; Cleves et al., 2014; O'Brown et al., 2015), and/or a variety of other changes in DNA sequence or content.

Despite the dominance of gene-centric thinking, an increasing number of biologists have begun asking if environmentally induced phenotypic change—that is, phenotypic plasticity (hereafter just plasticity and used synonymously with developmental plasticity)—might also play a leading role in the development and evolution of novelty (e.g., Price et al., 2003; West-Eberhard, 2003; Moczek et al., 2011; Badyaev, 2005; Pigliucci et al., 2006; Pfennig et al., 2010; Laland et al., 2015; Susoy et al., 2015; Schlichting and Wund, 2014; Lande, 2009). This line of reasoning stems, in part, from the observation that plasticity can trigger phenotypic divergence *within* species as great as that which is seen *between* species (Liem and Kaufman, 1984), suggesting that it might even contribute to *large-scale* evolutionary change or macroevolution (Pigliucci and Murren, 2003; Jablonski, 2017).

In this chapter, we discuss plasticity's possible contribution to the origins of complex, phenotypes. We begin by introducing the theory of how plasticity might promote novelty. Next, we discuss key predictions of this theory as well as a general scheme for evaluating these predictions. We do so with a particular focus on

Phenotypic Switching. DOI: https://doi.org/10.1016/B978-0-12-817996-3.00017-7

nonmodel, natural systems. We then present a case study that utilizes this scheme. We close with a brief discussion of the implications of this theory for understanding the causes of macroevolutionary innovation and diversification.

Plasticity-led evolution

Plasticity is ubiquitous across the tree of life (reviewed in Gilbert and Epel, 2015), and thinking about its role in fostering evolutionary innovation has a long history (West-Eberhard, 2003; Weismann, 1882; West-Eberhard, 1989; Waddington, 1953). Here we focus on a prevailing pathway that has been dubbed plasticity-led evolution (hereafter, PLE; Schwander and Leimar, 2011; Levis and Pfennig, 2016). We note, however, that plasticity can promote novelty and adaptation in a number of other ways (West-Eberhard, 2003, 1989).

Plasticity often enhances fitness under stressful conditions (Badyaev, 2005; Ghalambor et al., 2015; Schmalhausen, 1949), and when an individual encounters a novel environment (stressful or otherwise) that induces a phenotypic change, developmental processes can stabilize the expression of the altered phenotype, such that the individual can persist in the new environment [i.e., phenotypic accommodation occurs; sensu (West-Eberhard, 2003)]. If there is genetic variation in the tendency or manner in which individuals respond to the environment [i.e., if different genotypes exhibit different reaction norms (Schlichting and Pigliucci, 1998)], then selection can act on such variation and improve this new phenotype's functionality and adaptive value in the novel environment. Moreover, depending on whether or not plasticity is favored (Berrigan and Scheiner, 2004; Scheiner, 1993; Scheiner et al., 2017; Lande, 2015; Lande, 2014; Nijhout, 2003), selection can also respectively promote either *increased* environmental sensitivity—which might produce a polyphenism in which two or more distinct phenotypes exist and are induced by alternative environmental stimuli [sensu (Michener, 1961)]—or *decreased* environmental sensitivity—in which plasticity is lost and the phenotype becomes fixed. The latter outcome might occur for at least two, nonmutually exclusive reasons. First, plasticity can be lost through mutational degradation or genetic drift (Masel et al., 2007), as might occur when non-favored phenotypes are seldom expressed and thereby experience relaxed selection (Kawecki, 1994; Whitlock, 1996; Van Dyken and Wade, 2010). Second, when plasticity is costly (Murren et al., 2015; Snell-Rood et al., 2010), selection can actively eliminate it, leading to the canalization of the favored phenotype [i.e., genetic assimilation occurs; sensu (Waddington, 1953)].

Regardless of whether the outcome is a novel polyphenism, a new trait canalized through genetic assimilation, or something in between these two extremes, selection can act on an initially environmentally induced phenotype and promote an adaptive change in the form and/or regulation of that phenotype [this process is known as genetic accommodation; sensu (West-Eberhard, 2003)]. Thus, by jump starting an adaptive evolutionary sequence, plasticity could precede—and facilitate—the origin of a novel, complex phenotype (Fig. 16.1).

Although we described PLE earlier as starting with environmental induction, it could also start with a new genetic mutation (West-Eberhard, 2003). Essentially, developmental plasticity

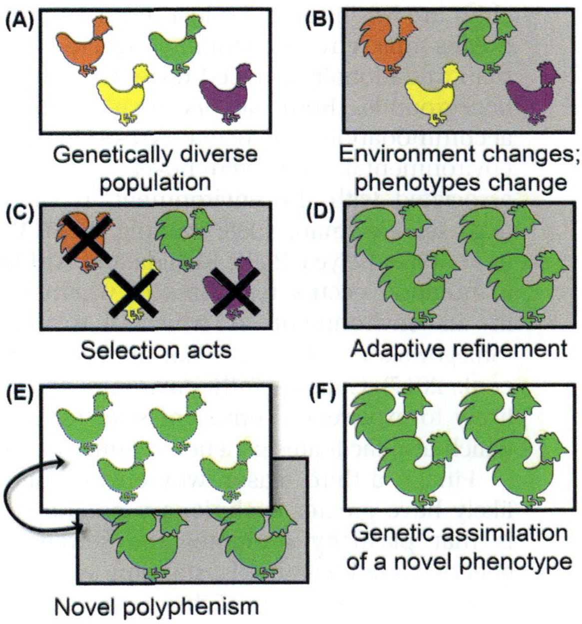

Figure 16.1 How plasticity can facilitate the evolution of a novel, complex phenotype. (A) A genetically diverse population (different colors: different genotypes) (B) experiences a novel environment *(shading)*, which induces novel phenotypes *(dashed lines)*, but genotypes differ in whether and how they respond (different head and/or tail feathers). (C) Selection acts on this formerly cryptic genetic variation (revealed by a change in environment) and disfavors genotypes that produce poorly adapted phenotypes ("X"). (D) This leads to the adaptive refinement of the favored phenotype (enlargement of the green chicken). (E) If individuals produce either this novel phenotype or the ancestral phenotype depending on their environment, then the result is a novel polyphenism.
(F) Alternatively, selection might favor the loss of plasticity (i.e., genetic assimilation), resulting in a novel phenotype that is produced regardless of the environment *(indicated by the loss of dashed lines)*.

must also accommodate any phenotypic changes wrought by a new mutation (i.e., mutationally induced changes do not occur in the absence of other developmental processes). Any such accommodation can then be acted on by selection to stabilize, refine, and/or extend the induced phenotype. Because a novel trait—whether created by a new mutation or a change in the environment—will nearly always involve some sort of adjustment by the existing developmental system, genetic and environmental inputs during development may often be interchangeable (West-Eberhard, 2003). However, there are at least three reasons to think that environmentally triggered novelties could have greater evolutionary potential than genetic mutations (West-Eberhard, 2003).

First, in contrast to most genetic mutations (which initially affect only one individual and its descendants), changes in the environment often impact *many* individuals simultaneously. This widespread induction means that environmentally triggered novelties are less likely to be lost through random genetic drift. It also means that environmentally triggered novelties are more likely than mutationally induced novelties to be tested in diverse genetic backgrounds. Both factors increase the chances that genetic accommodation will occur (West-Eberhard, 2003). Second, with environmentally triggered novelties, the inducing cue(s) is often associated with the environment in which the trait is adaptive (e.g., see Newman, 1988; Pfennig, 1990; Warkentin, 1995; Schmitt et al., 1999; Relyea, 2001; Kishida and Nishimura, 2004). In contrast, a mutation occurs regardless of whether or not the organism is in an environment in which that mutation is advantageous (Sniegowski and Lenski, 1995, but see Piacentini et al., 2014; Fanti et al., 2017). Consequently, environmentally induced traits are more likely to experience consistent selective pressures when expressed, which also facilitates genetic accommodation.

Finally, a third reason why environmentally triggered novelties likely have greater evolutionary potential than genetic mutations is that plasticity promotes the storage and release of cryptic genetic variation—that is, standing variation that is expressed only under new conditions (Gibson and Dworkin, 2004; Ledon-Rettig et al., 2014; Schlichting, 2008; Paaby and Rockman, 2014). When portions of the genome are not expressed (as is often the case with alleles associated with environmentally triggered traits) variation accrues in these regions because it is not exposed to, and removed by, selection (Van Dyken and Wade, 2010; Gibson and Dworkin, 2004; Schlichting, 2008). However, when the environment changes, formerly cryptic genetic variation that contributes to the development of novel environmentally induced phenotypes becomes exposed to selection. Genetic variation influencing the regulation of plastic traits is plentiful (Scheiner, 1993; Scheiner, 2002); the expression of such variation often increases when populations encounter novel environments (Lardies and Bozinovic, 2008); and the release of such variation fuels and facilitates genetic accommodation (Moczek, 2007) (Fig. 16.1B).

Recent reviews have confirmed that PLE likely plays an important role in the evolution of novelty and adaptation in both laboratory and natural systems (Schlichting and Wund, 2014; Levis and Pfennig, 2016, 2019a; Schneider and Meyer, 2017). Yet, skepticism surrounding PLE persists, primarily because of a perceived lack of evidence from the wild (Futuyma, 2015; Wray et al., 2014) as well as the difficulty in identifying a molecular signature that distinguishes PLE from other evolutionary routes (Kovaka, 2018; Levis and Pfennig, 2020). This skepticism may be further fueled by biases in the traits and taxa that have been used to study PLE: morphological traits have received the most attention, and research programs are dominated by fish and insects (Levis and Pfennig, 2020). In order to gain a more balanced trait and taxonomic sampling of potential cases of PLE, we previously described key criteria and empirical approaches for evaluating these criteria in natural systems (Levis and Pfennig, 2016). In the next section, we give an overview of these criteria and methods.

Empirically evaluating plasticity-led evolution

Often the most tractable approach for evaluating PLE in natural populations involves comparing plasticity in ancestral and derived lineages, such as when derived populations have undergone a recent range expansion from progenitor ancestral populations (e.g., Scoville and Pfrender, 2010). When ancestors and descendants cannot be evaluated directly, however, phylogenetic inference allows indirect tests of PLE (Schlichting and Wund, 2014; Wund, 2012). Alternatively, a widely used approach is to compare the reaction norms (i.e., the curve describing trait values across environments) of at least two different types of lineages: ones that possesses the focal, potentially canalized, trait (representing the derived condition) and ones that are closely related to the former lineage(s) but that lack the focal trait (representing the ancestral condition and that can therefore serve as an ancestor-proxy lineage). Reaction norms of these two types of lineages can be compared when both lineages are reared in both the derived environment (i.e., the environment in which the focal trait is associated) and the ancestral environment (Ledón-Rettig et al., 2008). This approach can be effective in evaluating the following critical predictions of PLE (Levis and Pfennig, 2016, 2019a):

Prediction 1—The focal trait can be environmentally induced in lineages showing the ancestral state. The most basic prediction of PLE is that the novel trait should exhibit ancestral plasticity. In other words, the derived trait (or components thereof) should be environmentally induced in lineages that normally lack the trait, but only when they experience the derived environment (West-Eberhard, 2003).

Prediction 2—Cryptic genetic variation underlying the focal trait should be uncovered when lineages with the ancestral state are exposed to the derived environment. As noted in the previous section, cryptic genetic variation can accrue when genetic variants are not expressed phenotypically. This could occur if lineages still showing the ancestral phenotypic state remain in the ancestral environment. However, once these lineages encounter a new environment, this formerly cryptic genetic variation might be expressed (and exposed to selection). Uncovering of cryptic genetic variation could be manifest as an increase in the trait's heritability (Ledón-Rettig et al., 2010). Observing such an increase would confirm the presence of cryptic genetic variation upon which selection could act (Moczek et al., 2011; Moczek, 2007). This confirmation is necessary to eliminate the possibility that the novel trait arose solely through lineage-specific mutations (as in Santos et al., 2017).

Prediction 3—The focal trait should exhibit evidence of having undergone an evolutionary change in its degree of plasticity and/or form in lineages with the derived trait. During genetic accommodation, a trait that is initially environmentally induced undergoes an evolutionary shift in degree of plasticity or form (West-Eberhard, 2003). These shifts can be detected as changes to the slope, curvature, and/or elevation of the reaction norm (Crispo, 2007; Murren et al., 2014). Finding that selection has led to the complete loss of plasticity in the derived lineage (i.e., no change in phenotype across environments) would imply that the trait has been genetically assimilated [i.e., its expression is now canalized (Waddington, 1953; Crispo, 2007)].

Prediction 4—The focal trait should exhibit evidence of having undergone adaptive refinement as it is induced and exposed to selection repeatedly. During genetic accommodation, selection improves the functionality of a trait. Thus, as a novel trait is expressed (and exposed to selection) more frequently, it should experience greater and more rapid refinement (West-Eberhard, 2003; Snell-Rood et al., 2010). Moreover, with environmentally induced (i.e., conditionally expressed) traits, adaptation should occur like a ratchet such that the more often a lineage expresses the focal trait, the greater degree of its refinement (West-Eberhard, 2003; Levis and Pfennig, 2016, 2019a, 2020). Said another way, there should be evidence of frequency-dependent adaptation (sensu Levis and Pfennig, 2019b).

Note that validation of any one of the above four predictions, by itself, is insufficient to establish that PLE has occurred. For instance, although ancestral plasticity has been documented in many systems [Prediction 1 (Schlichting and Wund, 2014; Schwander and Leimar, 2011)], such documentation (by itself) is insufficient to demonstrate PLE because the PLE hypothesis specifically requires that *selection* favored an evolutionary change in the degree of plasticity (West-Eberhard, 2003). Nonselective processes (e.g., drift, mutation) can also promote such shifts (Masel et al., 2007; Levis and Pfennig, 2019c) and must therefore be ruled out to demonstrate PLE. Thus, verification of all four predictions provides the strongest support for PLE. In the following text, we review such verification in spadefoot toads.

Plasticity-led evolution in spadefoot toads

Spadefoot toad tadpoles in the genus *Spea* have evolved a novel polyphenism not seen in other genera (Fig. 16.2; Ledón-Rettig et al., 2008). This polyphenism consists of an omnivore ecomorph, which primarily eats detritus, and a morphologically and behaviorally distinctive carnivore ecomorph, which specializes on shrimp, and which is characterized by a suite of unique, complex traits (Pfennig, 1990). Omnivores are the default morph; carnivores are induced when a young omnivore eats shrimp or other tadpoles (Pfennig, 1992; Levis et al., 2015). However, most populations harbor heritable variation in the propensity to produce carnivores (Pfennig, 1992). Carnivores arise developmentally from an omnivore-like form via accelerated growth of features (Pfennig, 1992), and frequency dependent, disruptive selection—stemming from resource competition—maintains both ecomorphs within most populations (Pfennig et al., 2007). In the following text, we describe how research has found support for all four of PLE's predictions, suggesting that this novel form likely arose via PLE.

First, several studies have found support for prediction 1 (Ledón-Rettig et al., 2008; Levis et al., 2018). By using species in the sister genus of *Spea* (*Scaphiopus couchii* and *Scaphiopus holbrookii*; Fig. 16.2) that do not regularly consume shrimp and do not produce carnivores, it was possible to test for ancestral plasticity in trophic traits. When fed a shrimp diet, these species showed changes in their trophic morphology (e.g., gut length and mouthparts), and some of these changes were in same direction as that seen in *Spea* carnivores.

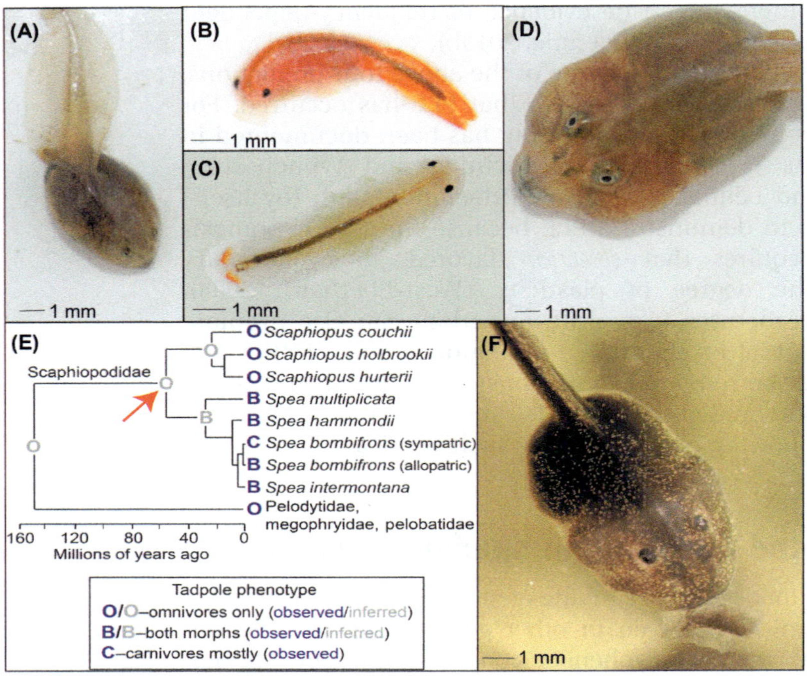

Figure 16.2 North American spadefoot toads of the genus *Spea* have evolved a unique resource polyphenism. Like most anuran tadpoles, spadefoots normally develop into a typical "omnivore" morph (A) by default. However, if a young tadpole ingests large animal prey, such as Anostracan fairy shrimp (B and C), it might develop into a novel carnivore morph (D). (E) In contrast to *Spea*, most frogs produce omnivores only, and it is therefore likely that the ancestor of *Spea* also did so. To study the evolutionary origin of the carnivore morph, Levis et al. (2018) used an omnivore-only producer, *Scaphiopus holbrookii*, as a proxy for the last common ancestor of *Scaphiopus* and *Spea (red arrow)*. (F) When *Sc. holbrookii* tadpoles were fed large animal prey, the tadpoles of this species exhibited diet-induced plasticity—in morphology and gene expression—suggesting that the ancestors of *Spea* likely possessed preexisting plasticity in these features as well. Levis et al. (2018) hypothesized that when an ancestral population began consuming large animal prey (fairy shrimp and other tadpoles), this novel diet uncovered selectable variation in morphology and gene expression. Because some of this variation was adaptive (e.g., producing a shorter gut is adaptive when consuming a protein-rich diet), selection presumably favored further refinement of the carnivore morph. As a footnote to this story, many animals produce a short gut when they eat meat. Thus, this observation suggests that preexisting phenotypic plasticity likely played a role in the evolution of the distinctive carnivore morph. *Source*: All photos taken by David Pfennig.

In support of prediction 2, cryptic genetic variation in *Sc. couchii* was detected when tadpoles were fed a shrimp diet and when they were exposed to corticosterone (a stress hormone). These *Sc. couchii* tadpoles developed and grew more slowly, had increased corticosterone levels, and exhibited greater heritability in size, development rate, and gut length when fed shrimp than when fed detritus. Additionally, *Sc. couchii* tadpoles exposed to corticosterone had greater heritability in development rate and gut length. Similarly, *Sc. holbrookii* tadpoles reared on a shrimp diet showed greater heritable variation in morphology than tadpoles reared on shrimp. Moreover, *Sc. holbrookii* tadpoles with greater levels of plasticity in carnivore traits achieved greater growth and development during intense competition for alternative resources. Taken together, these studies suggest that when *Spea*'s ancestor first began consuming shrimp, cryptic genetic variation was likely uncovered and that those individuals with greater levels of plasticity potentially had a fitness advantage.

Prediction 3 has support from intra- and inter-specific comparisons among spadefoot lineages. First, whereas the morphological and gene expression plasticity of *Sp. multiplicata* tadpoles is coordinated and adaptive, that of *Sc. holbrookii* is not (Levis et al., 2018). Namely, the slope of the reaction norms in *Spea* has evolved differently from those of *Scaphiopus*—which suggests that they have evolved differently from their common ancestor as well. Looking intraspecifically, (Levis and Pfennig, 2019b) found that derived populations of *Sp. bombifrons* have lost trophic plasticity due to selection imposed by *Sp. multiplicata*. That is, *Sp. bombifrons* tadpoles from a region of sympatry (where they cooccur with *Sp. multiplicata*) do not show the diet-dependent changes in gene expression seen in their allopatric (i.e., without the competitor) counterparts (Levis et al., 2017a). Moreover, these sympatric tadpoles are more carnivore-like at hatching prior to ever experiencing the carnivore-inducing cue and some of their features (i.e., jaw muscle width) is carnivore-like even on a detritus diet (i.e., the reaction norm is flat) (Levis and Pfennig, 2019b). Thus, changes in the form and regulation of plasticity likely occurred at the outset of this evolutionary sequence and are still occurring among populations of *Spea*.

Finally, prediction 4 posits that the focal trait should exhibit evidence of adaptive refinement. Initial evidence for this prediction comes from a comparison of gut cell proliferation (a measure of gut performance) among *Sc. couchii, Sp. multiplicata*, and *Sp. bombifrons*. Whereas *Sc. couchii* tadpoles did not exhibit increased gut cell proliferation when fed shrimp, both *Spea* species did, suggesting that shrimp digestion has been improved in *Spea*. In addition, various trophic features have been exaggerated in *Spea* compared to *Scaphiopus*, and the *Spea* lineages that most frequently produce carnivores also produce the most carnivore-like carnivores (Levis et al., 2018). Moreover, these carnivores do not just have more exaggerated carnivore features, they are actually *better* carnivores. Specifically,

Sp. bombifrons from populations that express the carnivore morph frequently are superior competitors for shrimp compared to *Sp. bombifrons* from populations that express the carnivore morph less frequently (i.e., sympatric *Sp. bombifrons* 'win' competition for shrimp) (Levis and Pfennig, 2019b). Thus, frequency-dependent adaptation has driven the refinement of a previously environmentally induced trait such that the trait now exhibits reduced plasticity, but greater functionality. In this way, a novel, coordinated, and well-adapted phenotype that predominates in some populations likely had its origins as an unrefined plastic response to a novel resource.

Plasticity and macroevolution

Although the spadefoot example focuses on two alternative phenotypes, plasticity generally, and PLE specifically, could also be responsible for greater levels of diversification (Pfennig et al., 2010). PLE is similar to the flexible stem hypothesis (West-Eberhard, 2003), in which the phenotypic plasticity of an ancestor generates phenotypically (Wund et al., 2008) and genetically (Gibert, 2017) divergent derived lineages via differential fixation of alternative morphs or phenotypes (Schneider and Meyer, 2017). Indeed, plasticity appears to have facilitated the evolution of a wide array of complex phenotypes from a new mode of reproduction in newts and salamanders (Levis and Pfennig, 2019a) to a diversity of trophic morphologies in fishes (Schneider and Meyer, 2017) to a novel resource-use phenotype in spadefoot toads (see above). These observations therefore lend credence to the claim that plasticity instigates novelty. However, there is another reason for suggesting that plasticity fosters evolutionary innovation: it can generate alternative phenotypes within species that are as divergent from each other—in morphology, behavior, and/or physiology—as are different species or even higher taxonomic categories.

Consider, for instance, the novel carnivore morph in spadefoot toads (Fig. 16.2D). This environmentally induced phenotype is as morphologically distinct from the default omnivore morph—produced by the *same* species—as the omnivore morph is to tadpoles from an entirely different *genus* (Fig. 16.3; for other examples, see Liem and Kaufman, 1984; Calsbeek et al., 2007; Susoy et al., 2016). In fact, for decades, herpetologists thought that these two morphs belonged to separate species (Bragg, 1965). Furthermore, the processes that are generally considered to foster adaptive radiation (competition, innovation, and ecological

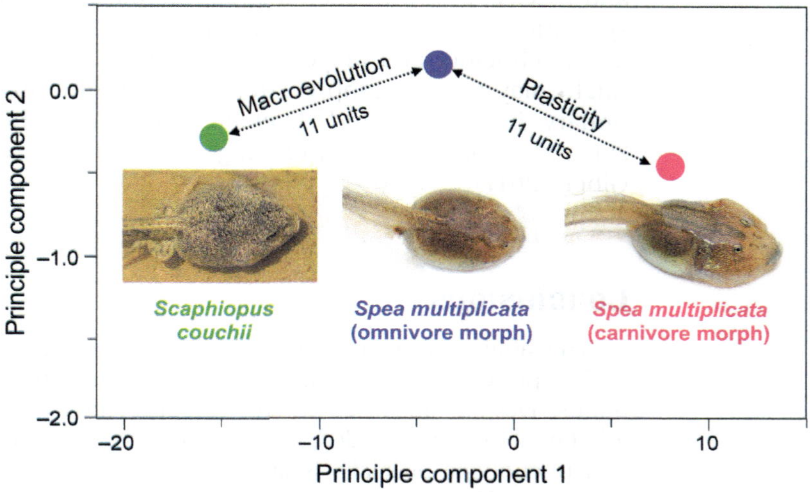

Figure 16.3 Plasticity can generate phenotypic divergence within species as great as that between species. Depending on their diet, spadefoot toad tadpoles in the genus *Spea* develop into either an omnivore morph or a carnivore morph (see Fig. 16.2). An analysis of body shape reveals that these two morphs (in this case, within *Sp. multiplicata*) are as divergent as are the tadpoles of different *genera* of spadefoot toads (numbers denote least squares mean differences between morphs/species in principle component space). *Source*: Modified from Levis, N.A., Pfennig, D.W., 2019a. Phenotypic plasticity, canalization, and the origins of novelty: evidence and mechanisms from amphibians. Semin. Cell Dev. Biol. 88, 80–90.

opportunity), can also drive intraspecific radiation via plasticity and contribute to morphological diversity within species (Levis et al., 2017b).

For these reasons, such extreme plasticity (specifically, polyphenism) has long intrigued evolutionary biologists, at least as far back as Weismann (Weismann, 1882), Goldschmidt (Goldschmidt, 1940), and Waddington (Waddington, 1957). They theorized that environmentally triggered morphs could be informative about the evolution of novelties and differences between species. Indeed, polyphenism has been dubbed 'intraspecific macroevolution' (West-Eberhard, 2003; Liem and Kaufman, 1984). Plasticity might ultimately accelerate macroevolutionary change because it enables alternative phenotypes to evolve semi-independently toward major new adaptive peaks without the negative fitness consequences that usually accompany such large-scale evolutionary shifts (reviewed in West-Eberhard, 2003; Snell-Rood et al., 2010; Gibert, 2017).

Moreover, because phenotypic alternatives can undergo canalization via genetic assimilation (see above), they might eventually become reproductively isolated from each other (Pfennig and McGee, 2010). In such a situation, plasticity could facilitate speciation and adaptive radiation (reviewed in Pfennig et al., 2010; Schneider and Meyer, 2017; Wund, 2012). Thus, plasticity might play an underappreciated role in promoting both evolutionary innovation *and* diversification (West-Eberhard, 1989, 2003; Moczek et al., 2011; Pfennig et al., 2010; Wund et al., 2008; Gibert, 2017).

Conclusions

The ability of organisms to alter their behavior, morphology, and/or physiology in response to environmental conditions is ubiquitous. Despite increased interest in plasticity's role in evolution, additional tests are needed. However, the current evidence suggests that phenotypic plasticity is likely an important contributor to the evolution of innovation, adaptation, and diversification.

References

Badyaev, A.V., 2005. Stress-induced variation in evolution: from behavioural plasticity to genetic assimilation. Proc. R. Soc. B: Biol. Sci. 272, 877–886.

Berrigan, D., Scheiner, S.M., 2004. Modeling the evolution of phenotypic plasticity. In: DeWitt, T.J., Scheiner, S.M. (Eds.), Phenotypic Plasticity: Functional and Conceptual Approaches. Oxford University Press, New York, pp. 82–97.

Bragg, A.N., 1965. Gnomes of the Night: The Spadefoot Toads. University of Pennsylvania Press, Philadelphia, PA.

Calsbeek, R., Smith, T.B., Bardeleben, C., 2007. Intraspecific variation in *Anolis sagrei* mirrors the adaptive radiation of Greater Antillean anoles. Biol. J. Linn. Soc. 90, 189–199.

Carroll, S.B., 2008. Evo-devo and an expanding evolutionary synthesis: a genetic theory of morphological evolution. Cell 134, 25–36.

Chan, Y.F., Marks, M.E., Jones, F.C., Villareal Jr., G., Shapiro, M.D., Brady, S.D., et al., 2010. Adaptive evolution of pelvic reduction in sticklebacks by recurrent deletion of *Pitx1* enhancer. Science 327, 302–305.

Cleves, P.A., Ellis, N.A., Jimenez, M.T., Nunez, S.M., Schluter, D., Kingsley, D.M., et al., 2014. Evolved tooth gain in sticklebacks is associated with a cis-regulatory allele of *Bmp6*. Proc. Natl. Acad. Sci. U.S.A. 111 (38), 13912–13917.

Colosimo, P.F., Peichel, C.L., Nereng, K., Blackman, B.K., Shapiro, M.D., Schluter, D., et al., 2004. The genetic architecture of parallel armor plate reduction in threespine sticklebacks. PLoS Biol. 2, e109.

Conant, G.C., Wolfe, K.H., 2008. Turning a hobby into a job: how duplicated genes find new functions. Nat. Rev. Genet. 9, 938–950.

Crispo, E., 2007. The Baldwin effect and genetic assimilation: revisiting two mechanisms of evolutionary change mediated by phenotypic plasticity. Evolution 61, 2469–2479.

Fanti, L., Piacentini, L., Cappucci, U., Casale, A.M., Pimpinelli, S., 2017. Canalization by selection of *de novo* induced mutations. Genetics 206 (4), 1995–2006.
Flagel, L.E., Wendel, J.F., 2009. Gene duplication and evolutionary novelty in plants. N. Phytol. 183, 557–564.
Futuyma, D.J., 2015. Can modern evolutionary theory explain macroevolution? In: Serrelli, E., Gontier, N. (Eds.), Macroevolution: Explanation, Interpretation and Evidence. Springer, New York, pp. 29–85.
Ghalambor, C.K., Hoke, K.L., Ruell, E.W., Fischer, E.K., Reznick, D.N., Hughes, K.A., 2015. Non-adaptive plasticity potentiates rapid adaptive evolution of gene expression in nature. Nature 525 (7569), 372–375.
Gibert, J.-M., 2017. The flexible stem hypothesis: evidence from genetic data. Dev. Genes. Evol. 227, 297–307.
Gibson, G., Dworkin, I., 2004. Uncovering cryptic genetic variation. Nat. Rev. Genet. 5, 1199–1212.
Gilbert, S.F., Epel, D., 2015. Ecological Developmental Biology: Integrating Epigenetics, Medicine, and Evolution, second ed. Sinauer, Sunderland, MA.
Goldschmidt, R., 1940. The Material Basis of Evolution. Yale University Press, New Haven, CT.
Hoekstra, H.E., Hirschmann, R.J., Bundley, R.A., Insel, P.A., Crossland, J.P., 2006. A single amino acid mutation contributes to adaptive beach mouse color pattern. Science 313, 101–104.
Jablonski, D., 2017. Approaches to macroevolution: 1. General concepts and origin of variation. Evol. Biol. 44 (4), 427–450. Available from: https://doi.org/10.1007/s11692-017-9420.
Kawecki, T.J., 1994. Accumulation of deleterious mutations and the evolutionary cost of being a generalist. Am. Nat. 144, 833–838.
Kishida, O., Nishimura, K., 2004. Bulgy tadpoles: inducible defense morph. Oecologia 140 (3), 414–421.
Kovaka, K., 2018. Underdetermination and evidence in the developmental plasticity debate. Br. J. Philos. Sci. 0, 1–26.
Laland, K.N., Uller, T., Feldman, M.W., Sterelny, K., Müller, G.B., Moczek, A.P., et al., 2015. The extended evolutionary synthesis: its structure, assumptions and predictions. Proc. R. Soc. B 282, 20151019.
Lande, R., 2009. Adaptation to an extraordinary environment by evolution of phenotypic plasticity and genetic assimilation. J. Evol. Biol. 22 (7), 1435–1446.
Lande, R., 2014. Evolution of phenotypic plasticity and environmental tolerance of a labile quantitative character in a fluctuating environment. J. Evol. Biol. 27 (5), 866–875.
Lande, R., 2015. Evolution of phenotypic plasticity in colonizing species. Mol. Ecol. 24, 2038–2045.
Lardies, M.A., Bozinovic, F., 2008. Genetic variation for plasticity in physiological and life-history traits among populations of an invasive species, the terrestrial isopod *Porcellio laevis*. Evol. Ecol. Res. 10, 747–762.
Ledon-Rettig, C.C., Pfennig, D.W., Chunco, A.J., Dworkin, I., 2014. Cryptic genetic variation in natural populations: a predictive framework. Integr. Comp. Biol. 54 (5), 783–793.
Ledón-Rettig, C.C., Pfennig, D.W., Nascone-Yoder, N., 2008. Ancestral variation and the potential for genetic accommodation in larval amphibians: implications for the evolution of novel feeding strategies. Evol Dev. 10 (3), 316–325.

Ledón-Rettig, C.C., Pfennig, D.W., Crespi, E.J., 2010. Diet and hormone manipulations reveal cryptic genetic variation: implications for the evolution of novel feeding strategies. Proc. R. Soc. B: Biol. Sci. 277, 3569–3578.

Levis, N.A., Pfennig, D.W., 2016. Evaluating 'plasticity-first' evolution in nature: key criteria and empirical approaches. Trends Ecol. Evol. 31 (7), 563–574.

Levis, N.A., Pfennig, D.W., 2019a. Phenotypic plasticity, canalization, and the origins of novelty: evidence and mechanisms from amphibians. Semin. Cell Dev. Biol. 88, 80–90.

Levis, N.A., Pfennig, D.W., 2019b. Plasticity-led evolution: evaluating the key prediction of frequency-dependent adaptation. Proc. R. Soc. B 20182754.

Levis, N.A., Pfennig, D.W., 2019c. How stabilizing selection and nongenetic inheritance combine to shape the evolution of phenotypic plasticity. J. Evol. Biol. 32, 706–716.

Levis, N.A., Pfennig, D.W., 2020. Plasticity-led evolution: a survey of developmental mechanisms and empirical tests. Evol Dev. 22, 71–87.

Levis, N.A., de la Serna Buzon, S., Pfennig, D.W., 2015. An inducible offense: carnivore morph tadpoles induced by tadpole carnivory. Ecol. Evol. 5, 1405–1411.

Levis, N.A., Serrato-Capuchina, A., Pfennig, D.W., 2017a. Genetic accommodation in the wild: evolution of gene expression plasticity during character displacement. J. Evol. Biol. 30, 1712–1723.

Levis, N.A., Martin, R.A., O'Donnell, K.A., Pfennig, D.W., 2017b. Intraspecific adaptive radiation: competition, ecological opportunity, and phenotypic diversification within species. Evolution 71, 2496–2509.

Levis, N.A., Isdaner, A.J., Pfennig, D.W., 2018. Morphological novelty emerges from pre-existing phenotypic plasticity. Nat. Ecol. Evol. 2 (8), 1289–1297.

Liem, K.F., Kaufman, L.S., 1984. Intraspecific macroevolution: functional biology of the polymorphic cichlid species *Cichlasoma minckleyi*. In: Echelle, A.A., Kornfield, I. (Eds.), Evolution of Fish Species Flocks. University of Maine Press, Orono, ME, pp. 203–215.

Masel, J., King, O.D., Maughan, H., 2007. The loss of adaptive plasticity during long periods of environmental stasis. Am. Nat. 169 (1), 38–46.

Michener, C.D., 1961. Social polymorphism in hymenoptera. Symp. R. Entomol. Soc. Lond. 1, 43–56.

Moczek, A.P., 2007. Developmental capacitance, genetic accommodation, and adaptive evolution. Evol. Dev. 9, 299–305.

Moczek, A.P., 2008. On the origins of novelty in development and evolution. Bioessays 30, 432–447.

Moczek, A.P., Sultan, S.E., Foster, S., Ledon-Rettig, C., Dworkin, I., Nijhout, H.F., et al., 2011. The role of developmental plasticity in evolutionary innovation. Proc. R. Soc. B: Biol. Sci. 278, 2705–2713.

Murren, C.J., Maclean, H.J., Diamond, S.E., Steiner, U.K., Heskel, M.A., Handelsman, C.A., et al., 2014. Evolutionary change in continuous reaction norms. Am. Nat. 183, 453–467.

Murren, C.J., Auld, J.R., Callahan, H., Ghalambor, C.K., Handelsman, C.A., Heskel, M.A., et al., 2015. Constraints on the evolution of phenotypic plasticity: limits and costs of phenotype and plasticity. Heredity 115, 293–301.

Newman, R.A., 1988. Adaptive plasticity in development of *Scaphiopus couchii* tadpoles in desert ponds. Evolution 42 (4), 774–783.

Nijhout, H.F., 2003. Development and evolution of adaptive polyphenisms. Evol. Dev. 5, 9–18.

O'Brown, N.M., Summers, B.R., Jones, F.C., Brady, S.D., Kingsley, D.M., 2015. A recurrent regulatory change underlying altered expression and *Wnt* response of the stickleback armor plates gene *EDA*. Elife 4, e05290.
Paaby, A.B., Rockman, M.V., 2014. Cyptic genetic variation: evolution's hidden substrate. Nat. Rev. Genet. 15, 247–258.
Pfennig, D.W., 1990. The adaptive significance of an environmentally-cued developmental switch in an anuran tadpole. Oecologia 85, 101–107.
Pfennig, D.W., 1992. Proximate and functional causes of polyphenism in an anuran tadpole. Funct. Ecol. 6, 167–174.
Pfennig, D.W., McGee, M., 2010. Resource polyphenism increases species richness: a test of the hypothesis. Phil. Trans. R. Soc. Lond. Ser. B. 365, 577–591.
Pfennig, D.W., Rice, A.M., Martin, R.A., 2007. Field and experimental evidence for competition's role in phenotypic divergence. Evolution 61, 257–271.
Pfennig, D.W., Wund, M.A., Snell-Rood, E.C., Cruickshank, T., Schlichting, C.D., Moczek, A.P., 2010. Phenotypic plasticity's impacts on diversification and speciation. Trends Ecol. Evol. 25, 459–467.
Piacentini, L., Fanti, L., Specchia, V., Bozzetti, M.P., Berloco, M., Palumbo, G., et al., 2014. Transposons, environmental changes, and heritable induced phenotypic variability. Chromosoma 123, 345–354.
Pigliucci, M., Murren, C.J., 2003. Genetic assimilation and a possible evolutionary paradox: can macroevolution sometimes be so fast as to pass us by? Evolution 57, 1455–1464.
Pigliucci, M., Murren, C.J., Schlichting, C.D., 2006. Phenotypic plasticity and evolution by genetic assimilation. J. Exp. Biol. 209, 2362–2367.
Price, T.D., Qvarnstrom, A., Irwin, D.E., 2003. The role of phenotypic plasticity in driving genetic evolution. Proc. R. Soc. Lond. Ser. B 270, 1433–1440.
Relyea, R.A., 2001. Morphological and behavioral plasticity of larval anurans in response to different predators. Ecology 82 (2), 523–540.
Santos, M.E., Le Bouquin, A., Crumière, A.J.J., Khila, A., 2017. Taxon-restricted genes at the origin of a novel trait allowing access to a new environment. Science 358, 386–390.
Scheiner, S.M., 1993. Genetics and evolution of phenotypic plasticity. Annu. Rev. Ecol. Syst. 24, 35–68.
Scheiner, S.M., 2002. Selection experiments and the study of phenotypic plasticity. J. Evol. Biol. 15, 889–898.
Scheiner, S.M., Barfield, M., Holt, R.D., 2017. The genetics of phenotypic plasticity. XV. Genetic assimilation, the Baldwin effect, and evolutionary rescue. Ecol. Evol. 7, 8788–8803.
Schlichting, C.D., 2008. Hidden reaction norms, cryptic genetic variation, and evolvability. Ann. N. Y. Acad. Sci. 1133, 187–203.
Schlichting, C.D., Pigliucci, M., 1998. Phenotypic Evolution: A Reaction Norm Perspective. Sinauer, Sunderland, MA.
Schlichting, C.D., Wund, M.A., 2014. Phenotypic plasticity and epigenetic marking: an assessment of evidence for genetic accommodation. Evolution 68 (3), 656–672.
Schmalhausen, I.I., 1949. Factors of Evolution: The Theory of Stabilizing Selection. University of Chicago Press, Chicago, IL [1986].
Schmitt, J., Dudley, S.A., Pigliucci, M., 1999. Manipulative approaches to testing adaptive plasticity: phytochrome-mediated shade-avoidance responses in plants. Am. Nat. 154, S43–S54.

Schneider, R.F., Meyer, A., 2017. How plasticity, genetic assimilation and cryptic genetic variation may contribute to adaptive radiations. Mol. Ecol. 26 (1), 330–350.

Schwander, T., Leimar, O., 2011. Genes as leaders and followers in evolution. Trends Ecol. Evol. 26 (3), 143–151.

Scoville, A.G., Pfrender, M.E., 2010. Phenotypic plasticity facilitates recurrent rapid adaptation to introduced predators. Proc. Natl. Acad. Sci. U.S.A. 107 (9), 4260–4263.

Snell-Rood, E.C., Van Dyken, J.D., Cruickshank, T., Wade, M.J., Moczek, A.P., 2010. Toward a population genetic framework of developmental evolution: the costs, limits, and consequences of phenotypic plasticity. Bioessays 32, 71–81.

Sniegowski, P.D., Lenski, R.E., 1995. Mutation and adaptation: the directed mutation controversy in evolutionary perspective. Annu. Rev. Ecol. Syst. 26, 553–578.

Susoy, V., Ragsdale, E.J., Kanzaki, N., Sommer, R.J., 2015. Rapid diversification associated with a macroevolutionary pulse of developmental plasticity. eLife 4, e05463.

Susoy, V., Herrmann, M., Kanzaki, N., Kruger, M., Nguyen, C.N., Rödelsperger, C., et al., 2016. Large-scale diversification without genetic isolation in nematode symbionts of figs. Sci. Adv. 2 (1), e1501031.

Van Dyken, J.D., Wade, M.J., 2010. The genetic signature of conditional expression. Genetics 184, 557–570.

Waddington, C.H., 1953. Genetic assimilation of an acquired character. Evolution 7, 118–126.

Waddington, C.H., 1957. The Strategy of the Genes. George Allen & Unwin, London.

Wagner, G.P., Lynch, V.J., 2010. Evolutionary novelties. Curr. Biol. 20, R48–R52.

Warkentin, K.M., 1995. Adaptive plasticity in hatching age: a response to predation risk trade-offs. Proc. Natl. Acad. Sci. U.S.A. 92 (8), 3507–3510.

Weismann, A., 1882. Studies in the theory of descent. With notes and additions by the author. Translated and edited, with notes, by Raphael Meldola, Sampson, Low, Marston, Searle, & Rivington, London.

West-Eberhard, M.J., 1989. Phenotypic plasticity and the origins of diversity. Annu. Rev. Ecol. Syst. 20, 249–278.

West-Eberhard, M.J., 2003. Developmental Plasticity and Evolution. Oxford University Press, New York.

Whitlock, M.C., 1996. The red queen beats the jack-of-all-trades: the limitations on the evolution of phenotypic plasticity and niche breadth. Am. Nat. 148, S65–S77.

Wray, G.A., Hoekstra, H.E., Futuyma, D.J., Lenski, R.E., Mackay, T.F.C., Schluter, D., et al., 2014. Does evolutionary theory need a rethink? No, all is well. Nature 514, 161–164.

Wund, M.A., 2012. Assessing the impacts of phenotypic plasticity on evolution. Integr. Comp. Biol. 52 (1), 5–15.

Wund, M.A., Baker, J.A., Clancy, B., Golub, J.L., Foster, S.A., 2008. A test of the "flexible stem" model of evolution: ancestral plasticity, genetic accommodation, and morphological divergence in the three spine stickleback radiation. Am. Nat. 172, 449–462.

Niche construction and the transition to herbivory: Phenotype switching and the organization of new nutritional modes

Lynn Chiu[1,2] and Scott F. Gilbert[3]
[1]*School of Biology, University of St. Andrews, Scotland, United Kingdom*
[2]*ImmunoConcept Lab, University of Bordeaux, Bordeaux, France*
[3]*Swarthmore College, Swarthmore, PA, United States*

Introduction

Eukaryotic organisms are not autonomous individuals. Rather, we are holobionts. The term "holobiont" refers to the consortium of zygote-derived cells plus their numerous associated microorganisms, including protists, archaeans, bacteria, fungi, and viruses. In the past decades, advances in microbiome research have shown that microbiomes are often essential for animals to develop, function, or reproduce normally (Funkhouser and Bordenstein, 2013; Gilbert et al., 2012; McFall-Ngai et al., 2013). The diverse ways animals and microbes shape and form symbiotic relationships with each other have important implications for the nature and boundaries of holobionts as biological individuals. What kinds of relationships imply that an animal is a "part" of a developing, living, evolving holobiont "organism"

Much work has centered on mutually beneficial, obligatory relationships within the holobiont. The reasoning is that holobionts are multispecies individuals when microbes are essential for host development and physiological functioning. The classic example is reef-building coral that survive only through the photosynthesis of their algal symbiont, *Symbiodinium*, which enters into the ectoderm of its host (zoon) and transports over 90% of its photosynthetically derived carbon compounds to the host cells (Muscatine et al., 1984). In exchange, the coral gives

Phenotypic Switching. DOI: https://doi.org/10.1016/B978-0-12-817996-3.00015-3

those endosymbionts critical nutrients and a safe, sunlit habitat in an otherwise nutrient-poor habitat (Roth, 2014). Without these symbionts, the coral, a keystone species for reef ecosystems, loses its color (bleaches) and dies. In animals, bacteria critical to the construction of the vertebrate immune system as well as gut capillaries and epithelia (Bates et al., 2006; Camp et al., 2014; Crabtree et al., 2007; Lee and Mazmanian, 2010; Stappenbeck et al., 2002). They also appear to be critical for the normal development of the vertebrate enteric and cerebral nervous systems (Cussotto et al., 2018; De Vadder et al., 2018).

Other arguments for holobiont organismality go beyond the complementarity of animal and microbes. Holobionts are individuals when host and microbes co-instigate developmentally novel and evolutionarily selectable phenotypes (Gilbert, 2016; Gilbert et al., 2010; Sudakaran et al., 2017). A remarkable example is the development and regulation of the mammalian immune system. The microbial colonization of the animal is facilitated by both maternal physiology and the animal's own immature immune system (Round et al., 2011; Chu and Mazmanian, 2013; Chiu and Gilbert, 2015). The recruited microbes in turn help induce the necessary development and functioning of host immune tissue, and these lifelong immune activities are well-regulated only in the continuous presence of microbes, which in turn constantly regulate which microbes stay with the animal (Chiu et al., 2017; Eberl, 2010; Eberl and Pradeu, 2017; Pradeu, 2012; Tauber, 2017). The immune system is a continuously coconstructed property of the holobiont.

In this paper, we develop a new line of reasoning for holobiont individuality—that animal and microbes coconstruct the ecological niches of the holobiont ("reciprocal niche construction"). Mounting evidence suggests that microorganisms can affect the niches of their host animals by changing the latter's diets, supplying defense systems against host predators, or by altering how hosts behave, determining a variety of niche parameters such as habitat space, diet, and ecological relationships with other species (Borges, 2017). Symbiotic microbes have been critical in determining nutritive niches. For instance, the endosymbiont *Buchnera aphidocola* enables the pea aphid *Acyrthosiphon pisum* to sustain itself on plant sap (Bennett and Moran, 2015). This bacterial symbiont is inherited vertically as a cytoplasmic contribution of the mother. One of us (Gilbert, 2019) has recently argued that horizontally transmitted microbial communities have also facilitated evolutionary innovations into new nutritive modes, specifically, from nonherbivory to herbivory diets (see also Vermeij and Lindberg, 2000). Consider the cow. It is an herbivore that occupies a plant-eating ecological niche. Yet, the animal—the multicellular organism developed from a single zygote—is not an herbivore. The animal, in fact, has no gene in her genome that encodes enzymes that facilitate the degradation of the cellulose or hemicellulose of grasses. Without its community of cellulose-digesting enzymes within its specialized rumen stomach, the cow cannot digest plant material. Cows occupy a plant dietary niche only *as* holobionts. The ability to eat living plant material opened numerous new evolutionary trajectories, giving rise to adaptations in animal digestive systems, mandibles, and locomotor organs.

In this chapter, we will focus on ruminants and elaborate niche construction frameworks to explain how evolutionary phenotypic switching occurred between drastically different nutritive modes as a function of gut microorganisms. It may be tempting to think of symbiont microbiota as building blocks that supply the animal with extended phenotypes and expanded niches. On this view, microorganisms harbor specific, functional traits that are recruited and added to the host organism. However, the transition from carnivory to herbivory is not just the gradual addition and removal of adaptive traits. It is also a dramatic shift in the significance and relevancy of the environments the microorganisms and host organisms find themselves in, mediated by their plasticity and abilities to construct their environments. The transition to herbivory involves various forms of niche construction at different levels of organization that facilitate transitions in niche space and adaptations. A holobiont perspective is crucial to capture, within a single explanatory unit, the important niche construction processes occurring at these different levels and between multiple species. We propose an eco-evo-devo (ecological evolutionary developmental) approach that focuses on changes in the developmental capacities of the holobiont that direct the dynamic construction of environmental opportunities and challenges.

Let us start by telling a microbe-filled story of how the cow gets its rumen.

How the bovine got its stomach

Herbivores typically arise from carnivores, making herbivory a derived condition. This runs counter to the ecological condition of extant ecosystems, wherein herbivores eat the herbs, and the carnivores eat the herbivores. However, genomic and paleontological evidence both show that the basal state of both Ecdysozoa and Vertebrata is carnivory, and that herbivores are late additions to the tree of life (Román-Palacios et al., 2019). In the sea, "macrophagous primary consumers" were not found in early marine ecosystems. Instead, the ocean fauna was dominated by filter feeders, detritovores, and the carnivores that ate them (Vermeij and Lindberg, 2000, p. 200).

Ruminants are an extremely successful mammalian lineage, and they are characterized by a multichambered stomach (including the rumen, which can digest plant fiber), cranial headgear (that include horns and antlers), and specialized hypsodont dentition for grinding plants. Representative species include cattle, buffalo, deer, sheep, goats, and yaks. Crown group ruminants begin to appear in the late Oligocene, about 35 million years ago. The horned ruminants appear about twelve million years later (Chen et al., 2019).

One of the classic examples of developmental and nutritive symbioses concerns the relationship between the microbial symbionts and the four-chambered stomach characteristic of the ruminants. In the following, we examine three critically important symbioses between the gut microbes and the bovine organism. In each case, microorganisms are proactively involved in facilitating the transition of the holobiont and its constituents into an herbivory niche. The first set of symbioses involves the roles of microbes in constructing the rumen of the gut. This new region of the stomach, from which "ruminants" get their name, creates a suitable environment for further bacterial proliferation while initiating developmental processes that mature the previously nonfunctional rumen organ. The second set of symbioses involves the roles of these rumen bacteria in digesting the plant fibers, thereby enabling the calf survive by digesting grass and plant fibers. The third set of symbioses involves the microbe-dependent neutralization of plant defense chemicals, allowing the calf to continue to utilize plants as a food source.

As we will see, these symbioses scaffold two types of niche construction. The first is "perturbational niche construction" (Odling-Smee et al., 2003) by microbes and their by-products that alter the host rumen environment. Here, the bacteria build their residence from the host tissue of their environment, which in turn deeply alters the digestive organs of the host. The second is "mediational niche construction" (Chiu, 2019) of the ecological niche experienced by the holobiont. Here, through the microbe-induced developmental and physiological processes, the holobiont gradually enters and maintains an herbivorous ecological niche whereby plants have significance as nutritive sources. This developmental process suggests an evolutionary sequence of niche transition also mediated by microorganisms.

Developmental symbiosis: the microbial-dependent development of the ruminant stomach

In the newborn calf, the four chambers of the stomach are already evident. The largest chamber of this organ is the abomasum, the "true stomach" that is homologous to other mammalian stomachs. This comprises about 60 % of the gastric volume. Another 15% of the volume is found in the omasum and reticulum chambers, and the rumen of the newborn contains about 25% of the stomach volume (Tamate et al., 1962).

Newborn calves have sterile rumens, and the digestive tube becomes colonized by microbes as the calf pass through the birth canal. Within 2 days of birth, the area of the rumen is seen to have microbes within it. These microbes are capable of degrading cellulose plant fibers (Moraïs and Mizrahi, 2019). However, the baby calf does not receive grass or grain to eat until it is weaned. Before that time, it receives milk from the mother cow, and the rumen, although containing bacteria, is small and nonfunctional. The milk does not go into the rumen, but is shunted by an esophageal groove into the abomasum. Upon weaning, the esophageal groove flattens, and the masticated grain proceeds directly into the rumen (Baldwin and Connor, 2017; Daniels and Yohe, 2014). There, bacteria such as *Ruminociccus flavefaciens* produce "cellulosomes," plant wall-digesting enzyme complexes that are bound to the bacterial cell surface, which efficiently metabolize the complex polysaccharides.

The bacteria in the rumen multiply when given this plant food, and as they proliferate, they produce volatile fatty acids, including butyrate. Butyrate causes the dramatic growth of the rumen as well as the differentiation of the ruminal papillae and musculature. Indeed, butyrate will cause the premature growth and differentiation of the calf rumen when it experimentally infused directly into the immature rumen or when it is placed into the milk drunk by the calf (Baldwin and Connor, 2017; Sander et al., 1959). Numerous genes encoding transcription factors, especially those associated with cell proliferation, are induced by the infusion of butyric acid into the rumen (Baldwin et al., 2018; Chen et al., 2019), and butyrate produced by gut bacteria is known to regulate gene expression through its activity as a regulator of histone deacetylase (Wu et al., 2012;

Yuille et al., 2018). The regulation of rumen development is thought to be achieved through bacterially driven host genome transcription through DNA methylation changes and microRNA production (Li et al., 2019; Malmuthuge et al., 2019). By three months, the rumen comprises about 85% of the calf's stomach volume. It is thusly that the gut bacteria construct their own niche, the rumen.[1]

Nutritional symbiosis: the microbe-dependent digestion of plant fiber

The second symbiotic effect of the ruminal bacteria is the digestion of plant fiber. The bacteria have helped build the rumen, and the rumen will now serve as a place of residence for these microbes and the calf's body will be modified to supply them and the calf with food. Plant cell walls, rich in pectin and cellulose, are the largest reservoir of organic carbon on earth (Gilbert, 2010), and the architecture and physiology of the rumen have evolved for millions of years to obtain the solar energy stored in these plant fibers (Mackie et al., 2002). Over 70% of the cow's energy comes from this microbial digestion of plant fiber (Flint et al., 2008; La Reau and Suen, 2018), demonstrating the primacy of microbes in a cow's herbivorous feeding strategy. Most animals (with the exception of certain plant-eating beetles that acquired their genes for digesting plant fiber by lateral gene transfer from microbes) do not have genes encoding the enzymes that digest these plant polysaccharides (Calderón-Cortés et al., 2012; Kirsch et al., 2014). Bacteria, protists and fungi, however, evolving as plant pathogens or saprophytic detritovores, often have genomes that do synthesize and secrete such enzymes.

Most of these plant cell wall-digesting enzymes belong to various glycoside hydrolase families, constituting cellulases, hemicellulases, and pectinases (Moraïs and Mizrahi, 2019). The enzymes can be secreted by bacteria, protists, fungi, and archea. Bacteria comprise the dominant proportion of the ruminal

[1]Butyrate can also function as a nutritional substrate, RNA splicing regulator, or as a G-protein activator, and microbiome-produced butyrate appears to be critical for normal human intestinal homeostasis (Dowhaniuk et al., 2019). Moreover, this short-chain fatty acid plays several roles in activating dormant development throughout the animal kingdom, including the reactivation of tick development through the smell of bovine butyrate (von Uexküll, 1934).

microbes (about 95% of the microbial population), with the archaea comprising 2%–5%. The eukaryotic component of the ruminal ecosystem is very small (Flint, 1997; Mizrahi, 2013). The microbial cellulose degraders (primarily bacteria and fungi) often produce several types of cellulases, including those that attack the ends of the fiber and those that cleave internally (Lombard et al., 2014). Hemicellulose has fewer carbohydrate units, and often contain xylose. Again, many different hemicellulose-digesting enzymes can come from a single microbe, and numerous species of microbes secrete such enzymes. Most of the cellulases and hemicellulases are secreted from bacteria, with some species producing both types of enzymes. Cellulose-degrading enzymes are produced by genera such as *Ruminococcus* and *Fibrobacter.* Many of the hemicellulases originate in those genera and in *Prevotella* (Dai et al., 2015). Interestingly, the giant panda (*Ailuropoda melanoleuca*), which only eats the new shoots of bamboo trees, comes from a carnivorous lineage and seems to be "learning" herbivory. It has a symbiont that can digest hemicellulose, and so its diet is fixed to the new shoots that have not yet made their cellulose and lignin-containing stalk. Giant pandas cannot eat mature branches (Zhang et al., 2018).

Within the rumen ecosystem, there are interactions at several levels. Synergy between the hemicellulases and cellulases is essential for plant fiber digestion, and different microorganisms contribute enzymes with overlapping functions for digesting these plant fibers (Artzi et al., 2017; Bayer et al., 2013). Moreover, the products of fiber digestion appear to regulate interactive networks that sustain the rumen ecosystem. The digestion of complex fibers into simple sugars supports the growth of a second set of microbes that can persist on the sugar monomers. The digestion of simple sugars, in turn, not only generates new metabolites for other microbes, but also generates hydrogen that can be used by methanogens. The presence of such methanogens accelerates the cellulose hydrolysis. Fungi and protists probably assist the bacteria in supporting this ring of symbiosis (Morvan et al., 1996; Newbold et al., 2015). Competitive interactions between bacteria, and between bacteria and protists, also contribute to the stability of the ruminal ecosystem (Allesina and Levine, 2011; Chen and Weimer, 2001; Moraïs and Mizrahi, 2019). The host genome also plays some role in determining which of the keystone cellulose-degrading and hemicellulose-degrading bacteria will thrive and proliferate in the rumen

(Sasson et al., 2017; Wallace et al., 2019). Thus, the symbiotic microbiome is stabilized through interactions between the microbes and between the microbes and the host environment.

Protective Symbiosis: the microbe-dependent detoxification of plant defense chemicals

Herbivory can be maintained only if the herbivore is capable of neutralizing those toxic secondary compounds that are made either by the plant or by the plant's symbionts. Each plant makes its own set of secondary compounds, and ingestion of these chemicals can lead to the death or severe impairment of the herbivore that eats them. Herbivory would not be possible unless animals had ways of degrading or excreting these poisons (Freeland and Janzen, 1974).

Dominguez-Bello (1996, p. 323) has characterized the rumen "as a detoxification chamber," where detoxification can be done by evolutionary changes in the animal's genome (Malenke et al., 2014) or by the acquisition of particular symbionts (Smith, 1992). For instance, ruminants need certain tryptophan-utilizing bacteria (such as *Clostridium sporogenes*) to degrade the alkaloid toxin ergovaline found in the seeds of tall fescue seeds (Harlow et al., 2017). The ergovaline, itself, is made by an endocytic fungus that lives between the cells of the plant. Indeed, there is a "the three-way interaction" between plants, herbivores, and microbes (Wielkopolan and Obrepalska-Steplowska, 2016), and Smith (1992, p. 25) notes that "it has been suggested that rumen microbial detoxification of poisonous plants might have been as important as rumen microbial degradation of cellulose for the evolutionary development and ecological expansion of ruminants as herbivores."

Thus, symbiotic microbes help construct the rumen, digest the fiber into short-chain fatty acids that can serve as nutrition for the organism, and detoxify the poisons that plants and their symbionts produce in the evolutionary defenses against such herbivory. In return, the microbes get room, board, and roommates—a residence, masticated plant material, lipids for their cell membranes, and other microbes that integrate them into a stable community. The bovine ruminant is completely dependent upon the symbionts after it has been weaned. Such vertebrates go from a set of maternal symbiotic relationships to the nutrition produced by the plant fiber-digesting ruminal ecosystem comprised of symbiotic microbes. As Moraïs and Mizrahi

(2019, p. 5) conclude, "Bacteria are obligate inhabitants of the rumen; without them, the host animal would not survive."

Perturbational and mediational niche construction

It is tempting to think of these three types of symbiotic relationships as microbe-induced adaptive phenotypes of the animal and the microbes to a plant dietary niche. The ruminant animal and microbes have benefited from these coconstructed traits and were thus likely selected for their continued alliance. Yet there are several[2] explanatory problems with this approach. One main issue is that it assumes a process of adaptation that gradually matches traits to environments through natural selection. Microbes helped induce or provide selectable, adaptive traits better adapted to a previously unexplored part of the environment—an empty niche. This way of thinking assumes that an herbivory niche preexisted in the environment ready for the ruminant ancestor to adapt into (eventually, through microbes).

The notion of an empty niche existing independently from organisms, however, is explanatorily problematic (Lewontin, 1982, 1983, 2000). Organisms—through their physiology—define which aspects of the external world are relevant and how they are significant. They also actively construct and change the physical features of the environment surrounding them. Ecological niches do not preexist the organisms that "occupy" them. If the "empty niche" does not exist, then niches cannot be used as a prior cause to explain evolutionary outcomes.

We argue that the rumen symbionts helped *construct* a different niche from the same environment. The characteristic niche of a cow—being an herbivore—is in large part determined by its gut microbiota. The microbes did not modify the external niche of the host (the plants outside of the cow). Rather,

[2]An issue with perceiving the holobiont as extended phenotype comes from *whose* phenotype is being extended. The default view is to privilege the multicellular eukaryote and describe microbes as its extensions. However, as Gould (1994) has noted, the prokaryotic microbes preceded eukaryotes by billions of years, and the eukaryote, itself, is derived from the symbiosis of archaean and bacterial prokaryotes. Richard Dawkins (1976) has claimed that "the individual organism is a survival machine for its genes." The "extended phenotype" principle would insist that the eukaryotic organism is an extended phenotype of the microbes and a survival machine for the microbial genomes. Animal and plant reproduction and evolution would become the generation of more niches for the proliferation and diversification of microbes (see Gilbert, in press). Such a fusion of Dawkins and Gould is a precarious undertaking.

vegetation, which had always been in the environment of the carnivores, has now been transformed into a nutritive resource. This evolutionary innovation switched organisms from one type of niche to another, creating a fundamental change in the types of relationships between them and the environment. Vegetation that was once not edible becomes edible to the organism; plant parts that were not nutritious became nutritious.

Perturbational niche construction

In standard niche construction theory, two types of niche construction have been identified (Laland et al., 2017; Odling-Smee et al., 2003): (1) *perturbational niche construction*, whereby modifications organisms make to the environment results in a part of the environment acquiring a different property (e.g., built structures), (2) *relocational niche construction*, whereby the organism surrounds itself with a different environment, leaving its original environment. Both perturbational and relocational niche construction processes determine the intrinsic, physical properties of the environment surrounding an organism. In the first case, the properties in the environment are directly modified. In the second case, the organism enters an environment with a different set of properties.

In the switch to an herbivore niche, perturbational niche construction occurred within the cow. Microorganisms created host-internal niches that induce complex innovations in host development, morphology, and function. Here, niche construction and developmental plasticity meet as microbes engage in a "developmental niche construction[3]" that alters the development of the host (Gilbert, 2016; Laland et al., 2008; Stotz, 2017; Sudakaran et al., 2017). Hosts in turn engage in perturbational niche construction that determine and alter the niches of its microorganisms (Bevins and Salzman, 2011; Donaldson et al., 2015). The rumen codevelops and coevolved with the microbial communities that can flourish within.

Mediational niche construction

There is a third type of niche construction that occurs when the organism experiences a different environment despite no changes to the latter's intrinsic, physical properties. The organism instead alters which aspects of the environment are

[3]See (Flynn et al., 2013; Stotz, 2017) for discussions on developmental niche construction and how it differs from selective niche construction.

relevant to it or change the significance of the environment to the organism. The environment *affords* a different opportunity for exploitation (Walsh, 2015).

This "experiential" type of niche construction was described by Richard Lewontin (1982, 1983, 2000), who noticed that organisms do not just alter their environments, but also determine which components of the environments are relevant and in what way. The organism's phenotype could determine how an environment is experienced. A fluctuating environment can be experienced as constant if the organism has ways of compensating for the changes through mechanisms such as fat storage or temperature regulation. Jakob von Uexküll (1934) and his notion of *Umwelt* also emphasized the interpretation of the environment by the organism. While most organisms would be unaffected or slightly irritated by the faint odor of butyrate coming from mammalian sweat, to the aforementioned tick, it is the pivotal signal that reactivates its dormant life cycle.

This third mode of niche construction has been coined "*mediational niche construction*," since the parts of the environment that are relevant (in what way) are mediated by the organism's makeup (Chiu, 2019). It has only recently been explicitly treated under the lens of phenotypic plasticity and organismal agency (Sultan, 2015; Walsh, 2015). Sonia Sultan (2015) has compiled a large collection of morphological, physiological, and behavioral changes in microbes, animals, and especially plants that alter whether an organism experiences its environment as plentiful, warm, fluctuating, threatening, or competitive. For instance, an environment that has more predators can be experienced as less "threatening" when plastic changes in the organism decreases the activities or impact of the predators.

To Sultan, the relevance of the environment is mediated through the perceptual and information transduction networks of the organism. These include experiences of the environment's temperature, threat levels, and the richness of its resources. When plastic responses to environmental cues occur in the organism, the phenotypic changes in these networks can alter the way the environment is experienced. Since the selective environment is determined by the experience of the environment, plastic, developmental responses to environmental cues can have evolutionary consequences mediational niche construction. Applying these perspectives to the ruminant holobiont, we can find an alternative interpretation of the three symbiotic relationships. The three symbiotic events between cows and microbes—microbe-dependent development, digestion, and toxin neutralization in and of the rumen—are features that determine the ecological niche of the cow holobiont. The cow, as a holobiont, can experience plants as edible and, furthermore, a safe, nontoxic source of nutrition, only because microorganisms helped *mediate* a plant-based nutritive niche (mediational niche construction). This niche was also possible only due to the microbe-induced development of the rumen, which was part of the microbiota's efforts to modify the rumen into its own ecological niche (perturbational niche construction). The transition into herbivory thus involved multiple types of niche construction.

Niche construction, plasticity, and developmental scaffolding

To better understand how developmental plasticity figures into the evolutionary transition between nutritive modes, we can make use of another conceptual framework that has been developed for holobiont systems—the idea that biological systems form "hybrid" life cycles of scaffolded development and reproduction (Chiu and Gilbert, 2015; Griesemer, 2014a,b). The basic idea is that each developmental stage or phase of a biological system is reached only by engaging with "scaffolds," catalytic entities and processes that allow novel processes to occur at lower difficulties and costs. Developmental scaffolds remove difficulties that would otherwise prevent some developmental process from occurring. The scaffold and scaffolded systems are temporary and distinct from each other, but when interacting is integrated into a larger, chimeric, hybrid system. This hybrid system acquires and realizes the developmental capacities to move on into the next stage only because of the interactions between the scaffold and the scaffolded. The hybrid then confers developmental capacities to the following phase, which may consist of a different scaffold-scaffolded hybrid.

In order to transform in and out of different hybrid states, the scaffold and scaffolded remain sensitive and responsive to each other. They are inherently plastic and readily engage in interactions that form hybrid systems. As in mammalian childbirth (Chiu and Gilbert, 2015), there is a mutual scaffolding between the animal and the microbial community. The microbial community scaffolds the development and procreation of the animal, while the animal scaffolds the organization and procreation of the microbes. The mammals create the environmental niches for the microbes; the microbes make the environmental niches for the mammals. In this coconstruction, they are each other's niches. This is reciprocal niche construction.

Indeed, in the ruminants, this is critical. Birth is a holobiont event, wherein symbionts are transferred from one generation to another, and this is when the calf receives the microbiome that will reside in its rumen. Then, there is an intermediate stage where the calf is fed only its mother's milk. This boosts its immunity, alters its microbiome, and nourishes the calf. Finally, there is weaning and the ingestion of grain that enables the production of the butyrate that induces ruminal growth and differentiation. The boundaries between the scaffolds, the scaffolded, and the perceived niche are a matter of perspective, and the same object or process can move between these categories.

Thus, we are able to provide a novel conceptual interpretation and explanation for the opening of herbivory as an evolutionary opportunity. We argue that microbes helped construct new physical niches by altering bodily development (developing the rumen) and generating a new mediational niche (plants as food) by altering the experience of the environment.

The coconstruction of niches in a symbiotic relationship is not a uniquely holobiont phenomenon. Coniche construction can be an instigator of mutualism between a microorganism and microorganisms when one party has an ecological niche (e.g., a nutritive source) that can only be constructed by another (or if both parties require a coconstructed niche) (Buser et al., 2014). For instance, unlike their evolutionarily cousins, bark beetles are unique in their abilities to live on nutrition-poor outer bark instead of the nutrition-rich phloem. This evolutionary shift in diet was made possible by symbiotic associations with fungi that transform the properties of wood (Six, 2019). These beetles kill the tree and transport these specific fungi to these trees, modifying the woody surface for the fungi. The fungi then digest the tree tissues, providing nutrients for the beetles. The fungi are obligate niche constructors for the beetles and vice versa. The beetles harbor highly selective mycangia that carry these specific fungi species and thus maintain their symbiotic relationship across generations.

However, unlike these cases where the modified niche is external to both symbionts, for the holobiont, microbes are modifying niches within the host, with consequences for niches experienced by the host and holobiont.

Implications for holobiont individuality

What kind of individual is a holobiont? Are they units of selection? In current debates, the answer hinges on whether the relationships between macro and micro actors and satisfy traditional criteria of evolutionary individuality. On the one hand, many have argued on the basis of functional integration, especially through obligatory symbiosis, that holobionts do satisfy criteria of anatomical, metabolic, developmental, immunological, and evolutionary units (Gilbert et al., 2012; Lloyd and Wade, 2019; McFall-Ngai et al., 2013; Rosenberg and Zilber-Rosenberg, 2016; Roughgarden et al., 2017; Zilber-Rosenberg and Rosenberg, 2008). Yet skeptics argue that such evidence is insufficient to establish the evolutionary individuality of holobionts as the organisms involved have conflicting evolutionary interests and their ecological relationships are similar to macroscale ecosystems (Bourrat and Griffiths, 2018; Christian et al., 2015; Costello et al., 2012; Douglas and Werren, 2016; Moran and Sloan, 2015; Skillings, 2016).

Neglected in these discussions is the importance of constructed environments and developmental plasticity for biological individuality. The examples we have described in this paper provide support for a different way of thinking about evolutionary units. The holobiont is an evolutionary unit or individual because evolutionary opportunities such as the plant dietary niche are made available for all constituents of the holobiont by the holobiont as a whole. The cow animal and its gut bacteria construct and solidify an herbivory ecological niche through perturbational (changes to the gut rumen) and mediational (changes to the significance of the plant environment) niche construction at a developmental time scale. This developmental process likely opened up evolutionary opportunities that placed the entire ensemble under the selective pressures specific to the plant-eating life.

Our proposal shifts attention from "what kinds of *individuals* are holobionts" to "what kinds of *processes* make holobionts." We examined niche-constructing processes. Symbiosis is critical in producing the niche construction processes that bias selective environments, as well as the developmental scaffolding that biases developmental variations. An eco-evo-devo approach asks how these processes determine developmental and evolutionary trajectories. Following Sultan and Walsh, we propose to focus on how these hybrid systems create and alter experienced niches. Following Griesemer, we propose to look at the way host-microbiota hybrid systems dynamically alter through time as they come to acquire, realize, and pass on developmental capacities to reproduce.

Organisms and their environments (including other organisms) are commingled. Changes in organisms—their agency and plasticity—entail changes in their experienced environments, which alters the way they develop, are ecologically organized, and evolve. Host microbiota is involved in these types of changes. Developmental plasticity is a key element to the host-microbiota coconstruction of a plant-based niche that is relevant, accessible, and easier to live in. Animals are inherently plastic and symbiotic microbes are mediators of developmental plasticity, which enables the holobiont organism to develop in particular directions. Symbiosis-inducible plasticity can be adaptive, with the symbionts facilitating specific trajectories of development (Dunbar et al., 2007; Kikuchi et al., 2012; Oliver et al., 2009; Tsuchida et al., 2010).

Hypothetically, the original members of the ruminant fiber-eating ecosystem were detritovores that found residence in a vertebrate stomach. However, this community has evolved enormously. While it is not known whether there are "phylo-symbiotic" associations of microbes and ruminants, such that those in hydra and primates, there appears to be a "core microbiome" in cow rumens. Sequencing the rumen microbiota of 1016 cows in four separated European countries. Wallace et al. (2019) found a core rumen microbiome of some 454 prokaryotes, 12 protists, and 46 fungi. A subset of 39 bacteria species was found to be linked physiologically to each other and to the host genomes. Differences within the core microbiomes correlated well with specific phenotypic traits such as methane production and milk quality, indicating the importance of the microbiome for holobiont phenotype (Wallace et al., 2019; Lima et al., 2019). It is also known that diet and housing regimen also contribute the diversity of the microbial populations of the rumen, and that genera between cows were more common than species (Hagey et al., 2019; Jami and Mizrahi, 2012). In this manner, the rumen appears to be similar to the conditions of gut microbes in humans (Bäckhed et al., 2012), suggesting that continuity of the species matters less than continuity of function.

Recently, some researchers (Doolittle and Booth, 2016; see also Suárez, 2018; Taxis et al., 2015) have proposed that the genes of the holobiont (the hologenome) constitutes a functional replicator, a network of genetic interactions that are instantiated across different generations of holobionts. In this view, it is the *functions* of the bacteria that matter, not the *species*, i.e., what

Doolittle and Booth (2016) call, "the song and not the singer." This metabolic, physiological, view of organism reflects some of the basic ideas of inheritance that were popular before the field of heredity became limited to genetics. Indeed, in one of his prescient paragraphs, Wilson (1896, p. 431) writes, "In its physiological aspect, therefore, inheritance is the recurrence in successive generations, of like forms of metabolism..." The holobiont has similarly been seen in this regard. Suárez and Triviño (2019) note that holobionts are units of selection, even if they acquire their microbiota from the environment. What is critical for inheritance is transgenerational *trait-recurrence*. In this case, the trait is herbivory, and the host would die if the transfer of the plant fiber-digesting consortium of bacteria was not made.

The continuity of symbiotic species might not be critical in holobiont continuity, since (1) the same sets of functions can be contributed by different species of bacteria, and (2) horizontal gene transfer is probably common among symbionts (J. Suarez, personal communication).[4] Indeed, horizontal gene transfer, mediated through bacteriophages, overrides mutation as the source of variation for *Escherichia coli* colonizing the human gut (Frazão et al., 2019), and intra-species transfers of particular genes that provide the holobiont with a new source of nutrition have been identified (Hehemann et al., 2010). Certainly, vertical transmission is not the *sine qua non* for being a unit of evolution or natural selection. The cow receives its symbiotic communities horizontally, and these microbes are critical for the nutrition and the existence of the cow. There is probably a spectrum of phylosymbiotic compatibility. In some organisms, such as Hydra (Fraune and Bosch, 2007), different species may have vastly different symbionts than other species; while in others animals, any number of different species might fulfill a core metabolic function. In some cases, the song can be performed well by numerous singers, while in other cases, the singer may be critical to the quality of the song. After all, any rocker can belt out *Satisfaction*; but no one sings it like Mick Jagger. This is one of the many research projects that are opened by viewing organisms as holobiont functions.

[4]Here we may be seeing the symbiotic equivalent of developmental systems drift, where the same trait may be accomplished by different genes in different species (Krol et al., 2011; True and Haag, 2001).

Conclusion

This essay has explored certain changes in our understanding of biological individuality that may have to be made if a holobiont view of organism means more than just a microbial extension of host phenotype. Using the cow as an example, we have presented a view of the holobiont that strengthens the claim that holobionts are developmental and evolutionary units.

Developmental plasticity and two modes of niche construction—perturbational and mediational niche construction—enable the host and microbial community to coconstruct an herbivory ecological niche experienced by the holobiont as a whole. Microbes are not just extended niche constructors of the *host,* but are coniche constructors of holobiont ecological and evolutionary niches. A cow is a holobiont, but is *Bos taurus* the holobiont or merely the animal component of it We would argue that it is the holobiont.

The exploration of the symbiotic origins of herbivory highlights the critical importance of developmental plasticity in making arguments in evolutionary developmental biology, for we are talking about holobiont evo-devo. To put it another way: If (1) evolution is brought about by inherited changes of development (a principle of evo-devo) and if (2) development is caused, in part, by symbiotic interactions with microbes; then changes in the interactions of host and microbial communities may cause changes in holobiont evolution.

In addition to the facilitation of developmental stages, this holobiont perspective also entails the transition between niches. The induced developmental capacities must be seen as capacities to exploit niches. Microbes can help secure certain selective ecological niches (maintaining access to niches and the ease and ability to properly utilize these niches) through scaffolding at the developmental level. Evolution and development are thus integrated not only in the production of new variants upon which natural selection can act; they are also integrated in the formation of new niches and the integration of the organism into the habitat it has helped to create.

References

Allesina, S., Levine, J., 2011. A competitive network theory of species diversity. Proc. Natl. Acad. Sci. U. S. A. 5638–5642.

Artzi, L., Bayer, E.A., Moraïs, S., 2017. Cellulosomes: bacterial nanomachines for dismantling plant polysaccharides. Nat. Rev. Microbiol. 15, 83–95. Available from: https://doi.org/10.1038/nrmicro.2016.164.

Bäckhed, F., Fraser, C.M., Ringel, Y., Sanders, M.E., Sartor, R.B., Sherman, P.M., et al., 2012. Defining a healthy human gut microbiome: current concepts, future directions, and clinical applications. Cell Host Microbe 12, 611–622. Available from: https://doi.org/10.1016/j.chom.2012.10.012.

Baldwin, R.L., Connor, E.E., 2017. Rumen function and development. Vet. Clin. North. Am. Food Anim. Pract., Digestive Disord. Forestomach 33, 427–439. Available from: https://doi.org/10.1016/j.cvfa.2017.06.001.

Baldwin, R.L., Li, R.W., Jia, Y., Li, C.-J., 2018. Transcriptomic impacts of rumen epithelium induced by butyrate infusion in dairy cattle in dry period. Gene Regul. Syst. Biol. 12, 1177625018774798. Available from: https://doi.org/10.1177/1177625018774798.

Bates, J., Mittge, E., Kuhlman, J., 2006. Distinct signals from the microbiota promote different aspects of zebrafish gut differentiation. Dev. Biol. 297 (2), 374–386.

Bayer, E.A., Shoham, Y., Lamed, R., 2013. Lignocellulose-decomposing bacteria and their enzyme systems. In: Rosenberg, E., DeLong, E.F., Lory, S., Stackebrandt, E., Thompson, F. (Eds.), The Prokaryotes: Prokaryotic Physiology and Biochemistry. Springer Berlin Heidelberg, Berlin, Heidelberg, pp. 215–266. Available from: https://doi.org/10.1007/978-3-642-30141-4_67.

Bennett, G.M., Moran, N.A., 2015. Hereditable symbiosis: the advantages and perils of an evolutionary rabbit hole. Proc. Natl. Acad. Sci. USA 112, 10169–10176. Available from: https://doi.org/10.1073/pnas.1421388112.

Bevins, C.L., Salzman, N.H., 2011. The potter's wheel: the host's role in sculpting its microbiota. Cell. Mol. Life Sci. 68, 3675–3685. Available from: https://doi.org/10.1007/s00018-011-0830-3.

Borges, R.M., 2017. Co-niche construction between hosts and symbionts: ideas and evidence. J. Genet. 96, 483–489.

Bourrat, P., Griffiths, P.E., 2018. Multispecies individuals. Hist. Philos. Life Sci. 40, 33. Available from: https://doi.org/10.1007/s40656-018-0194-1.

Buser, C.C., Newcomb, R.D., Gaskett, A.C., Goddard, M.R., 2014. Niche construction initiates the evolution of mutualistic interactions. Ecol. Lett. 17, 1257–1264. Available from: https://doi.org/10.1111/ele.12331.

Calderón-Cortés, N., Quesada, M., Watanabe, H., Cano-Camacho, H., Oyama, K., 2012. Endogenous plant cell wall digestion: a key mechanism in insect evolution. Annu. Rev. Ecol. Evol. Syst. 43, 45–71. Available from: https://doi.org/10.1146/annurev-ecolsys-110411-160312.

Camp, J., Frank, C., Lickwar, C., 2014. Microbiota modulate transcription in the intestinal epithelium without remodeling the accessible chromatin landscape. Genome Res. 1504–15016.

Chen, J., Weimer, P., 2001. Competition among three predominant ruminal cellulolytic bacteria in the absence or presence of non-cellulolytic bacteria. Microbiol. Read. Engl. 147, 21–30. Available from: https://doi.org/10.1099/00221287-147-1-21.

Chen, L., Qiu, Q., Jiang, Y., Wang, K., Lin, Z., Li, Z., et al., 2019. Large-scale ruminant genome sequencing provides insights into their evolution and distinct traits. Science 364, eaav6202. Available from: https://doi.org/10.1126/science.aav6202.

Chiu, L., 2019. Decoupling, commingling, and the evolutionary significance of experiential niche construction. In: Uller, T., Laland, K.N. (Eds.), Evolutionary Causation: Biological and Philosophical Reflections. MIT Press, p. 299.

Chiu, L., Bazin, T., Truchetet, M.-E., Schaeverbeke, T., Delhaes, L., Pradeu, T., 2017. Protective microbiota: from localized to long-reaching co-immunity. Front. Immunol. 8, 1678. Available from: https://doi.org/10.3389/fimmu.2017.01678.

Chiu, L., Gilbert, S.F., 2015. The birth of the holobiont: multi-species birthing through mutual scaffolding and niche construction. Biosemiotics 8, 191–210. Available from: https://doi.org/10.1007/s12304-015-9232-5.

Chu, H., Mazmanian, S.K., 2013. Innate immune recognition of the microbiota promotes host-microbial symbiosis. Nature Immunol. 14, 668–675. Available from: https://doi.org/10.1038/ni.2635.

Christian, N., Whitaker, B., Clay, K., 2015. Microbiomes: unifying animal and plant systems through the lens of community ecology theory. Front. Microbiol. 6, 869. Available from: https://doi.org/10.3389/fmicb.2015.00869.

Costello, E.K., Stagaman, K., Dethlefsen, L., Bohannan, B.J.M., Relman, D.A., 2012. The application of ecological theory toward an understanding of the human microbiome. Science 336, 1255–1262.

Crabtree, B., Holloway, D.E., Baker, M.D., Acharya, K.R., Subramanian, V., 2007. Biological and structural features of murine angiogenin-4, an angiogenic protein. Biochemistry 46, 2431–2443. Available from: https://doi.org/10.1021/bi062158n.

Cussotto, S., Sandhu, K.V., Dinan, T.G., Cryan, J.F., 2018. The neuroendocrinology of the microbiota-gut-brain axis: a behavioural perspective. Front. Neuroendocrinol. 51, 80–101. Available from: https://doi.org/10.1016/j.yfrne.2018.04.002.

Dai, X., Tian, Y., Li, J., Su, X., Wang, X., Zhao, S., et al., 2015. Metatranscriptomic analyses of plant cell wall polysaccharide degradation by microorganisms in the cow rumen. Appl. Environ. Microbiol. 81, 1375–1386. Available from: https://doi.org/10.1128/AEM.03682-14.

Daniels, K.M., Yohe, T.T., 2014. What do we know about rumen development, in: Tri State Dairy Nutrition Conference.

Dawkins, R., 1976. The Selfish Gene. Oxford University Press, Oxford.

De Vadder, F., Grasset, E., Mannerås Holm, L., Karsenty, G., Macpherson, A.J., Olofsson, L.E., et al., 2018. Gut microbiota regulates maturation of the adult enteric nervous system via enteric serotonin networks. Proc. Natl. Acad. Sci. U. S. A. 115, 6458–6463. Available from: https://doi.org/10.1073/pnas.1720017115.

Dominguez-Bello, M.G., 1996. Detoxification in the rumen. Ann. Zootech. 45, 323–327. Available from: https://doi.org/10.1051/animres:19960663.

Donaldson, G.P., Lee, S.M., Mazmanian, S.K., 2015. Gut biogeography of the bacterial microbiota. Nat. Rev. Microbiol. Advance online publication. Available from: https://doi.org/10.1038/nrmicro3552.

Doolittle, W.F., Booth, A., 2016. It's the song, not the singer: an exploration of holobiosis and evolutionary theory. Biol. Philos. 1–20. Available from: https://doi.org/10.1007/s10539-016-9542-2.

Douglas, A.E., Werren, J.H., 2016. Holes in the hologenome: why host-microbe symbioses are not holobionts. mBio 7, e02099-15. Available from: https://doi.org/10.1128/mBio.02099-15.

Dowhaniuk, J.K., Szamosi, J., Chorlton, S., Owens, J., Mileski, H., Clause, R.-F., et al., 2019. Starving the gut: a deficit of butyrate in the intestinal ecosystem of children with intestinal failure. J. Parenter. Enter. Nutr. n/a. Available from: https://doi.org/10.1002/jpen.1715.

Dunbar, H.E., Wilson, A.C.C., Ferguson, N.R., Moran, N.A., 2007. Aphid thermal tolerance is governed by a point mutation in bacterial symbionts. PLoS Biol. 5, 1006–1015.

Eberl, G., 2010. A new vision of immunity: homeostasis of the superorganism. Mucosal Immunol. 3, 450–460.
Eberl, G., Pradeu, T., 2017. Towards a general theory of immunity? Trends Immunol. 39 (4), 261–263.
Flint, H.J., 1997. The rumen microbial ecosystem--some recent developments. Trends Microbiol. 5, 483–488. Available from: https://doi.org/10.1016/S0966-842X(97)01159-1.
Flint, H.J., Bayer, E.A., Rincon, M.T., Lamed, R., White, B.A., 2008. Polysaccharide utilization by gut bacteria: potential for new insights from genomic analysis. Nat. Rev. Microbiol. 6, 121–131. Available from: https://doi.org/10.1038/nrmicro1817.
Flynn, E.G., Laland, K.N., Kendal, R.L., Kendal, J.R., 2013. Target article with commentaries: developmental niche construction. Dev. Sci. 16, 296–313. Available from: https://doi.org/10.1111/desc.12030.
Fraune, S., Bosch, T.C.G., 2007. Long-term maintenance of species-specific bacterial microbiota in the basal metazoan Hydra. Proc. Natl. Acad. Sci. 104, 13146–13151. Available from: https://doi.org/10.1073/pnas.0703375104.
Frazão, N., Sousa, A., Lässig, M., Gordo, I., 2019. Horizontal gene transfer overrides mutation in Escherichia coli colonizing the mammalian gut. Proc. Natl. Acad. Sci. 116 (36), 17906–17915201906958. Available from: https://doi.org/10.1073/pnas.1906958116.
Freeland, W.J., Janzen, D.H., 1974. Strategies in herbivory by mammals: the role of plant secondary compounds. Am. Nat. 269–289.
Funkhouser, L.J., Bordenstein, S.R., 2013. Mom knows best: the universality of maternal microbial transmission. PLoS Biol. 11 (8), e1001631.
Gilbert, H.J., 2010. The biochemistry and structural biology of plant cell wall deconstruction. Plant. Physiol. 153, 444–455. Available from: https://doi.org/10.1104/pp.110.156646.
Gilbert, S.F., 2016. Developmental plasticity and developmental symbiosis: the return of eco-devo. In: Wassarman, P.M. (Ed.), Curr. Top. Developmental Biol. B. Academic Press, Boca Raton, pp. 415–433.
Gilbert, S.F., 2019. Developmental symbiosis facilitates the multiple origins of herbivory. Evol. Dev. 0, e12291. Available from: https://doi.org/10.1111/ede.12291.
Gilbert, S.F. (in press) *Microhabitable.* (Lucia Pietroiusti, Fernando Garcia Dory, ed.).
Gilbert, S.F., McDonald, E., Boyle, N., Buttino, N., Gyi, L., Mai, M., et al., 2010. Symbiosis as a source of selectable epigenetic variation: taking the heat for the big guy. Philos. Trans. R. Soc. B Biol. Sci. 365, 671–678. Available from: https://doi.org/10.1098/rstb.2009.0245.
Gilbert, S.F., Sapp, J., Tauber, A.I., 2012. A symbiotic view of life: we have never been individuals. Q. Rev. Biol. 87, 325–341.
Gould, S.J., 1994. The evolution of life on earth. Sci. Am. 271 (4), 84–911994. Available from: http://brembs.net/gould.html.
Griesemer, J.R., 2014a. Reproduction and scaffolded developmental processes: an integrated evolutionary perspective. In: Minelli, A., Pradeu, T. (Eds.), Towards a Theory of Development. Oxford University Press, pp. 183–202.
Griesemer, J.R., 2014b. Reproduction and the scaffolded development of hybrids. In: Caporael, L.R., Griesemer, J.R., Wimsatt, W.C. (Eds.), Developing Scaffolds in Evolution, Culture, and Cognition. MIT Press, pp. 23–55.
Hagey, J.V., Bhatnagar, S., Heguy, J.M., Karle, B.M., Price, P.L., Meyer, D., et al., 2019. Fecal microbial communities in a large representative cohort of

California dairy cows. Front. Microbiol. 10, 1093. Available from: https://doi.org/10.3389/fmicb.2019.01093.
Harlow, B.E., Goodman, J.P., Lynn, B.C., Flythe, M.D., Ji, H., Aiken, G.E., 2017. Ruminal tryptophan-utilizing bacteria degrade ergovaline from tall fescue seed extract. J. Anim. Sci. 95, 980–988.
Hehemann, J.-H., Correc, G., Barbeyron, T., Helbert, W., Czjzek, M., Michel, G., 2010. Transfer of carbohydrate-active enzymes from marine bacteria to Japanese gut microbiota. Nature 464, 908.
Jami, E., Mizrahi, I., 2012. Similarity of the ruminal bacteria across individual lactating cows. Anaerobe 18, 338–343. Available from: https://doi.org/10.1016/j.anaerobe.2012.04.003.
Kikuchi, Y., Hayatsu, M., Hosokawa, T., Nagayama, A., Tago, K., Fukatsu, T., 2012. Symbiont-mediated insecticide resistance. Proc. Natl. Acad. Sci. U. S. A. 109, 8618–8622. Available from: https://doi.org/10.1073/pnas.1200231109.
Kirsch, R., Gramzow, L., Theißen, G., Siegfried, B.D., ffrench-Constant, R.H., Heckel, D.G., et al., 2014. Horizontal gene transfer and functional diversification of plant cell wall degrading polygalacturonases: key events in the evolution of herbivory in beetles. Insect Biochem. Mol. Biol. 52, 33–50. Available from: https://doi.org/10.1016/j.ibmb.2014.06.008.
Krol, A.J., Roellig, D., Dequéant, M.-L., Tassy, O., Glynn, E., Hattem, G., et al., 2011. Evolutionary plasticity of segmentation clock networks. Dev. Camb. Engl. 138, 2783–2792. Available from: https://doi.org/10.1242/dev.063834.
La Reau, A.J., Suen, G., 2018. The Ruminococci: key symbionts of the gut ecosystem. J. Microbiol. 56, 199–208. Available from: https://doi.org/10.1007/s12275-018-8024-4.
Laland, K.N., Odling-Smee, F.J., Gilbert, S.F., 2008. EvoDevo and niche construction: building bridges. J. Exp. Zoolog. B Mol. Dev. Evol. 310B, 549–566. Available from: https://doi.org/10.1002/jez.b.21232.
Laland, K.N., Odling-Smee, J., Endler, J., 2017. Niche construction, sources of selection and trait coevolution. Interface Focus. 7, 20160147. Available from: https://doi.org/10.1098/rsfs.2016.0147.
Lee, Y.K., Mazmanian, S.K., 2010. Has the microbiota played a critical role in the evolution of the adaptive immune system? Science 330, 1768–1773.
Lewontin, R., 1982. Organism and environment. In: Plotkin, H.C. (Ed.), Learning, Development and Culture. Wiley, Chichester, pp. 151–170.
Lewontin, R., 1983. The organism as the subject and object of evolution. Scientia 118, 63–82.
Lewontin, R., 2000. The Triple Helix: Gene, Organism and Environment. Harvard University Press, Cambridge, MA.
Li, Y., Carrillo, J.A., Ding, Y., He, Y., Zhao, C., Liu, J., et al., 2019. DNA methylation, microRNA expression profiles and their relationships with transcriptome in grass-fed and grain-fed cattle rumen tissue. PLoS One 14, e0214559. Available from: https://doi.org/10.1371/journal.pone.0214559.
Lima, J., Auffret, M.D., Stewart, R.D., Dewhurst, R.J., Duthie, C.A., Snelling, T.J., et al., 2019. Identification of rumen microbial genes involved in pathways linked to appetite, growth, and feed conversion efficiency in cattle. Frontiers Genet 10, 701. Available from: https://doi.org/10.3389/fgene.2019.00701.
Lloyd, E.A., Wade, M.J., 2019. Criteria for holobionts from community genetics. Biol. Theory. 14, 151–170. Available from: https://doi.org/10.1007/s13752-019-00322-w.
Lombard, V., Golaconda Ramulu, H., Drula, E., Coutinho, P.M., Henrissat, B., 2014. The carbohydrate-active enzymes database (CAZy) in 2013. Nucleic

Acids Res. 42, D490–D495. Available from: https://doi.org/10.1093/nar/gkt1178.

Mackie, R.I., McSweeney, C.S., Klieve, A.V., 2002. Microbial ecology of the ovine rumen. In: Freer, M., Dove, H. (Eds.), Sheep Nutr. CABI Publishing, Wallingford, Oxon, pp. 71–94.

Malenke, J.R., Skopec, M.M., Dearing, M.D., 2014. Evidence for functional convergence in genes upregulated by herbivores ingesting plant secondary compounds. BMC Ecol. 14, 23. Available from: https://doi.org/10.1186/1472-6785-14-23.

Malmuthuge, N., Liang, G., Guan, L.L., 2019. Regulation of rumen development in neonatal rumninants through microbial metagenomes and host transcriptomes. Genome Biol. 20, 172. Available from: https://doi.org/10.1186/s13059-019-1786-0.

McFall-Ngai, M., Hadfield, M.G., Bosch, T.C., Carey, H.V., Domazet-Lošo, T., Douglas, A.E., et al., 2013. Animals in a bacterial world, a new imperative for the life sciences. Proc. Natl. Acad. Sci. 110, 3229–3236.

Mizrahi, I., 2013. Rumen symbioses. In: Rosenberg, E., DeLong, E.F., Lory, S., Stackebrandt, E., Thompson, F. (Eds.), The Prokaryotes: Prokaryotic Biology and Symbiotic Associations. Springer Berlin Heidelberg, Berlin, Heidelberg, pp. 533–544. Available from: https://doi.org/10.1007/978-3-642-30194-0_1.

Moraïs, S., Mizrahi, I., 2019. The road not taken: the rumen microbiome, functional groups, and community states. Trends Microbiol. 27, 538–549. Available from: https://doi.org/10.1016/j.tim.2018.12.011.

Moran, N.A., Sloan, D.B., 2015. The hologenome concept: helpful or hollow? PLoS Biol. 13, e1002311. Available from: https://doi.org/10.1371/journal.pbio.1002311.

Morvan, B., Bonnemoy, F., Fonty, G., Gouet, P., 1996. Quantitative determination of H2-utilizing acetogenic and sulfate-reducing bacteria and methanogenic archaea from digestive tract of different mammals. Curr. Microbiol. 32, 129–133. Available from: https://doi.org/10.1007/s002849900023.

Muscatine, L., Falkowski, P.G., Porter, J.W., Dubinsky, Z., 1984. Fate of photosynthetic fixed carbon in light- and shade-adapted colonies of the symbiotic coral *Stylophora pistillata*. Proc. R. Soc. Lond. B Biol. Sci. 222, 181–202.

Newbold, C.J., de la Fuente, G., Belanche, A., Ramos-Morales, E., McEwan, N.R., 2015. The role of ciliate protozoa in the rumen. Front. Microbiol. 6, 1313. Available from: https://doi.org/10.3389/fmicb.2015.01313.

Odling-Smee, F.J., Laland, K.N., Feldman, M.W., 2003. Niche Construction: The Neglected Process in Evolution. Princeton University Press, Princeton, NJ.

Oliver, K.M., Degnan, P.H., Hunter, M.S., Moran, N.A., 2009. Bacteriophages encode factors required for protection in a symbiotic mutualism. Science 325, 992–994.

Pradeu, T., 2012. The Limits of the Self: Immunology and Biological Identity. Oxford University Press, New York.

Román-Palacios, C., Scholl, J.P., Wiens, J.J., 2019. Evolution of diet across the animal tree of life. Evol. Lett. 3, 339–347. Available from: https://doi.org/10.1002/evl3.127.

Rosenberg, E., Zilber-Rosenberg, I., 2016. Microbes drive evolution of animals and plants: the hologenome concept. mBio 7, e01395-15. Available from: https://doi.org/10.1128/mBio.01395-15.

Roth, M.S., 2014. The engine of the reef: photobiology of the coral-algal symbiosis. Front. Microbiol. 5, 422. Available from: https://doi.org/10.3389/fmicb.2014.00422.

Roughgarden, J., Gilbert, S.F., Rosenberg, E., Zilber-Rosenberg, I., Lloyd, E.A., 2017. Holobionts as units of selection and a model of their population dynamics and evolution. Biol. Theory 1–22. Available from: https://doi.org/10.1007/s13752-017-0287-1.

Round, J.L., Lee, S.M., Li, J., Tran, G., Chatila, T.M., Mazmanian, S.K., 2011. The Toll-like receptor 2 pathway establishes colonization by a commensal of the human microbiota. Science 332, 974–977. Available from: https://doi.org/10.1126/science.1206095.

Sander, E.G., Warner, R.G., Harrison, H.N., Loosli, J.K., 1959. The stimulatory effect of sodium butyrate and sodium propionate on the development of rumen mucosa in the young calf. J. Dairy. Sci. 42, 1600–1605. Available from: https://doi.org/10.3168/jds.S0022-0302(59)90772-6.

Sasson, G., Ben-Shabat, S.K., Seroussi, E., Doron-Faigenboim, A., Shterzer, N., Yaacoby, S., et al., 2017. Heritable bovine rumen bacteria are phylogenetically related and correlated with the cow's capacity to harvest energy from its feed. mBio 8, e00703–e00717. Available from: https://doi.org/10.1128/mBio.00703-17.

Six, D.L., 2019. A major symbiont shift supports a major niche shift in a clade of tree-killing bark beetles. Ecol. Entomol. . Available from: https://doi.org/10.1111/een.12786.

Skillings, D., 2016. Holobionts and the ecology of organisms: multi-species communities or integrated individuals? Biol. Philos. 31, 875–892. Available from: https://doi.org/10.1007/s10539-016-9544-0.

Smith, G.S., 1992. Toxification and detoxification of plant compounds by ruminants: an overview. Rangel. Ecol. Manag. Range Manag. Arch. 45, 25–30.

Stappenbeck, T.S., Hooper, L.V., Gordon, J.I., 2002. Developmental regulation of intestinal angiogenesis by indigenous microbes via Paneth cells. Proc. Natl. Acad. Sci. U. S. A. 99, 15451–15455. Available from: https://doi.org/10.1073/pnas.202604299.

Stotz, K., 2017. Why developmental niche construction is not selective niche construction: and why it matters. Interface Focus. 7, 20160157. Available from: https://doi.org/10.1098/rsfs.2016.0157.

Suárez, J., 2018. The importance of symbiosis in philosophy of biology: an analysis of the current debate on biological individuality and its historical roots. Symbiosis 76, 77–96. Available from: https://doi.org/10.1007/s13199-018-0556-1.

Suárez, J., Triviño, V., 2019. A metaphysical approach to holobiont individuality: holobionts as emergent individuals. Quad. Filos. 6, 59. Available from: https://doi.org/10.7203/qfia.6.1.14825.

Sudakaran, S., Kost, C., Kaltenpoth, M., 2017. Symbiont acquisition and replacement as a source of ecological innovation. Trends Microbiol. 25, 375–390. Available from: https://doi.org/10.1016/j.tim.2017.02.014.

Sultan, S.E., 2015. Organism and Environment: Ecological Development, Niche Construction, and Adaptation. Oxford University Press, New York.

Tamate, H., McGilliard, A.D., Jacobson, N.L., Getty, R., 1962. Effect of various dietaries on the anatomical development of the stomach in the calf. J. Dairy. Sci. 45, 408–420. Available from: https://doi.org/10.3168/jds.S0022-0302(62)89406-5.

Tauber, A.I., 2017. Immunity: The Evolution of an Idea. Oxford University Press, Oxford, New York.

Taxis, T.M., Wolff, S., Gregg, S.J., Minton, N.O., Zhang, C., Dai, J., et al., 2015. The players may change but the game remains: network analyses of ruminal microbiomes suggest taxonomic differences mask functional similarity.

Nucleic Acids Res. 43, 9600–9612. Available from: https://doi.org/10.1093/nar/gkv973.

True, J.R., Haag, E.S., 2001. Developmental system drift and flexibility in evolutionary trajectories. Evol. Dev. 3, 109–119. Available from: https://doi.org/10.1046/j.1525-142x.2001.003002109.x.

Tsuchida, T., Koga, R., Horikawa, M., Tsunoda, T., Maoka, T., Matsumoto, S., et al., 2010. Symbiotic bacterium modifies aphid body color. Science 330, 1102–1104. Available from: https://doi.org/10.1126/science.1195463.

Vermeij, G.J., Lindberg, D.R., 2000. Delayed herbivory and the assembly of marine benthic ecosystems. Paleobiology 26, 419–430. Available from: https://doi.org/10.1666/0094-8373(2000)026 < 0419:DHATAO > 2.0.CO;2.

von Uexküll, J., 1934. A Stroll Through the Worlds of Animals and Men: A Picture Book of Invisible Worlds, 2010th ed. University of Minnesota Press, Minneapolis, MN.

Wallace, R.J., Sasson, G., Garnsworthy, P.C., Tapio, I., Gregson, E., Bani, P., et al., 2019. A heritable subset of the core rumen microbiome dictates dairy cow productivity and emissions. Sci. Adv. 5, eaav8391. Available from: https://doi.org/10.1126/sciadv.aav8391.

Walsh, D.M., 2015. Organisms, Agency, and Evolution. Cambridge University Press, Cambridge.

Wielkopolan, Obrepalska-Steplowska, 2016. Three-way interaction among plants, bacteria, and coleopteran insects. Planta 313–332.

Wilson, E.B., *The Cell in Development and Inheritance*. Macmillan, NY.

Wu, S., Li, R.W., Li, W., Li, C.-J., 2012. Transcriptome characterization by RNA-seq unravels the mechanisms of butyrate-induced epigenomic regulation in bovine cells. PLoS One 7, e36940. Available from: https://doi.org/10.1371/journal.pone.0036940.

Yuille, S., Reichardt, N., Panda, S., Dunbar, H., Mulder, I.E., 2018. Human gut bacteria as potent class I histone deacetylase inhibitors in vitro through production of butyric acid and valeric acid. PLoS One 13, e0201073. Available from: https://doi.org/10.1371/journal.pone.0201073.

Zhang, W., Liu, W., Hou, R., Zhang, L., Schmitz-Esser, S., Sun, H., et al., 2018. Age-associated microbiome shows the giant panda lives on hemicelluloses, not on cellulose. ISME J. 12, 1319–1328. Available from: https://doi.org/10.1038/s41396-018-0051-y.

Zilber-Rosenberg, I., Rosenberg, E., 2008. Role of microorganisms in the evolution of animals and plants: the hologenome theory of evolution. FEMS Microbiol. Rev. 32, 723–735. Available from: https://doi.org/10.1111/j.1574-6976.2008.00123.x.

18

Nature, nurture, and noise in bird song ontogeny as determinants of phenotypic and functional variation among dialects

Jill A. Soha
Duke University, Durham, NC, United States

Introduction: Song function, ontogeny, and phenotypic variation

Functions of song

Songbirds use a variety of vocalizations to communicate with conspecifics. These vocalizations vary in acoustic structure and in the types of information they convey. By convention, the terms *call* and *song* denote functionally distinct vocalizations. Calls serve predominantly nonmating functions such as maintaining flock cohesion, indicating alarm, and signaling hunger. Songs, in contrast, serve functions directly related to mating, including mate attraction, pair bonding, and territory defense. To fully understand song dialect variation, it helps first to understand these functions of song.

To reproduce successfully, territorial songbirds must defend an area sufficient to provide the resources needed to rear young. Singing constitutes much of the behavior performed toward this end. As a territorial signal, song indicates to conspecifics that a given area is occupied by an individual (or pair) that is motivated to defend that territory. Potential rivals should benefit from using this information when deciding whether to move into an area. This function of song, long suspected from behavioral observation, has been confirmed by two types of experiments (Catchpole and Slater, 2008). Muting experiments show that males whose ability to sing is temporarily removed

Phenotypic Switching. DOI: https://doi.org/10.1016/B978-0-12-817996-3.00013-X

experience greater rates of territorial invasion by rival males (Smith, 1979; McDonald, 1989). Speaker replacement experiments show that after a male is removed from his territory, recordings of his song broadcast from speakers will delay invasion by rivals (Krebs, 1977; Falls, 1987; Nowicki et al., 1998). To learn more about how birds use song during territorial interactions, playback experiments are used to examine phenomena such as the matching of song types (Beecher et al., 2000) or temporal overlapping of songs (Naguib and Kipper, 2006) between rivals, and the use of low-amplitude song to signal aggressive motivation (Akçay et al., 2015).

Mate attraction is another function of song. To some extent, both territory defense and mate attraction may be expected to involve the same information (encoded in song) about the resource holding potential of the singer. In species in which males claim territories prior to pairing, females should benefit by choosing males occupying high-quality territories. It is hypothesized that females benefit further by mating with males possessing individual qualities that increase the female's own reproductive success. Many studies have aimed to define these qualities and investigate the extent to which song conveys information about them. For example, male longevity—which might indicate good genes as well as greater parental experience—is encoded in the song of blue tits (*Cyanistes caeruleus*), as older males begin singing earlier in the morning than do younger ones (Poesel et al., 2006). As another example, male swamp sparrow (*Melospiza georgiana*) song conveys information about individual developmental history, and females prefer the songs of males with higher early growth rates over those of males that have grown more slowly due to the impacts of nutritional stress (Searcy et al., 2010).

Regardless of the information that females obtain from male song, its basic function in mate attraction has also been confirmed in field experiments. Male seaside sparrows (*Ammodramus maritimus peninsulae*) that are temporarily muted fail to attract females, but they regain the ability to do so once their voice recovers (McDonald, 1989). Females of multiple species are more readily attracted to nest boxes from which male song is broadcast through a speaker than to nest boxes without song (Eriksson and Wallin, 1986; Mountjoy and Lemon, 1991; Johnson and Searcy, 1996). In addition to attracting females, male song serves to stimulate their reproductive behavior. In captive canaries, playback of male song triggers female nest-building behavior (Hinde and Steele, 1976), and in a number of species, song elicits copulation solicitation displays in females with sufficiently high levels of estradiol (King and West, 1977, Searcy and Marler, 1981).

The above functions of song—territory defense, mate attraction, and mate stimulation—apply to male song, and the experiments described earlier have been conducted in species in which males are the main singers. Females of most songbird species worldwide also sing, and singing in both sexes is ancestral (Odom et al., 2014). The functions of female song in various species include territory defense, defense of nesting sites or mates against other females, coordination of breeding and parental care, and less commonly, mate attraction (Langmore, 1998). Although female song function is increasingly well documented, we know very little about female song ontogeny or phenotypic variation in female song. The rest of this chapter is therefore primarily based on what we know about male song, but many of the points made are likely to apply to female song as well.

Song ontogeny

Young songbirds are predisposed to learn the song of their own species, but in almost all species studied to date, they must hear models in order to do so. For example, in New World sparrows of the family *Passerellidae*, males that hear no song models when young develop idiosyncratic songs containing only a subset of the normal species-typical features (song sparrow *Melospiza melodia* and swamp sparrow: Marler, 1984, grasshopper sparrow *Ammodramus savannarum*: Soha et al., 2009, white-crowned sparrow *Zonotrichia leucophrys*: Soha, 2017). Thus some properties of song are genetically encoded in the individual whereas others must be obtained from the environment. These and other internal and external factors affecting song ontogeny are discussed later in the chapter. The key point here is that although the relative influence of genes and environment on song ontogeny vary across species, both are known to be important factors.

The process of song learning includes two components: sensory learning, in which a bird hears and memorizes songs produced by adults of its species, and sensorimotor learning, in which the bird rehearses its own vocalizations and gradually matches them to internal models using auditory feedback. Normal song development fails to occur if a young bird either hears no song models or cannot hear itself vocalize (Konishi, 1965). During sensorimotor learning, a young bird progresses from subsong, which bears no resemblance to adult song, through plastic song containing increasingly recognizable species-typical notes and phrases in variable sequences, to crystallized adult song that is consistent from one rendition to the next. This progression has been documented in several distantly related species (Hultsch and Todt, 2004) and particularly well described in swamp sparrows (Marler and Peters, 1982) and zebra finches *Taeniopygia guttata* (Tchernichovski et al., 2001). Song development also often involves overproduction and selective attrition, such that more song types are memorized and rehearsed than are retained in the final repertoire (Peters and Nowicki, 2017).

The neural mechanisms underlying song learning are well studied. The primary model species for these studies has been the zebra finch, although the specialized neural circuit referred to as the song system was first discovered in the canary (*Serinus canaria*; Nottebohm et al., 1976) and has been documented at least to some extent in dozens of other species (DeVoogd et al., 1993, Brenowitz, 1997). The song system includes several discrete nuclei (clusters of neurons) in two interconnected functional pathways. The motor pathway receives input from auditory areas and controls the muscles of the vocal organ (the syrinx), and the anterior forebrain pathway modulates output based on feedback from the bird's own vocalizations. During song learning, changes occur in several anatomical, neurochemical, and electrophysiological properties of the song system (Ölveczky and Gardner, 2011). For example, the acoustic selectivity of neurons in multiple song nuclei changes during development (Volman, 1993; Solis and Doupe, 1997) as does the relative strength of projections from the two pathways into nucleus RA, where the pathways converge (Mooney, 2009). The extent, timing and exact nature of changes in the song system presumably vary across species, but both correlative and manipulative studies in model species indicate that these changes underlie sensory and sensorimotor development (Brainard and Doupe, 2002). The genetic underpinnings of the song system and its ontogeny are less well understood but are increasingly under study (Scharff and Adam, 2013).

Patterns of phenotypic variation

The process of song learning results in songs that vary among species, among individuals within a species, and often even within individuals (Fig. 18.1). Within-individual variation occurs most obviously in species with song repertoires, in which each individual sings multiple songs that differ in acoustic structure but typically serve the same set of functions (Fig. 18.1A–D). Repertoire sizes vary across species: repertoires of fewer than 10 songs are common, but in some species such as Bachman's sparrows (*Peucaea aestivalis*) and European robins (*Erithacus rubecula*), individuals sing dozens or even hundreds of songs. In a few species, songs not only differ in acoustic structure but also appear to fall into two functional categories. These categories might serve intrasexual and intersexual functions, respectively, or one might serve both territory defense and mate attraction functions while the other facilitates pair bonding or mate stimulation (Lohr et al., 2013). The number of songs within each category can also vary across species: grasshopper sparrows sing one song of each category (Fig. 18.1E and F; Lohr et al., 2013), whereas chestnut-sided warblers sing up to 4 songs of one category and 10 of the other (Lein, 1978).

Among species, the acoustic structure of song varies consistently enough that song encodes information about species identity, crucial in both intrasexual and intersexual communication. From an evolutionary perspective, both sexual selection and other forms of natural selection contribute to these species differences (e.g., Mahler and Gil, 2009). Mechanistically, anatomical and physiological features broadly limit song structure (Podos et al., 2004), but species differences in song are not fully attributable to such constraints: many species naturally mimic others or have done so in laboratory conditions (e.g., white-crowned sparrows: Soha and Marler, 2000). Developmental factors including genetically encoded learning predispositions (Marler, 1984, Wheatcroft and Qvarnstrom, 2017) and the acoustic and social environments experienced by young birds more strongly affect song structure and its variation across species. Divergence in song structure has been theorized to serve as a mechanism of reproductive isolation between incipient species (Irwin and Irwin, 2004; Podos, 2010), and song variation reflects phylogeny to some extent (e.g. Price and Lanyon, 2002). However, because cultural evolution facilitates rapid divergence, song is of limited use as a taxonomic character. Freeman and Montgomery (2017) found that song divergence between allopatric populations does not necessarily reflect the extent of discrimination by birds themselves, suggesting that song divergence is not always a reliable indicator of reproductive isolation.

Between the interspecies and intraindividual levels of phenotypic variation is variation in song among individuals within a species (Fig. 18.1G–J). In one sense, this is ubiquitous: even

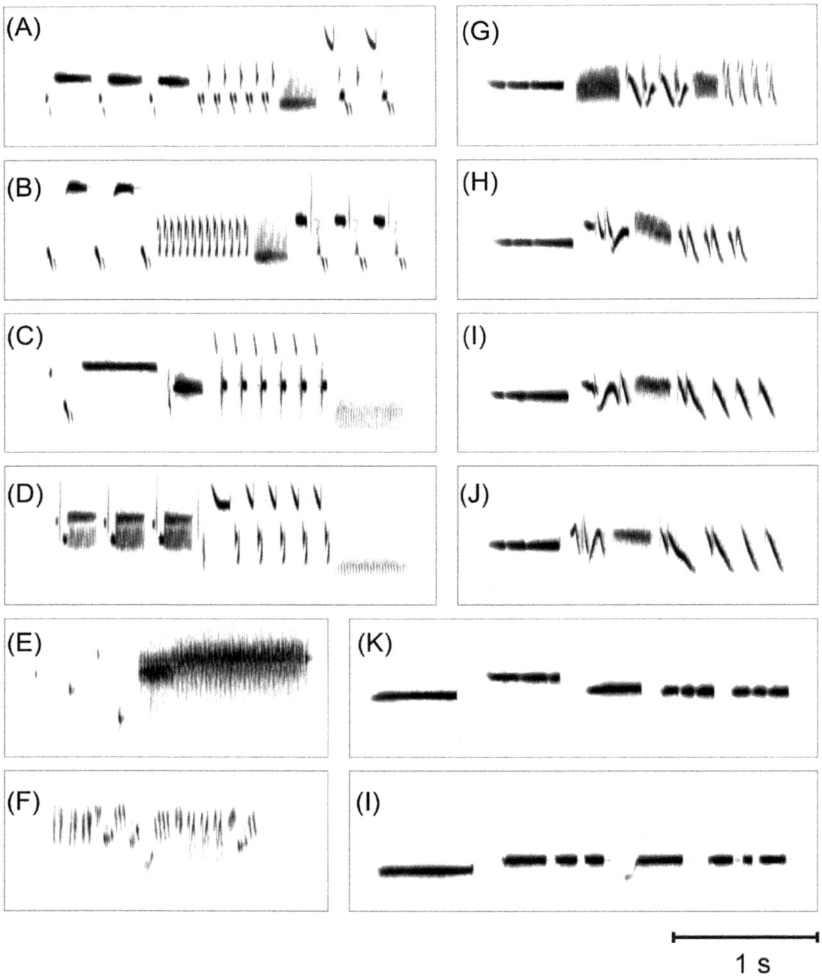

Figure 18.1 Spectrograms illustrating phenotypic variation in songs of New World sparrows (family Passerellidae): (A–D) four songs of one male song sparrow; (E) buzz song and (F) warble song of grasshopper sparrow; (G–I) three Puget Sound white-crowned sparrow song dialects, each from a different bird; (J) variant from a fourth bird of the dialect in (I) with a different complex syllable; (K–L) songs of two different white-throated sparrows. Vertical axis represents frequency: panels (E) and (F) show 2–12 kHz; all others show 0–10 kHz. Songs are excerpted from recordings in the Borror Laboratory of Bioacoustics archive, https://blb.osu.edu/archive: (A–D) BLB8142, (E) BLB28558, (F) BLB28274, (G) BLB25952, (H) BLB25707, (I) BLB24944, (J) BLB24927, (K) BLB5956, and (L) BLB27848.

when two individuals sing the same song, small differences exist in precise structure such that this song encodes individual identity. Typically, however, not all individuals of a species sing the same song. Within a population, variation among individuals can take different forms. Chipping sparrow (*Spizella passerina*) males each sing only one of several songs that exist in a patchwork distribution in the population (Liu and Kroodsma, 2006). A more common form of within-population variation, seen in species with repertoires of multiple songs, is sharing among individuals of some songs and not others. Song sparrow males each sing several songs, and at least in western North America, a portion of these songs is shared between neighboring males (Hill et al., 1999). This enables males to use either matching or nonmatching songs when countersinging, and this choice can convey information about aggressive motivation (Akçay et al., 2013).

Songs typically vary among populations of a given species, and this variation too can take different forms. Song features can change gradually with geographic distance, or stepwise transitions in song structure can occur between one area and another. The term *dialect* denotes regionally distinct songs, each sung by multiple individuals, separated by such stepwise transitions. Boundaries between dialects might reflect areas of unsuitable habitat, but this is not always the case. Dialects occur in some species that sing only a single song and in some with multisong repertoires (Podos and Warren, 2007). The communicative functions of the songs in each dialect are the same; within the local population, these songs serve the territorial and mating functions described earlier. In the next section, I describe how song ontogeny gives rise to phenotypic variation in song within a species. In the third section, I discuss song dialects in particular, examining how these can be considered functional polymorphisms from a receiver perspective and how these polymorphisms depend on receiver ontogeny.

How phenotypic variation arises: plasticity in song development

Extent of plasticity

The extent of plasticity in their vocal development distinguishes songbirds from most other nonhuman animals. The capacity for song learning gives rise to phenomena such as song matching between individuals, dialect formation, increased vocal complexity, and larger repertoire size, but whether or not these phenomena currently serve adaptive functions, they are unlikely to explain the initial evolution of plasticity in song development. Nottebohm (1991) speculated that this plasticity evolved from a more basic auditory feedback mechanism, present in animals that both produce loud sounds and have sensitive hearing, that prevents the production of sounds damaging to the cochlear cells of the producer. Hansen (1979) suggested that the ability to match those songs that most effectively penetrate the local environment might have been the initial adaptive benefit of imitative song learning. Lachlan and Slater (1999) argued that once song learning has evolved, it could be maintained by an evolutionary trap resulting from the interaction between culturally transmitted songs and the genes underlying the learning and recognition of those songs. Regardless of its evolutionary history and current functions, plasticity in song development is not unlimited. Songbirds are more or less restricted in what they will learn and when.

Restrictions on the acoustic parameters of songs learned by individuals of each species maintain species distinctiveness in song. These restrictions are not equally narrow in all species, however. For example, the acoustic diversity of learned song is lower in white-throated sparrows (*Zonotrichia atricapillus*), which sing only whistles (Fig. 18.1K and L), than in song sparrows, which sing a variety of trills, buzzes and note complexes (Fig. 18.1A–D). I will refer to the extent of variation in the acoustic parameters learned by a given species as *broadness*. Broadness is likely to be higher in species with larger repertoires and species such as Northern mockingbirds (*Mimus polyglottus*) that regularly mimic other species or environmental sounds. Broadness varies even among mimics, however. Northern mockingbirds preferentially mimic sounds that are similar to the nonimitative songs in this species' repertoire (Gammon, 2013), whereas European starlings (*Sturnus vulgaris*) imitate a wide variety of sounds in their warbled song (Hausberger et al., 1991). As heterospecific mimicry illustrates, the acoustic parameters of sounds that different species will learn can overlap. Restrictions on how mimicked songs are delivered within a repertoire, as well as the inclusion of species-specific songs, maintain species distinctiveness in these cases.

In addition to being limited in the acoustic dimension, plasticity in song ontogeny can be restricted temporally. Many species learn songs prior to sexual maturity and do not alter their repertoires thereafter (e.g., chaffinch *Fringilla coelebs*: Thorpe, 1958; zebra finch: Immelmann, 1969; song sparrow: Marler and Peters, 1987). Song learning in these species is described as age-limited or closed-ended. In other species, learning is open-ended and can occur into adulthood (brown-headed cowbirds *Molothrus ater*: Rothstein and Fleischer, 1987; European starlings: Eens et al., 1992). The rate of learning may decrease with increasing age in these cases, however (Eens et al., 1992). Interestingly, even in closed-ended learners, both components of the song learning process—sensory learning and sensorimotor learning—continue to function at some level after song crystallization. Even birds that do not produce new songs in adulthood may continue to memorize songs produced by others and use them in individual recognition (McGregor and Avery, 1986). Adult songbirds also use auditory feedback to maintain the structure of their songs, although the extent to which this occurs varies across species and decreases with age (Konishi, 2004). In addition, in closed-ended learners, the sensitive period can be extended somewhat by insufficient exposure to appropriate song models (Doupe and Kuhl, 1999, p. 617). Temporal limitations on vocal plasticity should therefore be seen as graded rather than absolute.

External influences

A key external influence on song learning is the set of song models that a young bird hears during the sensitive period. When young birds accurately imitate songs of adult conspecifics, as is common, the songs learned directly reflect those heard. Even if young birds do not copy song models exactly, songs heard early in life can still influence learning. Grasshopper sparrows do not learn their species-typical buzz songs through accurate imitation, but in a laboratory tutoring experiment, the buzz songs learned by young birds were more similar on average to the set of tutor songs they had heard than to a set of novel songs (Soha et al., 2009). In species with repertoires of multiple songs, birds that hear a larger variety of songs when young might learn more songs, as has been demonstrated in marsh wrens (*Cistothorus palustris*; Brenowitz et al., 1995). Independent of repertoire size, young birds might also preferentially imitate models of which they hear more different renditions (as suggested by results in song sparrows: Nowicki et al., 1999), as these might represent songs that are shared by multiple adults in the local population. Puget Sound white-crowned sparrows (*Z. l. pugetensis*) tend to learn the most common song variant in an area (Nelson and Poesel, 2014), and this could reflect such a mechanism, or it might simply result from imitating the song variant that is heard most often overall.

Other sounds, as well as the effects of the environment on sound transmission, might have more subtle effects on song learning. Songs or song components that are overlapped by environmental noise might be more difficult for young birds to hear, or to hear accurately, and therefore might be learned incorrectly if at all. A shift to higher frequencies has been documented in the songs of many songbird species in urban environments (Brumm and Zollinger, 2013), and an effect of low-frequency anthropogenic noise on song learning might partially explain this shift (but see Zollinger et al., 2017). That sound transmission through the environment can affect song learning has been demonstrated in a laboratory tutoring experiment with swamp sparrows. Young males tutored with both undegraded songs and songs that had been degraded by transmission through typical swamp sparrow breeding habitat selectively imitated the undegraded songs (Peters et al., 2012). Over generations, environmental effects on song transmission are hypothesized to cause the evolution of differences in song structure among species, subspecies, or even dialects (Morton, 1975, Nottebohm, 1975, Boncoraglio and Saino, 2007).

Social interaction with conspecifics is an important external influence that can affect song learning in multiple ways. In some species, males preferentially imitate models sung by tutors with whom they have more frequent or more aggressive interactions (zebra finches: Clayton, 1987; song sparrows: Akçay et al., 2017). Interactions with territorial neighbors during territory establishment can guide selective attrition at the end of song development, affecting which song or songs a first-year male retains in his repertoire, as documented in field sparrows (Nelson, 1992) and white-crowned sparrows (Nelson, 2000). Interactions among young birds can affect song ontogeny, as imitation of complete song in zebra finches is inhibited by male siblings (Tchernichovski and Nottebohm, 1998). Social interaction with adult females can also affect song development: visual feedback from females influences song rehearsal in young male brown-headed cowbirds (West and King, 1988) and zebra finches (Caruso-Peck and Goldstein, 2019). The extent to which each of these effects occurs likely varies across species.

Another potential external influence on song development is stress imposed by environmental conditions. Insufficient provisioning of nestlings or fledglings can result in nutritional stress during the period of song learning (Nowicki et al., 2002). The neural processes that underlie song learning use metabolic energy, and other physiological processes more crucial for survival, such as the immunological response (Spencer et al., 2005), are expected to take precedence for support if energetic resources are limited. If this occurs, song learning might suffer (Nowicki et al., 2002). Another mechanism by which environmental stressors might affect song development is through increased corticosterone (Buchanan et al., 2003). Laboratory studies have found detrimental effects of developmental stress on aspects of learned song including imitation accuracy (Holveck et al., 2008), repertoire size (Spencer et al., 2005, Schmidt et al., 2013), and song bout length (Buchanan et al., 2003).

Internal influences

Young songbirds are predisposed to learn species-typical song, and the genes underlying this predisposition constitute perhaps the most significant internal influence on the trajectory of song development. The possibility that selective song learning is guided by exposure to conspecific song early in life, rather than by genetic predispositions, has been refuted. Wheatcroft and Qvarnstrom (2017) found that nestling pied flycatchers (*Ficedula hypoleuca*) and collared flycatchers (*F. albicollis*) preferentially respond to conspecific song even after being cross-fostered as 1-to-3-day-old embryos, several days before hatching, and thereafter raised by heterospecific parents. This provides strong evidence for genetic control of the predisposition to learn conspecific song. The identity of the genes involved and the mechanisms by which they influence song learning in any given species remain unknown. However, a first step has been made toward answering these questions in Bengalese finches (*Lonchura striata domestica*), as genes underlying heritable variation in preferred song tempo are being identified in this species (D. Mets and M. Brainard, personal communication). Species-typical songs are described not only by tempo, of course, but also by frequency and amplitude and the patterns of their modulation, and thus presumably many genes together encode song learning predispositions.

Despite this expected complexity, genes do not fully encode songs in oscine songbirds. If they did, learning would be unnecessary. Rather, in these species, genes encode rudimentary neural representations of conspecific song, referred to as song templates (Marler, 1984). These templates direct young birds to pay attention to conspecific songs during the sensitive period, facilitate memorization of such songs, guide rehearsal and refinement during sensorimotor learning, and contribute to selective retention and crystallization of certain songs in the final repertoire (Marler, 1984, Soha, 2017). By thus limiting what a young bird is likely to learn, templates restrict plasticity. Importantly, however, these templates do not specify precise song structure, but allow a bird to develop any of a range of species-typical songs. The degree of potential variation in song structure allowed by templates represents the dimension of plasticity referred to above as broadness.

In addition to genetically encoded neural substrates, dynamic physiological factors influence the trajectory of song development and limit its plasticity. Steroid hormones have received particular attention as factors that restrict plasticity by affecting when each phase of song learning begins and ends. Androgen and estrogen receptors are present in some song system nuclei, enzymes that metabolize circulating androgens are present in the brain, and some stereoidogenesis occurs in the brain itself (Schlinger and Brenowitz, 2017). Estradiol appears to facilitate early sensory learning (i.e., song memorization) and increased levels of testosterone contribute to song crystallization at the end of sensorimotor learning (Bottjer and Johnson, 1997). It is simplistic to assume that estradiol alone controls the timing of the sensory period and that testosterone is the sole constraint on the sensorimotor period, but other influences on the timing of song learning, such as social factors, adequate exposure to song models, and the experience of song rehearsal itself (Doupe and Kuhl, 1999) might exert their effects in part by modulating hormone levels.

The physiological mechanisms underlying motivation, attention, and reward are also important internal factors in song development. Marler (1970) hypothesized that the process of song learning is intrinsically rewarding, as young white-crowned sparrows can learn from audio playback of recordings with no live tutor or other conspecifics present. Zebra finches, by contrast, require social interaction with a tutor in order to learn songs well. The mechanism by which social interaction facilitates learning in this species might involve external reward in the form of social feedback (Caruso-Peck and Goldstein, 2019) but likely also involves increased arousal and attention. The importance of attention was confirmed by Chen et al. (2016), who documented better imitation of tutors in young zebra finches that paid more attention during tutoring. The mechanisms of motivation and endogenous reward in song ontogeny require further study, but song practice in both young and adult birds might involve endogenous opioid and cannabinoid rewards (Riters et al., 2019). In adults, dopamine is thought to regulate the motivation to sing (Riters, 2012). Dopamine might also contribute to reinforcement learning and song plasticity during development. A dopaminergic projection exists from the ventral tegmental area to the song system nucleus

Area X (Lewis et al., 1981), and this projection develops over the age range during which song learning occurs in zebra finches (Soha et al., 1996, Harding et al., 1998). Fee and Goldberg (2011) hypothesized that reinforcement signals transmitted via this projection contribute to modulating subsequent vocal output. According to this hypothesis, fine-scale plasticity during song ontogeny depends on endogenous reward signaling in the song system.

Stochasticity

The external and internal influences described earlier act not in isolation but in combination during song development. The interactions between these factors can be complex, making the precise course of song learning unpredictable. Even a relatively simple example illustrates how song ontogeny depends on these interactions. Appropriate conditions of motivation and attention, reflecting the internal state of a young bird, must coincide with environmental conditions that facilitate song learning, including the occurrence of appropriate song models. It is unlikely that the precise time course of fluctuations in motivation and attention is genetically and/or environmentally determined. For this reason alone, we would not necessarily expect genetically identical individuals raised under identical environmental conditions, presented with the same array of song models to choose from, to learn the same songs.

At a reductionist level, this is due to stochasticity in physiological processes such as cellular metabolism and neural activity, both in the song system and in other areas of the brain. In other organisms, stochasticity has been documented in subcellular processes including gene expression and protein conformation change, as well as in broader-scale processes such as stem cell differentiation, axon growth, and intercellular communication including synaptic transmission (Balázsi et al., 2011; Clarke, 2012). Such stochasticity is increasingly believed to represent a third source of variation in organismal development, along with genetic and environmental influences. In the realm of behavior, differences in behavioral tendencies have been observed among cloned or parthogenetic animals raised in standardized environments (Archer et al., 2003; Vogt et al., 2008; Schuett et al., 2011), and stochastic developmental variation is believed to contribute to these differences (Vogt, 2015).

In the case of bird song development, stochastic processes could potentially have different effects during the sensory and the sensorimotor phases of learning. Stochastic variation during sensory learning might affect which of the available song models a bird memorizes, leading hypothetical songbird clones in identical environments to commit different models to memory. Stochastic variation during sensorimotor learning might in turn affect which of the memorized song models are rehearsed and retained, such that even clones that memorize the same set of songs might end up with different songs in their adult repertoires. Even if two clones were to memorize, rehearse, and retain the same song types, individual differences in their renditions of these songs would be expected to result from differences in the accuracy of memorization and differences in production and auditory feedback processing during rehearsal, also due to stochastic physiological effects.

Birds have not yet been cloned, due to the difficulty of implanting a partially developed embryo into an avian egg that is itself at the correct developmental stage. Twins, and in particular identical twins, can occur in birds but are uncommon (Bassett et al., 1999). The closest we have come to empirically addressing the question of whether songbird clones would learn the same songs has been to ask whether nestmates raised in a controlled environment do so. Nelson and Marler (2005) analyzed data from 35 pairs of white-crowned sparrow nestmates and found that even when raised in similar laboratory conditions and exposed to the same set of tutor songs, nestmates learned the same songs no more often than did nonnestmates. Nestmates are not genetically identical, however, and if extra-pair fertilizations occur (as is true in white-crowned sparrows and many other species), they might not even be full siblings. If it becomes possible in the future to study song ontogeny in genetically identical songbirds, we can expect that stochastic processes will result in variation in developmental outcomes, both for the reasons given here and also because stochastic developmental variation has already been documented in a variety of other taxa (reviewed in Vogt, 2015).

Results of plasticity

Sensitivity of song ontogeny to the influences described earlier, along with the likely contribution of developmental stochasticity, results in variation among individuals in a few dimensions of the song phenotype. The most obvious of these is the particular songs that are learned. This is especially evident in species with multisong repertoires, but even in species in which each individual sings only a single song type, multiple song types or variants might occur in the environment experienced by a young bird and thus serve as potential models for imitation. Puget Sound white-crowned sparrow dialects are identified by distinct terminal trill phrases, but within each dialect, different note complexes occur (Fig. 18.1I and J; Nelson et al., 2004). Even if a young male of this subspecies hears only one dialect during sensory learning, he is likely to hear multiple versions of that dialect containing different note complexes and might learn any of them.

Repertoire size is another feature that can vary among individuals due to plasticity in ontogeny. This too is most evident in species with multisong repertoires, such as song sparrows (5–13 songs each: Arcese et al., 2002) or marsh wrens (about 30–70 songs each in Eastern US populations: Kroodsma and Verner, 2013). However, even in species in which most individuals sing just one song, variation can occur. A small percentage of Puget Sound white-crowned sparrows retain two or more songs in their final repertoire. Interestingly, Nelson and Poesel (2014) found an inverse correlation between repertoire size and imitation accuracy, suggesting a tradeoff between the two in this subspecies.

In other species too, imitation accuracy can vary among individuals as a result of developmental plasticity. Nelson and Poesel (2014) suggested that the amount of time spent rehearsing a song during the sensorimotor phase might influence imitation accuracy. Developmental stress before or during the sensory phase has been found to affect imitation accuracy in swamp sparrows (Nowicki et al., 2002) and zebra finches (Holveck et al., 2008). Imitation accuracy might therefore signal male quality in some cases (Lachlan et al., 2014). In theory, however, lower imitation accuracy increases the utility of song in individual recognition. These conflicting pressures might explain why in Puget Sound white-crowned sparrows, different parts of song are imitated with different levels of accuracy (Nelson et al., 2004). More generally, to the extent that individual male quality is reflected in song features that vary as a result of plasticity during development (e.g., in response to developmental stress) and if these song features affect female mate choice, then sexual selection should promote the evolution of mechanisms that increase resilience of these features to stochastic developmental effects. Aspects of song that do not reflect male quality, such as (presumably) the particular subset of species-typical songs that constitute an individual repertoire, might be more sensitive to stochastic effects.

Song dialects as functional polymorphisms

How do dialects arise?

Above I described the mechanisms of plasticity that result in song variation among individuals. At the population level, this variation will take the form of dialects if cultural evolution results in two things: similarity in song among birds within an area, and the existence of different song forms in different areas. Mechanistically, similarity in songs within an area requires that birds learn accurate imitations of song models and that they do so within a limited distance from where they eventually set up their own breeding territory. The existence of multiple dialects, by contrast, requires some deviation from these criteria to have occurred at some point. The establishment of new dialects is presumably infrequent, as it requires the cultural transmission of inaccurate copies (new variants) to subsequent generations. This could happen if a few first-year birds settle together in an area that is isolated from the existing population, if at least one of these birds develops an inaccurate copy of the song heard at its natal site, and if breeding among these few birds results in offspring that imitate the new song variant. Alternatively, gradual dialect divergence through acoustic adaptation to different habitats might occur.

Several hypotheses have been proposed regarding the possible adaptive function of dialects. The earliest of these suggested that dialects serve as a signal of genetic adaptation to a local area (Marler and Tamura, 1962, Nottebohm, 1969). According to this hypothesis, individuals benefit from breeding in their natal area with a mate from the same natal area because this increases the likelihood that their offspring will be optimally adapted to the local environment. Subsequent hypotheses focused on social rather than genetic adaptation, proposing that locally typical songs indicate successful navigation of the local social environment by the singer. Adaptation to the social environment might mean, for example, that an individual is less likely to experience aggression from dominant individuals (Payne, 1981). If social adaptation means that an individual is more likely to successfully defend a territory, this information is useful to potential mates as well. A more recent hypothesis, falling somewhat between the genetic and social adaptation hypotheses, is that dialects indicate adaptation to local parasites (MacDougall-Shackleton et al., 2002). Site-specific immunological adaptation could be genetic, evolving over generations, or it could be based on individual exposure, developing within the lifetime of an individual. Either way, if birds singing the local dialect are better adapted to local parasites, these birds might be healthier and more likely to reproduce successfully (MacDougall-Shackleton et al., 2002). All three of these hypotheses propose that individuals (females, in temperate zone dialect species such as the white-crowned sparrow) should benefit by choosing mates (males) singing the local dialect.

In contrast to the above functional hypotheses, the "null" hypothesis is that song dialects are a functionless epiphenomenon. This hypothesis states that dialects are merely a byproduct of song learning and dispersal under certain conditions. Either of two combinations of song learning and dispersal parameters can result in dialects: (1) imitation of song models heard early in life and limited dispersal from the natal site, or (2) dispersal from the natal area and the ability to learn or match songs heard after dispersing. In either case, the song learning and dispersal parameters could be selected for reasons that are unrelated to any putative advantage of singing a local dialect rather than a foreign one (i.e., dialect function). As noted by Podos and Warren (2007), the epiphenomenon hypothesis does not require that birds singing foreign dialects experience any negative selection, and this distinguishes it from the functional hypotheses described earlier.

In a comprehensive literature survey, Podos and Warren (2007) assessed support for the various hypotheses of dialect evolution. They reviewed information about song dialect systems in 32 species or subspecies of birds. Among the features examined are the spatial scale of dialects, their temporal stability, individual repertoire size, timing of song learning, and whether the birds are migratory or sedentary. As the authors pointed out, song dialects appear to be uncommon, and over half of the species included in their analysis are from one family (Fringillidae). Despite this, variation was observed in the features of the song dialect systems documented to date. Both the rarity of dialect systems and the variation in their features suggest that no one functional hypothesis explains dialect evolution in all cases. Podos and Warren (2007) conclude that the epiphenomenon hypothesis appears most broadly applicable to all forms of geographic variation in song, including dialect variation.

Dialects and gene flow

From an evolutionary standpoint, another key question about dialects concerns their potential relationship to genetic divergence between populations. Do song dialects reflect such genetic divergence, reinforce it, or both, potentially facilitating speciation? The extent to which dialect populations differ genetically has so far been investigated in only a few species. In rufous-collared sparrows (*Zonotrichia capensis*), no correlation was found between song dialects and differences in allozyme frequencies (Lougheed and Handford, 1992) or in mtDNA sequences (Lougheed et al., 1993). In brown-headed cowbirds, the distribution of body size and beak flange color indicates gene flow between populations with different learned flight whistle dialects (Fleischer and Rothstein, 1988). The finding that allozyme frequencies differ between dialect populations in sedentary Nuttall's white-crowned sparrows (*Z. l. nuttalli*; Baker et al., 1982) was later refuted by reanalysis of the original data with geographic distance taken into account (Soha et al., 2004). Results from studies using microsatellite allele frequencies in two migratory subspecies of white-crowned sparrows, *Zonotrichia leucophrys oriantha* (MacDougall-Shackleton & MacDougall-Shackleton, 2001) and *Z. l. pugetensis* (Poesel et al., 2017), likewise indicate that dialects do not reflect significant genetic differences between populations (the original conclusion in the *Z. l. oriantha* study that dialect populations exhibit significant, if small, genetic differences is questionable based on issues of sampling and acoustic analysis; see Soha et al., 2004).

In the few species studied so far, then, learned vocal dialects appear not to correspond to genetic divergence between populations. Until this is examined in more species, we cannot conclude that this is a general property of dialect systems. It remains possible that dialects represent genetically distinct populations in some species and might facilitate speciation under certain conditions. However, at least in some cases, vocal dialects persist despite ongoing gene flow among dialect areas. How does this occur? In the case of brown-headed cowbird flight whistles, vocal learning after dispersal can easily explain the maintenance of dialects despite gene flow, as flight whistles can be learned into adulthood (Rothstein and Fleischer, 1987). In the case of white-crowned sparrow song dialects, the explanation is more complex. In this species, the sensitive period for song acquisition ends several months before the first breeding season. However, males can memorize and rehearse songs of multiple dialects, if exposed to them during the sensitive period. If a first-year male settles to breed in an area characterized by any one of these dialects, he can selectively retain it, matching his song to those of one or more territorial neighbors during repertoire attrition at the end of song ontogeny (Nelson, 2000). Not all birds do sing the local dialect, however. In each of the three subspecies of white-crowned sparrows mentioned above, about 10% of adult males sing nonlocal songs (Soha et al., 2004). Thus, matching of the local dialect during selective attrition does not occur in all individuals. To the extent that it does occur, however, it helps explain maintenance of dialects with ongoing gene flow in this species. These mechanisms mean that song dialects do not necessarily reflect genetic differences at the population level, just as the mechanisms of song ontogeny mean that song variation between individuals does not necessarily reflect genetic variation.

Functional variation among dialects

As described above, song dialects can be maintained by certain patterns of learning and dispersal—that is, cultural evolution—and do not necessarily represent genetically distinct populations, nor does geographic variation in the particular form of song dialects necessarily serve an adaptive function. In addition, the functions of song itself in mating and territorial defense, as described in the Introduction, are the same from dialect to dialect. Despite these observations, song dialects can be seen not only as phenotypic variants but also, from the receiver perspective, as functional variants. In other words, the effectiveness of a signaler's phenotype depends on the phenotype of the receiver. In species with song dialects, how an individual responds to conspecific song often depends on the dialect of that song. Playback studies in the wild have shown that males respond more aggressively to the local dialect than to foreign dialects in redwings (*Turdus iliacus*; Bjerke, 1984), corn buntings (*Miliaria calandra*; McGregor, 1983), and Puget Sound white-crowned sparrows (Nelson and Soha, 2004a). Females tested in captivity respond with more copulation solicitation displays to songs of their natal dialect than those of a foreign dialect in white-crowned sparrows (*Z. l. nuttalli*: Baker et al., 1981; *Z. l. oriantha*: MacDougall-Shackleton et al., 2001), rufous-collared sparrows (Danner et al., 2011), and yellowhammers (*Emberiza citrinella*; Baker et al., 1987).

These biases are not restricted to species in which geographic song variation takes the form of dialects. Stronger territorial responses by males to local songs have been documented in several species with more continuous geographic variation in song (e.g., northern cardinals *Cardinalis cardinalis*: Lemon, 1967; song sparrows: Harris and Lemon, 1974; Searcy et al., 1997). Stronger response to local song is not universal, however. In a few nondialect species (or subspecies), male responses are stronger to foreign songs (swamp sparrow: Balaban, 1988a) or similar to local songs and foreign songs (Gambel's white-crowned sparrow *Z. l. gambelii*: Nelson, 1998; grasshopper sparrow: Soha et al., 2016). Asymmetric discrimination, in which stronger responses to local song occur in one population but not in another, has also been documented in some species (black-throated blue warbler *Dendroica caerulescens*: Colbeck et al., 2010; Japanese tits *Parus minor* and varied tits *Poecile varius*: Hamao, 2016). Overall, however, higher male aggression to local song is the most common pattern observed so far, in both dialect and nondialect species.

Female preferences for local song, as evidenced by copulation solicitation displays in captivity, also occur in nondialect species. Females display more to local song in red-winged blackbirds (*Agelaius phoeniceus*; Searcy, 1990), song sparrows (Searcy et al., 2002), and swamp sparrows (Balaban, 1988b, Anderson, 2009). It has been hypothesized that females should be even more selective than males in their responses to song, under the premise that mating with the wrong male is more costly than needlessly engaging in territorial defense behavior (Ratcliffe and Otter, 1996). However, the results of a field playback test in Puget Sound white-crowned sparrows suggest that females are not equally selective in all contexts. Nelson and Soha (2004b) temporarily removed males from their territories and measured the responses of resident females to playback of three types of songs: conspecific songs with local phrases, conspecific song with foreign phrases, and heterospecific (song sparrow or junco *Junco hyemalis*) song. Females ignored heterospecific song but responded equally strongly to both types of conspecific song. (By contrast, males tested with the same stimuli responded more strongly to conspecific song containing local phrases: Nelson and Soha, 2004a.) It is possible that these field playback results reflect the social mate choice preferences of females and that females are less selective in this context than in choosing extra-pair mates (Nelson and Soha, 2004b). Extra-pair mating is common in songbirds, and Searcy (1992) hypothesized that the copulation solicitation display assay reveals extra-pair mate choice preferences. Regardless of why females might be choosier about extra-pair mates than social mates, the copulation solicitation display assay indicates that females can distinguish between local and foreign songs and that these do elicit different responses in some contexts.

In species with song dialects, therefore, both males and females respond to local and foreign dialects as though they are functionally distinct, at least in some contexts. The function of a given dialect, from the receiver perspective, depends on receiver experience. In males, many of the factors that affect song production learning also affect the ontogeny of song recognition and discrimination. Even after song production learning is complete, song perception learning can continue, as evidenced by the ability of males to distinguish territorial neighbors from strangers based on song (e.g., Brooks and Falls, 1975). In females, it has been found that the preferences revealed in the copulation solicitation assay are influenced by early exposure to song (Nuttall's white-crowned sparrows: Baker et al., 1981; swamp sparrows: Anderson et al., 2014). Perceptual learning also continues in adulthood in females, enabling females to recognize the songs of their mates (e.g., Lind et al., 1996). The experiences that shape differential responses to different song dialects are independent of genotype, such that two individuals for whom a particular dialect serves distinct functions need not differ genetically. Thus as is true of phenotypic variation in song among dialects, functional variation among dialects, from the receiver perspective, does not require genetic differences among individuals and might be influenced by stochastic processes during development.

Conclusion

Bird song learning has long provided a clear example of the interaction of genetics and environmental influences in behavioral development, but the contribution of stochasticity has yet to be investigated. The internal and external influences on song ontogeny, the extent to which these act, and the resulting intraspecific variation in song phenotypes have been well documented. In species with song dialects, phenotypic polymorphisms (song variants) are also, from the receiver perspective, functional polymorphisms. Whereas song phenotype depends on the developmental history of the signaler, functional variation among dialects depends on that of the receiver. The extent to which stochastic developmental effects influence both phenotypic and functional variation in bird song is an interesting question for future research.

References

Akçay, C., Anderson, R.C., Nowicki, S., Beecher, M.D., Searcy, W.A., 2015. Quiet threats: soft song as an aggressive signal in birds. Anim. Behav. 105, 267–274.

Akçay, C., Campbell, S.E., Beecher, M.D., 2017. Good tutors are not dear enemies in song sparrows. Anim. Behav. 129, 223–228.

Akçay, C., Tom, M.E., Campbell, S.E., Beecher, M.D., 2013. Song type matching is an honest early threat signal in a hierarchical animal communication system. Proc. R. Soc. B 280, 20122517.

Anderson, R.C., Peters, S., Nowicki, S., 2014. Effects of early auditory experience on the development of local song preference in female swamp sparrows. Behav. Ecol. Sociobiol. 68, 437–447.

Anderson, R.C., 2009. Operant conditioning and copulation solicitation display assays reveal a stable preference for local song by female swamp sparrows *Melospiza georgiana*. Behav. Ecol. Sociobiol. 64, 215–223.

Arcese, P., Sogge, M.K., Marr, A.B., Patten, M.A., 2002. Song sparrow (*Melospiza melodia*), version 2.0. In: Poole, A.F., Gill, F.B. (Eds.), The Birds of North America. Cornell Lab of Ornithology, Ithaca.

Archer, G.S., Friend, T.H., Piedrahita, J., Nevill, C.H., Walker, S., 2003. Behavioral variation among cloned pigs. Appl. Anim. Behav. Sci. 82, 151–161.

Baker, M.C., Bjerke, T.K., Lampe, H.U., Espmark, Y.O., 1987. Sexual response of female yellowhammers to differences in regional song dialects and repertoire sizes. Anim. Behav. 35, 395–401.

Baker, M.C., Spitler-Nabors, K.J., Bradley, D.C., 1981. Early experience determines song dialect responsiveness of female sparrows. Science 241, 819–821.

Baker, M.C., Thompson, D.B., Sherman, G.L., Cunningham, M.A., Tomback, D.F., 1982. Allozyme frequencies in a linear series of song dialect populations. Evolution 36, 1020–1029.

Balaban, E., 1988a. Bird song syntax: learned intraspecific variation is meaningful. Proc. Natl. Acad. Sci. USA 85, 3657–3660.

Balaban, E., 1988b. Cultural and genetic variation in swamp sparrows (*Melospiza georgiana*) 2. Behavioral salience of geographic song variants. Behaviour 105, 292–322.

Balázsi, G., van Oudenaarden, A., Collins, J.J., 2011. Cellular decision-making and biological noise: from microbes to mammals. Cell 144, 910–925.

Bassett, S.M., Potter, M.A., Fordham, R.A., Johnston, E.V., 1999. Genetically identical avian twins. J. Zool. Lond. 247, 475–478.

Beecher, M.D., Campbell, S.E., Burt, J.M., Hill, C.E., Nordby, J.C., 2000. Song-type matching between neighboring song sparrows. Anim. Behav. 59, 21–27.

Bjerke, T.K., 1984. The response of male redwings *Turdus iliacus* to playback of different dialects. Fauna Norv Ser. C. Cinclus 7, 24–27.

Boncoraglio, G., Saino, N., 2007. Habitat structure and the evolution of bird song: a meta-analysis of the evidence for the acoustic adaptation hypothesis. Funct. Ecol. 21, 134–142.

Bottjer, S.W., Johnson, F., 1997. Circuits, hormones, and learning: vocal behavior in songbirds. J. Neurobiol. 33, 602–618.

Brainard, M.S., Doupe, A.J., 2002. What songbirds teach us about learning. Nature 417, 351–358.

Brenowitz, E.A., Lent, K., Kroodsma, D.E., 1995. Brain space for learned song in birds develops independently of song learning. J. Neurosci. 15, 6281–6286.

Brenowitz, E.A., 1997. Comparative approaches to the avian song system. J. Neurobiol. 33, 517–531.

Brooks, R.J., Falls, J.B., 1975. Individual recognition by song in white-throated sparrows. I. Discrimination of songs of neighbors and strangers. Can. J. Zool. 53, 879–888.

Brumm, H., Zollinger, S.A., 2013. Avian vocal production in noise. In: Brumm, H. (Ed.), Animal Communication and Noise. Springer, Berlin, pp. 187–227.

Buchanan, K.L., Spencer, K.A., Goldsmith, A.R., Catchpole, C.K., 2003. Song as an honest signal of past developmental stress in the European starling (*Sturnus vulgaris*). Proc. R. Soc. Lond. B 270, 1149–1156.

Caruso-Peck, S., Goldstein, M.H., 2019. Female social feedback reveals non-imitative mechanisms of vocal learning in zebra finches. Curr. Biol. 29, 631–636.

Catchpole, C.K., Slater, P.J.B., 2008. Bird Song: Biological Themes and Variations. Cambridge University Press, Cambridge, UK.

Chen, Y., Matheson, L.E., Sakata, J.T., 2016. Mechanisms underlying the social enhancement of vocal learning in songbirds. Proc. Nat. Acad. Sci. USA 113, 6641–6646.

Clarke, P.G.H., 2012. The limits of brain determinacy. Proc. R. Soc. B 279, 1665–1674.

Clayton, N.S., 1987. Song tutor choice in zebra finches. Anim. Behav. 35, 714–721.

Colbeck, G.J., Sillett, T.S., Webster, M.S., 2010. Asymmetric discrimination of geographical variation in song in a migratory passerine. Anim. Behav. 80, 311–318.

Danner, J.E., Danner, R.M., Bonier, F., Martin, P.R., Small, T.W., Moore, I.T., 2011. Female, but not male, tropical sparrows respond more strongly to the local song dialect: implications for population divergence. Am. Nat. 178, 53–63.

DeVoogd, T.J., Krebs, J.R., Healy, S.D., Purvis, A., 1993. Relations between song repertoire size and the volume of brain nuclei related to song: comparative evolutionary analyses amongst oscine birds. Proc. R. Soc. Lond. B 254, 75–82.

Doupe, A.J., Kuhl, P.K., 1999. Birdsong and human speech: common themes and mechanisms. Ann. Rev. Neurosci. 22, 567–631.

Eens, M., Pinxten, R., Verheyen, R.F., 1992. Song learning in captive European starlings, *Sturnus vulgaris*. Anim. Behav. 44, 1131–1143.

Eriksson, D., Wallin, L., 1986. Male bird song attracts females – a field experiment. Behav. Ecol. Sociobiol. 19, 197–199.

Falls, J.B., 1987. Does song deter intrusion in white-throated sparrows *Zonotrichia albicollis*? Can. J. Zool. 66, 206–211.

Fee, M.S., Goldberg, J.H., 2011. A hypothesis for basal ganglia-dependent reinforcement learning in the songbird. Neurosci 198, 152–170.

Fleischer, R.C., Rothstein, S.I., 1988. Known secondary contact and rapid gene flow among subspecies and dialects in the brown-headed cowbird. Evolution 42, 1146–1158.

Freeman, B.G., Montgomery, G.A., 2017. Using song playbacks to measure species recognition between geographically isolated populations: A comparison with acoustic trait analysis. Auk 134, 857–870.

Gammon, D.E., 2013. How is model selection determined in a vocal mimic?: Tests of five hypotheses. Behav 1375–1397.

Hamao, S., 2016. Asymmetric response to song dialects among bird populations: the effect of sympatric related species. Anim. Behav. 119, 143–150.

Hansen, P., 1979. Vocal learning: its role in adapting sounds structures to long-distance propagation and a hypothesis on its evolution. Anim. Behav. 27, 1270–1271.

Harding, C.F., Barclay, S.R., Waterman, S.A., 1998. Changes in catecholamine levels and turnover rates in hypothalamic, vocal control, and auditory nuclei in male zebra finches during development. J. Neurobiol. 34, 329–346.
Harris, M.A., Lemon, R.E., 1974. Songs of song sparrows: reactions of males to songs of different localities. Condor 76, 33–44.
Hausberger, M., Jenkins, P.F., Keene, J., 1991. Species-specificity and mimicry in bird song: are they paradoxes? Behaviour 117, 53–81.
Hill, C.E., Campbell, S.E., Nordby, J.C., Burt, J.M., Beecher, M.D., 1999. Song sharing in two populations of song sparrows (*Melospiza melodia*). Behav. Ecol. Sociobiol. 46, 341–349.
Hinde, R.A., Steele, E., 1976. The effect of male song on an oestrogen-dependent behaviour in the female canary *Serinus canarius*. Horm. Behav. 7, 293–304.
Holveck, M.-J., Vieira de Castro, A.C., Lachlan, R.F., ten Cate, C., Riebel, K., 2008. Accuracy of song syntax learning and singing consistency signal early condition in zebra finches. Behav. Ecol. 19, 1267–1281.
Hultsch, H., Todt, D., 2004. Learning to sing. In: Marler, P., Slabbekoorn, H. (Eds.), Nature's Music: The Science of Birdsong. Elsevier Academic Press, London, pp. 80–107.
Immelmann, K., 1969. Song development in the zebra finch and other estrildid finches. In: Hinde, R.A. (Ed.), Bird Vocalizations. Cambridge University Press, Cambridge, pp. 61–74.
Irwin, D.E., Irwin, J.H., 2004. Speciation in a ring: the role of song. In: Marler, P., Slabbekoorn, H. (Eds.), Nature's Music: The Science of Birdsong. Elsevier Academic Press, London, p. 204.
Johnson, L.S., Searcy, W.A., 1996. Female attraction to male song in house wrens (*Troglodytes aedon*). Behaviour 133, 357–366.
King, A.P., West, M.J., 1977. Species identification in the North American cowbird: appropriate responses to abnormal song. Science 195, 1002–1004.
Konishi, M., 1965. The role of auditory feedback in the control of vocalization in the white-crowned sparrow. Z. Tierpsychol. 22, 770–783.
Konishi, M., 2004. The role of auditory feedback in birdsong. Ann. NY. Acad. Sci. 1016, 463–475.
Krebs, J.R., 1977. The significance of song repertoires: the Beau Geste hypothesis. Anim. Behav. 25, 475–478.
Kroodsma, D.E., Verner, J., 2013. Marsh Wren (*Cistothorus palustris*), version 2.0. In: Poole, A.F. (Ed.), The Birds of North America. Cornell Lab of Ornithology, Ithaca.
Lachlan, R.F., Anderson, R.C., Peters, S., Searcy, W.A., Nowicki, S., 2014. Typical versions of learned swamp sparrow song types are more effective signals than are less typical versions. Proc. R. Soc. B 281, 20140252.
Lachlan, R.F., Slater, P.J.B., 1999. The maintenance of vocal learning by gene–culture interaction: the cultural trap hypothesis. Proc. R. Soc. Lond. B 266, 701–706.
Langmore, N.E., 1998. Functions of duet and solo songs of female birds. Trends Ecol. Evol. 13, 136–140.
Lein, M.R., 1978. Song variation in a population of chestnut-sided warblers (*Dendroica pensylvanica*): its nature and suggested significance. Can. J. Zool. 56, 1266–1283.
Lemon, R.E., 1967. The response of cardinals to songs of different dialects. Anim. Behav. 15, 538–545.
Lewis, J.W., Ryan, S.M., Arnold, A.P., Butcher, L.L., 1981. Evidence for a catecholaminergic projection to Area X in the zebra finch. J. Comp. Neurol. 196, 347–354.

Lind, H., Dabelsteen, T., McGregor, P.K., 1996. Female great tits can identify mates by song. Anim. Behav. 52, 667–671.

Liu, W.-C., Kroodsma, D.E., 2006. Song learning by chipping sparrows: when, where and from whom. Condor 108, 509–517.

Lohr, B., Ashby, S., Wakamiya, S.M., 2013. The function of song types and song components in Grasshopper Sparrows (*Ammodramus savannarum*). Behav 150, 1085–1106.

Lougheed, S.C., Handford, P., Baker, A.J., 1993. Mitochondrial DNA hyperdiversity and vocal dialects in a subspecies transition of the rufous-collared sparrow. Condor 95, 889–895.

Lougheed, S.C., Handford, P., 1992. Vocal dialects and the structure of geographic variation in morphological and allozymic characters in the rufous-collared sparrow, *Zonotrichia capensis*. Evolution 46, 1443–1456.

MacDougall-Shackleton, E.A., Derryberry, E.P., Hahn, T.P., 2002. Nonlocal male mountain white-crowned sparrows have lower paternity and higher parasite loads than males singing local dialect. Behav. Ecol. 13, 682–689.

MacDougall-Shackleton, E.A., MacDougall-Shackleton, S.A., 2001. Cultural and genetic evolution in mountain white-crowned sparrows: song dialects are associated with population structure. Evolution 55, 2568–2575.

MacDougall-Shackleton, S.A., MacDougall-Shackleton, E.A., Hahn, T.P., 2001. Physiological and behavioural responses of female mountain white-crowned sparrows to natal- and foreign-dialect songs. Can. J. Zool. 79, 325–333.

Mahler, B., Gil, D., 2009. The evolution of song in the *Phylloscopus* leaf warblers (Aves: Sylviidae): A tale of sexual selection, habitat adaptation, and morphological constraints. Adv. Study Behav. 40, 35–66.

Marler, P., Peters, S., 1982. Structural changes in song ontogeny in the swamp sparrow *Melospiza georgiana*. Auk 99, 446–458.

Marler, P., Peters, S., 1987. A sensitive period for song acquisition in the song sparrow, *Melospiza melodia*: a case of age-limited learning. Ethology 76, 89–100.

Marler, P., Tamura, M., 1962. Song "dialects" in three populations of white-crowned sparrows. Condor 64, 368–377.

Marler, P., 1970. A comparative approach to vocal learning: song development in white-crowned sparrows. J. Comp. Physiol. Psychol. 71, 1–25.

Marler, P., 1984. Song learning: innate species differences in the learning process. In: Marler, P., Terrace, H.S. (Eds.), The Biology of Learning. Springer-Verlag, Berlin, pp. 289–309.

McDonald, M.V., 1989. Function of song in Scott's seaside sparrow. Anim. Behav. 38, 468–485.

McGregor, P.K., Avery, M.I., 1986. The unsung songs of great tits (*Parus major*): learning neighbours' songs for discrimination. Behav. Ecol. Sociobiol. 18, 311–316.

McGregor, P.K., 1983. The responses of corn buntings to playback of dialects. Z. Tierpsychol. 62, 256–260.

Mooney, R., 2009. Neural mechanisms for learned birdsong. Learn. Mem. 16, 655–669.

Morton, E.S., 1975. Ecological sources of selection on avian sounds. Am. Nat. 109, 17–34.

Mountjoy, D.J., Lemon, R.E., 1991. Song as an attractant for male and female European starlings, and the influence of song complexity on their response. Behav. Ecol. Sociobiol. 28, 97–100.

Naguib, M., Kipper, S., 2006. Effects of different levels of song overlapping on singing behaviour in male territorial nightingales (*Luscinia megarhynchos*). Behav. Ecol. Sociobiol. 59, 419–426.

Nelson, D.A., Hallberg, K.I., Soha, J.A., 2004. Cultural evolution of Puget Sound white-crowned sparrow song dialects. Ethology 110, 879–908.

Nelson, D.A., Marler, P., 2005. Do bird nestmates learn the same songs? Anim. Behav. 69, 1007–1010.

Nelson, D.A., Poesel, A., 2014. Tutor choice and imitation accuracy during song learning in a wild population of the Puget Sound white-crowned sparrow. Behav. Ecol. Sociobiol. 68, 1741–1752.

Nelson, D.A., Soha, J.A., 2004a. Perception of geographical variation in song by male Puget Sound white-crowned sparrows, *Zonotrichia leucophrys pugetensis*. Anim. Behav. 68, 395–405.

Nelson, D.A., Soha, J.A., 2004b. Male and female white-crowned sparrows respond differently to geographic variation in song. Behaviour 141, 53–69.

Nelson, D.A., 1992. Song overproduction and selective attrition lead to song sharing in the field sparrow (*Spizella pusilla*). Behav. Ecol. Sociobiol. 30, 415–424.

Nelson, D.A., 1998. Geographic variation in song of Gambel's white-crowned sparrow. Behaviour 135, 321–342.

Nelson, D.A., 2000. Song overproduction, selective attrition and song dialects in the white-crowned sparrow. Anim. Behav. 60, 887–898.

Nottebohm, F., Stokes, T.M., Leonard, C.M., 1976. Central control of song in the canary *Serinus canarius*. J. Comp. Neurol. 165, 457–486.

Nottebohm, F., 1969. The song of the chingolo, *Zonotrichia capensis*, in Argentina: Description and evaluation of a system of dialects. Condor 71, 299–315.

Nottebohm, F., 1975. Continental patterns of song variability in *Zonotrichia capensis*: some possible ecological correlates. Am. Nat. 109, 605–624.

Nottebohm, F., 1991. Reassessing the mechanisms and origins of vocal learning in birds. Trends Neurosci. 14, 206–211.

Nowicki, S., Peters, S., Searcy, W.A., Clayton, C., 1999. The development of within-song type variation in song sparrows. Anim. Behav. 57, 1257–1264.

Nowicki, S., Searcy, W.A., Hughes, M., 1998. The territory defense function of song in song sparrows: a test with the speaker occupation design. Behav 135, 615–628.

Nowicki, S., Searcy, W.A., Peters, S., 2002. Brain development, song learning and mate choice in birds: a review and experimental test of the "nutritional stress hypothesis". J. Comp. Physiol. A 188, 1003–1014.

Odom, K.J., Hall, M.L., Riebel, K., Omland, K.E., Langmore, N.E., 2014. Female song is widespread and ancestral in songbirds. Nat. Comm. 5, 3379.

Ölveczky, B.P., Gardner, T.J., 2011. A bird's eye view of neural circuit formation. Curr. Opin. Neurobiol. 21, 124–131.

Payne, R.B., 1981. Population structure and social behavior: Models for testing the ecological significance of song dialects in birds. In: Alexander, R.D., Tinkle, D.W. (Eds.), Natural Selection and Social Behavior: Recent Research and New Theory. Chiron Press, New York, pp. 108–120.

Peters, S., Derryberry, E.P., Nowicki, S., 2012. Songbirds learn songs least degraded by environmental transmission. Biol. Lett. 8, 736–739.

Peters, S., Nowicki, S., 2017. Overproduction and attrition: the fates of songs memorized during song learning in songbirds. Anim. Behav. 124, 255–261.

Podos, J., Huber, S.K., Taft, B., 2004. Bird song: the interface of evolution and mechanism. Annu. Rev. Ecol. Evol. Syst. 35, 55–87.

Podos, J., Warren, P.S., 2007. The evolution of geographic variation in birdsong. Adv. Stud. Behav. 37, 403–458.
Podos, J., 2010. Acoustic discrimination of sympatric morphs in Darwin's finches: a behavioural mechanism for assortative mating? Philos. Trans. R. Soc. Lond. B 365, 1031–1039.
Poesel, A., Fries, A.C., Miller, L., Gibbs, H.L., Soha, J.A., Nelson, D.A., 2017. High levels of gene flow among song dialect populations of the Puget Sound white-crowned sparrow. Ethology 123, 581–592.
Poesel, A., Kunc, H.P., Foerster, K., Johnsen, A., Kempenaers, B., 2006. Early birds are sexy: male age, dawn song and extrapair paternity in blue tits, *Cyanistes* (formerly *Parus*) *caeruleus*. Anim. Behav. 72, 531–538.
Price, J.J., Lanyon, S.M., 2002. Reconstructing the evolution of complex bird song in the oropendolas. Evolution 56, 1514–1530.
Ratcliffe, L., Otter, K., 1996. Sex differences in song recognition. In: Kroodsma, D. E., Miller, E.H. (Eds.), Ecology and Evolution of Acoustic Communication in Birds. Comstock Publishing Assoc, Ithaca, pp. 339–355.
Riters, L.V., Spool, J.A., Merullo, D.P., Hahn, A.H., 2019. Song practice as a rewarding form of play in songbirds. Behav. Process. 163, 91–98.
Riters, L.V., 2012. The role of motivation and reward neural systems in vocal communication in songbirds. Front. Neuroendocrinol. 33, 194–209.
Rothstein, S.I., Fleischer, R.C., 1987. Brown-headed cowbirds learn flight whistles after the juvenile period. Auk 104, 512–516.
Scharff, C., Adam, I., 2013. Neurogenetics of birdsong. Curr. Opin. Neurobiol. 23, 29–36.
Schlinger, B.A., Brenowitz, E.A., 2017. Neural and hormonal control of birdsong. In: Pfaff, D.W., Joels, M. (Eds.), Hormones, Brain and Behavior. Elsevier Science, pp. 255–290.
Schmidt, K.L., Moore, S.D., MacDougall-Shackleton, E.A., MacDougall-Shackleton, S.A., 2013. Early-life stress affects song complexity, song learning and volume of the brain nucleus RA in adult male song sparrows. Anim. Behav. 86, 25–35.
Schuett, W., Dall, S.R.X., Baeumer, J., Kloesener, M.H., Nakagawa, S., Beinlich, F., et al., 2011. 'Personality' variation in a clonal insect: the pea aphid, *Acyrthosiphon pisum*. Dev. Psychobiol. 53, 631–640.
Searcy, W.A., Marler, P., 1981. A test for responsiveness to song structure and programming in female sparrows. Science 213, 926–928.
Searcy, W.A., Nowicki, S., Hughes, M., Peters, S., 2002. Geographic song discrimination in relation to dispersal distances in song sparrows. Am. Nat. 159, 221–230.
Searcy, W.A., Nowicki, S., Hughes, M., 1997. The response of male and female song sparrows to geographic variation in song. Condor 99, 651–657.
Searcy, W.A., Peters, S., Kipper, S., Nowicki, S., 2010. Female response to song reflects male developmental history in swamp sparrows. Behav. Ecol. Sociobiol. 64, 1343–1349.
Searcy, W.A., 1990. Species recognition of song by female red-winged blackbirds. Anim. Behav. 40, 1119–1127.
Searcy, W.A., 1992. Song repertoire and mate choice in birds. Am. Zool. 32, 71–80.
Smith, D.G., 1979. Male singing ability and territory integrity in red-winged blackbirds *Agelaius pheoniceus*. Behav 68, 193–206.
Soha, J.A., Lohr, B., Gill, D.E., 2009. Song development in the grasshopper sparrow, *Ammodramus savannarum*. Anim. Behav. 77, 1479–1489.
Soha, J.A., Marler, P., 2000. A species-specific acoustic cue for selective song learning in the white-crowned sparrow. Anim. Behav. 60, 297–306.

Soha, J.A., Nelson, D.A., Parker, P., 2004. Genetic analysis of song dialect populations in Puget Sound white-crowned sparrows. Behav. Ecol. 15, 636–646.

Soha, J.A., Poesel, A., Nelson, D.A., Lohr, B., 2016. Non-salient geographic variation in birdsong in a species that learns by improvisation. Ethology 122, 343–353.

Soha, J.A., Shimizu, T., Doupe, A.J., 1996. Development of the catecholaminergic innervation of the song system of the male zebra finch. J. Neurobiol. 29, 473–489.

Soha, J.A., 2017. The auditory template hypothesis: a review and comparative perspective. Anim. Behav. 124, 247–254.

Solis, M.M., Doupe, A.J., 1997. Anterior forebrain neurons develop selectivity by an intermediate stage of birdsong learning. J. Neurosci. 17, 6447–6462.

Spencer, K.A., Buchanan, K.L., Leitner, S., Goldsmith, A.R., Catchpole, C.K., 2005. Parasites affect song complexity and neural development in a songbird. Proc. R. Soc. Lond. B 272, 2037–2043.

Tchernichovski, O., Mitra, P.P., Lints, T., Nottebohm, F., 2001. Dynamics of the vocal imitation process: how a zebra finch learns its song. Science 291, 2564–2569.

Tchernichovski, O., Nottebohm, F., 1998. Social inhibition of song imitation among sibling male zebra finches. Proc. Natl. Acad. Sci. USA 95, 8951–8956.

Thorpe, W.H., 1958. The learning of song patterns by birds, with especial reference to the song of the chaffinch *Fringilla coelebs*. Ibis 100, 535–570.

Vogt, G., Huber, M., Thiemann, M., van den Boogaart, G., Schmitz, O.J., Schubart, C.D., 2008. Production of different phenotypes from the same genotype in the same environment by developmental variation. J. Exp. Biol. 211, 510–523.

Vogt, G., 2015. Stochastic developmental variation, an epigenetic source of phenotypic diversity with far-reaching biological consequences. J. Biosci. 40, 159–204.

Volman, S.F., 1993. Development of neural selectivity for birdsong during vocal learning. J. Neurosci. 13, 4737–4747.

West, M.J., King, A.P., 1988. Female visual displays affect the development of male song in the cowbird. Nature 334, 244–246.

Wheatcroft, D., Qvarnstrom, A., 2017. Genetic divergence of early song discrimination between two young songbird species. Nat. Ecol. Evol. 1, 0192.

Zollinger, S.A., Slater, P.J.B., Nemeth, E., Brumm, H., 2017. Higher songs of city birds may not be an individual response to noise. Proc. R. Soc. B 284, 20170602.

19

Domestication as a process generating phenotypic diversity

Lyudmila N. Trut and Anastasiya V. Kharlamova
The Federal Research Center Institute of Cytology and Genetics of Siberian Branch of the Russian Academy of Sciences (IC&G SB RAS), Novosibirsk, Russia

Domestication in evolutionary terms

Variability and sources of its origin are central problem of evolutionary biology. Since Darwin up to present days the problem of domestication emerges in evolutionary terms (Darwin, 1859, 1868; Belyaev, 1979; Chen et al., 2013; Wilkins et al., 2014; Pendleton et al., 2018). The mainstream scientific research has been focused on the magnitude and pace of variability in domestic animals.

For illustration we will use the dog (*Canis familiaris*). It is not only the first domesticated species—the beginning of dog domestication estimated by different researches as 15,000–40,000 years ago (Savolainen et al., 2002; Germonpré et al., 2009), but also, by view of many researchers, is the pinnacle of changes under domestication.

Most astonishing is the evolutionary transformation of dog behavior. The changes triggered by domestication include not only the loss of the complex of species-specific aggressive responses to humans, but also the development of amiability and attachment to human, the ability to 'read' human gestures, gazes, words and to make use of such social cues in the process of adaptation to the environment. In addition, significant variation in other forms of dog behavior appeared, owing to which dogs nowadays serve in the police, as shepherds, security guards, hunters, and assist people with disabilities to lead a more active social life (Fig. 19.1).

What made a dog like this? What happened during a thousand-year coevolution of dog and human? From a scientific point of view, these are not idle questions.

In the process of dog domestication, its reproductive physiology underwent deep transformations, which resulted in the loss of reproductive seasonality. Seasonal changes in the length of daylight hours, which persisted throughout geological eras, represent

Phenotypic Switching. DOI: https://doi.org/10.1016/B978-0-12-817996-3.00014-1

Figure 19.1 Behavior of dog has transformed a lot during domestication. Now dog not only serves to human in different areas—she often is an equal companion.

the main environmental factor for the synchronization of reproductive function with environmental temperatures and feeding conditions that are optimal for raising offspring in most species of the temperate latitudes. This factor has lost its regulatory function. As a result, the offspring of dogs, like that of many other domesticates, but not of their wild ancestors, are born in any season of the year and more often than once a year.

Domestic dogs display an extraordinary range of variation in morphological parameters, which is most obvious and striking—it includes variation in overall body size, body proportions, skin, and hair. The dog has been classified as a distinct species (*Canis familiaris*). However, this is a unique species with no natural counterpart. None of the species displays such a magnitude of intraspecies phenotypic variation, for some traits it exceeds that among the species of the *Canidae* family. In addition, no species has such a huge number of forms (several hundred breeds). Range of dog's breed variability is so huge, that it can create the physical barrier for mating by natural way between breeds-dwarf and breeds-gain. However, as now confirmed by numerous independent studies, all such miscellaneous dog breeds originated from typical, uniform individuals of the same species, namely the wolf *Canis lupus* (Wayne and O'Brien, 1987; Vilà et al., 1997; Lindblad-Toh et al., 2005; vonHoldt et al., 2010).

Traditional view on possible mechanisms arisen of phenotypic diversity

The fundamental question that remains is: which factors (mechanisms) underlie such an incredible diversity displayed by the descendants of the wolf during their transformation into the dog?

At one time, accumulation and stochastic fixation of mutations was widely discussed. The historical age of *Canis familiaris* is, as was mentioned earlier, not less than 12,000–15,000 years. Undoubtedly, this is too short an evolutionary period for accumulating all the mutations that would be necessary to realize such diversity by means of artificial selection due to random events. Rate of spontaneous mutation has been estimated for mammals (at the example of mice) at least 1.8×10^{-10} mutations per base pair per cell division, with mutation rate per genome per replication being around 0.49 (Drake et al., 1998). As D. Belyaev wrote, frequency of aberrant phenotypes in the population of foxes selected for tame behavior was 10^{-2}–10^{-3}, that is, two to three orders higher than the expected frequency of spontaneous mutations (Belyaev, 1979). When discussing the nature of the variety of domestic animals and, above all, the variety of dogs, much has been said and written about the role of stochastic processes and inbreeding. However, some researchers believe that inferences about the role of inbreeding and genetic drift in the process of domestication are superficial (Vila et al., 2005; Wayne and O'Brien,

1987). In dogs, a great variety of alleles at nuclear loci has been detected. Based on these data, the authors believe that the genetic pool of modern dogs has evolved from a diverse gene pool, which in turn may indicate multiple independent wolf domestication events in different places and at different times and the absence of inbreeding in the process of domestication. In populations of the earliest dogs, inbreeding was also hampered by regular back-crossing to the wild wolf, which probably never stopped. For example, extensive admixture between dogs and wolves has been found in modern wolf populations—up to 25% of Eurasian wolf genomes showing signs of dog ancestry (Fan et al., 2016). Therefore, some researchers of domestication do believe that random drift could not make a significant contribution to the morphophysiological transformation seen in domestic dogs. Stochastic processes did take place in the evolution of domestic animals, but not at the earliest stage of domestication. Strongly inbred individuals were not the first ancestors of domestic animals, but the ancestors of different breeds. It was at the stage of breed formation that the dog populations went through population bottlenecks. The loss of nucleotide diversity during the early steps of domestication was much less (0.05) than that attributable to intensive breed formation (0.35) (Gray et al., 2009).

New insights into the nature of changes under domestication

It should be stressed that from the earliest stages of domestication animals were selected for certain characteristics of domesticated behavior. Therefore, the search for patterns of evolutionary transformation of animals during domestication should probably proceed from the postulate that for the processes of individual development the consequences of such selection are exactly opposite to those of natural selection. As shown by Schmalhausen (1949), the stabilizing role of natural selection is mainly in the stabilization of ontogenesis, that is, the creation of ontogenesis types providing better protection from disturbing factors and morphogenesis stabilization under given conditions. The creative role of stabilizing selection is to form ontogenetic patterns with more or less autonomous regulatory mechanisms of individual development, which ensure stabilization, that is, maximum stability of morphogenesis in changing environment.

In general, according to Lerner (1954) "a Mendelian population tends to retain its genetic composition arrived at by previous evolutionary history." However, any systematic selection, and especially selection for behavior, has been shown to lead to the destruction of the characters, reactions, and correlations created by stabilizing selection. As a result of a revolutionary change or accumulation of new changes the morphogenesis processes become more vulnerable and unstable giving rise to organisms with increased general reactivity to disturbing factors, that is, reduced stability of normal morphogenesis. This phenomenon was repeatedly emphasized by Schmalhausen (1949, 1982). However, the process of domestication is accompanied by the revolutionary changes that encompass the hormonal mechanisms of ontogenesis regulation, the machinery of homeostasis and ontogenesis stabilization, which is extremely sensitive to environmental influences. The break of developmental homeostasis can lead to the appearance of new phenotypic variants, that is, the variability previously masked by the regulatory systems can reveal itself, and therefore a break of the development-stabilizing machinery can be a factor accelerating evolutionary processes. The aforementioned conceptual framework was employed by Belyaev (1979) for creation of ideas about destabilizing selection, a particular case of which is selection for behavior during the domestication of animals. This selection affects the machinery of neurohormonal regulation of the development, which in nature has been created and maintained by stabilizing selection.

It is not only the high rate of phenotypic diversity that is surprising. An even more surprising fact is that in domesticates belonging to different systematic groups, the extremely high rates of emerging phenotypic diversity are combined with collateral (homologous) diversity patterns. Domesticates exhibit similar changes not only in some behavioral traits and physiology, but also in some morphological features.

In domestic animals, a similar complex of morphophysiological changes has been described in the literature as the "domestication syndrome" (Wilkins et al., 2014; Wilkins, 2017). Despite the fact that the discussion about specific evolutionary aspects of domestication is ongoing, their nature still remains unclear.

Farm-fox experiment

In the mid-50s of the last century, the Russian geneticist-evolutionist D. Belyaev began the Farm-fox experiment, which was based on his completely new insight into the nature of domestication. He hypothesized that the outcome of domestication could be predetermined already in its earliest phase, or during so-called "proto-domestication." During this period, a species takes first steps on a new evolutionary path, when for the first time in its evolutionary history it encounters human and anthropogenic environments and becomes involved in the formation of a new form of animal–human relationships. Belyaev believed that the key genetic factor acting during this period of animal domestication was natural selection for tolerance toward human, that is, for the ability to coexist with human (Belyaev, 1979). Gradually, the historical process of natural selection was replaced by artificial selection, which could be directed, probably unconsciously and unintentionally at the beginning, by primitive humans. Then a period of conscious, systematic, directed selection ultimately began. It is difficult to say when in the history of domestication one form of selection was replaced by another. Unfortunately, the documentary evidence of this process is lost in the historical past.

The gist of D. Belyaev's idea was again "scrolling through the documentary scenario" of the initial stage of domestication in an experiment, that is, to reproduce the absence of this crucial biological process in the present. This would provide an opportunity to trace dynamically, how this process spawned the striking variability, that is, form that is now apparent (Belyaev, 1979). Of course, it was impossible even in a rude form to reproduce the conditions of the earliest phase of historical domestication carried out by primitive humans. The purpose of our experiment was to iterate the action of the factor that appears to play a principal role in this stage—the strongest selection pressure for behavior, for the ability to adapt to ecological niches close to humans, that is, for the ability to be domesticated.

As already noted, this evolutionary experiment was launched in the mid-1950s on one of the objects of industrial breeding, the silver fox *Vulpes vulpes*.

Phenotypic changes of experimental foxes

Transformations during selection for the behavior of foxes toward tameability are beyond the scope of this review—they have been described in publications (Trut, 1999; Trut et al., 2004, 2009, 2012), but rather we focus on the effect of domestication on the phenotypic diversity.

Selection-induced changes in the behavior of foxes were accompanied by the appearance in some tame foxes of the morphological changes typical of many domestic animals (Belyaev, 1979; Trut et al., 2009). Fig. 19.2 illustrates a regular phenotype (Fig. 19.2A) and a gallery of phenotypic changes occurring in silver fox during its selection for domesticated behavior. Earliest of all,

Figure 19.2 Some phenotypic changes, appeared in domesticated population of silver-black foxes, with example of appropriate them similar dogs phenotypes: (A) Standard (normal) silver-black phenotype, usual for unselected foxes; (B) specific piebald spotting, determined by autosome semidominant gene—Star (fox, heterozygous for Star gene); (C) phenotype of fox, homozygous for Star gene; (D) 1-month-old puppies: left has floppy ears, right—with normal upright position of ears; (E) adult tame fox, carried two aberrations—Star on a head and floppy ear; (F) changes of cranial shape; 3-month-old tame fox pups: left has shortened skull (extremely rare case of visual shortened), right—with usual skull proportions; (G) tame fox with elongated mandible—"underbite"; (H) tame fox with shortened tail; (I) tame fox with curly tail, as in hasky.

there appeared changes in normal silver-black fur color. For instance, in the ninth selection generation (1969) the first three cubs born in different families had specific piebald star spotting on the head (Fig. 19.2B and C). Subsequently, such white patches on the head repeatedly appeared *de novo* in different families. In the next 10 generations (1969–79), the frequency of occurrence of this marker among domesticated foxes varied from one to nine per thousand descendants (1.1×10^{-3}–9.4×10^{-3}).

In the period from 1991 to 2001, that is, from the 31st to 41st selection generation, the frequency with which this change arose increased by two orders of magnitude; on average one out of every ten cubs had a white mark on the head; and subsequently there were several star marking carriers per each hundred descendants.

Floppy ears were also one of the first changes under domestication. Normally, all farm-bred fox cubs have their ears raised up (i.e., take the position as in adult animals) by the age of 2–3 weeks. In the cubs of tame foxes, this occurs somewhat later, but up to the age of 4 weeks their ears also up. However, from the ninth selection generation there appear descendants whose ears do not stay upright during the first month of life (Fig. 19.2D). They rise up significantly later than it normally occur in foxes unselected for behavior—at the age of two and rarely 3 months. In very few individuals the ears remain floppy all their lives. The cubs with this morphological marker are born much more seldom than the cubs with star marking. After the birth of the first flap-eared cub in the ninth generation (1969), 22 individuals with this aberration appeared in the period from 1969 to 1979. They appeared with a frequency of several descendants per thousand, and in all subsequent years, this frequency almost did not change.

Surprisingly, more than one aberration can be observed on some individuals at the same time—for example, a hanging ear and a star (Fig. 19.2E). As it was shown, one of the aberrations (brown mottling) induced by selection of foxes for domestication, can favor the appearance another one ("star" depigmentation) (Belyaev and Trut, 1986).

A special group of morphological changes observed during selection in tame foxes consists of the changes in some cranial dimensions affecting the skull shape, in particular shortening of its face region (Fig. 19.2F). Significant shortening for some elements of facial skull was detected in tame foxes at earlier stages of the experiment, as well as also widening of skull for tame males (Trut et al., 1991).

We should pay attention to such aberrations as an elongated lower jaw ("underbite"—Fig. 19.2G), which is revealed in the experimental foxes. It is very important also that reduced length and width of the canine and carnassial teeth were found in tame foxes (Wood et al., 2018). Reduction of tooth size is also listed as one of the specific changes noted in historically domesticated species such as pigs and dogs (Sánchez-Villagra et al., 2016).

Constraints of development and its focus

Despite the fact that the domestication of animals generates enormous phenotypic diversity, the most of domesticates, as already noted, display similar phenotypic deviations from standard phenotype. In this connection, the existence of barriers and constraints for realizing the variability, caused due to the integrity between ontogenetic processes and morphogenetic interactions, has been widely discussed. Many believe that variability is not unlimited and that the epigenetic mechanisms involved in the regulation and integration of morphogenetic processes might channel natural stochastic processes (Vorobjeva, 1980; Tatarinov, 1985; Shishkin, 1986; Mednikov, 1987). It has been suggested that each morphological system is characterized by a certain range of developmental pathways, the probability of realization of which depends on epigenetic and environmental interactions. In other words, the ontogenetic impact of both diverse random mutational changes and changes in developmental environment is always limited by the historically established properties of the regulatory systems that control development as a whole (Shishkin, 1984, 1986). A similar point of view has been espoused by other researchers (Schmalhausen, 1949; Mednikov, 1987), who introduced a concept of discreteness of the adaptive norms of the morphophysiological traits: depending on the conditions of development one and the same genotype can give rise to the phenotypes that differ by a number of morphophysiological characters (Mednikov, 1987). As J. Haldane wrote, "What is inherited is not a set of characters, but the capacity for reacting to the environment in such a way that, in a particular environment, particular characters are developed" (Haldane, 1949). The trigger for switching the system on searching for a new equilibrium is loss of the effective regulation of normal development in some critical situation (Shishkin, 2018). Regulatory systems play an important role in switching from one development program to another. An ample example of developmental bias is switching from exploratory behavior to fear response induced by the elevation of corticosteroid levels in the peripheral blood (Fig. 19.3). It is emphasized that minor quantitative changes can play an important role in developmental bias. It is also noted that minor quantitative changes in the temporal parameters of development can change

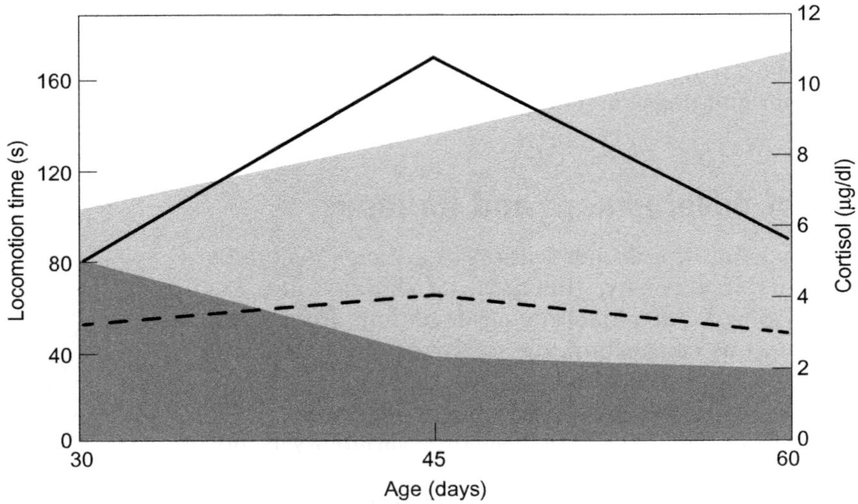

Figure 19.3 Total time of locomotion (as an indicator of exploratory behavior) and plasma cortisol levels in domesticated and farm-bred foxes at the age of 30–60 days: (*dotted line*) plasma cortisol levels in tame foxes. (*solid line*) Plasma cortisol levels in unselected foxes. (*light-grey field*) Locomotion of tame foxes. (*dark-grey field*) Locomotion of unselected foxes.

morphogenetic interactions and serve as development channel switches. An important evolutionary role of heterochronic processes has been underlined by many researchers, of whom first of all special mention should be given to Schmalhausen (1982).

Is there any relationship between the barriers and constraints of development and its direction observed in some evolutionary situations? As Smith (1983) rightly points out, the presence of certain developmental barriers related to the historically established regulatory correlation systems can only explain why organisms do not change in any conceivable directions, that is, why the rank of possible phenotypes of a species is limited. However, such development constraints fail to explain why organisms often change in one and the same direction, that is, cannot explain the phenomenon of morphophysiological convergence. For example, one of the cause of piebald spotting is the absence of melanocytes in hair follicles due to late melanoblasts migration to a certain area of the body (Rawles, 1953). In animals, depigmentation (white) spots most often appear on the body areas that are most distant from the neural crest, a starting point of melanoblasts migration. If, for some genetic or epigenetic reasons, the speed of melanoblasts migration slows down, then they will fail to reach the epidermis in the parts of the body that are most distant from the starting point of their migration (ventral area of the body, tips of the paws, snout, or tail). Therefore, the fact that in different animals the piebald spotting appears predominantly in the same places is fully explained by the existence of certain developmental features. However, how can the frequent occurrence of piebald spotting in animals under domestication be interpreted in terms of embryological constraints? Why does domestication naturally induce correlated changes in the rate of melanoblast migration, which results in the regular appearance of piebald spotting in almost all domestic animals? Moreover, in general, why does domestication inevitably lead to the emergence of similar morphological and complex physiological changes in the representatives of the same and many different systematic groups under their repeated involvement into the sphere of domestication?

Phenotypic changes of domesticated foxes and developmental rate

The association between the newly generated traits and changes in developmental rates in domestic foxes is noteworthy. Some behavioral and aberrant morphological changes are essentially correlates of the delayed development of normal somatic traits. Floppy ears, curly tails, widened skulls, and shortened snouts are all the juvenile traits that in some individuals are retained into adulthood. The emotional manifestation of affiliative behavior toward human, including desire for communication, inhibition of independent behavior in the presence of humans, is usually referred to behavioral neoteny, that is, the retention of juvenile behavior in adults (Udell et al., 2010).

A delay in developmental rates was observed as early as during embryonic morphogenesis. As already noted, specifically located depigmentation patch on the head (star) is a particular marker of domestication (Fig. 19.2B and C). It was shown that delayed development (proliferation and migration) of the embryonic melanocytes or primary melanoblasts is the mechanism underlying this depigmentation (Prasolova and Trut, 1993). The slowdown of the development ultimately results in the absence of melanocytes in specific areas of the skin and, consequently, in their depigmentation.

The role of changes in the development rate on the formation of morphological changes in domesticated foxes is manifested through craniological changes. As already noted, a shortened snout is a characteristic of many domesticates (Sánchez-Villagra et al., 2016). The trend toward similar changes in the conformation of tame foxes, as already noted, has been demonstrated at an earlier stage of our experiment (Trut et al., 1991). At the current stage we observe a slowdown of the growth rates of facial skull in tame foxes puppies during the first months of postnatal ontogenesis, which results in a shortening of the skull of tame foxes at the age of 2–3 months (unpublished data).

The selection of foxes for tameness and sensitive period of socialization.

Our work with foxes has demonstrated that a change in the rate of development of certain traits may arise as a correlated consequence of selection for tameability, that is, amenability to domestication. It is known that there is a particular period in early postnatal development of all mammals when young animals are unusually sensitive to socialization and exhibit active exploratory behavior, that is, get acquainted with the environment and adapt to one or another of its social factors (Scott, 1962; Serpell and Jagoe, 1997). In the life of mature-born animals, imprinting plays the huge role in process of adaptation. It extremely reduces the long of socialization period. However, in the early ontogenesis of immature-born animals, which includes the fox, this period begins with the onset of functioning of the sense organs and formation of locomotor activity, which allows an animal to explore its environment. The process of social adaptation in the period of early postnatal development becomes significantly more complex with the maturation of the neurophysiological substrate for fear response. After the first manifestation of fear, the puppies no longer actively explore the environment. The fear response causes them to avoid the action of social factors and is the sign of the end of the most sensitive period of socialization.

This period is critical not only for shaping behavior during ontogenesis, but also for its evolutionary modification. In this regard, of interest is the behavior characteristic of foxes such as tail wagging. At the outset, it occurred as a result of positive tactile sensations from petting and stroking provided by humans in early ontogenesis of animals. In the selection for tameness tail wagging was one of the behaviors which, along with others, served as a selection criterion. Over the course of selection the expression of this behavior became more and more intense, and started to appear more and more earlier in ontogenesis. Eventually, at a certain stage, the physiological threshold for the occurrence of this behavior decreased to such an extent that it began to occur and manifest itself even in the absence of outward trigger stimuli. In Waddington (1953) this process was called "genetic assimilation."

It turned out that the selection of foxes for domestication slows down the rate of fear response development and significantly delays the onset of its manifestation, prolonging period of juvenile behavior. In unselected foxes, the level of exploratory behavior, assessed by the locomotor activity, drops sharply at the age of 45 days, when a sharp rise in the level of corticosteroids is observed in peripheral blood accompanied by initial fear response (Fig. 19.3). In tame individuals this, as well as a rise in corticosteroids levels and a decrease in exploratory behavior, does not occur until the age of 4 months (Belyaev et al., 1984/85; Trut et al., 2004). It is quite possible that a slowdown in the development of the neurophysiological substrate for fear response is only one of the manifestations of the regulatory action on the development of genes involved in the selection for domestication. In this regard, of great interest are the data on the underlying genetic basis for some aspects of dog behavior. Specifically, it has been shown that structural variants in the *GTF2I* and *GTF2IRD1* genes contained within the Williams-Beuren syndrome locus are implicated in hypersocial behavior (vonHoldt et al., 2017). Such a behavior is the expression of neoteny, which is the retention of juvenile behavior into adulthood and increasing the ability of dogs to form primary attachments to human (Udell et al., 2010). As shown, this locus is under positive selection in dogs (vonHoldt et al., 2010).

It can be suggested that rapid changes in the behavior of foxes during domestication experiment are attributable to similar genetic mechanisms.

For tamed foxes changes in both allelic frequencies and in the expression of genes in prefrontal cortex and forebrain were shown (Wang et al., 2018). The two most reliable candidates out of significant genes that demonstrate their upregulation in tame foxes (*DKKL1* and *PCDHGA1*) might be involved in the development of behavioral phenotypes (Garafola and Henn, 2014; Bagot et al., 2016). Encouraging also is that some of genes with different allele frequencies between tame and unselected foxes are important in neural crest cell function (Simões-Costa and Bronner, 2015). This embryonic structure plays an important role in ontogenesis and in the formation of the domestication syndrome (Wilkins et al., 2014). It is important to note that whole-genome SNP analyses of village dogs and wolves identified candidate domestication regions, important for early steps of dog domestication, which also contained numerous genes linked to neural crest and central nervous system development. These genes, as now shown, can be important in forming of specific phenotypes distinguishing early dogs from wolves (Pendleton et al., 2018).

Therefore, the available literature data on dog domestication, as well as the data produced during the fox domestication experiment suggest that in the early stages of animal domestication, the selection might be targeted on a unique suite of behavior-related genes with a large phenotypic effect.

Acknowledgment

This work was supported by grant # AAAA-A17-117072710029-7.

References

Bagot, R.C., Cates, H.M., Purushothaman, I., Lorsch, Z.S., Walker, D.M., et al., 2016. Circuit-wide transcriptional profiling reveals brain region-specific gene networks regulating depression susceptibility. Neuron. 90 (5), 969–983. Available from: https://doi.org/10.1016/j.neuron.2016.04.015.

Belyaev, D.K., Plyusnina, I.Z., Trut, L.N., 1984. Domestication in the silver fox (*Vulpes vulpes*): changes in physiological boundaries of the sensitive period of primary socialization. Appl. Anim. Behav. Sci. 13, 359–370.

Belyaev, D.K., Trut, L.N., 1986. Genetic interrelations of specific changes in standard coat color of silver foxes ("singes" and "star") arising during domestication. Russian J. Genet. 22 (10), 119–128.

Belyaev, D.K., 1979. Destabilizing selection as a factor in domestication. J. Hered. 70, 301–308.

Chen, S., Krinsky, B.H., Long, M., 2013. New genes as drivers of phenotypic evolution. Nat. Rev. Genet. 14 (9), 645–660. Available from: https://doi.org/10.1038/nrg3521.

Darwin, C.R., 1859. On the Origin of the Species. Facsimile of the First Edition. Harvard Univ. Press, Cambridge, MA.

Darwin, C.R., 1868. The Variation of Animals and Plants under Domestication. Murray, John, London.

Drake, J.W., Charlesworth, B., Charlesworth, D., Crow, J.F., 1998. Rates of spontaneous mutation. Genetics. 148, 1667–1686.

Fan, Z., Silva, P., Gronau, I., Wang, S., Armero, A.S., Schweizer, R.M., et al., 2016. Worldwide patterns of genomic variation and admixture in gray wolves. Genome Res. 26, 163–173.

Garafola, C.S., Henn, F.A., 2014. A change in hippocampal protocadherin gamma expression in a learned helpless rat. Brain Res. 1593, 55–64.

Germonpré, M., Sablin, M.V., Stevens, R.E., Hedges, R.E.M., Hofreiter, M., Stiller, M., et al., 2009. Fossil dogs and wolves from Palaeolithic sites in Belgium, the Ukraine and Russia: osteometry, ancient DNA and stable isotopes. J. Archaeological Sci. 36, 473–490.

Gray, M.M., Granka, J.M., Bustamante, C.D., Sutter, N.B., Boyko, A.R., et al., 2009. Linkage disequilibrium and demographic history of wild and domestic canids. Genetics. 181, 1493–1505.

Haldane, J.B.S., 1949. In defence of genetics. Mod. Q. (NS) 4, 194–198.

Lerner, I.M., 1954. Genetic Homeostasis. Wiley, New York.

Lindblad-Toh, K., Wade, C.M., Mikkelsen, T.S., Karlsson, E.K., Jaffe, D.B., et al., 2005. Genome sequence, comparative analysis and haplotype structure of the domestic dog. Nature 438 (8), 803–819. Available from: https://doi.org/10.1038/nature04338.

Mednikov, B.M., 1987. Problem of species formation and adaptive norm. J. Gen. Biol. XLVIII (1), 15–26 (in Russian).

Pendleton, A.L., Shen, F., Taravella, A.M., Emery, S., Veeramah, K.R., Boyko, A.R., et al., 2018. Comparison of village dog and wolf genomes highlights the role of the neural crest in dog domestication. BMC Biology. 16, 64. Available from: https://doi.org/10.1186/s12915-018-0535-2.

Prasolova, L.A., Trut, L.N., 1993. Effect of gene Star on the migration rate of melanoblasts in silver foxes embryos (*Vulpes-vulpes*). Doklady Akademii Nauk 329 (6), 787–789 (in Russian).

Rawles, M.E., 1953. Pigment cell growth. Acad. Press, NY, p. 353.

Sánchez-Villagra, M., Geiger, M., Schneider, R.A., 2016. The taming of the neural crest: a developmental perspective on the origins of morphological covariation in domesticated mammals. R. Soc. open. sci. 3, 160107.

Savolainen, P., Zhang, Y., Luo, J., Lundeberg, J., Leitner, T., 2002. Genetic evidence for an East Asian origin of domestic dogs. Science. 298 (22), 1610–1613.

Schmalhausen, I.I., 1982. Organism as Whole in Individual and Historical Development: Proceedings. Publishing "Science" (in Russian).

Schmalhausen, I.I., 1949. Factors of Evolution. University of Chicago Press, Chicago.

Scott, J.P., 1962. Critical period in behavioral development. Science 138, 949–958.

Serpell, J., Jagoe, J., 1997. Early experience and the development of behavior. In: Serpell, J. (Ed.), The Domestic Dog: Its Evolution, Behavior and Interactions with People. Cambridge University Press, Cambridge, pp. 79–102.

Shishkin, M.A., 1986. Epigenetic system as an object of selective transformation. Morphology and evolution of animals. Publishing "Science", Moscow, pp. 63–74 (in Russian).

Shishkin, M.A., 2018. Evolution as a search for organizational equilibrium. BioSystems. 173, 174–180.

Shishkin, M.A., 1984. Onthogenesis and natural selection. Ontogenez. 15 (2), 115–136 (in Russian).

Simões-Costa, M., Bronner, M.E., 2015. Establishing neural crest identity: a gene regulatory recipe. Development. 142, 242–257.

Smith, J.M., 1983. The genetics of stasis and punctuation. Annu. Rev. Genet. 17, 11–55.

Tatarinov, L.P., 1985. The directionality of phylogenetic process and evolution forecasting. J Gen Biol XLVI (1), 3–19 (in Russian).

Trut, L., Oskina, I., Kharlamova, A., 2009. Animal evolution during domestication: the domesticated fox as a model. BioEssays. 31, 349–360.

Trut, L., Oskina, I., Kharlamova, A., 2012. Experimental studies of early canid domestication. CAB international. In: Ostrander, E.A., Ruvinsky, A. (Eds.), The Genetics of the Dog, second ed. pp. 12–37.

Trut, L.N., 1999. Early canid domestication: the farm-fox experiment. Am. Sci. 87 (2), 160–169.

Trut, L.N., Dzerzhinsky, F.Ya, Nikolsky, V.S., 1991. The intracranial allometry and morphological changes in silver foxes (*Vuples fulvus Desm.*) under domestication. Russ. J. Genet. 27 (9), 1605–1611.

Trut, L.N., Plyusnina, I.Z., Oskina, I.N., 2004. An experiment on fox domestication and debatable issues of evolution of the dog. Russ. J. Genet. 40, 644–655. Available from: https://doi.org/10.1023/B:RUGE.0000033312.92773.c1.

Udell, M.A.R., Dorey, N.R., Wynne, C.D.L., 2010. What did domestication do to dogs? A new account of dogs' sensitivity to human actions. Biol. Rev. Camb. Philos. Soc. 85, 327–345.

Vilà, C., Savolainen, P., Maldonado, J.E., Amorim, I.R., Rice, J.E., Honeycutt, R.L., et al., 1997. Multiple and ancient origins of the domestic dog. Science 276 (5319), 1687–1689.

Vila, C., Seddon, J., Ellegren, H., 2005. Genes of domestic mammals augmented by backcrossing with wild ancestors. Trends Genet. 21, 214–218.

vonHoldt, B.M., Pollinger, J.P., Lohmueller, K.E., Han, E., Parker, H.G., Quignon, P., et al., 2010. Genome-wide SNP and haplotype analyses reveal a rich history underlying dog domestication. Nature. 464 (8), 898–903. Available from: https://doi.org/10.1038/nature08837.
vonHoldt, B.M., Shuldiner, E., Koch, I.J., Kartzinel, R.Y., Hogan, A., Brubaker, L., et al., 2017. Structural variants in genes associated with human Williams-Beuren syndrome underlie stereotypical hypersociability in domestic dogs. Sci. Adv. 3 (7), e1700398. Available from: https://doi.org/10.1126/sciadv.1700398. 19.
Vorobjeva, E.I., 1980. The parallelism and convergence in evolution on *Crossopterygii*. In: Morphological Aspects of Evolution, Moscow, pp. 7–28 (in Russian).
Waddington, C.H., 1953. Genetic assimilation of an acquired character. Evolution. 7, 118–126.
Wang, X., Pipes, L., Trut, L.N., Herbeck, Y., Vladimirova, A.V., Gulevich, R.G., et al., 2018. Genomic responses to selection for tame/aggressive behaviors in the silver fox (*Vulpes vulpes*). PNAS 115 (41), 10398–10403. Available from: www.pnas.org/cgi/doi/10.1073/pnas.1800889115.
Wayne, R.K., O'Brien, S.J., 1987. Allozyme divergence within the Canidae. Syst. Zool. 36, 339–355.
Wilkins, A.S., 2017. Revisiting two hypotheses on the "domestication syndrome" in light of genomic data. Vavilovskii Zhurnal Genetiki i Selektsii = Vavilov J. Genet. Breed. 21 (4), 435–442. Available from: https://doi.org/10.18699/VJ17.26.
Wilkins, A.S., Wrangham, R.W., Fitch, W.T., 2014. The "domestication syndrome" in mammals: a unified explanation based on neural crest cell behavior and genetics. Genetics. 197 (3), 795–808.
Wood, E.K., Maddux, S.D., Southard, T.E., Kharlamova, A.V., Trut, L.N., Franciscus, R.G., 2018. Dental size reduction in tame Russian silver foxes and its implications for the 'self-domestication' model in hominin evolution. Am. J. Phys. Anthropol. 165, 305.

Further reading

Belyaev, D.K., Ruvinsky, A.O., Trut, L.N., 1975. Genetics of the W locus in foxes and expression of its lethal effects. J. Heredity 66, 331–338.
Belyaev, D.K., Ruvinsky, A.O., Trut, L.N., 1981. Inherited activation-inactivation of the star gene in foxes. J. Heredity 72, 267–274.
Freedman, A.H., Wayne, R.K., 2017. Deciphering the origin of dogs: from fossils to genomes. Annu. Rev. Anim. Biosci. 5, 281–307.
Freedman, A.H., Gronau, I., Schweizer, R.M., Ortega-Del Vecchyo, D., Han, E., et al., 2014. Genome sequencing highlights the dynamic early history of dogs. PLOS Genet. 10, e1004016.
Wayne, R.K., Ostrander, E., 1999. Origin, genetic diversity, and genome structure of the domestic dog. BioEssays. 21, 247–257.

The glycobiology of ovarian cancer progression: phenotypic switches and microenvironmental influences

Jimpi Langthasa, Purba Sarkar and Ramray Bhat
Department of Molecular Reproduction, Development and Genetics, Indian Institute of Science, Bangalore, India

Introduction

In an insightful lecture delivered at the Middlesex Hospital during the beginning of the last century, Sir John Bland-Sutton, later to be the President of the Royal Colleges of Physicians and Surgeons, discussed the difficulties in ascertaining the sites of origin of a neoplasm that had spread through the abdomen impacting not just the ovary but the intestines as well. To quote from his lecture, "We are apt to forget that the lethal effects of a carcinoma are not so much due to the primary tumour as to the accidents which arise from it. Cancer will attack an organ and cause no symptoms until it interferes with the function of the organ affected-as, for example, the colon or the pylorus. In many instances it pursues its course unnoticed, and even unsuspected, until it becomes infected with microorganisms; then sepsis and sloughing follow, and in due course produce the general toxic conditions which our seniors used to call cachexia, and regarded as something characteristic of cancer, whereas it is merely an indication of the onset of the terminal affections by which the lives of so many cancerous patients are terminated" (Bland-Sutton, 1908).

Even after about a century, and despite decades of research on early diagnosis and screening of ovarian cancer, the difficulties in treating ovarian cancer and especially its abdominal metastasis, persist. A report from a special conference on ovarian cancer held in 2018 as part of the American Association for Cancer Research emphasizes its bane: when it is detected at

Phenotypic Switching. DOI: https://doi.org/10.1016/B978-0-12-817996-3.00016-5

early stages in a very small fraction of patients, a combination of cytoreductive surgery and chemotherapy is able to formidably stretch the overall survival (OS). However, in most cases, the disease is diagnosed late and therefore results in very poor prognosis (Bast et al., 2019). The reason for this asymmetry can be found in the nature of progression of ovarian cancer. Beginning from the fallopian tubes and ovaries, it spreads to the peritoneal cavity, impacting ultimately the abdominal organs and obstructing their functions. This leads to the patient presenting to the outdoor clinic with broad symptoms such as pelvic discomfort, intestinal obstruction, altered habits of passage of stools and bloating. Clinical, serological, and radiological investigations subsequently help diagnose the disease, which by then has already metastasized and is difficult to treat (Vaughan et al., 2011). Recurrence of ovarian cancer is often coincident with resistance to chemotherapy (Thibault et al., 2014). Properties such as resistance may be acquired not just through a process of evolution within populations of variant cancer cells but also through exploration of the reaction norm of the plastic cancer cells as they migrate into different environments. Morbidity associated with ovarian cancer is more a consequence of the failure of functions of the metastasized abdominal organs, making it essential for experimental oncologists to focus on understanding the cellular-molecular mechanisms underlying its spread. Moreover, the contours of the treatment: cytoreduction and administration of taxane- and platinum-based drugs such as paclitaxel and carboplatin respectively, have remained largely unchanged over the last couple of decades, necessitating research into identification of newer targets, especially those that can be used to attack the micro- and macrometastases.

In this chapter we review the role of glycoconjugates, their binding partners and the enzymes involved in synthesis and degradation of expressed glycans in ovarian cancer. The rationale behind such an exercise is twofold: (1) to achieve an updated and historicist account of the glycobiological mechanisms of ovarian cancer, which we believe to the best of our knowledge, has not been hitherto chronicled in a formal manner and (2) an extensive knowledge of the changes occurring in the 'first sensorial layer' of the cells, the glycocalyx, provides us with a unique understanding of the phenotypic transitions that correlate with metastasis. Before delving into the same, we present, in the following section, a brief bird's-eye overview of the glycobiological mechanisms shown to prominently figure in the etiology of carcinogenesis in general.

Glycobiology and cancer

The observation of altered glycan diversities on the surface of malignant cells in comparison with their untransformed counterparts historically coincides with early and fundamental discoveries of genetic aberrations that are associated with, and causative to, malignant transformation. Therefore, changes in surface glycans or the glycocalyx of neoplastic cells can be considered one of the earliest observed hallmarks of cancer (Varki, 2017). Plant lectins, known for their ability to bind specific sugars on cell surfaces were found to display differential affinity in agglutinating transformed and untransformed cells (Aub et al., 1965a,b). The discovery of altered cell glycocalyces led to long-term and fruitful efforts to decipher its mechanism: differential expression of enzymes involved in transferring these sugars to existent glyco-templates have since been shown to have a profound effect on the phenotype of cancer cells. An oft-used example is the expression and function of an enzyme known as Mgat5, a glycosyltransferase that mediates the addition of β-1,6-*N*-acetylglucosamine to the α-linked mannose of biantennary *N*-glycans. Upregulated in cancers due to the regulatory influence of oncogenic transcription factors, Mgat5 extends the branching of N-linked glycans and enhances the invasive and metastatic properties of breast and mammary cancer cells (Buckhaults et al., 1997; Granovsky et al., 2000). Similarly, fucosylation and sialylation of the core and terminal portions of N-linked glycans, respectively mediated by upregulated glycosyltransferases increases the invasion and metastasis of a host of cancer epithelia (Pearce and Laubli, 2016; Shan et al., 2019; Tu et al., 2017). The literature on cancer sialylation is too vast to be summarized in this section and the reader is directed to the following excellent reviews on the subject (Pearce and Laubli, 2016) (Bull et al., 2014). O-linked glycans conjugated to serines of glycoproteins such as the mucin family, extensively populate the surface of epithelia. Prolific glycosylations of mucins bring about charge-dependent and steric changes on apical surfaces of epithelia leading to altered physicochemical properties of the tissue microenvironment. This has led to exciting advances at the interface of mechano-glycobiology (Gandhi et al., 2019). Paszek and coworkers have shown that bulky glycans constituting the glycocalyces can order the integrin receptors on the cell surfaces to enhance mechanotransductive signaling that is specific to metastatic tumor niches (Paszek et al., 2014). Shurer and coworkers show how the composition and topology of glycans on the surface of cells alter the physical properties of the membranes: the presence of mucin biopolymers and long chains of glycan polymers modulate the ability of membranes to conform to intracellular forces that guide the formation of cellular projections. Such projections are cognate to shape changes occurring within cells during invasion and migration (Shurer et al., 2018, 2019).

A special class of glycoconjugates, known as proteoglycans, are characterized by core proteins conjugated through their serines to long polymeric chains of iterated disaccharide units consisting of an amino sugar (either N-acetylated or N-sulfated glucosamine or N-acetylgalactosamine) and a uronic acid (glucuronic or iduronic acid) or galactose. These chains are known as glycosaminoglycans (GAGs) (Varki, 2017). The monosaccharides are often sulfated making these macromolecules among the most anionic of biomaterials (Yanagishita, 1993). Proteoglycans are dispersed throughout tissue matrices and provide the tissues with hydration and resistance to compression (Mouw et al., 2014). The largest class of proteoglycans is that of small leucine-rich proteoglycans (SLRPs) (Iozzo and Murdoch, 1996). Several SLRPs share a hallmark of the ability to regulate collagen fibrillogenesis (Chen and Birk, 2013). They are known to modulate collagen fibril growth, fibril organization, and extracellular matrix assembly (Ameye and Young, 2002). They also have a role in mediating cell-matrix interactions. Therefore, SLRPs have the ability to influence cell functionality including differentiation, apoptosis, proliferation, and migration through multiple mechanisms (Kadler et al., 2008). The GAGs that construct proteoglycans can be classified into a number of types based on the chemical backbone of the GAG. These include dermatan sulfate (DS), chondroitin sulfate (CS), heparan sulfate (HS)/heparin, and keratan sulfate (KS) (a fifth subclass of GAGs known as hyaluronan (HA) is not covalently conjugated to any proteins) (Vallen et al., 2014). Several cancers are known to exhibit increase in levels of HS proteoglycans (HSPGs) and HA. HSPGs are known to sequester growth factors and regulate their availability to cells: increased heparanase expression by cancer epithelia serve to release these growth factors and provide themselves with additional proliferative cues (Ilan et al., 2006). HA on the other hand has been shown to interact with cell surface receptors such as CD44 and CD168, altering the transductive dynamics of key signaling pathways such as those stimulated by receptor tyrosine kinases (Theocharis et al., 2019).

Along with the alterations in glycosylations associated with cancer (including, but certainly not limited to those mentioned above), there are also marked changes in the expression of proteins that bind to them: the endogenously expressed lectins. These include the most well-studied family of lectins known as galectins, which are known to bind β-galactoside containing glycoconjugates. Extensive research by several groups show galectins as being expressed to greater levels in cancer cells (Lahm et al., 2000) and the role of galectins in cancers is extremely context-dependent: when present in the extracellular milieu they modulate interactions between the cells and their surrounding matrices [glycan-dependent binding of Galectin-8 to leukocyte cell adhesion molecule (CD166) results in the trapping of the latter on the surface of breast cancer cells (Fernández et al., 2016)]. However, within nuclei, they are known to alter transcriptional responses through interactions with splicing modulators, and transcriptional factors in breast cancer epithelia (Bhat et al., 2016; Gao et al., 2018). Selectins, cell surface glycoproteins that require calcium ions to bind to sugars (and hence are known as C-lectins) are expressed on the surface of endothelia and white blood cells and may modulate their interaction with cancer cells, influencing in turn, tumor evasion. In line with this notion, injection of human tumor cells subcutaneously within mice that lack the expression of selectins results in enhanced oncogenic progression (Bhat et al., 2016). Richard Hyne's group has also shown that the recruitment of natural killer (NK) cells is required for the recognition of cancer cells by the immune system: this is however dependent on selectin levels. In selectin-null mice, the recruitment of NK cells to the tumor site is impaired leading to their exacerbated growth and progression (Sobolev et al., 2009). Another member of the family, the mannose receptor has also been implicated in metastasis. Generally known to be expressed on the surface of dendritic and other immune cells, its expression is elevated in gastric and colorectal cancer cells and is associated with poor prognosis.

This suggests that the dynamical interplay between the aberrantly regulated and transcribed genome of cancer cells and physicochemical cues from the tissue microenvironment often involves an 'intermediate filter': the dynamical glycocalyx. The latter serves both to contextualize the transduction of mechanical cues from the exterior to the cytosol, and to modulate the presentation of the secretome of the transformed cells. The heterogeneity of cancer cell phenotypes within tumor populations may also be reflected in the diversity of glycan-glycan binding protein levels at the surface of distinct tumor epithelia. In addition, examination of the glycans on transitory metastatic cells is bound to throw up several surprises: such alterations are potentially responsible for the signal that imparts survival and quiescence within such niches.

In the following sections, we will briefly review the current literature on the glycobiological dimension associated with ovarian cancer metastasis. In the first section, we will build a temporal narrative of ovarian cancer progression from a cellular-molecular perspective. The second and third sections will be devoted to elucidating the roles of N- and O-linked glycoproteins, and proteoglycans in metastasis. In the fourth, we will discuss how glycan-binding proteins affect the phenotype of ovarian cancer. The concluding section will elaborate open questions in the field in a bid to advocate a greater focus on glycobiological investigations of this disease and ensuing potential future diagnostic-therapeutic possibilities.

Ovarian cancer progression

How it begins

Ovarian cancer can be considered to be a "catch-all" phrase for a host of carcinomas that may differ in their oncogenic mutation signatures and histopathological origins (Kim et al., 2018). Typically, the most common and aggressive form of ovarian cancers are epithelial in origin, serous in histology and high-grade (hence high-grade serous ovarian cancer or HGSOC). Epithelial ovarian cancers arise from the epithelial lining of the peritoneum, which is derived from the mesoderm and is termed mesothelium. This lining therefore includes epithelia that cover the peritoneal side of the ovary [where the cells are known as ovarian surface epithelium (OSE)] and the fallopian tube (Kim et al., 2018).

The majority of ovarian cancers arise from epithelial cells. They are further subdivided into serous, mucinous, endometrioid, clear cell, transitional, and mixed epithelial type of tumors. Of these, serous tumors are most common presenting as uni- or multilocular cysts filled with serous fluid. Nonepithelial ovarian cancers may arise from the germ cells or the ovarian mesenchymal stroma (Kaku et al., 2003). Based on clinical and genetic observations, ovarian cancers can be classified into two types. Type 1 includes low-grade serous, endometrioid, clear cell, and mucinous cancers. These are indolent in their progression with typically better patient outcomes. Type 2 cancers consist of HGSOC, carcinosarcomas, and undifferentiated high-grade carcinomas with poor outcomes.

The site of origin of HGSOCs has been a matter of heated debate. The OSE has been traditionally thought to seed the growth of HGSOC. A histopathological phenomenon associated with OSE is the occurrence of the cortical inclusion cyst, the invagination of OSE in an ovulation-dependent or -independent manner into the cortical stroma. Inclusion cysts are thought to prefigure cancers that originate from the OSE (Banet and Kurman, 2015). However, the manipulations of murine OSE through mutations that are typical of HGSOCs result in slow kinetics of dissemination and inefficient metastasis unlike what is seen in the human cases. One possible way out of the paradox could be an alternative site of origin: this had led to a strong advocacy of a fallopian epithelial origin of HGSOC. In fact, in woman carrying a germline BRCA mutation, who had opted for a prophylactic removal of their ovaries and fallopian tubes, the latter was found to have several sites of epithelial dysplasia, known as serous tubal intraepithelial carcinoma (STIC) (Mingels et al., 2012). STIC is considered to be a premalignant state, such as ductal carcinoma in situ of breast with cells that do not proliferate but display pleomorphic darkly colored nuclei with prominent nucleoli (Voltaggio et al., 2016). Current investigations show however that the occurrences of STIC and HGSOC are not completely correlative (Eckert et al., 2016). Nevertheless, Type 2 ovarian cancers are increasingly thought to be of a fallopian origin with distinct genomic signatures (TP53 mutations) compared with OSE-derived Type 1 ovarian cancers with mutations in PTEN, BRAF, and KRAS (Kurman and Shih, 2011).

Considerable progress has also been made in the identification of molecular pathways involved in the regulation and progression of ovarian cancer epithelia within the primary tumor foci (Fig. 20.1). Wang and coworkers have recently shown that the miR-21, highly expressed in ovarian cancer cells activates Wnt signaling to regulate the expression of CD44v6 which increases the proliferation and migration of tumor epithelia (Wang et al., 2019). Hou and coworkers have demonstrated interesting dynamics of ABCB9, an important member of the multidrug resistance transporter family, which is upregulated in early stages of ovarian cancer but decreases in advanced stages, when pathways related to epithelial-to-mesenchymal transition (EMT), and TGF-beta-driven transduction are enriched (Hou et al., 2019). Delta-like 1 homolog, a noncanonical Notch ligand has also been shown to be overexpressed and participates in ovarian carcinogenesis through Notch activation and induction of EMT (Huang et al., 2019). Recent treatments with the inhibitor of DNA methyltransferase (DNMT), 5-Azacytidine show that M1 macrophages are likely suppressed during ovarian cancer progression and if activated, by DNMT inhibition, may increase patient survival and decrease burden (Travers et al., 2019).

How it transitions

Ovarian cancer metastasizes predominantly via the coelomic route wherein disseminated cells form spheroids: multicellular

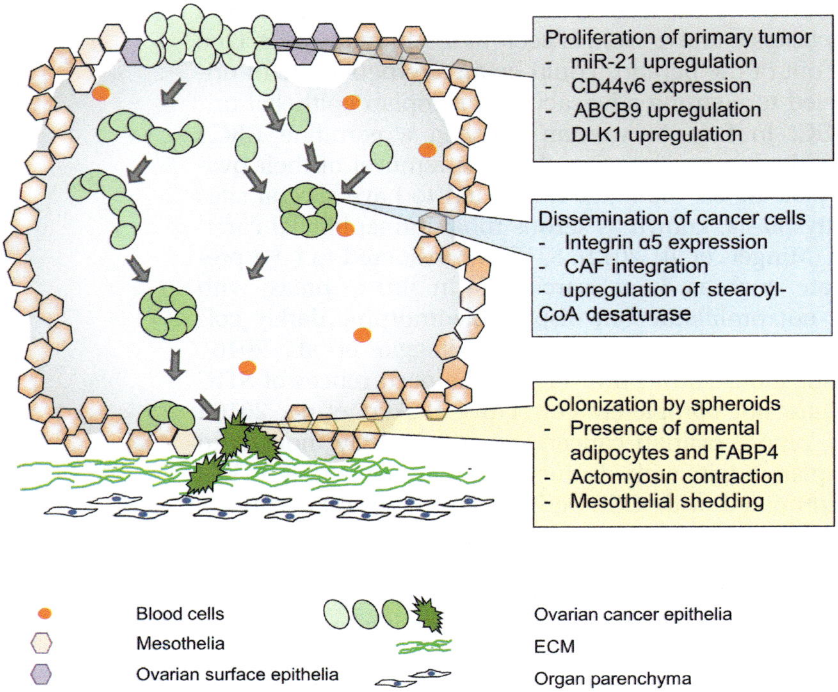

Figure 20.1 Schematic depiction of ovarian cancer progression. Text bubbles denote specific experimental observations on cellular-molecular phenomena underlying the proliferation of cancer cells in the primary ovarian tumor (*top*), peritoneal dissemination of cancer cells and the formation of spheroids (*middle*) and the colonization of underlying tissues by spheroids (*bottom*).

clusters of epithelial carcinoma cells that are found in the peritoneal fluid of patients with advanced ovarian cancer (Lengyel, 2010). The occurrence of spheroids correlates strongly with the recurrence of the disease as well as the acquisition of resistance to the first line of platinum-based chemotherapeutic drugs (Latifi et al., 2012). Spheroids are also known to harbor cellular niches that are dormant and have stem cell-like properties (Peart et al., 2015).

The exact mechanism of spheroid formation in patients is still under active investigation but two competing conceptual models exist. According to the first model, spheroids form via cell-cell adhesion after the ovarian cancer cells have shed from the primary foci (Boylan et al., 2016). The expression of vimentin and a concomitant decrease in cortical E-cadherin has been implicated to play a role in dissemination (Pease et al., 2012). According to the second, cancer cells detach not as single cells but as clusters or sheets that achieve a spheroidal geometry through unknown mechanisms in the fluid milieu (Al Habyan et al., 2018) (Fig. 20.1). Whichever may be the mode of dissemination, multicellular spheroids are more resistant to anoikis and presumably have a survival advantage over single cells. They also preserve, and may amplify, the intratumoral heterogeneity from the primary tumor with niches that can form intraabdominal metastases (Al Habyan et al., 2018; Pease et al., 2012). In fact, ovarian cancer spheroids may have distinct cell populations: spheroids constituted of disseminated cancer cells expressing high integrin α5 and cancer-associated fibroblasts have been proposed to be particularly suited to peritoneal invasion (Gao et al., 2019). Metabolic changes occur within spheroids as well: upregulation of stearoyl-CoA desaturase gene, which affects lipid homeostasis and suppresses ER stress enhances spheroid formation (Zhang et al., 2019).

Given the high potential for dissemination into a fluid microenvironment for ovarian cancers, the disease is an exemplar for understanding how the microenvironment may change cell states. For example, anchorage-independence, the regimes of flow, and its associated shear forces have been proposed to be responsible for increase in properties of resistance (Ip et al., 2016).

How it colonizes

The progression of metastasis continues with the homing of cancer cells and spheroids to the thin mesothelial layer that covers the abdominal organs. The mesothelia are then "cleared" by the proliferating and invading cancer epithelia. Iwanicki and coworkers observed an increased contractility of myosin II in cancer cells. Downregulation of α5 integrin, talin, or myosin II decreased contractility and mesothelial clearance by ovarian cancer spheroids. These observations suggest that mesothelial clearance may be dependent on cellular contractile forces from cancer cells (Davidowitz et al., 2012; Iwanicki et al., 2011) (Fig. 20.1). The final step is the colonization of underlying organs, whereby the liver, stomach, transverse colon, greater omentum, and the small intestine act as potential sites for the formation of metastasis (Davidowitz et al., 2014). The concept of preferential metastasis dates back to 1889, when Steven Paget published his "seed and soil" hypothesis where he stated that the pattern of metastatic spread of cancer is not random and that cancer cells exhibit preferences when metastasizing to organs. He proposed that the spread of tumor cells is governed by interaction and cooperation between the cancer cells (seed) and the host organ (soil) (Mikula-Pietrasik et al., 2018; Paget, 1889).

Ovarian cancer cells have a predisposition for metastasis to specific peritoneal locales, like the omentum, an organ largely made of adipocytes. At present, the mechanisms underlying the preferential homing of tumor cells to the omentum are not fully understood. Lengyel's group has shown that primary omental adipocytes promote homing, migration, and invasion of ovarian cancer cells. They demonstrate in cocultures of adipocytes and ovarian cancer cells that lipids are directly transferred from adipocytes to ovarian cancer cells, which led to growth of tumor in vitro and in vivo. Transferred fatty acids lead to rapid tumor growth in the secondary site. The role of fatty acid-binding protein 4 in omental metastasis of ovarian cancer is particularly noteworthy: its deficiency diminishes metastatic tumor growth in mice. Therefore, adipocytes constitute a major component of the "seed" microenvironment supporting the growth of cancer cells (Lengyel, 2010).

Li and coworkers have demonstrated that a glycoform of the protein IGF-1R of ovarian metastatic cancer stem cells, bears sialyl-lewis X, which acts as a ligand for P-selectin, present on human peritoneal mesothelial cells. The interaction between P-selectin and sialyl-lewis X mediates the adhesion of the cancer stem cells to the peritoneal mesothelial cells under ascitic shear stress. Their findings also bring out the critical role played by glycosyltransferases, particularly, fucosyltransferase-5, in the development and subsequent maintenance of metastasis (Li et al., 2019).

In the next three sections, we will showcase the relevant glycobiological molecules that contribute to the structure of this framework through modulation of behaviors of cells that are part of distinct metastatic niches of ovarian cancer.

N- and O-linked glycoproteins

A plethora of diseases not excluding ovarian cancer are associated with aberrant N-linked glycosylations (An et al., 2006; Saldova et al., 2007). Abnormal glycosylations have been proposed to contribute to cancer metastasis (Stowell et al., 2015). One of the earliest formal studies on the glycoforms secreted by ovarian cancer cells has been conducted by Zhang and coworkers. Performing Matrix-assisted laser desorption/ionization (MALDI)-based quantitative estimations of glycoforms secreted by the commonly employed cell line SKOV3 and a more metastatic variant SKOV3-ip, they show that N-glycan geometries with bisecting GlcNAc are relatively depleted in the more metastatic cell line (Fig. 20.2). This has been subsequently confirmed through the determination in expression of the enzyme involved in the synthesis of the bisecting topology, MGAT3 (Zhang et al., 2014b). Curiously though, Karen Abbott's group have verified that the mRNA levels of Mgat3 GnT-III, are elevated in epithelial ovarian carcinomas leading to bisection by GlcNAc of a plethora of vital proteins involved in signaling, including but not limited to, Notch, TGF-β and Wnt (Kurman and Shih, 2011). Subsequently, the same group has gone on to show that the suppression of Mgat3-encoded enzyme GnT-III, leads to decreased growth and metastasis of ovarian cancer in a xenograft murine

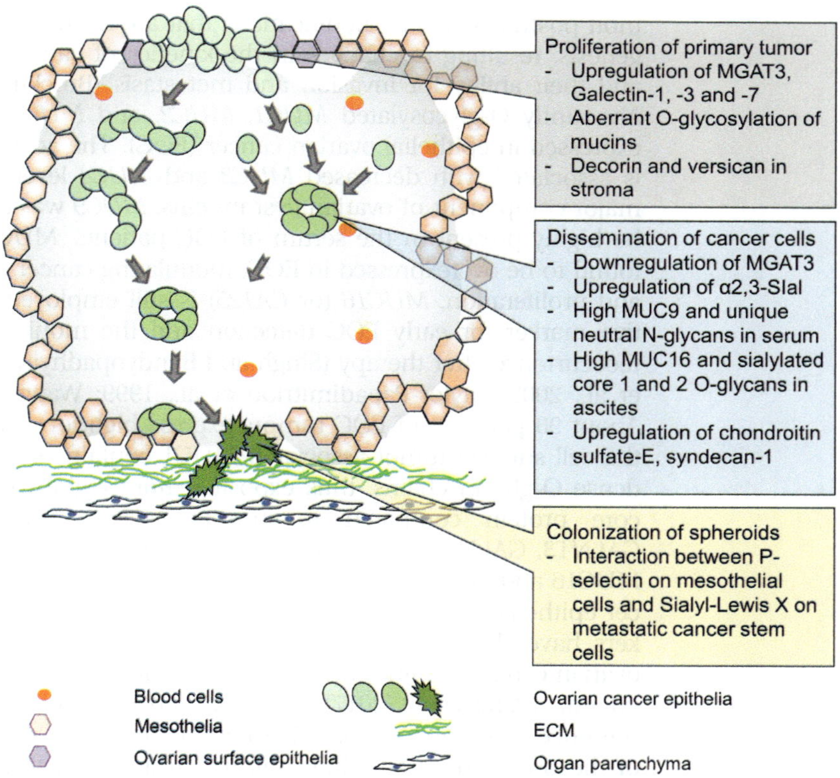

Figure 20.2 Schematic depiction of the glycobiological changes in ovarian cancer progression. Text bubbles denote specific observations on the alterations in expression and functions of N- and O-linked glycoproteins, proteoglycans, and lectins during the proliferation of cancer cells in the primary tumor (*top*), peritoneal dissemination of cancer cells and the formation of spheroids (*middle*), and the colonization of underlying tissues by spheroids (*bottom*).

model (Allam et al., 2017). This may be explained by the role of GnT-III in the glycosylation and activation of Notch: relocation of Notch receptor to the lysosome as a consequence of GnT-III suppression reduces its levels and impairs its activation to a greater extent than γ-secretase inhibition. The seemingly confusing relation between GnT-III and invasion has been partially addressed by Lu and coworkers who posit that the overexpression of GnT-III inhibits migration through a downregulation of α2,3-sialylation: however, this inhibition is manifested in cells, which have low levels of α2,6-sialic acid levels (Lu et al., 2016). Therefore, the effect of GnT-III expression on the phenotypes of ovarian cancer cells at different spatial and temporal stages of their progression is contextualized on the expression of distinct sialic acid linkages.

Aberrant mucin-type O-glycosylation is one of the most common posttranslational modifications that are misregulated in oncogenesis, resulting in changes in the kinetics of cancer cell survival and their ability for invasion and metastasis (Brockhausen, 1999). Aberrantly O-glycosylated *MUC1, MUC2,* and *MUC5AC* are overexpressed in epithelial ovarian cancer (EOC). The advanced disease is associated with decreased *MUC3* and *MUC4* levels. *MUC6* is a major component of ovarian cyst mucins. *MUC9* was also found to be highly present in the serum of EOC patients. MUC13 was also found to be overexpressed in EOC, modulating cancer cell adhesion and proliferation. *MUC16* (or *CA125*) is still employed as an effective marker for early EOC detection and the monitoring of EOC reoccurrence after therapy (Singh and Bandyopadhyay, 2007; Swartz et al., 2002; Taylor-Papadimitriou et al., 1999; Wang et al., 2014). About 90 per cent of EOC tumors express high levels of MUC1 on the cell surface: tumor-associated MUC1 contains shorter and less dense O-glycan chains, thus exposing more regions of the MUC1 core protein compared to normal MUC1. Transferases like GALNT3, GALNT14 contribute to the overexpression of MUC1 and MUC16 and other *O*-glycosylated mucin-like targets in ovarian cancer epithelia (Park et al., 2010; Wang et al., 2014). Yang and coworkers have shown that the inhibition of O-linked glycosylation in ovarian cancer cell lines brings about an increase in N-linked glycosylation (Yang et al., 2017). Kailemia and coworkers observed the presence of unique glycans like neutral structures composed only of hexose and *N*-acetylhexosamine residues in the serum of patients and conditioned medium from OC cell lines (Kailemia et al., 2017). After cleaving O-glycans from conditioned media and patient sera, profiling was done using high resolution MALDI-Fourier-transform ion cyclotron resonance-mass spectrometry. Structural identification of glycans was then performed by tandem mass spectrometry using infrared multiphoton dissociation (Kailemia et al., 2017). Within glycoproteins from ascites from ovarian cancer patients, O-linked glycans were observed to comprise mostly sialylated core 1 and 2 structures, and to a lesser extent sulfated core 2 structures (Karlsson and McGuckin, 2012).

Proteoglycans in ovarian cancer

Large amounts of DS are observed in normal ovarian surface epithelia as well as in the stroma of ovarian tumors and transformed epithelia (Ricciardelli and Rodgers, 2006; Ten Dam et al., 2009). Using a novel antibody GDA321, Ten Dam and coworkers observed DS colocalization with decorin, versican, and Type 1 collagen within cancer tissues. Another study identifies versican as being upregulated in the stromal regions of high-grade serous ovarian cancer wherein its expression correlates with invasion and platinum resistance. Versican is traditionally a CS proteoglycan although its colocalization in the previous study with DS along with the demonstration that the epimerase that converts CS to DS is also elevated in cancers may suggest that versican may have DS GAGs upon malignant transformation (Ghosh et al., 2010) (Fig. 20.2). CSs are further subdivided into various classes due to the presence of differently sulfated disaccharide units. These include CS-A (4-sulfated), CS-C (6-sulfated), CS-D (2, 6-sulfated), and CS-E (4, 6-disulfated). Various studies on cancerous tissues have demonstrated alterations in the sulfation level and type of disaccharide units (Vallen et al., 2012; Weyers et al., 2012, 2013). More invasive characteristics have been seen to be associated with the ability of ovarian cancer cells to form compact spheroids (Burleson et al., 2006; Sodek et al., 2009). Kuppevelt's group has shown that CS, especially CS-E has a very potent role in ovarian carcinogenesis. Using exogenous CS-E as well as with genetic perturbation of the enzyme GalNAc4S-6 sulfotransferase, their group shows that the CS promoted cell adhesion and forms more compact and bigger spheroids (Vallen et al., 2014). In their analyses of O-linked glycoconjugates in ovarian cancer ascites, Karlsson and McGuckin observed much of these to be extended KS proteoglycans (Karlsson and McGuckin, 2012). The role of these KS remains to be elucidated. The role of HSPGs in ovarian cancer is unsurprisingly context-dependent. Whereas syndecan-1, not expressed in ovarian tissues but upregulated in stroma of both primary and metastasized ovarian cancer cells, potentiates their growth, the downregulation of glypican-3, another HSPG, is associated with higher proliferation and metastasis (Guo et al., 2015; Liu et al., 2015).

Lectins

Among endogenously expressed glycan binding proteins (or lectins), C-type lectins (which require Ca^{2+} for binding) play a role in promoting tumor growth and metastasis. Carroll and coworkers have investigated the role of a C-type lectin known as P-selectin in ovarian cancer where they show that the interactions between tumor cells, mesothelial cells, and alternatively activated macrophages (AAM) influences the adhesion of tumor cells to mesothelial cells (Carroll et al., 2018). The study revealed AAMs secrete macrophage inflammatory protein MIP-1β, which activates CCR5/PI3K signaling in mesothelial cells, resulting in expression of P-selectin on the mesothelial cell surface. Tumor cells subsequently attached to this de novo P-selectin through CD24, resulting in increased tumor cell adhesion in static and flowed conditions (Carroll et al., 2018).

Galectins, S-type lectins are endogenously expressed proteins with a common jelly-roll fold and an affinity for binding lactose- or galactose-containing glycoconjugates (Barondes et al., 1994). Galectins bind to sugars through a conserved domain known as carbohydrate recognition domain. Of about fifteen galectins, Galectin-1 (Gal-1) has been shown to be expressed in the primary focus of serous ovarian adenocarcinoma (Zhang et al., 2014a). High Gal-1 expression in primary tumors correlates with poor prognosis for patients (Fig. 20.2). It relates to the long-range dissemination of tumoral cells (metastasis), dissemination into the adjacent tissue, and tumor immune escape (Zhang et al., 2014a). Downregulation of Gal-1 restores the sensitivity of ovarian cancer cells to cisplatin, whereas upregulation of Gal-1 results in chemoresistance (Zhang et al., 2014a). It implies that Galectin-1 functions may nucleate potential therapeutic studies contributing to the treatment of cisplatin-resistant cancer cells (Zhang et al., 2014a). In addition, Zucchetti and coworkers observed that OTX008, a calixarene-based Galectin-1 inhibitor decreases the growth of ovarian cancer cell lines (Zucchetti et al., 2013).

Apart from Gal-1, Gal-3 is also overexpressed in, and localized to, the cytoplasm and nucleus of serous epithelial ovarian cancer tissues (Brustmann, 2008; Liu et al., 2018). High expression levels of Gal-3 were associated with advanced clinical stage, platinum resistance, recurrence, high histological grade, and poor survival of the patients. β-catenin and Gal-3 expression were also found to be closely correlated with each other in advanced stage- and grade-resistant ovarian cancer (Liu et al., 2018). In addition, downregulation of Gal-3 showed decreased expression of the downstream molecules of Wnt/β-catenin pathway: c-myc and cyclin D1, which suggests that Gal-3 might activate Wnt/β-catenin signaling pathway in serous ovarian cancer (Liu et al., 2018). In addition, high serum levels of Gal-3 in ovarian cancer patients is also known to be associated with paclitaxel resistance and prone to disease recurrence, especially with the presence of local recurrence, carcinomatosis, lymphadenopathies, and distant metastasis (Schulz et al., 2017). This was further supported by the study, where a truncated, dominant-negative form of Galectin-3, known as Galectic-3C, was applied to ovarian cancer cell lines and primary cells established from ovarian cancer patients (Mirandola et al., 2014). Galectic-3C significantly reduced the growth, motility, invasion, and angiogenic potential of cultured ovarian cancer cell lines and primary cells, indicating its promise for further exploration as treatment for ovarian cancer (Mirandola et al., 2014).

Whereas Gal-7 expression is not seen in ovarian surface epithelia, its cytoplasmic localization can be appreciated in malignant cells. High Gal-7 expression is associated with more advanced grade tumors, metastasis, high mortality, and poor OS outcome in patients with EOC (Labrie et al., 2014). An in vitro study done by Labrie et al. suggests that p53 mutation induces Gal-7 expression and increases invasiveness of ovarian cancer cells by inducing MMP9 (Labrie et al., 2014). Higher expression of Gal-8 and Gal-9 showed significantly better overall- and disease-free survival in OC patients. However, moderate expression of Gal-9 in the cytoplasm of ovarian cancer cells showed a decreased progression-free survival and OS when compared to galectin-negative cases with Kaplan-Meier analysis (Schulz et al., 2018).

Cancer evolution and glycan-driven plasticity

The role of the microenvironment in the evolution of cancer cell survival and chemoresistance can be twofold. On the one hand, the microenvironment works as a filter for selection of specific cellular variants, resulting in the modulation of intratumoral heterogeneity. This may ensure that the cells that are disseminated into the ascites are resistant by selection to shear and anoikis. If selection were to act on a given trait that shares its underlying mechanism with other traits through convergent signal transduction, etc., such as resistance to shear and dormancy, then additional trait variation may be acquired in an exaptive way without necessarily exposing cells to simultaneous selective pressures. The second role of the tissue microenvironment is a more active one: instead of a selective filter, it transforms the cell phenotype that is inherently plastic, leading to the exhibition of traits that are concomitant with the environmental cues. Extensive demonstration by theorists has established that plasticity may also act as a major driver for the fixation of traits that persist even after the environmental cue is no longer present (Behera and Nanjundiah, 2004; Kaneko, 2009; Newman et al., 2009; West-Eberhard, 2005).

Notably, the alteration in the glycan repertoire in association with evolution of chemoresistance or invasiveness of cancer cells has been experimentally observed as described earlier. However, it is not clear whether such changes occur through selection of specific glycophenotypes, or whether phenotypic switching happens immediately over a few generations of cell division. Cancer cells are known to transition between different morphological states that are in turn underpinned by specific expression-based and biochemical signatures. Future studies will help us understand whether the plasticity of glycophenotypes and populational heterogeneity of glycophenotypes act in synergistic ways to exacerbate the challenges in therapy of ovarian cancer.

Conclusion

Our understanding of the glycobiological changes in ovarian cancer potentially holds clues to the development of novel future diagnostic and therapeutic strategies. Both the well-known Cancer antigen 125 (CA125) and human epididymis protein 4 (HE4) (which is being increasingly introduced currently for diagnosis) are two FDA-approved glycoprotein biomarkers for ovarian cancer. Targeting CA125 glycan alterations with *Vicia villosa* lectin, Chen and coworkers distinguished between benign and invasive ovarian neoplasms (Chen et al., 2013). Because CA125 glycan alterations include an increase in the fucosylation of their biantennary glycans, the *Lens culinaris* agglutinin, and *Ulex europaeus* agglutinin I that specifically recognize α1,6- and α1,2-linked fucose, have been investigated as diagnostic markers for ovarian cancer. The GAG HA has been considered a promising therapeutic target as a consequence of its ability to mediate adhesion of ovarian cancer cells to the peritoneum and confer them with resistance to chemotherapeutics in a CD44-dependent manner (Burleson et al., 2004; Catterall et al., 1999; Gardner et al., 1996; Yeo et al., 1996).

CD44 and β1 integrins mediate the migration of ovarian cancer cells within extracellular matrix proteins (Casey and Skubitz, 2000). Utilizing the HA-CD44 interaction, HA-paclitaxel conjugates have been employed to achieve better internalization of the taxanes into the metastasized cancer cells (Auzenne et al., 2007).

It is pertinent as we conclude our brief survey of ovarian cancer glycobiology to arrive at a few vital take-home principles: the dysregulation of glycoconjugates on the cell surfaces upon malignant transformation of ovarian cells is not an 'anarchic' one: specific chemical and evolutionary principles govern which glycosylations, and in what kind of arrangements eventually comprise the cancer glycocalyces as cells metastasize. This also implies that the formation of the glycocalyx is a multitrait systems property, which needs to be dissected with frameworks that incorporate experiments embedded in appropriate mathematical-computational efforts. A second corollary is that metastatic cells progress through multiple microenvironments and such systems properties may be cognate to specific microenvironments and hence change. Therefore, it is germane to analyze the expression and functions of glycosylations on cells within specific niches, as well as in a continuum across different spatial compartments. We conclude with the hope that future endeavors will flesh out such principles that are unique to ovarian cancer as also those that are universal across other neoplasia.

Acknowledgments

We would like to thank Shyamili Goutham for valuable inputs while writing the review. This work was supported by the DBT [BT/PR26526/GET/119/92/2017] SERB [ECR/2015/000280] and the Wellcome Trust/DBT India Alliance Fellowship [IA/I/17/2/503312] awarded to RB. PS would like to acknowledge funding support from DST INSPIRE fellowship (IF170197, IVR number-201600020331).

References

Al Habyan, S., Kalos, C., Szymborski, J., McCaffrey, L., 2018. Multicellular detachment generates metastatic spheroids during intra-abdominal dissemination in epithelial ovarian cancer. Oncogene 37 (37), 5127–5135. Available from: https://doi.org/10.1038/s41388-018-0317-x.

Allam, H., Johnson, B.P., Zhang, M., Lu, Z., Cannon, M.J., Abbott, K.L., 2017. The glycosyltransferase GnT-III activates Notch signaling and drives stem cell expansion to promote the growth and invasion of ovarian cancer. J. Biol. Chem. 292 (39), 16351–16359. Available from: https://doi.org/10.1074/jbc.M117.783936.

Ameye, L., Young, M.F., 2002. Mice deficient in small leucine-rich proteoglycans: novel in vivo models for osteoporosis, osteoarthritis, Ehlers-Danlos syndrome, muscular dystrophy, and corneal diseases. Glycobiology 12 (9), 107R–116R. Available from: https://doi.org/10.1093/glycob/cwf065.

An, H.J., Miyamoto, S., Lancaster, K.S., Kirmiz, C., Li, B., Lam, K.S., et al., 2006. Profiling of glycans in serum for the discovery of potential biomarkers for ovarian cancer. J. Proteome Res. 5 (7), 1626–1635. Available from: https://doi.org/10.1021/pr060010k.

Aub, J.C., Sanford, B.H., Cote, M.N., 1965a. Studies on reactivity of tumor and normal cells to a wheat germ agglutinin. Proc. Natl. Acad. Sci. U S A 54 (2), 396–399.

Aub, J.C., Sanford, B.H., Wang, L.H., 1965b. Reactions of normal and leukemic cell surfaces to a wheat germ agglutinin. Proc. Natl. Acad. Sci. U S A 54 (2), 400–402.

Auzenne, E., Ghosh, S.C., Khodadadian, M., Rivera, B., Farquhar, D., Price, R.E., et al., 2007. Hyaluronic acid-paclitaxel: antitumor efficacy against CD44(+) human ovarian carcinoma xenografts. Neoplasia 9 (6), 479–486. Available from: https://doi.org/10.1593/neo.07229.

Banet, N., Kurman, R.J., 2015. Two types of ovarian cortical inclusion cysts: proposed origin and possible role in ovarian serous carcinogenesis. Int. J. Gynecol. Pathol. 34 (1), 3–8. Available from: https://doi.org/10.1097/PGP.0000000000000120.

Barondes, S.H., Cooper, D.N., Gitt, M.A., Leffler, H., 1994. Structure and function of a large family of animal lectins. J. Biol. Chem. 269 (33), 20807–20810.

Bast Jr., R.C., Matulonis, U.A., Sood, A.K., Ahmed, A.A., Amobi, A.E., Balkwill, F. R., et al., 2019. Critical questions in ovarian cancer research and treatment: report of an American Association for Cancer Research Special Conference. Cancer 125 (12), 1963–1972. Available from: https://doi.org/10.1002/cncr.32004.

Behera, N., Nanjundiah, V., 2004. Phenotypic plasticity can potentiate rapid evolutionary change. J. Theor. Biol. 226 (2), 177–184. Available from: https://doi.org/10.1016/j.jtbi.2003.08.011.

Bhat, R., Belardi, B., Mori, H., Kuo, P., Tam, A., Hines, W.C., et al., 2016. Nuclear repartitioning of galectin-1 by an extracellular glycan switch regulates mammary morphogenesis. Proc. Natl. Acad. Sci. 113 (33), E4820–E4827. Available from: https://doi.org/10.1073/pnas.1609135113.

Bland-Sutton, J., 1908. A clinical lecture on cancer of the ovary: delivered at middlesex hospital. Br. Med. J. 1 (2453), 5–7. Available from: https://doi.org/10.1136/bmj.1.2453.5.

Boylan, K.L.M., Buchanan, P.C., Manion, R.D., Shukla, D.M., Braumberger, K., Bruggemeyer, C., et al., 2016. The expression of Nectin-4 on the surface of ovarian cancer cells alters their ability to adhere, migrate, aggregate, and proliferate. Oncotarget 8 (6), 9717–9738. Available from: https://doi.org/10.18632/oncotarget.14206.

Brockhausen, I., 1999. Pathways of O-glycan biosynthesis in cancer cells. Biochim. Biophys. Acta 1473 (1), 67–95. Available from: https://doi.org/10.1016/S0304-4165(99)00170-1.

Brustmann, H., 2008. Epidermal growth factor receptor expression in serous ovarian carcinoma: an immunohistochemical study with galectin-3 and cyclin D1 and outcome. Int. J. Gynecol. Pathol. 27 (3), 380–389. Available from: https://doi.org/10.1097/PGP.0b013e31815d060d.

Buckhaults, P., Chen, L., Fregien, N., Pierce, M., 1997. Transcriptional regulation of N-acetylglucosaminyltransferase V by the src oncogene.

J. Biol. Chem. 272 (31), 19575–19581. Available from: https://doi.org/10.1074/jbc.272.31.19575.

Bull, C., Stoel, M.A., den Brok, M.H., Adema, G.J., 2014. Sialic acids sweeten a tumor's life. Cancer Res. 74 (12), 3199–3204. Available from: https://doi.org/10.1158/0008-5472.CAN-14-0728.

Burleson, K.M., Casey, R.C., Skubitz, K.M., Pambuccian, S.E., Oegema, T.R., Skubitz, A.P.N., 2004. Ovarian carcinoma ascites spheroids adhere to extracellular matrix components and mesothelial cell monolayers. Gynecol. Oncol. 93 (1), 170–181. Available from: https://doi.org/10.1016/j.ygyno.2003.12.034.

Burleson, K.M., Boente, M.P., Pambuccian, S.E., Skubitz, A.P., 2006. Disaggregation and invasion of ovarian carcinoma ascites spheroids. J. Transl. Med. 4, 6. Available from: https://doi.org/10.1186/1479-5876-4-6.

Carroll, M.J., Fogg, K.C., Patel, H.A., Krause, H.B., Mancha, A.-S., Patankar, M.S., et al., 2018. Alternatively activated macrophages upregulate mesothelial expression of P-selectin to enhance adhesion of ovarian cancer cells. Cancer Res, canres 3341, 2017. Available from: https://doi.org/10.1158/0008-5472.can-17-3341.

Casey, R.C., Skubitz, A.P., 2000. CD44 and beta1 integrins mediate ovarian carcinoma cell migration toward extracellular matrix proteins. Clin. Exp. Metastasis 18 (1), 67–75.

Catterall, J.B., Jones, L.M., Turner, G.A., 1999. Membrane protein glycosylation and CD44 content in the adhesion of human ovarian cancer cells to hyaluronan. Clin. Exp. Metastasis 17 (7), 583–591.

Chen, S., Birk, D.E., 2013. The regulatory roles of small leucine-rich proteoglycans in extracellular matrix assembly. FEBS J. 280 (10), 2120–2137. Available from: https://doi.org/10.1111/febs.12136.

Chen, K., Gentry-Maharaj, A., Burnell, M., Steentoft, C., Marcos-Silva, L., Mandel, U., et al., 2013. Microarray glycoprofiling of CA125 improves differential diagnosis of ovarian cancer. J. Proteome Res. 12 (3), 1408–1418. Available from: https://doi.org/10.1021/pr3010474.

Davidowitz, R.A., Iwanicki, M.P., Brugge, J.S., 2012. In vitro mesothelial clearance assay that models the early steps of ovarian cancer metastasis. J. Vis. Exp. (60), . Available from: https://doi.org/10.3791/3888.

Davidowitz, R.A., Selfors, L.M., Iwanicki, M.P., Elias, K.M., Karst, A., Piao, H., et al., 2014. Mesenchymal gene program-expressing ovarian cancer spheroids exhibit enhanced mesothelial clearance. J. Clin. Invest. 124 (6), 2611–2625. Available from: https://doi.org/10.1172/JCI69815.

Eckert, M.A., Pan, S., Hernandez, K.M., Loth, R.M., Andrade, J., Volchenboum, S. L., et al., 2016. Genomics of ovarian cancer progression reveals diverse metastatic trajectories including intraepithelial metastasis to the fallopian tube. Cancer Discov. 6 (12), 1342–1351. Available from: https://doi.org/10.1158/2159-8290.CD-16-0607.

Fernández, M.M., Ferragut, F., Cárdenas Delgado, V.M., Bracalente, C., Bravo, A. I., Cagnoni, A.J., et al., 2016. Glycosylation-dependent binding of galectin-8 to activated leukocyte cell adhesion molecule (ALCAM/CD166) promotes its surface segregation on breast cancer cells. Biochim. Biophys. Acta 1860 (10), 2255–2268. Available from: https://doi.org/10.1016/j.bbagen.2016.04.019.

Gandhi, J.G., Koch, D.L., Paszek, M.J., 2019. Equilibrium modeling of the mechanics and structure of the cancer glycocalyx. Biophys. J. 116 (4), 694–708. Available from: https://doi.org/10.1016/j.bpj.2018.12.023.

Gao, Y., Li, X., Shu, Z., Zhang, K., Xue, X., Li, W., et al., 2018. Nuclear galectin-1-FOXP3 interaction dampens the tumor-suppressive properties of FOXP3 in breast cancer. Cell Death Dis. 9 (4), 416. Available from: https://doi.org/10.1038/s41419-018-0448-6.

Gao, Q., Yang, Z., Xu, S., Li, X., Yang, X., Jin, P., et al., 2019. Heterotypic CAF-tumor spheroids promote early peritoneal metastatis of ovarian cancer. J. Exp. Med. 216 (3), 688–703. Available from: https://doi.org/10.1084/jem.20180765.

Gardner, M.J., Catterall, J.B., Jones, L.M., Turner, G.A., 1996. Human ovarian tumour cells can bind hyaluronic acid via membrane CD44: a possible step in peritoneal metastasis. Clin. Exp. Metastasis 14 (4), 325–334.

Ghosh, S., Albitar, L., LeBaron, R., Welch, W.R., Samimi, G., Birrer, M.J., et al., 2010. Up-regulation of stromal versican expression in advanced stage serous ovarian cancer. Gynecol. Oncol. 119 (1), 114–120. Available from: https://doi.org/10.1016/j.ygyno.2010.05.029.

Granovsky, M., Fata, J., Pawling, J., Muller, W.J., Khokha, R., Dennis, J.W., 2000. Suppression of tumor growth and metastasis in Mgat5-deficient mice. Nat. Med. 6 (3), 306–312. Available from: https://doi.org/10.1038/73163.

Guo, Q., Yang, X., Ma, Y., Ma, L., 2015. Syndecan-1 serves as a marker for the progression of epithelial ovarian carcinoma. Eur. J. Gynaecol. Oncol. 36 (5), 506–513.

Hou, L., Zhang, X., Jiao, Y., Li, Y., Zhao, Y., Guan, Y., et al., 2019. ATP binding cassette subfamily B member 9 (ABCB9) is a prognostic indicator of overall survival in ovarian cancer. Medicine 98 (19), e15698. Available from: https://doi.org/10.1097/md.0000000000015698.

Huang, C.-C., Cheng, S.-H., Wu, C.-H., Li, W.-Y., Wang, J.-S., Kung, M.-L., et al., 2019. Delta-like 1 homologue promotes tumorigenesis and epithelial-mesenchymal transition of ovarian high-grade serous carcinoma through activation of Notch signaling. Oncogene 38 (17), 3201–3215. Available from: https://doi.org/10.1038/s41388-018-0658-5.

Ilan, N., Elkin, M., Vlodavsky, I., 2006. Regulation, function and clinical significance of heparanase in cancer metastasis and angiogenesis. Int. J. Biochem. Cell Biol. 38 (12), 2018–2039. Available from: https://doi.org/10.1016/j.biocel.2006.06.004.

Iozzo, R.V., Murdoch, A.D., 1996. Proteoglycans of the extracellular environment: clues from the gene and protein side offer novel perspectives in molecular diversity and function. FASEB J. 10 (5), 598–614.

Ip, C.K., Li, S.S., Tang, M.Y., Sy, S.K., Ren, Y., Shum, H.C., et al., 2016. Stemness and chemoresistance in epithelial ovarian carcinoma cells under shear stress. Sci. Rep. 6, 26788. Available from: https://doi.org/10.1038/srep26788.

Iwanicki, M.P., Davidowitz, R.A., Ng, M.R., Besser, A., Muranen, T., Merritt, M., et al., 2011. Ovarian cancer spheroids use myosin-generated force to clear the mesothelium. Cancer Discov. 1 (2), 144–157. Available from: https://doi.org/10.1158/2159-8274.CD-11-0010.

Kadler, K.E., Hill, A., Canty-Laird, E.G., 2008. Collagen fibrillogenesis: fibronectin, integrins, and minor collagens as organizers and nucleators. Curr. Opin. Cell Biol. 20 (5), 495–501. Available from: https://doi.org/10.1016/j.ceb.2008.06.008.

Kailemia, M.J., Park, D., Lebrilla, C.B., 2017. Glycans and glycoproteins as specific biomarkers for cancer. Anal. Bioanal. Chem. 409 (2), 395–410. Available from: https://doi.org/10.1007/s00216-016-9880-6.

Kaku, T., Ogawa, S., Kawano, Y., Ohishi, Y., Kobayashi, H., Hirakawa, T., et al., 2003. Histological classification of ovarian cancer. Med. Electron. Microsc. 36 (1), 9–17. Available from: https://doi.org/10.1007/s007950300002.

Kaneko, K., 2009. Relationship among phenotypic plasticity, phenotypic fluctuations, robustness, and evolvability; Waddington's legacy revisited under the spirit of Einstein. J. Biosci. 34 (4), 529–542.

Karlsson, N.G., McGuckin, M.A., 2012. O-Linked glycome and proteome of high-molecular-mass proteins in human ovarian cancer ascites: identification of sulfation, disialic acid and O-linked fucose. Glycobiology 22 (7), 918–929. Available from: https://doi.org/10.1093/glycob/cws060.

Kim, J., Park, E.Y., Kim, O., Schilder, J.M., Coffey, D.M., Cho, C.-H., et al., 2018. Cell origins of high-grade serous ovarian cancer. Cancers (Basel) 10 (11), 433. Available from: https://doi.org/10.3390/cancers10110433.

Kurman, R.J., Shih, I.-M., 2011. Molecular pathogenesis and extraovarian origin of epithelial ovarian cancer--shifting the paradigm. Hum. Pathol. 42 (7), 918–931. Available from: https://doi.org/10.1016/j.humpath.2011.03.003.

Labrie, M., Vladoiu, M.C., Grosset, A.-A., Gaboury, L., St-Pierre, Y., 2014. Expression and functions of galectin-7 in ovarian cancer. Oncotarget 5 (17), 7705–7721. Available from: https://doi.org/10.18632/oncotarget.2299.

Lahm, H., Hoeflich, A., Andre, S., Sordat, B., Kaltner, H., Wolf, E., et al., 2000. Gene expression of galectin-9/ecalectin, a potent eosinophil chemoattractant, and/or the insertional isoform in human colorectal carcinoma cell lines and detection of frame-shift mutations for protein sequence truncations in the second functional lectin domain. Int. J. Oncol. 17 (3), 519–524. Available from: https://doi.org/10.3892/ijo.17.3.519.

Latifi, A., Luwor, R.B., Bilandzic, M., Nazaretian, S., Stenvers, K., Pyman, J., et al., 2012. Isolation and characterization of tumor cells from the ascites of ovarian cancer patients: molecular phenotype of chemoresistant ovarian tumors. PLoS One 7 (10), . Available from: https://doi.org/10.1371/journal.pone.0046858e46858.

Lengyel, E., 2010. Ovarian cancer development and metastasis. Am. J. Pathol. 177 (3), 1053–1064. Available from: https://doi.org/10.2353/ajpath.2010.100105.

Li, S.S., Ip, C.K.M., Tang, M.Y.H., Tang, M.K.S., Tong, Y., Zhang, J., et al., 2019. Sialyl Lewis(x)-P-selectin cascade mediates tumor-mesothelial adhesion in ascitic fluid shear flow. Nat. Commun. 10 (1), 2406. Available from: https://doi.org/10.1038/s41467-019-10334-6.

Liu, Y., Zheng, D., Liu, M., Bai, J., Zhou, X., Gong, B., et al., 2015. Downregulation of glypican-3 expression increases migration, invasion, and tumorigenicity of human ovarian cancer cells. Tumour Biol. 36 (10), 7997–8006. Available from: https://doi.org/10.1007/s13277-015-3528-6.

Liu, Y., Xie, L., Wang, D., Li, D., Xu, G., Wang, L., et al., 2018. Galectin-3 and β-catenin are associated with a poor prognosis in serous epithelial ovarian cancer. Cancer Manag. Res. 10, 3963–3971. Available from: https://doi.org/10.2147/cmar.s171146.

Lu, J., Isaji, T., Im, S., Fukuda, T., Kameyama, A., Gu, J., 2016. Expression of N-acetylglucosaminyltransferase III suppresses alpha2,3-sialylation, and its distinctive functions in cell migration are attributed to alpha2,6-sialylation levels. J. Biol. Chem. 291 (11), 5708–5720. Available from: https://doi.org/10.1074/jbc.M115.712836.

Mikula-Pietrasik, J., Uruski, P., Tykarski, A., Ksiazek, K., 2018. The peritoneal "soil" for a cancerous "seed": a comprehensive review of the pathogenesis of intraperitoneal cancer metastases. Cell Mol. Life Sci. 75 (3), 509–525. Available from: https://doi.org/10.1007/s00018-017-2663-1.

Mingels, M.J., Roelofsen, T., van der Laak, J.A., de Hullu, J.A., van Ham, M.A., Massuger, L.F., et al., 2012. Tubal epithelial lesions in salpingo-oophorectomy specimens of BRCA-mutation carriers and controls. Gynecol. Oncol. 127 (1), 88–93. Available from: https://doi.org/10.1016/j.ygyno.2012.06.015.

Mirandola, L., Yu, Y., Cannon, M.J., Jenkins, M.R., Rahman, R.L., Nguyen, D.D., et al., 2014. Galectin-3 inhibition suppresses drug resistance, motility, invasion and angiogenic potential in ovarian cancer. Gynecol. Oncol. 135 (3), 573–579. Available from: https://doi.org/10.1016/j.ygyno.2014.09.021.

Mouw, J.K., Ou, G., Weaver, V.M., 2014. Extracellular matrix assembly: a multiscale deconstruction. Nat. Rev. Mol. Cell Biol. 15 (12), 771–785. Available from: https://doi.org/10.1038/nrm3902.

Newman, S.A., Bhat, R., Mezentseva, N.V., 2009. Cell state switching factors and dynamical patterning modules: complementary mediators of plasticity in development and evolution. J. Biosci. 34 (4), 553–572.

Paget, S., 1889. The distribution of secondary growths in cancer of the breast. Lancet 133 (3421), 571–573.

Park, J.-H., Nishidate, T., Kijima, K., Ohashi, T., Takegawa, K., Fujikane, T., et al., 2010. Critical roles of mucin 1 glycosylation by transactivated polypeptide N-acetylgalactosaminyltransferase 6 in mammary carcinogenesis. Cancer Res. 70 (7), 2759–2769. Available from: https://doi.org/10.1158/0008-5472.can-09-3911.

Paszek, M.J., DuFort, C.C., Rossier, O., Bainer, R., Mouw, J.K., Godula, K., et al., 2014. The cancer glycocalyx mechanically primes integrin-mediated growth and survival. Nature 511 (7509), 319–325. Available from: https://doi.org/10.1038/nature13535.

Pearce, O.M., Laubli, H., 2016. Sialic acids in cancer biology and immunity. Glycobiology 26 (2), 111–128. Available from: https://doi.org/10.1093/glycob/cwv097.

Peart, T., Ramos Valdes, Y., Correa, R.J.M., Fazio, E., Bertrand, M., McGee, J., et al., 2015. Intact LKB1 activity is required for survival of dormant ovarian cancer spheroids. Oncotarget 6 (26), 22424–22438. Available from: https://doi.org/10.18632/oncotarget.4211.

Pease, J.C., Brewer, M., Tirnauer, J.S., 2012. Spontaneous spheroid budding from monolayers: a potential contribution to ovarian cancer dissemination. Biol. open. 1 (7), 622–628. Available from: https://doi.org/10.1242/bio.2012653.

Ricciardelli, C., Rodgers, R.J., 2006. Extracellular matrix of ovarian tumors. Semin. Reprod. Med. 24 (04), 270–282. Available from: https://doi.org/10.1055/s-2006-948556.

Saldova, R., Royle, L., Radcliffe, C.M., Abd Hamid, U.M., Evans, R., Arnold, J.N., et al., 2007. Ovarian cancer is associated with changes in glycosylation in both acute-phase proteins and IgG. Glycobiology 17 (12), 1344–1356. Available from: https://doi.org/10.1093/glycob/cwm100.

Schulz, H., Schmoeckel, E., Kuhn, C., Hofmann, S., Mayr, D., Mahner, S., et al., 2017. Galectins-1, -3, and -7 are prognostic markers for survival of ovarian cancer patients. Int. J. Mol. Sci. 18 (6), 1230. Available from: https://doi.org/10.3390/ijms18061230.

Schulz, H., Kuhn, C., Hofmann, S., Mayr, D., Mahner, S., Jeschke, U., et al., 2018. Overall survival of ovarian cancer patients is determined by expression of galectins-8 and -9. Int. J. Mol. Sci. 19 (1), 323. Available from: https://doi.org/10.3390/ijms19010323.

Shan, M., Yang, D., Dou, H., Zhang, L., 2019. Fucosylation in cancer biology and its clinical applications. Prog. Mol. Biol. Transl. Sci. 162, 93–119. Available from: https://doi.org/10.1016/bs.pmbts.2019.01.002.

Shurer, C.R., Colville, M.J., Gupta, V.K., Head, S.E., Kai, F., Lakins, J.N., et al., 2018. Genetically encoded toolbox for glycocalyx engineering: tunable control of cell adhesion, survival, and cancer cell behaviors. ACS Biomater. Sci. Eng. 4 (2), 388–399. Available from: https://doi.org/10.1021/acsbiomaterials.7b00037.

Shurer, C.R., Kuo, J.C., Roberts, L.M., Gandhi, J.G., Colville, M.J., Enoki, T.A., et al., 2019. Physical principles of membrane shape regulation by the glycocalyx. Cell 177 (7), 1757–1770. Available from: https://doi.org/10.1016/j.cell.2019.04.017. e1721.

Singh, R., Bandyopadhyay, D., 2007. MUC1: a target molecule for cancer therapy. Cancer Biol. & Ther. 6 (4), 481–486. Available from: https://doi.org/10.4161/cbt.6.4.4201.

Sobolev, O., Stern, P., Lacy-Hulbert, A., Hynes, R.O., 2009. Natural killer cells require selectins for suppression of subcutaneous tumors. Cancer Res. 69 (6), 2531–2539. Available from: https://doi.org/10.1158/0008-5472.can-08-3126.

Sodek, K.L., Ringuette, M.J., Brown, T.J., 2009. Compact spheroid formation by ovarian cancer cells is associated with contractile behavior and an invasive phenotype. Int. J. Cancer 124 (9), 2060–2070. Available from: https://doi.org/10.1002/ijc.24188.

Stowell, S.R., Ju, T., Cummings, R.D., 2015. Protein glycosylation in cancer. Annu. Rev. Pathol. 10, 473–510. Available from: https://doi.org/10.1146/annurev-pathol-012414-040438.

Swartz, M.J., Batra, S.K., Varshney, G.C., Hollingsworth, M.A., Yeo, C.J., Cameron, J.L., et al., 2002. MUC4 expression increases progressively in pancreatic intraepithelial neoplasia. Am. J. Clin. Pathol. 117 (5), 791–796. Available from: https://doi.org/10.1309/7y7n-m1wm-r0yk-m2va.

Taylor-Papadimitriou, J., Burchell, J., Miles, D.W., Dalziel, M., 1999. MUC1 and cancer. Biochim. Biophys. Acta 1455 (2), 301–313. Available from: https://doi.org/10.1016/S0925-4439(99)00055-1.

Ten Dam, G.B., Yamada, S., Kobayashi, F., Purushothaman, A., van de Westerlo, E.M., Bulten, J., et al., 2009. Dermatan sulfate domains defined by the novel antibody GD3A12, in normal tissues and ovarian adenocarcinomas. Histochem. Cell Biol. 132 (1), 117–127. Available from: https://doi.org/10.1007/s00418-009-0592-2.

Theocharis, A.D., Manou, D., Karamanos, N.K., 2019. The extracellular matrix as a multitasking player in disease. FEBS J. Available from: https://doi.org/10.1111/febs.14818.

Thibault, B., Castells, M., Delord, J.P., Couderc, B., 2014. Ovarian cancer microenvironment: implications for cancer dissemination and chemoresistance acquisition. Cancer Metastasis Rev. 33 (1), 17–39. Available from: https://doi.org/10.1007/s10555-013-9456-2.

Travers, M., Brown, S.M., Dunworth, M., Holbert, C.E., Wiehagen, K.R., Bachman, K.E., et al., 2019. DFMO and 5-azacytidine increase M1 macrophages in the tumor microenvironment of murine ovarian cancer. Cancer Res. . Available from: https://doi.org/10.1158/0008-5472.can-18-4018.

Tu, C.-F., Wu, M.-Y., Lin, Y.-C., Kannagi, R., Yang, R.-B., 2017. FUT8 promotes breast cancer cell invasiveness by remodeling TGF-β receptor core fucosylation. Breast Cancer Res. 19 (1), 111. Available from: https://doi.org/10.1186/s13058-017-0904-8.

Vallen, M.J., Massuger, L.F., ten Dam, G.B., Bulten, J., van Kuppevelt, T.H., 2012. Highly sulfated chondroitin sulfates, a novel class of prognostic biomarkers in ovarian cancer tissue. Gynecol. Oncol. 127 (1), 202–209. Available from: https://doi.org/10.1016/j.ygyno.2012.06.022.

Vallen, M.J., Schmidt, S., Oosterhof, A., Bulten, J., Massuger, L.F., van Kuppevelt, T.H., 2014. Primary ovarian carcinomas and abdominal metastasis contain 4,6-disulfated chondroitin sulfate rich regions, which provide adhesive properties to tumour cells. PLoS One 9 (11), e111806. Available from: https://doi.org/10.1371/journal.pone.0111806.

Varki, A., 2017. Essentials of Glycobiology, third ed. Cold Spring Harbor Laboratory Press, Cold Spring Harbor, NY.

Vaughan, S., Coward, J.I., Bast Jr., R.C., Berchuck, A., Berek, J.S., Brenton, J.D., et al., 2011. Rethinking ovarian cancer: recommendations for improving outcomes. Nat. Rev. Cancer 11 (10), 719–725. Available from: https://doi.org/10.1038/nrc3144.

Voltaggio, L., Cimino-Mathews, A., Bishop, J.A., Argani, P., Cuda, J.D., Epstein, J.I., et al., 2016. Current concepts in the diagnosis and pathobiology of intraepithelial neoplasia: A review by organ system. CA Cancer J. Clin. 66 (5), 408–436. Available from: https://doi.org/10.3322/caac.21350.

Wang, Z.-Q., Bachvarova, M., Morin, C., Plante, M., Gregoire, J., Renaud, M.-C., et al., 2014. Role of the polypeptide N-acetylgalactosaminyltransferase 3 in ovarian cancer progression: possible implications in abnormal mucin O-glycosylation. Oncotarget 5 (2), 544–560. Available from: https://doi.org/10.18632/oncotarget.1652.

Wang, Y., Yang, X., Yuan, M., Xian, S., Zhang, L., Yang, D., et al., 2019. Promotion of ovarian cancer cell invasion, migration and colony formation by the miR-21/Wnt/CD44v6 pathway. Oncol. Rep. 42 (1), 91–102. Available from: https://doi.org/10.3892/or.2019.7153.

West-Eberhard, M.J., 2005. Phenotypic accommodation: adaptive innovation due to developmental plasticity. J. Exp. Zool. B Mol. Dev. Evol. 304 (6), 610–618. Available from: https://doi.org/10.1002/jez.b.21071.

Weyers, A., Yang, B., Yoon, D.S., Park, J.H., Zhang, F., Lee, K.B., et al., 2012. A structural analysis of glycosaminoglycans from lethal and nonlethal breast cancer tissues: toward a novel class of theragnostics for personalized medicine in oncology? OMICS 16 (3), 79–89. Available from: https://doi.org/10.1089/omi.2011.0102.

Weyers, A., Yang, B., Park, J.H., Kim, Y.S., Kim, S.M., Lee, S.E., et al., 2013. Microanalysis of stomach cancer glycosaminoglycans. Glycoconj. J. 30 (7), 701–707. Available from: https://doi.org/10.1007/s10719-013-9476-8.

Yanagishita, M., 1993. Function of proteoglycans in the extracellular matrix. Acta Pathol. Jpn. 43 (6), 283–293.

Yang, S., Höti, N., Yang, W., Liu, Y., Chen, L., Li, S., et al., 2017. Simultaneous analyses of N-linked and O-linked glycans of ovarian cancer cells using solid-phase chemoenzymatic method. Clin. Proteom. 14 (1), 3. Available from: https://doi.org/10.1186/s12014-017-9137-1.

Yeo, T.K., Nagy, J.A., Yeo, K.T., Dvorak, H.F., Toole, B.P., 1996. Increased hyaluronan at sites of attachment to mesentery by CD44-positive mouse ovarian and breast tumor cells. Am. J. Pathol. 148 (6), 1733–1740.

Zhang, P., Shi, B., Zhou, M., Jiang, H., Zhang, H., Pan, X., et al., 2014a. Galectin-1 overexpression promotes progression and chemoresistance to cisplatin in epithelial ovarian cancer. Cell Death Dis. 5 (1), e991. Available from: https://doi.org/10.1038/cddis.2013.526.

Zhang, X., Wang, Y., Qian, Y., Wu, X., Zhang, Z., Liu, X., et al., 2014b. Discovery of specific metastasis-related N-glycan alterations in epithelial ovarian cancer

based on quantitative glycomics. PLoS One 9 (2), e87978. Available from: https://doi.org/10.1371/journal.pone.0087978.
Zhang, Q., Yu, S., Lam, M.M.T., Poon, T.C.W., Sun, L., Jiao, Y., et al., 2019. Angiotensin II promotes ovarian cancer spheroid formation and metastasis by upregulation of lipid desaturation and suppression of endoplasmic reticulum stress. J. Exp. & Clin. Cancer Res. 38 (1), 116. Available from: https://doi.org/10.1186/s13046-019-1127-x.
Zucchetti, M., Bonezzi, K., Frapolli, R., Sala, F., Borsotti, P., Zangarini, M., et al., 2013. Pharmacokinetics and antineoplastic activity of galectin-1-targeting OTX008 in combination with sunitinib. Cancer Chemother. Pharmacol. 72 (4), 879–887. Available from: https://doi.org/10.1007/s00280-013-2270-2.

21

Epithelial-mesenchymal transition in cancer

Dongya Jia[1], Mohit Kumar Jolly[2], Herbert Levine[1,3,4] and José N. Onuchic[1,5,6,7]

[1]Center for Theoretical Biological Physics, Rice University, Houston, TX, United States [2]Center for BioSystems Science and Engineering, Indian Institute of Science, Bangalore, India [3]Department of Bioengineering, Northeastern University, Boston, MA, United States [4]Department of Physics, Northeastern University, Boston, MA, United States [5]Department of Chemistry, Rice University, Houston, TX, United States [6]Department of Biosciences, Rice University, Houston, TX, United States [7]Department of Physics and Astronomy, Rice University, Houston, TX, United States

Introduction

Cancer cells exhibit abnormal phenotypes characterized by hallmarks such as sustained proliferative signaling, deregulated cellular energetics, and the activation of invasion and metastasis pathways (Hanahan and Weinberg, 2011). Recent investigations have shown that there exists significant heterogeneity at the single-cell level even within a single tumor. This inherent cancer cell heterogeneity exists both at the genetic level due to genomic instability (Burrell et al., 2013), and at the nongenetic level due to cellular plasticity, that is, the ability of cells to switch between different phenotypes even at fixed genome (Brock et al., 2009). It is commonly believed that a major impediment to the efficacy of therapies is the presence of one or more subpopulations of cancer cells that are drug tolerant or drug resistant (Easwaran et al., 2014).

Here we focus on a specific example of nongenetic heterogeneity related to the epithelial-mesenchymal transition (EMT) (Polyak and Weinberg, 2009). This phenotypic axis is important in its own right due to its relevance for motility and invasiveness. In addition, EMT is connected to the acquisition of the stem-like properties (Wicha et al., 2006) and to the phenomenon of metabolic reprogramming (Porporato et al., 2014). The decision-making of these different cellular traits are indeed coupled so that varying the EMT

Phenotypic Switching. DOI: https://doi.org/10.1016/B978-0-12-817996-3.00018-9

status can tune cancer stemness and metabolic profiles and vice versa (Jolly et al., 2015; Luo et al., 2018; Sciacovelli and Frezza, 2017). Phenotypic transitions in cancer can occur in varying degrees so that cells can acquire "hybrid" phenotypes which combine features of paradoxically exclusive phenotypes of normal cells (Lu et al., 2013; Kröger et al., 2019; Jolly et al., 2015; Boareto et al., 2015; Jia et al., 2019; Yu et al., 2017; Jia et al., 2018). Due to the combined features of different phenotypes, the hybrid phenotypes often exhibit enhanced plasticity so as to enable cancer cells to better adapt to hostile environments (Porporato et al., 2014; Jolly et al., 2015; Goldman et al., 2019). In addition, the hybrid phenotypes may occupy "transition sweet spots" so that they can flexibly switch to other phenotypes for better survival (Li et al., 2016; Ruscetti et al., 2016). Consequently, the "hybrid" phenotypes contribute to cancer cell heterogeneity and therapy resistance. A better understanding of mechanisms underlying phenotypic transitions in cancer and the emergence of "hybrid" phenotypes can imply critical clues for better therapeutic strategies.

Epithelial-mesenchymal transition

EMT is a trans-differentiation program by which cells lose their epithelial features such as cell-cell adhesion, apico-basal polarity, and cobblestone morphology, and gain mesenchymal features such as enhanced motility and a spindle-like morphology. EMT is reversible and cells entering the mesenchymal phenotype can undergo mesenchymal-epithelial transition (MET) to regain epithelial features. EMT and MET constitute a fundamental process that plays a critical role during embryonic development, for example, neural crest delamination, gastrulation and myogenesis. EMT that occurs during embryonic development is often referred to as type I EMT (Kalluri and Weinberg, 2009). The concept of EMT was originally proposed by Greenburg and Hay who showed that epithelia from embryonic anterior lens lose their polarity when suspended in collagen gels, detach from the explants, elongate and migrate as individual cells, thus exhibiting mesenchymal-associated characteristics (Greenburg and Hay, 1982). Later, Nieto and colleagues reported an essential role for EMT in the formation of mesoderm and the migration of the neutral crest during chicken embryonic development (Nieto et al., 1994). Interruption of normal EMT program can cause severe developmental problems (Nieto et al., 1994; Carver et al., 2001). In addition, EMT occurs during physiological wound healing and tissue regeneration as well as during pathological fibrosis (Thiery et al., 2009). Repair-associated EMT is referred to as type II EMT (Kalluri and Weinberg, 2009). In this scenario,

EMT is triggered by inflammation and ceases when inflammation subsides. Upon persistent inflammation, EMT can be maintained and eventually results in organ destruction. Last but not least, EMT can also be utilized by cancer cells to metastasize. The cancer-associated EMT is also referred to as type III EMT (Kalluri and Weinberg, 2009).

In the context of cancer, EMT functions as a central regulator of metastasis (Ye and Weinberg, 2015). EMT significantly contributes to metastasis, which accounts for more than 90% of cancer-related deaths (Gupta and Massagué, 2006). The epithelial cells in a primary tumor can undergo EMT and thereby lose cell-cell adhesion, detach from the primary tumor, invade through the stromal tissue and enter the blood or lymphatic vessels (intravasation), and migrate there as circulating tumor cells (CTCs). These CTCs can exit from the blood or lymphatic vessels (extravasation) and form the secondary tumor in a different organ (Nieto, 2013). Interestingly, the EMT-inducing transcription factors (EMT-TFs), functioning as the executors of EMT, have been shown to be key regulators of metastasis even in the context of noncarcinomas, such as melanoma, leukemia, sarcoma, and glioblastoma (Kahlert et al., 2017; Somarelli et al., 2016; Lee et al., 2014; Li et al., 2015; Caramel et al., 2013).

In addition to the contribution to metastasis, EMT plays a critical role in the acquisition of cancer therapy resistance. For example, the paclitaxel-resistant DU145 prostate cancer cells exhibits downregulation of the epithelial markers keratins and upregulation of the mesenchymal marker vimentin (Vim) and EMT-TFs ZEB1 and SNAIL relative to their parental cells both in vitro and in vivo (Kim et al., 2013). Consequently, the paclitaxel-resistant DU145 cells show an increased migratory and invasive capacity. Another example is the sensitivity of EGFR-containing NSCLC cells to the EGFR inhibitor erlotinib depends on their EMT status (Thomson et al., 2005). The epithelial NSCLC cell lines such as H292 and H322 exhibit greater sensitivity to erlotinib relative to the mesenchymal NSCLC cells lines such as Calu6 and H1703 both in vitro and in xenografts. Interesting, as the A549 cell line has been characterized as a mixture of epithelial and mesenchymal cells (George et al., 2017); it exhibits intermediate sensitivity to erlotinib. Even in the scenarios where the necessity of EMT for metastasis needs further investigation, EMT has been shown to significantly induce chemoresistance in mouse models of both spontaneous breast-to-lung metastasis and pancreatic adenocarcinoma (PDAC) (Zheng et al., 2015; Fischer et al., 2015). Notably, therapy exposure also promotes EMT. For example, exposure of A549 cells to the EGFR inhibitor gefitinib can lead to a downregulation of the epithelial adhesion molecule E-cadherin (E-cad), cytokeratin 8/18 and an upregulation of Vim, suggesting the acquisition of a more mesenchymal phenotype of A549, referred to as A549/GR (Rho et al., 2009). Interestingly, these A549/GR cells exhibit cross resistance to other EGFR inhibitors such as erlotinib and CL-387,758. Last but not least, the EMT status of tumor cells also affects the infiltration of cancer-killing immune cells, indicating a possible mechanism for cancer cells to avoid immune destruction (Jia et al., 2019). In summary, these studies are consistent with the idea that cancer cells undergoing EMT typically exhibit increased resistance to therapies and suppression of immune attack, thus accelerating their malignancy (Shibue and Weinberg, 2017).

Regulation of epithelial-mesenchymal transition

EMT is regulated by multilayered regulatory networks containing EMT-TFs, microRNAs (miRNAs), alternative splicing factors, and epigenetic modifiers (De Craene and Berx, 2013). On one hand, EMT can be triggered by multiple signaling pathways such as TGF-β, EGF, and WNT, by oncometabolites such as fumarate (Sciacovelli et al., 2016) and D-2 hydroxyglutarate (Colvin et al., 2016), as well as by the change of density and stiffness of extracellular matrix (ECM) and endogenous mechanical stress (Gomez et al., 2010; Kumar et al., 2014) (Fig. 21.1). These pathways often function through activating EMT-TFs such as SNAIL, TWIST, and ZEB. The activation of these EMT-TFs often results in the downregulation of E-cad and thus weakening cell-cell adhesion and consequently triggering EMT. On the other hand, EMT can be inhibited or reverted by various molecular players including tumor suppressor genes such as p53, miRNAs such as miR-34 and miR-200, and the MET-inducing transcription factors (MET-TFs) such as OVOL and GRHL2 (Chang et al., 2011; Cieply et al., 2013; Li and Yang, 2014; Zaravinos, 2015). Indeed, these molecular players often repress EMT by directly inhibiting the EMT-TFs.

Despite the complexity of EMT regulation, there appears to be a "hub" in the regulatory network that functions as the decision-making module of EMT. The "hub" consists two mutually inhibitory feedback loops, one between the miR-34 family and the SNAIL family and the other between the miR-200 family and the ZEB family (Lu et al., 2013; Park et al., 2008; Siemens et al., 2011) (Fig. 21.1). High levels of miR-34 and miR-200 maintain an epithelial phenotype while high levels of SNAIL and ZEB promote EMT. Many of the aforementioned EMT-inducing or -repressing signaling pathways regulate EMT through this core EMT

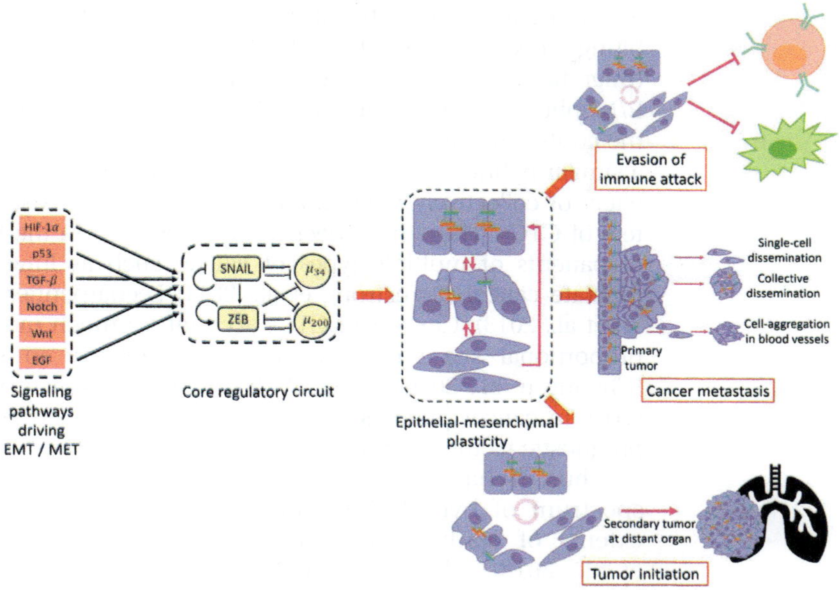

Figure 21.1 An overview of EMT—causes and consequences. EMT is regulated by multiple signaling pathways, such as TGF-β, HIF-1α, and Notch. These signaling pathways often regulate EMT via a core regulatory circuit consisting of two mutually inhibitory feedback loops—miR-34/SNAIL and miR-200/ZEB. The miR-34/SNAIL/miR-200/ZEB functions as a three-way decision-making circuit regulating the transitions among epithelial, mesenchymal and hybrid E/M phenotypes. EMT is closely connected with tumor metastasis, acquisition of stem-like properties and evasion of immune attack. *Source*: The figure is adapted from Jia, D., Li, X., Bocci, F., Tripathi, S., Deng, Y., Jolly, M.K., et al., 2019. Quantifying cancer epithelial-mesenchymal plasticity and its association with stemness and immune response, J. Clin. Med. 8 (5), 725. Reproduced with permission from the authors; published by MDPI, 2019.

regulatory circuit. Notably, even though both the SNAIL family and the ZEB family can promote EMT, these EMT-TFs have non-redundant roles in regulating metastasis. As shown in a mouse model of PDAC, while deletion of SNAIL does not lead to significant change in tumor metastasis (Zheng et al., 2015), deletion of ZEB1 significantly reduces the formation of metastases and the tumor colonization capacity (Krebs et al., 2017). Mathematical modeling approaches have been applied to identify the operating principles of the core EMT regulatory circuit. This will be discussed in detail in the next section.

Hybrid epithelial/mesenchymal phenotypes

In the developmental context of embryogenesis and the physiological context of wound healing, cells can undergo a partial EMT to acquire a hybrid epithelial/mesenchymal (E/M) phenotype exhibiting mixed epithelial (cell-cell adhesion) and mesenchymal (migration) features (Micalizzi et al., 2010). Cells in such a hybrid E/M phenotype can migrate collectively instead of individually. In the context of cancer, cells often undergo a partial EMT and acquire a similar hybrid E/M phenotype (Jolly et al., 2015). The downstream effect of collectively migrating cancer cells is the existence as clusters of CTCs. These have been widely observed in the bloodstream of patients of multiple types of cancer such as lung, breast and prostate cancer (Armstrong et al., 2011; Lecharpentier et al., 2011; Yu et al., 2013). Clusters of CTCs account for much more than their proportional share of total metastases relative to single CTCs (Cheung et al., 2016; Liu et al., 2019). The presence of clusters of CTCs in cancer patients has already been identified as an adverse prognostic factor (Fabisiewicz and Grzybowska, 2017).

Thus, it has become clear that cancer cells can acquire a spectrum of hybrid E/M phenotypes characterized by varying extents of epithelial and mesenchymal features (Pastushenko et al., 2018). The hybrid E/M phenotypes were proposed by mathematical modeling of gene regulatory networks governing EMT. A mechanism-based mathematical model focusing on the dynamics of the core EMT regulatory circuit—miR-34/SNAIL/miR-200/ZEB—showed that this core circuit can give rise to three stable steady states, corresponding to an epithelial phenotype, a mesenchymal phenotype and an additional hybrid E/M phenotype (Lu et al., 2013). Furthermore, this study shows that the miR-200/ZEB circuit tends to function as a three-way decision-making circuit regulating the transitions among epithelial, mesenchymal and hybrid E/M phenotypes, and the miR-34/SNAIL circuit mainly functions as a noise buffer to integrate the input signals. Alternatively, another mathematical modeling study analyzing the same EMT regulatory circuit argued that both the miR-34/SNAIL and miR-200/ZEB circuits function as bistable circuits, accounting for the transition from epithelial to hybrid E/M and the transition from hybrid E/M to mesenchymal, respectively. Based on this second model, the emergence of a hybrid E/M phenotype is mainly due to the miR-34/SNAIL circuit switching from miR-34^{high}/$SNAIL^{low}$ to miR-34^{low}/$SNAIL^{high}$ while the miR-200/ZEB circuit remains miR-200^{high}/ZEB^{low} (Zhang et al., 2014). Despite the different detailed mechanisms accounting for the emergence of the hybrid E/M phenotypes, both studies propose that EMT is not a binary process and instead, a stable hybrid E/M phenotype can be the endpoint of the transition (Jia et al., 2017). The hybrid E/M phenotypes have been further characterized by analyzing several extended versions of the core EMT regulatory circuit (Huang et al., 2017; Biswas et al. 2019; Jolly et al., 2018; Mooney et al., 2017; Steinway et al., 2015). The emergence of the hybrid E/M phenotypes has been further demonstrated to be a robust dynamical feature determined mainly by the topologies of EMT regulatory networks and not necessarily very sensitive to the kinetic parameters (Huang et al., 2017; Font-Clos et al., 2018). In summary, hybrid E/M phenotypes that exhibit a combination of epithelial and mesenchymal markers should be expected based on EMT regulatory networks.

In agreement with these analyses, hybrid E/M phenotypes have been recently characterized both *in vitro* and *in vivo*. Several non-small cell lung cancer (NSCLC) cell lines that coexpress the canonical epithelial marker E-cad and mesenchymal marker Vim have been identified, which indicates a hybrid E/M phenotype at the cell population level (Schliekelman et al., 2015). Among these hybrid NSCLC cell lines, the H1975 cells coexpress E-cad and Vim at a single-cell level, thus being composed of single hybrid E/M cells. H1975 cells can stably maintain their hybrid E/M phenotype over multiple passages for at least two months (Jolly et al., 2016). Consistently, the H1975 cells exhibit collective migration, a typical feature of the hybrid E/M phenotype, during a scratch assay. Another study uses the epithelial markers E-cad and pan-cytokeratin (Pan-CK) and the mesenchymal marker Vim to identify two subpopulations of the ovarian carcinoma cell lines that tentatively exhibit hybrid E/M phenotypes. One subpopulation is characterized by E-cad$^+$/Pan-CK$^+$/Vim$^+$, referred to as intermediate E and the other is characterized by E-cad$^-$/Pan-CK$^+$/Vim$^-$, referred to as intermediate M. Among these hybrid E/M cell lines, the intermediate M SKOV3 cells exhibit much higher spheroidogenic efficiency and metastatic potential relative to other ovarian carcinoma cell lines. Last but not least, the RT4 human bladder cancer cells exhibit coexpression of E-cad and ZEB1 at a single-cell level and exhibit collective migration in vitro, thus being composed of single hybrid E/M cells (Bocci et al., 2019). In summary, these studies show that cells can exhibit a hybrid E/M phenotype at both a population level and a single-cell level in vitro (Huang et al., 2013).

The hybrid E/M phenotype has also been identified in vivo. Using a genetically engineered mouse model of skin squamous cell carcinoma (SCC), three subpopulations of SCC cells characterized by CD61$^-$/CD51$^-$/CD106$^-$, CD51$^+$, and CD106$^+$ exhibit coexpression of the epithelial marker keratin 14 (K14) and the mesenchymal marker Vim, and thus can be considered to be hybrid E/M cells (Pastushenko et al., 2018). These hybrid E/M SCC cells, especially the CD61$^-$/CD51$^-$/CD106$^-$ and CD106$^+$

ones, can generate more metastases relative to other subpopulations of SCC, which tend more towards an epithelial or a mesenchymal phenotype. Another study using lineage tracing in a mouse model of PDAC demonstrated that PDAC cells can undergo a partial EMT to achieve a hybrid E/M phenotype as the endpoint of their transitions via the relocation of the membrane E-cad and claudin-7 to intracellular loci (Aiello et al., 2018). In addition, head and neck squamous cell carcinoma (HNSCC) cells from patients can also exhibit a hybrid E/M phenotype characterized by the coexpression of the epithelial markers such as KRT17 and EpCAM and the mesenchymal markers such as Vim and TGF-β-induced (TGFBi), as elucidated by single-cell transcriptomics analysis (Puram et al., 2017). Last but not least, the HMLER cells characterized by $CD104^{+}/CD44^{high}$ also exhibit a hybrid E/M phenotype which can be stably maintained *in vivo* (Kröger et al., 2019). Intriguingly, the tumors formed by the implanted hybrid E/M HMLER cells can be ten times larger than those formed by implanted epithelial or mesenchymal HMLER cells. The high tumorigenic capacity of hybrid E/M HMLER cells in vivo cannot be achieved by a mixture of epithelial and mesenchymal cells. In summary, these studies validate the existence and importance of hybrid E/M cells in vivo.

Due to the aggressiveness of the hybrid E/M phenotypes and their contribution to metastasis, identifying the molecular players that promote and stabilize the hybrid E/M phenotypes can suggest important therapeutic strategies targeting metastasis. To that end, mathematical modeling approaches have been applied. The transcription factors OVOL, GRHL2, NUMB, NRF2, and ΔNP63α and phosphorylated SP1/NFAT and JAG1 have all been proposed as phenotypic stability factors (PSFs) that can promote and stabilize the hybrid E/M phenotypes (Jolly et al., 2016; Gould et al., 2016; Jia et al., 2015; Marcelo et al., 2016; Bocci et al., 2017). These PSFs function in two manners. On one hand, the PSFs expand the parameter space and thereby the expected physiological conditions in which the hybrid E/M states can exist. On the other hand, these PSFs increase the mean residence time of the hybrid E/M states and thereby their expected percentages in a cell population are enhanced (Biswas et al., 2019). Experimental validation of the PSF concept comes from the observation that knocking down the PSFs in the hybrid E/M H1975 cells disrupts their collective cell migration behavior, downregulates the expression of E-cad, upregulates the expression of Vim, thus promoting a complete EMT as indicated by individual cell migration (Jolly et al., 2016; Bocci et al., 2017; Jolly et al., 2017).

As discussed earlier, during metastasis, cells are able to transition into/out of epithelial, mesenchymal, and multiple hybrid E/M phenotypes. Such a switching ability of cancer cells during EMT is referred to as epithelial-mesenchymal plasticity. It is worth noting that even without external stimuli, cancer cells can spontaneously undergo transitions between epithelial, mesenchymal and hybrid E/M phenotypes. For example, when plating in culture, the subpopulation of the hybrid E/M prostate tumor cells characterized by coexpression of EpCAM and Vim is sufficient to give rise to both epithelial and mesenchymal subpopulations (Ruscetti et al., 2016). Another study shows that the fluorescence activated cell sorting (FACS)-sorted EpCAM^{high} epithelial subpopulation of the breast cancer PMC42-LA cells can generate within three weeks EpCAM^{low} mesenchymal cells with a similar ratio of EpCAM^{high}/ EpCAM^{low} as parental cells (Bhatia et al., 2019). Similarly, the EpCAM^{low} mesenchymal subpopulation can generate EpCAM^{high} epithelial cells but with a relatively slow pace. Last but not least, when implanting epithelial, mesenchymal and hybrid E/M HMLER cells respectively into the mammary fat pad of mice, the tumors formed by either epithelial or mesenchymal HMLER cells contain a very high percentage of hybrid E/M HMLER cells while the tumors formed by hybrid E/M HMLER cells remain mostly in a hybrid phenotype (Kröger et al., 2019). If the plasticity of epithelial or mesenchymal HMLER cells is blocked by ZEB1 knockout or ZEB1 overexpression respectively, their tumor formation capacity is significantly reduced. It is worth noting that there exists cooperation among cells with varying degrees of EMT. For example, the disseminated tumor cells of mouse 4T1 breast cancer which are characterized by primarily epithelial-like subpopulation with a moderate size of mesenchymal-like subpopulation exhibit highest metastasis formation ability relative to both the epithelial parental 4T1 cells and the mesenchymal CTCs in vivo (Liu et al., 2019). In summary, these studies suggest that epithelial, mesenchymal and hybrid E/M phenotypes can exhibit varying plasticity in various contexts. As we develop therapeutic strategies targeting metastasis, a better understanding of the operating principles underlying cancer epithelial-mesenchymal plasticity is necessary to improve therapeutic design (Jia et al., 2019).

Therapeutic strategies targeting epithelial-mesenchymal plasticity

When focusing on epithelial-mesenchymal plasticity, therapeutic strategies should consider directly targeting the hybrid

E/M phenotypes or blocking the transition ability of cells from epithelial or mesenchymal to hybrid phenotypes. By doing this, epithelial-mesenchymal plasticity may be eliminated, and consequently metastasis may be halted. To target the hybrid E/M phenotypes, the PSFs, such as the TFs OVOL, GRHL2, and NUMB, should be considered since their presence promotes the hybrid E/M phenotypes (Jolly et al., 2016; Jia et al., 2015; Bocci et al., 2017). Intriguingly, a recent study shows that the EMT and MET processes may not go through the same hybrid E/M phenotypes, indicating the potentially different roles of hybrid E/M phenotypes in EMT and MET (Karacosta et al., 2019). Future studies to characterize the hybrid E/M phenotypes specifically associated with EMT versus MET will be important to better understand cancer epithelial-mesenchymal plasticity.

From the perspective of dynamical systems theory, each cell phenotype can be considered to be an "attractor" on a genetic landscape that is determined by the interactions of multiple molecular players and is robust to environmental fluctuations. This brings to mind the famous metaphor of developmental biology, Waddington's epigenetic landscape, which is used to describe the differentiation processes of stem cells. Cancer cells are regarded as "abnormal attractors" on the landscape (Fig. 21.2) as modified by genomic changes, "attractors" that are not accessible to normal cells (Jia et al., 2017). The ultimate goal of cancer therapies is to destabilize and eliminate these "cancer attractors." To that end, therapeutic strategies that contain stepwise approaches—first destabilizing the relatively aggressive "cancer attractors," such as the ones representing hybrid E/M phenotypes, followed by the induction of transitions of the less aggressive "cancer attractors" to "normal attractors" and completed by deepening the basin of attractions of "normal attractors" to prevent tumor relapse, can be developed more effectively than conventional therapies targeting individual molecules/pathways (Jia et al., 2017). Considering that EMT is a multidimensional spatiotemporal program containing alterations of mRNA expression, protein concentration, and relocation which changes cell junction, apico-basal polarity, and cell morphology, leading to adaptation of cell migratory and invasive capacity (Jolly et al., 2017), future studies should combine these different aspects to better characterize the basins of attractions corresponding to various EMT statuses on the landscape.

The characterization of the hybrid E/M phenotypes, proposed by mathematical modeling studies and verified by both in vitro and in vivo experiments, serves as an example of the power of a combination of computational and experimental approaches in elucidating cancer epithelial-mesenchymal plasticity. As Dr. Hanahan and

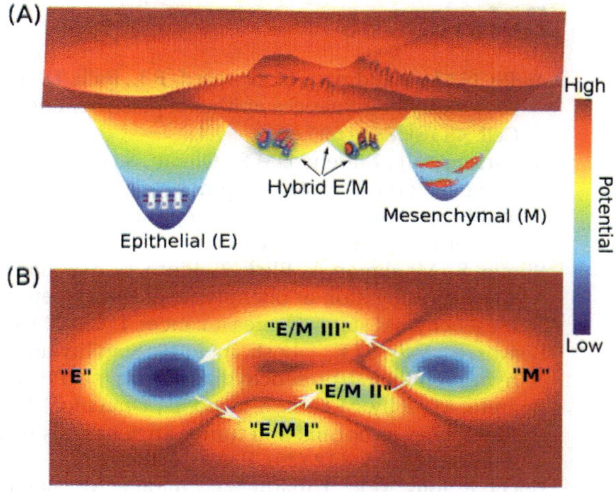

Figure 21.2 Schematic illustration of the quasi-potential landscape for EMT in three-dimensional space (A) and 2-dimensional projection (B). In (A), the basins of attractions representing the attractors "E," "M," and "hybrid E/M" are labeled along with the cartoons representing the corresponding epithelial (cobble-stone shaped with tight cell-cell adhesion), mesenchymal (spindle-shaped with no cell-cell adhesion), and hybrid E/M (combined epithelial and mesenchymal features) phenotypes. The stability of each attractor—"E," "M," and "hybrid E/M" is represented by the quasi-potential which is derived from the probability of finding cells in that attractor. Lower potential represents higher stability of an attractor. *Source*: The figure is adapted from Jia, D., Jolly, M.K., Kulkarni, P., Levine, H., 2017. Phenotypic plasticity and cell fate decisions in cancer: insights from dynamical systems theory, Cancers (Basel) 9 (7). Reproduced with permission from the authors; published by MDPI, 2017.

Dr. Weinberg concluded in their seminal review article "Hallmarks of Cancer: The Next Generation"—"as before, we continue to foresee cancer research as an increasingly logical science, in which myriad phenotypic complexities are manifestations of a small set of underlying organizing principles" (Hanahan and Weinberg, 2011), we should continue performing systems biology studies to identify the organizing principles underlying EMT and other cancer phenotypic complexities, through integrating mathematical modeling, bioinformatics with in vitro and in vivo experiments as well as clinical insights.

Acknowledgments

This work is supported by the National Science Foundation (NSF) grant for the Center for Theoretical Biological Physics NSF PHY-1427654. D.J. is supported by a training fellowship from the Gulf Coast Consortia, on the Computational Cancer Biology Training Program (CPRIT Grant No. RP170593). J.N.O. is a CPRIT Scholar in Cancer Research sponsored by the Cancer Prevention and Research Institute of Texas. M.K.J. is supported by Ramanujan Fellowship awarded by SERB, DST, Government of India (SB/S2/RJN-049/2018).

References

Aiello, N.M., Maddipati, R., Norgard, R.J., Balli, D., Li, J., Yuan, S., et al., 2018. EMT subtype influences epithelial plasticity and mode of cell migration. Dev. Cell. 45 (6), 681–695.e4.

Armstrong, A.J., Marengo, M.S., Oltean, S., Kemeny, G., Bitting, R.L., Turnbull, J. D., et al., 2011. Circulating tumor cells from patients with advanced prostate and breast cancer display both epithelial and mesenchymal markers. Mol. Cancer Res. 9 (8), 997–1007.

Bhatia, S., Monkman, J., Blick, T., Pinto, C., Waltham, M., Nagaraj, S.H., et al., 2019. Interrogation of phenotypic plasticity between epithelial and mesenchymal states in breast cancer. J. Clin. Med. 8 (6), 893.

Biswas, K., Jolly, M.K., Ghosh, A., 2019. Stability and mean residence times for hybrid epithelial/mesenchymal phenotype. Phys. Biol. 16 (2), 025003.

Boareto, M., Jolly, M.K., Ben-Jacob, E., Onuchic, J.N., 2015. Jagged mediates differences in normal and tumor angiogenesis by affecting tip-stalk fate decision. PNAS. 112 (29), E3836–E3844.

Bocci, F., Jolly, M.K., Tripathi, S.C., Aguilar, M., Hanash, S.M., Levine, H., et al., 2017. Numb prevents a complete epithelial-mesenchymal transition by modulating Notch signalling. J. R. Soc. Interface 14 (136).

Bocci, F., Tripathi, S.C., Vilchez Mercedes, S.A., George, J.T., Casabar, J.P., Wong, P.K., et al., 2019. NRF2 activates a partial epithelial-mesenchymal transition and is maximally present in a hybrid epithelial/mesenchymal phenotype. Integr. Biol. (Camb.) 11 (6), 251–263.

Brock, A., Chang, H., Huang, S., 2009. Non-genetic heterogeneity--a mutation-independent driving force for the somatic evolution of tumours. Nat. Rev. Genet. 10 (5), 336–342.

Burrell, R.A., McGranahan, N., Bartek, J., Swanton, C., 2013. The causes and consequences of genetic heterogeneity in cancer evolution. Nature. 501 (7467), 338–345.

Caramel, J., Papadogeorgakis, E., Hill, L., Browne, G.J., Richard, G., Wierinckx, A., et al., 2013. A switch in the expression of embryonic EMT-inducers drives the development of malignant melanoma. Cancer Cell. 24 (4), 466–480.

Carver, E.A., Jiang, R., Lan, Y., Oram, K.F., Gridley, T., 2001. The mouse snail gene encodes a key regulator of the epithelial-mesenchymal transition. Mol. Cell Biol. 21 (23), 8184–8188.

Chang, C.-J., Chao, C.-H., Xia, W., Yang, J.-Y., Xiong, Y., Li, C.-W., et al., 2011. p53 regulates epithelial-mesenchymal transition and stem cell properties through modulating miRNAs. Nat. Cell Biol. 13 (3), 317–323.

Cheung, K.J., Padmanaban, V., Silvestri, V., Schipper, K., Cohen, J.D., Fairchild, A.N., et al., 2016. Polyclonal breast cancer metastases arise from collective dissemination of keratin 14-expressing tumor cell clusters. Proc. Natl Acad. Sci. USA 113 (7), E854–E863.

Cieply, B., Farris, J., Denvir, J., Ford, H.L., Frisch, S.M., 2013. Epithelial-mesenchymal transition and tumor suppression are controlled by a reciprocal feedback loop between ZEB1 and Grainyhead-like-2. Cancer Res. 73 (20), 6299–6309.

Colvin, H., Nishida, N., Konno, M., Haraguchi, N., Takahashi, H., Nishimura, J., et al., 2016. Oncometabolite D-2-hydroxyglurate directly induces epithelial-mesenchymal transition and is associated with distant metastasis in colorectal cancer. Sci. Rep. 6, 36289.

De Craene, B., Berx, G., 2013. Regulatory networks defining EMT during cancer initiation and progression. Nat. Rev. Cancer 13 (2), 97–110.

Easwaran, H., Tsai, H.-C., Baylin, S.B., 2014. Cancer epigenetics: tumor heterogeneity, plasticity of stem-like states, and drug resistance. Mol. Cell. 54 (5), 716–727.

Fabisiewicz, A., Grzybowska, E., 2017. CTC clusters in cancer progression and metastasis. Med. Oncol. 34 (1), 12.

Fischer, K.R., Durrans, A., Lee, S., Sheng, J., Li, F., Wong, S.T.C., et al., 2015. Epithelial-to-mesenchymal transition is not required for lung metastasis but contributes to chemoresistance. Nature. 527 (7579), 472–476.

Font-Clos, F., Zapperi, S., Porta, C.A.M.L., 2018. Topography of epithelial–mesenchymal plasticity. PNAS. 115 (23), 5902–5907.

George, J.T., Jolly, M.K., Xu, S., Somarelli, J.A., Levine, H., 2017. Survival outcomes in cancer patients predicted by a partial EMT gene expression scoring metric. Cancer Res. 77 (22), 6415–6428.

Goldman, A., Khiste, S., Freinkman, E., Dhawan, A., Majumder, B., Mondal, J., et al., 2019. Targeting tumor phenotypic plasticity and metabolic remodeling in adaptive cross-drug tolerance. Sci. Signal. 12 (595), eaas8779.

Gomez, E.W., Chen, Q.K., Gjorevski, N., Nelson, C.M., 2010. Tissue geometry patterns epithelial-mesenchymal transition via intercellular mechanotransduction. J. Cell Biochem. 110 (1), 44–51.

Gould, R., Bassen, D.M., Chakrabarti, A., Varner, J.D., Butcher, J., 2016. Population heterogeneity in the epithelial to mesenchymal transition is controlled by NFAT and phosphorylated Sp1. PLoS Comput. Biol. 12 (12), e1005251.

Greenburg, G., Hay, E.D., 1982. Epithelia suspended in collagen gels can lose polarity and express characteristics of migrating mesenchymal cells. J. Cell Biol. 95 (1), 333–339.

Gupta, G.P., Massagué, J., 2006. Cancer metastasis: building a framework. Cell. 127 (4), 679–695.

Hanahan, D., Weinberg, R.A., 2011. Hallmarks of cancer: the next generation. Cell. 144 (5), 646–674.

Huang, B., Lu, M., Jia, D., Ben-Jacob, E., Levine, H., Onuchic, J.N., 2017. Interrogating the topological robustness of gene regulatory circuits by randomization. PLoS Comput. Biol. 13 (3), e1005456.

Huang, R.Y.-J., Wong, M.K., Tan, T.Z., Kuay, K.T., Ng, A.H.C., Chung, V.Y., et al., 2013. An EMT spectrum defines an anoikis-resistant and spheroidogenic intermediate mesenchymal state that is sensitive to e-cadherin restoration by a src-kinase inhibitor, saracatinib (AZD0530). Cell Death Dis. 4, e915.

Jia, D., Jolly, M.K., Boareto, M., Parsana, P., Mooney, S.M., Pienta, K.J., et al., 2015. OVOL guides the epithelial-hybrid-mesenchymal transition. Oncotarget. 6 (17), 15436–15448.

Jia, D., Jolly, M.K., Kulkarni, P., Levine, H., 2017. Phenotypic plasticity and cell fate decisions in cancer: insights from dynamical systems theory. Cancers (Basel) 9 (7).

Jia, D., Jolly, M.K., Tripathi, S.C., Den Hollander, P., Huang, B., Lu, M., et al., 2017. Distinguishing mechanisms underlying EMT tristability. Cancer Converg. 1 (1), 2.

Jia, D., Li, X., Bocci, F., Tripathi, S., Deng, Y., Jolly, M.K., et al., 2019. Quantifying cancer epithelial-mesenchymal plasticity and its association with stemness and immune response. J. Clin. Med. 8 (5), 725.

Jia, D., Lu, M., Jung, K.H., Park, J.H., Yu, L., Onuchic, J.N., et al., 2019. Elucidating cancer metabolic plasticity by coupling gene regulation with metabolic pathways. Proc. Natl Acad. Sci. USA 116 (9), 3909–3918.

Jia, D., Park, J.H., Jung, K.H., Levine, H., Kaipparettu, B.A., 2018. Elucidating the metabolic plasticity of cancer: mitochondrial reprogramming and hybrid metabolic states. Cells. 7 (3), pii: E21.

Jolly, M.K., Boareto, M., Debeb, B.G., Aceto, N., Farach-Carson, M.C., Woodward, W.A., et al., 2017. Inflammatory breast cancer: a model for investigating cluster-based dissemination. NPJ Breast Cancer 3, 21.

Jolly, M.K., Boareto, M., Huang, B., Jia, D., Lu, M., Ben-Jacob, E., et al., 2015. Implications of the hybrid epithelial/mesenchymal phenotype in metastasis. Front. Oncol. 5, 155.

Jolly, M.K., Jia, D., Boareto, M., Mani, S.A., Pienta, K.J., Ben-Jacob, E., et al., 2015. Coupling the modules of EMT and stemness: a tunable 'stemness window' model. Oncotarget. 6 (28), 25161–25174.

Jolly, M.K., Preca, B.-T., Tripathi, S.C., Jia, D., George, J.T., Hanash, S.M., et al., 2018. Interconnected feedback loops among ESRP1, HAS2, and CD44 regulate epithelial-mesenchymal plasticity in cancer. APL. Bioeng. 2 (3), 031908.

Jolly, M.K., Tripathi, S.C., Jia, D., Mooney, S.M., Celiktas, M., Hanash, S.M., et al., 2016. Stability of the hybrid epithelial/mesenchymal phenotype. Oncotarget. 7 (19), 27067–27084.

Jolly, M.K., Ware, K.E., Gilja, S., Somarelli, J.A., Levine, H., 2017. EMT and MET: necessary or permissive for metastasis? Mol. Oncology. 11 (7), 755–769.

Kahlert, U.D., Joseph, J.V., Kruyt, F.A.E., 2017. EMT- and MET-related processes in nonepithelial tumors: importance for disease progression, prognosis, and therapeutic opportunities. Mol. Oncol. 11 (7), 860–877.

Kalluri, R., Weinberg, R.A., 2009. The basics of epithelial-mesenchymal transition. J. Clin. Invest. 119 (6), 1420–1428.

Karacosta, L.G., Anchang, B., Ignatiadis, N., Kimmey, S.C., Benson, J.A., Shrager, J.B., et al., 2019. Mapping lung cancer epithelial-mesenchymal transition states and trajectories with single-cell resolution. bioRxiv 570341. Mar.

Kim, J.J., Yin, B., Christudass, C.S., Terada, N., Rajagopalan, K., Fabry, B., et al., 2013. Acquisition of paclitaxel resistance is associated with a more aggressive and invasive phenotype in prostate cancer. J. Cell Biochem. 114 (6), 1286–1293.

Krebs, A.M., Mitschke, J., Lasierra Losada, M., Schmalhofer, O., Boerries, M., Busch, H., et al., 2017. The EMT-activator Zeb1 is a key factor for cell plasticity and promotes metastasis in pancreatic cancer. Nat. Cell Biol. 19 (5), 518–529.

Kröger, C., Afeyan, A., Mraz, J., Eaton, E.N., Reinhardt, F., Khodor, Y.L., et al., 2019. Acquisition of a hybrid E/M state is essential for tumorigenicity of basal breast cancer cells. PNAS. 116 (15), 7353–7362.

Kumar, S., Das, A., Sen, S., 2014. Extracellular matrix density promotes EMT by weakening cell-cell adhesions. Mol. Biosyst. 10 (4), 838–850.

Lecharpentier, A., Vielh, P., Perez-Moreno, P., Planchard, D., Soria, J.C., Farace, F., 2011. Detection of circulating tumour cells with a hybrid (epithelial/mesenchymal) phenotype in patients with metastatic non-small cell lung cancer. Br. J. Cancer 105 (9), 1338–1341.

Lee, J.-K., Joo, K.M., Lee, J., Yoon, Y., Nam, D.-H., 2014. Targeting the epithelial to mesenchymal transition in glioblastoma: the emerging role of MET signaling. Onco Targets Ther. 7, 1933–1944.

Li, C., Hong, T., Nie, Q., 2016. Quantifying the landscape and kinetic paths for epithelial–mesenchymal transition from a core circuit. Phys. Chem. Chem Phys 18 (27), 17949–17956.

Li, F.Z., Dhillon, A.S., Anderson, R.L., McArthur, G., Ferrao, P.T., 2015. Phenotype switching in melanoma: implications for progression and therapy. Front. Oncol. 5, 31.

Li, S., Yang, J., 2014. Ovol proteins: guardians against EMT during epithelial differentiation. Developmental Cell. 29 (1), 1–2.

Liu, X., Li, J., Cadilha, B.L., Markota, A., Voigt, C., Huang, Z., et al., 2019. Epithelial-type systemic breast carcinoma cells with a restricted mesenchymal transition are a major source of metastasis. Sci. Adv. 5 (6), eaav4275.

Liu, X., Taftaf, R., Kawaguchi, M., Chang, Y.-F., Chen, W., Entenberg, D., et al., 2019. Homophilic CD44 interactions mediate tumor cell aggregation and polyclonal metastasis in patient-derived breast cancer models. Cancer Discov. 9 (1), 96–113.

Lu, M., Jolly, M.K., Levine, H., Onuchic, J.N., Ben-Jacob, E., 2013. MicroRNA-based regulation of epithelial–hybrid–mesenchymal fate determination. PNAS. 110 (45), 18144–18149.

Luo, M., Shang, L., Brooks, M.D., Jiagge, E., Zhu, Y., Buschhaus, J.M., et al., 2018. Targeting breast cancer stem cell state equilibrium through modulation of redox signaling. Cell Metab. 28 (1), 69–86.e6.

Marcelo, B., Kumar, J.M., Aaron, G., Mika, P., Mani Sendurai, A., Shiladitya, Sengupta, et al., 2016. Notch-Jagged signalling can give rise to clusters of cells exhibiting a hybrid epithelial/mesenchymal phenotype. J. R. Soc. Interface 13 (118), 20151106.

Micalizzi, D.S., Farabaugh, S.M., Ford, H.L., 2010. Epithelial-mesenchymal transition in cancer: parallels between normal development and tumor progression. J. Mammary Gland. Biol. Neoplasia 15 (2), 117–134.

Mooney, S.M., Talebian, V., Jolly, M.K., Jia, D., Gromala, M., Levine, H., et al., 2017. The GRHL2/ZEB feedback loop—a key axis in the regulation of EMT in breast cancer. J. Cell. Biochem. 118 (9), 2559–2570.

Nieto, M.A., 2013. Epithelial plasticity: a common theme in embryonic and cancer cells. Science. 342 (6159), 1234850.

Nieto, M.A., Sargent, M.G., Wilkinson, D.G., Cooke, J., 1994. Control of cell behavior during vertebrate development by Slug, a zinc finger gene. Science. 264 (5160), 835–839.

Park, S.-M., Gaur, A.B., Lengyel, E., Peter, M.E., 2008. The miR-200 family determines the epithelial phenotype of cancer cells by targeting the E-cadherin repressors ZEB1 and ZEB2. Genes. Dev. 22 (7), 894–907.

Pastushenko, I., Brisebarre, A., Sifrim, A., Fioramonti, M., Revenco, T., Boumahdi, S., et al., 2018. Identification of the tumour transition states occurring during EMT. Nature. 556 (7702), 463.

Polyak, K., Weinberg, R.A., 2009. Transitions between epithelial and mesenchymal states: acquisition of malignant and stem cell traits. Nat. Rev. Cancer. 9 (4), 265–273.

Porporato, P.E., Payen, V.L., Pérez-Escuredo, J., De Saedeleer, C.J., Danhier, P., Copetti, T., et al., 2014. A mitochondrial switch promotes tumor metastasis. Cell Rep. 8 (3), 754–766.

Puram, S.V., Tirosh, I., Parikh, A.S., Patel, A.P., Yizhak, K., Gillespie, S., et al., 2017. Single-cell transcriptomic analysis of primary and metastatic tumor ecosystems in head and neck cancer. Cell. 171 (7), 1611–1624.e24.

Rho, J.K., Choi, Y.J., Lee, J.K., Ryoo, B.-Y., Na, I.I., Yang, S.H., et al., 2009. Epithelial to mesenchymal transition derived from repeated exposure to gefitinib determines the sensitivity to EGFR inhibitors in A549, a non-small cell lung cancer cell line. Lung Cancer. 63 (2), 219–226.

Ruscetti, M., Dadashian, E.L., Guo, W., Quach, B., Mulholland, D.J., Park, J.W., et al., 2016. HDAC inhibition impedes epithelial-mesenchymal plasticity and suppresses metastatic, castration-resistant prostate cancer. Oncogene. 35 (29), 3781–3795.

Schliekelman, M.J., Taguchi, A., Zhu, J., Dai, X., Rodriguez, J., Celiktas, M., et al., 2015. Molecular portraits of epithelial, mesenchymal, and hybrid states in lung adenocarcinoma and their relevance to survival. Cancer Res. 75 (9), 1789–1800.

Sciacovelli, M., Frezza, C., 2017. Metabolic reprogramming and epithelial-to-mesenchymal transition in cancer. FEBS J. 284 (19), 3132–3144.

Sciacovelli, M., Gonçalves, E., Johnson, T.I., Zecchini, V.R., da Costa, A.S.H., Gaude, E., et al., 2016. Fumarate is an epigenetic modifier that elicits epithelial-to-mesenchymal transition. Nature. 537 (7621), 544–547.

Shibue, T., Weinberg, R.A., 2017. EMT, CSCs, and drug resistance: the mechanistic link and clinical implications. Nat. Rev. Clin. Oncol. 14 (10), 611–629.

Siemens, H., Jackstadt, R., Hünten, S., Kaller, M., Menssen, A., Götz, U., et al., 2011. miR-34 and SNAIL form a double-negative feedback loop to regulate epithelial-mesenchymal transitions. Cell Cycle. 10 (24), 4256–4271.

Somarelli, J.A., Shetler, S., Jolly, M.K., Wang, X., Bartholf Dewitt, S., Hish, A.J., et al., 2016. Mesenchymal-epithelial transition in sarcomas is controlled by the combinatorial expression of microRNA 200s and GRHL2. Mol. Cell Biol. 36 (19), 2503–2513.

Steinway, S.N., Zañudo, J.G.T., Michel, P.J., Feith, D.J., Loughran, T.P., Albert, R., 2015. Combinatorial interventions inhibit TGFβ-driven epithelial-to-mesenchymal transition and support hybrid cellular phenotypes. NPJ Syst. Biol. Appl. 1, 15014.

Thiery, J.P., Acloque, H., Huang, R.Y.J., Nieto, M.A., 2009. Epithelial-mesenchymal transitions in development and disease. Cell. 139 (5), 871–890.

Thomson, S., Buck, E., Petti, F., Griffin, G., Brown, E., Ramnarine, N., et al., 2005. Epithelial to mesenchymal transition is a determinant of sensitivity of non-small-cell lung carcinoma cell lines and xenografts to epidermal growth factor receptor inhibition. Cancer Res. 65 (20), 9455–9462.

Wicha, M.S., Liu, S., Dontu, G., 2006. Cancer stem cells: an old idea–a paradigm shift. Cancer Res. 66 (4), 1883–1890. discussion 1895-1896.

Ye, X., Weinberg, R.A., 2015. Epithelial-mesenchymal plasticity: a central regulator of cancer progression. Trends Cell Biol. 25 (11), 675–686.

Yu, L., Lu, M., Jia, D., Ma, J., Ben-Jacob, E., Levine, H., et al., 2017. Modeling the genetic regulation of cancer metabolism: interplay between glycolysis and oxidative phosphorylation. Cancer Res. 77 (7), 1564–1574.

Yu, M., Bardia, A., Wittner, B.S., Stott, S.L., Smas, M.E., Ting, D.T., et al., 2013. Circulating breast tumor cells exhibit dynamic changes in epithelial and mesenchymal composition. Science 339 (6119), 580–584.

Zaravinos, A., 2015. The regulatory role of microRNAs in EMT and cancer. J. Oncol. 2015, 865816.

Zhang, J., Tian, X.-J., Zhang, H., Teng, Y., Li, R., Bai, F., et al., 2014. TGF-β-induced epithelial-to-mesenchymal transition proceeds through stepwise activation of multiple feedback loops. Sci. Signal. 7 (345), ra91.

Zheng, X., Carstens, J.L., Kim, J., Scheible, M., Kaye, J., Sugimoto, H., et al., 2015. Epithelial-to-mesenchymal transition is dispensable for metastasis but induces chemoresistance in pancreatic cancer. Nature. 527 (7579), 525–530.

22

Phenotypic switching and prostate diseases: a model proposing a causal link between benign prostatic hyperplasia and prostate cancer

Prakash Kulkarni[1], Anusha Nathan[1], Ravi Salgia[1], Govindan Rangarajan[2,3] and Mohit Kumar Jolly[4]

[1]Department of Medical Oncology and Therapeutics Research, City of Hope National Medical Center, Duarte, CA, United States [2]Department of Mathematics, Indian Institute of Science, Bangalore, India [3]Center for Neuroscience, Indian Institute of Science, Bangalore, India [4]Center for BioSystems Science and Engineering, Indian Institute of Science, Bangalore, India

The difficulty lies, not in the new ideas, but in escaping from the old ones.

John Maynard Keynes

Introduction

Diseases of the prostate are some of the most highly prevalent and devastating ailments in men especially, as they age. In this chapter we focus on the two most common prostate diseases namely benign prostatic hyperplasia (BPH) and prostate cancer (PCa). Consistent with the theme of this book, we endeavor to highlight how phenotypic switching plays an important role in these diseases but remained unappreciated until the last decade. We present a model that points to a potential causal link between BPH and PCa. This new thinking not only sheds new light on this important medical conundrum but can also have a huge impact on how we treat and manage these burgeoning maladies.

Phenotypic Switching. DOI: https://doi.org/10.1016/B978-0-12-817996-3.00020-7

Although BPH is generally assumed to be an age-related disease, histologic (asymptomatic) BPH lesions are encountered in men in their 20s and 30s (Wolin et al., 2015). Histologic BPH affects 50% of all men by the time they turn 50 and escalates to >90% in men 85 years or older (Oesterling, 1996). Further, a significant fraction of these cases progress to symptomatic forms of the disease, which can result in lower urinary tract symptoms and other complications (Getzenberg and Kulkarni, 2014).

Similarly, although PCa is typically detected in men who are in their late 60s (Bostwick et al., 1992), the disease has been detected in young adults who are barely in their 20s (Haas et al., 2008; Baade et al., 2009). It is therefore not surprising that BPH and PCa can coexist in the same patient. Indeed, in an autopsy study of cadavers with and without PCa, BPH was identified in 80% and 45% of cadavers with and without PCa, respectively (Sommers, 1957). Consistent with these observations, Ørsted et al. (2011) who evaluated >3,000,000 Danish men found that BPH was associated with an increased incidence of PCa. Therefore, it may be inferred that most men diagnosed with BPH, particularly symptomatic BPH, have PCa within their prostates. Furthermore, in addition to the practically ubiquitous histologic association, both diseases share many commonalities in their etiologies including androgen-dependence, smoking, a diet high in fat (Ørsted and Bojesen, 2013), and inflammation (De Nunzio et al., 2011; Sfanos and De Marzo, 2012; Gurel et al., 2014; Sfanos et al., 2018). Notwithstanding these similarities and the high statistical likelihood, the prevailing wisdom enunciates that BPH and PCa are distinct and unrelated pathological states (De Marzo et al., 1998; Schenk et al., 2011; Chang et al., 2012), or at best, the link, if there is one, is unclear or controversial (Guess, 2001; De Nunzio et al., 2011; Ørsted and Bojesen, 2013).

Several reasons appear to aid these conclusions. In the majority of the cases, there is a zonal distinction in the occurrence of the two diseases in the prostate gland. Thus, while BPH is typically associated with the transition zone (TZ), PCa is predominantly (70%–80%) detected in the peripheral zone (PZ) (McNeal et al., 1988). Of note, in addition to the anatomical differences, there also appear to be molecular (Sinnott et al., 2015) and biological (Lee et al., 2015) differences between the prostate zones. Indeed, these differences suggest that subtypes potentially resulting from different etiologic pathways might arise in these zones (Sinnott et al., 2015). Consistently, TZ PCa is associated with favorable pathologic features and better recurrence-free survival despite being diagnosed with larger cancers and higher PSA values (Lee et al., 2015). Additionally, while PCa is thought to be a proliferative disease of the epithelial cells, BPH is often attributed to hyperplasia of the stromal cells. Finally, it is generally held that genetic instability and somatic mutations that are the hallmarks of cancer and are detected in PCa, are rarely observed in BPH.

However, mutations in genes such as p53 have been detected in BPH (Meyers et al., 1993; Schlechte et al., 1998; Mohamed et al., 2017), and so are chromosomal abnormalities (Dal Cin and Van Den Berghe, 1994; Aly et al., 1994; Altok et al., 2016). Furthermore, studies by Walsh and coworkers identified family history of BPH as a risk factor for clinical BPH and suggested the presence of a predisposing gene in patients with early onset BPH (Sanda et al., 1994). And more recently, a genome-wide association study by Gudmundsson et al. (2018) discovered 23 genome-wide significant variants, located at 14 loci with little or no overlap between the BPH variants and published PCa risk variants. Nonetheless, the dominant *weltanschauung* is that the two diseases are indeed very distinct. We refer to this traditional view as the Discrete Model (Fig. 22.1). Of note, the Discrete Model

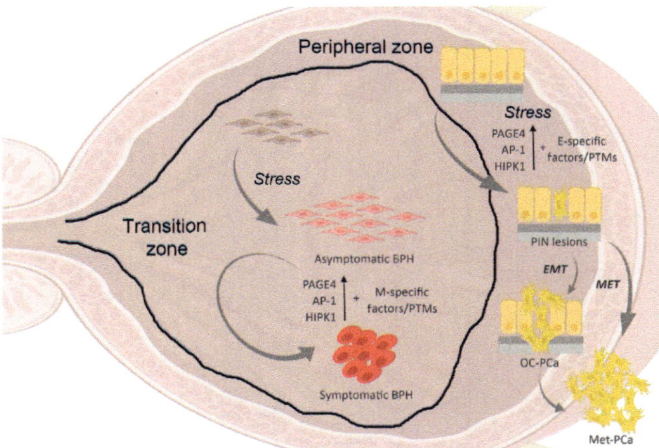

Figure 22.1 The discrete model. This model suggests that, prostatic epithelial cells give rise to prostate cancer in the PZ while the stromal cells give rise to BPH in the TZ. In the PZ, in response to stress, the activities of prostate-associated gene 4 (PAGE4), AP-1, and HIPK1 are upregulated and altered in response to epithelial (E)-specific factors and/or posttranslational modifications (PTMs) in prostatic intraepithelial neoplasia lesions and in organ-confined prostate cancer (OC-PCa). Malignant cells undergo an epithelial-to-mesenchymal transition (EMT), and migrate to distant locations. Here they undergo the reverse transition from mesenchymal to epithelial (MET) and give rise to metastatic disease (MET-PCa). More aggressive malignant cells with metastatic potential are thought to arise very early on (indicated by the arrow on the top) but lie dormant indefinitely. In response to stress, stromal cells proliferate in the TZ to give rise to asymptomatic BPH. Per the model, in response to additional factors that are turned on/off and/or PTMs specific to the mesenchymal (M) phenotype, the activities of PAGE4, AP-1, and HIPK1 are upregulated resulting in symptomatic BPH.

ignores the fact that because the two diseases can coexist, prognosis and selection of appropriate treatment options can be confusing and at times, misleading (Wilt et al., 2010; Roehrborn et al., 2011; Gudmundsson et al., 2018). Thus, there is an unmet need for a better understanding of the link between BPH and PCa.

From BPH to PCa via EMT/MET

Here, we shall discuss a different model that we refer to as the Integrated Model (Fig. 22.2) that underscores a plausible causality between the two diseases (Kulkarni and Getzenberg, 2016). The integrated model that embodies four steps and rests on a phenotypic switching underpinning, implies that in response to stress caused by factors such as oxidative damage and inflammation, phenotypic plasticity of the prostatic epithelial cell can be exploited by intrinsic disorder and promiscuous interactions of certain proteins to specify cell fate via EMT and MET. It is important to point out that while the

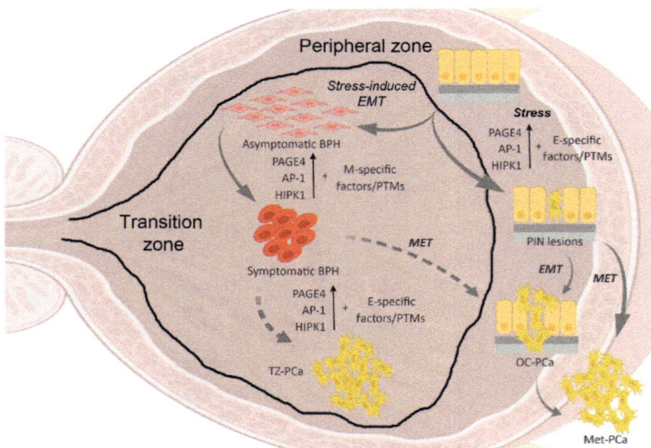

Figure 22.2 The integrated model. Per this model, prostatic epithelial cells give rise both to prostate cancer in the peripheral zone (PZ) as well as BPH in the transition zone (TZ). Prostate cancer in the PZ is envisaged to originate as in the case of the discrete model. However, the model states that in response to stress, epithelial cells undergo an epithelial-to-mesenchymal transition (EMT) and give rise to asymptomatic BPH per Alonso-Magdalena et al. (2009) and Tu et al. (2016). However, the model states that in response to additional factors that are turned on/off and/or PTMs specific to the mesenchymal (M) phenotype, the activities of PAGE4, AP-1, and HIPK1 are upregulated resulting in symptomatic BPH. Per the model, symptomatic BPH cells can undergo a reverse mesenchymal to epithelial transition (MET) and give rise to prostate cancer in the TZ (TZ-PCa), or these cells transcend the TZ and migrate to the PZ where they undergo and MET and give rise to organ-confined prostate cancer (OC-PCa). The dashed arrows indicate the phenotypic transitions envisaged in the integrated model.

Table 22.1 The four steps embodying in the Integrated Model.

Step	Event	Reference
1.	**a.** Field effect	Klein et al. (2013), Gabriel et al. (2016)
	b. EMT	Alonso-Magdalena et al. (2009), Shao et al. (2014), Hu et al. (2014), Shi et al. (2017)
2.	Asymptomatic to symptomatic BPH	Correlative evidence demonstrating selective upregulation of PAGE4 (Prakash et al., 2002) and AP-1 (Lin-Tsai et al., 2014), respectively in symptomatic BPH
3.	MET in transition zone	Assumption
4.	Migration of mesenchymal cells to peripheral zone from transition zone	Assumption

discrete model was inspired by experimental evidence for some aspects that are germane to it, some aspects are based on assumptions or are ideas that need to be experimentally verified. Table 22.1 lists the various aspects of the discrete model that are supported by data, and indicates those that are assumed or suggested as theoretical possibilities.

The Integrated Model

It is now becoming increasingly evident that the entire prostate can become unstable, a phenomenon often referred to as a "field effect" or changes in the normal adjacent areas of the prostate (Klein et al., 2013; Gabriel et al., 2016). Thus, molecular alterations in what appears to be morphologically (and histologically) normal prostate tissue are widespread. Among these are telomere dysfunction (Tu et al., 2016), the identification of molecular changes associated with aggressive disease within normal areas (Alinezhad et al., 2016), patterns of DNA methylation (Luo et al., 2013), as well as nuclear morphometric changes (Mairinger et al., 1999; Veltri et al., 2004). In fact, all of the hallmarks associated with instability and the origin of cancer are found within the normal areas of the diseased prostate (Nonn et al., 2009). Thus, the unstable prostate resulting from changes in the hormonal status, environmental insults, and inflammatory damage, is susceptible for the development of relatively hypeproliferative states and give rise to prostate diseases.

Anatomically, the human prostate gland comprises three zones namely, the central (CZ), transitional (TZ), and the peripheral (PZ) zone. The first step that is predicated by this instability invokes EMT in the PZ (Fig. 22.2). Although it is generally held that BPH is a so-called "stromal" disease, Alonso-Magdalena et al. (2009) showed that BPH may result due to an accumulation of mesenchymal-like cells derived from the prostatic epithelium and the endothelium. In their study, the authors found no evidence of proliferation in the stroma. However, in the epithelium of some ducts, they observed that 0.7% of the basal and 0.4% of luminal cells were positive for Ki67 and proliferating cell nuclear antigen, two key markers of proliferation. Furthermore, the regions of the ductal epithelium where the epithelial cells did not express E-cadherin, lost their polarization, became spindle shaped, and stained highly positively for the EMT markers Smad3 and Snail. These observations are entirely consistent with changes that occur when a cell transitions from an epithelial to a mesenchymal phenotype in the prostate (Fig. 22.3). Mesenchymal cells become elongated, lose cell-cell adhesions, and become motile. The most well-known target in epithelial cells is E-cadherin (CDH1), a cell membrane protein that allows for cell-cell attachment via a Ca^{2+}-dependent binding with E-cadherin molecules on adjacent cells. In a normal prostate epithelial cell, E-cadherin is also connected internally with actin filaments. Changes in the cytoskeleton are most evident within the intermediate filaments, which changes from keratin to vimentin providing the mesenchymal cells with their structure and contributing to motility (Mooney et al., 2016).

Subsequent work by Shao et al. (2014) who used both human BPH samples as well as a rodent model of BPH that was induced

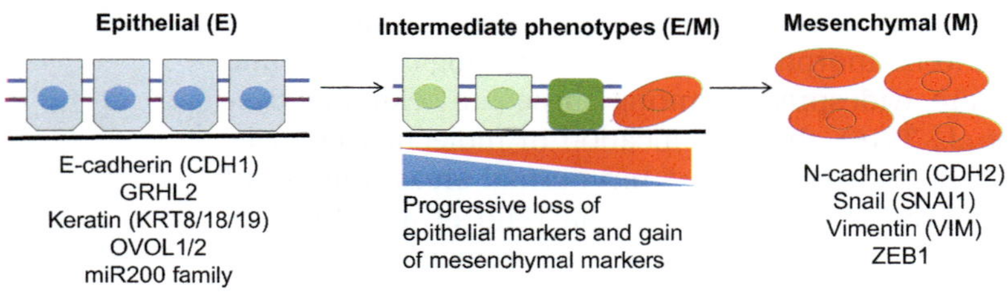

Figure 22.3 EMT progression. Epithelial (E) cells are well anchored to the basement membrane, connected to each other in sheets via E-cadherin junctions and express various markers, which may vary depending on the individual cancer cell. E cells can then transition into mesenchymal cells by first going through a series of intermediates (E/M) that are usually metastable. Mesenchymal (M) cells are fully motile and can exhibit various genes and properties including the ability to adapt to stress. *Source*: Reproduced with permission from Mooney, S.M., Jolly, M.K., Levine, H., Kulkarni, P., 2016. Phenotypic plasticity in prostate cancer: role of intrinsically disordered proteins. Asian J. Androl. 18 (5), 704–710.

by estrogen/androgen (E2:t, 1:100), also demonstrated that EMT and estrogen receptor α-mediated epithelial dedifferentiation mark the development of BPH. Additionally, the role of estrogen in inducing EMT in both two-dimensional and three-dimensional (3D) culture systems was also recently confirmed by Shi et al. (2017) lending further credence to role of E2 and EMT in the pathogenies of BPH in elderly men. Consistent with these observations López-Novoa and Nieto (2009) demonstrated that a chronic inflammatory microenvironment that is common to tumorigenesis is a decisive factor in the induction of the pathological EMT, and Hu et al. (2014) working with a cell culture model system showed that stromal cell supernatant was able to induce EMT in BPH-1 cells, possibly through secreting TGF-b1 to activate Smad signaling.

The second step per the Integrated Model entails another phenotypic switch and results in the transition from asymptomatic to symptomatic disease (Fig. 22.2). It should be emphasized that this step is based on the observations by Prakash et al. (2002) who employed gene expression profiling to discern differences between symptomatic and asymptomatic BPH. The authors found that, at the gene expression level, while asymptomatic BPH shares similarities with the normal prostate, symptomatic BPH resembles BPH patients who also had PCa elsewhere in the gland (BPH with cancer group) (Prakash et al., 2002). More importantly, both symptomatic BPH and BPH with cancer groups were enriched in genes associated with inflammation corroborating the role of stress, and those encoding cytokines and cell proliferation factors. Interestingly, the authors also found that the expression profile distinguishing the BPH with cancer group from all other groups was highly enriched in several oncogenes such as Myc, Jun, Fos, etc. even though the samples used were devoid of cancer (Prakash et al., 2002).

One of the genes from the panel that differentiated symptomatic and asymptomatic BPH in this study was the cancer/testis antigen (CTA) PAGE4. Subsequently it was found that PAGE4 is a highly intrinsically disordered protein (IDP) (Zeng et al., 2011), and is upregulated in response to stress including inflammatory stress (Zeng et al., 2013). Functional studies using cell-based reporter assays revealed that PAGE4 acts as a powerful potentiator of transactivation by the oncogene c-Jun (Rajagopalan et al., 2014) that heterodimerizes with members of the Fos family of transcription factors to constitute the AP-1 complex involved in several cellular processes such as differentiation, proliferation, and apoptosis (Shaulian and Karin, 2002; Hess et al., 2004; Ye et al., 2014). Biochemical studies revealed that the stress-response kinase HIPK1 phosphorylates PAGE4, and this site-specific phosphorylation that remodels its conformational ensemble, is critical for potentiation of Jun activity (Mooney et al., 2014). In addition to HIPK1, PAGE4 is phosphorylated by a second kinase namely, CDC-Like Kinase 2 (CLK2), a dual-specificity kinase that phosphorylates serine-rich proteins of the spliceosome complex. However in contrast to HIPK1, PAGE4 is hyperphosphorylated by CLK2 at multiple S/T residues. Hyperphosphorylation by CLK2 attenuates rather than potentiates c-Jun transactivation (Kulkarni et al., 2017).

In PCa cells, since AP-1 can inhibit the androgen receptor (AR) activity which in turn can repress expression of CLK2 (Kulkarni et al., 2017), it may be surmised that there is a negative feedback loop in the PAGE4/AP-1/AR circuit. Indeed, mathematical modeling suggests that this feedback loop can give rise to sustained or damped oscillations in the levels of AR activity, HIPK1-PAGE4 and CLK2-PAGE4 (Fig. 22.4), suggesting that androgen dependence of a PCa cell can be a dynamic trait (Kulkarni et al., 2017; Lin et al., 2018). Furthermore, PCa cells that are insensitive to androgen typically show increased AR activity as an adaptive auto-regulatory mechanism (Isaacs et al., 2012). Therefore, as the intracellular levels of HIPK1-PAGE4 and CLK2-PAGE4 vary, cells can go on phenotypic excursions with varying sensitivities to androgen. Additional interactions of these components could convert these oscillations into a multistable system. Thus, it follows that this heterogeneous population can better evade the effects of ADT, the

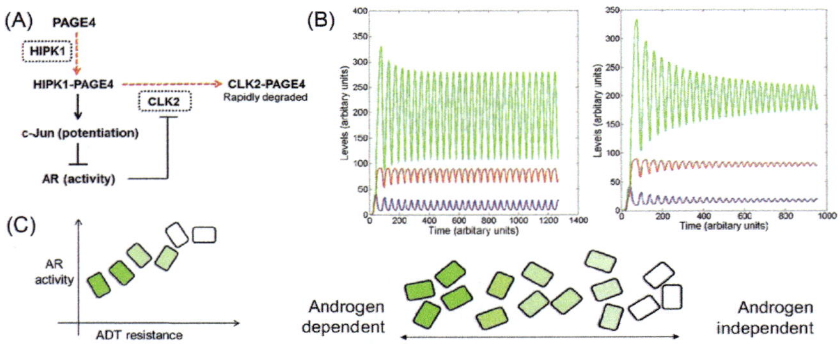

Figure 22.4 Modeling the PAGE4/AP-1/AR/CLK2 regulatory circuit. (A) Regulatory circuit for PAGE4/AP-1/AR/CLK2 interactions. Dashed red lines denote enzymatic reactions, and solid black lines denote nonenzymatic reactions. CLK2 and HIPK1, the two enzymes involved, are shown in dotted rectangles. (B) Dynamics of the circuit showing sustained and damped oscillations for HIPK1-PAGE4 (PAGE4M, shown in *blue*), CLK2-PAGE4 (PAGE4H, shown in *red*), and CLK2 (shown in *green*). (C) Distribution of androgen dependence for an isogenic population over a spectrum, as indicated by the shade of green. Dark green boxes denote highly androgen-dependent (i.e., ADT-sensitive) cells, and white boxes denote androgen-independent cells. *Source*: Reproduced with permission from Kulkarni, P., Jolly, M.K., Jia, D., Mooney, S.M., Bhargava, A., Kagohara, L.T., et al., 2017. Phosphorylation-induced conformational dynamics in an intrinsically disordered protein and potential role in phenotypic heterogeneity. Proc. Natl. Acad. Sci. U.S.A. 114 (13), E2644–E2653.

first line treatment option for PCa patients, as compared to a homogeneous PCa population. This nongenetic mechanism involving phenotypic switching is in contrast with the Darwinian clonal evolution model (Brock et al., 2009) which assumes the existence of mutually exclusive androgen-dependent and androgen-independent clones. Thus, in addition to genetic changes, phenotypic plasticity in PCa may be driven by underlying dynamics of the PAGE4/AP-1/AR circuit. Consistently, while CLK2, like PAGE4, is expressed in androgen-dependent PCa while HIPK1 is expressed both in androgen-dependent and androgen-independent PCa cells (Kulkarni et al., 2017).

Of note, the PAGE4 protein is dramatically upregulated in the proliferative inflammatory atrophy (PIA) (Zeng et al., 2013) and high-grade prostatic intraepithelial neoplasia (HG-PIN) lesions (Sampson et al., 2012) that are infiltrated with lymphocytes and are regarded as PCa precursors (Nelson et al., 2003). In addition, the protein is also upregulated in OC-PCa (Zeng et al., 2013) (Fig. 22.5). More importantly, AP-1 is also confirmed to be upregulated in PCa (Ouyang et al., 2008) and in symptomatic but not in asymptomatic BPH (Lin-Tsai et al., 2014). Together, these observations highlight the pathological relevance of the PAGE4/AP-1 interaction in regulating the switch from asymptomatic to symptomatic BPH. However, per the Integrated Model, specification of cell fate would depend on the PTMs that remodel the PAGE4 ensemble as well as the presence or absence of the factors with which PAGE4/AP-1 complex interacts (Mooney et al., 2014; He et al., 2015). Thus, the Integrated Model implies that additional factors that interact with PAGE4 and its PTMs specific to the mesenchymal phenotype drive the transition of asymptomatic to symptomatic BPH.

In the subsequent step, the integrated model envisages a MET to occur in the TZ resulting in PCa in the TZ that presents in the clinic albeit less frequently (20%–30%) (McNeal et al., 1988). TZ cancers or so-called incidental carcinomas, tend to follow a nonaggressive clinical course whereas cancers from the PZ are more aggressive, tend to invade the periprostatic tissues, and have higher biochemical recurrence rates (McNeal et al., 1988; Colombo et al., 2001). MET is postulated to be aided by PTMs and factors in the TZ local microenvironment (Fig. 22.2). It may be worth noting that, consistent with this hypothesis, classical work from Cunha and colleagues using tissue recombinant techniques demonstrated that alteration in the stromal microenvironment was sufficient to promote malignant transformation of human prostatic epithelial cells (Cunha et al., 2002).

In the last step, the Integrated Model suggests that, instead of, or in addition to, colonizing the TZ mesenchymal cells resulting

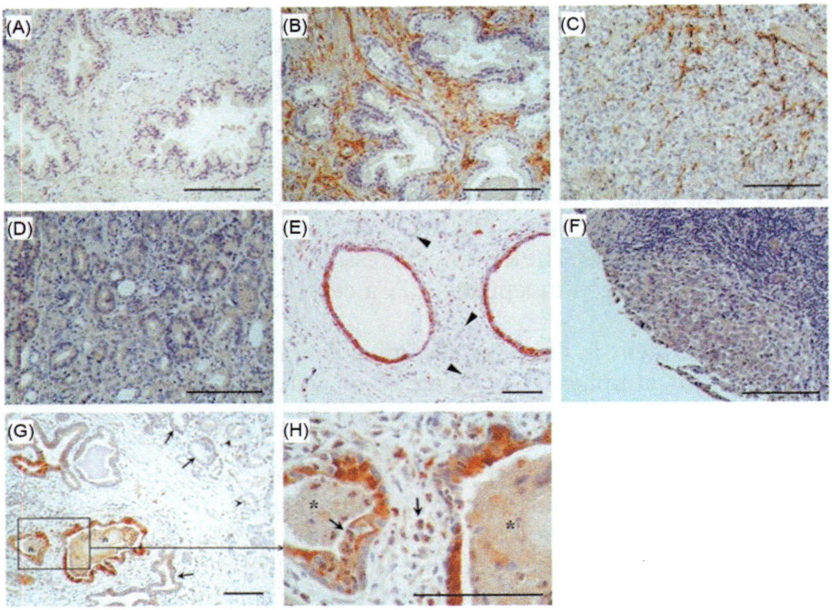

Figure 22.5 Immunohistochemistry analysis of Prostate-associated Gene 4 (PAGE4) in prostate cancer (PCa). (A) Negative staining in the normal prostate. (B) Intense staining shown in the stromal tissue in benign prostatic hyperplasia (BPH). (C) Positive staining in the stromal cells but negative in the cancer cells in some PCa specimens. (D) Moderate staining in the cancer cells but negative in the stromal cells in some PCa specimens. (E) Positive staining in the atrophic glands but negative in the cancer cells (*arrowhead*). (F) Negative staining in metastatic PCa. (G) Intense staining shown in cancer adjacent "normal" glands (*asterisk*) associated with inflammation but only moderate staining in the cancer cells (*arrowhead*). (H) High power view of boxed area in (G). Asterisk, PIA lesions; *arrows*, inflammatory cells. Scale bars in all panels, 100 μm. *Source*: From Kulkarni, P., Solomon, T.L., He, Y., Chen, Y., Bryan, P.N., Orban, J., 2018. Structural metamorphism and polymorphism in proteins on the brink of thermodynamic stability. Protein Sci. 27 (9), 1557–1567.

from step 3 that harbor elevated levels of PAGE4, AP-1, and HIPK1, particularly those that retain a mixed phenotype with aggressive tumor-initiating properties (see below section on The key drivers of EMT/MET in PCa are IDPs), migrate to the PZ. In the PZ, aided by PTMs and factors in the local microenvironment, they stochastically revert to an epithelial phenotype via MET and seed the epithelial compartment to give rise to PCa in the PZ, the most common site of PCa (Fig. 22.2).

The key drivers of EMT/MET in PCa are IDPs

The key transcription factors that regulate the EMT/MET phenotypic switching constitute a gene regulatory circuit (Fig. 22.6).

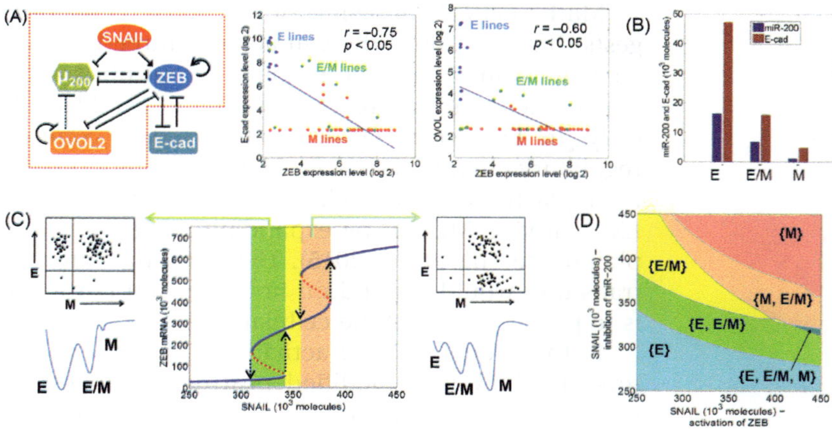

Figure 22.6 Modeling of PCa transitions. (A) (*Left*) miR-200, OVOL (Ovo-Like Zinc Finger 1/2), and ZEB (zinc finger E-box-binding homeobox 1/2) circuit with solid lines indicating direct transcriptional activation (*arrows*) or inhibition (*bar*). The dashed line is a miRNA-mediated translational inhibition whereas the dotted line is an indirect inhibition. Within the red dotted lines is the circuit considered by Jia et al. The graphs in the middle and right depict the correlation between the levels of ZEB1, CDH1, and OVOL in NCI-60 cell lines. The CellMiner website was used to query the corresponding gene expression levels. (B) Levels of CDH1 and miR-200 in three phenotypes, as predicted by the model considered in (A). (C) Graph indicates how the levels of SNAIL (which includes SNAI1/2 [Snail Family Zinc Finger 1/2]) protein change the levels of ZEB mRNA per cell. The yellow area is where only the E/M (Epithelial/Mesenchymal hybrid) state exists. Corresponding in silico predicted fluorescence-activated cell sorting (top) and "Waddington landscape" figures (below) have been included. (D) The phase diagram has the SNAIL signal split into two external inputs, which can either activate ZEB or repress miR-200. *Source*: (B and D) Modified from Jia, D., Jolly, M.K., Boareto, M., Parsana, P., Mooney, S.M., Pienta, K.J., et al., 2015. OVOL guides the epithelial-hybrid-mesenchymal transition. Oncotarget 6 (17), 15436–15448, while (A) is an extension of the model with the addition of E-cadherin and a detailed method provided in the supplemental information. Reproduced with permission from Mooney, S.M., Jolly, M.K., Levine, H., Kulkarni, P., 2016. Phenotypic plasticity in prostate cancer: role of intrinsically disordered proteins. Asian J. Androl. 18 (5), 704–710.

In PCa, ZEB1 plays a crucial role by acting as the trigger and ZEB1 is positively regulated by SNAIL1. Upregulation of ZEB1 represses genes critical to the epithelial phenotype but activates those genes that characterize a mesenchymal phenotype (Gemmill et al., 2011). Furthermore, ZEB1 inhibits, and is directly inhibited by, the transcription factors OVOL1/2 (OVOL) and micro-RNA 200c (miR200c), forming mutually inhibitory feedback loops. OVOL in turn can also inhibit its own transcription (Jia et al., 2015), and E-Cadherin can indirectly inhibit ZEB1 (Schmalhofer et al., 2009). Indeed, downregulation of ZEB1 induces MET validating the regulatory circuit. Therefore, while upregulation of ZEB1 and SNAIL1 and downregulation of miR200c, E-Cadherin, and OVOL delineate the mesenchymal phenotype, the opposite is true for the epithelial phenotype (Mooney et al., 2016).

Mathematical modeling of the EMT regulatory circuit suggests that miR-200/ZEB1 can act as a three-way switch that can facilitate transitions among the different phenotypes with varying plasticity and aggressiveness (Lu et al., 2013). More specifically, in PCa cells, OVOL that is a key component of this regulatory circuit, plays a crucial role in modulating the phenotypic plasticity of PCa cells in multiple ways such as restricting EMT, driving MET or by expanding the existence of the hybrid phenotype (Jia et al., 2015). However, the rates of phenotypic transition would depend on how robust a particular phenotype is to perturbations of the cellular PINs.

In this model, SNAIL1 activity can differ depending on cofactors and how it is modified posttranslationally and therefore, can produce two separate signals to repress miR200c or activate ZEB1. The resulting phase diagram demonstrates the existence of combinations of different distinct phenotypes. Of note, they may all coexist depending on the differential repressing or activating activities of SNAIL1 (Lu et al., 2013; Jia et al., 2015). Taken together, the model underscores the fact that phenotypic plasticity in the context of EMT/MET at least in PCa, is not dependent on a linear upregulation of transcription but rather a phase transition through a bifurcation indicative of a rewiring of the cell's PIN (Fig. 22.6).

In addition to the epithelial or mesenchymal phenotype a PCa cell can assume, some cells may display a hybrid phenotype (Lu et al., 2013; Jia et al., 2015). Interestingly, it has been observed that such cells have maximum cellular plasticity, possess heightened tumor-initiating properties (Grosse-Wilde et al., 2015), and metastatic potential. The Integrated model postulates that some hybrid metastable cells in which interactions between PAGE4 and the AP-1 complex can rewire the protein interaction network (PIN) to uncover the tumor-initiating properties such as dedifferentiation, transcend the TZ to give rise to PCa lesions in the PZ. Consistently, PAGE4, which is undetectable in terminally differentiated somatic cells, is dramatically upregulated when the somatic cells are reprogrammed to form induced pluripotent stem cells (Kulkarni et al., 2016).

Phenotypic plasticity is a manifestation of plasticity at the molecular level

A remarkable feature of the factors that actuate phenotypic switching is that many of them are, predicted to be and in some cases, experimentally confirmed, to be IDPs. For example, the products of most oncogenes such as JUN, FOS, MYC (Iakoucheva et al., 2002; Uversky, 2014), the factors that reprogramme somatic cells to pluripotent stem cells such as OCT3/4, SOX2, MYC, NANOG, and KLF4 (Xue et al., 2012) and the cancer/testis antigens including PAGE4 (Rajagopalan et al., 2011) several of which are implicated in EMT and the genesis of cancer stem cell-like cells (Yang et al., 2015; Gordeeva, 2018; Cao et al., 2018; Ardalan Khales et al., 2019) are IDPs. Furthermore, the key factors implicated in EMT/MET, namely, OVOL1/2, ZEB1, and SNAI1 are also strongly predicted to be IDPs (Mooney et al., 2016).

Two main characteristics of the IDPs may help explain why they may be the key drivers in phenotypic switching. First, IDPs lack rigid 3D structure and have a rugged energy landscape with many local minima separated with low-energy barriers (Turoverov et al., 2010). In contrast to energy landscapes of highly ordered proteins, this type of energy landscape enables stochastic IDP fluctuations between numerous conformational states in response to modest perturbations, such as the overexpression of its binding partners. Second, despite lack of structure, they do not exist as a random coil; they exist as ensembles and exhibit conformational preferences either upon interacting with a partner (Uversky and Dunker, 2010; Wright and Dyson, 2015) or in response to posttranslational modifications such as phosphorylation (Bah et al., 2015; Wrighton, 2015; Hausrath and Kingston, 2017; reviewed, DeForte and Uversky, 2016; Kulkarni et al., 2018; Lin et al., 2019). Furthermore, some IDPs stochastically undergo excursions between multiple conformations (Choi et al., 2011; Choi et al., 2019). Therefore, with multiple conformational states and rapid conformational dynamics, IDPs can engage in myriad "promiscuous" interactions and typically occupy hub positions in PINs. Such stochastic interactions can result in "conformational" noise (Mahmoudabadi et al., 2013). It is now widely accepted that noise in gene expression ("transcriptional" noise) underlies substantial phenotypic variations, resulting in genetically identical cells to switch phenotypes and behave differently (Taniguchi et al., 2010; Munsky et al., 2012; Urban and Johnston, 2018; Engl, 2019). In such systems, especially those driven by positive feedback loops, transcriptional noise allows cell fate decisions to be probabilistic and can lead to novel phenotypes (Eldar and Elowitz, 2010; Balázsi et al., 2011; Farquhar et al., 2019). Of note, the role of IDPs is not limited to the propagation of transcriptional noise. Rather, IDPs could likely relay, and even amplify, other intrinsic and extrinsic types of noise and perturbations in the system.

Conclusions and future directions

From the foregoing it follows that contrary to the prevailing deterministic view, stochasticity may be a confounding factor in specifying cell fate. Therefore, genetic instability that invariably occurs in PCa, may be considered as the result rather than the cause. In fact, it is now amply clear that contrary to the prevailing wisdom that mutations occur purely stochastically at constant, gradual rates, microbes, plants, flies, and human cancer cells possess mechanisms of mutagenesis that are upregulated by stress responses (Fitzgerald et al., 2017; Fitzgerald and Rosenberg, 2019). Consistent with this argument, it has also been shown that at least in the case of Myc-driven cancers, brief or partial suppression of Myc to its physiological levels on can result in sustained tumor regression with tumor cells undergoing proliferative arrest, differentiation, senescence, and apoptosis (Shachaf et al., 2004; Li et al., 2014; Gabay et al., 2014). Recently, it was demonstrated that cellular plasticity of cancer cells undergoing EMT can be exploited to force transdifferentiation of breast cancer cells into postmitotic and functional adipocytes, resulting in repression of primary tumor invasion and metastasis (Ishay-Ronen et al., 2019). Finally, a recent study by Yizhak et al. (2019) demonstrating that even normal human tissue can harbor cancer-associated mutations points to the critical role of factors other than mutations alone in cancer.

A causal link between BPH and PCa although suspected has thus far remained elusive. As far as we are aware, the model proposed here is the first to establish this causality. A fundamental tenet on which the model heavily relies is that molecular plasticity of the IDPs underlies the phenotypic plasticity. However, it has never been considered since it was tacitly assumed that structure defines all biological function and hence, by inference, the actuators are highly structured (rather than unstructured and highly plastic) molecules. The integrated model provides a conceptual basis to link plasticity at the molecular level (conformational dynamics of the IDPs) to phenotypic plasticity at the cellular level and provides a theoretical framework to illustrate how phenotypic plasticity of the impressionable prostate can give rise to both (symptomatic) BPH and PCa. Furthermore, it should be noted that the two models need not be mutually exclusive; the integrated model may be considered in addition to, not instead of, the discrete model.

The key aspects that eclipse symptomatic BPH and PCa including its precursors are stress, including inflammation and oxidative damage, upregulation of IDPs such as PAGE4 and the EMT markers OVOL1/2, ZEB1, and SNAI1, and the coexpression of PAGE4 and the AP-1 complex. Together these aspects that constitute the underpinnings of the integrated model provide compelling evidence linking this causality. Furthermore, the fact that the conformational ensembles of IDPs can be remodeled to populate different clusters that can promiscuously interact with different partners and hence, guide different functional outputs with large functional consequences such as state switching, underscores the virtues of the integrated model in demonstrating the causality. Of note, consistent with the "field effect" theory, a comparison of the expression of PAGE4 in cancerous and immediately adjacent "normal" samples obtained from the same patient by laser microdissection demonstrated that PAGE4 expression is much higher in the latter (Zeng et al., 2013) lending further credence to the Integrated Model. As additional evidence accumulates in the future, the Integrated Model will not only reaffirm the role of phenotypic switching in BPH and PCa but will also confirm the causal link between these two prostate diseases. Establishing this causality can improve the accuracy of disease prognostication and expedite intervention, perhaps even preemptively.

References

Alinezhad, S., Vaananen, R.M., Tallgren, T., Perez, I.M., Jambor, I., Aronen, H., et al., 2016. Stratification of aggressive prostate cancer from indolent disease—prospective controlled trial utilizing expression of 11 genes in apparently benign tissue. Urol. Oncol. S1078–S1439, 610–619.

Alonso-Magdalena, P., Brossner, C., Reiner, A., Cheng, G., Sugiyama, N., Warner, M., et al., 2009. A role for epithelial-mesenchymal transition in the etiology of benign prostatic hyperplasia. Proc. Natl. Acad. Sci. U.S.A. 106, 2859–2863.

Altok, M., Bağcı, Ö., Umul, M., Güneş, M., Akyüz, M., Uruç, F., et al., 2016. Chromosomal aberrations in benign prostatic hyperplasia patients. Investig. Clin. Urol. 57 (1), 45–49.

Aly, M.S., Dal Cin, P., Van de Voorde, W., van Poppel, H., Ameye, F., Baert, L., et al., 1994. Chromosome abnormalities in benign prostatic hyperplasia. Genes Chromosomes Cancer 9 (4), 227–233.

Ardalan Khales, S., Abbaszadegan, M.R., Majd, A., Forghanifard, M.M., 2019. Linkage between EMT and stemness state through molecular association between TWIST1 and NY-ESO1 in esophageal squamous cell carcinoma. Biochimie 163, 84–93.

Baade, P.D., Youlden, D.R., Krnjacki, L.J., 2009. International epidemiology of prostate cancer: geographical distribution and secular trends. Mol. Nutr. Food Res. 53 (2), 171–184.

Bah, A., Vernon, R.M., Siddiqui, Z., Krzeminski, M., Muhandiram, R., Zhao, C., et al., 2015. Folding of an intrinsically disordered protein by phosphorylation as a regulatory switch. Nature 519 (7541), 106–109.

Balázsi, G., van Oudenaarden, A., Collins, J.J., 2011. Cellular decision making and biological noise: from microbes to mammals. Cell 144, 910–925.

Bostwick, D.G., Cooner, W.H., Denis, L., Jones, G.W., Scardino, P.T., Murphy, G.P., 1992. The association of benign prostatic hyperplasia and cancer of the prostate. Cancer 70, 291–301.
Brock, A., Chang, H., Huang, S., 2009. Non-genetic heterogeneity—a mutation-independent driving force for the somatic evolution of tumours. Nat. Rev. Genet. 10, 336–342.
Cao, Q.H., Liu, F., Li, C.Z., Liu, N., Shu, M., Lin, Y., et al., 2018. Testes-specific protease 50 (TSP50) promotes invasion and metastasis by inducing EMT in gastric cancer. BMC Cancer 18 (1), 94.
Chang, R.T., Kirby, R., Challacombe, B.J., 2012. Is there a link between BPH and prostate cancer? Practitioner 256, 13–16.
Choi, U.B., McCann, J.J., Weninger, K.R., Bowen, M.E., 2011. Beyond the random coil: stochastic conformational switching in intrinsically disordered proteins. Structure 19 (4), 566–576.
Choi, U.B., Sanabria, H., Smirnova, T., Bowen, M.E., Weninger, K.R., 2019. Spontaneous switching among conformational ensembles in intrinsically disordered proteins. Biomolecules 9 (3), pii: E114.
Colombo, P., Patriarca, C., Alfano, R.M., Cassani, B., Ceva Grimaldi, G., Roncalli, M., et al., 2001. Molecular disorders in transitional vs. peripheral zone prostate adenocarcinoma. Int. J. Cancer 94 (3), 383–389.
Cunha, G.R., Hayward, S.W., Wang, Y.Z., 2002. Role of stroma in carcinogenesis of the prostate. Differentiation 70 (9–10), 473–485.
Dal Cin, P., Van Den Berghe, H., 1994. Chromosome abnormalities in benign prostatic hyperplasia. In: Petrovich, Z., Baert, L. (Eds.), Benign Prostatic Hyperplasia: Innovations in Management. Springer-Verlag, New York, pp. 49–56.
De Marzo, A.M., Meeker, A.K., Epstein, J.I., Coffey, D.S., 1998. Prostate stem cell compartments: Expression of the cell cycle inhibitor p27Kip1 in normal, hyperplastic, and neoplastic cells. Am. J. Pathol. 153, 911–919.
De Nunzio, C., Kramer, G., Marberger, M., Montironi, R., Nelson, W., Schroder, F., et al., 2011. The controversial relationship between benign prostatic hyperplasia and prostate cancer: the role of inflammation. Eur. Urol. 60, 106–117.
DeForte, S., Uversky, V.N., 2016. Order, disorder, and everything in between. Molecules 21 (8), pii: E1090.
Eldar, A., Elowitz, M.B., 2010. Functional roles for noise in genetic circuits. Nature 467 (7312), 167–173.
Engl, C., 2019. Noise in bacterial gene expression. Biochem. Soc. Trans. 47 (1), 209–217.
Farquhar, K.S., Charlebois, D.A., Szenk, M., Cohen, J., Nevozhay, D., Balázsi, G., 2019. Role of network-mediated stochasticity in mammalian drug resistance. Nat. Commun. 10 (1), 2766.
Fitzgerald, D.M., Hastings, P.J., Rosenberg, S.M., 2017. Stress-induced mutagenesis: implications in cancer and drug resistance. Annu. Rev. Cancer Biol. 1, 119–140.
Fitzgerald, D.M., Rosenberg, S.M., 2019. What is mutation? A chapter in the series: how microbes "jeopardize" the modern synthesis. PLoS Genet. 15 (4), e1007995.
Gabay, M., Li, Y., Felsher, D.W., 2014. MYC activation is a hallmark of cancer initiation and maintenance. Cold Spring Harb. Perspect. Med. 4 (6), a014241.
Gabriel, K.N., Jones, A.C., Nguyen, J.P., Antillon, K.S., Janos, S.N., Overton, H.N., et al., 2016. Association and regulation of protein factors of field effect in prostate tissues. Int. J. Oncol. 49 (4), 1541–1552.

Gemmill, R.M., Roche, J., Potiron, V.A., Nasarre, P., Mitas, M., Coldren, C.D., et al., 2011. ZEB1-responsive genes in non-small cell lung cancer. Cancer Lett. 300 (1), 66–78.

Getzenberg, R.H., Kulkarni, P., 2014. Etiology and pathogenesis. In: Kaplan, S.A., McVary, K.T. (Eds.), Male Lower Urinary Tract Symptoms and Benign Prostatic Hyperplasia. Wiley Blackwell, New York.

Gordeeva, O., 2018. Cancer-testis antigens: unique cancer stem cell biomarkers and targets for cancer therapy. Semin. Cancer Biol. 53, 75–89.

Grosse-Wilde, A., Fouquier d'Hérouël, A., McIntosh, E., Ertaylan, G., Skupin, A., Kuestner, R.E., et al., 2015. Stemness of the hybrid epithelial/mesenchymal state in breast cancer and its association with poor survival. PLoS One 10 (5), e0126522.

Gudmundsson, J., Sigurdsson, J.K., Stefansdottir, L., Agnarsson, B.A., Isaksson, H.J., Stefansson, O.A., et al., 2018. Genome-wide associations for benign prostatic hyperplasia reveal a genetic correlation with serum levels of PSA. Nat. Commun. 9 (1), 4568.

Guess, H.A., 2001. Benign prostatic hyperplasia and prostate cancer. Epidemiol. Rev. 23, 152–158.

Gurel, B., Lucia, M.S., Thompson, I.M., Goodman, P.J., Tangen, C.M., Kristal, A.R., et al., 2014. Chronic inflammation in benign prostate tissue is associated with high-grade prostate cancer in the placebo arm of the prostate cancer prevention trial. Cancer Epidemiol. Biomarkers Prev. 23, 847–856.

Haas, G.P., Delongchamps, N., Brawley, O.W., Wang, C.Y., de la Roza, G., 2008. The worldwide epidemiology of prostate cancer: Perspectives from autopsy studies. Can. J. Urol. 15, 3866–3871.

Hausrath, A.C., Kingston, R.L., 2017. Conditionally disordered proteins: bringing the environment back into the fold. Cell Mol. Life Sci. 74 (17), 3149–3162.

He, Y., Chen, Y., Mooney, S.M., Rajagopalan, K., Bhargava, A., Sacho, E., et al., 2015. Phosphorylation-induced conformational ensemble switching in an intrinsically disordered cancer/testis antigen. J. Biol. Chem. 290 (41), 25090–25102.

Hess, J., Angel, P., Schorpp-Kistner, M., 2004. AP-1 subunits: quarrel and harmony among siblings. J. Cell Sci. 117 (Pt 25), 5965–5973.

Hu, S., Yu, W., Lv, T.J., Chang, C.S., Li, X., Jin, J., 2014. Evidence of TGF-β1 mediated epithelial-mesenchymal transition in immortalized benign prostatic hyperplasia cells. Mol. Membr. Biol. 31 (2–3), 103–110.

Iakoucheva, L.M., Brown, C.J., Lawson, J.D., Obradović, Z., Dunker, A.K., 2002. Intrinsic disorder in cell-signaling and cancer-associated proteins. J. Mol. Biol. 323 (3), 573–584.

Isaacs, J.T., D'Antonio, J.M., Chen, S., Antony, L., Dalrymple, S.P., Ndikuyeze, G.H., et al., 2012. Adaptive auto-regulation of androgen receptor provides a paradigm shifting rationale for bipolar androgen therapy (BAT) for castrate resistant human prostate cancer. Prostate. 72 (14), 1491–1505.

Ishay-Ronen, D., Diepenbruck, M., Kalathur, R.K.R., Sugiyama, N., Tiede, S., Ivanek, R., et al., 2019. Gain fat-lose metastasis: converting invasive breast cancer cells into adipocytes inhibits cancer metastasis. Cancer Cell. 35 (1), 17–32.

Jia, D., Jolly, M.K., Boareto, M., Parsana, P., Mooney, S.M., Pienta, K.J., et al., 2015. OVOL guides the epithelial-hybrid-mesenchymal transition. Oncotarget. 6 (17), 15436–15448.

Klein, E.A., Moscovita Falzarano, S.M., Zhang, N., Knezevic, D., Maddala, T., Lawrence, H.J., et al., 2013. Evidence for a field effect in early prostate cancer (PCa): gene expression profiles in normal-appearing prostate tissue (NT) adjacent to tumor (T) as predictors of clinical outcome. J. Clin. Oncol. 15 (Suppl), 5029.

Kulkarni, P., Dunker, A.K., Weninger, K., Orban, J., 2016. Prostate-associated gene 4 (PAGE4), an intrinsically disordered cancer/testis antigen, is a novel therapeutic target for prostate cancer. Asian J. Androl. 18 (5), 695–703.

Kulkarni, P., Getzenberg, R.H., 2016. Disorder, promiscuous interactions, and stochasticity regulate state switching in the unstable prostate. J. Cell Biochem. 117 (10), 2235–2240.

Kulkarni, P., Jolly, M.K., Jia, D., Mooney, S.M., Bhargava, A., Kagohara, L.T., et al., 2017. Phosphorylation-induced conformational dynamics in an intrinsically disordered protein and potential role in phenotypic heterogeneity. Proc. Natl. Acad. Sci. U.S.A. 114 (13), E2644–E2653.

Kulkarni, P., Solomon, T.L., He, Y., Chen, Y., Bryan, P.N., Orban, J., 2018. Structural metamorphism and polymorphism in proteins on the brink of thermodynamic stability. Protein Sci. 27 (9), 1557–1567.

Lee, J.J., Thomas, I.C., Nolley, R., Ferrari, M., Brooks, J.D., Leppert, J.T., 2015. Biologic differences between peripheral and transition zone prostate cancer. Prostate. 75 (2), 183–190.

Li, Y., Casey, S.C., Felsher, D.W., 2014. Inactivation of MYC reverses tumorigenesis. J. Intern. Med. 276 (1), 52–60.

Lin, X., Roy, S., Jolly, M.K., Bocci, F., Schafer, N.P., Tsai, M.Y., et al., 2018. PAGE4 and conformational switching: insights from molecular dynamics simulations and implications for prostate cancer. J. Mol. Biol. 430 (16), 2422–2438.

Lin, X., Kulkarni, P., Bocci, F., Schafer, N.P., Roy, S., Tsai, M.Y., et al., 2019. Structural and dynamical order of a disordered protein: molecular insights into conformational switching of PAGE4 at the systems level. Biomolecules 9 (2), pii: E77.

Lin-Tsai, O., Clark, P.E., Miller, N.L., Fowke, J.H., Hameed, O., Hayward, S.W., et al., 2014. Surgical intervention for symptomatic benign prostatic hyperplasia is correlated with expression of the AP-1 transcription factor network. Prostate. 74 (6), 669–679.

López-Novoa, J.M., Nieto, M.A., 2009. Inflammation and EMT: an alliance towards organ fibrosis and cancer progression. EMBO Mol. Med. 1 (6-7), 303–314.

Lu, M., Jolly, M.K., Levine, H., Onuchic, J.N., Ben-Jacob, E., 2013. MicroRNA-based regulation of epithelial-hybrid-mesenchymal fate determination. Proc. Natl. Acad. Sci. U.S.A. 110 (45), 18144–18149.

Luo, J.H., Ding, Y., Chen, R., Michalopoulos, G., Nelson, J., Tseng, G., et al., 2013. Genome-wide methylation analysis of prostate tissues reveals global methylation patterns of prostate cancer. Am. J. Pathol. 182, 2028–2036.

Mahmoudabadi, G., Rajagopalan, K., Getzenberg, R.H., Hannenhalli, S., Rangarajan, G., Kulkarni, P., 2013. Intrinsically disordered proteins and conformational noise: implications in cancer. Cell Cycle 12 (1), 26–31.

Mairinger, T., Mikuz, G., Gschwendtner, A., 1999. Nuclear chromatin texture analysis of nonmalignant tissue can detect adjacent prostatic adenocarcinoma. Prostate 41, 12–19.

McNeal, J.E., Redwine, E.A., Freiha, F.S., Stamey, T.A., 1988. Zonal distribution of prostatic adenocarcinoma. Correlation with histologic pattern and direction of spread. Am. J. Surg. Pathol. 12, 897–906.

Meyers, F.J., Chi, S.G., Fishman, J.R., deVere White, R.W., Gumerlock, P.H., 1993. p53 mutations in benign prostatic hyperplasia. J. Natl. Cancer Inst. 85 (22), 1856–1858.

Mohamed, H.M., Aly, M.S., Hussein, T.D., 2017. Genetic alterations in benign prostatic hyperplasia patients. Ger. Med. Sci. 15, Doc16.
Mooney, S.M., Qiu, R., Kim, J.J., Sacho, E.J., Rajagopalan, K., Johng, D., et al., 2014. Cancer/testis antigen PAGE4, a regulator of c-Jun transactivation, is phosphorylated by homeodomain-interacting protein kinase 1, a component of the stress-response pathway. Biochemistry. 53 (10), 1670–1679.
Mooney, S.M., Jolly, M.K., Levine, H., Kulkarni, P., 2016. Phenotypic plasticity in prostate cancer: role of intrinsically disordered proteins. Asian J. Androl. 18 (5), 704–710.
Munsky, B., Neuert, G., van Oudenaarden, A., 2012. Using gene expression noise to understand gene regulation. Science. 336 (6078), 183–187.
Nelson, W.G., De Marzo, A.M., Isaacs, W.B., 2003. Prostate cancer. N. Engl. J. Med. 349 (4), 366–381.
Nonn, L., Ananthanarayanan, V., Gann, P.H., 2009. Evidence for field cancerization of the prostate. Prostate 69, 1470–1479.
Oesterling, J.E., 1996. Benign prostatic hyperplasia: a review of its histogenesis and natural history. Prostate 6, 67–73.
Ørsted, D.D., Bojesen, S.E., 2013. The link between benign prostatic hyperplasia and prostate cancer. Nat. Rev. Urol. 10, 49–54.
Ørsted, D.D., Bojesen, S.E., Nielsen, S.F., Nordestgaard, B.G., 2011. Association of clinical benign prostate hyperplasia with prostate cancer incidence and mortality revisited: a nationwide cohort study of 3,009,258 men. Eur. Urol. 60, 691–698.
Ouyang, X., Jessen, W.J., Al-Ahmadie, H., Serio, A.M., Lin, Y., Shih, W.J., et al., 2008. Activator protein-1 transcription factors are associated with progression and recurrence of prostate cancer. Cancer Res. 68 (7), 2132–2144.
Prakash, K., Pirozzi, G., Elashoff, M., Munger, W., Waga, I., Dhir, R., et al., 2002. Symptomatic and asymptomatic benign prostatic hyperplasia: molecular differentiation by using microarrays. Proc. Natl. Acad. Sci. U.S.A. 99, 7598–7603.
Rajagopalan, K., Mooney, S.M., Parekh, N., Getzenberg, R.H., Kulkarni, P., 2011. A majority of the cancer/testis antigens are intrinsically disordered proteins. J. Cell Biochem. 112 (11), 3256–3267.
Rajagopalan, K., Qiu, R., Mooney, S.M., Rao, S., Shiraishi, T., Sacho, E., et al., 2014. The Stress-response protein prostate-associated gene 4, interacts with c-Jun and potentiates its transactivation. Biochim. Biophys. Acta 1842 (2), 154–163.
Roehrborn, C.G., Andriole, G.L., Wilson, T.H., Castro, R., Rittmaster, R.S., 2011. Effect of dutasteride on prostate biopsy rates and the diagnosis of prostate cancer in men with lower urinary tract symptoms and enlarged prostates in the Combination of Avodart and Tamsulosin trial. Eur. Urol. 59, 244–249.
Sampson, N., Ruiz, C., Zenzmaier, C., Bubendorf, L., Berger, P., 2012. PAGE4 positivity is associated with attenuated AR signaling and predicts patient survival in hormone-naive prostate cancer. Am. J. Pathol. 181 (4), 1443–1454.
Sanda, M.G., Beaty, T.H., Stutzman, R.E., Childs, B., Walsh, P.C., 1994. Genetic susceptibility of benign prostatic hyperplasia. J. Urol. 152 (1), 115–119.
Schenk, J.M., Kristal, A.R., Arnold, K.B., Tangen, C.M., Neuhouser, M.L., Lin, D. W., et al., 2011. Association of symptomatic benign prostatic hyperplasia and prostate cancer: results from the prostate cancer prevention trial. Am. J. Epidemiol. 173, 1419–1428.
Schlechte, H., Lenk, S.V., Löning, T., Schnorr, D., Rudolph, B.D., Ditscherlein, G., et al., 1998. p53 tumour suppressor gene mutations in benign prostatic hyperplasia and prostate cancer. Eur. Urol. 34 (5), 433–440.

Schmalhofer, O., Brabletz, S., Brabletz, T., 2009. E-cadherin, beta-catenin, and ZEB1 in malignant progression of cancer. Cancer Metastasis Rev. 28 (1-2), 151–166.

Sfanos, K., De Marzo, A., 2012. Prostate cancer and inflammation. Histopathology 60, 199–215.

Sfanos, K.S., Yegnasubramaniam, S., Nelson, W.G., De Marzo, A.M., 2018. The inflammatory microenvironment and microbiome in prostate cancer development. Nat. Rev. Urol. 15, 11–24.

Shachaf, C.M., Kopelman, A.M., Arvanitis, C., Karlsson, A., Beer, S., Mandl, S., et al., 2004. MYC inactivation uncovers pluripotent differentiation and tumour dormancy in hepatocellular cancer. Nature 431 (7012), 1112–1117.

Shao, R., Shi, J., Liu, H., Shi, X., Du, X., Klocker, H., et al., 2014. Epithelial-to-mesenchymal transition and estrogen receptor α mediated epithelial dedifferentiation mark the development of benign prostatic hyperplasia. Prostate. 74 (9), 970–982.

Shaulian, E., Karin, M., 2002. AP-1 as a regulator of cell life and death. Nat. Cell Biol. 4 (5), E131–E136.

Shi, X., Peng, Y., Du, X., Liu, H., Klocker, H., Lin, Q., et al., 2017. Estradiol promotes epithelial-to-mesenchymal transition in human benign prostatic epithelial cells. Prostate. 77 (14), 1424–1437.

Sinnott, J.A., Rider, J.R., Carlsson, J., Gerke, T., Tyekucheva, S., Penney, K.L., et al., 2015. Molecular differences in transition zone and peripheral zone prostate tumors. Carcinogenesis. 36 (6), 632–638.

Sommers, S.C., 1957. Endocrine changes with prostatic carcinoma. Cancer. 10, 345–358.

Taniguchi, Y., Choi, P.J., Li, G.W., Chen, H., Babu, M., Hearn, J., et al., 2010. Quantifying E. coli proteome and transcriptome with single-molecule sensitivity in single cells. Science. 329 (5991), 533–538.

Tu, L., Huda, N., Grimes, B.R., Slee, R.B., Bates, A.M., Cheng, L., et al., 2016. Widespread telomere instability in prostatic lesions. Mol. Carcinog. 55, 842–852.

Turoverov, K.K., Kuznetsova, I.M., Uversky, V.N., 2010. The protein kingdom extended: ordered and intrinsically disordered proteins, their folding, supramolecular complex formation, and aggregation. Prog. Biophys. Mol. Biol. 102 (2-3), 73–84.

Urban, E.A., Johnston Jr., R.J., 2018. Buffering and amplifying transcriptional noise during cell fate specification. Front. Genet. 9, 591–604.

Uversky, V.N., 2014. Wrecked regulation of intrinsically disordered proteins in diseases: pathogenicity of deregulated regulators. Front. Mol. Biosci. 1, 6.

Uversky, V.N., Dunker, A.K., 2010. Understanding protein non-folding. Biochim. Biophys. Acta 1804 (6), 1231–1264.

Veltri, R.W., Khan, M.A., Miller, M.C., Epstein, J.I., Mangold, L.A., Walsh, P.C., et al., 2004. Ability to predict metastasis based on pathology findings and alterations in nuclear structure of normal-appearing and cancer peripheral zone epithelium in the prostate. Clin. Cancer Res. 10, 3465–3473.

Wilt, T.J., Macdonald, R., Hagerty, K., Schellhammer, P., Tacklind, J., Somerfield, M.R., et al., 2010. 5-a-Reductase inhibitors for prostate cancer chemoprevention: an updated Cochrane systematic review. BJU Int. 106, 1444–1451.

Wolin, K.Y., Grubb, R.L., Pakpahan, R., Ragard, L., Mabie, J., Andriole, G.L., et al., 2015. Physical activity and benign prostatic hyperplasia related outcomes and nocturia. Med. Sci. Sports Exerc. 47, 581–592.

Wright, P.E., Dyson, H.J., 2015. Intrinsically disordered proteins in cellular signalling and regulation. Nat. Rev. Mol. Cell Biol. 16 (1), 18–29.
Wrighton, K.H., 2015. Protein folding. Phosphorylation regulates IDP folding. Nat. Rev. Mol. Cell Biol. 16 (2), 66.
Xue, B., Oldfield, C.J., Van, Y.Y., Dunker, A.K., Uversky, V.N., 2012. Protein intrinsic disorder and induced pluripotent stem cells. Mol. Biosyst. 8 (1), 134–150.
Yang, P., Huo, Z., Liao, H., Zhou, Q., 2015. Cancer/testis antigens trigger epithelial-mesenchymal transition and genesis of cancer stem-like cells. Curr. Pharm. Des. 21 (10), 1292–1300.
Ye, N., Ding, Y., Wild, C., Shen, Q., Zhou, J., 2014. Small molecule inhibitors targeting activator protein 1 (AP-1). J. Med. Chem. 57 (16), 6930–6948.
Yizhak, K., Aguet, F., Kim, J., Hess, J.M., Kübler, K., Grimsby, J., et al., 2019. RNA sequence analysis reveals macroscopic somatic clonal expansion across normal tissues. Science. 364 (6444), pii: eaaw0726.
Zeng, Y., He, Y., Yang, F., Mooney, S.M., Getzenberg, R.H., Orban, J., et al., 2011. The cancer/testis antigen prostate-associated gene 4 (PAGE4) is a highly intrinsically disordered protein. J. Biol. Chem. 286 (16), 13985–13994.
Zeng, Y., Gao, D., Kim, J.J., Shiraishi, T., Terada, N., Kakehi, Y., et al., 2013. Prostate-associated gene 4 (PAGE4) protects cells against stress by elevating p21 and suppressing reactive oxygen species production. Am. J. Clin. Exp. Urol. 1 (1), 39–52.

23

Phenotypic plasticity and lineage switching in prostate cancer

Jason A. Somarelli[1,2], Andrew J. Armstrong[1,2], Maya U. Sheth[1,2], Kathryn E. Ware[1,2] and Mohit Kumar Jolly[3]

[1]Department of Medicine, Duke University Medical Center, Durham, NC, United States [2]Duke Cancer Institute, Durham, NC, United States [3]Centre for BioSystems Science and Engineering, Indian Institute of Science, Bangalore, India

Prostate development and the progression to prostate cancer

Prostate cancer is responsible for more than 70 deaths each day in the United States (Siegel et al., 2019). The cancer derives typically from the peripheral zone of the prostate gland, a male reproductive gland that resides below the bladder and produces fluids found in semen. Such seminal fluids and prostatic secretions have a normal function in promoting sperm motility, energy stores, and survival during transit through the female cervix and uterus to facilitate fertilization. The prostate forms embryologically in utero in a "castrate" host, the mother, but differentiates under the control of fetal testicular androgens from an endoderm lineage into the urogenital sinus (Wong et al., 2003; Toivanen and Shen, 2017). The urogenital sinus forms epithelial buds, which proliferate via androgen signaling into the surrounding urogenital mesenchyme (Toivanen and Shen, 2017; Kellokumpu-Lehtinen et al., 1980). These epithelial buds form epithelial cords that then undergo canalization and differentiate into basal and luminal cells, giving rise to the glandular structures of the prostate (Aaron et al., 2016). During this process of epithelial expansion and differentiation, urogenital mesenchyme also differentiates into fibroblasts and smooth muscle mesenchyme (Hayward et al., 1996). These epithelial and mesenchymal cells work coordinately during development of the prostate: the mesenchyme promotes differentiation of the epithelial cells into prostate epithelium while the epithelial cells induce differentiation of the

Phenotypic Switching. DOI: https://doi.org/10.1016/B978-0-12-817996-3.00021-9

prostate mesenchyme [reviewed in (Wilson, 2011)]. Proper development of the prostate gland requires a tightly-regulated cross talk between multiple developmental factors, including expression of signaling factors, such as fibroblast growth factors (FGF)7 and FGF10 (Donjacour et al., 2003), Wnt5a (Allgeier et al., 2008), β-catenin (Mehta et al., 2013), and the TGF-β repressor, Dickkopf 3 (Romero et al., 2013) as well as developmental transcription factors Nkx3.1 (Bhatia-Gaur et al., 1999) and SOX9 (Thomsen et al., 2008) and androgen receptor (AR) cofactors FOXA1 (Gao et al., 2005) and HOXB13 (Economides and Capecchi, 2003). A comprehensive review of prostate embryogenesis and development is available from Toivanen and Shen (2017).

The mature prostate gland is comprised of basal epithelial cells on the exterior of the gland, luminal epithelial secretory cells lining the interior of the gland, and neuroendocrine cells scattered throughout the gland. Mesenchymal stromal cells, including fibroblasts, smooth muscle cells, endothelia, and leukocytes, surround the glandular structures (Henry et al., 2018). In addition to basal and luminal epithelial cells and mesenchyme, rare neuroendocrine cells also contribute to epithelial cell differentiation and identity (Szczyrba et al., 2017). The origin of these neuroendocrine cells is unclear; the cells are thought to derive from either neural crest or epithelial cells (Toivanen and Shen, 2017).

Targeting the androgen receptor signaling axis in prostate adenocarcinoma

As a male reproductive organ, the development of the normal prostate epithelium is driven by the AR (Wilson, 2011). The AR is a nuclear hormone receptor and transcription factor that binds testosterone, localizes to the nucleus, and activates gene expression programs important for cellular differentiation and survival (Davey and Grossmann, 2016). The AR regulates these programs as both a transcriptional activator and repressor by interacting with a variety of cofactors. There is also some evidence that the AR may function independently of its role in DNA binding as an activator of secondary signaling cascades, including the extracellular signal-regulated kinases (ERK), Ak strain transforming (AKT), and mitogen-activated protein kinase (MAPK) pathways (Davey and Grossmann, 2016), though the significance of these functions remains controversial. AR signaling is crucial for growth and survival of normal prostate cells. In most cases, when normal prostate cells transform to prostate cancer they remain dependent on AR signaling. Indeed, with the exception of several rare prostate cancer types, including pure small cell carcinoma of the prostate (Kumar et al., 2018), carcinosarcoma of the prostate (Somarelli et al., 2015), and prostate sarcoma (Tward et al., 2018), the vast majority of prostate cancer patients present with prostate adenocarcinoma, which is an epithelial cancer derived from the glandular epithelial cells of the prostate. A small fraction (<5%) of prostate cancers, both intraductal and ductal adenocarcinoma variants, likely arise from prostatic ducts, while the majority of tumors exhibit glandular/luminal cell differentiation and AR positivity. This is in contrast to breast cancer, where the majority of breast adenocarcinomas are ductal and a minority subset is lobular in origin. Exactly which type of glandular epithelial cell is the cell-of-origin for prostate adenocarcinoma remains controversial: Studies have found that both basal and luminal cells can act as the cell of origin (Zhang et al., 2017a; Liu et al., 2016; Park et al., 2016), though possibly to different extents (Wang and Shen, 2015). However, despite the remaining uncertainty around the cell of origin, it is clear that the vast majority of prostate adenocarcinomas are composed of an AR-positive epithelial cell type.

While the majority of men with localized prostate cancer can be cured with surgery or radiation, or safely observed for years during active surveillance, approximately one of seven men with prostate cancer will die of their disease due to relapse and metastatic spread (Miller et al., 2019). Some men present clinically with metastatic disease, with no opportunity for local approaches, and the prevalence of metastatic prostate cancer varies widely across the globe, ranging from 5%–10% in the United States to 30%–50% in regions of the world where screening is not performed (Rawla, 2019). For men with metastatic or relapsed disease, hormonal therapy has been the mainstay of therapy since the Nobel Prize winning discoveries of the hormonal dependence of prostate cancer in 1941 by Huggins and Hodges (Huggins and Hodges, 1972). As demonstrated by Huggins and Hodges, oncogenic addiction to AR signaling creates a therapeutic vulnerability for prostate adenocarcinoma. Based on this oncogenic addiction to AR signaling, standard-of-care treatment for recurrent and metastatic prostate adenocarcinoma includes androgen deprivation therapy (Crawford et al., 2017). Yet, although androgen deprivation therapy is often an effective treatment, most men develop progressive disease over time despite castrate levels of serum testosterone. This progression is determined by screening for rises in prostate-specific antigen (PSA), a well-defined target of AR signaling (Crawford et al., 2017).

The majority of men with metastatic prostate cancer, and a significant fraction of men with biochemically relapsed prostate cancer, defined by rising PSA levels after local therapy, will go on to develop metastatic, castration-resistant prostate cancer (mCRPC), in which AR signaling is either restored or bypassed (Dai et al., 2017). The early use of androgen deprivation therapy with docetaxel or more potent AR inhibition in men with metastatic hormone sensitive prostate cancer improves survival and delays the onset of mCRPC, but progression after 3–5 years is typically observed (Fizazi et al., 2017; James et al., 2017; Davis et al., 2019; Chi et al., 2019; Armstrong et al., 2019b). For patients treated initially with androgen deprivation therapy alone, next-generation hormonal therapies are generally used upon progression (Ryan et al., 2013; Beer et al., 2014; de Bono et al., 2011; Scher et al., 2012). These therapies include abiraterone acetate and enzalutamide, both of which potently inhibit AR signaling (Crawford et al., 2017). A major concern in the field presently is that the early use of these potent AR signaling inhibitors, while improving the survival and quality of life of patients, has created a bottleneck of resistance in which new forms of prostate cancer and new resistance mechanisms are emerging. These novel molecular subtypes and resistance mechanisms have few treatments presently available other than taxane chemotherapy (Tannock et al., 2004), sipuleucel-T immunotherapy (Kantoff et al., 2010), and radium-223 radiopharmaceutical therapy (Parker et al., 2013).

Treatment-induced phenotypic plasticity overcomes androgen receptor blockade

At autopsy, the majority of prostate cancers in the pre-AR inhibition era were typical adenocarcinomas that demonstrate AR and PSA positivity, while a small subset (<5%) demonstrated small cell or neuroendocrine histology (Shah et al., 2004). However, recent autopsy series in the abiraterone/enzalutamide treatment era demonstrate an increase in neuroendocrine prostate cancer (NEPC) variants as well as aggressive anaplastic histologist variants that lack AR and NEPC biomarkers, the so-called double negative tumors (Bluemn et al., 2017) (Fig. 23.1). Similarly, while most prostate cancers are initially AR-positive and epithelial-like, approximately 1% of prostate cancers are small cell *de novo*, while approximately 20% of prostate cancers exhibit NEPC biomarker expression after exposure to potent AR inhibitors (Aggarwal et al., 2018). This lineage plasticity or phenotypic plasticity is most-often observed during the emergence of treatment resistance (Aggarwal et al., 2019) and metastatic progression (Jadaan et al., 2015; Kadakia et al., 2015) and can include phenotypic switches to become more dedifferentiated or mesenchymal-like [reviewed in

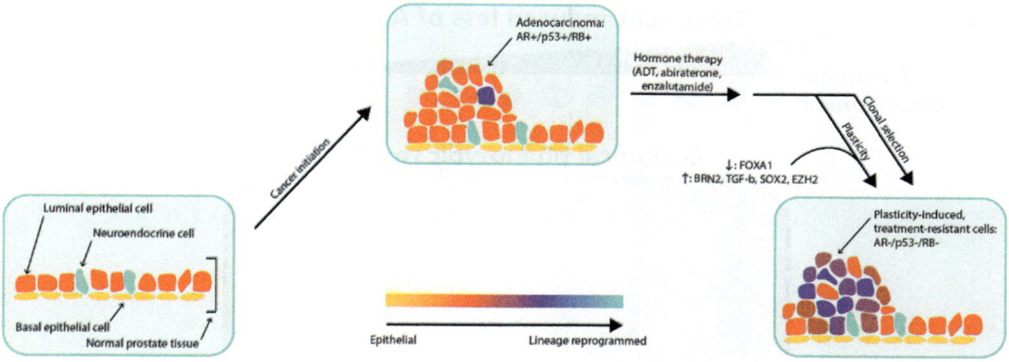

Figure 23.1 Phenotypic plasticity in prostate cancer progression. Prostate cancer most often arises from an epithelial-like cell type with oncogenic dependence on AR signaling. Hormone therapy promotes clonal selection and/or phenotypic plasticity, resulting in neuroendocrine and mesenchymal-like features. These phenotypes are driven by genomic losses in RB, p53, and FOXA1, and upregulation of lineage plasticity factors BRN2, TGF-β, SOX2, EZH2, and others.

(Bitting et al., 2014), squamous (Stoyanova et al., 2013), bone-like (Armstrong et al., 2019a; Gonzalez et al., 2017), or neuroendocrine-like (Mu et al., 2017; Ku et al., 2017)]. Aside from plasticity, such histologic transformations of typical adenocarcinoma to new histologic variants may also be due to clonal selection and divergent evolution related to the acquisition of new genetic and epigenetic alterations. These phenotypic transitions impart upon prostate cancer cells several key features for survival, including the abilities to (1) overcome the resource depletion encountered during treatment by altering gene expression pathways and drug targets, (2) disseminate and colonize new sites throughout the body, and (3) evade the immune system (Fig. 23.2). Here we will discuss emerging data around these three concepts connecting lineage plasticity with survival, dissemination, and immune evasion.

Overcoming resource depletion

The AR normally activates a cellular program of differentiation important to normal glandular formation rather than proliferation and invasion/metastasis. However, during prostate cancer progression, the AR is reprogrammed to occupy novel sites on the DNA through pioneering factors, such as HOXB13 (Nerlakanti et al., 2018; Chen et al., 2018; Pomerantz et al., 2015), FOXA1 (Pomerantz et al., 2015), CHD1 (Augello et al., 2019), and other cofactors that rewire AR biology to facilitate tumor progression. Androgen deprivation therapy inhibits AR signaling, which

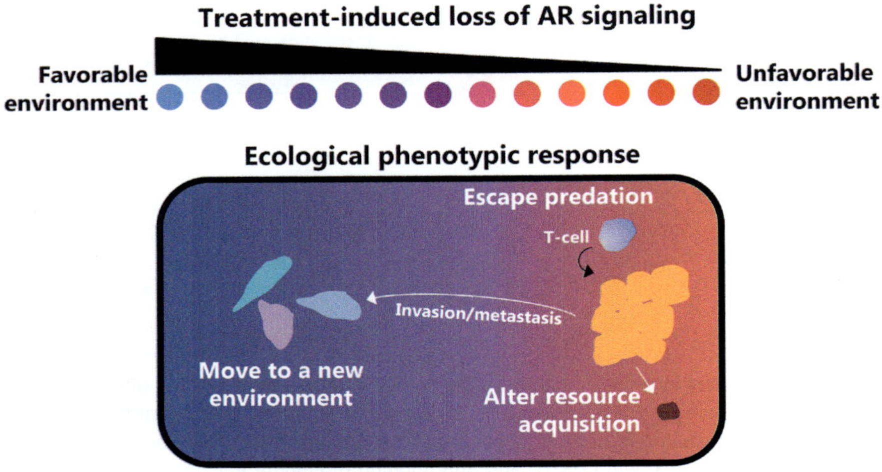

Figure 23.2 Hormone therapy induces convergent phenotypic responses. When faced with therapy-induced resource depletion, prostate cancer cells undergo phenotypic responses necessary to survive, including: (1) overcoming resource depletion through metabolic reprogramming or dormancy, (2) migrating to a new environment, and (3) evading predation by the immune system.

is the most critical resource of the prostate cancer cell. Cancer cells respond to this loss of resources by adapting and rewiring their gene expression programs through changes in cofactor levels, genetic selection, epigenetics [reviewed in (Ruggero et al., 2018)], microRNA expression [reviewed in (Kojima et al., 2017)], transcriptional reprogramming (Robinson et al., 2015a), alternative splicing and mRNA processing (Sowalsky et al., 2015), and translation (Jaiswal et al., 2018; D'Abronzo et al., 2017). This rewiring in response to androgen deprivation can induce a spectrum of phenotypic transformations toward a more neuroendocrine or mesenchymal-like phenotype [reviewed in (Bitting et al., 2014) and (Lo et al., 2017)] (Fig. 23.1), which resembles the epithelial-mesenchymal transition observed during normal embryologic development [reviewed in (Nieto et al., 2016)]. Interestingly, epithelial-mesenchymal-like transitions (Alonso-Magdalena et al., 2009; Lu et al., 2012; Shao et al., 2014) are observed even during progression from normal prostate epithelium to benign prostatic hyperplasia, suggesting that prostate cells are capable of undergoing these transitions even in the absence of androgen deprivation therapy. In addition, squamous differentiation is a well-appreciated phenotype of prostates treated with androgen deprivation therapy, indicating that morphology can be dynamic over time in the same patient and tumor depending on the cellular context and stressors applied. It is also worth noting that both androgens and antiandrogens can induce expression of factors important for promoting cellular plasticity depending on the model and cellular context (Colditz et al., 2016).

The increased expression of mesenchymal genes during androgen deprivation therapy imbues the cell with a multitude of aggressive features (Bitting et al., 2014) (Table 23.1). For one, induction of plasticity regulators induces changes in cellular signaling that enable cells to bypass AR signaling, including activation of ERK (Tang et al., 2018; Wu et al., 2016), MAPK (Tang et al., 2018; Randle et al., 2013), and phosphoinositide 3-kinase/AKT/mammalian target of rapamycin signaling (Seol et al., 2019; Shao et al., 2018; Wang et al., 2017b; Shi et al., 2018). In addition, epithelial plasticity can also enhance production of AR splice variants that lack the ligand binding domain for androgens (Cottard et al., 2013; Kong et al., 2015). These AR splice variants are well-established markers of resistance to abiraterone and enzalutamide (Antonarakis et al., 2014; Armstrong et al., 2019c). Moreover, the AR splice variants may reciprocally induce factors downstream that promote further epithelial plasticity, including N-cadherin (Cottard et al., 2017) and TGF-β (Sun et al., 2014). TGF-β, in particular, is a powerful and well-characterized plasticity-inducing ligand (Wendt et al., 2009). TGF-β activates a suite of master regulators of epithelial plasticity transcription factors, including members of the Snail/Zeb family (Smith et al., 2009).

Several of these plasticity-related transcription factors are implicated in hormone therapy resistance. One of the known drivers of both epithelial plasticity and therapy resistance is Snail. Snail is a master regulator of mesenchymal cell fate that has been associated with aggressive features across multiple cancers (Wang et al., 2013). Snail promotes enzalutamide resistance, in part through interactions with AR. Snail activity increases AR signaling and promotes nuclear localization, although the exact details of this relationship remain unclear (Ware et al., 2016; Miao et al., 2017). Another plasticity master regulator, Twist, is activated during androgen deprivation and enzalutamide resistance, and reversing Twist through small molecule inhibition of PKC induces sensitivity to enzalutamide (Shiota et al., 2014).

Another key regulator of plasticity-mediated treatment resistance is the Wingless and Int1 (WNT)/β-catenin pathway (Wu et al., 2018; Li et al., 2018; Sha et al., 2018; Liu et al., 2015). WNT also cross talks with the TGF-β pathway through expression of Frizzled 8 ligand, which can bind and activate both WNT and TGF-β complexes (Murillo-Garzon et al., 2018). Soluble WNT family members can be secreted from adjacent tumor associated stroma

Table 23.1 Molecular drivers of plasticity in prostate cancer.

Gene/ Protein	Mechanism	Link to EMT?	Link to NEPC?	References
AR	Reprogrammed to occupy novel sites on DNA through factors including HOXB13, FOXA1, CHD1 through changes in gene expression program	Yes	Yes	Nerlakanti et al., 2018; Chen et al., 2018; Pomerantz et al., 2015; Augello et al., 2019
AR	Bypassing of AR signaling by activation of other pathways and/or production of AR splice variants lacking the ligand binding domain	Yes	Yes	Tang et al., 2018; Wu et al., 2016; Randle et al., 2013; Seol et al., 2019; Shao et al., 2018; Wang et al., 2017a,b; Shi et al., 2018; Cottard et al., 2013; Kong et al., 2015
Snail	Induces gene expression program that increase migration and invasion; promotes enzalutamide resistance through interaction with AR	Yes		Ware et al., 2016; Miao et al., 2017
Twist	Activated during androgen deprivation and enzalutamide resistance	Yes		Shiota et al., 2014
WNT16B	Induces epithelial plasticity and treatment resistance through WNT/beta-catenin pathway and cross talk with TGF-B complexes	Yes		Sun et al., 2012
WNT5A	Production by tumor cells drives motility and AR therapy resistance	Yes		Sandsmark et al., 2017
WNT5A	Acts as immunosuppressive factor to create immunoevasive microenvironment	Yes		Holtzhausen et al., 2015; Zhao et al., 2018
TGF-beta	Regulates EMT, immunosuppressive ligand	Yes	Yes	Chen et al., 2015
SPDEF	Loss of SPDEF induces TGF-B and CCL2 expression, which in turn induce epithelial plasticity and cellular migration	Yes		Chen et al., 2017; Tsai et al., 2018; Cioni et al., 2018; Izumi et al., 2013
MYCN	Genomic gain contributes to lineage reprogramming		Yes	Beltran et al., 2016
AURKA	Genomic gain contributes to lineage reprogramming		Yes	Beltran et al., 2016
FOXA1	Genomic loss lessens repression of TGF-B and IL-8	EMT inhibitor	NEPC inhibitor	Song et al., 2019; Jin et al., 2013; Kim et al., 2017
PEG10	Upregulation in NEPC activates Snail through TGF-B	Yes	Yes	Akamatsu et al., 2015

(*Continued*)

Table 23.1 (Continued)

Gene/ Protein	Mechanism	Link to EMT?	Link to NEPC?	References
SOX2	Induces lineage plasticity with RB1/TP53 loss	Yes	Yes	Mu et al., 2017; He et al., 2017
EZH2	Upregulation induces epigenetic reprogramming through loss of DAB2IP, methylation of STAT3, and silencing of miRNA-338-5p/-421, which represses SPINK1	Yes	Yes	Zhang et al., 2018a,b; Luo et al., 2019; Bhatia et al., 2019; Min et al., 2010
RB1/ TP53	Enriched in NEPC tumors over mCRPC adenocarcinomas; may drive NEPC- and EMT-like reprogramming	Yes	Yes	Mu et al., 2017; Ku et al., 2017; Puca et al., 2019
BRN2	Repression by AR drives neuroendocrine transcription program with SOX2	Yes	Yes	Gritsina et al., 2019
STAT3	Activates EMT	Yes		Cho et al., 2014; Cho et al., 2013

during chemotherapy or radiation, such as WNT16B, inducing epithelial plasticity in tumor cells and treatment resistance (Sun et al., 2012). In addition, tumor cells can produce or utilize canonical and noncanonical WNT molecules, such as WNT5a, to drive motility and AR therapy resistance. New therapies that target both of these pathways have emerged with the potential to reprogram prostate cancer cells back to an epithelial-like phenotype and reestablish sensitivity to hormone therapy. The TGF-β inhibitor, galunisertib, has been shown to reverse epithelial plasticity and sensitize prostate cancer cells to enzalutamide (Paller et al., 2019; Song et al., 2019). Similarly, the drug, metformin, can reverse enzalutamide resistance by inhibiting TGF-β/STAT3-mediated epithelial-mesenchymal transition (Liu et al., 2017). In addition to TGF-β, the β-catenin inhibitor, ICG001, acted synergistically with enzalutamide to inhibit proliferation and tumor growth (Zhang et al., 2018b). In addition to activation of plasticity regulators through TGF-β and other mechanisms, loss of epithelial lineage transcription factors, such as FOXA1, activates epithelial plasticity programs through increased TGF-β (Song et al., 2019). It is also worth noting that activation of many of these lineage reprogramming pathways is initiated and sustained by the underlying genetics of the cell. For example, prostate cancer cells with TMPRRS2:ERG fusions induce a TGF-β-mediated plasticity gene expression program (Ratz et al., 2017), which may indicate that the TMPRRS2:ERG fusion background creates a permissive cell state that promotes epithelial plasticity.

Plasticity-driven metastasis

In the context of prostate cancer, hormone therapy induces phenotypic plasticity, which creates a cellular phenotype permissive for metastasis. While the extensive literature supporting a role for epithelial plasticity in prostate cancer metastasis has been reviewed elsewhere (Bitting et al., 2014; Lo et al., 2017; Culig, 2019; Soundararajan et al., 2018), studies have continued to unravel the molecular drivers of plasticity-mediated metastasis in prostate cancer. One key determinant of plasticity that has gained increased attention is the SAM Pointed Domain Containing ETS Transcription Factor (SPDEF). SPDEF represses transcription of TGF-β, and loss of SPDEF leads to increased metastasis through TGFB1 expression in xenograft models (Chen et al., 2017). Similarly, SPDEF downregulation leads to induction of epithelial plasticity through production of the soluble factor, CCL2 (Tsai et al., 2018), an inducer of prostate cancer cellular migration (Cioni et al., 2018) and metastasis (Izumi et al., 2013). Another important factor in promoting metastatic behavior in prostate cancer is Snail. Snail, in addition to its role in enzalutamide resistance, induces a gene expression program that increases migration and invasion (Ware et al., 2016). Consistent with this change in invasion, Snail is also strongly expressed in prostate cancer metastases as compared to localized disease (Ware et al., 2016).

Epithelial plasticity in immune evasion

Concomitant with the creation of a treatment-resistant, prosurvival, and metastatic phenotype, epithelial plasticity also mediates immune evasion (Chae et al., 2018; Liang et al., 2018; Tripathi et al., 2016). One of the mechanisms for this immune-evasive phenotype is by upregulating immune checkpoint molecules. This upregulation is observed across cancers and for multiple immune checkpoint proteins (Mak et al., 2016). Consistent with this study, in melanoma, loss of the epithelial specific splicing factor, ESRP1, correlates with an epithelial-mesenchymal-like signature and increased immune checkpoint activation (Yao et al., 2016). Although the mechanisms by which immune checkpoints are upregulated during epithelial-mesenchymal transition remain poorly understood, studies on the immune checkpoint molecule, PD-L1, have shown that PD-L1 is upregulated during epithelial-mesenchymal plasticity through activation of Zeb1, and siRNA-mediated knockdown of

Zeb1 represses PD-L1 expression (Noman et al., 2017), suggesting epithelial plasticity factors transcriptionally regulate immune checkpoint molecules. Notably, PD-L1 is also upregulated in enzalutamide-resistant cell lines and patients (Bishop et al., 2015), and its induction can be observed with multiple cellular stressors that result in interferon release (Mimura et al., 2018; Garcia-Diaz et al., 2017). This upregulation has therapeutic relevance, as treatment of enzalutamide-resistant patients with a combination of an anti-PD-1 agent, pembrolizumab, and enzalutamide has shown dramatic responses in about 20% of patients in early phase clinical trials (Graff et al., 2016), while anti-PD-1 therapy alone has limited activity in unselected men with mCRPC (Bono et al., 2018). While most prostate cancers are nonimmunogenic, certain subsets (5%–10%), such as those characterized by CDK12 biallelic loss and those with microsatellite instability, have a higher response rate to PD-1-directed therapies. Thus, understanding the plasticity of immune checkpoint biomarker expression during lineage plasticity, metastasis, and treatment resistance is critical to developing novel immunotherapy approaches for men with mCRPC.

In addition to immune checkpoint expression, epithelial plasticity also stimulates expression of secreted immunosuppressive factors, such as soluble WNTs and TGF-β. Indeed, nearly 20% of men with metastatic castration resistant prostate cancer have somatic alterations in key WNT pathway regulators (Robinson et al., 2015b). Among these WNT pathway regulators the noncanonical WNT ligand, WNT5A, correlates with epithelial-mesenchymal transition signatures and prostate cancer aggressiveness (Sandsmark et al., 2017). However, while it remains to be determined if WNT5A is inducing immune tolerance in prostate cancer, evidence from other cancers suggests that WNT5A creates a protumorigenic, immunoevasive microenvironment. For instance, in a model of nonsmall cell lung cancer, WNT5A knockdown inhibits tumor growth and metastasis (Wang et al., 2017a). Likewise, in melanoma, WNT5A stimulates production of the enzyme indoleamine 2,3-dioxygenase-1, which increases differentiation of T regulatory cells (Holtzhausen et al., 2015) and fosters immune tolerance. Importantly, the addition of a WNT pathway inhibitor increased the efficacy of immune checkpoint therapies (Holtzhausen et al., 2015; Zhao et al., 2018). Similarly, TGF-β is a potent immunosuppressive ligand that has been studied extensively (Yoshimura and Muto, 2011; Yang et al., 2010). In prostate cancer, Snail-mediated production of TGF-β induces epigenetic downregulation of

HLA-I, the loss of which promotes immune escape (Chen et al., 2015).

EMT and NEPC—common molecular drivers with unique phenotypic outputs?

The clinical implementation of sustained AR signaling blockade by abiraterone acetate and enzalutamide has significantly prolonged the lives of men with both non-metastatic (Hussain et al., 2018) and metastatic prostate cancer (Ryan et al., 2013; Beer et al., 2014; Scher et al., 2012). However, despite the substantial benefit to patients, the eventual development of resistance to these agents has been nearly universal. The mechanisms of resistance to these agents can broadly be categorized into two classes: AR signaling dependent and AR signaling independent (Crona and Whang, 2017). Among the AR signaling-independent mechanisms, the transformed small cell or NEPC subtype has emerged as a key phenotype of hormone therapy resistance (Boudadi and Antonarakis, 2016; Akamatsu et al., 2018).

Despite the importance of the neuroendocrine-like plasticity in resistance, defining NEPC has been challenging outside of the standard morphologic criteria for small cell histology on biopsy. For example, many patients with adenocarcinoma can have high serum markers of chromogranin A, a neuroendocrine biomarker, but can have highly variable histologic expression of chromogranin A, synaptophysin, neuron-specific enolase, and CD56. In addition, some tumors express both AR and NEPC biomarkers (Aggarwal et al., 2018), suggesting that many tumors exist on a continuum of differentiation. While AR and PSA can be expressed in NEPC, the canonical AR transcriptome is typically reduced or altered significantly in these tumors, suggesting a broad rewiring of transcription away from traditional AR signaling and dependence. The classical clinical phenotype associated with NEPC or small cell tumors, also termed aggressive-variant prostate cancer, is that of high disease burden and visceral metastases, but without significant PSA elevations. However, we now know that NEPC-like features can be present on biopsy of men with mCRPC even with high PSA levels, in lymph nodes or in bone metastases, and cannot always be predicted by the clinical phenotype, which limits our ability to recognize these variants in the clinic.

NEPC displays a unique gene expression signature (Beltran et al., 2016) that is driven by distinct underlying genomic features, including genomic gains in MYCN and AURKA (Beltran et al., 2011), and genomic loss of FOXA1 (Song et al., 2019;

Adams et al., 2019; Parolia et al., 2019) and RB1/TP53 (Mu et al., 2017; Ku et al., 2017). In the context of these genetic alterations, lineage reprogramming to neuroendocrine-like prostate cancer is mediated by loss of AR signaling and increases in the transcription factors FOXA1 (Song et al., 2019), FOXA2 (Park et al., 2017), PEG10 (Akamatsu et al., 2015), FOXC2 (Paranjape et al., 2016), and SOX2 (Mu et al., 2017), and the epigenetic modifier, EZH2 (Zhang et al., 2018a; Luo et al., 2019). Loss of RB1 may underlie much of these changes, given that TP53/RB1 loss is clearly enriched in NEPC tumors over prostate adenocarcinomas. RB1 loss may lead to upregulation of DLL3, a critical NEPC signaling molecule in the NOTCH family, and currently a target for therapeutic development (Puca et al., 2019).

Loss of these critical tumor suppressors, such as TP53/RB1 in NEPC can lead to phenotypic convergence with gene expression pathways common to epithelial-mesenchymal plasticity. FOXA1 is an epithelial transcription factor that inhibits both epithelial-mesenchymal transition (Jin et al., 2013) and neuroendocrine-like differentiation (Kim et al., 2017). Lineage regulation by FOXA1 is mediated by repressing secretion of TGF-β and IL-8 (Song et al., 2019; Kim et al., 2017). Another lineage reprogramming factor, PEG10, which is normally responsible for control of placental development (Ono et al., 2006), becomes upregulated in neuroendocrine-like prostate cancer and activates Snail through activation of TGF-β (Akamatsu et al., 2015). Expression is also sufficient to induce epithelial-mesenchymal plasticity in hepatocellular carcinoma by upregulating TGF-β (Zhang et al., 2017b). BRN2 is a neural transcription factor repressed by AR and can drive a neuroendocrine transcriptional program in cooperation with SOX2 (Gritsina et al., 2019). SOX2, another driver of epithelial-mesenchymal plasticity (Gao et al., 2015; Liu et al., 2018; Wang et al., 2018; Li et al., 2013), promotes enzalutamide resistance by inducing lineage plasticity toward a neuroendocrine-like phenotype in the genetic background of Rb1 and p53 loss (Mu et al., 2017). In addition to transcriptional control, epigenetic reprogramming through EZH2 activation converges on both epithelial-mesenchymal and neuroendocrine-like plasticity. In the context of epithelial-mesenchymal plasticity, EZH2 upregulation induces loss of DAB2IP, a repressor of epithelial-mesenchymal transition (Min et al., 2010), and EZH2 mRNA and protein expression has been associated with a subgroup of patients with more aggressive disease (Labbe et al., 2017). EZH2 also methylates STAT3 through a non-histone methyltransferase activity to drive neuroendocrine-like differentiation (Luo et al., 2019). STAT3 is a known activator of epithelial-mesenchymal transition in prostate cancer (Cho et al., 2014; Cho et al., 2013).

EZH2 also epigenetically silences miRNA-338-5p/-421, which represses the metastasis-promoting gene, SPINK1 (Bhatia et al., 2019).

Despite the extensive cross talk between gene regulatory networks in epithelial-mesenchymal and neuroendocrine-like lineage reprogramming, the exact interaction between these types of plasticity remains unclear; however, the role of SOX2 in neuronal differentiation may help explain the relationship between epithelial-mesenchymal plasticity and neuroendocrine-like signatures in prostate cancer. Elegant studies in mouse embryonic fibroblasts showed that early epithelial-mesenchymal transition was required for expression of SOX2 (He et al., 2017). SOX2 expression subsequently induced mesenchymal-epithelial transition en route to neuronal differentiation (He et al., 2017). Based on this study, it is plausible that dysfunctional prostate cancer cells existing in a quasi-epithelial-mesenchymal or epigenetically poised state (Chaffer et al., 2013) are pushed toward a neuronal-like lineage by loss of Rb1 and further induction of SOX2 expression and selected upon during hormonal therapies. In this scenario, an epithelial-mesenchymal plasticity signature would be enriched in Rb1 + /p53 + cells while Rb1-/p53- cells exhibiting upregulation of SOX2 may pseudo-differentiate toward a more neuronal-like lineage. The consistent loss of Rb1 and p53 throughout neuroendocrine-like cancers supports this idea (Akeno et al., 2017; Konukiewitz et al., 2017; Meder et al., 2016). In addition, the top gene sets associated with NEPC histology and subtype in patients are related to EMT (Beltran et al., 2016), and many EMT-related factors are also coexpressed in NEPC tumors, such as N-cadherin. Furthermore, the role of Rb1 as a driver of neuroendocrine-like cancers is likely dependent on p53 (Sage, 2012) and/or other mutations, since Rb1 loss promotes apoptosis of p53 wild-type neural progenitor cells (Matsui et al., 2017). Contrary to this, however, both Rb1 and p53 loss are associated with epithelial-mesenchymal plasticity features in breast cancer (Jiang et al., 2011), which supports a more consistent genetic background between epithelial-mesenchymal and neuroendocrine-like reprogramming.

Future integrated computational-experimental efforts can be valuable in decoding the underlying networks linking these two axes of phenotypic plasticity: (1) epithelial-mesenchymal plasticity, and (2) neuroendocrine-like differentiation. Computational models have been valuable in identifying the association of epithelial-mesenchymal plasticity with another axis of plasticity: tumor-initiation capacity or "stem-like" properties (Jolly et al., 2015; Jolly et al., 2014), an association that has been since validated extensively in vitro and in vivo (Kroger et al., 2019; Bierie et al., 2017;

Grosse-Wilde et al., 2015). Such coupled cell-fate decisions can arise due to interconnected regulatory networks that are often "multistable" (Jia et al., 2017). Multistability refers to the feature of a system in which more than one phenotype can coexist within a population, even for isogenic/clonal populations. Thus, given enough "biological noise," cells may switch from one phenotype to another. This "noise" can arise due to various factors—cell-to-cell variation in kinetic rates of biochemical reactions and in concentrations of various biomolecules, varying transcription and translation rates, and even promiscuous interactions among intrinsically disordered proteins involved in cellular signaling (Kulkarni et al., 2017; Balazsi et al., 2011). Such variability may facilitate phenotypic heterogeneity and consequent higher adaptability to a cellular population through enabling a diverse set of responses; such "bet-hedging" may even contribute to the emergence of drug resistance (Inde and Dixon, 2018; Jolly et al., 2018). Thus, to elucidate the observed associations between therapy resistance, epithelial-mesenchymal plasticity, and neuroendocrine-like differentiation, investigating the bidirectional connections among these networks may help us to arrive at revised therapeutic strategies to curb disease aggressiveness.

Phenotypic plasticity is a dynamic driver of prostate cancer aggression

Improvements in AR signaling blockade have had substantial benefit for prostate cancer patients, but these new therapies have also led to the emergence of resistant disease with aggressive features. Potent resource depletion in the background of a dysfunctional cancer genome has rewired gene expression and epigenetic programs to select for and/or induce a population of cancer cells with a dynamic, shifting cellular fate. These cellular fates are not binary, but rather exist as a spectrum of phenotypes, often coexisting within the same tumor. Indeed, elements of small cell or neuroendocrine-like cancer, some of which have sustained AR expression, are often observed admixed with adenocarcinomatous features (Tsai et al., 2017; Wang and Epstein, 2008).

There are many unanswered questions in the field of NEPC biology and lineage plasticity. Some of these questions include: (1) How are neural cells and neuronal signaling in the prostate and those that originate from the central nervous system involved in this lineage plasticity and drug resistance? (2) Which transcription factors are critical to promoting this lineage plasticity in patients? (3) Do master regulators of EMT cooperate with neural transcription factors or promote neuroendocrine-like gene expression programs? (4) Given the connections between fibroblast growth factor receptor (FGFR) signaling and EMT (Acevedo et al., 2007; Corn et al., 2013), what is the relationship of EMT to subtypes of prostate cancer that lack both AR signaling and a neuroendocrine-like phenotype? (5) How redundant are neural transcription factors, such as BRN2, and what are the critical steps to reprogramming? (6) Why do NEPCs generally lack DNA homologous repair defects (Aggarwal et al., 2018) despite being platinum sensitive? Continued innovation in single-cell sequencing and genomics technologies are sure to help further unravel the complex genetic and gene regulatory diversity within and between prostate tumors. In addition to these many questions, one of the most critical barriers is the identification of the most promising targets for therapy in these highly adaptable, plastic tumors. While the standard of care for NEPC remains platinum-based chemotherapy, these responses are short lived (Aparicio et al., 2013), and outcomes remain poor. Continued efforts to understand this complex biology will hopefully result in new targeted therapies directed at these plasticity phenotypes to prolong the lives of men with metastatic prostate cancer.

References

Aaron, L., Franco, O.E., Hayward, S.W., 2016. Review of prostate anatomy and embryology and the etiology of benign prostatic hyperplasia. Urol. Clin. North. Am. 43 (3), 279–288.

Acevedo, V.D., et al., 2007. Inducible FGFR-1 activation leads to irreversible prostate adenocarcinoma and an epithelial-to-mesenchymal transition. Cancer Cell 12 (6), 559–571.

Adams, E.J., et al., 2019. FOXA1 mutations alter pioneering activity, differentiation and prostate cancer phenotypes. Nature 571 (7765), 408–412.

Aggarwal, R., et al., 2018. Clinical and genomic characterization of treatment-emergent small-cell neuroendocrine prostate cancer: a multi-institutional prospective study. J. Clin. Oncol. 36 (24), 2492–2503.

Aggarwal, R.R., et al., 2019. Whole-genome and transcriptional analysis of treatment-emergent small-cell neuroendocrine prostate cancer demonstrates intraclass heterogeneity. Mol. Cancer Res. 17 (6), 1235–1240.

Akamatsu, S., et al., 2015. The placental gene PEG10 promotes progression of neuroendocrine prostate cancer. Cell Rep. 12 (6), 922–936.

Akamatsu, S., et al., 2018. Clinical and molecular features of treatment-related neuroendocrine prostate cancer. Int. J. Urol. 25 (4), 345–351.

Akeno, N., et al., 2017. TRP53 mutants drive neuroendocrine lung cancer through loss-of-function mechanisms with gain-of-function effects on chemotherapy response. Mol. Cancer Ther. 16 (12), 2913–2926.

Allgeier, S.H., et al., 2008. WNT5A selectively inhibits mouse ventral prostate development. Dev. Biol. 324 (1), 10–17.
Alonso-Magdalena, P., et al., 2009. A role for epithelial-mesenchymal transition in the etiology of benign prostatic hyperplasia. Proc. Natl. Acad. Sci. U.S.A. 106 (8), 2859–2863.
Antonarakis, E.S., et al., 2014. AR-V7 and resistance to enzalutamide and abiraterone in prostate cancer. N. Engl. J. Med. 371 (11), 1028–1038.
Aparicio, A.M., et al., 2013. Platinum-based chemotherapy for variant castrate-resistant prostate cancer. Clin. Cancer Res. 19 (13), 3621–3630.
Armstrong, A.J., et al., 2019a. Pharmacodynamic study of radium-223 in men with bone metastatic castration resistant prostate cancer. PLoS One 14 (5), e0216934.
Armstrong, A.J., et al., 2019b. Phase III study of androgen deprivation therapy (ADT) with enzalutamide (ENZA) or placebo (PBO) in metastatic hormone-sensitive prostate cancer (mHSPC): The ARCHES trial. J. Clin. Oncol. 37 (7_Suppl), 687–687.
Armstrong, A.J., et al., 2019c. Prospective multicenter validation of androgen receptor splice variant 7 and hormone therapy resistance in high-risk castration-resistant prostate cancer: the PROPHECY study. J. Clin. Oncol. 37 (13), 1120–1129.
Augello, M.A., et al., 2019. CHD1 loss alters AR binding at lineage-specific enhancers and modulates distinct transcriptional programs to drive prostate tumorigenesis. Cancer Cell 35 (4), 603–617. e8.
Balazsi, G., van Oudenaarden, A., Collins, J.J., 2011. Cellular decision making and biological noise: from microbes to mammals. Cell 144 (6), 910–925.
Beer, T.M., et al., 2014. Enzalutamide in metastatic prostate cancer before chemotherapy. N. Engl. J. Med. 371 (5), 424–433.
Beltran, H., et al., 2011. Molecular characterization of neuroendocrine prostate cancer and identification of new drug targets. Cancer Discov. 1 (6), 487–495.
Beltran, H., et al., 2016. Divergent clonal evolution of castration-resistant neuroendocrine prostate cancer. Nat. Med. 22 (3), 298–305.
Bhatia, V., et al., 2019. Epigenetic silencing of miRNA-338-5p and miRNA-421 drives SPINK1-positive prostate cancer. Clin. Cancer Res. 25 (9), 2755–2768.
Bhatia-Gaur, R., et al., 1999. *Roles for Nkx3.1 in prostate development and cancer.* Genes. Dev. 13 (8), 966–977.
Bierie, B., et al., 2017. Integrin-beta4 identifies cancer stem cell-enriched populations of partially mesenchymal carcinoma cells. Proc. Natl. Acad. Sci. U.S.A. 114 (12), E2337–E2346.
Bishop, J.L., et al., 2015. PD-L1 is highly expressed in Enzalutamide resistant prostate cancer. Oncotarget 6 (1), 234–242.
Bitting, R.L., et al., 2014. The role of epithelial plasticity in prostate cancer dissemination and treatment resistance. Cancer Metastasis Rev. 33 (2–3), 441–468.
Bluemn, E.G., et al., 2017. Androgen receptor pathway-independent prostate cancer is sustained through FGF signaling. Cancer Cell 32 (4), 474–489. e6.
Bono, J.S.D., et al., *KEYNOTE-199: Pembrolizumab (pembro) for docetaxel-refractory metastatic castration-resistant prostate cancer (mCRPC).* 2018. **36** (15_suppl): pp. 5007–5007.
Boudadi, K., Antonarakis, E.S., 2016. Resistance to novel antiandrogen therapies in metastatic castration-resistant prostate cancer. Clin. Med. Insights Oncol. 10 (Suppl. 1), 1–9.

Chae, Y.K., et al., 2018. Epithelial-mesenchymal transition (EMT) signature is inversely associated with T-cell infiltration in non-small cell lung cancer (NSCLC). Sci. Rep. 8 (1), 2918.

Chaffer, C.L., et al., 2013. Poised chromatin at the ZEB1 promoter enables breast cancer cell plasticity and enhances tumorigenicity. Cell 154 (1), 61–74.

Chen, X.H., et al., 2015. TGF-beta and EGF induced HLA-I downregulation is associated with epithelial-mesenchymal transition (EMT) through upregulation of snail in prostate cancer cells. Mol. Immunol. 65 (1), 34–42.

Chen, W.Y., et al., 2017. Loss of SPDEF and gain of TGFBI activity after androgen deprivation therapy promote EMT and bone metastasis of prostate cancer. Sci. Signal. 10 (492), pii: eaam6826.

Chen, Z., et al., 2018. Diverse AR-V7 cistromes in castration-resistant prostate cancer are governed by HoxB13. Proc. Natl. Acad. Sci. U.S.A. 115 (26), 6810–6815.

Chi, K.N., et al., 2019. Apalutamide for metastatic, castration-sensitive prostate cancer. N. Engl. J. Med. 381 (1), 13–24.

Cho, K.H., et al., 2013. STAT3 mediates TGF-beta1-induced TWIST1 expression and prostate cancer invasion. Cancer Lett. 336 (1), 167–173.

Cho, K.H., et al., 2014. A ROS/STAT3/HIF-1alpha signaling cascade mediates EGF-induced TWIST1 expression and prostate cancer cell invasion. Prostate 74 (5), 528–536.

Cioni, B., et al., 2018. Loss of androgen receptor signaling in prostate cancer-associated fibroblasts (CAFs) promotes CCL2- and CXCL8-mediated cancer cell migration. Mol. Oncol. 12 (8), 1308–1323.

Colditz, J., et al., 2016. Androgens induce a distinct response of epithelial-mesenchymal transition factors in human prostate cancer cells. Mol. Cell Biochem. 421 (1–2), 139–147.

Corn, P.G., et al., 2013. Targeting fibroblast growth factor pathways in prostate cancer. Clin. Cancer Res. 19 (21), 5856–5866.

Cottard, F., et al., 2013. Constitutively active androgen receptor variants upregulate expression of mesenchymal markers in prostate cancer cells. PLoS One 8 (5), e63466.

Cottard, F., et al., 2017. Dual effects of constitutively active androgen receptor and full-length androgen receptor for N-cadherin regulation in prostate cancer. Oncotarget 8 (42), 72008–72020.

Crawford, E.D., Petrylak, D., Sartor, O., 2017. Navigating the evolving therapeutic landscape in advanced prostate cancer. Urol. Oncol. 35S, S1–S13.

Crona, D.J., Whang, Y.E., 2017. Androgen receptor-dependent and -independent mechanisms involved in prostate cancer therapy resistance. Cancers (Basel) 9 (6), 209–216.

Culig, Z., 2019. Epithelial mesenchymal transition and resistance in endocrine-related cancers. Biochim. Biophys. Acta Mol. Cell Res. 1866 (9), 1368–1375.

D'Abronzo, L.S., et al., 2017. The androgen receptor is a negative regulator of eIF4E phosphorylation at S209: implications for the use of mTOR inhibitors in advanced prostate cancer. Oncogene 36 (46), 6359–6373.

Dai, C., Heemers, H., Sharifi, N., 2017. Androgen signaling in prostate cancer. Cold Spring Harb. Perspect. Med. 7 (9), pii: a030452.

Davey, R.A., Grossmann, M., 2016. Androgen receptor structure, function and biology: from bench to bedside. Clin. Biochem. Rev. 37 (1), 3–15.

Davis, I.D., et al., 2019. Enzalutamide with standard first-line therapy in metastatic prostate cancer. N. Engl. J. Med. 381 (2), 121–131.

de Bono, J.S., et al., 2011. Abiraterone and increased survival in metastatic prostate cancer. N. Engl. J. Med. 364 (21), 1995–2005.

Donjacour, A.A., Thomson, A.A., Cunha, G.R., 2003. FGF-10 plays an essential role in the growth of the fetal prostate. Dev. Biol. 261 (1), 39–54.
Economides, K.D., Capecchi, M.R., 2003. Hoxb13 is required for normal differentiation and secretory function of the ventral prostate. Development 130 (10), 2061–2069.
Fizazi, K., et al., 2017. Abiraterone plus prednisone in metastatic, castration-sensitive prostate cancer. N. Engl. J. Med. 377 (4), 352–360.
Gao, N., et al., 2005. Forkhead box A1 regulates prostate ductal morphogenesis and promotes epithelial cell maturation. Development 132 (15), 3431–3443.
Gao, H., et al., 2015. SOX2 promotes the epithelial to mesenchymal transition of esophageal squamous cells by modulating slug expression through the activation of STAT3/HIF-alpha signaling. Int. J. Mol. Sci. 16 (9), 21643–21657.
Garcia-Diaz, A., et al., 2017. Interferon receptor signaling pathways regulating PD-L1 and PD-L2 expression. Cell Rep. 19 (6), 1189–1201.
Gonzalez, A., et al., 2017. Distinct osteomimetic response of androgen-dependent and independent human prostate cancer cells to mechanical action of fluid flow: prometastatic implications. Prostate 77 (3), 321–333.
Graff, J.N., et al., 2016. Early evidence of anti-PD-1 activity in enzalutamide-resistant prostate cancer. Oncotarget 7 (33), 52810–52817.
Gritsina, G., Gao, W.Q., Yu, J., 2019. Transcriptional repression by androgen receptor: roles in castration-resistant prostate cancer. Asian J. Androl. 21 (3), 215–223.
Grosse-Wilde, A., et al., 2015. Stemness of the hybrid epithelial/mesenchymal state in breast cancer and its association with poor survival. PLoS One 10 (5), e0126522.
Hayward, S.W., et al., 1996. Stromal development in the ventral prostate, anterior prostate and seminal vesicle of the rat. Acta Anat. (Basel) 155 (2), 94–103.
He, S., et al., 2017. Sequential EMT-MET induces neuronal conversion through Sox2. Cell Discov. 3, 17017.
Henry, G.H., et al., 2018. A cellular anatomy of the normal adult human prostate and prostatic urethra. Cell Rep. 25 (12), 3530–3542. e5.
Holtzhausen, A., et al., 2015. *Melanoma-derived Wnt5a promotes local dendritic-cell expression of IDO and immunotolerance: opportunities for pharmacologic enhancement of immunotherapy.* Cancer Immunol. Res. 3 (9), 1082–1095.
Huggins, C., Hodges, C.V., 1972. *Studies on prostatic cancer. I. The effect of castration, of estrogen and androgen injection on serum phosphatases in metastatic carcinoma of the prostate.* CA Cancer J. Clin. 22 (4), 232–240.
Hussain, M., et al., 2018. Enzalutamide in men with nonmetastatic, castration-resistant prostate cancer. N. Engl. J. Med. 378 (26), 2465–2474.
Inde, Z., Dixon, S.J., 2018. The impact of non-genetic heterogeneity on cancer cell death. Crit. Rev. Biochem. Mol. Biol. 53 (1), 99–114.
Izumi, K., et al., 2013. Targeting the androgen receptor with siRNA promotes prostate cancer metastasis through enhanced macrophage recruitment via CCL2/CCR2-induced STAT3 activation. EMBO Mol. Med. 5 (9), 1383–1401.
Jadaan, D.Y., Jadaan, M.M., McCabe, J.P., 2015. Cellular plasticity in prostate cancer bone metastasis. Prostate Cancer 2015, 651580.
Jaiswal, P.K., et al., 2018. Eukaryotic translation initiation factor 4 gamma 1 (eIF4G1) is upregulated during prostate cancer progression and modulates cell growth and metastasis. Sci. Rep. 8 (1), 7459.
James, N.D., et al., 2017. Abiraterone for prostate cancer not previously treated with hormone therapy. N. Engl. J. Med. 377 (4), 338–351.

Jia, D., et al., 2017. Phenotypic plasticity and cell fate decisions in cancer: insights from dynamical systems theory. Cancers (Basel) 9 (7), pii: E70.

Jiang, Z., et al., 2011. RB1 and p53 at the crossroad of EMT and triple-negative breast cancer. Cell Cycle 10 (10), 1563–1570.

Jin, H.J., et al., 2013. Androgen receptor-independent function of FoxA1 in prostate cancer metastasis. Cancer Res. 73 (12), 3725–3736.

Jolly, M.K., et al., 2014. Towards elucidating the connection between epithelial-mesenchymal transitions and stemness. J. R. Soc. Interface 11 (101), 20140962.

Jolly, M.K., et al., 2015. Coupling the modules of EMT and stemness: a tunable 'stemness window' model. Oncotarget 6 (28), 25161–25174.

Jolly, M.K., et al., 2018. Phenotypic plasticity, bet-hedging, and androgen independence in prostate cancer: role of non-genetic heterogeneity. Front. Oncol. 8, 50.

Kadakia, K.C., et al., 2015. Comprehensive serial molecular profiling of an "N of 1" exceptional non-responder with metastatic prostate cancer progressing to small cell carcinoma on treatment. J. Hematol. Oncol. 8, 109.

Kantoff, P.W., et al., 2010. Sipuleucel-T immunotherapy for castration-resistant prostate cancer. N. Engl. J. Med. 363 (5), 411–422.

Kellokumpu-Lehtinen, P., Santti, R., Pelliniemi, L.J., 1980. Correlation of early cytodifferentiation of the human fetal prostate and Leydig cells. Anat. Rec. 196 (3), 263–273.

Kim, J., et al., 2017. FOXA1 inhibits prostate cancer neuroendocrine differentiation. Oncogene 36 (28), 4072–4080.

Kojima, S., Goto, Y., Naya, Y., 2017. The roles of microRNAs in the progression of castration-resistant prostate cancer. J. Hum. Genet. 62 (1), 25–31.

Kong, D., et al., 2015. Androgen receptor splice variants contribute to prostate cancer aggressiveness through induction of EMT and expression of stem cell marker genes. Prostate 75 (2), 161–174.

Konukiewitz, B., et al., 2017. Somatostatin receptor expression related to TP53 and RB1 alterations in pancreatic and extrapancreatic neuroendocrine neoplasms with a Ki67-index above 20. Mod. Pathol. 30 (4), 587–598.

Kroger, C., et al., 2019. Acquisition of a hybrid E/M state is essential for tumorigenicity of basal breast cancer cells. Proc. Natl Acad. Sci. U.S.A. 116 (15), 7353–7362.

Ku, S.Y., et al., 2017. Rb1 and Trp53 cooperate to suppress prostate cancer lineage plasticity, metastasis, and antiandrogen resistance. Science 355 (6320), 78–83.

Kulkarni, P., et al., 2017. Phosphorylation-induced conformational dynamics in an intrinsically disordered protein and potential role in phenotypic heterogeneity. Proc. Natl Acad. Sci. U.S.A. 114 (13), E2644–E2653.

Kumar, K., et al., 2018. Poorly differentiated small-cell-type neuroendocrine carcinoma of the prostate: a case report and literature review. Case Rep. Oncol. 11 (3), 676–681.

Labbe, D.P., et al., 2017. TOP2A and EZH2 provide early detection of an aggressive prostate cancer subgroup. Clin. Cancer Res. 23 (22), 7072–7083.

Li, X., et al., 2013. SOX2 promotes tumor metastasis by stimulating epithelial-to-mesenchymal transition via regulation of WNT/beta-catenin signal network. Cancer Lett. 336 (2), 379–389.

Li, K., et al., 2018. Oldhamianoside II inhibits prostate cancer progression via regulation of EMT and the Wnt/beta-catenin signaling pathway. Oncol. Lett. 15 (6), 9457–9463.

Liang, J., et al., 2018. The correlation between the immune and epithelial-mesenchymal transition signatures suggests potential therapeutic targets and prognosis prediction approaches in kidney cancer. Sci. Rep. 8 (1), 6570.
Liu, H., et al., 2015. FOXO3a modulates WNT/beta-catenin signaling and suppresses epithelial-to-mesenchymal transition in prostate cancer cells. Cell Signal. 27 (3), 510–518.
Liu, X., et al., 2016. Low CD38 identifies progenitor-like inflammation-associated luminal cells that can initiate human prostate cancer and predict poor outcome. Cell Rep. 17 (10), 2596–2606.
Liu, Q., et al., 2017. Metformin reverses prostate cancer resistance to enzalutamide by targeting TGF-beta1/STAT3 axis-regulated EMT. Cell Death Dis. 8 (8), e3007.
Liu, X., et al., 2018. Sox2 promotes tumor aggressiveness and epithelialmesenchymal transition in tongue squamous cell carcinoma. Int. J. Mol. Med. 42 (3), 1418–1426.
Lo, U.G., et al., 2017. The role and mechanism of epithelial-to-mesenchymal transition in prostate cancer progression. Int. J. Mol. Sci. 18 (10), pii: E2079.
Lu, T., et al., 2012. Targeting androgen receptor to suppress macrophage-induced EMT and benign prostatic hyperplasia (BPH) development. Mol. Endocrinol. 26 (10), 1707–1715.
Luo, J., et al., 2019. LncRNA-p21 alters the antiandrogen enzalutamide-induced prostate cancer neuroendocrine differentiation via modulating the EZH2/STAT3 signaling. Nat. Commun. 10 (1), 2571.
Mak, M.P., et al., 2016. A patient-derived, pan-cancer EMT signature identifies global molecular alterations and immune target enrichment following epithelial-to-mesenchymal transition. Clin. Cancer Res. 22 (3), 609–620.
Matsui, T., et al., 2017. Retinoblastoma protein controls growth, survival and neuronal migration in human cerebral organoids. Development 144 (6), 1025–1034.
Meder, L., et al., 2016. NOTCH, ASCL1, p53 and RB alterations define an alternative pathway driving neuroendocrine and small cell lung carcinomas. Int. J. Cancer 138 (4), 927–938.
Mehta, V., et al., 2013. Beta-catenin (CTNNB1) induces Bmp expression in urogenital sinus epithelium and participates in prostatic bud initiation and patterning. Dev. Biol. 376 (2), 125–135.
Miao, L., et al., 2017. Disrupting androgen receptor signaling induces Snail-mediated epithelial-mesenchymal plasticity in prostate cancer. Cancer Res. 77 (11), 3101–3112.
Miller, K.D., et al., 2019. Cancer treatment and survivorship statistics, 2019. CA Cancer J. Clin. 69 (5), 363–385.
Mimura, K., et al., 2018. PD-L1 expression is mainly regulated by interferon gamma associated with JAK-STAT pathway in gastric cancer. Cancer Sci. 109 (1), 43–53.
Min, J., et al., 2010. An oncogene-tumor suppressor cascade drives metastatic prostate cancer by coordinately activating Ras and nuclear factor-kappaB. Nat. Med. 16 (3), 286–294.
Mu, P., et al., 2017. SOX2 promotes lineage plasticity and antiandrogen resistance in TP53- and RB1-deficient prostate cancer. Science 355 (6320), 84–88.
Murillo-Garzon, V., et al., 2018. Frizzled-8 integrates Wnt-11 and transforming growth factor-beta signaling in prostate cancer. Nat. Commun. 9 (1), 1747.
Nerlakanti, N., et al., 2018. Targeting the BRD4-HOXB13 coregulated transcriptional networks with bromodomain-kinase inhibitors to suppress

metastatic castration-resistant prostate cancer. Mol. Cancer Ther. 17 (12), 2796–2810.
Nieto, M.A., et al., 2016. Emt: 2016. Cell 166 (1), 21–45.
Noman, M.Z., et al., 2017. The immune checkpoint ligand PD-L1 is upregulated in EMT-activated human breast cancer cells by a mechanism involving ZEB-1 and miR-200. Oncoimmunology 6 (1), e1263412.
Ono, R., et al., 2006. Deletion of Peg10, an imprinted gene acquired from a retrotransposon, causes early embryonic lethality. Nat. Genet. 38 (1), 101–106.
Paller, C., et al., 2019. TGF-beta receptor I inhibitor enhances response to enzalutamide in a pre-clinical model of advanced prostate cancer. Prostate 79 (1), 31–43.
Paranjape, A.N., et al., 2016. Inhibition of FOXC2 restores epithelial phenotype and drug sensitivity in prostate cancer cells with stem-cell properties. Oncogene 35 (46), 5963–5976.
Park, J.W., et al., 2016. Prostate epithelial cell of origin determines cancer differentiation state in an organoid transformation assay. Proc. Natl Acad. Sci. U.S.A. 113 (16), 4482–4487.
Park, J.W., et al., 2017. FOXA2 is a sensitive and specific marker for small cell neuroendocrine carcinoma of the prostate. Mod. Pathol. 30 (9), 1262–1272.
Parker, C., et al., 2013. Alpha emitter radium-223 and survival in metastatic prostate cancer. N. Engl. J. Med. 369 (3), 213–223.
Parolia, A., et al., 2019. Distinct structural classes of activating FOXA1 alterations in advanced prostate cancer. Nature 571 (7765), 413–418.
Pomerantz, M.M., et al., 2015. The androgen receptor cistrome is extensively reprogrammed in human prostate tumorigenesis. Nat. Genet. 47 (11), 1346–1351.
Puca, L., et al., 2019. Delta-like protein 3 expression and therapeutic targeting in neuroendocrine prostate cancer. Sci. Transl. Med. 11 (484), pii: eaav0891.
Randle, D.D., et al., 2013. Snail mediates invasion through uPA/uPAR and the MAPK signaling pathway in prostate cancer cells. Oncol. Lett. 6 (6), 1767–1773.
Ratz, L., et al., 2017. TMPRSS2:ERG gene fusion variants induce TGF-beta signaling and epithelial to mesenchymal transition in human prostate cancer cells. Oncotarget 8 (15), 25115–25130.
Rawla, P., 2019. Epidemiology of prostate cancer. World J. Oncol. 10 (2), 63–89.
Robinson, D., et al., 2015a. Integrative clinical genomics of advanced prostate cancer. Cell 161 (5), 1215–1228.
Robinson, D., et al., 2015b. Integrative clinical genomics of advanced prostate cancer. Cell 162 (2), 454.
Romero, D., et al., 2013. Downregulation of Dickkopf-3 disrupts prostate acinar morphogenesis through TGF-beta/Smad signalling. J. Cell Sci. 126 (Pt 8), 1858–1867.
Ruggero, K., et al., 2018. Epigenetic regulation in prostate cancer progression. Curr. Mol. Biol. Rep. 4 (2), 101–115.
Ryan, C.J., et al., 2013. Abiraterone in metastatic prostate cancer without previous chemotherapy. N. Engl. J. Med. 368 (2), 138–148.
Sage, J., 2012. The retinoblastoma tumor suppressor and stem cell biology. Genes. Dev. 26 (13), 1409–1420.
Sandsmark, E., et al., 2017. A novel non-canonical Wnt signature for prostate cancer aggressiveness. Oncotarget 8 (6), 9572–9586.
Scher, H.I., et al., 2012. Increased survival with enzalutamide in prostate cancer after chemotherapy. N. Engl. J. Med. 367 (13), 1187–1197.

Seol, M.A., et al., 2019. Interleukin-7 contributes to the invasiveness of prostate cancer cells by promoting epithelial-mesenchymal transition. Sci. Rep. 9 (1), 6917.

Sha, J., et al., 2018. PRKAR2B promotes prostate cancer metastasis by activating Wnt/beta-catenin and inducing epithelial-mesenchymal transition. J. Cell Biochem. 119 (9), 7319–7327.

Shah, R.B., et al., 2004. Androgen-independent prostate cancer is a heterogeneous group of diseases: lessons from a rapid autopsy program. Cancer Res. 64 (24), 9209–9216.

Shao, R., et al., 2014. Epithelial-to-mesenchymal transition and estrogen receptor alpha mediated epithelial dedifferentiation mark the development of benign prostatic hyperplasia. Prostate 74 (9), 970–982.

Shao, G., et al., 2018. GCN5 inhibition prevents IL-6-induced prostate cancer metastases through PI3K/PTEN/Akt signaling by inactivating Egr-1. Biosci. Rep. 38 (6), pii: BSR20180816.

Shi, J., et al., 2018. Tumor microenvironment promotes prostate cancer cell dissemination via the Akt/mTOR pathway. Oncotarget 9 (10), 9206–9218.

Shiota, M., et al., 2014. Inhibition of protein kinase C/Twist1 signaling augments anticancer effects of androgen deprivation and enzalutamide in prostate cancer. Clin. Cancer Res. 20 (4), 951–961.

Siegel, R.L., Miller, K.D., Jemal, A., 2019. Cancer statistics, 2019. CA Cancer J. Clin. 69 (1), 7–34.

Smith, A.P., et al., 2009. A positive role for Myc in TGFbeta-induced Snail transcription and epithelial-to-mesenchymal transition. Oncogene 28 (3), 422–430.

Somarelli, J.A., et al., 2015. Carcinosarcomas: tumors in transition? Histol. Histopathol. 30 (6), 673–687.

Song, B., et al., 2019. Targeting FOXA1-mediated repression of TGF-beta signaling suppresses castration-resistant prostate cancer progression. J. Clin. Invest. 129 (2), 569–582.

Soundararajan, R., et al., 2018. EMT, stemness and tumor plasticity in aggressive variant neuroendocrine prostate cancers. Biochim. Biophys. Acta Rev. Cancer 1870 (2), 229–238.

Sowalsky, A.G., et al., 2015. Whole transcriptome sequencing reveals extensive unspliced mRNA in metastatic castration-resistant prostate cancer. Mol. Cancer Res. 13 (1), 98–106.

Stoyanova, T., et al., 2013. Prostate cancer originating in basal cells progresses to adenocarcinoma propagated by luminal-like cells. Proc. Natl Acad. Sci. U.S.A. 110 (50), 20111–20116.

Sun, Y., et al., 2012. Treatment-induced damage to the tumor microenvironment promotes prostate cancer therapy resistance through WNT16B. Nat. Med. 18 (9), 1359–1368.

Sun, F., et al., 2014. Androgen receptor splice variant AR3 promotes prostate cancer via modulating expression of autocrine/paracrine factors. J. Biol. Chem. 289 (3), 1529–1539.

Szczyrba, J., et al., 2017. Neuroendocrine cells of the prostate derive from the neural crest. J. Biol. Chem. 292 (5), 2021–2031.

Tang, G., et al., 2018. MiRNALet-7a mediates prostate cancer PC-3 cell invasion, migration by inducing epithelial-mesenchymal transition through CCR7/MAPK pathway. J. Cell Biochem. 119 (4), 3725–3731.

Tannock, I.F., et al., 2004. Docetaxel plus prednisone or mitoxantrone plus prednisone for advanced prostate cancer. N. Engl. J. Med. 351 (15), 1502–1512.

Thomsen, M.K., et al., 2008. Sox9 is required for prostate development. Dev. Biol. 316 (2), 302–311.
Toivanen, R., Shen, M.M., 2017. Prostate organogenesis: tissue induction, hormonal regulation and cell type specification. Development 144 (8), 1382–1398.
Tripathi, S.C., et al., 2016. Immunoproteasome deficiency is a feature of non-small cell lung cancer with a mesenchymal phenotype and is associated with a poor outcome. Proc. Natl Acad. Sci. U.S.A. 113 (11), E1555–E1564.
Tsai, H.K., et al., 2017. Gene expression signatures of neuroendocrine prostate cancer and primary small cell prostatic carcinoma. BMC Cancer 17 (1), 759.
Tsai, Y.C., et al., 2018. Androgen deprivation therapy-induced epithelial-mesenchymal transition of prostate cancer through downregulating SPDEF and activating CCL2. Biochim. Biophys. Acta Mol. Basis Dis. 1864 (5 Pt A), 1717–1727.
Tward, J.D., et al., 2018. Demographics, stage distribution, and relative roles of surgery and radiotherapy on survival of persons with primary prostate sarcomas. Cancer Med. 7 (12), 6030–6039.
Wang, W., Epstein, J.I., 2008. *Small cell carcinoma of the prostate. A morphologic and immunohistochemical study of 95 cases.* Am. J. Surg. Pathol. 32 (1), 65–71.
Wang, Z.A., Shen, M.M., 2015. Comparative lineage tracing reveals cellular preferences for prostate cancer initiation. Mol. Cell Oncol. 2 (3), e985548.
Wang, Y., et al., 2013. The role of Snail in EMT and tumorigenesis. Curr. Cancer Drug. Targets 13 (9), 963–972.
Wang, B., et al., 2017a. Wnt5a promotes epithelial-to-mesenchymal transition and metastasis in non-small-cell lung cancer. Biosci. Rep. 37 (6), , p. BSR20171092.
Wang, S., et al., 2017b. c-Met, CREB1 and EGFR are involved in miR-493-5p inhibition of EMT via AKT/GSK-3beta/Snail signaling in prostate cancer. Oncotarget 8 (47), 82303–82313.
Wang, K., et al., 2018. FGFR1-ERK1/2-SOX2 axis promotes cell proliferation, epithelial-mesenchymal transition, and metastasis in FGFR1-amplified lung cancer. Oncogene 37 (39), 5340–5354.
Ware, K.E., et al., 2016. Snail promotes resistance to enzalutamide through regulation of androgen receptor activity in prostate cancer. Oncotarget 7 (31), 50507–50521.
Wendt, M.K., Allington, T.M., Schiemann, W.P., 2009. Mechanisms of the epithelial-mesenchymal transition by TGF-beta. Future Oncol. 5 (8), 1145–1168.
Wilson, J.D., 2011. The critical role of androgens in prostate development. Endocrinol. Metab. Clin. North. Am. 40 (3), 577–590. ix.
Wong, Y.C., Wang, X.H., Ling, M.T., 2003. Prostate development and carcinogenesis. Int. Rev. Cytol. 227, 65–130.
Wu, J., et al., 2016. Polo-like kinase 1 induces epithelial-to-mesenchymal transition and promotes epithelial cell motility by activating CRAF/ERK signaling. Elife 5, pii: e10734.
Wu, S., et al., 2018. 2'Hydroxyflavanone inhibits epithelialmesenchymal transition, and cell migration and invasion via suppression of the Wnt/betacatenin signaling pathway in prostate cancer. Oncol. Rep. 40 (5), 2836–2843.
Yang, L., Pang, Y., Moses, H.L., 2010. TGF-beta and immune cells: an important regulatory axis in the tumor microenvironment and progression. Trends Immunol. 31 (6), 220–227.

Yao, J., et al., 2016. Altered expression and splicing of ESRP1 in malignant melanoma correlates with epithelial-mesenchymal status and tumor-associated immune cytolytic activity. Cancer Immunol. Res. 4 (6), 552–561.
Yoshimura, A., Muto, G., 2011. TGF-beta function in immune suppression. Curr. Top. Microbiol. Immunol. 350, 127–147.
Zhang, D., et al., 2017a. Developing a novel two-dimensional culture system to enrich human prostate luminal progenitors that can function as a cell of origin for prostate cancer. Stem Cell Transl. Med. 6 (3), 748–760.
Zhang, M., et al., 2017b. PEG10 is imperative for TGF-beta1-induced epithelialmesenchymal transition in hepatocellular carcinoma. Oncol. Rep. 37 (1), 510–518.
Zhang, Y., et al., 2018a. Androgen deprivation promotes neuroendocrine differentiation and angiogenesis through CREB-EZH2-TSP1 pathway in prostate cancers. Nat. Commun. 9 (1), 4080.
Zhang, Z., et al., 2018b. Inhibition of the Wnt/beta-catenin pathway overcomes resistance to enzalutamide in castration-resistant prostate cancer. Cancer Res. 78 (12), 3147–3162.
Zhao, F., et al., 2018. Paracrine Wnt5a-beta-catenin signaling triggers a metabolic program that drives dendritic cell tolerization. Immunity 48 (1), 147–160. e7.

24

Implications of non-genetic heterogeneity in cancer drug resistance and malignant progression

Kaitlyn Johnson[1], Eric Brenner[2] and Amy Brock[1,2]

[1]Department of Biomedical Engineering, The University of Texas at Austin, Austin, TX, United States [2]Institute for Cellular and Molecular Biology, The University of Texas at Austin, Austin, TX, United States

Introduction

Intratumoral heterogeneity is widely recognized as a critical factor in tumor progression, adaptation, and treatment response. Broadly speaking, intratumoral heterogeneity can be defined as the presence of distinct cellular phenotypes within a tumor cell population. The diversity of a cancer cell population can be examined at multiple spatial scales ranging from single nucleotide mutations in the genome to broad functional cellular behaviors such as growth rate or drug sensitivity. Numerous studies have demonstrated that increased heterogeneity is correlated with increased resistance to treatment and poorer patient prognosis (Fillmore and Kuperwasser, 2008b; Maley et al., 2017; Brock et al., 2009; Pisco and Huang, 2015). While the importance of understanding heterogeneity in cancer is well acknowledged by the field, the adoption of multiple different definitions of heterogeneity can cause complications, potentially preventing a common language between basic discovery and clinical measures (Maley et al., 2017). Heterogeneity may manifest as clonal, or genetic, heterogeneity, in which distinct subclones of cells harbor genetic mutations that can confer phenotypic diversity to daughter cells; or nongenetic heterogeneity, which describes multistability in gene expression dynamics whereby one genome produces multiple stable or metastable phenotypic states. Observations of individual cells able to reversibly transition into different phenotypic states

Phenotypic Switching. DOI: https://doi.org/10.1016/B978-0-12-817996-3.00022-0

either spontaneously (Thanos et al., 2018; Gupta et al., 2011) or due to an environmental stimulus (Pisco and Huang, 2015; Hardeman et al., 2017; Chen et al., 2018) imply that a cell does not need to harbor a permanent genetic mutation to exhibit multiple cellular phenotypes.

While heterogeneity in observed cellular phenotypes is likely due to a combination of genetic and nongenetic diversity, in this chapter, we focus on the implications of nongenetic heterogeneity and its role in subpopulation dynamics relevant to cancer progression and treatment response. Even at the nongenetic level, heterogeneous cell states have been historically defined in numerous ways using a variety of tools to study and classify distinct subpopulations. Recently, single-cell RNA sequencing (scRNA-seq) has enabled a tremendous increase in the throughput and molecular level detail at which single cells can be characterized (Klein et al., 2015; Cao et al., 2017; Kolodziejczyk et al., 2015), opening up the possibility for higher dimensional, complex definitions of cell subpopulations. This exciting new technology presents a number of advantages over historical cell state definitions used to study subpopulation dynamics, however, the utility of these cell states still lies in their relevance to the biological question under investigation. We assert the validity of more traditional, functional characterizations of nongenetic heterogeneity, and propose ways in which the field of cancer research can integrate high-dimensional scRNA-seq data with current mechanistic understanding of distinct cell phenotypes to best study subpopulation dynamics related to progression and treatment.

Theory of cell states in a phenotypic landscape

The presence of distinct subpopulations of cells able to transition between cell states to form a heterogeneous cell population is commonly described using the analogy of a phenotypic landscape. The concept was first introduced by Waddington (Waddington, 1940) to model differentiation and development. Stem cells occupy the top level of the landscape and as cells differentiate, they descend into valleys and assume stable discrete phenotypes represented by "basins"; these are defined by their characteristic gene expression profiles and the resulting phenotype. In this framework, cells are more likely to equilibrate into stable states represented by a "potential well" in the landscape, but reverse transitions back toward stemness are theoretically possible. The probability of transitioning between states is directly proportional to the energy barrier between states, that is, the height of the basins. This framework has recently been extended to understanding cancer cell fates (Huang, 2013; Brock et al., 2009; Paudel et al., 2018). In cancer, a clear hierarchy of cell types is not generally believed to exist, but instead multiple metastable phenotypic states can coexist, consistent with the observations of nongenetic heterogeneity in cancer, in which the landscape represents all theoretically possible physiological cell states. A population of cancer cells may spread out across these available phenotypes, and subpopulation compositions of cancer cell types represent a quasi-equilibrium of the landscape. The effect of a perturbation on this landscape can come in the form of a drug treatment, where in theory a treatment can have the effect of temporarily altering the topography of the landscape, resulting in temporary changes in phenotypic composition followed by a return to initial proportions. Alternatively, a drug can act to permanently alter the landscape by changing the relative depth of the wells that represent cell states, resulting in reequilibration followed by stable changes in phenotypic composition of a cell population (Brock et al., 2009; Zhou et al., 2014; Li et al., 2016). These stable changes could be achieved through either permanent environmental stimuli, or mutations and epigenetic alterations in the cancer cells themselves that may make available cell states that were previously inaccessible or change the stability of existing states relative to one another.

While the phenotypic landscape concept can in some ways be seen as more of an analogy than a true biological phenomenon to be exploited and tested, its utility lies in the framework it provides for understanding complex subpopulation dynamics. Heterogeneity in cancer cells not only encompasses the presence of distinct subpopulations in different amounts present throughout a tumor, but also that these subpopulations exhibit temporal variation. Under the landscape framework, one can model changing subpopulation compositions in cancer cell populations by defining cells into distinct cell states and allowing cells to grow, be killed by drug, and transition into other states either stochastically or induced by a stimulus. Simple mathematical models can be developed under the guiding principle of cell states as attractor states, and can be used to describe and predict how subpopulations of cancer cells will change based on the topography of the landscape at any given time (Pisco and Huang, 2015). Studying the effect of perturbation through these modeling methods can provide evidence for the critical role of cell state transitions, rather than differential growth rates, as the driver of observed subpopulation levels in time (Gupta et al., 2011; Pisco and Huang, 2015). These models can help predict the implications of different environmental stimuli and treatment strategies on the population composition over time. Models of phenotypic state switching can be employed to help us understand the dynamic processes related to cancer progression and treatment response by integrating theoretical models with experimental quantification of levels of relevant subpopulations over time.

Experimental evidence for nongenetic heterogeneity

What experimental findings have been used to demonstrate the presence and utility of nongenetic heterogeneous cell states in cancer? To begin, in order to measure heterogeneity, one must first define the unit that is being measured. It turns out this is not quite so simple. While the field has sought to develop a universal definition of diversity, cell states, and heterogeneity (Maley et al., 2017), we argue here that cell states should be defined not universally, but in the context of the relevant biological question. For example, this means if we are interested in examining heterogeneity in response to treatment, then heterogeneity in cell-cycle phase may be less relevant, and may not need to be incorporated into the model of heterogeneity. In this overview we discuss heterogeneity defined at a breadth of levels—from molecular quantification of gene expression levels in a single cell to observations of diversity in cell morphology, functional behavior, and behavior related to proliferation rates, metastatic potential, and responsiveness to therapies. A few key cell state definitions used as critical evidence for nongenetic heterogeneity in cancer are described in Table 24.1. Here, we only consider observations of stable changes in cellular phenotypes as cell states, as opposed to the quantification of cell states defined by rapid stochastic fluctuations in gene expression levels. In an analogy to thermodynamics, we are interested in the ability of cells to transition from one macrostate to another, rather than focusing on transitions at the level of microstates. Even at this higher level, the definition of a state can vary widely based on context. For example, cell states can be classified by their molecular characterizations (Gupta et al., 2011), where cells are categorized as basal-like, luminal-like, and stem-like based on their surface markers that are commonly used to characterize these distinct cell types. Other times, cell states may be classified by their proliferative capacity (Paudel et al., 2018), in which cells are assigned states based on doubling time, or based

Table 24.1 Examples of different categories of cancer cell state definitions used in the context of heterogeneous cancer cell populations in the literature, identified by function and molecular characterization.

Cell state definition	Functional characterization	Molecular characterization	Source
Epithelial/ mesenchymal	Epithelial: cell-to-cell adhesion, less mobile, polygonal and cobble stone-like, apical-basal polarity Mesenchymal: lack of cell-to-cell adhesions, elongated and spindle-like invasive, mobile, front-back polarity, elevated resistance to apoptosis	Epithelial: high E-cadherin, low vimentin, cytokeratins Mesenchymal: high N-cadherin, high vimentin, production of ECM degrading enzymes, FSP1, desmin Transcription factors associated with transition to mesenchymal: SNAIL. TWIST, ZEB, FOXC2, and YAP families	Elosegui-artola et al. (2017), Wei et al. (2015), Yu et al. (2015), Ren et al. (2016), Ai et al. (2014), Chaffer et al. (2016), Ye and Weinberg (2015), Kalluri et al. (2010), Mani et al. (2008), Brabletz et al. (2018)
Drug sensitive/ resistant	Sensitive: higher growth rate, quicker death in response to drug Resistant: slower growth rate, slower/less death in presence and after drug exposure, mechanisms include: evading targeted treatment, multidrug resistance pump, persistence, changes in morphology	Sensitive: high Ki67 (proliferation marker), mutations dependent (i.e., high expression of BRAF/ Kras/GPX4 others) Resistant: MDR1 expression high, low expression of Kras, BRAF, upregulated DNA repair	Pisco and Huang (2015), Chen et al. (2018), Howard et al. (2018), Hangauer et al. (2017)
OxPhos/ glycolysis metabolism	OxPhos: oxidative phosphorylation as primary mechanism of ATP production, use oxygen, sensitive to OxPhos inhibition Glycolysis: glucose used to produce ATP	OxPhos: upregulation of PCG1 and transcription factor MITF Glycolysis: upregulation of RAS-RAF-MEK-ERK signaling axis, glutamine transporter ASCT2	Hardeman et al. (2017), Deberardinis et al. (2008), Deberardinis and Chandel (2016), Heiden et al. (2009), Davies et al. (2015), Vazquez et al. (2013)
Hypoxic/ vascular	Hypoxic: deoxygenated cell state Vascular: well-oxygenated and nutrient rich environment, characterized by presence of blood vessels in three-dimensional tumor environment	Hypoxia: H&E high Vascular: CD31 high	Syed and Woodall (2019), Sorace et al. (2017), Junttila and De Sauvage (2013), Goel et al. (2011)
Carcinomatous/ sarcomatoid		Carcinomatous: k7 positive, upregulated in cell-cell	

(*Continued*)

Table 24.1 (Continued)

Cell state definition	Functional characterization	Molecular characterization	Source
	Carcinomatous: round, less invasive, better cell-cell junctions Sarcomatoid: spindle-like, more aggressive, invasive/migration	junction genes and epithelial related genes Sarcomatoid: k7 negative, upregulated EMT related genes (TWIST1, TGFB, ZEB1), upregulated stem cell genes	Thanos et al. (2018), Wang et al. (2012), Miettinen et al. (1999)
Basal/luminal/stem	Basal: spherical, mesenchymal-like, stem-like, upregulated TGF-β Luminal: differentiated, upregulated E-cadherin Stem: ability to generate other phenotypes, self-renewal	Basal: CD44 high, CD24 neg, EpCam neg Luminal: CD44 low, CD24 high, EpCam high Stem: CD44 high, CD24 neg, EpCam Lo	Gupta et al. (2011), Fillmore and Kuperwasser (2008b), Shipitsin et al. (2007)

on sensitivity to drug (Howard et al., 2018). In some studies, gene expression states may be more completely characterized by the entire gene expression state using single-cell RNA sequencing and dimensionality reduction (Stumpf et al., 2017; Patel, 2014; Mojtahedi et al., 2016). While none of these methods of categorizing cell states can perfectly capture the most parsimonious definition and binning of cell states, a diverse set of experimental measures may further our understanding of the dynamic composition of cancer cell populations.

Well-characterized cell states: epithelial-to-mesenchymal transition

The epithelial-to-mesenchymal cell state transition (EMT) provides an illustrative example of how an environmental stimulus can induce a phenotypic transition that leads to altered subpopulation composition. In particular, EMT has been observed to occur on time scales that nearly eliminate the plausibility that cell population composition changes are due to differences in proliferation rate (Kalluri et al., 2010; Zhang et al., 2015; Li and Balazsi, 2018), and thus must be due to genome-independent phenotypic plasticity. Both the epithelial and mesenchymal cell states have been characterized in detail at the molecular and function level, with many known biomarkers used as evidence for the presence of each cell type. The epithelial cell phenotype is defined by increased cell-to-cell adhesions, less motility, and increased sensitivity to chemotherapy treatment (Maclean et al., 2014). Epithelial cells tend to be polygonal and cobblestone like in morphology (Ye and Weinberg, 2015) and their proliferation rate is usually higher than mesenchymal cells (Chen et al., 2012). The epithelial phenotype is associated with molecular characterizations of high levels of E-cadherin to form the cell-to-cell adhesions, and low levels of vimentin and N-cadherin (Ai et al., 2014; Wei et al., 2015; Yu et al., 2015; Ren et al., 2016). In contrast, mesenchymal cells are often characterized by their motile phenotype, elongated and spindle-like shape, and lack of cell-to-cell adhesions (Kalluri et al., 2010; Ye and Weinberg, 2015; Chaffer et al., 2016; Brabletz et al., 2018). These characteristics implicate them in relation to metastasis and invasive potential via upregulation of ECM degrading enzymes and an increased expression of N-cadherin and vimentin (Kalluri et al., 2010; Ye and Weinberg, 2015). The mesenchymal phenotype is generally associated with aggressiveness, and they tend to be more resistant to cytotoxic agents (Li and Balazsi, 2018). Additionally, transcription factor families such as SNAIL, and TWIST, and ZEB (Kalluri et al., 2010; Ye and Weinberg, 2015) are known to drive the EMT phenotypic transition, with YAP, Snail, and Twist1 nuclear localization associated with downstream mesenchymal-like behavior of cells (Ai et al., 2014; Lu et al., 2014; Wei et al., 2015; Yu et al., 2015; Ren et al., 2016; Jia et al., 2017).

EMT and its reverse, mesenchymal-epithelial transition (MET), are two of the most prominent naturally occurring transdifferentiation programs that ensure interconversion of cells to form various cell types (Ye and Weinberg, 2015). Foundational studies in a variety of biological processes have demonstrated that EMT is activated rapidly in response to a stimulus- for example in gastrulation, wound healing, organ development, and tissue regeneration (Kalluri et al., 2010; Ye and Weinberg, 2015). Because the EMT process has been well studied in development and tissue repair, epithelial and mesenchymal cells are well characterized by distinct molecular and functional properties, and these can be used to study the phenotypic plasticity of epithelial and mesenchymal cells in response to different stimuli.

In both normal cells and cancerous cells, certain environmental conditions have been demonstrated to induce EMT and its reverse, MET. Specifically, in vitro studies have demonstrated that TGF-β (Zhang et al., 2015) induce EMT in certain types of cancer cells (Kalluri et al., 2010; Ye and Weinberg, 2015; Zhang et al., 2015), and that the phenotype can be reversed back via MET when TGF-β is removed from culture. TGF-β-induced EMT has been shown to be mediated by either Smad proteins via the ALK-5 receptor or through p38 MAPK and RhoA pathways (Kalluri et al., 2010). Additionally, studies have demonstrated that the stiffness of the microenvironment of a cell can induce EMT through mechanotransduction signaling pathways mediated through TWIST1 (Wei et al., 2015) and directly through forced nuclear localization of the transcription factor YAP (Elosegui-artola et al., 2017). The effect of environmental stimuli of TGF-β and stiffness provide direct evidence that cancer cells are able to adapt to their environment by inducing changes in phenotype in response to an external perturbation. EMT serves as one well-characterized example in which nongenetic heterogeneity is present within cancer and results in subpopulation compositions that are able to dynamically respond to the environment via phenotypic plasticity. EMT provides a model system that demonstrates how the diverse means of defining the epithelial and mesenchymal cell states—through morphology, growth rate, invasive potential, and molecular characterization—can all be useful in specific contexts and do not represent competing definitions of cell states but rather demonstrate that various cell state definitions give functional meaning to molecular characterizations and vice versa.

Observations of drug naïve cell states

Just as distinct subpopulations of epithelial and mesenchymal cell states can be observed in cancer cell populations, there exists an abundance of evidence for the presence of distinct subpopulations of cells within a whole population, even in the absence of environmental stimuli that might drive phenotypic adaptation. The existence of phenotypic diversity does not necessitate the presence of nongenetic heterogeneity, as observed phenotypic composition could be obtained through differential growth rate of distinct subclones with different genomes that give rise to different phenotypes. Thus, this often leads to the question, are the observed proportions of subpopulations of cells maintained through differential proliferation rates of distinct subtypes or the interconversion between different cell states to maintain equilibrium proportions? While it is likely true that in cancer, both differential growth rates of distinct subclones and phenotypic plasticity both contribute to subpopulation composition, there exists an abundance of evidence for the idea that subpopulation proportions are maintained through phenotypic transitions between distinct cell states.

Recent studies have demonstrated that phenotypic transitions are the most likely mechanism for maintaining equilibrium proportions of cell states within a cancer cell line that might otherwise be considered homogenous (Gupta et al., 2011). Researchers quantified the baseline proportions of distinct cell states in SUM159 and SUM149 breast cancer cell lines using previously defined and characterized cell-surface markers corresponding to phenotypes of: stem-like (CD44 high CD24 neg EpCam lo), basal (CD44 high, CD24 negative, EpCam neg), and luminal (CD44 low CD24 high EpCam high) (Shipitsin et al., 2007; Fillmore and Kuperwasser, 2008a). Using fluorescence activated cell sorting (FACS), the baseline proportions of each cell phenotypes, defined by the cell-surface marker levels states above, were measured in each cell line (Gupta et al., 2011). The cell-surface markers and FACS were used to isolate pure subpopulations, and the resulting proportions of cells in each cell state were sampled over time following isolation (Gupta et al., 2011). The results revealed a rapid progression back toward the equilibrium proportions of the original cell line, and based on the short time it took to recapitulate initial proportions, cell state transitions were more likely to achieve the observed proportions than differential proliferation rates, demonstrating the role of phenotypic state switching in maintaining the observed nongenetic heterogeneity in cancer cell lines (Gupta et al., 2011).

In the phenotypic landscape model (Huang, 2011), cells may overcome the stability of their gene expression configuration to exit an attractor state in one direction, but the mechanisms by which this occurs are not well characterized. The question of mechanism of "escape" is addressed in blood cells by perturbing differentiated erythroid (red) and myeloid (white) blood cell lineages and observing the dynamic response to perturbation by quantifying cell state composition in time (Mojtahedi et al., 2016). Stimulation with cytokines of erythropoietin and/or IL-3/GM-CSF trigger this transition the resulting changes in subpopulation composition over time were measured at the single-cell resolution using both FACS sorting on Sca1 and c-kit surface protein expression markers and single-cell qPCR analysis of 19 genes. Transition was characterized by a transition period at day 3 in which cells exhibited a higher diversity in cell state space, consistent with the analogy of the landscape temporarily flattening. This was followed by coalescence into two distinct cell clusters, corresponding to committed erythroid *(red)* and myeloid *(white)* cell states. These empirical data relating changes in phenotypic compositions over time provide the basis for a mathematical modeling framework to describe the probability of transitioning from one attractor state to another as functions of the relative depth of the well (Mojtahedi et al., 2016). These modeling frameworks allow us to better understand how subpopulation dynamics are maintained through phenotypic plasticity (Gupta et al., 2011; Mojtahedi et al., 2016).

Since these landmark studies describing the dynamics of cell state transitions, additional work has demonstrated the important role of drug-naïve phenotypic state switching for different definitions of cell state. For example, in human liver cancer, heterogeneity in histopathology has demonstrated that sarcomatoid cholangiocarcinoma is characterized by distinct stable phenotypes classified as "sarcomatoid" and "carcinomatous" cells (Miettinen et al., 1999; Wang et al., 2016; Thanos et al., 2018). The sarcomatoid cell type is characterized morphologically as spindle-shaped and functionally are known to be more invasive and motile, similar to mesenchymal cells. Transcriptional profiles of the two cell types indicates that sarcomatoid cells are down regulated in cell-to-cell junction related genes and are upregulated in invasion/migration related genes, EMT related genes such as TWIST1, TGFB, and ZEB1, and stem cell genes, compared to the carcinomatous cells. While morphology, invasivity, motility, and molecular characterizations were used to "define" distinct subtypes of cells in this context, a single marker, keratin-7 expression, was used to isolate cell phenotypes via antibody staining (Thanos et al., 2018). To determine whether observed heterogeneity in cell states is due to the ability of cells to transition from one state to another or due to independently growing subclones of sarcomatoid and carcinomatous cells, k7 positive (carcinomatous), k7 negative (sarcomatoid), cells were grown separately as single-cell-derived subclones. The results revealed that k7 positive cells (carcinomas) were able to give rise to k7 heterogeneous and k7 negative daughter cells, indicating the k7 positive cells have the ability to transition into different phenotypes. The researchers propose that the k7 heterogeneous cells represent an unstable transition state between the two phenotypes, and demonstrate that tumors seeded with the k7 heterogeneous cells, with the ability to transition between states more easily, had a higher "take rate" in mice than isolated k7 positive and k7 negative subpopulations (Thanos et al., 2018). This work demonstrates that phenotypic plasticity plays a significant role in maintaining heterogeneous subpopulations of cancer cells even in the absence of any environmental stimulus.

Additional instances of drug-naive phenotypic state switching have been described and characterized in cancer in a number of ways. For example, in the field of EMT, stochastic cell state transitions between epithelial and mesenchymal cells have been observed in the absence of any transition driving perturbation (Tian et al., 2013). Additionally, cell states have been characterized functionally by the observed growth rate, with changes in the observed population growth rate over time and passage number explained due to changing subpopulations of cells in different phenotypic states (Paudel et al., 2018). Thus, even in the absence of treatment or severe environmental pressures, cancer cells exhibit the capacity to overcome energy barriers between "basins" in the landscape to convert with some nonzero probability into alternative cell states and explore state-space. Although normal healthy cells likely exhibit these capabilities as well, it is possible that cancer cells exist in a "flatter" landscape (Li et al., 2016; Mojtahedi et al., 2016), making the breadth of state space that is explored more likely and resulting in nongenetic heterogeneity and the ability to evade environmental pressures that is characteristic of cancer cell populations.

Adaptive cell states that are induced in response to cancer treatment

The development of drug resistance in cancer is often explained by the presence of rare genetic mutations in the tumor population that allow for resistant subclones to expand in the presence of the selective pressure of treatment via Darwinian selection. However, recently there has been an increased interest in an alternative mechanism, in which the treatment itself induces an altered, drug resistant or tolerant, phenotypic state (Pisco and Huang, 2015; Greene and Gevertz, 2017). The ability of cells within a population to transition from a drug sensitive to a drug resistant state is demonstrated in a number of cancer types and in response to both targeted and cytotoxic therapies in cancer (Brock et al., 2009; Zhou et al., 2014; Pisco and Huang, 2015; Greene and Gevertz, 2017; Hardeman et al., 2017). In this section, we will highlight the evidence for drug-induced phenotypic switching in cancer and overview the many ways in which these altered phenotypic states have been characterized in the context relevant to understanding mechanisms of observed drug resistance.

Targeted therapies are designed to specifically target and kill or inhibit the growth of cancer cells exhibiting a characteristic not present in high abundance on normal healthy cells. This could include anything from cell-surface receptors often overexpressed in cancer to oncogene addictions in certain cancer types. Some examples of these targets for which known therapies have been developed include: Kras addicted pancreatic ductal adenocarcinoma (PDAC) (Chen et al., 2018), HER2+ breast cancer (Hangauer et al., 2017), EGFR positive nonsmall cell lung cancer, and BRAF mutated melanoma (Hardeman et al., 2017; Paudel et al., 2018). The promise of targeted therapies is based on the assumption that cancer cells are dependent on activation of the specific target, and variants with preexisting resistance to that target are rare. However, a number of studies of targeted drug treatments on oncogene addicted cancers have demonstrated that exposure to targeted treatment can enable cells to adapt to an alternative, potentially reversible, cell state. For example, in Kras addicted PDAC, Kras inhibition treatment results in the induction of a drug-tolerant cell state characterized by differences in cell morphology, proliferative kinetics, and tumor-initiating capacity (Chen et al., 2018). This drug-induced tolerant state is demonstrated to be reversible, resulting in no significant mutational or transcriptional changes but changes in gene expression related to cell signaling and focal adhesion pathways that cause the cells to have an increased dependence on adhesion for viability *in vitro* (Chen et al., 2018). Similarly, in HER2+ breast cancer, lapatinib treatment induces a nonmutational "persister" cell state that is characterized by a transient dependency on GPX4 (Hangauer et al., 2017). In this work, the ability to capitalize on nongenetic heterogeneity and phenotypic plasticity is demonstrated by attacking the drug-induced GPX4 dependent state with a GPX4 inhibitor, resulting in cell death in vitro that is not observed on the parental cell line alone with GPX4 inhibitor (Hangauer et al., 2017).

In addition to targeting drug-induced states with new targeted treatments based on the drug-induced states molecular dependencies, recent work has also shed light on characterizing the drug-induced state via broader phenotypic changes. For example, BRAF mutated melanoma cells states have been defined in terms of their tumor metabolic phenotype (Hardeman et al., 2017). In this work, it is observed that cells fall along a spectrum of sensitivity to BRAF inhibition, with cell states resistant to BRAF inhibition characterized by a metabolic phenotype of oxidative phosphorylation instead of glycolysis. Thus, to overcome resistance to BRAF inhibition treatment, melanoma cells were treated with zalcitabine, a drug that suppresses normal oxidative phosphorylation and forces cells into glycolysis, and found that cells could be resensitized to BRAF inhibition, thus demonstrating the reversibility of the drug resistant phenotype via driving the alternative metabolic phenotype (Hardeman et al., 2017). In this context, knowing both the phenotype of the BRAF sensitivity and the metabolic phenotype allowed researchers to probe whether or not the phenotypic switching observed in response to drug was directly linked to an alternate phenotype. These types of relationships to characterize drug-induced phenotypes can be useful in developing targeted treatment strategies aimed at rationally modifying the landscape in favor of cell states with greater drug susceptibility.

The idea that cancer cells might evade attack from targeted treatment by utilizing alternative pathways for survival is quite rational, however, how does a specific multidrug resistant phenotype emerge in response to broad-based chemotherapeutic agents? Again, observed drug resistance is classically explained by selection of resistant mutant cancer cells, however recent work has demonstrated that a multidrug resistant state, characterized by expression of the MDR1 drug-pumping family of genes, is directly induced via Lamarckian induction, following exposure to chemotherapy in HL60 leukemic cells (Pisco and Huang, 2015). This landmark paper not only demonstrated a drug-induced phenotypic state characterized by the functional ability to efflux drug, but also that this drug-induced phenotype was a result of cell-autonomous gene induction that was independent of fitness benefit, as the elevated levels of expression of MDR1 proteins were still observed even in the inhibition of the functional drug-pumping mechanism. These findings introduced the idea that drug exposure can "instruct" a cell to switch between attractor states in a directed manner. Not only does this finding indicate that drug-induced attractor states should be examined to identify novel targets of overcoming resistance, but also implies that drug treatment schedules must take resistance induction dynamics into consideration when developing optimal treatment strategies, as has been explored by a number of recent theoretical works (Greene and Gevertz, 2017; Wood et al., 2012; Gatenby, 1991; Gatenby et al., 2009). In addition, the observation of drug-induced resistance lends itself to categorizing cell states based on the most relevant variable of interest—drug sensitivity. By defining cell states in terms of drug tolerance via an LD50 metric, the subpopulation composition of cancer cells over time in response to a pulse treatment were quantified in order to characterize the dynamic drug resistance response (Howard et al., 2018). While understanding the dynamics of drug resistant phenotypes over time is important, it is equally important to integrate this knowledge into a mechanistic understanding of the biological process that drives drug-induced resistance. While the drug-induced resistant state has been characterized by a specific resistance mechanism, the MDR1 high state (Pisco and Huang, 2015), it is quite likely that multidrug resistance is due to multiple resistance mechanisms. To understand the processes driving resistance induction, it is sometimes necessary to identify and quantify nongenetic heterogeneity through much higher throughput molecular and physical measurements.

Future of cell state characterization: defining cell states via relevant heterogeneity

As we have demonstrated, the definition of what warrants a distinct cell state varies—from the highest level, phenotypes can be defined by the degree at which the cell can perform a specific function (i.e., grow) (Paudel et al., 2018), efflux drug (Pisco and Huang, 2015), adhere to a surface (Hangauer et al., 2017) to the lowest level of molecular detail in which an individual cell state is defined by all of its transcript levels at a specific time. Because of the wide variability in classifying cell states, the degree of heterogeneity, diversity, and plasticity can be equally difficult to quantify (Maley et al., 2017). In this consensus statement (Maley et al., 2017), the authors call for a universal classification system to allow for a common language and conceptual categories for diversity and heterogeneity. The temptation here is that more is better, and this may lead us to believe that scRNA-seq technologies provide the most complete means of defining cell states in cancer. While a universal set of guidelines may be useful for clinical translation, in studies of the fundamental roles of phenotypic plasticity and nongenetic heterogeneity, it is important to consider the biological question being addressed when determining the relevance of cell state definitions that define heterogeneity. While we acknowledge enormous power of scRNA-seq in describing high-dimensional cell states, we caution that the importance of this level of detail lies in the ability to connect gene expression states to the mechanistic underpinnings of the processes of interest to understanding cancer progression.

Many of the most foundational pieces of evidence for nongenetic heterogeneity in cancer have been identified in the absence of a "complete" quantification of gene expression state.

In earlier studies (Pisco and Huang, 2015; Gupta et al., 2011), *a priori* knowledge of a few relevant surface markers informed the choice of readout for measuring cell population composition over time. While these definitions of cell state may seem rudimentary compared to the high-throughput capabilities of scRNA-seq, they were well suited to estimating the dynamics of cell state transitions in vitro in response to perturbations. In the clinic, scRNA-seq is not yet feasible broadly, in real time, at the patient-specific level. However, imaging of patient tumors can be used to quantify the phenotypic states in response to treatment, and may be performed on a voxel-to-voxel basis to capture phenotypic and spatial heterogeneity of the patient tumor (Syed and Woodall, 2019). In this work, imaging metrics related to vascularity, hypoxia, and cell number were used to demonstrate the effect of treatment on these relevant aspects of tumor progression and overall heterogeneity (Syed and Woodall, 2019). While cell states could have been defined by a number of imaging metrics available at each voxel, this work quantified cell states based on metrics that could be tied back to their function (Syed and Woodall, 2019). The strategy to identify relevant metrics in the vast amount of imaging data can also be applied to extract meaning from the immense amount of data obtained from scRNA-seq technology. Here, we discuss how scRNA-seq technologies may advance our understanding of cell states and nongenetic heterogeneity, but also acknowledge inherent limitations in the technology and analysis.

Traditionally, cell states have been characterized in a number of ways, many of which are presented in this chapter and include morphology and specific protein markers. Recent developments in single-cell RNA sequencing technology allow for high-throughput identification of both prevalent and rare cell types and states in a neoplasm. Numerous studies have used scRNA-seq to reveal novel aspects of tumor heterogeneity and implications for chemotherapy resistance (Bartoschek et al., 2018; Gan et al., 2018; Karaayvaz et al., 2018; Kim et al., 2018; Kumar et al., 2018; Levitin et al., 2018; Poirion et al., 2018; Rodriguez-Meira et al., 2019). Despite the large breadth of information provided by scRNA-seq, the technology does have some disadvantages. First, in the interest of financial feasibility, scRNA-seq studies typically include data from only a few samples. This may limit the scope of studies using human tumor samples where larger sample sizes may be important due to a high level of variation. It also limits the ability of the technology to capture relevant dynamics of cell state compositions, as the vast majority of scRNA-seq studies are performed at a single static time point due to the high cost of additional samples (Stumpf et al., 2017). Second, standard protocols for performing scRNA-seq on tissues require chemical and/or mechanical dissociation to produce a single-cell suspension. This processing may introduce a bias toward cells able to survive the stress of dissociation. In addition, the protocol removes cells from their native environment, meaning that spatial information is lost. Methods have been developed to profile gene expression at the single-cell level without losing spatial information (Shah et al., 2017; Wang et al., 2018) however, these methods are limited to measuring approximately 1000 or fewer target transcripts. A further consideration is that scRNA-seq measurements do not include RNA molecules that are outside of cells. These may be contained in extracellular vesicles, ribonucleoprotein particles, and virions—all of which have been implicated as relevant in control of cancer progression (Redzic et al., 2014; Xu et al., 2018).

In addition to scRNA-seq, several other single-cell technologies have been developed to evaluate cellular heterogeneity and these may complement or replace scRNA-seq in some applications. For example, single-cell epigenomics techniques can be used to better understand gene regulatory mechanisms and to reveal phenotypes that persist over a large temporal range (Kelsey et al., 2017; Shema et al., 2019). Another emerging field of technology is single-cell proteomics. Given that proteins are the final products in the central dogma and are the actual effectors of cellular function, measuring levels of proteins can offer valuable information about the phenotypic state of a cell. Unfortunately, proteomic assays (such as mass spectrometry) struggle to detect low abundance proteins, an issue that is even more problematic in single-cell measurements. However, new technologies have increased the sensitivity of single-cell mass spectrometry, suggesting that single-cell proteomics may have growing applications in the near future (Budnik et al., 2018; Specht and Slavov, 2018; Wang and Zhang, 2018; Doerr, 2019).

While scRNA-seq provides a significant increase in the magnitude of gene expression space sampled at single-cell resolution, these shortcomings must be taken into consideration when evaluating scRNA-seq data—it is not all encompassing, and it is limited in its ability to capture spatial and dynamic data. For this reason, we think scRNA-seq data should not be critical to identifying cell states and heterogeneity, but rather an additional tool used to understand a process of interest. Allowing for a definition of heterogeneity that is tailored to the biological context is already used in some aspects of scRNA-seq analysis. For example, the removal of cell-cycle genes is often an initial step in scRNA-seq data analysis, as these changes may overshadow other differences in transcript levels, which are more critical to the biological processes of interest. Sifting through scRNA-seq data is analogous to focusing on a plane of interest that allows cell states to be characterized differently in different contexts to identify relevant functional features. Regardless of its advantages and disadvantages, single-cell RNA sequencing provides the most extensive molecular characterization of phenotypic cell states currently available. While the bread of cell state characterizations are important (Kolodziejczyk et al., 2015), in cancer, we argue that the relevance of cell state definitions lies in how these cell state compositions change in time and respond to environmental pressures such as treatment or the immune system. Thus, new insights may come from the integration of dynamic single-cell gene expression data to assign functional states to cells over time, as has been performed in single-cell resolutions studies of differentiation (Stumpf et al., 2017). In this pivotal work, gene expression state information is used to develop and test hypothesis for models of how subpopulation composition of embryonic stem cells transitions into fully differentiated neural progenitor cells (Stumpf et al., 2017). In cancer, cell state clusters (Patel, 2014) may be integrated with the functional information they correspond to in order to develop models of cell state transitions that describe tumor progression and drug response. Pulling from the principles applied in the field of differentiation would provide a better understanding of how the dynamic processes characterized by changes in phenotypic composition are driven by different environmental cues in cancer, and allow us to develop strategies to combat tumor progression, drug resistance, and immune suppression by manipulating subpopulation dynamics.

In conclusion, cell state heterogeneity that is independent of genetic identity is frequently observed in cancer, and studies of these changing cell state compositions have the potential to improve our understanding of disease progression and treatment response. By allowing for an infinite combination of cell state definitions via integrating single-cell RNA sequencing with functional and molecular level characterizations, the possibilities of exploiting phenotypic plasticity for developing new therapeutic strategies and understanding the dynamics of certain processes are enormous.

References

Ai, L., et al., 2014. TRIM29 suppresses TWIST1 and invasive breast cancer behavior. Cancer Res. 4875–4888. Available from: https://doi.org/10.1158/0008-5472.CAN-13-3579.

Bartoschek, M., et al., 2018. Spatially and functionally distinct subclasses of breast cancer-associated fibroblasts revealed by single cell RNA sequencing. Nat. Commun. 9, 5150. Available from: https://doi.org/10.1038/s41467-018-07582-3.

Brabletz, T., et al., 2018. EMT in cancer. Nature 18, 128–134. Available from: https://doi.org/10.1038/nrc.2017.118.

Brock, A., Chang, H., Huang, S., 2009. Non-genetic heterogeneity—a mutation-independent driving force for the somatic evolution of tumours. Nat. Rev. Genet. 10 (5), 336–342. Available from: https://doi.org/10.1038/nrg2556.

Budnik, B., et al., 2018. SCoPE-MS: mass spectrometry of single mammalian cells quantifies proteome heterogeneity during cell differentiation. Genome Biol. 19, 1–12.

Cao, J., et al., 2017. Comprehensive single cell transcriptional profiling of a multicellular organism by combinatorial indexing. bioRxiv 667 (August), 1–35. Available from: https://doi.org/10.1101/104844.

Chaffer, C.L., et al., 2016. EMT, cell plasticity and metastasis. Cancer Metastasis Rev. 35, 645–654. Available from: https://doi.org/10.1007/s10555-016-9648-7.

Chen, L., et al., 2012. The enhancement of cancer stem cell properties of MCF-7 cells in 3D collagen scaffolds for modeling of cancer and anti-cancer drugs. Biomaterials 33 (5), 1437–1444. Available from: https://doi.org/10.1016/j.biomaterials.2011.10.056.

Chen, P., et al., 2018. Adaptive and reversible resistance to Kras inhibition in pancreatic cancer cells. Cancer Res. 78, 985–1003. Available from: https://doi.org/10.1158/0008-5472.CAN-17-2129.

Davies, J.E.H., et al., 2015. Vemurafenib resistance reprograms melanoma cells towards glutamine dependence. J. Transl. Med. 1–11. Available from: https://doi.org/10.1186/s12967-015-0581-2.

Deberardinis, R.J., Chandel, N.S., 2016. Fundamentals of cancer metabolism. Sci Adv. 2 (5), e1600200. Available from: https://doi.org/10.1126/sciadv.1600200.

Deberardinis, R.J., et al., 2008. Review the biology of cancer: metabolic reprogramming fuels cell growth and proliferation. Cell 11–20. Available from: https://doi.org/10.1016/j.cmet.2007.10.002.

Doerr, A., 2019. Single-cell proteomics. Nat. Methods 16 (1), 20. Available from: https://doi.org/10.1038/s41592-018-0273-y.

Elosegui-artola, A., et al., 2017. Force triggers YAP nuclear entry by regulating transport across nuclear pores article force triggers YAP nuclear entry by regulating transport across nuclear pores. Cell 171 (6), 1397–1410. Available from: https://doi.org/10.1016/j.cell.2017.10.008. e14.

Fillmore, C.M., Kuperwasser, C., 2008a. Human breast cancer cell lines contain stem-like cells that self-renew, give rise to phenotypically diverse progeny and survive chemotherapy. Breast Cancer Res. 10 (2), 1–13. Available from: https://doi.org/10.1186/bcr1982.

Fillmore, C.M., Kuperwasser, C., 2008b. Research article human breast cancer cell lines contain stem-like cells that self-renew, give rise to phenotypically diverse progeny and survive chemotherapy. Breast Cancer Res. 10 (2), 1–13. Available from: https://doi.org/10.1186/bcr1982.

Gan, Y., et al., 2018. Identification of cancer subtypes from single-cell RNA-seq data using a consensus clustering method. BMC Med. Genomics 11 (Suppl. 6), 117. Available from: https://doi.org/10.1186/s12920-018-0433-z.

Gatenby, R.A., 1991. Population ecology issues in tumor growth. Cancer Res. 2, 2542–2548.

Gatenby, R.A., et al., 2009. Adaptive therapy. Cancer Res. 69 (11), 4894–4903. Available from: https://doi.org/10.1158/0008-5472.CAN-08-3658.

Goel, S., et al., 2011. Normalization of the vasculature for treatment of cancer and other diseases. Physiol. Rev. 91, 1071–1121. Available from: https://doi.org/10.1152/physrev.00038.2010.

Greene, J.M., Gevertz, J.L., 2017. A mathematical approach to differentiate spontaneous and induced evolution to drug resistance during cancer treatment. bioRxiv 1–30.

Gupta, P.B., et al., 2011. Stochastic state transitions give rise to phenotypic equilibrium in populations of cancer cells. Cell 146 (4), 633–644. Available from: https://doi.org/10.1016/j.cell.2011.07.026.

Hangauer, M.J., et al., 2017. To GPX4 inhibition. Nat. Publ. Group 551 (7679), 247–250. Available from: https://doi.org/10.1038/nature24297.

Hardeman, K.N., et al., 2017. Dependence on glycolysis sensitizes BRAF-mutated melanomas for increased response to targeted BRAF inhibition. Nat. Publ. Group. 42604, 1–9. Available from: https://doi.org/10.1038/srep42604.

Heiden, M.G.V., et al., 2009. Understanding the Warburg effect: cell proliferation. Science 324 (May), 1029–1034.

Howard, G.R., et al., 2018. A multi-state model of chemoresistance to characterize phenotypic dynamics in breast cancer. Sci. Rep. 1–11. Available from: https://doi.org/10.1038/s41598-018-30467-w.

Huang, S., 2011. The molecular and mathematical basis of Waddington's epigenetic landscape: a framework for post-Darwinian biology? Bioessays 34, 149–157. Available from: https://doi.org/10.1002/bies.201100031.

Huang, S., 2013. Genetic and non-genetic instability in tumor progression: Link between the fitness landscape and the epigenetic landscape of cancer cells. Cancer Metastasis Rev. 32 (3–4), 423–448. Available from: https://doi.org/10.1007/s10555-013-9435-7.

Jia, D., et al., 2017. Distinguishing mechanisms underlying EMT tristability. Cancer Converg. 1 (1), 2. Available from: https://doi.org/10.1186/s41236-017-0005-8.

Junttila, M.R., De Sauvage, F.J., 2013. Influence of tumour micro-environment heterogeneity on therapeutic response. Nature 501, 346–354. Available from: https://doi.org/10.1038/nature12626.

Kalluri, R., et al., 2010. The basics of epithelial-mesenchymal transition. J. Clin. Invest. 119 (May), 1420–1428. Available from: https://doi.org/10.1172/JCI39104.1420.

Karaayvaz, M., et al., 2018. Unravelling subclonal heterogeneity and aggressive disease states in TNBC through single-cell RNA-seq. Nat. Commun. 9, 3588. Available from: https://doi.org/10.1038/s41467-018-06052-0.

Kelsey, G., Stegle, O., Reik, W., 2017. Single-cell epigenomics: recording the past and predicting the future. Science 358 (6359), 69–75. Available from: https://doi.org/10.1126/science.aan6826.

Kim, C., et al., 2018. Chemoresistance evolution in triple-negative breast cancer delineated by single-cell sequencing. Cell 173 (4), 879–893. Available from: https://doi.org/10.1016/j.cell.2018.03.041. e13.

Klein, A.M., et al., 2015. Droplet barcoding for single-cell transcriptomics applied to embryonic stem cells. Cell 161 (5), 1187–1201. Available from: https://doi.org/10.1016/j.cell.2015.04.044.

Kolodziejczyk, A.A., et al., 2015. The technology and biology of single-cell RNA sequencing. Mol. Cell 58 (4), 610–620. Available from: https://doi.org/10.1016/j.molcel.2015.04.005.

Kumar, M.P., et al., 2018. Analysis of single-cell RNA-Seq identifies cell-cell communication associated with tumor characteristics. Cell Rep. 25 (6), 1458–1468. Available from: https://doi.org/10.1016/j.celrep.2018.10.047. e4.

Levitin, H.M., Yuan, J., Sims, P.A., 2018. Single-cell transcriptomic analysis of tumor heterogeneity. Trends Cancer 4 (4), 264–268. Available from: https://doi.org/10.1016/j.trecan.2018.02.003.

Li, C., Balazsi, G., 2018. A landscape view on the interplay between EMT and cancer. NPJ Syst. Biol. Appl. 4 (34).

Li, Q., Wennborg, A., Aurell, E., Dekel, E., Zou, J.-Z., et al., 2016. Dynamics inside the cancer cell attractor reveal cell heterogeneity, limits of stability, and escape. Proc. Natl. Acad. Sci. U.S.A. 113 (10), 2672–2677. Available from: https://doi.org/10.1073/pnas.1519210113.

Lu, M., et al., 2014. Toward decoding the principles of cancer metastasis circuits. Cancer Res. 74 (17), 4574–4587. Available from: https://doi.org/10.1158/0008-5472.CAN-13-3367.

Maclean, A.L., et al., 2014. Epithelial-mesenchymal transition in metastatic cancer cell populations affects tumor dormancy in a simple mathematical model. Biomedicines 2 (4), 384–402. Available from: https://doi.org/10.3390/biomedicines2040384.

Maley, C.C., et al., 2017. Classifying the evolutionary and ecological features of neoplasms. Nat. Publ. Group 17 (10), 605–619. Available from: https://doi.org/10.1038/nrc.2017.69.

Mani, S.A., et al., 2008. The epithelial-mesenchymal transition generates cells with properties of stem cells. Cell 133, 704–715. Available from: https://doi.org/10.1016/j.cell.2008.03.027.

Miettinen, M., et al., 1999. Epithelioid sarcoma: an immunohistochemical analysis of 112 classical and variant cases and a discussion of the differential diagnosis. Hum. Pathol. 934–942.

Mojtahedi, M., et al., 2016. Cell fate decision as high-dimensional critical state transition. PLoS Biol. 14, 1–28. Available from: https://doi.org/10.1371/journal.pbio.2000640.

Patel, A.P., 2014. Single-cell RNA-seq highlights intratumoral heterogeneity in primary glioblastoma. Science 1396 (June), 1–10. Available from: https://doi.org/10.1126/science.1254257.

Paudel, B.B., et al., 2018. A nonquiescent "idling" population state in drug-treated, BRAF-mutated melanoma. Biophysical J. 114 (6), 1499–1511. Available from: https://doi.org/10.1016/j.bpj.2018.01.016.

Pisco, A.O., Huang, S., 2015. Non-genetic cancer cell plasticity and therapy-induced stemness in tumour relapse: 'What does not kill me strengthens me'. Br. J. Cancer 112 (11), 1725–1732. Available from: https://doi.org/10.1038/bjc.2015.146.

Poirion, O., et al., 2018. Using single nucleotide variations in single-cell RNA-seq to identify subpopulations and genotype-phenotype linkage. Nat. Commun. 9 (4892), 1–13. Available from: https://doi.org/10.1038/s41467-018-07170-5.

Redzic, J.S., et al., 2014. Extracellular RNA mediates and marks cancer progression. Semin. Cancer Biol. 28, 14–23. Available from: https://doi.org/10.1016/j.semcancer.2014.04.010.

Ren, H., et al., 2016. TWIST1 and BMI1 in cancer metastasis and chemoresistance. J. Cancer 7 (9), 1074–1080. Available from: https://doi.org/10.7150/jca.14031.

Rodriguez-Meira, A., et al., 2019. Unravelling intratumoral heterogeneity through high-sensitivity single-cell mutational analysis and parallel RNA sequencing. Mol. Cell. 73, 1292–1305.e8. Available from: https://doi.org/10.1016/j.molcel.2019.01.009.

Shah, S., et al., 2017. seqFISH accurately detects transcripts in single cells and reveals robust spatial organization in the hippocampus. Neuron 94 (4), 752–758. Available from: https://doi.org/10.1016/j.neuron.2017.05.008. e1.

Shema, E., Bernstein, B.E., Buenrostro, J.D., 2019. Single-cell and single-molecule epigenomics to uncover genome regulation at unprecedented resolution. Nat. Genet. 51 (1), 19–25. Available from: https://doi.org/10.1038/s41588-018-0290-x.

Shipitsin, M., et al., 2007. Molecular definition of breast tumor heterogeneity. Cancer Cell 11 (March), 259–273. Available from: https://doi.org/10.1016/j.ccr.2007.01.013.

Sorace, A.G., et al., 2017. Quantitative F-FMISO-PET imaging shows reduction of hypoxia following trastuzumab in a murine model of HER2 + breast cancer. Mol. Imaging Biol. 19 (1), 130–137. Available from: https://doi.org/10.1007/s11307-016-0994-1.Quantitative.

Specht, H., Slavov, N., 2018. Transformative opportunities for single-cell proteomics. J. proteome Res. 17 (8), 2565–2571. Available from: https://doi.org/10.1021/acs.jproteome.8b00257.

Stumpf, P.S., et al., 2017. Stem cell differentiation as a non-Markov stochastic process. Cell Syst. 5 (3), 268–282. Available from: https://doi.org/10.1016/j.cels.2017.08.009. e7.

Syed, A.K., Woodall, R., 2019. Characterizing trastuzumab-induced alterations in intratumoral heterogeneity with quantitative imaging and immunohistochemistry in HER2 + breast cancer. Neoplasia 21 (1), 17–29. Available from: https://doi.org/10.1016/j.neo.2018.10.008.

Thanos, D., et al., 2018. Stochastic phenotype switching leads to intratumor heterogeneity in human liver cancer. Hematology 68 (3), 1–16. Available from: https://doi.org/10.1002/hep.29679.

Tian, X., Zhang, H., Xing, J., 2013. Coupled reversible and irreversible bistable switches underlying TGFβ-induced epithelial to mesenchymal transition. Biophys. J. 105 (4), 1079–1089. Available from: https://doi.org/10.1016/j.bpj.2013.07.011.

Vazquez, F., et al., 2013. Article PGC1 a expression defines a subset of human melanoma tumors with increased mitochondrial capacity and resistance to oxidative stress. Cancer Cell 23 (3), 287–301. Available from: https://doi.org/10.1016/j.ccr.2012.11.020. Elsevier Inc.

Waddington, C.H., 1940. Organisers and Genes. Cambridge University Press, Cambridge.

Wang, Z., Zhang, X., 2018. Single cell proteomics for molecular targets in lung cancer: high-dimensional data acquisition and analysis. Adv. Exp. Med. Biol. 1068, 73–87. Available from: https://doi.org/10.1007/978-981-13-0502-3_7.

Wang, Q., Cui, B., Weng, J., 2012. Clinicopathological characteristics and outcome of primary sarcomatoid carcinoma and carcinosarcoma of the liver. J. Gastrointest. Surg. 16, 1715–1726. Available from: https://doi.org/10.1007/s11605-012-1946-y.

Wang, K., et al., 2016. Hidden in the mist no more: physical force in cell biology. Nat. Methods 13 (2), 124–125. Available from: https://doi.org/10.1038/nmeth.3744.Hidden.

Wang, X., et al., 2018. Three-dimensional intact-tissue sequencing of single-cell transcriptional states. Sci. (N. York, N.Y.) 361 (6400). Available from: https://doi.org/10.1126/science.aat5691.

Wei, S.C., et al., 2015. Matrix stiffness drives epithelial-mesenchymal transition and tumour metastasis through a TWIST1-G3BP2 mechanotransduction pathway. Nat. Cell Biol. 17 (5), 678–688. Available from: https://doi.org/10.1038/ncb3157.

Wood, K., et al., 2012. Mechanism-independent method for predicting response to multidrug combinations in bacteria. Proc. Natl. Acad. Sci. 109 (30), 12254–12259. Available from: https://doi.org/10.1073/pnas.1201281109.

Xu, R., et al., 2018. Extracellular vesicles in cancer - implications for future improvements in cancer care. Nat. Rev. 15 (10), 617–638. Available from: https://doi.org/10.1038/s41571-018-0036-9.

Ye, X., Weinberg, R.A., 2015. Epithelial–mesenchymal plasticity: a central regulator of cancer progression. Trends Cell Biol. 25 (11), 675–686. Available from: https://doi.org/10.1016/j.tcb.2015.07.012.

Yu, Y., et al., 2015. Down-regulation of miR-129-5p via the Twist1-Snail feedback loop stimulates the epithelial-mesenchymal transition and is associated with poor prognosis in breast cancer. Oncotarget 6 (33), 34423–34436.

Zhang, J., et al., 2015. TGF-β–induced epithelial-to-mesenchymal transition proceeds through stepwise activation of multiple feedback loops. Sci. Signal. 7 (345), 1–12.

Zhou, J.X., et al., 2014. Nonequilibrium population dynamics of phenotype conversion of cancer cells. PLoS One 9 (12), 1–19. Available from: https://doi.org/10.1371/journal.pone.0110714.

25

Phenotypic plasticity: the emergence of cancer stem cells and collective cell migration

Caterina A.M. La Porta[1,2] and Stefano Zapperi[3,4]

[1]Center for Complexity and Biosystems, University of Milan, Milano, Italy [2]Department of Environmental Science and Policy, University of Milan, Milano, Italy [3]Department of Physics, University of Milan, Milano, Italy [4]CNR - Consiglio Nazionale delle Ricerche, ICMATE, Milano, Italy

Phenotypic plasticity and cancer stem cells

Cancer is highly heterogeneous. This fact brings to many important consequences: there is a profound variation between different individuals with the same cancer and, on the other hand, the cells of a cancer of a specific subject show many significant differences between each other. The different phenotypes of tumor cells are due not only to epigenetic intratumor heterogeneity but also to epigenetic changes due to the impact of the environment. It has been reported that genetically homogeneous tumor cells show a remarkable diversity with respect to therapy response or other environmental stimuli (Kreso et al., 2013; Sharma et al., 2010). Epigenetic gene regulation at molecular level to DNA methylation, posttranslational modification of histones, noncoding RNAs, chromatin remodeling are the most common mechanisms contributing to cellular epigenetic heterogeneity. In cancer, the robusteness of the system is the capability to small cells to adapt to another environment, having the possibility to evolve new cellular ecosystems. These capabilities to adapt to every possible environment lead to the difficulty to find a successful strategy to bite it. The genetic mechanisms that contribute mainly to the ability of cancer to adapt to different microenvironment is genetic destabilization (Luoto et al., 2013; Gillies et al., 2012; Klein and Glazer, 2010).

Moreover, epigenetic change is a second important mechanism that helps tumor heterogeneity (Easwaran et al., 2014). These changes can be modified during cell division, giving the opportunity at the cells to acute, oscillating, intracellular, or changes in

Phenotypic Switching. DOI: https://doi.org/10.1016/B978-0-12-817996-3.00023-2

the microenvironment that requires cycle of adaptation for survival (Gatenby and Gillies, 2008). Furthermore, differences in tumor cell metabolism as direct impact of genetic muations and/or altered microenvironment, can directly impact epigenetic changes (Kinnaird et al., 2016). Interesting data showed a change in mammary stem cell activity during development (Wahl and Spike, 2017). In tumors, our group showed recently the capability of human melanoma cells to overshoot aggressive markers during their growth (Sellerio et al., 2015) (Fig. 25.1A). This mechanism was demonstrated to be controlled by a complex network of miRNAs (Fig. 25.1B and C), confirming the tight control by the cell of this process. The direct and more important consequence of these findings is that the cells have an intrinsic capability to change in dependence of the environment (Sellerio et al., 2015). Similar results were published recently in breast cancer are reviewed in (Wahl and Spike, 2017).

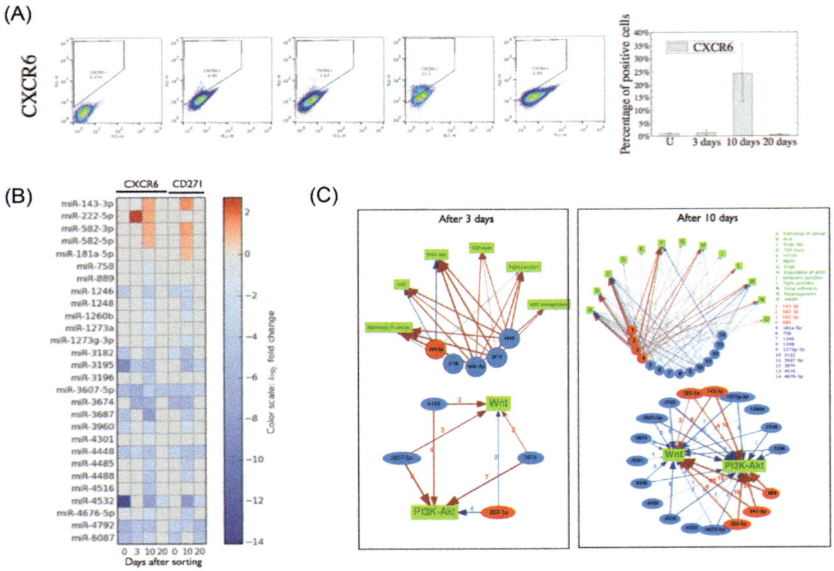

Figure 25.1 Overshoot of CSC population in melanoma cells. (A) The level of expression of a CSC marker (CXCR6) in unsorted cells (U) and at different times after sorting is reported as flow cytometric analysis. Notice the overshoot at 10 days. (B) The list of miRNAs that are differentially expressed with respect to unsorted cells for CXCR6- or CD271- negative cells at 3, 10, and 20 days after sorting. The cells are also analyzed immediately after sorting (0). (C) The network formed by differentially expressed miRNAs in CXCR6-negative (3 and 10 days after sorting) and pathways identified to be significantly targeted by those miRNAs. *Source*: Adapted from Sellerio A.L., Ciusani E., Ben-Moshe N.B., Coco S., Piccinini A., Myers C.R., et al., 2015. Overshoot during phenotypic switching of cancer cell populations. Sci. Rep. 5, 15464.

The landscape of epithelial-mesenchymal plasticity

The aggressive phenotype of tumor cells shows a mesenchymal-like phenotype which is characterized by many differences including an amoeboid shape, the expression of N-cadherin instead of E-cadherin, all these characteristics lead to a more mobile cell, less attached to the others (Beerling et al., 2016; La Porta and Zapperi, 2018). Normal and cancer cells can switch from epithelial-to-mesenchymal state (EMT) or vice versa (MET) in dependence by the environment. There are many evidences of hybrid states E/M which show a more aggressive phenotype (Jolly et al., 2014; Grosse-Wilde et al., 2015; Jolly et al., 2016; George et al., 2017). The same hybrid states have been found in embryo development and wound healing (Jolly et al., 2014).

In a recent work, we have analyzed the emergence of hybrid E/M states by studying the gene regulatory network underlying EMT/MET (Font-Clos et al., 2018), building on previous work on Boolean networks (Steinway et al., 2014, 2015) (see Fig. 25.2A). The distinctive feature of our approach is to focus on the topographic landscape describing the stability of E and M states and their transition, following the old idea of Waddington (Waddington, 1957) who idealized a phenotypic transition as a marble rolling over a landscape. By studying and classifying the kinetic attractors of the EMT/MET gene regulatory networks, we were able to reconstruct the landscape (see Fig. 25.2B) and study the effect of external perturbation on the attractor states. The landscape displays a very large variety of states with stable E or M states, separated by a multitude of metastable hybrid E/M states.

Using the model, we could perturb the state with switching on or off selected genes and follow the ensuing trajectory. Due to the complex hierarchical organization of the states, the result of these perturbations is difficult to predict. In general, however, metastable hybrid states are more likely to undergo complete EMT after a perturbation than stable epithelial states (see Fig. 25.2C). More interestingly, the reconstructed landscape can be used to map gene expression data, both for bulk samples and for single cells, and follow their trajectory during phenotypic transformation. As an illustration, we report in Fig. 25.2D the mapped trajectory of a Transforming growth factor beta (TGF-β)-treated lung adenocarcinoma cell line undergoing EMT (Abnaof et al., 2014).

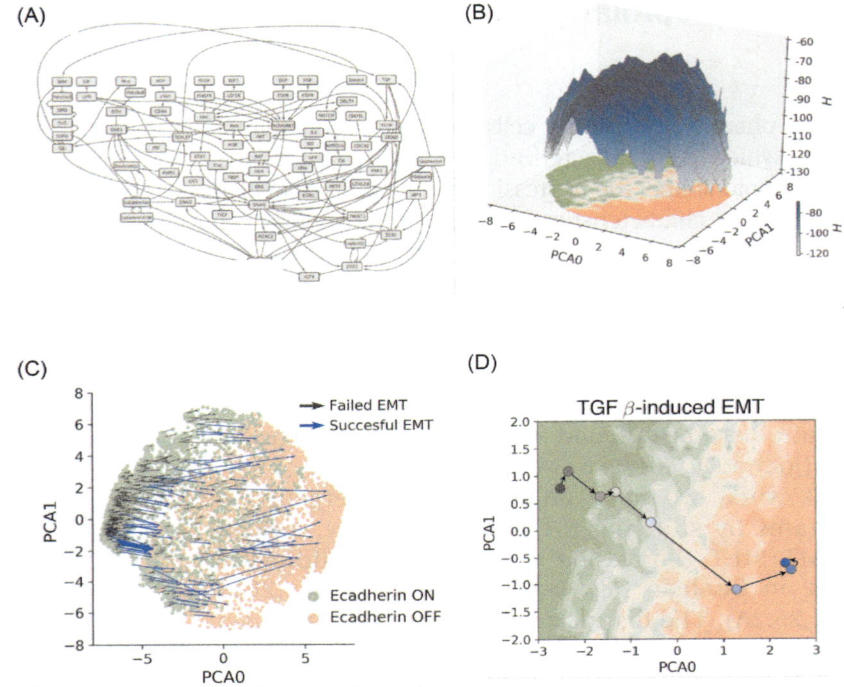

Figure 25.2 Landscape of the EMT. (A) The Boolean network used to investigate the EMT landscape. (B) Three-dimensional reconstruction of topography of EMT. The xy-planes report the principal component analysis (PCA) projection of 10^6 steady states. Color corresponds to the ratio of steady states that express E-cadherin. The z-axis corresponds to the value of H, a measure of the stability of the states, showing that high-H states *(colored in darker blue shades)* coincide with hybrid states. (C) Transition map simulating SNAI1 overexpression. The model displays different forms of SNAI1-induced EMT. (D) Data from TGF-β-treated lung adenocarcinoma cell lines (Abnaof et al., 2014) can be mapped on the landscape yielding a trajectory moving from the E to the M region. *Source*: Adapted from Font-Clos F., Zapperi S., La Porta C.A.M., 2018. Topography of epithelial–mesenchymal plasticity. Proc. Natl. Acad. Sci. U.S.A. 115 (23), 5902–5907.

The consequence of these findings is that from E to M state, there are multiple metastable states where the cell goes through to reach the more stable state E or M. The direction is due to the environment that controls the molecular machinery of the cell. The so-called "cell niche" represents the cell–cell interaction, cell-matrix, and paracrine factors that are present in the microenvironments and that control the plasticity of the cell. It is also possible that a certain mutational background may increase the probability of reprogramming in response to extracellular stimuli including inflammation (Schwitalla et al., 2013; Westphalen et al., 2014). An important consequence that we discussed recently is that we can modify the state of the cells acting on the microenvironment (La Porta and Zapperi, 2018).

Senescence and cancer cell plasticity

Cellular senescence is a physiological mechanism involved in many biological processes ranging from development to cancer. In contrast to apoptotic cells, senescent cells remain viable for an extended period of time exhibiting a senescence-associated secretory phenotype (SASP). SAPS includes proinflammatory cytokines, chemokines, and other factors highlighting the complex interactions between these cells and the environment (Kuilman and Peeper, 2009; Kuilman et al., 2010). In the past few years, emerging literature shows the crucial role of senescent cells in cellular homeostasis and in affecting the immune network responses (Chandra et al., 2015). In general, senescent cells are considered a static end point of the life cycle of the cell, because of the signs of terminal differentiation that can be detected in this cells (Seshadri and Campisi, 1990). Some studies later showed the reversibility of these cells and indicating the presence of presenescence and early-senescent cells that can reenter into the cell cycle (Beausejour et al., 2003; Sage et al., 2003; Milanovic et al., 2018; Yu et al., 2018). This kind of senescence reprogramming appears to be related to cancer stem cells plasticity, as discussed in recent papers (Milanovic et al., 2018; Yu et al., 2018; Latella et al., 2017; Ritschka et al., 2017). Since it is not easy to define rigorously senescent cells, it is still unclear if these cells are really dormant instead of senescent. On the other hand, evidence seems to suggest a more dynamic state for senescent cells. Interestingly, the dynamic nature of senescent cells implies that the cell arrest is not necessary terminal (Milanovic et al., 2018). For example, single cell tracking experiments, using a fluorescent based SA-β-galactosidase assay showed senescent cells at the transition to actively replicating DNA and dividing (Debacq-Chainiaux et al., 2009).

Another interesting study showed that it is possible to enhance the reprogramming of senescent cells using "Yamanaka factors" (Oct4, Sox2, Klf4, Myc) in primary neighbor cells through the SASP (Ritschka et al., 2017) The overlap between the cancer stemness and senescence pathway suggests the idea of a senescence-associated reprogramming and cancer stemness hypothesis (Milanovic et al., 2018). In particular, in nonstem leukemia cells, it appears that the passage through a temporary senescent state helps inducing a reprogramming state (Milanovic et al., 2018). It has been also been postulated that a profoundly remodeled chromatin state, cocontrolling both the senescence and the stemness capacity, allows the cell to overcome Waddington-type landscape barrier toward a stem cell state (Chandra et al., 2015). Under stress or damage, this double capability might be lost.

Our group investigated the population dynamics of cancer stem cells (CSC) and senescent cells by combining mathematical models and experiments in melanoma cells (La Porta et al., 2012). Our experiments showed that populations of melanoma cells in vitro became senescent after a few months, but then the population rapidly reversed into a steady state where the number of senescent cells was reduced (see Fig. 25.3). In our theoretical approach, we considered the hierarchically organized transitions from cancer stem cells to cancer cells and finally into senescent cells without considering explicitly phenotypic transformation from cancer cells or senescent cells back into cancer stem cells. Yet, the model was able to show reversal of senescent cells at the population level (La Porta et al., 2012). Similar results were also obtained simulating therapeutic intervention that would induce senescence in a cancer cell population: the population rapidly restores the steady state reducing the number of senescent cells.

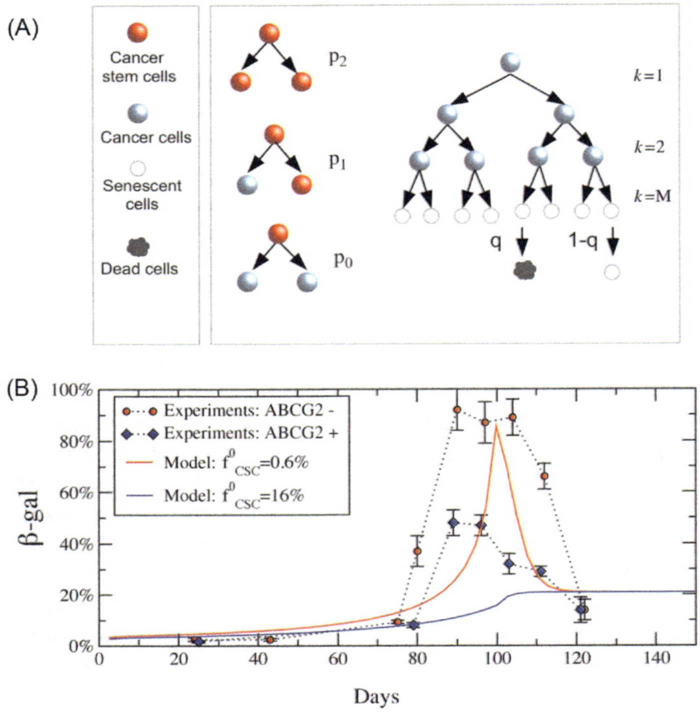

Figure 25.3 Senescence and cancer stem cells. (A) Branching processes for cancer growth. At each generation, CSCs can divide symmetrically or asymmetrically with different probability. Cancer cells divide a finite number of times before turning senescent. (B) Experiments in vitro show that after sorting for CSC markers, cancer cells turn massively turn senescent, as shown by staining with β-gal. Senescence is, however, only transient in the population. *Source*: Images from La Porta C.A. M., Zapperi S., Sethna J.P., 2012. Senescent cells in growing tumors: population dynamics and cancer stem cells. PLoS Comput. Biol. 8 (1), e1002316 (CC license).

As discussed in our paper (La Porta et al., 2012), it is clear from all this evidence that targeting senescence cells could be more complicated than expected. The simplest view is that enhancing senescence could have an opposite effect as the one intended. On the other hand, inhibiting senescence could be a strategy to affect the balance between stem cells and senescence. In the literature there are indeed contrasting results on this issue (Demaria et al., 2017; Farr et al., 2017; Zhu et al., 2015).

Phenotypic transformations and collective cell migration

Cell plasticity controlled by EMT is often considered in the context of metastasis because it provides a general mechanism by which epithelial cells can detach from the bulk and migrate. Yet, it is becoming apparent that cancer cells can also move as a cohesive and coordinated group, in a process known as collective cell migration (Friedl and Gilmour, 2009). In addition to biochemical mechanisms (Malinverno et al., 2017), collective cell migration is affected by physical interactions among cells and with the substrate (Tambe et al., 2011; Brugues et al., 2014; Haeger et al., 2014; Lange and Fabry, 2013; Koch et al., 2012). In particular, when cell proliferation leads to larger cell density, collective fluid-like motion gives way to slowing down and dynamical arrest (Angelini et al., 2011; Park et al., 2015) in a way that is reminiscent of jamming in glasses and soft solids (Liu et al., 2010).

As for inanimate matter, cell jamming can be induced by different physical mechanism such as the increase of cell density or reduction of cell motility (Doxzen et al., 2013). Cell adhesion should also play a role, but the results in the literature are so far controversial with some experiments suggesting counterintuitively that increased intracellular adhesion may lead to unjamming (Park et al., 2015) while other experiments support the idea that adhesion promotes jamming (Garcia et al., 2015).

A vivid illustration of the effect due to environmental changes in triggering a transformation into a collectively migrating phenotype is provided by wound healing experiments in which a cell monolayer is scratched. In a recent paper (Chepizhko et al., 2018), we showed that the presence of the wound triggers a front of collective activity starting from the wound and then spreading toward the bulk of the layer. We showed by using different cell lines that cells in confluent monolayers are typically in a jammed state characterized by limited motility and confined diffusion. Once an edge is created, however, cells sense the edge and transform into a highly mobile and collective phenotype (Chepizhko et al., 2018). We reached this conclusion by analyzing time-lapse images and comparing the results with simulations of an active particle model. The model was shown to quantitatively reproduce experimentally measured velocity distributions allowing for a systematic investigation of the role of key physical parameters. In this way, it was possible to map each experimental condition into a jamming-unjamming phase diagram (Chepizhko et al., 2018).

Concluding remarks

Developing novel therapeutic strategies for cancer that would take into account the heterogeneity between the patients is already considered a fundamental goal of cancer research. Improved understanding of tumor plasticity represents a key step in the ongoing quest for precision medicine. To this end, we would need to elucidate the complex network of miRNAs and other factors regulating phenotypic plasticity of cancer cells. An interdisciplinary combination of computational analysis and experiments represents in our opinion the way forward to reach these goals.

References

Abnaof, K., Mallela, N., Walenda, G., Meurer, S.K., Seré, K., Lin, Q., et al., 2014. TGF-beta stimulation in human and murine cells reveals commonly affected biological processes and pathways at transcription level. BMC Syst. Biol. 8 (1), 55. Available from: https://doi.org/10.1186/1752-0509-855.

Angelini, T.E., Hannezo, E., Trepat, X., Marquez, M., Fredberg, J.J., Weitz, D.A., 2011. Glass-like dynamics of collective cell migration. Proc. Natl Acad. Sci. U.S.A. 108 (12), 4714–4719. Available from: https://doi.org/10.1073/pnas.1010059108.

Beausejour, C.M., Krtolica, A., Galimi, F., Narita, M., Lowe, S.W., Yaswen, P., et al., 2003. Reversal of human cellular senescence: roles of the p53 and p16 pathways. EMBO J. 22 (16), 4212–4222. Available from: https://doi.org/10.1093/emboj/cdg417.

Beerling, E., Seinstra, D., de Wit, E., Kester, L., van der Velden, D., Maynard, C., et al., 2016. Plasticity between epithelial and mesenchymal states unlinks EMT from metastasis-enhancing stem cell capacity. Cell Rep. 14 (10), 2281–2288. Available from: https://doi.org/10.1016/j.celrep.2016.02.034.

Brugues, A., Anon, E., Conte, V., Veldhuis, J.H., Gupta, M., Colombelli, J., et al., 2014. Forces driving epithelial wound healing. Nat. Phys. 10 (9), 683–690. Available from: https://doi.org/10.1038/nphys3040.

Chandra, T., Ewels, P.A., Schoenfelder, S., Furlan-Magaril, M., Wingett, S.W., Kirschner, K., et al., 2015. Global reorganization of the nuclear landscape in senescent cells. Cell Rep. 10 (4), 471–483. Available from: https://doi.org/10.1016/j.celrep.2014.12.055.

Chepizhko, O., Lionetti, M.C., Malinverno, C., Giampietro, C., Scita, G., Zapperi, S., et al., 2018. From jamming to collective cell migration through a boundary induced transition. Soft Matter. 14 (19), 3774–3782. Available from: https://doi.org/10.1039/c8sm00128f.

Debacq-Chainiaux, F., Erusalimsky, J.D., Campisi, J., Toussaint, O., 2009. Protocols to detect senescence-associated beta-galactosidase (SA-betagal) activity, a biomarker of senescent cells in culture and in vivo. Nat. Protoc. 4 (12), 1798–1806. Available from: https://doi.org/10.1038/nprot.2009.191.

Demaria, M., O'Leary, M.N., Chang, J., Shao, L., Liu, S., Alimirah, F., et al., 2017. Cellular senescence promotes adverse effects of chemotherapy and cancer relapse. Cancer Discov. 7 (2), 165–176. Available from: https://doi.org/10.1158/2159-8290.CD-16-0241.

Doxzen, K., Vedula, S.R.K., Leong, M.C., Hirata, H., Gov, N.S., Kabla, A.J., et al., 2013. Guidance of collective cell migration by substrate geometry. Integr. Biol. (Camb.) 5 (8), 1026–1035. Available from: https://doi.org/10.1039/c3ib40054a.

Easwaran, H., Tsai, H.-C., Baylin, S.B., 2014. Cancer epigenetics: tumor heterogeneity, plasticity of stem-like states, and drug resistance. Mol. Cell 54 (5), 716–727. Available from: https://doi.org/10.1016/j.molcel.2014.05.015.

Farr, J.N., Xu, M., Weivoda, M.M., Monroe, D.G., Fraser, D.G., Onken, J.L., et al., 2017. Targeting cellular senescence prevents age-related bone loss in mice. Nat. Med. 23 (9), 1072–1079. Available from: https://doi.org/10.1038/nm.4385.

Font-Clos, F., Zapperi, S., La Porta, C.A.M., 2018. Topography of epithelial–mesenchymal plasticity. Proc. Natl. Acad. Sci. U.S.A. 115 (23), 5902–5907. Available from: https://doi.org/10.1073/pnas.1722609115.

Friedl, P., Gilmour, D., 2009. Collective cell migration in morphogenesis, regeneration and cancer. Nat. Rev. Mol. Cell Biol. 10 (7), 445–457. Available from: https://doi.org/10.1038/nrm2720.

Garcia, S., Hannezo, E., Elgeti, J., Joanny, J.-F., Silberzan, P., Gov, N.S., 2015. Physics of active jamming during collective cellular motion in a monolayer. Proc. Natl. Acad. Sci. U.S.A. 112 (50), 15314–15319. Available from: https://doi.org/10.1073/pnas.1510973112.

Gatenby, R.A., Gillies, R.J., 2008. A microenvironmental model of carcinogenesis. Nat. Rev. Cancer 8 (1), 56–61. Available from: https://doi.org/10.1038/nrc2255.

George, J.T., Jolly, M.K., Xu, S., Somarelli, J.A., Levine, H., 2017. Survival outcomes in cancer patients predicted by a partial EMT gene expression scoring metric. Cancer Res. 77 (22), 6415–6428. Available from: https://doi.org/10.1158/00085472.CAN-16-3521.

Gillies, R.J., Verduzco, D., Gatenby, R.A., 2012. Evolutionary dynamics of carcinogenesis and why targeted therapy does not work. Nat. Rev. Cancer 12 (7), 487–493. Available from: https://doi.org/10.1038/nrc3298.

Grosse-Wilde, A., Fouquier d'Heroüel, A., McIntosh, E., Ertaylan, G., Skupin, A., Kuestner, R.E., et al., 2015. Stemness of the hybrid epithelial/mesenchymal state in breast cancer and its association with poor survival. PLoS One 10 (5), e0126522. Available from: https://doi.org/10.1371/journal.pone.0126522.

Haeger, A., Krause, M., Wolf, K., Friedl, P., 2014. Cell jamming: collective invasion of mesenchymal tumor cells imposed by tissue confinement. Biochim. Biophys. Acta 1840 (8), 2386–2395. Available from: https://doi.org/10.1016/j.bbagen.2014.03.020.

Jolly, M.K., Huang, B., Lu, M., Mani, S.A., Levine, H., Ben-Jacob, E., 2014. Towards elucidating the connection between epithelial-mesenchymal transitions and stemness. J. R. Soc. Interface 11 (101), 20140962. Available from: https://doi.org/10.1098/rsif.2014.0962.

Jolly, M.K., Tripathi, S.C., Jia, D., Mooney, S.M., Celiktas, M., Hanash, S.M., et al., 2016. Stability of the hybrid epithelial/mesenchymal phenotype. Oncotarget 7 (19), 27067–27084. Available from: https://doi.org/10.18632/oncotarget.8166.

Kinnaird, A., Zhao, S., Wellen, K.E., Michelakis, E.D., 2016. Metabolic control of epigenetics in cancer. Nat. Rev. Cancer 16 (11), 694–707. Available from: https://doi.org/10.1038/nrc.2016.82.

Klein, T.J., Glazer, P.M., 2010. The tumor microenvironment and DNA repair. Semin. Radiat. Oncol. 20 (4), 282–287. Available from: https://doi.org/10.1016/j.semradonc.2010.05.006.

Koch, T.M., Münster, S., Bonakdar, N., Butler, J.P., Fabry, B., 2012. 3d traction forces in cancer cell invasion. PLoS One 7 (3), e33476. Available from: https://doi.org/10.1371/journal.pone.0033476.

Kreso, A., O'Brien, C.A., van Galen, P., Gan, O.I., Notta, F., Brown, A.M.K., et al., 2013. Variable clonal repopulation dynamics influence chemotherapy response in colorectal cancer. Science 339 (6119), 543–548. Available from: https://doi.org/10.1126/science.1227670.

Kuilman, T., Peeper, D.S., 2009. Senescence-messaging secretome: SMS-ing cellular stress. Nat. Rev. Cancer 9 (2), 81–94. Available from: https://doi.org/10.1038/nrc2560.

Kuilman, T., Michaloglou, C., Mooi, W.J., Peeper, D.S., 2010. The essence of senescence. Genes. Dev. 24 (22), 2463–2479. Available from: https://doi.org/10.1101/gad.1971610.

La Porta, C.A.M., Zapperi, S., 2018. Explaining the dynamics of tumor aggressiveness: at the crossroads between biology, artificial intelligence and complex systems. Semin. Cancer Biol. 53, 42–47. Available from: https://doi.org/10.1016/j.semcancer.2018.07.003.

La Porta, C.A.M., Zapperi, S., Sethna, J.P., 2012. Senescent cells in growing tumors: population dynamics and cancer stem cells. PLoS Comput. Biol. 8 (1), e1002316. Available from: https://doi.org/10.1371/journal.pcbi.1002316.

Lange, J.R., Fabry, B., 2013. Cell and tissue mechanics in cell migration. Exp. Cell Res. 319 (16), 2418–2423. Available from: https://doi.org/10.1016/j.yexcr.2013.04.023.

Latella, L., Dall'Agnese, A., Boscolo, F.S., Nardoni, C., Cosentino, M., Lahm, A., et al., 2017. DNA damage signaling mediates the functional antagonism between replicative senescence and terminal muscle differentiation. Genes. Dev. 31 (7), 648–659. Available from: https://doi.org/10.1101/gad.293266.116.

Liu, A.J., Nagel, S.R., Langer, J.S., 2010. The jamming transition and the marginally jammed solid. Annu. Rev. Condens. Matter Phys. 1, 347–369. Available from: https://doi.org/10.1146/annurev-conmatphys-070909-104045.

Luoto, K.R., Kumareswaran, R., Bristow, R.G., 2013. Tumor hypoxia as a driving force in genetic instability. Genome Integr. 4 (1), 5. Available from: https://doi.org/10.1186/2041-9414-4-5.

Malinverno, C., Corallino, S., Giavazzi, F., Bergert, M., Li, Q., Leoni, M., et al., 2017. Endocytic reawakening of motility in jammed epithelia. Nat. Mater. 16 (5), 587–596. Available from: https://doi.org/10.1038/nmat4848.

Milanovic, M., Fan, D.N.Y., Belenki, D., Däbritz, J.H.M., Zhao, Z., Yu, Y., et al., 2018. Senescence-associated reprogramming promotes cancer stemness. Nature 553 (7686), 96–100. Available from: https://doi.org/10.1038/nature25167.

Park, J.-A., Kim, J.H., Bi, D., Mitchel, J.A., Qazvini, N.T., Tantisira, K., et al., 2015. Unjamming and cell shape in the asthmatic airway epithelium. Nat. Mater. 14, 1040–1048. Available from: https://doi.org/10.1038/nmat4357.

Ritschka, B., Storer, M., Mas, A., Heinzmann, F., Ortells, M.C., Morton, J.P., et al., 2017. The senescence-associated secretory phenotype induces cellular plasticity and tissue regeneration. Genes. Dev. 31 (2), 172–183. Available from: https://doi.org/10.1101/gad.290635.116.

Sage, J., Miller, A.L., Perez-Mancera, P.A., Wysocki, J.M., Jacks, T., 2003. Acute mutation of retinoblastoma gene function is sufficient for cell cycle re-entry. Nature 424 (6945), 223–228. Available from: https://doi.org/10.1038/nature01764.

Schwitalla, S., Fingerle, A.A., Cammareri, P., Nebelsiek, T., Göktuna, S.I., Ziegler, P.K., et al., 2013. Intestinal tumorigenesis initiated by dedifferentiation and acquisition of stem-cell-like properties. Cell 152 (1–2), 25–38. Available from: https://doi.org/10.1016/j.cell.2012.12.012.

Sellerio, A.L., Ciusani, E., Ben-Moshe, N.B., Coco, S., Piccinini, A., Myers, C.R., et al., 2015. Overshoot during phenotypic switching of cancer cell populations. Sci. Rep. 5, 15464. Available from: https://doi.org/10.1038/srep15464.

Seshadri, T., Campisi, J., 1990. Repression of c-fos transcription and an altered genetic program in senescent human fibroblasts. Science 247 (4939), 205–209. Available from: https://doi.org/10.1126/science.2104680.

Sharma, S.V., Lee, D.Y., Li, B., Quinlan, M.P., Takahashi, F., Maheswaran, S., et al., 2010. A chromatin-mediated reversible drug-tolerant state in cancer cell subpopulations. Cell 141 (1), 69–80. Available from: https://doi.org/10.1016/j.cell.2010.02.027.

Steinway, S.N., Zañudo, J.G., Ding, W., Rountree, C.B., Feith, D.J., Loughran, T.P., et al., 2014. Network modeling of TGF-beta signaling in hepatocellular carcinoma epithelial-tomesenchymal transition reveals joint sonic hedgehog and Wnt pathway activation. Cancer Res. 74 (21), 5963–5977. Available from: https://doi.org/10.1158/0008-5472.CAN-14-0225. arXiv:http://cancerres.aacrjournals.org/content/74/21/5963.full.pdf, <http://cancerres.aacrjournals.org/content/74/21/5963>.

Steinway, S.N., Zañudo, J.G.T., Michel, P.J., Feith, D.J., Loughran, T.P., Albert, R., 2015. Combinatorial interventions inhibit TGFβ-driven epithelial-to-mesenchymal transition and support hybrid cellular phenotypes. NPJ Syst. Biol. Appl. 1, 15014. Available from: https://doi.org/10.1038/npjsba.2015.14.

Tambe, D.T., Hardin, C.C., Angelini, T.E., Rajendran, K., Park, C.Y., Serra-Picamal, X., et al., 2011. Collective cell guidance by cooperative intercellular forces. Nat. Mater. 10 (6), 469–475. Available from: https://doi.org/10.1038/nmat3025.

Waddington, C., 1957. The Strategy of the Genes. Allen & Unwin, London.

Wahl, G.M., Spike, B.T., 2017. Cell state plasticity, stem cells, EMT, and the generation of intra-tumoral heterogeneity. NPJ Breast Cancer 3, 14. Available from: https://doi.org/10.1038/s41523-017-0012-z.

Westphalen, C.B., Asfaha, S., Hayakawa, Y., Takemoto, Y., Lukin, D.J., Nuber, A. H., et al., 2014. Long-lived intestinal tuft cells serve as colon cancer-initiating cells. J. Clin. Invest. 124 (3), 1283–1295. Available from: https://doi.org/10.1172/JCI73434.

Yu, Y., Schleich, K., Yue, B., Ji, S., Lohneis, P., Kemper, K., et al., 2018. Targeting the senescence-overriding cooperative activity of structurally unrelated h3k9 demethylases in melanoma. Cancer Cell 33 (4), 785. Available from: https://doi.org/10.1016/j.ccell.2018.03.009.

Zhu, Y., Tchkonia, T., Pirtskhalava, T., Gower, A.C., Ding, H., Giorgadze, N., et al., 2015. The achilles' heel of senescent cells: from transcriptome to senolytic drugs. Aging Cell 14 (4), 644–658. Available from: https://doi.org/10.1111/acel.12344.

26

Adaptive phenotypic switching in breast cancer in response to matrix deprivation

Ananthalakshmy Sundararaman[1], Saurav Kumar[2] and Annapoorni Rangarajan[2]

[1]*School of Biochemistry, Faculty of Life Sciences, University of Bristol, Bristol, United Kingdom* [2]*Department of Molecular Reproduction, Development and Genetics, Indian Institute of Science, Bangalore, India*

Cancer cells invade surrounding tissues and metastasize to distant sites through the blood or lymphatic vascular systems. In a glandular organ like the breast, initiated cancer cells proliferate into the lumen of the duct or alveoli that pushes them away from the basement membrane, leading to matrix detachment. This is classically recognized as luminal filling upon histological analysis of early stage breast cancer tissues (Pradeep et al., 2012). Further, during the process of metastasis, while in transit within the circulation, cancer cells experience matrix deprivation, which is estimated to be in the order of a few minutes (Labelle and Hynes, 2012). This event of complete matrix deprivation triggers a set of signaling events that cause dramatic phenotype switching. This rapid event is followed by transient adhesion to endothelial cells of the vasculature, and extravasation at the secondary site. The stroma at the secondary site may not be conducive for attachment and therefore cancer cells may face longer durations of matrix deprivation ranging from hours to days (Debnath, 2008). Thus, breast cancer progression involves multiple steps where the cells face a lack of matrix attachment. Survival in a matrix-deprived state requires cancer cells to undergo reversible phenotype switching since ultimately the matrix-detached cancer cells reattach at the secondary site to give rise to metastasis. One of the major events in phenotype switching under detachment is the reprogramming of cellular metabolism. The metabolic plasticity allows cancer cells to fine tune their energy requirements to match the availability of extracellular matrix (ECM)-dependent signals and resources like

Phenotypic Switching. DOI: https://doi.org/10.1016/B978-0-12-817996-3.00025-6

glucose. This adaptive switch not only prevents apoptosis but can also help circumvent nonapoptotic forms of cell death under detachment while simultaneously bringing about energy homeostasis. In this chapter, we will explore the cellular responses to matrix deprivation and the role of phenotype switching in regulating cancer cell plasticity and metastasis.

Heterogeneity and phenotypic plasticity in cancer

Cancer progression is generally associated with increasing heterogeneity. The bulk tumor includes cells with different molecular signatures. This gives rise to different tumor subpopulations, which could be genetically distinct within and across different sites (spatial heterogeneity) (Dagogo-Jack and Shaw, 2018). The same set of cells could also vary their molecular make-up over time (temporal heterogeneity) (Dagogo-Jack and Shaw, 2018). The heterogeneity of tumors of the same histological type between different patients (intertumoral heterogeneity) arises from patient-specific factors like the patient genetic profile and somatic mutation landscape. On the other hand, the heterogeneity within the different tumor cells in a single patient (intratumoral heterogeneity) could arise not only from the genetic variations between cells, but also due to the ability of cancer cells to access diverse phenotypes with the same genetic profile. The role of genomic instability in driving genetic heterogeneity is well appreciated. The ability of the same genotype/clone of cancer cells to display different phenotypes, termed phenotypic plasticity, is less appreciated as a driver of cancer progression. Our ability to sample genetic variations, through single cell sequencing, has developed rapidly, but studying nongenetic heterogeneity is proving to be more challenging, particularly at a single cell resolution. The complexity of regulatory networks that govern the phenotype of cells is immense. It is this very same complex network that allows a single genome within the normal cells in our body to give rise to a multitude of distinct, stable phenotypic states in different cell types during development. We need to bear in mind that the cancer cells can also access many of these diverse phenotypes within their existing genetic make-up (Pisco and Huang, 2015). Therefore, phenotype switching is of great relevance in cancer progression. The different microenvironments that the cancer cells face are strong drivers of phenotype switching and heterogeneity (Meacham and Morrison, 2013). Our focus would be one such phenotype, the metabolic phenotype, and the phenotypic switching in response to matrix deprivation. The phenotype switching in this context gives us insights into the adaptive significance of accessing different metabolic phenotypes during cancer progression in general.

Sensing matrix deprivation

Cell-matrix interactions are critical determinants of cell fate during normal growth and development. Cells sense their position in tissues through integrin receptors that link the cells to the ECM. Attachment to the ECM triggers signaling through downstream nonreceptor tyrosine kinases (RTKs), such as Src, FAK or ILK, that confers survival signals and regulate cell shape and mobility (Nagaprashantha et al., 2011). Cross talk between integrins and growth factor receptors involving RTKs, such as, epidermal growth factor receptor (EGFR), plays a major role in cell fate determination to negotiate proliferation, differentiation, migration, invasion, and apoptosis during normal development and pathological conditions. The lack of attachment-mediated signaling, and the associated cytoskeletal disruption, causes the release of proapoptotic factors from the cytoskeleton, like Bim and BMF (BCl2 modifying factor) (Nagaprashantha et al., 2011). These then tip the balance between pro and anti-apoptotic factors triggering cell death. Cell death triggered by lack of matrix attachment was first observed in epithelial and endothelial cells and termed as "anoikis"—a Greek word meaning "homelessness" (Frisch and Francis, 1994; Frisch and Ruoslahti, 1997). Anoikis maintains normal tissue homeostasis and prevents abnormal migration of normally adherent cells. However, cancer cells must overcome anoikis in order to traverse through the circulation and cause metastasis. They do so through multiple different means: (1) cancer cells harbor cellular alterations that allow them to behave as if they are in the right environment, for example, through mutations in the integrin signaling cascade, (2) hyperactivation of survival signals that counterbalance the proapoptotic signaling, for example, through the overexpression or activating mutations of RTKs (Guadamillas et al., 2011), and (3) entering a dormant state, often through general translational repression (Sequeira et al., 2007; Ng et al., 2012) and the induction of autophagy (Lock and Debnath, 2008; Avivar-Valderas et al., 2011) to tide over the period of detachment until favorable environment is available for attachment.

In addition to integrins, two additional signals allow cancer cells to sense and adapt to matrix detachment. One of them is the increase in the reactive oxygen species (ROS) (Schafer et al., 2009), which could be detrimental or favorable to cancer progression depending on the genetic context. The other signal is that of an increase in intracellular calcium that is encountered transiently by cells undergoing detachment (Sundararaman et al., 2016). Calcium ions are intracellular second messengers that fine tune a variety of cellular processes. Although endoplasmic reticulum is the major reservoir of calcium within the cells, several studies indicate the role of mitochondrial calcium in regulating the metabolic phenotype of cancer cells (Cannino et al., 2018). Mitochondrial calcium can change the amplitude and frequency of cytosolic calcium signals, regulate the redox, and metabolic phenotype of cancer cells, and play a role in cellular response to stress conditions like matrix deprivation. Preventing an elevation of calcium in the cytosol of detached cells improves anchorage-independent sphere formation (Sundararaman et al., 2016).

In both normal and cancer cells, while the sensing of matrix detachment is quite similar, their adaptive responses vary dramatically. While normal cells mount a very unsuccessful adaptive response ending in cell death, cancer cells often improve the degree of response as well as access novel molecular pathways through genetic and nongenetic alterations to survive and thrive under matrix deprivation. It is therefore imperative to understand the plasticity in cancer cells that enables them to access novel molecular phenotypes.

Signaling pathways regulating metabolic plasticity

As we focus on the metabolic plasticity in this chapter, let us briefly consider two key signaling pathways that influence the metabolic phenotype of cells.

The PI3K/Akt-mTOR pathway

The phosphoinositide-3 kinase (PI3K)/protein kinase B/PKB (Akt)-mechanistic target of rapamycin (mTOR) pathway is frequently deregulated in cancer through activating mutations in PI3K or inactivating mutations in PTEN, a suppressor of the pathway (Tan et al., 2015). Activation of the lipid kinase PI3K by extracellular stimuli, such as growth factor signaling, leads to Akt activation. Akt activates the mTOR and coordinates cell growth and metabolism with external cues (Saxton and Sabatini, 2017). The PI3K/Akt-mTOR pathway is largely anabolic, promoting the synthesis of proteins, lipid, and nucleotides, while inhibiting the catabolic process of autophagy, thereby satisfying the growing demands of cancer cells (Lien et al., 2016). This signaling pathway rewires metabolism in cancer cells to overcome the physiological limits on nutrient utilization and to support the requirement for enhanced proliferation in cancer cells.

AMPK pathway

AMP-activated protein kinase (AMPK) is a serine/threonine kinase responsive to cellular AMP/ATP ratio, making it an ideal sensor of cellular bio-energetic and metabolic status. The enzyme is allosterically activated by AMP. AMPK is rapidly activated under a host of stress conditions such as hypoxia, nutrient deprivation, including a loss of attachment to the matrix (Sundararaman et al., 2016). Upon activation, AMPK phosphorylates many target proteins that bring about catabolic reactions to enable cells to increase ATP generation while inhibiting biosynthetic enzymes to decrease ATP consumption. In matrix-deprived cells, AMPK activation in turn inhibits Akt (Saha et al., 2018) and mTOR (Ng et al., 2012), thereby reducing the energy demand. Thus, AMPK reprograms metabolism from anabolism to catabolism (Herzig and Shaw, 2018). Crucially this enzyme protects cells from anoikis by rewiring metabolism (Ng et al., 2012; Saha et al., 2018; Herzig and Shaw, 2018; Jeon et al., 2012). As is evident, AMPK and Akt-mTOR pathways drive the cell to distinct metabolic fates. Recent work from our laboratory revealed that the metabolic plasticity under matrix deprivation is dependent on a switch between these two pathways (Saha et al., 2018).

Role of metabolic plasticity in cancer cell survival under matrix deprivation

Matrix detachment causes bio-energetic stress within cells (Schafer et al., 2009). Metabolic plasticity allows cancer cells to fine tune their energetic requirements. By entering a more quiescent state, detached cancer cells avoid energy crisis and imminent death. Several studies have examined the role of metabolic reprogramming in matrix-deprived cells.

The major challenges faced by cells under anchorage deprivation are (1) a reduction in glucose uptake and consequent glycolytic flux and ATP synthesis, and (2) a robust increase in

ROS. Normal cells often succumb to these challenges by undergoing anoikis or nonapoptotic forms of cell death (Ishikawa et al., 2015; Overholtzer et al., 2007; Hawk et al., 2018; Brown et al., 2018). Cancer cells circumvent these obstacles by utilizing the oncogenic mutations that enable them to have higher glucose uptake and an increased flux through the pentose phosphate pathway (PPP) pathway to produce more reducing equivalents.

Cancer cells utilize the plasticity in metabolic signaling and modulate the flux through major metabolic pathways to adapt to matrix deprivation. Metabolic plasticity is therefore a key adaptive feature in cancer cells that enables them to survive the metastatic cascade. We will explore the key studies delineating the plasticity in metabolism that arises in response to matrix deprivation.

Plasticity in glucose metabolism

Cancer cells can vary the glycolytic rate over a 100-fold (Levine and Puzio-Kuter, 2010). In fact the aberration in glucose metabolism is one of the first identified and most distinctive properties of cancer cells. The term "Warburg effect" refers to the increased uptake of glucose and its nonoxidative metabolism in cancer cells. This addiction of cancer cells to glucose is utilized in positron emission tomography with 18 fluorodeoxyglucose, which preferentially accumulates in tumor cells after its phosphorylation. Cancer cells display considerable plasticity and robustness in their glucose utilization when facing matrix deprivation. Glucose is a versatile precursor that can be turned into energy by complete oxidative phosphorylation while intermediates of the glycolytic pathway can serve as biosynthetic precursors for amino acids and nucleotides. When normal cells detach from substratum, they fail to take up glucose adequately and, consequently, fail to maintain cellular ATP levels (Schafer et al., 2009), leading to cell death. Cancer cells possess alterations that allow them to compensate for this process. They either upregulate glucose transporters to take up more glucose or depend on alternate carbon sources to drive tricarboxylic acid cycle (TCA) cycle to maintain ATP levels. Let us look at specific examples of this plasticity under matrix deprivation.

Cancer cells express increased levels of the glucose transporter GLUT1. If GLUT1 is depleted, it can reduce the survival of mammary tumor cells under matrix-deprived conditions in soft agar (Young et al., 2011). This suggests an increased expression and/or translocation of transporters to the cell surface to compensate for loss of glucose uptake under detachment. An NMR-based study showed Osteopontin-a expression led to increased cellular glucose levels in non-adherent breast cancer cells compared to control cells (Shi et al., 2014a). The same group also demonstrated that Osteopontin-c, an isoform exclusively expressed in cancerous breast cells and not in normal breast cells, promotes anchorage-independent growth by upregulating gene products and metabolites involved in glycolysis, hexose monophosphate shunt and TCA cycle and thus coordinating energy metabolism (Shi et al., 2014b). Thus the expression of cancer-specific splice variants of osteopontin could be a mechanism to compensate for any reduction in glucose uptake and glycolysis caused by detachment. Anchorage-deprived cancer cells could replenish the TCA cycle intermediates through pyruvate uptake. This has been demonstrated in invasive ovarian cancer cells under detached conditions, that compensate for a reduction in glycolysis by increasing pyruvate uptake and mitochondrial ATP generation (Caneba et al., 2012).

Anchorage deprivation rapidly activates AMPK as described previously (Sundararaman et al., 2016). This is achieved preemptively, even before a change in AMP/ATP ratio is detectable (AS and AR, unpublished observations). Recent studies suggest that fructose-1,6 bisphosphate (FBP), produced by the first irreversible step in glycolysis catalyzed by phosphofructokinase, is a key metabolite whose levels are sensed by AMPK. This is achieved through the ability of FBP bound aldolase to act as a sensor of glycolytic flux. In case of glucose deprivation, the aldolase without FBP occupancy triggers AMPK activation (Hardie, 2018; Zhang et al., 2017). From these results, we could postulate that, the reduced glycolytic flux under matrix deprivation might also rapidly activate AMPK through aldolase before the cells experience energetic stress. AMPK activation can then improve mitochondrial biogenesis restoring or even preventing any fall in ATP levels in cancer cells under matrix deprivation.

Hypoxia-induced factor 1α (HIF1α) is a key regulator of glycolytic metabolism in cancer cells. This oxygen-labile protein pairs with HIF1β to drive the transcription of glycolytic enzymes and transporters (Marin-Hernandez et al., 2009). Although classically known to be stabilized under low oxygen tension, cancer cells express increased levels of HIF1α under normoxic conditions downstream of several oncogenic pathways. HIF1α confers anoikis resistance to breast cancer cells downstream of oncogenic ErBB2 (Whelan et al., 2013). Hypoxia per se confers anoikis resistance to cells through the transcriptional activity of HIF1α (Whelan and Reginato, 2011). HIF1α regulates the transcription of 9 out of the 10 glycolytic enzymes (Semenza et al., 1994). Thus stabilization of HIF1α enables cancer cells to overcome defects in glucose uptake under detachment and therefore maintain ATP levels. AMPK and HIF1α are master regulators of oxidative phosphorylation (OXPHOS) and glycolysis respectively. Their balance in activity determines the relative utilization of the two pathways to break down glucose. Anchorage deprivation activates AMPK and increases OXPHOS in cells. Recent studies indicate that cancer cells can utilize hybrid metabolic phenotypes of upregulating both OXPHOS and glycolysis, in addition to switching from one to another in response to stimuli (Jia et al., 2019). We can envisage oncogene driven HIF1α stabilization and matrix deprivation triggered AMPK activation coexisting in cancer cells in suspension. These intermediate phenotypic states are thus easily accessible only to cancer cells and not their normal counterparts. Thus, enhanced plasticity in the utilization of glucose aids cell survival under matrix detachment.

Some of the oncogenic mutations, like the oncogenic Ras, signal through different effectors under attached versus detached conditions (Mason et al., 2016). This study demonstrates that oncogenic Ras supports ATP production preferentially in matrix-deprived cells. This is achieved through the canonical PI3K effector but surprisingly in an Akt independent manner. Therefore, while matrix-attached cells utilize PI3K-Akt pathway, matrix-detached cells with Ras mutation utilize PI3K-SGK1 for ATP production (Mason et al., 2016). The same genetic make-up is leveraged differently under matrix deprivation to contribute to plasticity.

Additional examples include expression of an isoform of pyruvate kinase, PKM2, the enzyme that catalyzes the last rate-limiting step in glycolysis. The switch to PKM2 drives biosynthesis of complex macromolecules by funneling the glycolytic intermediates toward glycerol synthesis and PPP. This protein exists in a dimeric, low catalytically active state, especially when oxidized on its cysteine residues that can happen due to the high ROS conditions under detachment. As discussed later, this allows cancer cells to produce more reducing equivalents through the PPP to survive under matrix deprivation. Thus, PKM2 allows metabolic phenotype switching and plasticity in cancer cells while the normal PKM1 counterpart is constitutively active preventing normal cells from regulating their glycolytic flux efficiently in response to matrix deprivation (Mitchell et al., 2018; Dong et al., 2016). Indeed, the plasticity in the metabolite flow through branches of glycolytic pathway is a characteristic adaptive feature in cancer cells that allow them to divide and proliferate under matrix deprivation.

Yet another classic example of plasticity in metabolic flux in cancer cells is the regulation of pyruvate entry into the mitochondria. Untransformed mammary epithelial cells upregulate pyruvate dehydrogenase (PDH) kinase (PDK1 and/or PDK4) under detachment, which phosphorylates and inactivates PDH (Grassian et al., 2011). This reduces pyruvate entry into mitochondria and ROS generation. However, for normal cells this proves to be too little, too late to prevent cell death in suspension. In contrast, cancer cells express elevated levels of PDK4 conferring them with a much greater ability to reduce pyruvate entry specifically under detached conditions (Kamarajugadda et al., 2012). In support of this, it was observed that depleting PDK in cancer cells sensitized them to anoikis and decreased their metastatic potential (Kamarajugadda et al., 2012). Metabolic switching to control the utilization of glucose at various steps, starting from its uptake, to diversion through various branch pathways of glycolysis, through to controlling mitochondrial entry of pyruvate and dependence on OXPHOS provide cancer cells with a survival advantage that normal cells do not have under matrix deprivation.

Plasticity in redox signaling

Increase in ROS is one of the key challenges and the biggest threats to cell survival under detachment. Indeed, the survival of even normal cells is improved by antioxidants under detachment (Schafer et al., 2009). Cancer cells have developed immense plasticity in handling ROS and some cancer types even use the increase in ROS to aid aggressive growth and metastasis. Let us look at how cancer cells handle ROS increase under detachment. Work from our group (Sundararaman et al., 2016) and others (Li et al., 1999) (reviewed in (Hawk et al., 2018)) suggest that the ROS increase during detachment is mainly from the mitochondria. The increased ROS levels signal to AMPK through the upstream kinase LKB1 to trigger metabolic adaptation (Cannino et al., 2018). The outcome of ROS increase (cell death vs survival) depends on the ability of cells to scavenge the ROS levels in a timely manner through the antioxidant defense system. Cancer cells employ multiple mechanisms to counter the increase in ROS levels under detachment including (1) decreased ROS production, (2) increased antioxidant enzyme production, and (3) increased production of reducing equivalents through PPP pathway (Hawk et al., 2018).

Metabolic adaptation to increased ROS levels under detachment is critical for cancer cell survival and metastasis. Many studies have investigated the effects of antioxidants in metastasis, which requires cells to evade anoikis. Studies show that the use of antioxidants increase metastasis in a lung cancer mouse model (Sayin et al., 2014) and in malignant melanoma model (Le Gal et al., 2015). This is consistent with oxidative stress leading to inhibition of metastatic capacity of cells (Hawk et al., 2016). Further, small molecule inhibitors that counter the ability of cancer cells to handle increased ROS levels show significant antitumor activity in mouse models (Raj et al., 2011).

Work from the Schafer group indicates that antioxidant enzymes, catalase and superoxide dismutase (SOD), are utilized by breast cancer cells to quench ROS under detachment. Catalase expression additionally triggers AMPK signaling by increasing the total AMPK levels (Davison et al., 2013). AMPK activation further promotes the restoration of ATP synthesis in anchorage-deprived cells thereby aiding in their survival. Intriguingly, the inhibition of these antioxidant enzymes does not compromise survival in attached cells, suggesting that antioxidant defenses of the cells become much more significant in the context of detachment. This is reflective in metastatic efficiency of cancer cells as well, wherein the inhibition of catalase was found to reduce tumor burden in lungs of mice after tail vein injection (Davison et al., 2013).

Cancer cells also exhibit an increase in MnSOD (SOD2) that counters mitochondrial superoxide production by converting it into H_2O_2 that can then be reduced to water by catalase. The induction of this enzyme, under detachment, is dependent on the transcription factor NFκB (Kamarajugadda et al., 2013). Interestingly, in attached cells, MnSOD mediated conversion of O_2^- to H_2O_2 could directly activate AMPK and promote glycolysis (Hart et al., 2015). The MnSOD-AMPK axis is strongly correlated to tumor progression (Hart et al., 2015). This work agrees with our observations that ROS can activate AMPK upon detachment (Sundararaman et al., 2016). Taken together, ROS levels spike under detachment and the antioxidant enzyme MnSOD plays a crucial role in generating H_2O_2 that signals from the mitochondria to AMPK. AMPK in turn can mount an effective redox response through the transcription factor Nrf2 that transcriptionally upregulates many antioxidant enzymes (DeNicola et al., 2011).

Studies with the mammary acinar model have delineated the importance of PPP in restoring redox balance in matrix-deprived cells (Schafer et al., 2009). PPP-derived nicotinamide adenine dinucleotide phosphate (NADPH) serves as a significant contributor to a cell's antioxidant capacity (Hawk et al., 2018). Among the pathways that help cancer cells adapt to high ROS levels under matrix deprivation, AMPK signaling stands out in both being able to sense ROS increase and being able to respond to it by coordinating an increase in the redox capacity of the cell. AMPK orchestrates NADPH production in cells through the inhibition of ACC1 and ACC2 (Jeon et al., 2012). Through these enzymes AMPK conserves NADPH that would otherwise be utilized in fatty acid synthesis and makes more NADPH available through fatty acid oxidation. Indeed, this axis was shown to be important in solid tumor formation in vivo (Overholtzer et al., 2007).

Several studies have delineated ways in which increased ROS in cancer cells allows them to have a survival advantage mainly in attached conditions. However, few studies also show an advantage to increasing ROS even in detached conditions. In prostate cancer cells, the persistent increase in ROS under detached conditions keeps Src in an oxidized and active state. This maintains EGFR phosphorylation in a ligand independent manner conferring anoikis-resistance through the degradation of the proapoptotic protein Bim. Antioxidants are shown to cause anoikis sensitivity in this case (Giannoni et al., 2009). Another example in which increased ROS levels aid cancer progression even in detachment is seen in K-Ras transformed cancers. These cells show increased conversion of glutamine to αKG to increase TCA flux and ROS production (Weinberg et al., 2010). ROS from the mitochondria activates ERK signaling to promote anchorage-independent growth. Mitochondrial function inhibition (using TFAM knockout mouse model) compromises tumor growth in oncogenic K-Ras driven mouse model of lung cancer (Weinberg et al., 2010).

These studies are in contradiction to the earlier studies showing that mitochondrial ROS is detrimental to anchorage-independent growth (Jiang et al., 2016; Rohwer et al., 2008). The contradiction can be explained by the fact that metabolic plasticity in cancers is linked to their genotype and certain mutations cause a larger range of possible phenotypes that the cancer cells can display. In these cases, the increased ROS is no longer a threat but an ally to drive proliferation. This depends on the inherent antioxidant capacity that would allow cells to tolerate higher amounts of ROS if it can be quenched soon after the ROS mediated signaling is completed. Ras transformed cells show a higher antioxidant capacity (Kimmelman, 2015) and therefore can funnel the mitochondrial ROS for anchorage-independent growth and survival. Cells utilize mitochondrial ROS signaling to activate AMPK among other responses. AMPK can in turn regulate the levels of cytosolic and mitochondrial ROS to levels that sustain survival and metastasis (Jeon et al., 2012; Rabinovitch et al., 2017). ROS signaling to metabolic effectors like AMPK and HIF-1α could be specifically targeted to prevent cancer progression.

Plasticity in lipid metabolism

Matrix-deprived cells undergo a switch from fatty acid synthesis (FAS) to a predominantly fatty acid oxidizing (FAO) phenotype. This is critical as fatty acid synthesis requires ATP and reducing equivalents (NADH and FADH2), and as discussed before, these are in short supply under matrix detachment. Cells funnel reducing equivalents toward countering the increased ROS levels under detachment. To do so they undergo fatty acid oxidation (β oxidation) that generates ATP and reducing equivalents by shutting down the key regulatory enzyme Acetyl CoA carboxylase (ACC1 and ACC2). ACC is a bonafide substrate of AMPK and this switch from FAS to FAO is orchestrated by AMPK signaling. ACC1 is a cytosolic protein catalyzing the rate-limiting step of malonyl CoA synthesis from acetyl CoA for fatty acid synthesis. ACC2 on the other hand is associated with mitochondrial membranes and allosterically inhibits Carnitin palmitoyl transferase CPT1, the enzyme responsible for import of long chain fatty acids into the mitochondria. By phosphorylation and inhibition of ACC1 and ACC2, AMPK inhibits fatty acid synthesis while simultaneously promoting entry of fatty acids into mitochondria for oxidation (Fullerton et al., 2013).

Under attached, growth factor replete conditions, the PI3K/Akt-mTORC1 pathways keeps β oxidation under check and promotes lipogenesis through increasing the transcriptional expression of several proteins via sterol regulatory element-binding proteins (Lamming and Sabatini, 2013). The switch to fatty acid oxidation in cancer cells is therefore an adaptive response to matrix deprivation. Transcriptomics data on matrix-deprived MDA-MB-231 breast cancer cells from our laboratory also supports elevated FAO in suspension (Saha et al., 2018). In non-transformed MCF10A cells, reduced FAO under detached conditions triggers anoikis due to increased ROS levels (Schafer et al., 2009). Indeed, promoting fatty acid oxidation through the upregulation of CPT1A in suspension confers anoikis resistance in colon cancer cells (Wang et al., 2018). Taken together, the adaptive switch in lipid metabolism toward catabolism allows cells to produce ATP and reducing equivalents that can enable survival under matrix deprivation.

Plasticity in amino acid metabolism

In order to understand the rewiring of amino acid metabolism under detachment we can take glutamine metabolism as a well-studied case in point. Cancer cells are known to be highly reliant on glutamine for their increased proliferation as glutamine serves as a precursor for macromolecules. Glutamine is an important carbon and nitrogen source for cancer cells. Glutamine gets converted to α-ketoglutarate (αKG) in 2 steps catalyzed by glutaminase (that converts glutamine to glutamate) and glutamate dehydrogenase (GDH) (that converts glutamate to αKG), respectively. This therefore helps replenish TCA cycle intermediates. Glutaminase inhibition hinders anchorage-independent growth of cancer cell lines HepG2 and A549 (Lee et al., 2014). The second step enzyme, GDH, is transcriptionally upregulated under matrix deprivation in lung cancer cells (Jin et al., 2018). The GDH1-derived αKG enhances CaMKK2-mediated AMPK activation and is therefore especially critical in cancers lacking the AMPK upstream kinase LKB1 (Jin et al., 2018). AMPK, as discussed previously, rewires metabolism to aid in anchorage-independent survival. Thus, the increase in glutaminolysis is an adaptive metabolic switch tailored to survive matrix deprivation in cancers.

The glutamine-derived αKG has two fates, the first fate, progressing along the TCA cycle, is the generation of succinate catalyzed by αketoglutarate dehydrogenase. As the name suggests, it is oxidative in nature and generates NADH. The second is reductive carboxylation, essentially going backward on the TCA cycle. This uses reducing equivalents from NAD(P)H to generate citrate catalyzed by mitochondrial IDH2 (isocitrate dehydrogenase). This citrate can then generate oxaloacetate and acetyl CoA to provide for lipid synthesis in attached cells (Mullen et al., 2014). Either way, glutamine replenishes carbon in the TCA cycle, a term called anaplerosis. A third fate of αKG, quite characteristic of matrix-deprived cells, is its transport to cytosol and conversion to citrate by cytosolic IDH1. This citrate reenters mitochondria and gives rise to αKG catalyzed by mitochondrial IDH2 (note the same reaction in reverse catalyzed by the same isozymes) to generate mitochondrial NAD(P)H and alleviating mitochondrial ROS. Therefore, in this case the use of cytosolic IDH1 essentially takes reducing equivalents from the cytosol to regenerate them in the mitochondria to quench ROS at the site of production. Blocking citrate entry into mitochondria or IDH1 or IDH2 reduces spheroid sizes under matrix deprivation without altering growth of cancer cells in monolayer (Jiang et al., 2016). This provides a good example of plasticity in metabolic flux through different branches (starting from αKG) that enables cancer cells to adapt to matrix deprivation.

Role of autophagy in the adaptive phenotypic switch

Autophagy is a tightly regulated process that allows for orderly degradation and recycling of cellular components. In most contexts, autophagy promotes tumorigenesis in response to microenvironmental stresses (White, 2015). Long-term metabolic adaptation involves a switch to autophagy, which allows the cells to supplement their nutritional resources through degradation of endogenous proteins to provide a pool of essential amino acids. Autophagy was initially identified as playing a role in the context of lumen clearing during development. Using mammary acinar model in three-dimensional culture, it was demonstrated that autophagy is selectively triggered in the cells at the center lacking matrix attachment (Fung et al., 2008). In cancer cells, upon detachment, PERK signals through LKB1 to increase ATG6 and ATG8 to switch on autophagy. Autophagy is at the cross roads of signaling downstream of matrix-attached and -detached conditions. It is maintained at basal levels under attached conditions. The PI3K/Akt/mTOR signaling acts as a gatekeeper for autophagy, inhibiting it under matrix-attached conditions. Matrix deprivation promptly activates autophagy by turning off mTOR signaling through its negative regulator AMPK (Ng et al., 2012). The switch in signaling from PI3K-Akt-mTOR axis to AMPK under detachment serves as a major metabolic switch that turns the primarily anabolic nature of cancer cell metabolism to a catabolic one. Autophagy is also tightly regulated by inputs from these key kinases. Several feed forward and feedback loops from AMPK and Akt/mTOR kinases to various components of the autophagy pathway ensures that the cellular response is robust. For example, the autophagy initiating kinase ULK1 (ATG1) is a direct target of both AMPK and mTOR. While AMPK directly phosphorylates and activates ULK1 on Ser 317 and Ser777, mTOR inhibits the ULK1-AMPK interaction by phosphorylating it at Ser 757(Kim et al., 2011). Another example of a protein simultaneously targeted by these two pathways is

TSC2. Active Akt phosphorylates TSC2 at S939 position and inhibits its activity while AMPK phosphorylates and activates it (Inoki et al., 2003). Similarly, AMPK phosphorylates Beclin1 to increase autophagy while Akt phosphorylates a different residue on Beclin1 to inhibit it (Zhang et al., 2016; Wang et al., 2012). Therefore, through several independent means, AMPK and Akt signals ensure that the autophagic response is a near all or none behavior in each context. Indeed, autophagy is required for the survival of dormant disseminated breast cancer cells (Vera-Ramirez et al., 2018). Therefore, upregulating autophagy is an important metabolic phenotype switch under matrix deprivation.

Bistable systems and plasticity

Bistability is the property by which a dynamic system exists in one of the two stable states depending on the external condition. One example is the differentiation of human mesenchymal progenitors into myeloid or osteogenic lineage (Wang et al., 2009). Often times, double negative feedback loops regulate the switch between the two states. For example, the Wnt–β-catenin signaling and HNF4α operative in a double negative feedback loop to regulate epithelial-to-mesenchymal transition (Yang et al., 2013; Venta et al., 2012). Each state is then reinforced with positive feedback loops. The phosphorylation-dephosphorylation cycle of CyclinB-CDK1 mediates regulation of mitotic entry, which constitutes another reversible bistable system that allows the cell to execute an all or none response. For metabolic regulation in matrix-deprived cells, a novel AMPK-Akt double negative feedback achieves the rapid switch between the catabolic and anabolic phenotypes (Saha et al., 2018).

AMPK-Akt double negative feedback loop confers bistability to the metabolic phenotype

Cancer cells primarily occupy two distinct metabolic phenotypes based on the availability of matrix attachment. We recently identified a nongenetic mechanism involving the AMPK-Akt double negative feedback loop between attached and detached cells that reversibly regulates this metabolic plasticity (Saha et al., 2018). Let us look at the dichotomy in the metabolic phenotype driven by these two enzymes in detail.

AMPK is a cellular energy sensor activated by an increase in AMP/ATP ratio. The first link between AMPK activation and matrix deprivation was made in 2008 (Fung et al., 2008), wherein AMPK dependent autophagy was shown to promote cell survival. A subsequent study placed PERK upstream of the LKB1-AMPK pathway in sensing matrix deprivation and triggering autophagy (Avivar-Valderas et al., 2011). The role of AMPK was further expanded under matrix deprivation from promoting autophagy to inhibiting protein synthesis by mTOR inhibition (Ng et al., 2012) and mediating redox homeostasis through NADPH synthesis (Jeon et al., 2012). Data from our laboratory revealed that AMPK activation directly prevents the initiation of apoptosis by phosphorylation of PEA15 (Hindupur et al., 2014). While one report showed that matrix detachment for 24 h leads to a decrease in ATP/ADP ratio in MCF10A cells (Schafer et al., 2009), and another report showed the same at 16 h for A549 cells (Jeon et al., 2012) that could trigger AMPK, work from our laboratory showed that even before a change in AMP/ATP ratio sets in, AMPK can be activated by an increase in calcium and ROS flux in detached cells (Sundararaman et al., 2016). Several studies have confirmed that AMPK activation switches off anabolic pathways and turns on catabolic pathways (Hardie, 2018). Recent investigations from our group revealed that matrix deprivation acts as a flip switch between two serine-threonine kinases AMPK and Akt to regulate the cellular metabolism to make it conducive for cancer cells to survive under the stress of matrix deprivation, at the expense of proliferation. We showed that matrix detachment-triggered AMPK mediates an increase in the phosphatase PHLPP2 that inactivates Akt under detached conditions, while Akt activation triggered by adhesion in turn facilitated PP2Cα mediated AMPK inactivation under attachment. This double negative feedback loop between these two kinases mediated by their respective phosphatases provides bistability to the metabolic phenotype of cancer cells (Saha et al., 2018). Further evidence for the double negative loop comes from the analysis of gene expression patterns that closely resemble an Akt signature under attached conditions and in primary and metastatic lesions but switched to an AMPK dependent expression pattern in circulating tumor cells (CTCs) (Saha et al., 2018). When Akt is active, as is often the case in multiple cancers, there is metabolic reprogramming toward anabolic processes. Cancer cells need to overcome the physiological barriers to nutrient utilization to drive the enhanced growth and proliferation. The context dependent utilization of AMPK and Akt to switch their metabolic phenotype contributes to cancer progression. When activated out of context, under attachment, AMPK can enforce a metabolic checkpoint on cell division causing cell cycle arrest (Jones et al., 2005). Similarly, forced

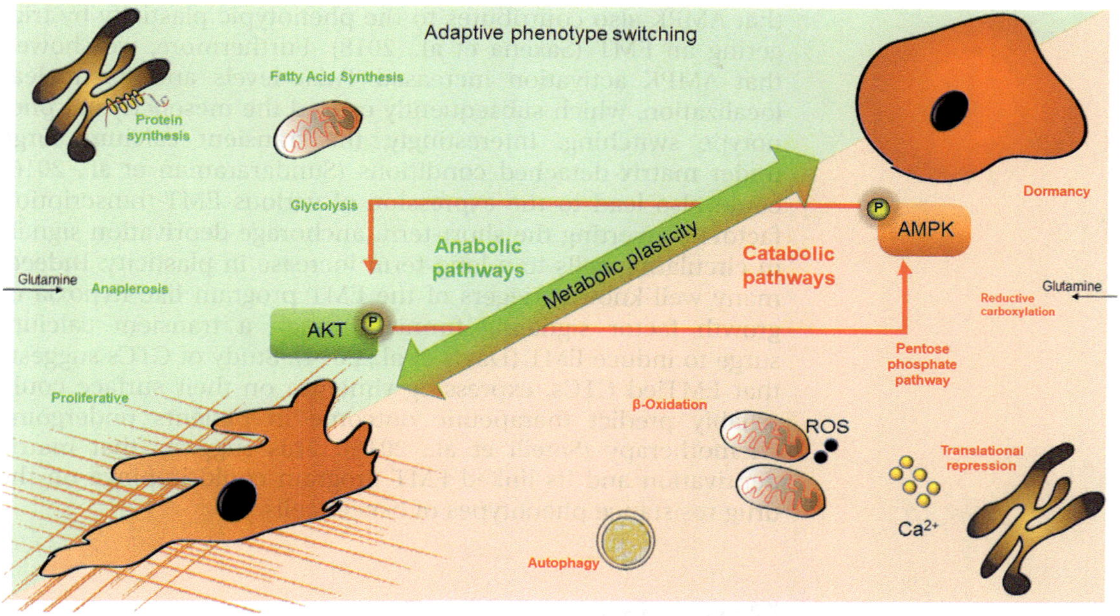

Figure 26.1 Cancer cells exploit different metabolic phenotypes to thrive in matrix attached versus matrix deprived states.

activation of Akt under detached conditions also triggers anoikis (Saha et al., 2018). This suggests that the bistability in the system is critical for phenotypic plasticity in cancer cells. The metabolic plasticity in cancers can be targeted by identifying metabolic pathways significantly upregulated under detached conditions that are dispensable for normal cell survival and tissue homeostasis (Fig. 26.1). In addition to the reversible signaling and metabolic plasticity, cancer cells also display reversible activation of the epithelial-to-mesenchymal transition (EMT) and stemness programs.

Matrix deprivation and EMT

The process of EMT, which confers migratory and invasive properties, is also associated with anoikis resistance. In fact, the molecular pathways that enable EMT are also involved in conferring survival properties to matrix-deprived cells. For example, twist is a master regulator of EMT (Yang et al., 2004), and twist overexpression also enables anoikis resistance (Smit et al., 2009). Likewise, in addition to regulating the metabolic plasticity of matrix-deprived cells, work from our laboratory revealed that AMPK also contributes to the phenotypic plasticity by triggering an EMT (Saxena et al., 2018). Furthermore, we showed that AMPK activation increased Twist levels and its nuclear localization, which subsequently caused the mesenchymal phenotype switching. Interestingly, the transient calcium surge under matrix-detached conditions (Sundararaman et al., 2016) could also lead to the expression of various EMT-transcription factors converting the short-term anchorage deprivation signals in circulating cells to a long-term increase in plasticity. Indeed, many well-known triggers of the EMT program like hypoxia or growth factor signaling funnel through a transient calcium surge to induce EMT (Davis et al., 2014). Study of CTCs suggests that EMTied CTCs, expressing vimentin on their surface could reliably predict therapeutic outcome in patients undergoing chemotherapy (Satelli et al., 2015). This suggests that matrix deprivation and its linked EMT program could impinge on the drug resistance phenotypes of cancer cells.

Matrix deprivation and stemness

Matrix deprivation expands the limited plasticity of normal mammary epithelium

The mammary gland is a highly branched structure composed of a double layer of cells; the outer myoepithelial cells encase the inner luminal epithelial cells. The mammary gland is unique in its ability to undergo remodeling postnatally during puberty, pregnancy, lactation and weaning. This suggests a hierarchical organization of cells into stem, progenitor, and differentiated pools in adult mammary gland. Initial studies to identify these stem cells relied on transplanting cells, isolated based on specific surface markers, to regenerate the entire mammary glandular structure. This suggested various markers could distinguish multipotent stem cells within the adult mammary gland. However, recent in situ lineage tracing studies reveal that multipotent stem cells exist only in embryonic mammary placode, and the postnatal mammary gland contains only unipotent progenitors (Seldin et al., 2017). If so, how could we isolate cells and regenerate the entire mammary tree? This conundrum is resolved by recognizing that there is an expansion of the plasticity with regard to cell fate in the in vitro cultured cells. Thus, the removal of cells from their microenvironment and matrix deprivation in the process of cell isolation contributed to expansion of plasticity. Indeed, long-term culture of cells in anchorage-deprived states, in the form of mammospheres, has been used as a tool to characterize normal breast stem/progenitor cells (Dey et al., 2009). This ability to show plasticity across lineages by initially unipotent progenitors in ex vivo culture reflects the properties hijacked and expanded on by breast cancers. These historic developments in the stem-cell field are a strong indicator that matrix deprivation can cause an expansion in plasticity in cancer cells during metastasis.

Matrix deprivation enhances stemness properties in cancer cells

The presence or absence of a cellular hierarchy is one of the many factors contributing to phenotypic and functional heterogeneity in cancers. The cancer stem-cell (CSC) model postulates that a hierarchy among cancer cells exists wherein only a small subset of cells is capable of sustaining tumorigenesis (Visvader and Lindeman, 2012). Detachment of cancer cells from the ECM, during the course of metastasis, causes an enrichment of cells with stem-cell-like properties (Charpentier and Martin, 2013). Several factors contribute to the gain in stem-like properties in matrix-deprived condition. Mitochondrial biogenesis through ERRα-PGC1 signaling pathway governs stemness and anoikis resistance in MCF7 cells (De Luca et al., 2015). We have seen before that autophagy is characteristically upregulated under matrix deprivation. Autophagy is also reported to promote $CD44^{+}CD24^{-/low}$ stem-like progeny in breast cancer cells (Cufí et al., 2011). Another report supports a role for autophagy protein Beclin1 in the formation of mammospheres and CSC maintenance under detachment (Gong et al., 2013). Salinomycin causes a reduction in the percentage of $CD44^{+}/CD24^{-}$ stem-like cells and sensitizes MDA-MB-231 cells to anoikis suggesting that anoikis resistance and cancer stemness properties go hand in hand (An et al., 2015). Gaining stemness properties is an adaptive response of cancer cells to matrix deprivation.

EMT and stemness are associated with metabolic reprogramming in cancer

The induction of EMT can promote metabolic plasticity in the cancer cells (Ye and Weinberg, 2015). The EMT-transcription factor Snail (SNAI1) is reported to regulate glucose flux toward the PPP. Mechanistically, Snail represses phosphofructokinase that is involved in the regulation of the first rate-limiting step of glycolysis. Knockdown of Snail shows a reduction in the number of anchorage-independent colony formation by breast cancer cell lines (Kim et al., 2017). Another EMT-transcription factor Twist1 promotes upregulation of dihydropyrimidine dehydrogenase, enzyme involved in pyrimidine catabolism that promotes EMT (Shaul et al., 2014). Work done by Kondaveeti et al. have correlated CSC properties with the EMT by growing HER2-positive BT-474 breast cancer cells as anchorage-independent mammosphere cultures (Kondaveeti et al., 2015). They have observed that cells with $CD44^{+/high}/24^{low}$ marker display mesenchymal feature and have high glucose uptake and higher lactate production rate (Kondaveeti et al., 2015). Thus, EMT and metabolic plasticity have an important role in survival of cells in matrix-deprived condition. In addition, promotion of glucose metabolism can induce EMT. For example, phosphoglucose isomerase, converts glucose-6P to fructose 6-P and is secreted by cancer cells. This acts as a cytokine to induce NF-kB-dependent stabilization of Zeb1 and Zeb2 in breast cancer cells triggering EMT (Sciacovelli and Frezza, 2017). Increased glycolytic activity can promote partial EMT (Gaude and Frezza, 2016) leading to transition in E/M hybrid state (Jia et al., 2018). The E/M hybrid cells are more likely to metastasize and survive in matrix-deprived condition (Aceto et al., 2014). The metabolic phenotype of cells directly feed onto other forms of phenotypic plasticity across the epithelial-to-mesenchymal spectrum and the cell fate hierarchy and together represent the phenotypic landscape of anchorage-deprived cancer cells. While we understand these phenotypic states individually, their propensity for transition during metastasis, especially when faced with matrix deprivation is not understood well.

Future directions

While we have addressed phenotypic plasticity with respect to metabolism primarily in this chapter, our understanding of the global landscape of phenotypes favored during cancer progression is miniscule. Cancer heterogeneity is regulated by the other cell types in the microenvironment, mechanochemical inputs, oxygen, and nutrient availability, systemic factors like circulating estrogen and progesterone in breast cancer, chemotherapy, and others. Their own intrinsic genetic mutations, the cell type of origin, extent of dedifferentiation, epigenetic alterations, and various other stochastic state transitions make the study of cancer cell phenotypic plasticity a challenging task (Gupta et al., 2011; Gupta et al., 2019). Cancer cells are often in a state of phenotypic equilibrium between different subpopulations for a given microenvironmental condition (Gupta et al., 2011). Understanding these most accessed phenotypes and their potential to switch will guide future therapeutic interventions to prevent cancer related deaths.

References

Aceto, N., et al., 2014. Circulating tumor cell clusters are oligoclonal precursors of breast cancer metastasis. Cell 158, 1110–1122.

An, H., et al., 2015. Salinomycin promotes anoikis and decreases the CD44 + / CD24-stem-like population via inhibition of STAT3 activation in MDA-MB-231 cells. PLoS One 10, e0141919.

Avivar-Valderas, A., et al., 2011. PERK integrates autophagy and oxidative stress responses to promote survival during extracellular matrix detachment. Mol. Cell Biol. 31, 3616–3629. Available from: https://doi.org/10.1128/MCB.05164-11.

Brown, C.W., Amante, J.J., Mercurio, A.M., 2018. Cell clustering mediated by the adhesion protein PVRL4 is necessary for alpha6beta4 integrin-promoted ferroptosis resistance in matrix-detached cells. J. Biol. Chem. 293, 12741–12748. Available from: https://doi.org/10.1074/jbc.RA118.003017.

Caneba, C.A., Bellance, N., Yang, L., Pabst, L., Nagrath, D., 2012. Pyruvate uptake is increased in highly invasive ovarian cancer cells under anoikis conditions for anaplerosis, mitochondrial function, and migration. Am. J. Physiol. Endocrinol. Metab. 303, E1036–E1052. Available from: https://doi.org/10.1152/ajpendo.00151.2012.

Cannino, G., Ciscato, F., Masgras, I., Sanchez-Martin, C., Rasola, A., 2018. Metabolic Plasticity of Tumor Cell Mitochondria. Front. Oncol. 8, 333. Available from: https://doi.org/10.3389/fonc.2018.00333.

Charpentier, M., Martin, S., 2013. Interplay of stem cell characteristics, EMT, and microtentacles in circulating breast tumor cells. Cancers 5, 1545–1565.

Cufí, S., et al., 2011. Autophagy positively regulates the CD44 + CD24-/low breast cancer stem-like phenotype. Cell cycle 10, 3871–3885.

Dagogo-Jack, I., Shaw, A.T., 2018. Tumour heterogeneity and resistance to cancer therapies. Nat. Rev. Clin. Oncol. 15, 81–94. Available from: https://doi.org/10.1038/nrclinonc.2017.166.

Davis, F.M., et al., 2014. Induction of epithelial-mesenchymal transition (EMT) in breast cancer cells is calcium signal dependent. Oncogene 33, 2307–2316. Available from: https://doi.org/10.1038/onc.2013.187.

Davison, C.A., et al., 2013. Antioxidant enzymes mediate survival of breast cancer cells deprived of extracellular matrix. Cancer Res. 73, 3704–3715. Available from: https://doi.org/10.1158/0008-5472.CAN-12-2482.

De Luca, A., et al., 2015. Mitochondrial biogenesis is required for the anchorage-independent survival and propagation of stem-like cancer cells. Oncotarget 6, 14777.

Debnath, J., 2008. Detachment-induced autophagy during anoikis and lumen formation in epithelial acini. Autophagy 4, 351–353. Available from: https://doi.org/10.4161/auto.5523.

DeNicola, G.M., et al., 2011. Oncogene-induced Nrf2 transcription promotes ROS detoxification and tumorigenesis. Nature 475, 106–109. Available from: https://doi.org/10.1038/nature10189.

Dey, D., et al., 2009. Phenotypic and functional characterization of human mammary stem/progenitor cells in long term culture. PLoS One 4, e5329. Available from: https://doi.org/10.1371/journal.pone.0005329.

Dong, G., et al., 2016. PKM2 and cancer: the function of PKM2 beyond glycolysis. Oncol. Lett. 11, 1980–1986. Available from: https://doi.org/10.3892/ol.2016.4168.

Frisch, S.M., Francis, H., 1994. Disruption of epithelial cell-matrix interactions induces apoptosis. J. Cell Biol. 124, 619–626. Available from: https://doi.org/10.1083/jcb.124.4.619.

Frisch, S.M., Ruoslahti, E., 1997. Integrins and anoikis. Curr. Opin. Cell Biol. 9, 701–706. Available from: https://doi.org/10.1016/s0955-0674(97)80124-x.

Fullerton, M.D., et al., 2013. Single phosphorylation sites in Acc1 and Acc2 regulate lipid homeostasis and the insulin-sensitizing effects of metformin. Nat. Med. 19, 1649–1654. Available from: https://doi.org/10.1038/nm.3372.

Fung, C., Lock, R., Gao, S., Salas, E., Debnath, J., 2008. Induction of autophagy during extracellular matrix detachment promotes cell survival. Mol. Biol. Cell 19, 797–806.

Gaude, E., Frezza, C., 2016. Tissue-specific and convergent metabolic transformation of cancer correlates with metastatic potential and patient survival. Nat. Commun. 7, 13041.

Giannoni, E., Fiaschi, T., Ramponi, G., Chiarugi, P., 2009. Redox regulation of anoikis resistance of metastatic prostate cancer cells: key role for Src and EGFR-mediated pro-survival signals. Oncogene 28, 2074–2086. Available from: https://doi.org/10.1038/onc.2009.77.

Gong, C., et al., 2013. Beclin 1 and autophagy are required for the tumorigenicity of breast cancer stem-like/progenitor cells. Oncogene 32, 2261.

Grassian, A.R., Schafer, Z.T., Brugge, J.S., 2011. ErbB2 stabilizes epidermal growth factor receptor (EGFR) expression via Erk and Sprouty2 in extracellular matrix-detached cells. J. Biol. Chem. 286, 79–90.

Guadamillas, M.C., Cerezo, A., Del Pozo, M.A., 2011. Overcoming anoikis--pathways to anchorage-independent growth in cancer. J. Cell Sci. 124, 3189–3197. Available from: https://doi.org/10.1242/jcs.072165.

Gupta, P.B., et al., 2011. Stochastic state transitions give rise to phenotypic equilibrium in populations of cancer cells. Cell 146, 633–644. Available from: https://doi.org/10.1016/j.cell.2011.07.026.

Gupta, P.B., Pastushenko, I., Skibinski, A., Blanpain, C., Kuperwasser, C., 2019. Phenotypic plasticity: driver of cancer initiation, progression, and therapy resistance. Cell Stem Cell 24, 65–78. Available from: https://doi.org/10.1016/j.stem.2018.11.011.

Hardie, D.G., 2018. Keeping the home fires burning: AMP-activated protein kinase. J. R. Soc. Interface 15, . Available from: https://doi.org/10.1098/rsif.2017.0774. pii: 20170774.

Hart, P.C., et al., 2015. MnSOD upregulation sustains the Warburg effect via mitochondrial ROS and AMPK-dependent signalling in cancer. Nat. Commun. 6, 6053. Available from: https://doi.org/10.1038/ncomms7053.

Hawk, M.A., McCallister, C., Schafer, Z.T., 2016. Antioxidant activity during tumor progression: a necessity for the survival of cancer cells? Cancers (Basel) 8, 92. Available from: https://doi.org/10.3390/cancers8100092.

Hawk, M.A., et al., 2018. RIPK1-mediated induction of mitophagy compromises the viability of extracellular-matrix-detached cells. Nat. Cell Biol. 20, 272–284. Available from: https://doi.org/10.1038/s41556-018-0034-2.

Herzig, S., Shaw, R.J., 2018. AMPK: guardian of metabolism and mitochondrial homeostasis. Nat. Rev. Mol. Cell Biol. 19, 121–135. Available from: https://doi.org/10.1038/nrm.2017.95.

Hindupur, S.K., et al., 2014. Identification of a novel AMPK-PEA15 axis in the anoikis-resistant growth of mammary cells. Breast Cancer Res. 16, 420. Available from: https://doi.org/10.1186/s13058-014-0420-z.

Inoki, K., Zhu, T., Guan, K.L., 2003. TSC2 mediates cellular energy response to control cell growth and survival. Cell 115, 577–590.

Ishikawa, F., Ushida, K., Mori, K., Shibanuma, M., 2015. Loss of anchorage primarily induces non-apoptotic cell death in a human mammary epithelial cell line under atypical focal adhesion kinase signaling. Cell Death Dis. 6, e1619. Available from: https://doi.org/10.1038/cddis.2014.583.

Jeon, S.M., Chandel, N.S., Hay, N., 2012. AMPK regulates NADPH homeostasis to promote tumour cell survival during energy stress. Nature 485, 661–665. Available from: https://doi.org/10.1038/nature11066.

Jia, D., Park, J., Jung, K., Levine, H., Kaipparettu, B., 2018. Elucidating the metabolic plasticity of cancer: mitochondrial reprogramming and hybrid metabolic states. Cells 7, 21.

Jia, D., et al., 2019. Elucidating cancer metabolic plasticity by coupling gene regulation with metabolic pathways. Proc. Natl Acad. Sci. U S A 116, 3909–3918. Available from: https://doi.org/10.1073/pnas.1816391116.

Jiang, L., et al., 2016. Reductive carboxylation supports redox homeostasis during anchorage-independent growth. Nature 532, 255–258. Available from: https://doi.org/10.1038/nature17393.

Jin, L., et al., 2018. The PLAG1-GDH1 axis promotes anoikis resistance and tumor metastasis through CamKK2-AMPK signaling in LKB1-deficient lung cancer. Mol. Cell 69, 87–99. Available from: https://doi.org/10.1016/j.molcel.2017.11.025. e87.

Jones, R.G., et al., 2005. AMP-activated protein kinase induces a p53-dependent metabolic checkpoint. Mol. Cell 18, 283–293. Available from: https://doi.org/10.1016/j.molcel.2005.03.027.

Kamarajugadda, S., et al., 2012. Glucose oxidation modulates anoikis and tumor metastasis. Mol. Cell Biol. 32, 1893–1907. Available from: https://doi.org/10.1128/MCB.06248-11.

Kamarajugadda, S., et al., 2013. Manganese superoxide dismutase promotes anoikis resistance and tumor metastasis. Cell Death Dis. 4, e504. Available from: https://doi.org/10.1038/cddis.2013.20.

Kim, J., Kundu, M., Viollet, B., Guan, K.L., 2011. AMPK and mTOR regulate autophagy through direct phosphorylation of Ulk1. Nat. Cell Biol. 13, 132–141. Available from: https://doi.org/10.1038/ncb2152.

Kim, N.H., et al., 2017. Snail reprograms glucose metabolism by repressing phosphofructokinase PFKP allowing cancer cell survival under metabolic stress. Nat. Commun. 8, 14374.

Kimmelman, A.C., 2015. Metabolic dependencies in RAS-driven cancers. Clin. Cancer Res. 21, 1828–1834. Available from: https://doi.org/10.1158/1078-0432.CCR-14-2425.

Kondaveeti, Y., Reed, I.K.G., White, B.A., 2015. Epithelial–mesenchymal transition induces similar metabolic alterations in two independent breast cancer cell lines. Cancer Lett. 364, 44–58.

Labelle, M., Hynes, R.O., 2012. The initial hours of metastasis: the importance of cooperative host-tumor cell interactions during hematogenous dissemination. Cancer Discov. 2, 1091–1099. Available from: https://doi.org/10.1158/2159-8290.CD-12-0329.

Lamming, D.W., Sabatini, D.M., 2013. A Central role for mTOR in lipid homeostasis. Cell Metab. 18, 465–469. Available from: https://doi.org/10.1016/j.cmet.2013.08.002.

Le Gal, K., et al., 2015. Antioxidants can increase melanoma metastasis in mice. Sci. Transl. Med. 7, 308re308. Available from: https://doi.org/10.1126/scitranslmed.aad3740.

Lee, Y.Z., et al., 2014. Discovery of selective inhibitors of Glutaminase-2, which inhibit mTORC1, activate autophagy and inhibit proliferation in cancer cells. Oncotarget 5, 6087–6101. Available from: https://doi.org/10.18632/oncotarget.2173.

Levine, A.J., Puzio-Kuter, A.M., 2010. The control of the metabolic switch in cancers by oncogenes and tumor suppressor genes. Science 330, 1340–1344. Available from: https://doi.org/10.1126/science.1193494.

Li, A.E., et al., 1999. A role for reactive oxygen species in endothelial cell anoikis. Circ. Res. 85, 304–310.

Lien, E.C., Lyssiotis, C.A., Cantley, L.C., 2016. Metabolic reprogramming by the PI3K-Akt-mTOR pathway in cancer. Recent. Results Cancer Res. 207, 39–72. Available from: https://doi.org/10.1007/978-3-319-42118-6_3.

Lock, R., Debnath, J., 2008. Extracellular matrix regulation of autophagy. Curr. Opin. Cell Biol. 20, 583–588. Available from: https://doi.org/10.1016/j.ceb.2008.05.002.

Marin-Hernandez, A., Gallardo-Perez, J.C., Ralph, S.J., Rodriguez-Enriquez, S., Moreno-Sanchez, R., 2009. HIF-1alpha modulates energy metabolism in cancer cells by inducing over-expression of specific glycolytic isoforms. Mini Rev. Med. Chem. 9, 1084–1101.

Mason, J.A., et al., 2016. Oncogenic Ras differentially regulates metabolism and anoikis in extracellular matrix-detached cells. Cell Death Differ. 23, 1271–1282. Available from: https://doi.org/10.1038/cdd.2016.15.

Meacham, C.E., Morrison, S.J., 2013. Tumour heterogeneity and cancer cell plasticity. Nature 501, 328–337. Available from: https://doi.org/10.1038/nature12624.

Mitchell, A.R., et al., 2018. Redox regulation of pyruvate kinase M2 by cysteine oxidation and S-nitrosation. Biochem. J. 475, 3275–3291. Available from: https://doi.org/10.1042/BCJ20180556.

Mullen, A.R., et al., 2014. Oxidation of alpha-ketoglutarate is required for reductive carboxylation in cancer cells with mitochondrial defects. Cell Rep. 7, 1679–1690. Available from: https://doi.org/10.1016/j.celrep.2014.04.037.

Nagaprashantha, L.D., Vatsyayan, R., Lelsani, P.C., Awasthi, S., Singhal, S.S., 2011. The sensors and regulators of cell-matrix surveillance in anoikis resistance of tumors. Int. J. Cancer 128, 743–752. Available from: https://doi.org/10.1002/ijc.25725.

Ng, T.L., et al., 2012. The AMPK stress response pathway mediates anoikis resistance through inhibition of mTOR and suppression of protein synthesis. Cell Death Differ. 19, 501–510. Available from: https://doi.org/10.1038/cdd.2011.119.

Overholtzer, M., et al., 2007. A nonapoptotic cell death process, entosis, that occurs by cell-in-cell invasion. Cell 131, 966–979. Available from: https://doi.org/10.1016/j.cell.2007.10.040.

Pisco, A.O., Huang, S., 2015. Non-genetic cancer cell plasticity and therapy-induced stemness in tumour relapse: 'What does not kill me strengthens me'. Br. J. Cancer 112, 1725–1732. Available from: https://doi.org/10.1038/bjc.2015.146.

Pradeep, C.R., et al., 2012. Modeling ductal carcinoma in situ: a HER2-Notch3 collaboration enables luminal filling. Oncogene 31, 907–917. Available from: https://doi.org/10.1038/onc.2011.279.

Rabinovitch, R.C., et al., 2017. AMPK maintains cellular metabolic homeostasis through regulation of mitochondrial reactive oxygen species. Cell Rep. 21, 1–9. Available from: https://doi.org/10.1016/j.celrep.2017.09.026.

Raj, L., et al., 2011. Selective killing of cancer cells by a small molecule targeting the stress response to ROS. Nature 475, 231–234. Available from: https://doi.org/10.1038/nature10167.

Rohwer, N., et al., 2008. Hypoxia-inducible factor 1alpha mediates anoikis resistance via suppression of alpha5 integrin. Cancer Res. 68, 10113–10120. Available from: https://doi.org/10.1158/0008-5472.CAN-08-1839.

Saha, M., et al., 2018. AMPK-Akt double-negative feedback loop in breast cancer cells regulates their adaptation to matrix deprivation. Cancer research, canres. 78, 1497–1510. Available from: https://doi.org/10.1158/0008-5472.CAN-17-2090.

Satelli, A., et al., 2015. Epithelial-mesenchymal transitioned circulating tumor cells capture for detecting tumor progression. Clin. Cancer Res. 21, 899–906. Available from: https://doi.org/10.1158/1078-0432.CCR-14-0894.

Saxena, M., et al., 2018. AMP-activated protein kinase promotes epithelial-mesenchymal transition in cancer cells through Twist1 upregulation. J. Cell Sci. 131, . Available from: https://doi.org/10.1242/jcs.208314. pii: jcs208314.

Saxton, R.A., Sabatini, D.M., 2017. mTOR signaling in growth, metabolism, and disease. Cell 168, 960–976. Available from: https://doi.org/10.1016/j.cell.2017.02.004.

Sayin, V.I., et al., 2014. Antioxidants accelerate lung cancer progression in mice. Sci. Transl. Med. 6, . Available from: https://doi.org/10.1126/scitranslmed.3007653221-215.

Schafer, Z.T., et al., 2009. Antioxidant and oncogene rescue of metabolic defects caused by loss of matrix attachment. Nature 461, 109–113. Available from: https://doi.org/10.1038/nature08268.

Sciacovelli, M., Frezza, C., 2017. Metabolic reprogramming and epithelial-to-mesenchymal transition in cancer. FEBS J. 284, 3132–3144.

Seldin, L., Le Guelte, A., Macara, I.G., 2017. Epithelial plasticity in the mammary gland. Curr. Opin. Cell Biol. 49, 59–63. Available from: https://doi.org/10.1016/j.ceb.2017.11.012.

Semenza, G.L., Roth, P.H., Fang, H.M., Wang, G.L., 1994. Transcriptional regulation of genes encoding glycolytic enzymes by hypoxia-inducible factor 1. J. Biol. Chem. 269, 23757–23763.

Sequeira, S.J., et al., 2007. Inhibition of proliferation by PERK regulates mammary acinar morphogenesis and tumor formation. PLoS One 2, e615. Available from: https://doi.org/10.1371/journal.pone.0000615.

Shaul, Y.D., et al., 2014. Dihydropyrimidine accumulation is required for the epithelial-mesenchymal transition. Cell 158, 1094–1109.

Shi, Z., Mirza, M., Wang, B., Kennedy, M.A., Weber, G.F., 2014a. Osteopontin-a alters glucose homeostasis in anchorage-independent breast cancer cells. Cancer Lett. 344, 47–53. Available from: https://doi.org/10.1016/j.canlet.2013.10.008.

Shi, Z., Wang, B., Chihanga, T., Kennedy, M.A., Weber, G.F., 2014b. Energy metabolism during anchorage-independence. Induction by osteopontin-c. PLoS One 9, e105675. Available from: https://doi.org/10.1371/journal.pone.0105675.

Smit, M.A., Geiger, T.R., Song, J.Y., Gitelman, I., Peeper, D.S., 2009. A Twist-Snail axis critical for TrkB-induced epithelial-mesenchymal transition-like transformation, anoikis resistance, and metastasis. Mol. Cell Biol. 29, 3722–3737. Available from: https://doi.org/10.1128/MCB.01164-08.

Sundararaman, A., Amirtham, U., Rangarajan, A., 2016. Calcium-oxidant signaling network regulates AMP-activated protein kinase (AMPK) activation upon matrix deprivation. J. Biol. Chem. 291, 14410–14429. Available from: https://doi.org/10.1074/jbc.M116.731257.

Tan, H., Bao, J., Zhou, X., 2015. Genome-wide mutational spectra analysis reveals significant cancer-specific heterogeneity. Sci. Rep. 5, 12566. Available from: https://doi.org/10.1038/srep12566.

Venta, R., Valk, E., Koivomagi, M., Loog, M., 2012. Double-negative feedback between S-phase cyclin-CDK and CKI generates abruptness in the G1/S switch. Front. Physiol. 3, 459. Available from: https://doi.org/10.3389/fphys.2012.00459.

Vera-Ramirez, L., Vodnala, S.K., Nini, R., Hunter, K.W., Green, J.E., 2018. Autophagy promotes the survival of dormant breast cancer cells and metastatic tumour recurrence. Nat. Commun. 9, 1944. Available from: https://doi.org/10.1038/s41467-018-04070-6.

Visvader, J.E., Lindeman, G.J., 2012. Cancer stem cells: current status and evolving complexities. Cell stem Cell 10, 717–728.

Wang, L., et al., 2009. Bistable switches control memory and plasticity in cellular differentiation. Proc. Natl Acad. Sci. U S A 106, 6638–6643. Available from: https://doi.org/10.1073/pnas.0806137106.

Wang, R.C., et al., 2012. Akt-mediated regulation of autophagy and tumorigenesis through Beclin 1 phosphorylation. Science 338, 956–959. Available from: https://doi.org/10.1126/science.1225967.

Wang, Y.N., et al., 2018. CPT1A-mediated fatty acid oxidation promotes colorectal cancer cell metastasis by inhibiting anoikis. Oncogene 37, 6025–6040. Available from: https://doi.org/10.1038/s41388-018-0384-z.

Weinberg, F., et al., 2010. Mitochondrial metabolism and ROS generation are essential for Kras-mediated tumorigenicity. Proc. Natl Acad. Sci. U S A 107, 8788–8793. Available from: https://doi.org/10.1073/pnas.1003428107.

Whelan, K.A., Reginato, M.J., 2011. Surviving without oxygen: hypoxia regulation of mammary morphogenesis and anoikis. Cell Cycle 10, 2287–2294. Available from: https://doi.org/10.4161/cc.10.14.16532.

Whelan, K.A., et al., 2013. The oncogene HER2/neu (ERBB2) requires the hypoxia-inducible factor HIF-1 for mammary tumor growth and anoikis resistance. J. Biol. Chem. 288, 15865–15877. Available from: https://doi.org/10.1074/jbc.M112.426999.

White, E., 2015. The role for autophagy in cancer. J. Clin. Invest. 125, 42–46. Available from: https://doi.org/10.1172/JCI73941.

Yang, J., et al., 2004. Twist, a master regulator of morphogenesis, plays an essential role in tumor metastasis. Cell 117, 927–939. Available from: https://doi.org/10.1016/j.cell.2004.06.006.

Yang, M., et al., 2013. A double-negative feedback loop between Wnt-beta-catenin signaling and HNF4alpha regulates epithelial-mesenchymal transition in hepatocellular carcinoma. J. Cell Sci. 126, 5692–5703. Available from: https://doi.org/10.1242/jcs.135053.

Ye, X., Weinberg, R.A., 2015. Epithelial–mesenchymal plasticity: a central regulator of cancer progression. Trends Cell Biol. 25, 675–686.

Young, C.D., et al., 2011. Modulation of glucose transporter 1 (GLUT1) expression levels alters mouse mammary tumor cell growth in vitro and in vivo. PLoS One 6, e23205. Available from: https://doi.org/10.1371/journal.pone.0023205.

Zhang, D., et al., 2016. AMPK regulates autophagy by phosphorylating BECN1 at threonine 388. Autophagy 12, 1447–1459. Available from: https://doi.org/10.1080/15548627.2016.1185576.

Zhang, C.S., et al., 2017. Fructose-1,6-bisphosphate and aldolase mediate glucose sensing by AMPK. Nature 548, 112–116. Available from: https://doi.org/10.1038/nature23275.

27

Phenotypic instability induced by tissue disruption at the origin of cancer

Jean-Pascal Capp
INSA/University of Toulouse, Toulouse Biotechnology Institute, UMR CNRS 5504, UMR INRA 792, Toulouse, France

Introduction

The role of cellular phenotypic plasticity and heterogeneity in cancer has been highlighted in recent years by works at the single-cell level studying various phenomena ranging from the initiation steps to drug resistance (Lawson et al., 2018). Indeed interrogating the whole genome, transcriptome, epigenome, and proteome in single cells is now possible. Profiling thousands of cells from an individual tumor allow to characterize the extent of intratumoral heterogeneity in individual tumors at various regulatory levels, and to spatially localize cell populations within tumors. Thus key advances have been made to understand the function and effect of different cell subpopulations, especially in terms of plasticity, on tumorigenesis, including which features promote tumor initiation, progression, or drug resistance.

Especially, a subset of cancer cells called cancer stem cells (CSCs) harboring increased phenotypic plasticity is suspected to maintain tumor growth and confer therapeutic resistance (Capp, 2019). Disruption of normal epigenetic marks is often proposed to explain such behavior. Epigenetic event are even now considered has able to initiate cancer, as genetic events do (Flavahan et al., 2017). Nevertheless, the origin of this epigenetic instability remains elusive. The most common explanation is now that genetic alterations initiating cancer induce a "tumor reprogramming" that "resets" the epigenetic status and gene expression in initially healthy cells (Vicente-Duenas et al., 2015). Thus, oncogenes would contribute to the development of cancers less by inducing cell proliferation than by causing

Phenotypic Switching. DOI: https://doi.org/10.1016/B978-0-12-817996-3.00026-8

"developmental reprogramming" of the epigenome of the affected cells (Sanchez-Garcia, 2015) that would allow them to aberrantly and pathologically differentiate. Oncogenesis is indeed accompanied by loss of developmental information at the epigenetic level, and reversion to a "pseudoprimitive" state that combines regulatory DNA features of embryonic stem (ES) cells and of other developing lineages (Stergachis et al., 2013).

Nevertheless, a phenomenon more recently studied in the field has to be taken into consideration: the stochastic nature of gene expression (or gene expression noise) and its role in phenotypic diversification and cellular plasticity in cancer cell populations. Especially, the origin of early differentiation problems in cancer may be considered otherwise by focusing at the tissue level and by considering that tissue disruption can produce cells with increased stochastic gene expression (SGE) and phenotypic plasticity that behave abnormally, initial genetic alterations being present or not (Capp, 2005; Capp, 2012a; Capp, 2017; Capp and Bataille, 2018).

In this contribution, recent results on the role of cellular plasticity and SGE in cancer growth and drug resistance will first be briefly reviewed. Then the relationship between SGE and stemness will be analyzed so as to propose a framework where CSCs are mainly characterized by increased SGE that is associated to dedifferentiation (of differentiated cells) or non-differentiation (of normal stem cells), and that explains their high phenotypic instability. Finally, the origin of this instability will be envisaged as being induced by tissue disruption that would therefore be at the origin of cancer.

Phenotypic plasticity in cancer growth and drug resistance: role of cancer stem cells and stochastic gene expression

Single-cell technologies provide the opportunity to measure cell states in individual cells, especially in cancer cell populations. Apart from genetic diversity, tumor cells were shown to display inherent functional variability in tumor propagation potential, which contributes to both cancer growth and therapy tolerance (Kreso et al., 2013). Increased intratumoral non-genetic heterogeneity was associated with decreased survival for instance in glioblastoma (Patel et al., 2014) and drug resistance through selection of preexisting drug-resistant cells (Sharma et al., 2018). Basal cell carcinomas for instance contain a more stem-like tumor cell population that expresses Lgr5, a well-known marker of many epithelial stem cells, and a more differentiated population that does not express Lgr5. Treatment with vismodegib (GDC0449) mediates cancer regression by promoting the differentiation of tumor cells, but persisting, a slow-cycling tumor population that expresses Lgr5 persists and is responsible for tumor relapse following treatment discontinuation (Sanchez-Danes et al., 2018).

Apart from phenotypic heterogeneity *per se*, lineage plasticity seems especially important for cancer progression (Le Magnen et al., 2018). For instance, during the reversible phenotype switching that occurs in the progression from nonmetastatic to metastatic melanoma, melanoma cells acquire invasive and migratory properties that promote metastasis, suggesting a partial dedifferentiation (Quintana et al., 2008). Stochastic transitions between differentiation states were also shown to make subpopulations of cells purified for a given phenotypic state return towards equilibrium proportions over time (Gupta et al., 2011), highlighting the potential importance of this plasticity in the tumoral process. Another pioneering study using single-cell multiplex quantitative polymerase chain reaction (qPCR) approach showed that human colon cancers contain distinct cell populations that mirror the cellular lineages of the normal colon (Dalerba et al., 2011). This transcriptional heterogeneity was not due to underlying genetic heterogeneity, as injection of single CSCs into immune-deficient mice gave rise to monoclonal tumors as heterogeneous as the parental one. Also, both mouse (de Sousa e Melo et al., 2017) and human (Shimokawa et al., 2017) models showed that differentiated colon cells have the intrinsic capacity to dedifferentiate into CSCs expressing Lgr5 upon ablation of preexisting CSCs. These results support the fact that CSCs originate from differentiated cells that dedifferentiate rather than from a normal stem cell.

Phenotypic switching is frequently mentioned in oncogenesis through the concept of epithelial-to-mesenchymal transition (EMT) that generates mesenchymal-like cells and a variety of intermediate cell states between the epithelial and the mesenchymal states, which could function as CSC (Dongre and Weinberg, 2019). While phenotypic plasticity through EMT was supposed to be a rare event among cancer cells, it appears that this event frequently occurs at a clonal level in breast cancers (Rios et al., 2019), implying that the differentiation state of tumor cells is inherently unstable or plastic.

Consistent with its association with more aggressive cancer phenotypes, cellular plasticity has also emerged as an important mechanism of drug resistance (Pisco and Huang, 2015). Transdifferentiation, for instance, is a mechanism of treatment resistance in a mouse model of castration-resistant prostate cancer (Zou et al., 2017). Similarly, antiandrogen resistance in some prostate cancers is mediated by lineage plasticity (Mu et al., 2017). More surprising is the fact that, in the absence of preexisting phenotypic heterogeneity, stress-induced transdifferentiation can be a major driver of drug resistance and metastasis, a mechanism that ultimately relies on epigenetic mechanisms (Sharma et al., 2018).

From a mechanistic point of view, the contribution of various phenomena producing phenotypic diversification such as epigenetic regulation, SGE, or variability in the microenvironment can be evoked (Marusyk et al., 2012). Histone modifications are especially important for lineage plasticity in cancer (Flavahan et al., 2017; Liau et al., 2017; Pastore et al., 2019). For instance, a reversible epigenetic transition allows glioblastoma stem cells to transit between proliferative and slow-cycling states. This plasticity enables glioblastoma tumors to propagate, adapt, and persist in the face of environmental and therapeutic pressures (Liau et al., 2017). In chronic lymphocytic leukemia, intratumoral epigenetic diversity seems to allow leukemic cells to stochastically activate alternative gene regulatory programs, facilitating the emergence of novel cell states, ultimately facilitating tumor growth and drug tolerance (Pastore et al., 2019).

In addition, the role of SGE in the appearance of therapeutic resistance independently of genetic variations has been demonstrated recently (Shaffer et al., 2017; Hinohara et al., 2018). A first study showed that human melanoma cells can display profound transcriptional variability at the single-cell level that predicts which cells will ultimately resist to drug treatment (Shaffer et al., 2017). Indeed, SGE among these cells produces infrequent, semi-coordinated transcription of a number of resistance markers at high levels in a very small percentage of cells that finally can convert their transient transcriptional state to a stably resistant state. A similar phenomenon has been observed in breast cancer in relationship with epigenetic regulation. Indeed cellular transcriptomic heterogeneity in breast cancers seems to be due to SGE and is modulated by epigenetic regulators such as the KDM5 family of histone H3 lysine 4 demethylases (Hinohara et al., 2018). Especially, higher KDM5B activity may increase cellular transcriptomic heterogeneity because KDM5 inhibitor treatment in luminal ER+ breast cancers resulted in more uniform gene expression patterns (Hinohara et al., 2018). In addition to decreasing cell-to-cell transcriptomic heterogeneity, inhibition of KDM5 also increased sensitivity to fulvestrant, even in previously fulvestrant-resistant cells (Hinohara et al., 2018), suggesting a direct role of SGE in therapeutic resistance. Besides KDM5B, higher expression of several other histone demethylases also correlates with higher transcriptomic heterogeneity in human tumors, implying that histone demethylases in general may play a role in regulating transcriptomic heterogeneity (Hinohara and Polyak, 2019).

Cellular plasticity and stemness conferred by high stochastic gene expression

Both cellular plasticity and SGE can contribute to tumor growth and be responsible for drug resistance. Nevertheless, their intimate relationship, while intensively studied in stem cell biology, is rarely mentioned in cancer research. Since pioneering researches by the end of the 2000s (Efroni et al., 2008; Chang et al., 2008), SGE is acknowledged as a source of cellular heterogeneity and phenotypic plasticity that are both hallmarks of embryonic and adult stem cells (Graf and Stadtfeld, 2008). First, gene expression in mouse embryonic stem (ES) cells is generalized and widespread, and differentiation is associated with large-scale repression of gene expression at the genomic level (Efroni et al., 2008). Gene expression profiles clearly become more discontinuous in differentiated cells. The "hyperactive" gene expression in ES cells is associated with the presence of high levels of proteins involved in chromatin remodeling and gene transcription. Moreover almost all of the tissue-specific and differentiation genes that have been examined are sporadically expressed at a low level. Second, variation in gene expression arising from transcriptional noise and network fluctuation and the associated phenotypic heterogeneity account for stochasticity of cell fate decisions in stem and progenitor cells (Chang et al., 2008).

The whole genome is heterogeneously expressed in ES cells, thus creating early, stochastic and reversible commitments (also known as "priming") towards differentiation (Hough et al., 2009; Trott et al., 2012). To a certain extent, this heterogeneity seems to be related to the level of expression (itself stochastic) of some pluripotency factors such as Nanog (MacArthur et al., 2012). For example, cells that express less Nanog differentiate more easily (Kalmar et al., 2009), which is related to more frequent stochastic and reversible commitments to a cell type (MacArthur et al., 2012). Works on mouse ES cells have directly showed that increasing transcriptional heterogeneity by inhibition of histone acetyl-transferase, and candidate noise modulator, Kat2a (yeast ortholog Gcn5) reduced pluripotency and accelerated mesendodermal differentiation, with increased probability of transitions into lineage commitment (Moris et al., 2018). This is a direct relationship between transcriptional heterogeneity and cell fate transitions through manipulation of the histone acetylation landscape of mouse ES cells.

More generally, the sophistication of single-cell studies on fate decisions in developing tissues provides new insights into how differentiation is gained. It was thus shown, for example, that progenitors of mouse neural cells express with oscillations the three transcription factors characteristic of the three cell types they generate (neurons, astrocytes, and oligodendrocytes), whereas differentiation is associated with the stable and sustained expression of only one of these factors (Robson, 2014). Multiple developing systems have shown that multilineage potential consists in co-expressing genes typically associated with each of their alternative lineage fates (Laslo et al., 2006; Brunskill et al., 2014; Goolam et al., 2016). It is proposed that the varied expression of genes associated with each alternative fate in single cells can act as a substrate for selection by signals (Moris et al., 2016).

Thus, many results indicate that initial diversity is necessary for differentiation, and that differentiation genes must be first expressed by cells in order to make them able to initiate a differentiation process in response to environmental factors. Indeed, the expression of differentiation genes appears to be a prerequisite for differentiation in response to a signal, and integration into a tissue. In general, genes that are characteristic of differentiated cell types are expressed with more variably than the genes expressed ubiquitously in all the cells, which seems to indicate a functional role of this high variability (Padovan-Merhar et al., 2015). Moreover, expression remains highly probabilistic within a differentiated tissue such as the liver (Bahar Halpern et al., 2015), suggesting also a role in the functional plasticity of hepatocytes (Lam and Deans, 2015).

As mentioned earlier, the degree of SGE is modulated during development and differentiation: while gene expression is initially widespread and highly variable, cells progressively transit toward more homogeneous, coordinated and restricted gene expression profiles (Efroni et al., 2008; Richard et al., 2016; Moussy et al., 2017) associated with a more restrictive chromatin (Gaspar-Maia et al., 2011). But, of note, when hematopoietic stem or progenitors cells are induced to differentiate in vitro, a transient state characterized by an increase in gene expression variability occurs before its strong decrease (Richard et al., 2016; Moussy et al., 2017). Thus differentiation is clearly associated with the establishment of "constraints" that make SGE decrease and that stabilize and homogenize gene expression patterns from cell-to-cell in the population (MacArthur and Lemischka, 2013).

Cell−cell interactions as constraints decreasing cellular plasticity and stemness during differentiation

When considering that stemness is characterized by high SGE and that differentiation occurs mainly by a reduction of the expressed part of the genome and decreased SGE, one can ask which the constraints that stabilize phenotypes are. The literature about the phenomena that contribute to decrease SGE during development and differentiation is growing and tends to highlight the role of cellular interactions and the subsequently activated signaling pathways. Cell−cell interactions appear to be the main "constraints" leading to stable differentiated states. For instance, in Drosophila, signals received via the EGF receptor in eye cells are necessary for decreasing expression noise of a transcription factor important in the balance between multipotency and differentiation since genetic loss caused sustained noise (Pelaez et al., 2015). Indeed, loss of EGFR signaling prevents cells from differentiating and generates a prolonged noisy expression of this transcription factor, suggesting that its dynamic heterogeneity is a necessary element of cell state transitions in the eye, and that cell states are stabilized through noise reduction (Pelaez et al., 2015).

A similar mechanism was already described in *Caenorhabditis elegans* embryos where strong signaling is essential to maintain low expression variability and to ensure reliable neuroblast development (Ji et al., 2013). Following division of cells derived from the neuroblasts Q of *C. elegans* embryos (precursors of the future nervous system), two Q neuroblasts are positioned symmetrically and bilaterally in the embryo but then migrate in opposite directions, which are associated with the expression of different genes. In the "left" neuroblast, a gene called *mab-5* is needed and sufficient to ensure the migration of descending cells. In the "right" one, the absence of *mab-5* ensures the opposite migration.

The expression of *mab-5* is dependent on a signaling pathway called Wnt, which arises from the interaction of a cell surface ligand on the four Frizzled receptors of the Wnt pathway, which are therefore required for the expression of *mab-5* in the offspring of the "left" cell. Interestingly, embryos that do not have some of these receptors exhibit partial penetration migration defects, where some descending cells of the "left" cell migrate in the opposite direction. The authors showed that this phenotypic heterogeneity has its origin in the increased variability of *mab-5* expression in the absence of certain receptors. In cells that do not have some Frizzled receptors, the increased variability and decreased expression of *mab-5* is associated with a disrupted chromatin organization with more heterogeneous and reduced "transcription centers" (the nucleus locations of the cell where transcription events are concentrated). Thus the transcription of *mab-5* appears to be inherently highly stochastic, with a modulation of this variability obtained by Wnt signaling: strong signaling reduces its SGE and is needed for proper cell segregation.

These few examples show that perturbation of cell communication can enhance SGE and phenotypic heterogeneity among differentiating cells. Additional convincing examples are provided by studies of the developing mouse blastocyst where an initial phase of stochastic expression of individual genes precedes signal reinforcement through Fgf4 that segregates early lineages (Ohnishi et al., 2014), or by studies on the rat pituitary tissue where direct cell contacts through gap junctions spatially coordinate prolactin gene expression while enzymatic digestion of extracellular proteins or pharmacological inhibition of gap junctions reduce transcriptional coordination between cells (Featherstone et al., 2016).

Finally, the relationship between SGE and stemness and the role of cell–cell interactions in the acquisition of differentiated phenotypes lead to consider that disruption of the cellular interactions that stabilize phenotypes and homogenize gene expression patterns would produce CSCs defined as cells with increased SGE (Capp, 2019). Considering that dedifferentiation and acquisition of stem-like properties are key features of the tumoral process, disruption of cellular interactions is expected to be sufficient to produce phenotypic plasticity, less differentiated cancer cells, and ultimately oncogenesis (Capp, 2005; Capp, 2012a; Capp, 2017; Capp and Bataille, 2018). Oncogenesis seems to be possible through removal of the constraints established during differentiation at the tissue level, without the need to evoke oncogenic event at the origin of this epigenetic instability.

Tissue disruption at the origin of phenotypic instability and consequently cancer

The results mentioned in chapter I about the role of phenotypic switching in tumor growth and drug resistance encourage to reconsider the foundations of the stem-like properties of at least some cancer cells. Indeed, this phenotypic instability seems to be very close to phenomena observed in normal stem cells that have been associated to SGE (Capp, 2019). Transcriptional heterogeneity produced by SGE is even considered as necessary for cell fate decisions. The similarity with these normal processes makes probable the fact that increased SGE is also at the origin of phenotypic switching in cancer cells. Moreover, it should encourage to consider whether this increased SGE and subsequent phenotypic plasticity is a side-effect of other hallmarks of cancer or if it is one of the, or even the main, driving force in cancer (Capp, 2011).

Years of debate have discussed whether stemness can be acquired stochastically by any cell in the tumor depending on the environmental context and random intracellular events, or is restricted to a small cancer cell subpopulation that is at the top of a classical hierarchical model from where, when a cell leaves the stem state, the phenomenon is irreversible and none of the descendants can restore it. Recent data suggest that dedifferentiation is common and can occur in any cell of the cancerous population (Rios et al., 2019). This plasticity of the CSC state has also been recently demonstrated in glioblastoma (Dirkse et al., 2019). Conversely, glioblastoma CSCs can express various differentiation markers and still contribute to tumor initiation and self-renewal (Chen et al., 2010). Thus, it seems that, as previously suggested, stemness could be acquired by any cell at any time, rendering difficult any therapeutic intervention targeting CSCs (Dirks, 2010). Stochastic models of CSC have been proposed to explain how each cell of a tumor could act as CSC (Nguyen et al., 2012; Beck and Blanpain, 2013). Those models include the possibility that variations in this ability are due to intrinsic stochastic variations, especially in terms of gene expression. Considering that this property is both dependent on intrinsic stochastic events, and under environmental influences, leads to point the interplay between cell–cell interactions and SGE in the acquisition of the stem state. Especially, the phenotypic instability intrinsically associated to CSC might originate from an initial loss of environmental control by the normal tissue structure that could be considered as the sole required event to start cancerous transformation.

Thus differentiation and quiescence would no longer be maintained, and cancer-stem like properties and proliferation would appear, without any genetic requirements. Subsequently, genetic and epigenetic instabilities would necessary appear because of an increased cellular instability, increasing phenotypic instability and the risk of malignant transformation. If normal stem cells fail in their differentiation process because the cellular interactions that must be established are not possible, they would remain with their high phenotypic instability but "non-controlled" by the microenvironment. Thus they would also present the required phenotypic plasticity to produce tumors.

Genetic alterations have long been, and are still in most cases, considered as the required initial events for oncogenesis. However, among many arguments that can be formulated against this idea (Capp, 2005; Capp, 2012a; Capp, 2017; Capp and Bataille, 2018), the fact that oncogenic events currently considered as been able to produce cancer are widespread in normal tissues such as in the skin (Martincorena et al., 2015), esophagus (Yokoyama et al., 2019) or many others (Yizhak et al., 2019) encourages to moderate this view. This does not mean that genetic alterations have no role in cancer transformation, but they are at least not sufficient to initiate cancer. Nevertheless, they are obvious strong contributors in cancer progression. They are not sufficient but are they necessary to initiate cancer? Works showing the ability of the sole microenvironmental defect to produce cancer [reviewed in (Capp, 2017)] suggest that they might be dispensable, at least in early steps. Additionally, the ability of a healthy tissue to normalize cancer cells shows that even already transformed cells remain sensitive to normal environmental cues [also reviewed in (Capp, 2017)], suggesting that genetic alterations are not the main controller of cancer development.

Finally, while genetic alterations would be essential and inevitable contributors to tumor growth, they would not be its initiators and would exert their influence only if the normal tissue environment is also altered. Why does it make a difference? The ability of the normal microenvironment to control genetically altered cells and reduce phenotypic plasticity suggests that mimicking this normal microenvironment with molecules that would simulate the presence of normal interaction partners in the microenvironment, together with molecules stimulating the reexpression of the proteins necessary to interact with these elements, would be efficient in stopping cell proliferation and cancer evolution (Capp, 2012b; Capp and Bataille, 2018). This would correspond to an innovative therapeutic strategy purely based on the consideration that phenotypic instability and plasticity are the inherent cancer cell properties that should be first of all targeted in cancer.

References

Bahar Halpern, K., et al., 2015. Bursty gene expression in the intact mammalian liver. Mol. Cell 58 (1), 147–156.

Beck, B., Blanpain, C., 2013. Unravelling cancer stem cell potential. Nat. Rev. Cancer 13 (10), 727–738.

Brunskill, E.W., et al., 2014. Single cell dissection of early kidney development: multilineage priming. Development 141 (15), 3093–3101.

Capp, J.P., 2005. Stochastic gene expression, disruption of tissue averaging effects and cancer as a disease of development. Bioessays 27 (12), 1277–1285.

Capp, J.P., 2011. Stochastic gene expression is the driving force of cancer. Bioessays 33 (10), 781–782.

Capp, J.P., 2012a. Nouveau regard sur le cancer. Pour une révolution des traitements. Belin-Pour la science, Paris.

Capp, J.P., 2012b. Stochastic gene expression stabilization as a new therapeutic strategy for cancer. Bioessays 34 (3), 170–173.

Capp, J.P., 2017. Tissue disruption increases stochastic gene expression thus producing tumors: cancer initiation without driver mutation. Int. J. Cancer 140 (11), 2408–2413.

Capp, J.P., 2019. Cancer stem cells: from historical roots to a new perspective. J. Oncol. 5189232.

Capp, J.P., Bataille, R., 2018. Multiple myeloma exemplifies a model of cancer based on tissue disruption as the initiator event. Front. Oncol. 8, 355.

Chang, H.H., et al., 2008. Transcriptome-wide noise controls lineage choice in mammalian progenitor cells. Nature 453 (7194), 544–547.

Chen, R., et al., 2010. A hierarchy of self-renewing tumor-initiating cell types in glioblastoma. Cancer Cell 17 (4), 362–375.

Dalerba, P., et al., 2011. Single-cell dissection of transcriptional heterogeneity in human colon tumors. Nat. Biotechnol. 29 (12), 1120–1127.

de Sousa e Melo, F., et al., 2017. A distinct role for Lgr5(+) stem cells in primary and metastatic colon cancer. Nature 543 (7647), 676–680.

Dirks, P., 2010. Cancer stem cells: invitation to a second round. Nature 466 (7302), 40–41.

Dirkse, A., et al., 2019. Stem cell-associated heterogeneity in Glioblastoma results from intrinsic tumor plasticity shaped by the microenvironment. Nat. Commun. 10 (1), 1787.

Dongre, A., Weinberg, R.A., 2019. New insights into the mechanisms of epithelial-mesenchymal transition and implications for cancer. Nat. Rev. Mol. Cell Biol. 20 (2), 69–84.

Efroni, S., et al., 2008. Global transcription in pluripotent embryonic stem cells. Cell Stem Cell 2 (5), 437–447.
Featherstone, K., et al., 2016. Spatially coordinated dynamic gene transcription in living pituitary tissue. Elife 5, e08494.
Flavahan, W.A., Gaskell, E., Bernstein, B.E., 2017. Epigenetic plasticity and the hallmarks of cancer. Science 357 (6348), pii: eaal2380.
Gaspar-Maia, A., et al., 2011. Open chromatin in pluripotency and reprogramming. Nat. Rev. Mol. Cell Biol. 12 (1), 36–47.
Goolam, M., et al., 2016. Heterogeneity in Oct4 and Sox2 targets biases cell fate in 4-cell mouse embryos. Cell 165 (1), 61–74.
Graf, T., Stadtfeld, M., 2008. Heterogeneity of embryonic and adult stem cells. Cell Stem Cell 3 (5), 480–483.
Gupta, P.B., et al., 2011. Stochastic state transitions give rise to phenotypic equilibrium in populations of cancer cells. Cell 146 (4), 633–644.
Hinohara, K., Polyak, K., 2019. Intratumoral heterogeneity: more than just mutations. Trends Cell Biol. 29 (7), 569–579.
Hinohara, K., et al., 2018. KDM5 histone demethylase activity links cellular transcriptomic heterogeneity to therapeutic resistance. Cancer Cell 34 (6), 939–953. e9.
Hough, S.R., et al., 2009. A continuum of cell states spans pluripotency and lineage commitment in human embryonic stem cells. PLoS One 4 (11), e7708.
Ji, N., et al., 2013. Feedback control of gene expression variability in the *Caenorhabditis elegans* Wnt pathway. Cell 155 (4), 869–880.
Kalmar, T., et al., 2009. Regulated fluctuations in nanog expression mediate cell fate decisions in embryonic stem cells. PLoS Biol. 7 (7), e1000149.
Kreso, A., et al., 2013. Variable clonal repopulation dynamics influence chemotherapy response in colorectal cancer. Science 339 (6119), 543–548.
Lam, A., Deans, T.L., 2015. A noisy tug of war: the battle between transcript production and degradation in the liver. Dev. Cell 33 (1), 3–4.
Laslo, P., et al., 2006. Multilineage transcriptional priming and determination of alternate hematopoietic cell fates. Cell 126 (4), 755–766.
Lawson, D.A., et al., 2018. Tumour heterogeneity and metastasis at single-cell resolution. Nat. Cell Biol. 20 (12), 1349–1360.
Le Magnen, C., Shen, M.M., Abate-Shen, C., 2018. Lineage plasticity in cancer progression and treatment. Annu. Rev. Cancer Biol. 2, 271–289.
Liau, B.B., et al., 2017. Adaptive chromatin remodeling drives glioblastoma stem cell plasticity and drug tolerance. Cell Stem Cell 20 (2), 233–246. e7.
MacArthur, B.D., Lemischka, I.R., 2013. Statistical mechanics of pluripotency. Cell 154 (3), 484–489.
MacArthur, B.D., et al., 2012. Nanog-dependent feedback loops regulate murine embryonic stem cell heterogeneity. Nat. Cell Biol. 14 (11), 1139–1147.
Martincorena, I., et al., 2015. Tumor evolution. High burden and pervasive positive selection of somatic mutations in normal human skin. Science 348 (6237), 880–886.
Marusyk, A., Almendro, V., Polyak, K., 2012. Intra-tumour heterogeneity: a looking glass for cancer? Nat. Rev. Cancer 12 (5), 323–334.
Moris, N., Pina, C., Arias, A.M., 2016. Transition states and cell fate decisions in epigenetic landscapes. Nat. Rev. Genet. 17 (11), 693–703.
Moris, N., et al., 2018. Histone acetyltransferase KAT2A stabilizes pluripotency with control of transcriptional heterogeneity. Stem Cell 36 (12), 1828–1838.

Moussy, A., et al., 2017. Integrated time-lapse and single-cell transcription studies highlight the variable and dynamic nature of human hematopoietic cell fate commitment. PLoS Biol. 15 (7), e2001867.
Mu, P., et al., 2017. SOX2 promotes lineage plasticity and antiandrogen resistance in TP53- and RB1-deficient prostate cancer. Science 355 (6320), 84–88.
Nguyen, L.V., et al., 2012. Cancer stem cells: an evolving concept. Nat. Rev. Cancer 12 (2), 133–143.
Ohnishi, Y., et al., 2014. Cell-to-cell expression variability followed by signal reinforcement progressively segregates early mouse lineages. Nat. Cell Biol. 16 (1), 27–37.
Padovan-Merhar, O., et al., 2015. Single mammalian cells compensate for differences in cellular volume and DNA copy number through independent global transcriptional mechanisms. Mol. Cell 58 (2), 339–352.
Pastore, A., et al., 2019. Corrupted coordination of epigenetic modifications leads to diverging chromatin states and transcriptional heterogeneity in CLL. Nat. Commun. 10 (1), 1874.
Patel, A.P., et al., 2014. Single-cell RNA-seq highlights intratumoral heterogeneity in primary glioblastoma. Science 344 (6190), 1396–1401.
Pelaez, N., et al., 2015. Dynamics and heterogeneity of a fate determinant during transition towards cell differentiation. Elife 4, e08924.
Pisco, A.O., Huang, S., 2015. Non-genetic cancer cell plasticity and therapy-induced stemness in tumour relapse: 'What does not kill me strengthens me'. Br. J. Cancer 112 (11), 1725–1732.
Quintana, E., et al., 2008. Efficient tumour formation by single human melanoma cells. Nature 456 (7222), 593–598.
Richard, A., et al., 2016. Single-cell-based analysis highlights a surge in cell-to-cell molecular variability preceding irreversible commitment in a differentiation process. PLoS Biol. 14 (12), e1002585.
Rios, A.C., et al., 2019. Intraclonal plasticity in mammary tumors revealed through large-scale single-cell resolution 3D imaging. Cancer Cell 35 (4), 618–632. e6.
Robson, P., 2014. Deciphering developmental processes from single-cell transcriptomes. Dev. Cell 29 (3), 260–261.
Sanchez-Danes, A., et al., 2018. A slow-cycling LGR5 tumour population mediates basal cell carcinoma relapse after therapy. Nature 562 (7727), 434–438.
Sanchez-Garcia, I., 2015. How tumour cell identity is established? Semin. Cancer Biol. 32, 1–2.
Shaffer, S.M., et al., 2017. Rare cell variability and drug-induced reprogramming as a mode of cancer drug resistance. Nature 546 (7658), 431–435.
Sharma, A., et al., 2018. Longitudinal single-cell RNA sequencing of patient-derived primary cells reveals drug-induced infidelity in stem cell hierarchy. Nat. Commun. 9 (1), 4931.
Shimokawa, M., et al., 2017. Visualization and targeting of LGR5(+) human colon cancer stem cells. Nature 545 (7653), 187–192.
Stergachis, A.B., et al., 2013. Developmental fate and cellular maturity encoded in human regulatory DNA landscapes. Cell 154 (4), 888–903.
Trott, J., et al., 2012. Dissecting ensemble networks in ES cell populations reveals micro-heterogeneity underlying pluripotency. Mol. Biosyst. 8 (3), 744–752.
Vicente-Duenas, C., et al., 2015. Tumoral stem cell reprogramming as a driver of cancer: Theory, biological models, implications in cancer therapy. Semin. Cancer Biol. 32, 3–9.

Yizhak, K., et al., 2019. RNA sequence analysis reveals macroscopic somatic clonal expansion across normal tissues. Science 364 (6444), pii: eaaw0726.

Yokoyama, A., et al., 2019. Age-related remodelling of oesophageal epithelia by mutated cancer drivers. Nature 565 (7739), 312–317.

Zou, M., et al., 2017. Transdifferentiation as a mechanism of treatment resistance in a mouse model of castration-resistant prostate cancer. Cancer Discov. 7 (7), 736–749.

28

Evolutionary strategies to overcome cancer cell resistance to treatment

Pedro M. Enriquez-Navas and Robert A. Gatenby
Department of Cancer Physiology, Moffitt Cancer Center & Research Institute, Tampa, FL, United States

Introduction

The year 2018 marked the 15th anniversary of the completion of the Human Genome Project (https://homedna.com/blog/human-genome-project-celebrating-15th-anniversary; https://ghr.nlm.nih.gov/primer/hgp/accomplishments). This huge biomedical achievement permitted identification of genes that are altered in various cancers and to assess the potential risk for multiple diseases in individual patients. In the intervening years the cost to sequence the whole genome of a single person or tumor has steadily declined and personal genetic data can now be obtained for only a few hundred dollars.

These data have been valuable in cancer treatment primarily for identifying mutated genes that provide key fitness advantages to cancer cell in their Darwinian competition with each other in the context of the often harsh intratumoral environment (Dunbar et al., 2018). Many of these mutant proteins can now be altered by small molecules, and this "targeted therapy" has successfully treated a number of different cancer types. Similarly, many "gene signatures" have been identified and can be used to assess risk and identify the probability of sensitivity or resistance to a number of therapeutic strategies (Topol, 2014).

However, despite the "tsunami" of molecular data that has been generated in cancer biology, corresponding improvements in clinical outcomes have not occurred (Gabor Miklos, 2005; Katsios and Roukos, 2010). Thus, despite new strategies, such as targeted therapy, advances in older strategies, such as immunotherapy, and development of improved drugs in chemo- and hormonal-therapies, most metastatic cancers remain fatal. While

Phenotypic Switching. DOI: https://doi.org/10.1016/B978-0-12-817996-3.00028-1

many of these treatments are highly effective, nearly all of them ultimately fail, leading to tumor progression and ultimately death of the patient.

Evolution in cancer

Failure to eradicate cancer with both new and old treatments is primarily due to the long-recognized capacity of cancer cells to evolve resistance. However, "somatic evolution" is usually viewed as synonymous with the genetic model of malignancy. That is, Darwinian dynamics are typically thought to be a sequence of mutations that individually confer a fitness advantage. In contrast, some investigators (Vidi et al., 2013; Weaver, 2014) have argued from the primacy of environmental factors demonstrating, for example, that genetically identical cancer cells can behave very differently when placed in disparate tissue environments. Interestingly, these seemingly divergent views are fully integrated in classic Darwinian dynamics (Gatenby et al., 2009) wherein "evolution by natural selection" is governed by the interactions of local environmental selection factors and heritable phenotypic traits in the evolving organism. Here, genetics serves as the "mechanism of inheritance." Thus, genetic changes are a "molecular history" of the evolution of an individual or species but not the fundamental unit of selection. In other words, evolution selects phenotypes and not genotypes. Furthermore, since multiple genetic combinations can result in identical phenotypes [termed "supervenience"(Gatenby et al., 2014)], the genetic record will often be an imperfect and potentially confusing history of prior environmental selection forces and adaptive strategies.

Evolution, phenotypic switching, and cancer therapy

An important extension of the supervenience principle is "phenotypic switching" in which the cancer cell can alter the phenotypic expression of its genome in response to changes in its environment. Cancer cells, with access to the vast information stored in the human genome, have a remarkably wide range of evolutionarily achievable phenotypic properties. This plasticity, termed reaction norm, can "select" from all of the phenotypic properties found in any cell type that found throughout each individual's development from a fertilized egg through adulthood.

Perhaps the most obvious and important environmental change in cancers is the imposition of therapy. The optimal goal of systemic treatment applied to metastatic cancers is to eradicate the cancer population. However, this is a monumental task. Each cubic centimeter of cancer is estimated to contain up to one billion (10^9) cells and disseminated cancers do not usually become fatal in adults until the tumor burden exceeds about 500 g. Thus, treatment is often directed toward a cancer cell population that is larger than that of the current human number of humans on earth by more than an order of magnitude.

Despite these large numbers, cancer therapies are often very effective in reducing the tumor burden. In fact, some treatments can render the metastatic tumor sites undetectable by imaging, physical exam, or serum tumor biomarkers. Nevertheless, in most metastatic cancers, even highly successful treatment usually fails due to emergence of resistant populations. The associated population dynamics are diverse. In many cases, the large pretreatment population contains cells that are randomly resistant to treatment before it is applied. Alternatively, some cancer cells have such a high degree of phenotypically plasticity that they can switch to a resistant phenotype prior to induction of cell death by treatment.

Although the sheer size and diversity of a cancer population virtually assures the presence of resistant phenotypes, it is important to recognize this is necessary but not sufficient to result in treatment failure. The latter will occur only when the resistant cells proliferate sufficiently to produce a large, clinically significant population. As noted earlier, this requires a minimum population of several billion cells and is fundamentally governed by evolutionary dynamics that determine the fitness of each cell phenotype. Furthermore, the molecular machinery of resistance and the proliferative capacity of resistant cells are fundamentally connected by the complex cost/benefit trade-off of the resistance mechanism(s) in the context of an often substrate limited intratumoral environment.

Can evolution be integrated into cancer therapy?

Recognition that the population growth necessary for tumor progression is governed by eco-evolutionary principles has led to the hypothesis that perturbing these dynamics can delay or even prevent tumor progression following successful treatment (Gatenby, 2009). Investigating the Darwinian dynamics of resistance led to a conclusion that current principles of cancer therapy are often evolutionarily unwise. Traditional drug administration in metastatic cancer applies continuous, maximum tolerated dose (MTD). Treatment is modified or discontinued only when there is unambiguous progression of disease or if symptoms from toxicity are unacceptable. By applying continuous MTD of the same drug or drug combination, the therapist effectively kills all or nearly all of the treatment-sensitive cells leaving only those that are resistant. If the initial treatment was effective in significantly reducing the tumor burden, we can assume the resistant population prior to therapy was very small. Furthermore, using the principle of "evolutionary triage" (Gatenby et al., 2014) we can estimate the resistant cells were significantly less fit than the sensitive cells. Most likely, this reflects the fitness cost of the molecular machinery of resistance. In other words, whether resistance was present prior to treatment or arose from phenotypic switching, synthesis, maintenance, and operation of the molecular machinery necessary for resistance requires energy and substrate. This represents an evolutionary cost that must be balanced by a corresponding benefit. In the absence of treatment, there is no fitness benefit so the resistant population is less fit than the sensitive population and, therefore, smaller. Only in the presence of therapy is the resistant phenotype fitter than its sensitive competitors. Thus, continuous MTD treatment both intensely selects for resistance and eliminates virtually all of their potential competitors. This leaves only resistant cells to compete with each other to gain a proliferative advantage. This allows the resistant cells to evolve compensatory mechanisms for the cost of resistance so that they will proliferate at an increasing rate leading to progression by cells growing even faster than the pretreatment population. In evolutionary biology, these dynamics are commonly observed in management of invading species or agricultural pests. Termed "competitive release," application of high dose pesticides results in rapid emergence of resistance resulting in, for example, uncontrollable crop damage. Interestingly, pesticide manufacturers are required to generate a "resistance management plan" (Hemingway et al., 2013; Mnzava et al., 2015) to obtain approval for any new agent. In contrast, cancer medicine takes a more "fatalistic" view of resistance as inevitable and unstoppable.

Note, however, that the evolutionary dynamics that govern proliferation of resistant phenotype offer opportunities for intervention that can alter the population interaction. That is, emergence of a resistant phenotype probably *is* inevitable given the size, diversity, and spatial dispersion of metastatic cancers but rapid proliferation of these cells to form a clinically evident recurrence *is not*.

Evolution during cancer therapy

In evolutionary terms, the entire cancer population within an individual patient is a clade (arising from one cell or a small population of cells) that is divided into multiple species. Since populations typically achieve a fitness maximum to a stable environment within five generations, most speciation events will occur when there is a change in environmental selection pressures. The dynamics are readily visible in Fig. 28.1, in which breast cancer cells within a population stain for Ki67 (indicating proliferation) and cleaved caspase 3 (indicating cell death). This continuous cycle of cell birth and death permits Darwinian selection in which fitter cells (within the local environment) proliferate at the expense of less fit phenotypes. Importantly, fitness is entirely contextual. That is, it is determined by the local environmental conditions which, in cancers, can vary over time and space. In Fig. 28.2, the spatial gradient of expression of glucose transporter 1 (GLUT1) likely represents a decline in oxygen concentration as it diffuses into the intraductal tumor from blood vessels on the other side of the basement membrane. Here, the phenotype switching observed along the radius of the duct is not arbitrary or the result of mutations. Rather, it simply an acclimation to local conditions. In addition to these spatial variations, malignant cells in invasive

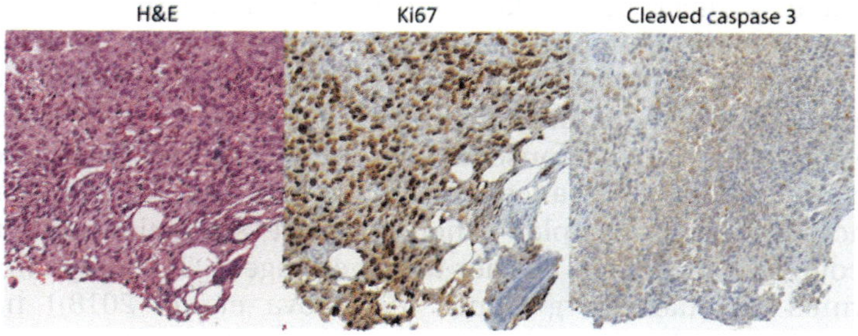

Figure 28.1 MDA-mb-231 tumor subsequence slices stained for *(left to right)*: H&E, Ki67, and cleaved caspase. Comparing Ki67 and cleaved caspase 3 demonstrate the presence of proliferative and dying cells, respectively.

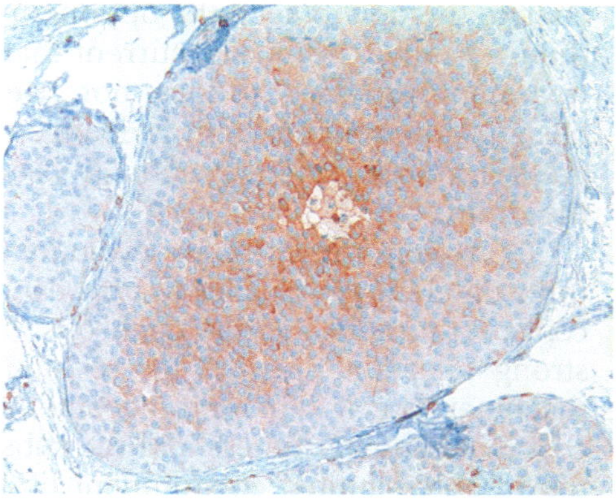

Figure 28.2 Tumor cells stained for GLUT1, a marker for glucose transporter. Note the gradient from the blood vessel to the outer rim of the tumor.

cancers may be subject to cyclical or stochastic temporal changes as the dysregulated blood vessels often result in chaotic perfusion dynamics. Should blood flow in an adjacent vessel stop or reverse, the cells will need to rapidly adjust to hypoxia or normoxia and then again deal with a burst of oxygen free radicals when flow resumes. This can impose novel selection forces that will, in general, increase the capacity of cancer cells to survive in a wide range of conditions facilitating invasion and metastases (Gillies et al., 2018).

Incorporating evolutionary principles into cancer treatment

Game theoretic analysis of cancer treatment views it as a competition between the treating physician who applies therapy and cancer population, which counters with resistant strategies. In theory, the game is heavily weighted in favor of the treating physician because he/she has two major competitive advantages (Stankova et al., 2018). First, only the oncologist is sentient. Cancer cells, like all evolving organisms, can only adapt to current or prior conditions and can *never* anticipate the future. In contrast, the treating physician can design current therapy to anticipate and exploit future therapy directed against the eco-evolutionary consequences of cancer cell adaptation. Second, the game has a leader-follower organization so that the cancer cells cannot "play" their adaptive strategies until the oncologist has applied his/her treatment agent. These dynamics [termed a "Stackelberg Game" (Stankova et al., 2018)] have been extensively investigated and consistently confer a significant advantage on the leader because he/she can continuously change therapy to keep the opponent constantly unprepared for the next move. Ironically, current therapy, by applying the same treatment repeatedly, does not use this advantage. In fact, because the treating agent is changed only when the tumor progresses, current therapy actually cedes control of the game to the tumor.

To exploit these tumor evolutionary dynamics, we present three different approaches to treat cancer based on three key concepts: (1) tumors are heterogeneous, (2) cancer treatment applies strong selection pressure for resistance, and (3) the molecular machinery a cell deploys to become resistant is associated with a cost that can alter fitness and the subsequent cell proliferation in the resource-limited environment of a clinical cancer. The latter is likely consistently found chemotherapy, hormone therapy, and immunotherapy. While a cost of resistance in targeted therapy is not necessarily obvious, the concept of "evolutionary triage (Gatenby et al., 2014)" presented above suggests the presence of rare resistance mutations, because it has failed to expand in the cancer environment, is less fit than the dominant mutation at the time of therapy. Some experimental studies seem to be consistent with this assumption.

Adaptive therapy

A second general strategy limits the application of a treatment drug to maintain a population of sensitive cells, which can then suppress proliferation of resistant cells. This typically requires an on/off strategy. The drug is applied until the tumor burden decreases by some fraction (e.g., half) of its pretreatment volume based on radiographic size measurements or serum biomarkers (e.g., prostate specific antigen (PSA)). At this point, treatment is discontinued and not started again until the pretreatment burden is again reached. While tumor regrows to its prior size, the proliferative phase occurs without application of treatment. Therefore, there is no selection for resistance. In fact, the sensitive cells, because they are fitter than the resistant phenotypes in the absence of therapy, will proliferate at the expense of the resistant population. Thus, when the tumor is again ready for treatment, the ratio of sensitive to resistant cells will be roughly the same as it was prior to the first treatment. The cycle is then repeated.

We have tested this concept in several preclinical models (Enriquez-Navas et al., 2016, 2019; Gatenby et al., 2009) and currently it is being applied in a clinical trial (Zhang et al., 2017) and different more clinical trials are accruing patients or waiting for approval.

In our preclinical studies, we have noted that there are two phases in adaptive therapy to control tumor growth. In rapidly growing exponentially, two or three initial doses of drug at MTD must be applied to control tumor growth. However, subsequently, the dose can be lowered and then given less frequently. In many cases, the mice could be given treatment "vacations" for weeks without losing tumor control. We hypothesize the "plateau" (undercontrol) phase is due to selection for a slower growing cancer cell "species" by the prior application of cytotoxic drugs that will tend to eliminate rapidly proliferating cells and have less effect on slower proliferation subpopulation. We have also noted that the controlled tumor consistently demonstrates increased vascularity compared to controls. Corroborated by mathematical models (Gallaher et al., 2018), these results have shown that the broad principles in adaptive therapy may be effective in controlling even heterogeneous and rapidly growing tumors. However, there are many opportunities to vary the specific protocols to optimize the treatment based on the parameters of each patient and each tumor.

The first clinical trial of Adaptive Therapy (Zhang et al., 2017) focused on abiraterone therapy for men with metastatic castration-resistant prostate cancer. A preliminary report an adaptive protocol of drug administration substantially increased (by a factor of about 2.5) the time to progression (compared to published data and a contemporaneous cohort) while decreasing the cumulative abiraterone dose to 40% of the amount the patients would have received with standard continuous dose at MTD until progression. Importantly, the pattern of response to Adaptive Therapy was highly variable and mathematical analysis of the patients who progressed (as of this writing 11 of 18 patients are still on treatment) demonstrated alternative strategies that are predicted to have prolonged tumor control (West et al., 2019).

A significant challenge in adaptive therapy is that it is highly demanding of tumor information. To make the correct treatment decision, the oncologists need metrics of ongoing tumor evolution during treatment including the size and locations of critical subpopulations (e.g., therapy sensitive and resistant). There are currently no technologies that can consistently and accurately provide this information. Ongoing research is focused on the potential roles of circulating tumor cells, serum cell-free DNA, and quantitative analysis of cross sectional imaging such as computerize axial tomography (CAT) or magnetic resonance imaging (MRI) scans.

Finally, we emphasize the role of mathematical modeling in trial design. Once a model of the specific treatment and underlying evolutionary dynamics has been established, thousands of simulations [termed "phase *i* trials" (Scott, 2012)] can be performed to define critical parameters as well as specific treatment protocol designs. Furthermore, because the treatment is based on a mathematical model, the observed outcome in each patient can be analyzed. That is, the mathematical model can be used in a time-reverse mode to move from the observed outcomes to the (unobserved) initial conditions (West et al., 2019). The model can take the initial conditions and move forward in time using different protocols to identify alternative treatment strategies that would have improved outcomes (West et al., 2019). Consistent application of this "after action analysis" (AAA) (Stankova et al., 2018) will also permit a more detailed analysis of the entire cohort. Finally, the AAA will identify key components in the initial conditions that need to be quantified in pre-therapy investigations. In this iterative process, our goal is to use the mathematical models to design patient-specific optimization treatment plans that provide decision support to assist in the specific application of individual agents (or combination of agents) as well as the optimal sequence of treatments.

Can the cost of resistance be increased?

The competition between sensitive and resistant cancer cells is governed by their relative fitness. Thus, any strategy that selectively decreases the fitness of resistant cells may prolong response and improve outcomes. One of the most common mechanisms that cells develop to survive chemotherapies is the upregulation of membrane pumps. For example, the multidrug resistant (MDR) complex will extrude cytotoxic drugs from the cytoplasm to the extracellular space (Enriquez-Navas et al., 2016). The cost of this synthesis, maintenance, and operation of these pumps has been investigate empirically and shown to consume up to 50% of the energy budget of resistant cells (Gatenby et al., 2009; Enriquez Navas et al., 2019). Notably, these MDRs (as their name implies) are nonspecific and usually act upon a wide range of agents. This suggested a strategy to further reduce the fitness of resistant cells by increasing the cost of the MDR operation through administration of a noncytotoxic drug that is nevertheless, extruded by the MDR membrane pumps. In effect, this forces the resistant cells to consume more energy while providing no gain as would is the case when they extrude cytotoxic drugs.

In pre-clinical studies, verapamil (a drug used cardiovascular diseases) has been shown to be an MDR-1 ligand. When administered with doxorubicin to breast cancer cells that had previously evolved resistance to doxorubicin through upregulation of MDR proteins (Zhang et al., 2017; Gallaher et al., 2018), increased efficacy was observed probably because it acted as a competitive inhibitor. Clinical trials using simultaneous administration of verapamil and a cytotoxic agent have found some success.

However, in the context of adaptive therapy, mathematical models demonstrate that verapamil or a similar agent will selectively inhibit proliferation of resistant cells if administered after the treatment drug has been withdrawn (Kam et al., 2015).

Double bind

As noted earlier, a major advantage accrued to the sentient treating physician is the ability to plan future events. In contrast, evolving cancer cells can never anticipate the future and adapt to a treatment that has not been applied. This advantage can be exploited by strategically linking two treatments over time so that the cancer cell adaptive strategy to the initial therapy renders it more vulnerable to a follow-up agent. This strategy has been termed "double bind" (West et al., 2019) or "sucker's gambit" (Merlo et al., 2006; Maley et al., 2004).

Some tumors are sensitive to several different drugs. If toxicity is not an issue, one therapeutic strategy is to simply apply all of the drugs at the same time. This, in fact, is often used in clinical oncology and, in general, multidrug therapy increases the rate and duration of response compared to monotherapy. However, there appears to be a limit to this benefit because adding additional drugs beyond a threshold of two or three usually does not significantly improve the response rate but does increase toxicity.

Thus, an alternative treatment plan using the "double bind" or "sucker's gambit" (Gatenby et al., 2009; Maley et al., 2004) strategy suggests that sequential rather than simultaneous combinations of drugs may be equally effective when evolutionary dynamics are used to design the protocol. For example, Antonia et al. (Antonia et al., 2006) investigated the efficacy of a p53 vaccine in a heavily pretreated cohort of patients with small cell lung cancer. Although an immune response was observed in nearly all of vaccine-treated patients, there was only one partial clinical response. However, when this group was subsequently treated with a standard chemotherapy regimen, the response rate was over 60% (compared to an expected rate of $<5\%$) and the patients who had the greatest immune response to the vaccine also were most likely to respond to chemotherapy. The results suggest that either the adaptive response to the vaccine increased the cells' vulnerability to chemotherapy or the latter decreased the efficacy of the response to the vaccine.

A general evolutionary model for this strategy is "predator facilitation." For example, if the goal is to control a destructive rodent population in a field, one can introduce owls. They will be very effective "treatment agents" but the rodents adapt by hiding under bushes. However, this adaptive strategy renders the rodents vulnerable to addition of a new agent—snakes. When the owl is present, mice will hide under the bushes, which will make them more sensitive to snakes and vice versa.

Conclusion

Most cancer patients die because of metastatic disease. While a large and growing number of systemic treatment agents are available to the treating oncologist, evolution of resistance remains a significant barrier to improving overall survival. In the past, investigations into the emergence of resistance focused on identifying and then blocking the molecular machinery that cells deployed as they evolved resistance. These approaches were generally unsuccessful probably because each cancer cell, with access to the vast information stored in the human genome, have the capacity to deploy a wide range of adaptive strategies. So, when one is blocked, the cell can use any number of alternatives.

An alternative approach to the problem of resistance to cancer therapy focuses on disrupting the evolutionary dynamics that govern proliferation of resistant cells. That is, the presence of few cells that possess the molecular machinery necessary for resistance is not clinically significant until that population expands through multiple generations into billions of individuals. By using more than a century of accumulated knowledge gained by evolutionary biologist, these evolutionary dynamics can be disrupted to delay or even prevent proliferation of a clinically important population of resistance cells. However, successful integration of Darwinian principles into cancer treatment requires a multidisciplinary team that includes oncologists, experimentalists, evolutionary biologists, and mathematicians.

Finally, this approach also requires changes in the research focus of cancer therapy. For the past century, improvements in cancer therapy have relied on new drug development. This, of course, should continue and there is a worldwide pharmaceutical industry that will do so. However, we propose that many currently available drugs are not administered in a way that optimizes patient outcomes. Thus, cancer treatment investigations also need to focus on *tactics*. The conventional view, that greatest benefit in patients with metastatic cancer is achieved by using the MTD to kill as many cancer cells as possible, is often evolutionarily unwise. In fact, in some models, these tactics maximally accelerate evolution of resistance. Thus, administering currently available drugs using evolutionarily enlightened strategies may improve outcomes without incurring the often enormous costs associated with new drug discovery.

References

Antonia, S.J., Mirza, N., Fricke, I., Chiappori, A., Thompson, P., Williams, N., et al., 2006. Combination of p53 cancer vaccine with chemotherapy in patients with extensive stage small cell lung cancer. Clin. Cancer Res. 12 (3 Pt 1), 878–887. Available from: https://doi.org/10.1158/1078-0432.CCR-05-2013. Epub 2006/02/10. PubMed PMID: 16467102.

Dunbar, C.E., High, K.A., Joung, J.K., Kohn, D.B., Ozawa, K., Sadelain, M., 2018. Gene therapy comes of age. Science 359 (6372), eaan4672. Available from: https://doi.org/10.1126/science.aan4672.

Enriquez-Navas, P.M., Kam, Y., Das, T., Hassan, S., Silva, A., Foroutan, P., et al., 2016. Exploiting evolutionary principles to prolong tumor control in preclinical models of breast cancer. Sci. Transl. Med. 8 (327), 327ra24. Available from: https://doi.org/10.1126/scitranslmed.aad7842. 327ra24. Epub 2016/02/26; PMCID: PMC4962860. PubMed PMID: 26912903.

Enriquez Navas, P.M., Garcia, L., Abdalah, M., Stringfield, O., Luddy, K., Hassan, S., et al., 2019. Intermittent hormonal therapy shows similar outcome than SOC in ER + breast cancer preclinical model. bioRxiv 509158. Available from: https://doi.org/10.1101/509158.

Gabor Miklos, G.L., 2005. The human cancer genome project—one more misstep in the war on cancer. Nat. Biotechnol. 23, 535. Available from: https://doi.org/10.1038/nbt0505-535.

Gallaher, J.A., Enriquez-Navas, P.M., Luddy, K.A., Gatenby, R.A., Anderson, A.R.A., 2018. Spatial heterogeneity and evolutionary dynamics modulate time to recurrence in continuous and adaptive cancer therapies. Cancer Res. 78 (8), 2127–2139. Available from: https://doi.org/10.1158/0008-5472.CAN-17-2649. Epub 2018/02/01; PMCID: PMC5899666. PubMed PMID: 29382708.

Gatenby, R.A., 2009. A change of strategy in the war on cancer. Nature 459 (7246), 508–509. Available from: https://doi.org/10.1038/459508a. Epub 2009/05/30. PubMed PMID: 19478766.

Gatenby, R.A., Brown, J., Vincent, T., 2009. Lessons from applied ecology: cancer control using an evolutionary double bind. Cancer Res. 69 (19), 7499–7502. Available from: https://doi.org/10.1158/0008-5472.can-09-1354. Epub 2009/09/16. PubMed PMID: 19752088.

Gatenby, R.A., Silva, A.S., Gillies, R.J., Frieden, B.R., 2009. Adaptive therapy. Cancer Res. 69 (11), 4894–4903. Available from: https://doi.org/10.1158/0008-5472.CAN-08-3658. PMCID: 3728826. PubMed PMID: 19487300.

Gatenby, R.A., Cunningham, J.J., Brown, J.S., 2014. Evolutionary triage governs fitness in driver and passenger mutations and suggests targeting never mutations. Nat. Commun. 5, 5499. Available from: https://doi.org/10.1038/ncomms6499. Epub 2014/11/20; PMCID: 4260773. PubMed PMID: 25407411.

Gillies, R.J., Brown, J.S., Anderson, A.R.A., Gatenby, R.A., 2018. Eco-evolutionary causes and consequences of temporal changes in intratumoural blood flow. Nat. Rev. Cancer 18 (9), 576–585. Available from: https://doi.org/10.1038/s41568-018-0030-7. Epub 2018/06/13. PubMed PMID: 29891961.

Hemingway, J., Vontas, J., Poupardin, R., Raman, J., Lines, J., Schwabe, C., et al., 2013. Country-level operational implementation of the Global Plan for Insecticide Resistance Management. Proc. Natl Acad. Sci. U.S.A. 110 (23), 9397–9402. Available from: https://doi.org/10.1073/pnas.1307656110. Epub 2013/05/23; PMCID: PMC3677419. PubMed PMID: 23696658.

https://ghr.nlm.nih.gov/primer/hgp/accomplishments.

https://homedna.com/blog/human-genome-project-celebrating-15th-anniversary. Available from: https://homedna.com/blog/human-genome-project-celebrating-15th-anniversary.

Kam, Y., Das, T., Tian, H., Foroutan, P., Ruiz, E., Martinez, G., et al., 2015. Sweat but no gain: inhibiting proliferation of multidrug resistant cancer cells with "ersatzdroges". Int. J. Cancer 136 (4), E188–E196. Available from: https://doi.org/10.1002/ijc.29158. Epub 2014/08/27; PMCID: 4262548. PubMed PMID: 25156304.

Katsios, C., Roukos, D.H., 2010. Individual genomes and personalized medicine: life diversity and complexity. Per Med. 7 (4), 347–350. Available from: https://doi.org/10.2217/pme.10.30. PubMed PMID: 29788639.

Maley, C.C., Reid, B.J., Forrest, S., 2004. Cancer prevention strategies that address the evolutionary dynamics of neoplastic cells: simulating benign cell boosters and selection for chemosensitivity. Cancer Epidemiol. Biomarkers Prev. 13 (8), 1375–1384. Epub 2004/08/10. PubMed PMID: 15298961.

Merlo, L.M., Pepper, J.W., Reid, B.J., Maley, C.C., 2006. Cancer as an evolutionary and ecological process. Nat. Rev. Cancer 6 (12), 924–935. Available from: https://doi.org/10.1038/nrc2013. Epub 2006/11/17. PubMed PMID: 17109012.

Mnzava, A.P., Knox, T.B., Temu, E.A., Trett, A., Fornadel, C., Hemingway, J., et al., 2015. Implementation of the global plan for insecticide resistance management intrun 1502 malaria vectors: progress, challenges and the way forward. Malar. J. 14, 173. Available from: https://doi.org/10.1186/s12936-015-0693-4. Epub 2015/04/23; PMCID: PMC4423491. PubMed PMID: 25899397.

Scott, J., 2012. Phase i trialist. Lancet Oncol. 13 (3), 236. Epub 2012/04/11. PubMed PMID: 22489289.

Stankova, K., Brown, J.S., Dalton, W.S., Gatenby, R.A., 2018. Optimizing cancer treatment using game theory: a review. JAMA Oncol. 5 (1), 96–103. Available from: https://doi.org/10.1001/jamaoncol.2018.3395. Epub 2018/08/12. PubMed PMID: 30098166.

Topol, E.J., 2014. Individualized medicine from prewomb to tomb. Cell. 157 (1), 241–253. Available from: https://doi.org/10.1016/j.cell.2014.02.012. Epub 2014/04/01; PMCID: PMC3995127. PubMed PMID: 24679539.

Vidi, P.A., Bissell, M.J., Lelievre, S.A., 2013. Three-dimensional culture of human breast epithelial cells: the how and the why. Methods Mol. Biol. 945, 193–219. Available from: https://doi.org/10.1007/978-1-62703-125-7_13. Epub 2012/10/26; PMCID: 3666567. PubMed PMID: 23097109.

Weaver, V.M., 2014. The microenvironment matters. Mol. Biol. Cell 25 (21), 3254–3258. Available from: https://doi.org/10.1091/mbc.E14-06-1080. Epub 2014/11/02; PMCID: PMC4214768. PubMed PMID: 25360044.

West, J.B., Dinh, M.N., Brown, J.S., Zhang, J., Anderson, A.R., Gatenby, R.A., 2019. Multidrug cancer therapy in metastatic castrate-resistant prostate cancer: an evolution-based strategy. Clin. Cancer Res 25 (14), 4413–4421. Available from: https://doi.org/10.1158/1078-0432.CCR-19-0006. Epub 2019/04/18. PubMed PMID: 30992299.

Zhang, J., Cunningham, J.J., Brown, J.S., Gatenby, R.A., 2017. Integrating evolutionary dynamics into treatment of metastatic castrate-resistant prostate cancer. Nat. Commun. 8 (1), 1816. Available from: https://doi.org/10.1038/s41467-017-01968-5.

Index

Note: Page numbers followed by "*f*" and "*t*" refer to figures and tables, respectively.

A

Aberrant mucin-type O-glycosylation, 538–539
Abnormal attractors, 562
Abnormal glycosylations, 536–538
Accommodation, genetic, 425–427
Acetylation, 259–260
  histone, 213
Acetyl-CoA, 352–353
  molecule, 68
    oxidation of, 66–67
Acetyl CoA carboxylase 1 (ACC1), 662
Acetyl CoA carboxylase 2 (ACC2), 662
Acidification, of mitochondrial matrix, 80–81
Adaptation, 313–320, 326–328
  of cell, 309–320
  dynamics, 322, 330–331
  environmental, 212*f*, 234, 239–240, 242
  metabolism in, 316
  plasticity as, 423–424
Adaptive cell states, 627–630
Adaptive phenotypic plasticity, 409
Adaptive phenotypic switch, autophagy role in, 664–665
Adaptive therapy, 697–699
Aerobic glycolysis, 11–12
Ak strain transforming (AKT), 592–593
Alcohol oxidase, 84
  oligomerization of, 84
Alkaline proteins, 82
Alleles, equilibrium frequency of, 285–286
Allometries, reaction norms and, 424
Alpha B-crystallins, 89–90
Alternative phenotypes, 404–405
Amino acid metabolism, plasticity in, 663–664
Amoeboid to mesenchymal transition (AMT), 7–8
AMP-activated protein kinase (AMPK), 11–12, 655, 657–658, 660–661, 665–667
Amphimedon queenslandica, 262
AMPK-Akt double negative feedback loop, 665–667, 667*f*
AMT. *See* Amoeboid to mesenchymal transition (AMT)
Amyloids, 107
Anchorage-deprived cancer cells, 656–657
Androgen receptor (AR), 591–592
Androgen receptor blockade, phenotypic plasticity overcomes, 594–602, 595*f*, 596*f*
Androgen receptor signaling, 592–594
  axis in prostate adenocarcinoma, 592–594
Anfinsen theory, 90–91
Angiopoietin 2 (Ang2), 385–386
Angiotensin II protein expression, 389–390
Ang2 protein expression, 385–386
Anoikis, 653, 655–659, 661–663
Anoikis resistance, 657–658, 661–663, 667–669
Antiandrogen resistance, 679–680
Apia mellifera, 214
Apoptosis, 366–367
AR. *See* Androgen receptor (AR)
Arabidopsis thaliana, 230
Araschnia levana, 411–412
Aromatase, 410
Artificial selection, 513–514
Attractors, 366
  abnormal, 562
Autophagy, role in adaptive phenotypic switch, 664–665
Autopsy, 594–595

B

Bacillus subtilis, 30–31, 31*f*, 32*f*
  "bet-hedging" property of, 33
Balanced equilibrium system, 162–163
Behavior of dog, 511, 512*f*
Benign prostatic hyperplasia (BPH), 569–572, 571*f*
  from BPH to PCa via EMT/MET, 572–573
Beta-catenin, 263–265
"Bet hedging" strategy, 5, 230, 241, 372, 386–387, 391, 394
Bicyclus anynana, 420

Biochemical traits, stochastic developmental variation (SDV) of, 223
Bioenergetics and chromatin, 68
Biological markers, fluorescent, 23–25
Biological noise, 196–197, 361–362, 604–605
 in developmental plasticity, 368–370, 369*f*
 dynamic vWF mosaicism in vitro, 378–379
 implications of, 388–394
 and phenotypic plasticity, 367–371
 plasticity in response to, 369*f*
 in response to environmental change, 370–371
Biological noise-mediated heterogeneity, 371–372
Biological regulatory networks, 365–366
Biological switch, 369*f*, 392*f*
Biomolecular condensate, 262, 269
Biosynthesis, ecdysone, 421–422
Bird song learning, 503
Bistability
 frustrated, 350*b*, 351*f*
 HCV-induced, 132
 in interferon signaling network, 133–137
 to metabolic phenotype, 665–667, 667*f*
Bistable systems and plasticity, 665–667
Blood brain barrier, 387–390
B lymphocyte, 34–35
Boltzmann law, 162–163
Boolean function, 254–256
Bovine ruminant, 466–467
BPH. *See* Benign prostatic hyperplasia (BPH)
BRAF inhibition, 628–629
Brain capillaries, loss of vWF mosaic expression in, 387–388
Brain-neuroendocrine axis, 410
Breast cancer progression, 651–652
Britten-Davidson MGRN program, 255
Britten-Davidson model, 254–256
Bumblebees, queen determination in, 412–413
Butyrate, 463–464

C

Cadherins, 256
CA125 glycan, 542–543
Calcium
 ions, 653–654
 mitochondrial, 653–654
Canalization and polyphenism, 431*f*
Cancer, 321 *See also specific types of cancer*
 drug resistance in, 627
 epithelial-mesenchymal transition in, 553–563
 evolution and glycan-driven plasticity, 541–542
 evolution in, 692
 glycobiology and, 529–532
 growth and drug resistance, 678–681
 heterogeneity in, 652–653, 677
 metabolic reprogramming in, 669–670
 ovarian, 527–528
  progression, 532–536, 534*f*, 537*f*
 phenotypic plasticity in, 652–653
 phenotypic transitions in, 553–554
 progression, 652–653
 prostate, 569–570
 therapy resistance, 554–556
Cancer cells
 invasion, pattern, 7–8, 9*f*
 plasticity, 643–645
 resistant, 699
 state definitions, 621*t*
 survival, metabolic plasticity, 655–665
Cancer stem cells (CSCs), 11, 644, 677–678
 model, 669
 phenotypic plasticity and, 639–640
 population in melanoma cells, 640*f*
 senescence and, 644*f*
Cancer therapy, 693–694
 evolutionary principles, 696–697
 evolution during, 694–696
Candida albicans, 12–13, 215
Capillary cells, 366–367
Capsaspora owczarzaki, 260
Cardiac phenotypic plasticity, 13
Cardiac remodeling, 13
CAT. *See* Collective to amoeboid transition (CAT)
Catechol-o-methyltransferase (COMT), 92
β-Catenin, 81, 263–265, 271–272
CDC-like kinase 2 (CLK2), 575–576
 hyperphosphorylation of PAGE4, 89
CD2 receptor, 82–83
CDSN. *See* Corneodesmin (CDSN)
Cdx2, 168–169, 168*f*
Cell adaptation, to unforeseen challenge, 310–320, 314*f*
Cell adhesion, 645
Cell biology, universality in, 200–202, 202*f*
Cell–cell interaction, 316–319
 as constraints, 683–685
Cell cycle, 312
Cell decision making, 161
Cell differentiation, 53–56, 62–63, 158–159, 158*f*, 169, 321

biological explanations in terms of chance, 42–46, 44*f*
cell types and, 196*f*
in immunology, 33–37
in multicellular organisms, 320–321
from noise to chance as explanatory, 30–33
positive view of chance, 37–42, 39*f*
stem, 174
Cell division, 264
Cell dynamics, 326–327
Cell fate decision making
in differentiation and development, 174
Waddington landscape and quantified landscape, 176–177
of differentiation and reprograming, 160–168
for differentiation and reprograming through regulations, 169–170
discrete paths *vs.* continuous paths, 182
dynamics of differentiation/ development, 175–176
gene regulatory network dynamics of, 166–167
intermediate states, 181–182
irreversibility and time arrow, 178
kinetic rates of differentiation/ development and reprograming, 166–168
network dynamics of, 165–166
optimal paths for quantifying differentiation/ development and reprograming process, 164–166
origins of bifurcations and phase transitions, 177
quantifications of transition states, speed, and optimal paths of, 179–181
transition states for, 180–182
Cell fate dynamics, physical laws of, 192–196, 196*f*
Cell-matrix interactions, 653
Cell migration, phenotypic transformations and collective, 645–646
Cell niche, 642–643
Cell phenotype, 53–54, 68–69
Cell plasticity, 645
Cell populations, universal protein distribution in, 323–325, 324*f*
Cell state, 620–622
adaptive, 627–630
characterization, 630–633
drug naive, 624–627
dynamics, statistical mechanics, 196–200
in phenotypic landscape, 618–620
stochastic dynamical equation for, 165
via relevant heterogeneity, 630–633
well-characterized, 622–624
Cell-state heterogeneity
in isogenic microbial populations, 335–337
of yeast in well-mixed chemostat, 346–353
Cell-state organization, 305–306, 309–310
exploratory dynamics of, 322–328
physical framework of, 323–325
Cell-to-cell nongenetic variability, 23–25
Cell-to-cell phenotypic variability, 23–25
Cellular differentiation, epigenetic landscape and implications in, 5–7, 6*f*
Cellular level, phenotypic plasticity at, 5–13
Cellular memory, stochastic phenotype switching and, 371
Cellular phenotypes, 62, 335
heterogeneity in, 618
Cellular phenotypic plasticity, 677
Cellular plasticity and stemness, 681–683
during differentiation, 683–685
Cellular senescence, 643
Central limit theorem, 201
Central nervous system and hormones, 409–410
CFTR gene, 94
CFTR protein, 94
CGV. *See* Cryptic genetic variation (CGV)
Chance
definition, 24*t*, 42–46
and reductionism/ antireductionism debate, 46–47
Chaperone, 84–85
Chemostat, 312–313, 347–348
adaptation dynamics in, 314*f*
cell-state heterogeneity of yeast in well-mixed, 346–353
experiments, 313–314, 316–319
Chlamydomonas, 298–299
Cholangiocarcinoma, sarcomatoid, 626–627
Chromatin, 64–68
bioenergetics and, 68
fluctuation, 65
modifications, 239–240
organization, effect of, 3–4
poised, 3–4
posttranslational modifications of, 67*f*
three-dimensional structure of, 3–4
Chromosomes, 285
Chronic lymphocytic leukemia, 680

Chrosomus eosneogaeus, 230–231
Circulating tumor cells (CTCs), 555, 665–667
CJD. *See* Creutzfeldt Jakob Disease (CJD)
C-lectins, 530–531
CLK2. *See* CDC-Like Kinase 2 (CLK2)
Clonal cell populations, phenotypic heterogeneity in, 337
Clonal microbial colonies, 338
Coarse-grained process, 323–325
Codons, 90–91
  sense, 114
Coefficients of variation (CV), 218–221
Coexpression networks, 267
Cofactor, 66
Collective to amoeboid transition (CAT), 7–8
Colonization, 429–430
comK gene, 30–31
ComK protein, 30–33
Computerize axial tomography (CAT), 698
COMT. *See* Catechol-o-methyltransferase (COMT)
Conformational diversity, generates phenotypic diversity, 120–122
Conformational noise, 3–4
Constraints of development, 519–521, 520*f*
Conventional nonlinear dynamics, 161–162
Core gene regulatory motif, 173
Corneodesmin (CDSN), 94–95
Corticosteroids, 522–523
Cortisol, 410
Cotranslational folding, synonymous mutations modulate, 95–97, 97*f*
CpG dinucleotides, DNA methylation of, 379–382
CpG dyads, 383–384
CpG methylation, 383–384
Creutzfeldt Jakob Disease (CJD), 106–107
Cryptic genetic variation (CGV), 427–428
  in genetic accommodation, 427–430
Crystallin, gamma-B, 95–96
CSCs. *See* Cancer stem cells (CSCs)
CTCs. *See* Circulating tumor cells (CTCs)
CTLs, 148–149
Curl flux, 163–168, 164*f*, 171–172, 175–176
cyp19a gene, 408
cyp19a methylation, 408
Cystic fibrosis, 94
Cytosine-to-thymine transition, 240

D

Daphnia, 406
  D. longicephala, 406
Darwinian dynamics, 692
Darwinian evolution, 63
Darwin's theory, 283
Decision making, 157, 159
  cell, 161
  cell fate, 159–160
Dedifferentiation, 7
Delta factors, 40–41
Deterministic dynamics, nonequilibrium landscape in, 175–176
Deterministic model, 55–57
Developmental niche construction, 468
Developmental noise, 231, 233*f*
Developmental plasticity, 403, 445–446, 497
  biological noise in, 368–370, 369*f*
  regulation of, 404–422
Developmental rates, 521–523
Dicentrarchus labrax, 408
Dictyostelium discoideum, 14
Differentially methylated DNA regions (DMRs), 229–230
Differentiation
  cellular plasticity and stemness during, 683–685
  inherent cell functions in origin of, 266–269
Diffusion coefficient, 199
Diffusion equation, Fokker-Planck, 162
Dihydrolipoamide dehydrogenase (DLD)
  crystal structure of human, 80–81, 81*f*
  impairment of DLD activity, 80–81
Direct-acting antiviral agents (DAAs) treatments, 131–132, 141–144, 142*f*
  mutational pathways of resistance to, 144–147
  potential considerations and strategies for optimizing, 144–150
Disctyostelium discoideum (DdS4), 80
Disentangling genetic variation
  in field studies, 229–231
  in laboratory animals, 227–229
  by mathematical modeling, 231–232
Diversification, isologous, 257–258
Diversity, 281–282
  phenotypic, 513–514
Division of labor, 337–338, 353–354
DLD. *See* Dihydrolipoamide dehydrogenase (DLD)
DMRs. *See* Differentially methylated DNA regions (DMRs)
DNA cytosine-5-ethyltransferases (DNMTs), 407
DNA methylation, 173, 213, 234–235, 288–289, 379–384, 380*f*, 407–408

of CpG dinucleotides, 379–382
Dnmt3, 407
DNMT-coding gene, 407
Dnmt3 RNAi, 407
DNMTs. *See* DNA cytosine-5-ethyltransferases (DNMTs)
Dog
behavior of, 511, 512*f*
domestic, 513
domestication, 511–512
Domain–domain interactions, 88–89
in multidomain proteins, 85–86
Domesticated foxes, phenotypic changes of, 521–523
Domestication
in evolutionary terms, 511–513
nature of changes under, 514–516
selection of foxes for, 522–523
Domestication syndrome, 515, 523
Domestic dogs, 513
Dopamine, 494–495
Dopamine receptor D2 (DRD2) gene, 92
Double bind, 699–700
Double-stranded RNA (dsRNA) complex, 133–135
Drosophila melanogaster, 428–429
Drug naive cell states, 624–627
Drug resistance
in cancer, 627
cancer growth and, 678–681
Dsx, 419
Dung beetles, 413
dynactin p62, 407
Dynamical systems theory, 365–366
Dynamic heterogeneity, of vWF expression in vivo, 372–378, 375*f*, 376*f*, 379*f*
Dynamic mosaic heterogeneity, 389*f*, 391–392
Dynamic VWF expression, 377–378
Dynamic vWF mosaicism in vitro, biological noise, 378–379

E

E-cadherin, 534–535, 574
Ecdysone, 411–412
biosynthesis, 421–422
ECM. *See* Extracellular matrix (ECM)
Eco-evo-devo approach, 461, 472–473
Ecosystem
rumen, 465–466
ruminant fiber-eating, 473
Effectors, 413–422
involved in plastic patterning, 418–420
of plastic growth, 414–418
EGFR. *See* Epidermal growth factor receptor (EGFR)
EGFR phosphorylation, 661
EIV. *See* Environmentally induced variation (EIV)
Embryonic stem cells (ESCs), 3–7, 157–158, 168–169, 391
EMT. *See* Epithelial-mesenchymal transition (EMT)
EMT-inducing transcription factors (EMT-TFs), 555
Endocrine integrators
of phenotypic plasticity
in insects, 411–413
in vertebrates, 410–411
Endocrine signals, 415*f*
Endogenous interferon, responsiveness, 141–144, 142*f*
Endogenous mechanical stress, 556
Endothelial cells, 385–386
Endothelial heterogeneity, 367
Endothelial progenitors, 368
Endothelial-specific molecule 1 (ESM1), 389–390
Energy barriers, 363–364
Enhancers, 262, 264–266
Environmental adaptation, 212*f*, 234, 239–240, 242
Environmental change, biological noise in response to, 370–371
Environmental effects, on epigenetic variation, 230–231, 232*f*
Environmental fluctuation, Gaussian probability distribution of, 165
Environmentally induced variation (EIV), 207–208, 211–213, 212*f*, 232–236, 241
in field studies, 229–231
in laboratory animals, 227–229
occurrence of, 215–218
Environmental sensors, 408
Environmental signals, 405, 410
plasticity in response to, 369*f*
Environmental triggers, multicellularity in response to, 116–117
Environmental variation, 209
Environment-sensitive period, 404–405
Enzalutamide, 593–594, 597–600
Enzymes
activities, 65–66
plant cell wall-digesting, 464–465
ephrin-B2, 368
Epidermal growth factor receptor (EGFR), 653
Epigenetic barriers, 393*f*
Epigenetic changes, 639–640
in clonal fish, 230–231
Epigenetic control, of vWF expression and heterogeneity, 379–384, 380*f*, 382*f*
Epigenetic gene regulation, 639

Epigenetic inheritance, 290–291, 294
Epigenetic landscape, 59–62, 363–364
  random walk across, 53–54
    concepts, 54–57
    conceptual framework, 63–68
    historical origins, 57–59, 58*f*
Epigenetic marks, 214
Epigenetic memory, of gene expression, 65–66
Epigenetic molecular process, 315
Epigenetic phenotypic switching, 294, 297*f*
Epigenetic reactions, 66
Epigenetic regulation, of polyphenisms, 407–408
Epigenetics, 59–60, 172–173, 213, 241
  heterogeneity from, 178–179
  and morphological diversity in parthenogenetic marbled crayfish, 238*f*
Epigenetic switching, 297–298
Epigenetic variation, 239–240
  in asexual fish, 232*f*
  environmental effects on, 230–231, 232*f*
Epigenetic watershed, 432–434
  model, 432–433
    of genetic accommodation, 432–433, 433*f*
Epimutations, 230
Epimutation-to-genetic mutation transition, 240
Epithelial cell phenotype, 622–623
Epithelial-mesenchymal plasticity, 603–604
  landscape of, 641–643
  therapeutic strategies targeting, 561–563, 563*f*
Epithelial-mesenchymal transition (EMT), 7–8, 533, 554–556, 557*f*, 595–597, 600–605, 622–624, 642*f*, 679
  in cancer, 553–563
  matrix deprivation and, 667–668
  regulation of, 556–557
  and stemness, 669–670
Epithelial ovarian cancer (EOC), 538–539
Epithelial plasticity, in immune evasion, 600–602
Equilibrium dynamics, 163–164
Equilibrium system, 166–167
  balanced, 162–163
Escherichia coli, 79, 338
ESCs. *See* Embryonic stem cells (ESCs)
ESM1. *See* Endothelial-specific molecule 1 (ESM1)
"Essence/existence" dualism, 57–59
Estradiol, 494
Euchromatin, 65
Eukaryotes, gene splicing, 91–92
Eukaryotic organisms, 459
Evolution, 255, 257, 260, 262, 307–309, 321
  in cancer, 692
  Lamarckian, 240
  "plasticity and epigenetics-first" concept of, 241
Evolutionary capacitors, 427–428
Evolutionary triage, 693–694
Exploratory dynamics, 307–308, 329
Extracellular matrix (ECM), 7–8, 556, 651–653, 669
Extracellular signal-regulated kinases (ERK), 592–593
Extrinsic signals, 53

F

FACS. *See* Fluorescence activated cell sorting (FACS)
FACs. *See* Function-amplifying centers (FACs)
Farm-fox experiment, 515–516
FAS. *See* Fatty acid synthesis (FAS)
Fatty acid oxidizing (FAO) phenotype, 662
Fatty acid synthesis (FAS), 662
Feeding plasticity, 234
Ferromagnetic materials, 363–364
Fibroblast growth factor receptor (FGFR), 605–606
Fibrosis, cystic, 94
FISH. *See* Fluorescence in situ hybridization (FISH)
Fixed-point attractors, 195
Flocculin, 116
FLO11 gene, 118
Fluctuating asymmetry (FA), 229
Fluorescence activated cell sorting (FACS), 561, 625
Fluorescence in situ hybridization (FISH), 378–379
Fluorescent biological markers, 23–25
Fokker-Planck diffusion equation, 162–164
Fokker-Planck equation, 169–170, 197–199
Fokker-Planck formalisms, 198–200
Foxes, domesticated
  phenotypic changes of, 521–523
FOXO activity, 414
Frizzled receptors, 683–684
Functional polymorphisms, song dialects as, 498–503
Function-amplifying centers (FACs), 266, 269–271, 272*f*
Fungal cell, prion propagation and transmission in, 110–112
Fungal prion-forming proteins, 112
Fungal prions, 118–119

G

GAGs. *See* Glycosaminoglycans (GAGs)
GAL80, 315
Galectins, 540
Gal-7 expression, 541
GAL metabolic system, 316–319
GAL system, 310–313, 311*f*
Gamma-B crystallin, 95–96
[GAR$^+$]/Pma1/Std1, 118–120
Gene-centric thinking, 443
Gene circuit dynamics, 172–173
Gene—environment interactions, 362–363
Gene expression, 2–3, 54–55, 70–71, 281–282, 281*f*, 287–289, 292–293, 306–307, 316–319, 682–683
  epigenetic memory of, 65–66
  microarray, 80
  noise in, 27–29
  phenotypes and, 316–319, 318*f*
  regulation by synonymous mutations, 96–97, 97*f*
  stochastic, 336–337
    effect of, 338
  stochasticity of, 63–64
Gene expression dynamics, 169–170
Gene networks, 161, 163
General systems biology model, 200
Gene regulation, 61, 63–64, 321–322
  epigenetic, 639
Gene regulatory mechanisms, 262
Gene regulatory motif circuit, determining differentiation, 168–169, 168*f*
Gene regulatory networks (GRNs), 161–162, 253–255
Gene splicing, synonymous mutations dictate, 91–92, 93*t*
Genetic accommodation, 425–427, 429–430, 445–446, 448–449
  cryptic genetic variation in, 427–430
  epigenetic watershed model of, 432–433, 433*f*
  model of, 430–434
Genetic assimilation, 445, 453–454, 522
Genetic determinism, 56–57
Genetic program, 36
Genetic roots, of phenotypic plasticity, 363–364, 363*f*
Genetic variation, 209, 446–447
  disentangling
    in field studies, 229–231
    in laboratory animals, 227–229
Gene transcription, 64–65
Genome, 1, 5–7, 53, 55–57, 61–62
  regulatory, 253–255
Genome-wide methylomes, 239–240
Genome-wide protein coding sequence analysis, 94–95
Genomic Evolutionary Rate Profiling, 96–97
Genotype-phenotype maps, 3–4
Genotypes, 1, 305–306
GFP, 324*f*
GGRD model. *See* Global genome regulatory dynamics (GGRD) model
Global genome regulatory dynamics (GGRD) model, 255, 257–259, 270
Global transcriptional regulation, 117–118
D-Glucan, 87–88
Glucosamine, 118–119
Glucose, 656
Glucose metabolism, plasticity in, 656–659
Glucose transporter 1 (GLUT1), 694–696, 695*f*
β-Glucuronidase, 86–87
  three-domain association of, 87–88
GLUT1, 656–657
Glutaminase inhibition hinders, 663
Glutamine, 663
Glycan
  CA125, 542–543
  O-linked, 529–530
  in ovarian cancer, 528
Glycan-glycan binding protein, 531
Glycobiology and cancer, 529–532
Glycoconjugates, 528
Glycolysis, aerobic, 11–12
P-Glycoprotein, 96
Glycoproteins, N- and O-linked, 536–539
Glycosaminoglycans (GAGs), 530
Glycosylation
  aberrant mucin-type O-, 538–539
  abnormal, 536–538
Godfrey-Smith's contextualism, 43
Gonadal development, 408
Gonadotropin-releasing hormone, 410
G-protein coupled receptors, 3–4
Grainyhead homolog, 265
Grainyhead-like, 259, 262–265
Graphium sarpedon nipponum, 405
GRNs. *See* Gene regulatory networks (GRNs)
GTF2I genes, 522–523
GTF2IRD1 genes, 522–523
GYF-domain, 82–83

H

Hansenula polymorpha, 84
HCV. *See* Hepatitis C virus (HCV)

Heart capillaries, loss of vWF mosaic expression in, 384–387
Heat shock protein 90 (Hsp90), 428–429
Hematopoietic progenitors, 391
Hemicellulose, 464–465
Hepatitis C virus (HCV)
  induces bistability in interferon signaling network, 133–137, 134*f*
  viral kinetics, 137–139
Hepatitis C virus (HCV) infection, 131–132
  interferons and, 133
  natural outcomes of, 149–150, 150*f*
Herbivores, 460–462, 466
Hermaphroditism, sequential, 404
Heterogeneity, 172–173
  biological noise-mediated, 371–372
  in cancer, 652–653, 677
  cell state
    in isogenic microbial populations, 335–337
    via relevant, 630–633
  in cellular phenotypes, 618
  dynamic mosaic, 389*f*
  endothelial, 367
  epigenetic control of, 379–384, 380*f*, 382*f*
  from epigenetics, 178–179
  intratumoral, 617–618
  metabolic, 337–341
  nongenetic, experimental evidence for, 620–630
  phenotypic, 63–64, 137–144, 335–336
  stress-induced, 336–337
Heterogeneous cells, in yeast colony, 339–341, 340*f*
Hexokinase-1, substrate affinity of, 88–89
HGSOC. *See* High-grade serous ovarian cancer (HGSOC)
Hha/YdgT-like proteins, 80
HIF1α. *See* Hypoxia-induced factor 1α (HIF1α)
High-dimensional systems, 325–326
High-grade serous ovarian cancer (HGSOC), 532–533
Hippo/Warts pathway, 418
Histone, 65, 82
  acetylation, 213
  methylation, 67–68
  modifications, inheritance, 290*f*
H3K9-trimethylated histones, 382–383
HL60 leukemic cells, 629–630
holE gene, 80
Holobiont, 459–461, 470
  individuality, implications, 472–474
Holometabolous insects, developmental stages of, 214
Holozoa, 259
Holozoans, 267
  unicellular, 267–268
Homeodomain-interacting protein kinase-1 (HIPK1) phosphorylates, 89, 90*f*
Homeodomain transcription factors, 419
Homeostasis, in speciation, 429–430
Hormone
  central nervous system and, 409–410
  receptor, 409–410
  secretion, 409
Hsp90. *See* Heat shock protein 90 (Hsp90)
HS proteoglycans (HSPGs), 530
Human Genome Project, 691
Hybrid epithelial/mesenchymal (E/M) phenotype, 557–563
Hymenopterans, 406
Hypoxia-induced factor 1α (HIF1α), 657–658
Hysteresis, 371

I

IDP. *See* Intrinsically disordered proteins (IDPs)
IDRs. *See* Intrinsically disordered protein regions (IDRs)
Immune evasion, epithelial plasticity in, 600–602
Immunity-related GTPase M (IRGM) gene, 95
Immunofluorescence assays, 378–379
Immunological hypothesis, viral kinetics on, 148*f*
Immunology, cell differentiation in, 33–37
Induced pluripotent stem cells (iPSCs), 2–3, 5–7
Induced variation
  environmentally
    in field studies, 229–231
    in laboratory animals, 227–229
Inherent cell functions, in origin of differentiation, 266–269
Inheritance
  epigenetic, 294
  evolution and principle of, 282–288, 286*f*, 290*f*
  of histone modifications, 290*f*
  importance of, 283
  Lamarckian, 57–59
  Lamarck's theory of, 283
  nongenomic, 305–306
  timescale of, 281*f*
Inhibitory pheromone, 412–413
InR. *See* Insulin receptor (InR)
Insects
  endocrine integrators of phenotypic plasticity in, 411–413
  nonlinear allometries in, 426*f*
Insulin receptor (InR), 414
Integrated Model, 572–578, 572*f*, 573*t*

Integrators, 409–413
Interferon and HCV infection, 133
Interferon-based treatment outcome, 137–141
Interferon responsiveness, phenotypic heterogeneity in, 137
Interferon signaling network, bistability in, 133–137
Interferon-stimulated genes (ISGs), 133
Intermediate states, cell fate decision making, 181–182
Intracellular level, phenotypic plasticity at, 3–5
Intracellular molecular regulatory networks, intrinsic complexity of, 192
Intratumoral heterogeneity, 617–618
Intrinsically disordered protein regions (IDRs), 256
Intrinsically disordered proteins (IDPs), 89, 322–323, 329, 575–576, 578–581
In vivo, noise-mediated endothelial cell heterogeneity, 372–388
iPSCs. *See* Induced pluripotent stem cells (iPSCs)
ISGs. *See* Interferon-stimulated genes (ISGs)
Isogenic microbial cells, 335–336, 349
Isogenic microbial populations, cell state heterogeneity in, 335–337
Isogenic populations, phenotypic heterogeneity in, 336
Isologous diversification, 257–258

J

Janus kinase-signal transducer and activator of transcription (JAK-STAT) pathway, 133
Johannsen, Wiliam, 284
Juvenile hormone (JH), 411–413, 421

K

Kairomones, 406
Kauffman-type models, 257–258
Kinetic rate, for differentiation, 171
Kramer's rate theory, 166–167
Krebs cycle, 66–68
Kruppel homolog 1 (Kr-h1), 421

L

Laboratory animals, phenotypic variation in, 227, 228*f*
LacZ expression, 375*f*
LacZ reporter gene, 375*f*
Lamarckian evolution, 240
Lamarckian inheritance, 57–59
Lamarck's theory of inheritance, 283
Landscape-flux theory, 169–170
Langevin equation, 197
Langevin formalisms, 198
LDTFs. *See* Lineage determining transcription factors (LDTFs)
Lectins, 540–541
Leukemia
  cells, nonstem, 643–644
  chronic lymphocytic, 680
Linaria vulgaris, pelorism in, 291*f*
Lineage determining transcription factors (LDTFs), 261, 266, 268
Lineage plasticity, 594–595, 595*f*, 603–606
Lipid metabolism, plasticity in, 662–663
Living cell, as sloppy dynamical system, 328–330
L-lactic acid, 119–120
Locusta migratoria, 406–408
Lyapunov function, 175–176
Lymphocytic leukemia, chronic, 680

M

Macroevolution, plasticity and, 452–454, 453*f*
Macrophages, 82
  plasticity, 8–11, 10*f*
Macrophagous primary consumers, 461–462
Macroscopic global dynamics, insensitivity of, 328–329
Magnetic resonance imaging (MRI), 698
Male seaside sparrows, 484–485
Mammalian cells, 391
Mammalian G-protein coupled receptors, 370–371
Mammary acinar model, 660–661
Mammary gland, 668–669
MAPK signaling, ultrasensitivity in, 370–371
Marbled crayfish, as promising model, 236–241, 238*f*, 239*f*
Marmorkrebs, as promising model, 236–241
MAT. *See* Mesenchymal to amoeboid transition (MAT)
Mathematical modeling, stochastic developmental variation by, 231–232
Matrix-assisted laser desorption/ionization (MALDI), 536–538
Matrix deprivation
  and EMT, 667–668
  enhances stemness properties in cancer cells, 669
  expands limited plasticity of normal mammary epithelium, 668–669
  and stemness, 668–669
Maximum tolerated dose (MTD), 693–694

mCRPC. *See* Metastatic, castration-resistant prostate cancer (mCRPC)
MDA-mb-231 tumor subsequence, 695*f*
Mediational niche construction, 462–463, 467–470
Medium-chain acyl-CoA dehydrogenase (MCAD) deficiency, 92
Melanoma cells, CSC population in, 640*f*
Mendel, Gregor, 284
Mesenchymal cells, 574, 622–623
Mesenchymal-epithelial transition (MET), 554–555, 623
Mesenchymal to amoeboid transition (MAT), 7–8
Mesenchymal to epithelial transition (MET), 7–8
Mesenchyme, 591–592
MET. *See* Mesenchymal-epithelial transition (MET)
Metabolic adaptation, 664–665
Metabolic cooperation, 71–72
Metabolic cycle, yeast, 347–348, 348*f*, 351–353
Metabolic heterogeneity, 337–341
  case study, 344–346
  frustrated bistability, 350–351, 351*f*
Metabolic phenotype, bistability to, 665–667, 667*f*
Metabolic plasticity
  in cancer cell survival, 655–665
  signaling pathways regulating, 654–655
Metabolic reprogramming, 11–12
  in cancer, 669–670
Metabolic states, 339–342, 347–348, 353–354
Metabolism
  in adaptation, 316
  amino acid, plasticity in, 663–664
  lipid, plasticity in, 662–663
Metabolites, 337–338
  sentinel, 66, 68–69
Metamorphosis, 404, 420–422
Metastasis, 555–557, 561–562, 645, 651–652
  plasticity-driven, 600
Metastatic, castration-resistant prostate cancer (mCRPC), 593–594
Metazoan differentiation system, 271
Metazoan evolutionary innovation, prolific cell differentiation as, 269–273
Metazoan-specific modes, of transcriptional regulation, 259–262
Methylation, 213, 259–260, 407
  cyp19a, 408
  DNA, 213, 234–235, 288–289, 380*f*, 407–408
  histone, 67–68
Methylomes, genome-wide, 239–240
Methylotrophic organisms, regulation of alcohol oxidase, 84, 85*f*
Methylotrophic yeast, 84
MGRN model. *See* Modular gene regulatory network (MGRN) model
Microarray gene expression, 80
Microbe-dependent
  detoxification of plant defense chemicals, 466–467
  digestion of plant fiber, 464–466
Microbes, 335–336, 467
  prototypical, 335
  symbiotic, 460–461, 466–467
Microbial cells, isogenic, 335–336
Microbial colonies, clonal, 338
Microbial colonization of animal, 460
Microbial communities, 335, 337–338
Microbial-dependent development, of ruminant stomach, 463–464
Microbial populations, cell state heterogeneity in isogenic, 335–337
miRNAs, 94–95
Mitochondrial calcium, 653–654
Mitochondrial matrix, acidification of, 80–81
Mitogen-activated protein kinase (MAPK), 592–593
Modular gene regulatory network (MGRN) model, 254–259
Molecular mechanisms, 367
  identification of, 232–236
Molecular signals, 82
Monophenic canalization and polyphenisms, 430
Monte Carlo sampling, 166
Moonlighting proteins, 78–81
Mosaic heterogeneity, 389–390, 392*f*, 394
  dynamic, 389*f*, 391–392
Mosaic vWF expression, 375–377
[$MOT3^+$]/Mot3, 116–117
Motor pathway, 486
Mouth-form plasticity, 234
mRNA
  synonymous mutations impair interactions of, 94–95
  synonymous mutations regulate folding of, 92–94
MTD. *See* Maximum tolerated dose (MTD)
Multicellularity, 259, 262
  in response to environmental triggers, 116–117

role in evolution of differentiation, 263–265
Multicellular organisms, 259
cell differentiation in, 320–321
development of, 253
Multidomain proteins, 85–86
domain–domain interactions in, 85–86
Multidrug resistance 1 (MDR1) gene, 96
Multidrug resistant (MDR) complex, 699
Multiple cellular states, at population level, 341–346, 343*f*
Multiscale models, 137–139, 143–145, 146*f*, 150–151
Multistablility, 363–364
Mutant tRNAs, 113–114
Mutations, 55, 284, 427–428
in prion-forming domain (PrD), 112
silent, 90–97
synonymous, 90–91
dictate gene splicing, 91–92, 93*t*
impact of, 91
impair interactions of mRNA, 94–95
modulate cotranslational folding, 95–97, 97*f*
regulate folding of mRNA secondary structure, 92–94
regulation of gene expression by, 96–97, 97*f*
Mycobacterium tuberculosis (Mtb), 12–13
Mycoviruses, impact of, 105–106

N

N- and O-linked glycoproteins, 536–539
Native cells, 311–312
Natural killer (NK) cells, 530–531
Natural selection, 258, 281–282, 513–516
NEPC. *See* Neuroendocrine prostate cancer (NEPC)
Neural crest, 520–521, 523
Neural sensors, 405–406, 408
Neurochemicals, 406
Neuroendocrine prostate cancer (NEPC), 594–595, 602–605
Neurotransmitters, 406
Niche construction, 470–471
developmental, 468
mediational, 467–470
perturbational, 462–463, 468
relocational, 468
Nicotinamide adenine dinucleotide phosphate (NADPH), 660–661
Nilaparvata lugens, 416
Nitrix oxide (NO), 10–11
N-linked glycans, 529–530
Noise, 29–30, 197–198
biological, 196–197
definition, 24*t*
in gene expression, 27–29
Noise-mediated endothelial cell heterogeneity in vivo, 372–388
Nonadaptive plasticity, and evolution of robustness, 422–423
Nonepithelial ovarian cancers, 532
Nonequilibrium dynamics, 163–164, 176, 180
Nonequilibrium systems, 163–164, 167–168
Nongenetic components, of phenotypic variation, 236–241
Nongenetic factors, 78
Nongenetic heterogeneity, experimental evidence for, 620–630
Nongenetic phenotypic variation, manifestations of, 207, 208*f*
Nongenetic variability, 23–25
Nongenomic inheritance, 305–306
Nonholozoan opisthokonts, 259
Nonlinear dynamics, conventional, 161–162
Non-small cell lung cancer (NSCLC), 559
Nonstem leukemia cells, 643–644
Notch factors, 40–41
Notch genes, expression of, 40–41
Notch receptors, 40–41
Novel polyphenism, 445
NRs. *See* Nuclear receptors (NRs)
NS5A inhibitors, RAVs associated with, 146*f*
NSCLC. *See* Non-small cell lung cancer (NSCLC)
Nuclear β-catenin, 264
Nuclear receptors (NRs), 260–261
Nutritional symbiosis, 464–466

O

OAZ1 gene, 115
Oct4, 168–169, 168*f*
Oligomerization
of alcohol oxidase, 84
modulation of protein function by, 84–85, 85*f*
O-linked glycans, 529–530
Ommatidia, differentiation of, 368–370
Omnivores, 449
On the Origin of Species (Darwin), 283
Onthophagus, 413
O. taurus, 415–416
Ontogenesis, 57
stabilization of, 514
Ontogenesis recapitulating phylogenesis, 59
Opisthokonts, nonholozoan, 259
"Order from disorder" rule, 57
Organismal level, phenotypic plasticity at, 13–15

OSE. *See* Ovarian surface epithelium (OSE)
Ovarian cancer, 527–528
glycans in, 528
nonepithelial, 532
progression, 532–536, 534*f*, 537*f*
proteoglycans in, 539–540
Ovarian surface epithelium (OSE), 532–533
Oxidation, of acetyl-CoA molecule, 66–67
Oxidative phosphorylation (OXPHOS) pathway, 11–12

P

PAGE4. *See* Prostate associated gene-4 (PAGE4)
Pancreatic adenocarcinoma (PDAC), 555–556
Pancreatic ductal adeno carcinoma (PDAC), 628
Pancreatic progenitor, 7
Papilio xuthus, 405
Passerellidae, 485, 488*f*
Path finders, 7–8
Path generators, 7–8
PDAC. *See* Pancreatic adenocarcinoma (PDAC)
Pelorism, in *Linaria vulgaris*, 291*f*
Pentose phosphate pathway (PPP), 655–656
Peripheral zone (PZ), 570–571, 571*f*
Peroxisomes, 84
Perturbational niche construction, 462–463, 467–470
PFPs. *See* Prion-forming proteins (PFPs)
Phenotypes, 1, 305–306
alternative, 404–405
cellular, 62, 335
and gene expression, 316–319, 318*f*
hybrid epithelial/mesenchymal (E/M), 557–561
of organism, 207
prion-mediated, 112–120
Phenotypic changes
of domesticated foxes, 521–523
of experimental foxes, 516–519, 517*f*
Phenotypic differentiation, 71–72
Phenotypic diversity, 513–514, 624–625
conformational diversity generates, 120–122
Phenotypic dynamics, 306–307
Phenotypic heterogeneity, 63–64, 137–144, 335–336
in clonal cell populations, 337
in interferon responsiveness, 137
in isogenic populations, 336
Phenotypic instability, tissue disruption at origin of, 685–687
Phenotypic landscape, theory of cell states in, 618–620
Phenotypic landscape model, 625–626
Phenotypic plasticity, 1–2, 15, 207–208, 208*f*, 210, 215, 227–230, 234, 403–404, 423, 580–581, 605–606
adaptive, 409
biological noise and, 367–371
in cancer, 652–653
growth and drug resistance, 678–681
and cancer stem cells, 639–640
cardiac, 13
at cellular level, 5–13
emergence of, 2–3
evolution via, 425–427
from genes to, 363–367
genetic roots of, 363–364, 363*f*
in insects, 411–413
at intracellular level, 3–5
at organismal level, 13–15
overcomes androgen receptor blockade, 594–602, 595*f*, 596*f*
overcoming resource depletion, 595–600
in prostate cancer progression, 595*f*
Waddington landscape as metaphor for, 2–3, 3*f*
Phenotypic stability factors (PSFs), 560
Phenotypic switching, 215
epigenetically determined, 288–295
evolutionary consequences of, 295–301
Phenotypic transformations and collective cell migration, 645–646
Phenotypic transitions, in cancer, 553–554
Phenotypic variability, 14, 106
Phenotypic variation, 209–210
genetic component of, 210
in laboratory animals, 227, 228*f*
manifestations of, 214
nongenetic components of, 207–208, 236–241
nongenetic, manifestations of, 207, 208*f*
nongenetic sources of, 241
patterns, songbirds, 486–489, 488*f*
plasticity in song development, 489–497
extent of plasticity, 489–491
external influences, 491–493
internal influences, 493–495
results of plasticity, 496–497
stochasticity, 495–496
in populations, 207

in wild populations, 229
Pheromone, 406, 412–413
inhibitory, 412–413
soldier-inhibiting, 412–413
Phosphorylation, 89
EGFR, 661
Photoreceptors, 368–370
Phylogenesis, ontogenesis recapitulating, 59
Physarum polycephalum, 14
Physical laws, of cell fate dynamics, 192–196, 196*f*
Physics of complex systems, 322, 330–331
PIA. *See* Proliferative inflammatory atrophy (PIA)
Pichia pastoris, 84
PI3K/Akt-mechanistic target of rapamycin (mTOR) pathway, 654–655
PI3K/Akt/mTOR signaling, 664–665
PIN. *See* Protein interaction network (PIN)
Plant cell wall-digesting enzymes, 464–465
Plant defense chemicals, microbe-dependent detoxification of, 466–467
Plant fiber, microbe-dependent digestion of, 464–466
Plant lectins, 529–530
Plastic growth, effectors of, 414–418
Plasticity, 172–173, 470–471
See also specific types of plasticity
as adaptation, 423–424
adaptive phenotypic, 409
in amino acid metabolism, 663–664
bistable systems and, 665–667
cancer evolution and glycan-driven, 541–542
developmental, 403, 445–446, 497
regulation of, 404–422
epithelial-mesenchymal, 603–604
feeding, 234
in glucose metabolism, 656–659
lineage, 594–595, 595*f*, 603–606
in lipid metabolism, 662–663
and macroevolution, 452–454, 453*f*
macrophages, 8–11, 10*f*
mouth-form, 234
phenotypic. *See* Phenotypic plasticity
in redox signaling, 659–662
of vWF expression, 377–378
"Plasticity and epigenetics-first" concept of evolution, 242
Plasticity-driven metastasis, 600
Plasticity-led evolution (PLE), 444–447, 445*f*
empirically evaluating, 447–449
in spadefoot toads, 449–452, 450*f*
Plasticity-related transcription factors, 597
Plastic patterning, effectors involved in, 418–420
Plastic seasonal adaptations, 409
PLE. *See* Plasticity-led evolution (PLE)
Polymorphisms, song dialects as functional, 498–503
Polypeptide chains, 77
Polyphenic traits, 404
Polyphenism, 214, 403–404, 406, 411, 435, 449, 453
canalization and, 431*f*
epigenetic regulation of, 407–408
evolution of, 432
monophenic canalization and, 430
reaction norms, 424–425
and robustness, 426*f*
seasonal, 411–412, 419–420
sequential, 420
Population dynamics, 314*f*, 316–319, 342–344
Population level, multiple cellular states at, 341–346, 343*f*
Posttranslational modifications, modulation of protein function by, 89–90, 90*f*
Potamopyrgus antipodarum, 229–230
PPP. *See* Pentose phosphate pathway (PPP)
PrD. *See* Prion-forming domain (PrD)
Precision medicine, 646
Predator facilitation, 700
Premetazoan opisthokonts, 267
Priming, 681–682
Prion, 106–107
formation and loss, 108–110
fungal, 118–119
propagation and transmission in fungal cell, 110–112
variants, 121–122
Prion-forming domain (PrD), 112
mutations in, 112
Prion-forming proteins (PFPs), 107–108
Prion-mediated phenotypes, 112–120
Pristionchus pacificus, 234
Procambarus fallax, 236
Procambarus virginalis, 236
Progenitors
endothelial, 368
hematopoietic, 391
pancreatic, 7
Proliferative inflammatory atrophy (PIA), 577
Prolific cell differentiation, as metazoan evolutionary innovation, 269–273
Promiscuous protein interactions, 323
Prostate
development, 591–592
diseases of, 569

Prostate adenocarcinoma, androgen receptor signaling axis in, 592–594
Prostate associated gene-4 (PAGE4), 89, 90*f*
  PAGE4/AP-1/AR/CLK2 regulatory circuit, 576*f*
  PAGE4 protein, 577
  in prostate cancer (PCa), 578*f*
Prostate cancer (PCa), 569–570, 598*t*
  aggression, 605–606
  EMT/MET in, 578–580
  progression to, 591–592
  prostate-associated Gene 4 (PAGE4) in, 578*f*
  transitions, 579*f*
Prostate specific antigen (PSA), 697
Prostate transition zone, 570–571, 571*f*, 572*f*, 577
Protective symbiosis, 466–467
Protein
  distribution in cell populations, 323–325, 324*f*
  expression, 308–309
    in single cells, 361–362
  secreted, 263–264
Protein function
  influence of domain association on, 85–89, 87*f*, 88*f*
  modulation of
    by oligomerization, 84–85, 85*f*
    by posttranslational modifications, 89–90, 90*f*
Protein interaction network (PIN), 580
Protein–protein interactions, 81–84, 83*f*
Proteins, 77, 97–98
  alkaline, 82
  amino acid sequence of, 77–78
  Hha/YdgT-like, 80
  moonlighting, 78–81
  multidomain, 85–86
  regulators, 172–173
  RNA-binding, 94–95
  TCF/LEF, 264
  3D structure of, 77
Proteoglycans, 530
  in ovarian cancer, 539–540
Proto-domestication, 515–516
Prototypical microbe, 335
PrP^{Sc} form, 107
PrP^{Sc} prion, 107–108
P-selectin, 386–387
PSFs. *See* Phenotypic stability factors (PSFs)
[PSI +] prion, 120–121, 121*f*
[PSI^{+}]/Sup35, 113–115
Pupal coloration, 405
Pyruvate dehydrogenase (PDH) kinase, 659

Q

Quantified landscape, for cell fate decision making in differentiation and development, 176–177
Quantitative polymerase chain reaction (qPCR), 679
Quasi-potential landscape, 62
Quorum sensing, 338–339

R

Randomness, definition, 24*t*
Ras protein, 81
RAVs. *See* Resistance-associated amino acid variants (RAVs)
Reaction-diffusion mechanism, 231
reaction norm genotype, 422–423
Reaction norms and allometries, 424
Reactive oxygen species (ROS), 653–654, 660–662
Redox signaling, plasticity in, 659–662
Reductionism, 46–47, 307
Regulatory genome, 253–255
Regulatory RNAs, 253–254, 289
Relocational niche construction, 468
Resistance-associated amino acid variants (RAVs), 141–142
  associated with NS5A inhibitors, 146*f*
Resistant cancer cells, 699
Reverse engineering, 307
Rhythmic behaviors, 365–366
RNA
  binding proteins, 94–95
  polymerase enzyme complex, 64–65
  regulatory, 253–254, 289
RNAseq, 180
Robo4, 370–371
Robust mechanism, 55
ROS. *See* Reactive oxygen species (ROS)
RTKs, 653
Rumen ecosystem, 465–466
Ruminant, 462
  fiber-eating ecosystem, 473
Ruminant stomach, microbial-dependent development of, 463–464

S

Saccharomyces cerevisiae, 105–106, 299–300
  life cycle in, 107*f*
S-adenosyl-methionine (SAM), 66–67
SAM. *See* S-adenosyl-methionine (SAM)
SAM Pointed Domain Containing ETS Transcription Factor (SPDEF), 600
Sarcomatoid cell type, 626–627
Sarcomatoid cholangiocarcinoma, 626–627
SASP. *See* Senescence-associated secretory phenotype (SASP)
Schistocerca gregaria, 214, 405–406

scRNA-seq. *See* Single-cell RNA sequencing (scRNA-seq)
SDV. *See* Stochastic developmental variation (SDV)
Seasonal polyphenism, 411–412, 419–420
Secreted protein, 263–264
Selection, 516–518
  artificial, 513–514
  of foxes for domestication, 522–523
  natural, 513–516
  stabilizing, 514–515
Self-organization, 307–308
Self-referencing process, 53
Seminal fluids, 591–592
Senescence
  and cancer cell plasticity, 643–645
  and cancer stem cells, 644*f*
  cellular, 643
Senescence-associated secretory phenotype (SASP), 643
Senescent cells, 644
Sense codons, 114
Sensing matrix deprivation, 653–654
Sensorimotor learning, 485–486
Sensors, 404–408
  environmental, 408
  neural, 405–406, 408
Sensory hairs, stimulation of, 405–406
Sentinel metabolites, 66, 68–69
Sequential hermaphroditism, 404
Sequential polyphenism, 420
Serotonin, 406
Serous tubal intraepithelial carcinoma (STIC), 532–533
Sex change, in fish, 410
Sex determination, 408
Sex switching, 409
SGE. *See* Stochastic gene expression (SGE)
sHSPs. *See* Small heat shock proteins (sHSPs)
Signaling pathways, regulating metabolic plasticity, 654–655
Silent mutation, 90–91
  tunes gene function, 90–97
Silent Variant Analyzer, 96–97
Single-cell RNA sequencing (scRNA-seq), 64–65, 618, 630–633
Single-cell technologies, 678–679
Single nucleotide polymorphisms, 105
Single nucleotide variants (SNVs), 237–239
Skepticism, 447
Ski8 protein, 83
Sloppy dynamical system, living cell as, 328–330
SLRPs. *See* Small leucine-rich proteoglycans (SLRPs)
Small heat shock proteins (sHSPs), 84–85, 89–90
  suboligomeric complexes of, 84–85
Small leucine-rich proteoglycans (SLRPs), 530
SNVs. *See* Single nucleotide variants (SNVs)
Social insects, caste determination in, 412–413
Soldier-inhibiting pheromone, 412–413
Somatic evolution, 692
Songbirds
  mate attraction, 484
  phenotypic variation patterns, 486–489, 488*f*
  song development. *See* Song development, songbirds
  song function, 483–485
  song ontogeny, 485–486
  territorial, 483–484
Song development, songbirds
  plasticity in, phenotypic variation, 489–497
    extent of plasticity, 489–491
    external influences, 491–493
    internal influences, 493–495
    results of plasticity, 496–497
    stochasticity, 495–496
Song dialects
  as functional polymorphisms, 498–503
    dialects and gene flow, 499–501
    dialects arise, 498–499
    functional variation among dialects, 501–503
Song function, songbirds, 483–485
SOX2 expression, 604
Spadefoot toads, plasticity-led evolution (PLE) in, 449–452, 450*f*
SPDEF. *See* SAM Pointed Domain Containing ETS Transcription Factor (SPDEF)
Speciation, role of robustness and homeostasis in, 429–430
Stabilization of ontogenesis, 514
Stabilizing selection, 514–515
STAT3, 603–604
Stem cells, 157–158
  differentiation, 174
  differentiation and development, 160–164, 164*f*
  embryonic, 157–158, 391
  transdifferentiation of adult, 11
Stemness
  cellular plasticity and, 681–683
  EMT and, 669–670

Stemness (*Continued*)
matrix deprivation and, 668–669
STIC. *See* Serous tubal intraepithelial carcinoma (STIC)
Stochastic developmental effects, 495–497
Stochastic developmental variation (SDV), 207–208, 210–213, 212*f*, 232–236, 241
of behavior, 225, 226*f*
of biochemical traits, 223
of coloration, 221–223, 222*f*
determination of, 218–226
in field studies, 229–231
in identically reared groups of isogenic animals, 218–221, 220*t*
in kingdoms of life, 215–218, 216*t*
in laboratory animals, 227–229
of life history traits, 223, 224*f*
of life span, 224–225
by mathematical modeling, 231–232
by mathematical simulation, 231, 233*f*
of morphological traits, 218, 219*f*
of reproductive traits, 223–224
Stochastic fluctuations, 25
Stochastic gene expression, 336–337
effect of, 338
Stochastic gene expression (SGE), 678–681
cellular plasticity and stemness conferred by, 681–683
Stochasticity, 33
definition, 24*t*
of gene expression, 63–64
in intracellular process, 25
song development, songbirds, 495–496
Stochastic phenotype switching, and cellular memory, 371
Stochastic system, irreversibility in, 178
Stop codons, 113–115
Stress-induced heterogeneity, 336–337
Stromal disease, 574
Sucker's gambit, 700
Sugar hydrolyzing enzyme, 86–87
Sustained virological response (SVR), 147
[SWI^{+}]/Swi1, 117–118
Symbiodinium, 459–460
Symbiosis, protective, 466–467
Symbiotic microbes, 460–461, 466–467
SynEx, 32–33
Synonymous mutations, 90–91
dictate gene splicing, 91–92, 93*t*
impact of, 91
impair interactions of mRNA, 94–95
modulate cotranslational folding, 95–97, 97*f*
regulate folding of mRNA secondary structure, 92–94
regulation of gene expression by, 96–97, 97*f*
Systems biology, 190–191, 562–563

T

TADs. *See* Topologically associating domains (TADs)
Taeniopygia guttata, 485–486
Tamed foxes, 523
Tameness, 522
Targeted therapies, 628
TCA cycle. *See* Tri-carboxylic acid (TCA) cycle
TCF/LEF proteins, 264
Territorial songbirds, 483–484
TFs. *See* Transcription factors (TFs)
Therapy resistance, cancer, 554–556
Thioredoxin, 79
Tissue disruption, at origin of phenotypic instability, 685–687
Topologically associating domains (TADs), 260–261
TOR activity, 414
Trachemys scripta, 408
Transcriptional regulation
global, 117–118
metazoan-specific modes of, 259–262
Transcription factors (TFs), 55–56, 61, 253–254, 264–268
enhancers, 262, 264–265
plasticity-related, 597
Transdifferentiation, 7–8, 321
of adult stem cells, 11
of cell types, 2–3
Transforming growth factor beta (TGF-β), 641
Trans-glycosidase, 3D structure and dynamics of, 86–87, 87*f*
Transition states
cell fate decision making, 181–182
for cell fate decision making, 180
Transition state theory, 166–167
Transition zone (TZ)
prostate, 570–571, 571*f*, 572*f*, 577
Tri-carboxylic acid (TCA) cycle, 352–353, 663–664
Trichoplax adhaerens, 261–262
tRNAs, mutant, 113–114
Tumor evolutionary dynamics, 696–697
Tumor heterogeneity, 631–632
Tumor reprogramming, 677–678
Tumors, heterogeneity of, 652–653

U

Ultrasensitivity, in *MAPK* signaling, 370–371
Unicellular holozoans, 267–268
Universality, in cell biology, 200–202, 202*f*
Urogenital sinus, 591–592
U5 snRNP 52K (U5 52K) protein, 82–83

V

VE-cadherin, 378–379
Vertebrates, endocrine integrators of phenotypic plasticity in, 410–411
Vimentin, 7–8
Viral kinetics, 137–144, 138*f*
  on immunological hypothesis, 148*f*
V(D)J genes, 35*f*
von Willebrand disease, 388–389
von Willebrand Factor (vWF) expression, 371
  epigenetic control of, 379–384, 380*f*, 382*f*
  mosaic expression, loss of
    in brain capillaries, 387–388
    in heart capillaries, 384–387
  plasticity of, 377–378
  regulation, dimensions of bistable, 392*f*
  stochastic toggle of, 384–385
  in vivo, dynamic heterogeneity, 372–378, 375*f*, 376*f*, 379*f*

W

Waddington landscape, 5–7
  for cell fate decision making in differentiation and development, 176–177
  as metaphor for phenotypic plasticity, 2–3, 3*f*
  and paths for differentiation/development, 168–174, 170*f*
Waddington metaphor, 363–364
Waddington's epigenetic landscape, 432–433
Waddington's formalism, 195
Waddington's notion of cell type, 203
Waddington's principle, 198
Warburg effect, 11–12, 656
WNT, 597–600
Wnt pathway, 81
Woodward's proportionality criteria, 44–45

Y

Yamanaka factors, 643–644
Yeast
  adaptation experiments, implications, 320–330
  cells, 311–312, 315–316
  genome, 347–348
  metabolic cycle, 347–348, 348*f*
  methylotrophic, 84
Yeast colony, 341
  heterogeneous cells in, 339–341, 340*f*
  spatial organization within, 337–341
Yeast metabolic cycle, 347, 348*f*, 351–353

Z

Zootermopsis nevadensis, 407
Zygote-derived cells, 459

9780128179963